ENVIRONMENTAL MODELING

ENVIRONMENTAL SCIENCE AND TECHNOLOGY

A Wiley-Interscience Series of Texts and Monographs

Edited by JERALD L. SCHNOOR, *University of Iowa*
ALEXANDER ZEHNDER, *Swiss Federal Institute for Water Resources and Water Pollution Control*

A complete list of the titles in this series appears at the end of this volume

ENVIRONMENTAL MODELING

Fate and Transport of Pollutants in Water, Air, and Soil

JERALD L. SCHNOOR

Department of Civil & Environmental Engineering
The University of Iowa
Iowa City, Iowa

A WILEY-INTERSCIENCE PUBLICATION
JOHN WILEY & SONS, INC.
New York • Chichester • Brisbane • Toronto • Singapore

This text is printed on acid-free paper.

Library of Congress Cataloging in Publication Data:

Schnoor, Jerald L.
Environmental modeling : fate and transport of pollutants in water, air, and soil / Jerald L. Schnoor.
p. cm. — (Environmental science and technology)
"A Wiley-Interscience publication."
Includes bibliographical references and index.
ISBN 0-471-12436-2 (cloth : alk. paper)
1. Water—Pollution—Mathematical models. 2. Pollution—Mathematical models. I. Title. II. Series.
TD423.S37 1996
628.5'2—dc20 96-6187

Printed in the United States of America

10 9 8

To Jana, Ben, and Britta
for making my days bright

SERIES PREFACE

Environmental Science and Technology

We are in the third decade of the Wiley-Interscience Series of texts and monographs in Environmental Science and Technology. It has a distinguished record of publishing outstanding reference texts on topics in the environmental sciences and engineering technology. Classic books have been published here, graduate students have benefited from the textbooks in this series, and the series has also provided for monographs on new developments in various environmental areas.

As new editors of this Series, we wish to continue the tradition of excellence and to emphasize the interdisciplinary nature of the field of environmental science. We publish texts and monographs in environmental science and technology as it is broadly defined from basic science (biology, chemistry, physics, toxicology) of the environment (air, water, soil) to engineering technology (water and wastewater treatment, air pollution control, solid, soil, and hazardous wastes). The series is dedicated to a scientific description of environmental processes, the prevention of environmental problems, and preservation and remediation technology.

There is a new clarion for the environment. No longer are our pollution problems only local. Rather, the scale has grown to the global level. There is no such place as "upwind" any longer; we are all "downwind" from somebody else in the global environment. We must take care to preserve our resources as never before and to learn how to internalize the cost to prevent environmental degradation into the product that we make. A new "industrial ecology" is emerging that will lessen the impact our way of life has on our surroundings.

In the next 50 years, our population will come close to doubling, and if the developing countries are to improve their standard of living as is needed, we will require a gross world product several times what we currently have. This will create new pressures on the environment, both locally and globally. But there are new opportunities also. The world's people are recognizing the need for sustainable development and leaving, a legacy of resources for future generations at least equal to what we had. The goal of this series is to help understand the environment, its functioning, and how problems can be overcome; the Series will also provide new insights and new sustainable technologies that will allow us to preserve and hand down an intact environment to future generations.

JERALD L. SCHNOOR
ALEXANDER J. B. ZEHNDER

PREFACE

From the turn of this century, the emphasis on water quality has changed markedly from public health and sanitation problems, to treatment of conventional pollutants in wastewater, to concern over toxic chemicals in the global environment today. This book is an attempt to wed elementary concepts of pollutant fate and transport with chemical principles in a modern text to assess environmental quality. Previous textbooks on environmental modeling have emphasized engineering transport fundamentals or equilibrium aquatic chemistry, but not both. To address chemical problems in the environment today, we need to be fully aware of transport, reactions, and chemical speciation that determine toxicological effects. It draws examples from a wide variety of water quality issues including conventional pollutants in rivers, eutrophication of lakes, and toxic organic chemicals and heavy metals in surface and groundwaters. Current concerns over atmospheric deposition, groundwater contamination, and global change are also included.

First, there is a crucial need to answer the deceptively simple question, "Where do all the chemical pollutants go?" For this, we must employ mathematical models. In a mathematical framework we can account for simultaneous chemical reactions and transport in the environment. Second, we need to consider the related question regarding effects, "What are the chemical metabolites, toxicity, and long-term consequences in the environment?" For that question, we must consider chemical speciation and exposure. Models of chemical fate are necessary to compute the dose (concentration), frequency, and duration of chemical exposure to aquatic biota and humans.

Environmental Modeling is designed to be an introductory textbook for senior undergraduate and graduate students in environmental sciences and engineering. Prerequisite studies might include undergraduate or introductory graduate courses in environmental chemistry, physicochemical processes, and calculus through differential equations. The objective of the book is to demonstrate how to develop and solve mathematical models for a wide variety of chemical pollutants. The emphasis is on natural waters. Students will gain an understanding of mathematical models such that they can read environmental journals from a critical viewpoint. They will be capable of building their own models from mass balance equations, and hopefully, they will gain an appreciation for related environmental disciplines (geochemistry, microbiology, ecology, toxicology, hydrology, chemistry, and physics).

Chapters follow in a pedagogical sequence: introduction, transport processes, chemical principles, specific applications in water and groundwater, and atmospheric deposition and global change. The first two chapters include introductory material on water quality modeling and transport processes. Chemical kinetics and equilibrium modeling in Chapters 3 and 4 are a necessary primer for subsequent chapters on water, groundwater, and atmosphere. Chapters 5–9 are detailed applications of mathematical models for various pollutants in lakes, rivers, estuaries, and groundwater.

A modern textbook on modeling of chemicals in the environment would be remiss without discussion of, potentially, our most serious environmental problem of the 21st century, that is, alteration of the atmosphere and global change. Chapter 10 seeks to lay a foundation for understanding these disturbances in our global cycles of oxidation and reduction from a chemical perspective. We seek to stretch the concepts of aquatic chemistry to the atmospheric aqueous phase (fogs and cloud waters) and to demonstrate its importance at the air–water interface, including deposition to water and land. Aquatic effects of acid deposition are one manifestation of disturbances in global cycles.

Chapter 11 concludes the book with a look at changes in global cycles of carbon, nitrogen, and sulfur in water, air, and soil. Time scales and remedies are discussed for the accumulation of trace gases in the atmosphere. It will be difficult to stabilize these gases, considering the need for many countries to develop, but there are possibilities for sustainable development suggested in Chapter 11. I hope that this book serves to advance the field and to improve the modeling of chemical pollutants in the environment.

Environmental Modeling would not have been possible without the inspiration of Werner Stumm, who provided collaboration and sabbatical support at the Swiss Federal Institute for Environmental Science and Technology (EAWAG) in 1988 when this project was first conceived. He was influential in the development of Chapters 4, 8, and 10, especially. The mentor who opened the world of mathematical modeling to me was Donald J. O'Connor, Manhattan College. I am most grateful to Nikolaos Nikolaidis, O-Yul Kwon, James Szydlik, Deborah Mossman, Chikashi Sato, Sijin Lee, Laura Sigg, Mary Bergs, Philippe Behra, Steven Chapra, Chip Levy, Thomas Voice, and David Wiggert for commenting on parts of the text; for the encouragement of Gene Parkin, Richard Valentine, Forrest Holly, and Pedro Alvarez and the Department of Civil and Environmental Engineering, University of Iowa; and to Alexander Zehnder, Director of EAWAG, where the book was completed during 1994–95. Earlier versions of the manuscript were used as textbooks at The University of Iowa, the University of Connecticut, Michigan State University, and Toledo University.

The expert assistance of Jane Frank, Center for Global and Regional Environmental Research at The University of Iowa, who processed the entire text, and Heidi Bolliger, EAWAG, who drafted many of the figures, was much appreciated. In the end, it was the students at The University of Iowa who "wrote" this book by virture of their many queries and corrections, challenges and cooperation, including Shyam Asolekar, Steven Banwart, R. Christopher Barry, Joel Burken, Louis Licht, James

Lin, Drew McAvoy, Dhileepan Nair, Kurtis Paterson, Todd Rees, Y. L. Yan, and Xiameng Zhong, who have gone on to distinguished careers in academia and industry. They were the principal reason for writing this book and their successors are the audience that will advance the field still further.

JERALD L. SCHNOOR

Iowa City, Iowa

CONTENTS

1

INTRODUCTION

If we are going to live so intimately with these chemicals, eating and drinking them into the very marrow of our bones, we had better know something about their nature and power.

—Rachel Carson, *Silent Spring*

1.1 SCOPE OF ENVIRONMENTAL MODELING

Why should we build mathematical models of environmental pollutants? Three reasons indicate the scope and rationale for this book:

- To gain a better understanding of the fate and transport of chemicals by quantifying their reactions, speciation, and movement.
- To determine chemical exposure concentrations to aquatic organisms and/or humans in the past, present, or future.
- To predict future conditions under various loading scenarios or management action alternatives.

In the first case, we want to know, "Where do all the chemicals go?" Are they with us forever? How rapidly are they degraded? Such issues pertain to the *fate, transport,* and *persistence* of chemicals in the environment. Classic models address conventional pollutants, eutrophication, toxic organic chemicals, and metals in surface waters and groundwater.[1–9] Recently, mathematical models have become more sophisticated in terms of their chemistry, and this book seeks to solidify the bond between water quality modeling and aquatic chemistry. Chemical speciation models[10,11] are coupled with kinetic transport models for determining fate and chemical speciation.

The second purpose of mathematical models, to determine chemical exposure concentrations, is one of growing importance. It pertains to assessing the *effects* of chemical pollutants. New water quality criteria are promulgated to account for acute and chronic effects levels using frequency and duration of exposure. These criteria result in water quality standards that are enforceable by law and require the application of mathematical models for waste load allocations, risk assessments, or environmental impact assessments. Toxic chemicals—ammonia, arsenic, cadmium, chlorine, chromium, copper, cyanide, lead, and mercury—have been regulated.[12] The criteria specify an acute threshold concentration and a chronic-no-effect concentration for each toxicant as well as tolerable durations and frequencies.

Water quality criteria[13] state that aquatic organisms and their uses should not be affected unacceptably if two conditions are met: (1) the 4-day average concentration of the toxicant does not exceed the recommended chronic criterion more than once every three years on the average and (2) the 1-hour average concentration does not exceed the recommended acute criterion more than once every three years on the average. New criteria recognize that toxic effects are a function both of the magnitude of a pollutant concentration and of the organism exposure time to that concentration. A very brief exposure to a relatively high concentration may be less harmful than a prolonged exposure to a lower concentration. It is only through the use of aquatic chemical models that such frequency–duration relationships can be developed. It is only with a detailed knowledge of aquatic chemistry that such models can be built, tested, and interpreted properly.

We must distinguish between acute and chronic toxicity thresholds, but also it is important to know the toxic chemical species. In the future we will need site-specific water quality criteria that also would require mathematical modeling to aid in determining water quality standards and excursions. For example, research has shown that copper exhibits much less toxicity and/or bioavailability in site-specific tests compared to single-species laboratory bioassay testing.[14] It is likely that aqueous copper forms strong complexes with ligands in situ (organic ligands in particular) that are not as toxic or bioavailable to aquatic organisms. Dissolved organic carbon concentrations vary widely in natural waters so site-specific criteria should be developed. Because copper exceeds water quality criteria in some locations naturally, this is a difficult issue. Publicly owned wastewater treatment plants often exceed water quality criteria for some toxic metals under extreme, low flow conditions. The likelihood of in situ toxicity, either acute or chronic, becomes the point of primary concern. To provide the proper perspective, we must have valid exposure assessments, which require the use of mathematical models. For example, development of a site-specific water quality criterion for pentachlorophenol has been established.[15]

The third major purpose of environmental models is to predict future chemical concentrations under various loading scenarios or management action alternatives. Waste load allocations and exposure models for risk assessment fall into this category. Regardless how much monitoring data are available, it will always be desirable to have an estimate of chemical concentrations under different conditions, results for a future waste loading scenario, a predicted "hindcast" or reconstructed history, or estimates at an alternate site where field data do not exist. For all these reasons we need chemical fate and transport models, and we need models that are increasingly sophisticated in their chemistry, as we move toward site-specific water quality standards and chemical speciation considerations in ecotoxicology. To model aquatic chemical systems, we begin with a simple mass balance based on the principle of continuity: matter is neither created nor destroyed in macroscopic chemical, physical, and biological interactions.

1.2 MASS BALANCES

Water quality may be defined as "something inherent or distinctive about water." These distinctive characteristics can be chemical, physical, or biological parameters. Most water quality parameters are measured in mass quantities or concentration units (mg, mg L^{-1}, moles $liter^{-1}$). Thus we frequently use a mass balance to determine the fate of these parameters in natural waters and to assess degree of pollution expected under various conditions.

The fate of chemicals in the aquatic environment is determined by two factors: their reactivity and the rate of their physical transport through the environment. All mathematical models of the fate of chemicals are simply useful accounting procedures for the calculation of these processes as they become quite detailed. To the extent that we can accurately predict the chemical, biological, and physical reactions and transport of chemical substances, we can "model" their fate and persistence and the inevitable exposure to aquatic organisms.

Figure 1.1 is a schematic of the mass balance modeling approach to the solution of mass transport problems with chemical reaction. Key elements in a mass balance are defined below:

(i) A clearly defined control volume.
(ii) A knowledge of inputs and outputs that cross the boundary of the control volume.
(iii) A knowledge of the transport characteristics within the control volume and across its boundaries.
(iv) A knowledge of the reaction kinetics within the control volume.

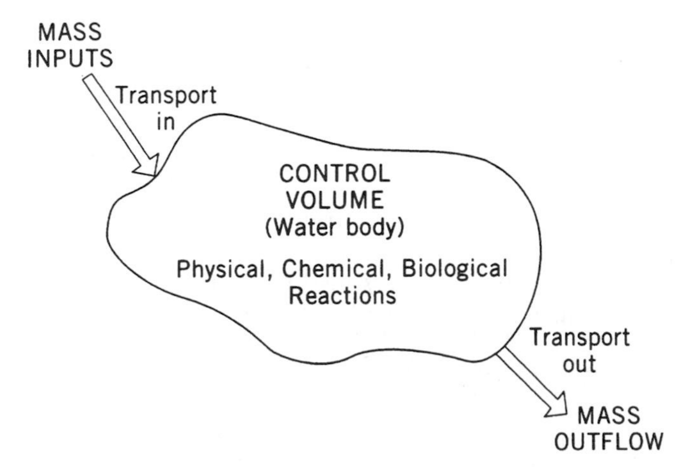

Figure 1.1 Generalized approach for mass balance models utilizing the control-volume concept and transport across boundaries.

A control volume can be as small as an infinitesimal thin slice of water in a swiftly flowing stream or as large as the entire body of oceans on the planet Earth. The important point is that the boundaries are clearly defined with respect to their location (element i) so that the volume is known and mass fluxes across the boundaries can be determined (element ii). Within the control volume, the transport characteristics (degree of mixing) must be known either by measurement or an estimate based on the hydrodynamics of the system. Likewise, the transport in adjacent or surrounding control volumes may contribute mass to the control volume (much as smoke can travel from another room to your room within a house), so transport across the boundaries of the control volume must be known or estimated (element iii).

A knowledge of the chemical, biological, and physical reactions that the substance can undergo within the control volume (element iv) is needed. If there were no degradation reactions taking place in aquatic ecosystems, every pollutant that was ever released to the environment would still be here to haunt us. Fortunately, there are natural purification processes that serve to assimilate some wastes and to ameliorate aquatic impacts. We must understand these reactions from a quantitative viewpoint in order to assess the potential damage to the environment from pollutant discharges and to allocate allowable limits for these discharges.

A mass balance is simply an accounting of mass inputs, outputs, reactions, and accumulation as described by the following equation:

$$\begin{array}{c}\text{Accumulation within} \\ \text{the control volume}\end{array} = \underbrace{\begin{array}{c}\text{Mass} \\ \text{inputs}\end{array} - \begin{array}{c}\text{Mass} \\ \text{outflows}\end{array}}_{\text{Transport}} \pm \text{Reactions} \qquad (1)$$

It is the subject of Chapter 2 to describe the mathematical formulation for the "Transport" terms in equation (1), and it is the subject of Chapters 3, 4, and 7 to describe environmental chemical models for the "Reactions" term in equation (1). Mass balances are based on first principles (continuity) and are the foundation for this entire book. Chapters 5–11 are applications and examples of the power and utility of this approach.

If a chemical is being formed within the control volume (such as the combination of two reactants to form a product, $A + B \rightarrow P$), then the algebraic sign in front of the "Reactions" term is positive when writing a mass balance for the product. If the chemical is being destroyed or degraded within the control volume, then the algebraic sign of the "Reactions" term is negative. If the chemical is conservative (i.e., nonreactive or inert), the "Reactions" term is zero.

$$\text{Accumulation} = \text{Inputs} - \text{Outflows} \pm \text{Reactions} \qquad (2)$$

A list of reactive and nonreactive chemicals are provided in Table 1.1, many of which are considered in later chapters.

If the system is at steady state (i.e., no change in concentration with respect to

Table 1.1 Classification of Substances Relative to Their Reactions in Water (Aqueous Phase)

Reaction Formation (+)	Reaction Degradation (–)	Conservative Substances
Products in chemical reactions	Reactants in chemical reactions	Rhodamine WT® dye
Algal growth	Biochemical oxygen demand (BOD)	Chloride, bromide
Bacterial growth	Radioisotope decay	Total dissolved solids (TDS)
Gas absorption	Particle sedimentation	Nonbiodegradable organics
Chemical desorption	Bacteria die-away	Total metal
	Organics degradations	Stable isotopes (^{15}N, ^{13}C)
	Gas stripping	
	Chemical adsorption	

time, $dC/dt = 0$), then there is no accumulation in the system and outflows are simply equal to inputs plus or minus reactions.

$$\text{Outflows} = \text{Inputs} \pm \text{Reactions} \tag{3}$$

The importance of an accurate mass balance for water cannot be overemphasized. Without a good water balance it is impossible to obtain an accurate mass balance for the aquatic chemical of interest. Water can be viewed as a conservative substance with numerous inputs and outflows from the water body. The accumulation of mass of water is termed the "change in storage." If the system is nearly isothermal, then the mass of storage is accounted for by the volume of inflows and outflows.

$$\Delta\text{Storage} = \sum\text{Inflows} - \sum\text{Outflows} + \text{Direct precipitation} - \text{Evaporation} \tag{4}$$

Inflows may include the volumetric inputs of tributaries and overland flow; outflows are all discharges from the water body; direct precipitation is the water that falls directly on the surface, while evaporation is the volume of water that leaves the surface of the water body to the atmosphere. ΔStorage can be measured in lakes or rivers by a change in elevation or stage. Inflows and outflows should be gaged or measured frequently during the period of investigation. Precipitation gages and evaporation pans can be utilized with sufficient accuracy to measure direct precipitation and evaporation. If the lake or stream basin is not sufficiently "tight" with respect to inputs or outflows to groundwater (GW), the piezometric surface of the groundwater adjacent to the water body must also be measured in order to determine the magnitude of the interaction.

$$\begin{aligned}\Delta\text{Storage} = {} & \sum\text{Inflows} + \text{GW Inputs} - \sum\text{Outflows} - \text{GW Outseepage} \\ & + \text{Direct precipitation} - \text{Evaporation}\end{aligned} \tag{5}$$

Equation (5) is represented schematically in Figure 1.2 showing various inflow and outflow routes in a hypothetical lake.

The interrelationship of groundwater and surface water and errors in terms of water budgets have been discussed in several papers by Winter.[16,17] In the best of situations it is possible to achieve an annual water balance within 5% (total inflows are within 5% of total outflows plus storage). A comprehensive water budget could be expressed by the following algebraic difference equation:

$$\Delta V = (\sum Q_{\text{in}} + Q_{\text{gw}} - \sum Q_{\text{out}} - Q_{\text{seep}} + IA - EA)\,\Delta t \tag{6}$$

where Q = flowrate, $m^3\ d^{-1}$
I = precipitation rate, $m\ d^{-1}$
A = surface area of water body, m^2
E = evaporation rate, $m\ d^{-1}$
Δt = time increment, days
ΔV = change in storage volume, m^3

The difference equation can be made into a differential equation if we divide both sides of the equation by Δt and take the limit as $\Delta t \to 0$.

$$\frac{dV}{dt} = \sum Q_{\text{in}} + Q_{\text{gw}} - \sum Q_{\text{out}} - Q_{\text{seep}} + IA - EA \tag{7}$$

Equation (7) may be integrated and solved for the volume as a function of all time.

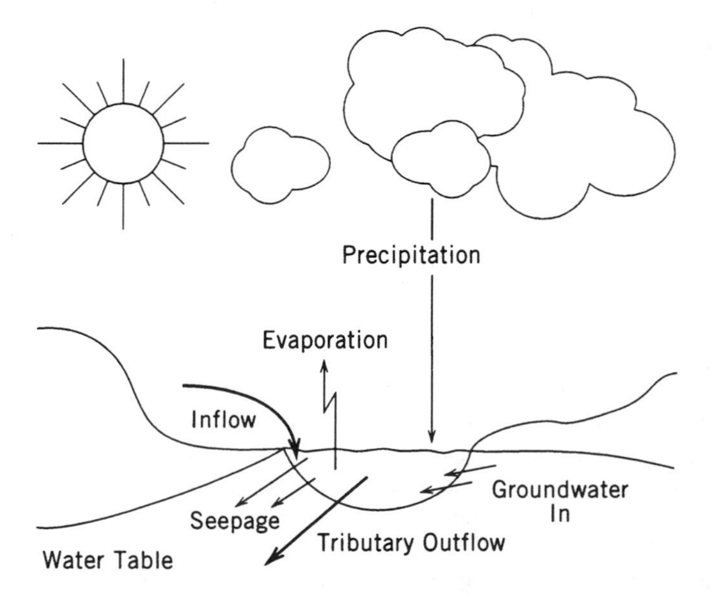

Figure 1.2 Schematic of a lake with inflows and outflows for computation of a water budget [equation (5)].

Example 1.1 Mass Balance of Water (Volume) in a Lake

Calculate the volume of a lake over time during a drought if the sum of all inputs is 100 m^3 s^{-1} and the outflows are 110 m^3 s^{-1} and increasing 1 m^3 s^{-1} every day due to evaporation and water demand. Initial volume of the lake is 1×10^9 m^3. See Figure 1.3. (*Note:* Convert all units from seconds to days.)

Solution:

$$\frac{dV}{dt} = \sum Q_{\text{in}} - \sum Q_{\text{out}}$$

$$\sum Q_{\text{in}} = 100 \text{ m}^3 \text{ s}^{-1} = 8.64 \times 10^6 \text{ m}^3 \text{ d}^{-1}$$

$$\sum Q_{\text{out}} = 110 \text{ m}^3 \text{ s}^{-1} + (1 \text{ m}^3 \text{ s}^{-1})\, t$$

$$= 9.50 \times 10^6 \text{ m}^3\text{d}^{-1} + (8.64 \times 10^4 \text{ m}^3\text{d}^{-1})\, t$$

$$\int_{V_0}^{V} dV = \int_0^t 8.64 \times 10^6 - (9.50 \times 10^6 + 8.64 \times 10^4\, t)\, dt$$

$$V - V_0 = \left| -0.864 \times 10^6\, t - \tfrac{1}{2}(8.64 \times 10^4\, t^2) \right|_0^t$$

$$V = 10^9 - 8.60 \times 10^5 t - 4.32 \times 10^4\, t^2$$

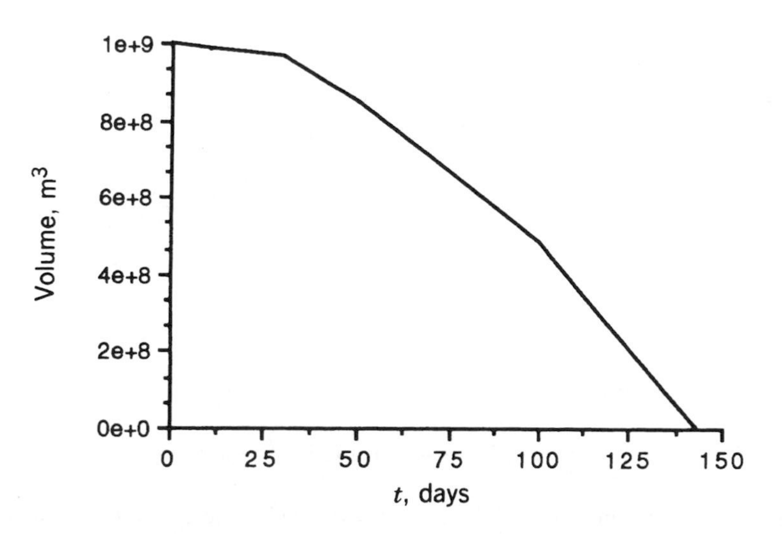

Figure 1.3 Plot of volume versus time for hypothetical lake during a drought period. This problem serves as a simple example of constructing a mass balance equation (water balance, in this case) and integrating to obtain the solution for the state variable (water volume as a function of time).

t, days	V, m^3
0	1×10^9
10	9.87×10^8
30	9.70×10^8
50	8.49×10^8
100	4.82×10^8
143	0

Example 1.2 Algebraic Mass Balance on Toxic Chemical in a Lake

Calculate the steady-state concentration of a toxic chemical in a lake under the following conditions. Assume steady state ($dC/dt = 0$) and constant volume ($Q_{in} = Q_{out}$) and a degradation rate of 50 kg d^{-1}.

Given

$$C_{in} = 100 \ \mu g \ L^{-1}$$
$$Q_{in} = Q_{out} = 10 \ m^3 \ s^{-1}$$
$$-\text{Rxn} = 50 \ kg \ d^{-1}$$

***Solution*:** Write the mass balance equation for the lake as a control volume [Equation (1)].

$$\text{Accumulation} = \text{Inputs} - \text{Outflows} \pm \text{Rxns}$$

$$\text{Accumulation} = 0 \text{ at steady state}$$

$$\text{Outflows} = \text{Inputs} - \text{Rxn (degradation)}$$

$$Q_{out} \times C_{out} = Q_{in} \times C_{in} - \text{Rxn}$$

$$(10 \ m^3 \ s^{-1}) \times C_{out} = (10 \ m^3 s^{-1})(100 \ \mu g \ L^{-1}) - 50 \ kg \ d^{-1}$$

$$Q_{in} \times C_{in} = \frac{10 \ m^3}{s} \ \frac{100 \ \mu g}{L} \ \frac{1000 \ L}{m^3} \ \frac{86{,}400 \ s}{d} \ \frac{g}{10^6 \ \mu g}$$

$$C_{out} = \frac{\text{Mass per day}}{Q_{out}} = \frac{36.4 \ kg \ d^{-1}}{10 \ m^3 \ s^{-1}}$$

Convert units into $\mu g \ L^{-1}$:

$$C_{out} = 42.1 \ \mu g \ L^{-1}$$

Simple mass balance models yield an expected concentration of chemical species. If the model is *steady state,* the answer is constant with time (e.g., Example 1.2). If the model is *time variable,* then the predicted state variable changes over time, as in Example 1.1. Spatially, mathematical models of aquatic chemicals may be one-, two-, or three-dimensional. Models can be homogeneous or heterogeneous in terms of the physical setting of the prototype (the natural system being modeled). For example, a groundwater aquifer may consist of sand deposits with a clay lense of different porosity and permeability.

Most of the models in this text are deterministic (i.e., they have one expected outcome for a given initial condition and model parameters). However, we will discuss uncertainty analysis using a probabilistic (stochastic) model that will allow prediction of not only the expected outcome (mean or best estimate) but also the variance of that estimate. In the future, mathematical models should provide decision-makers with a best estimate *and* a standard deviation (how certain one is of the results).

1.3 MODEL CALIBRATION AND VERIFICATION

To perform mathematical modeling of aquatic chemicals, four ingredients are necessary: (1) field data on chemical concentrations and mass discharge inputs, (2) a mathematical model formulation, (3) rate constants and equilibrium coefficients for the mathematical model, and (4) some performance criteria with which to judge the model.

Without field data, model calibration and verification are impossible. Depending on the ultimate use of the model, the amount of field reconnaissance varies. If the model is to be used for regulatory purposes, there should be enough field data to be confident of model results. Usually this requires two sets of field measurements, one for model calibration and one for verification under somewhat different circumstances (a different year of field measurements or an alternate site).

Model calibration involves a comparison between simulation results and field measurements. Model coefficients and rate constants should be chosen initially from literature or laboratory studies. Flow discharge rates are also needed as input to drive the model. After you run the model, a statistical comparison is made between model results for the state variables (chemical concentrations) and field measurements. If errors are within an acceptable tolerance level, the model is considered calibrated. If errors are not acceptable, rate constants and coefficients must be systematically varied (tuning the model) to obtain an acceptable simulation. The parameters should not be "tuned" outside the range of experimentally determined values reported in the literature. Thus the model is calibrated.

A few definitions may be helpful relating to model calibration and verification.

Mathematical model—a quantitative formulation of chemical, physical, and biological processes that simulates the system.

State variable—the dependent variable that is being modeled (in this context, usually a chemical concentration).

Model parameters—coefficients in the model that are used to formulate the mass balance equation (e.g., rate constants, equilibrium constants, stoichiometric ratios).

Model inputs—forcing functions or constants required to run the model (e.g., flowrate, input chemical concentrations, temperature, sunlight).

Calibration—a statistically acceptable comparison between model results and field measurements; adjustment or "tuning" of model parameters is allowed within the range of experimentally determined values reported in the literature.

Verification—a statistically acceptable comparison between model results and a second (independent) set of field data for another year or at an alternate site; model parameters are fixed and no further adjustment is allowed after the calibration step.

Simulation—use of the model with any input data set (even hypothetical input) and not requiring calibration or verification with field data.

Validation—scientific acceptance that (1) the model includes all major and salient processes, (2) the processes are formulated correctly, and (3) the model suitably describes observed phenomena for the use intended.

Robustness—utility of the model established after repeated applications under different circumstances and at different sites.

Post audit—a comparison of model predictions to future field measurements at that time.

Sensitivity analysis—determination of the effect of a small change in model parameters on the results (state variable), either by numerical simulation or mathematical techniques.

Uncertainty analysis—determination of the uncertainty (standard deviation) of the state variable expected value (mean) due to uncertainty in model parameters, inputs, or initial state via stochastic modeling techniques.

Statistical criteria for acceptance of model calibration and verification should be established *a priori,* before the simulations are begun. How "good" the model results are depends on desired use of the model or predictions. Likewise, criteria for acceptance of a calibration or verification depend on the intended use of the model. For example, a criterion for acceptance of a dissolved oxygen model calibration might be: "The prediction of dissolved oxygen concentration in the stream should be within $\pm$ 0.5 mg L^{-1} in at least 90% of the observations." There are several other types of statistical criteria that can be established.

- Statistical "goodness of fit" criteria using chi-square or Kolmogorov–Smirnov tests (tests of the sampling distribution of the variance).
- Paired *t*-tests of model results and field observations at the same time (a test of the means).
- Linear regression of paired data for model predictions and field observations at the same time.

- A comparison of model results to field observations and their standard deviation (or geometric deviation, if appropriate).
- Parameter estimation techniques such as nonlinear curve-fitting regressions (weighted or unweighted) or Kalman filters to determine model parameters in an optimal fashion (e.g., the sum of the square of the residuals is a minimum).

To verify the model, a statistical comparison between simulation results and a second set of field data is required. Coefficients and rate constants cannot be changed from the model calibration. This procedure provides some confidence that the model is performing acceptably. Performance criteria may be as simple as, "model results should be within one order of magnitude of the field concentrations at all times," or as stringent as, "the mean squared error of the residuals (difference between field measurements and model results) should be a minimum prescribed or optimal value."

Acceptance of a model calibration of verification does not necessarily imply that the model, itself, is validated. It is possible that the model works well under one set of circumstances but poorly under another. As the model is applied to different situations at various locations, we gain confidence in the model and its robustness. Exactly when the model is considered validated is a question akin to when does an adolescent become an adult. It is a gradual process. As the model is used, it is "flexed" in more and different ways, further testing its formulation and validity. Post audits of model results are an important test of the usefulness of the model. These audits are performed after model predictions have been made as data become available in the future. Very few post audits have been reported in the literature. More are needed.

Repeated model testing is crucial to gaining confidence in the model and to understanding its limitations. We have found that every model based on experience elsewhere fails miserably in Iowa!

Example 1.3 Calibration and Criteria Testing of a Dissolved Oxygen Model for a Hypothetical Stream

A waste discharge with biochemical oxygen demand (BOD) at km 0.0 causes a depletion in dissolved oxygen in a stream (D.O. sag curve). Model calibration results are tabulated below (D.O. model) together with field measurements (D.O. field) expressed in concentration units, mg L^{-1}. See Figure 1.4.

	Concentration, mg L^{-1}	
Distance, km	D.O. Model	D.O. Field
0	8.0	8.0
5	6.3	6.6
10	5.4	5.5
20	4.58	4.4

(continued)

Distance, km	Concentration, mg L^{-1} D.O. Model	D.O. Field
30	4.64	4.6
40	5.1	4.5
50	5.5	5.2
60	6.0	6.0
80	6.7	7.0
100	7.2	7.3

Problem: Determine if the model calibration is acceptable according to the following statistical criteria:

a. Chi-square goodness of fit at a 0.10 significance level (a 90% confidence level).
b. Paired t-test (difference between the mean and zero) at significance level $p = 0.10$.
c. Linear least-squares regression of model results (D.O. model on x-axis) versus observed data (D.O. field data on y-axis) with $r^2 > 0.8$.

Solution:

a. The chi-square goodness of fit test is based on the test statistic

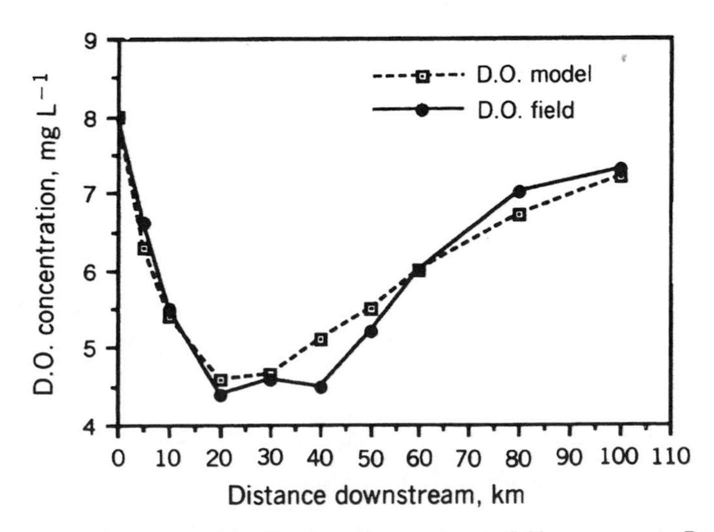

Figure 1.4 Dissolved oxygen model calibration and comparison to field measurements. Dashed line is the continuous model result, and the circled points are field measurements. A table of model results and field values at the point of measurement is given above.

$$\chi^2 = \sum_{i=1}^{n} \frac{(\text{observed value}_i - \text{expected value}_i)^2}{\text{expected value}}$$

where the observed values are the D.O. field data, and the expected values are the D.O. model results. In order to accept the model results as a good fit,

$$P(\chi^2 \le \chi_0^2) = 1 - \alpha$$

where α is the confidence level and $\le \chi_0^2$ is the chi-square distribution value for $n-1$ degrees of freedom. The criterion for the D.O. modeling effort is

$$P(\chi^2 \le 4.17) = 0.10$$

where $\chi_0^2 = 4.17$ for $n = 10$ and $\alpha = 0.9$. The value for $\chi_0^2 = 4.17$ was determined from a statistical table for the chi-square distribution with 9 degrees of freedom $(n-1)$ and $P = 0.10$.

The table below shows that

$$0.1254 \le 4.17$$

Therefore the model passes the goodness of fit test at a 0.10 significance level.

Distance	χ_i^2	d_i (obs$_i$ – expect$_i$)	d_i^2
0	0	0	0
5	0.0143	0.3	0.09
10	0.0019	0.1	0.01
20	0.0071	–0.18	0.0324
30	0.0003	–0.04	0.0016
40	0.0706	–0.6	0.36
50	0.0164	–0.3	0.09
60	0	0	0
80	0.0134	0.3	0.09
100	0.0014	0.1	0.01
	0.1254	–0.32	0.684

b. The paired t-test is used to test the difference between pairs of data at a specified confidence limit. The test statistic is

$$t = \frac{\bar{d}\sqrt{n}}{S_d}$$

where

$$\bar{d} = \frac{|\sum d_i|}{n}$$

$$d_i = \text{difference between values in data pair}_i$$

$$S_d = \sqrt{\frac{\sum d_i^2}{n-1} - \frac{n}{n-1}\,\bar{d}^2}$$

The acceptance criterion for the t-test is

$$P(t \leq t_0) = p$$

for $n - 1$ degrees of freedom. The criterion for the D.O. model is

$$P(t \leq 1.833) = 0.10$$

The value 1.833 was determined from a table of t-values with 9 degrees of freedom and $P = 0.10$.

The table above shows

$$\sum d_i = -0.32 \quad \text{and} \quad \sum d_i^2 = 0.684$$

The test statistic can be calculated:

$$\bar{d} = \frac{|-0.32|}{10} = 0.032$$

$$S_d = \sqrt{\frac{0.684}{9} - \frac{10}{9}\,(0.032)^2} = 0.2736$$

$$t = \frac{0.032\sqrt{10}}{0.2736} = 0.3699$$

The model results are found to be indistinguishable from the field data at a significance level of 0.10 from the paired t-test because

$$0.3699 \leq 1.833$$

There is less than a 10% probability that these two populations of data (model and field observations) could have been selected randomly from different distributions. The model meets the statistical criteria selected for means.

c. Perfect model predictions would yield

$$y = 1.0x + 0$$

Table 1.2 Table of Significant Reactions for Selected Priority Pollutant Organic Chemicals in Natural Waters[22]

	Biotrans-formations	Chemical Hydrolysis	Chemical Oxidation	Phototrans-formations	Volatili-zation	Sorption/ Bioconcen-tration
Pesticides						
Acrolein	x		x			
DDT—chlorinated hydrocarbon	x	x				x
Parathion-organo-P ester	x	x		x		
TCDD—tetra-chlorodibenzo-*p*-dioxin	x					x
Polychlorinated biphenyls (PCBs)						
Aroclor 1248	x				x	x
Halogenated aliphatic hydrocarbons						
Chloroform	x	x			x	
Halogenated ethers						
2-Chloroethyl vinyl ether	x	x	x		x	
Monocyclic aromatics						
2,4-Dimethylphenol	x					
Pentachlorophenol	x					x
Phthalate esters						
Bis(2-ethyl-hexyl)phthalate	x	x			x	x
Polycyclic aromatic hydrocarbons						
Anthracene	x		x	x	x	x
Benzo[*a*]pyrene	x		x	x		x
Nitrosamines and miscellaneous						
Benzidine	x		x	x		x
Dimethyl nitrosamine	x			x		

The D.O. model meets the linear regression criteria of $r^2 > 0.8$. It would have been better to have more observations (field data) to test the model. All three models become more powerful (in the statistical sense) as $n \rightarrow 30$ data points.

1.4 ENVIRONMENTAL MODELING AND ECOTOXICOLOGY

In environmental modeling, we seek to understand better the fate of chemicals in our environment and the role of humans in those chemical cycles. What distinguishes modern times from the past is not our negative impact on the environment, but rather the *magnitude* of that impact. We are tremendously powerful. The tools and industries of modern civilization can cause major change in a very short time. In industrialized nations, the anthropogenic energy flow per unit area exceeds photosynthesis by a factor of about 10. Organic chemicals produced by industry (~ 150 kg $capita^{-1}$ yr^{-1} or 40 g m^{-2} yr^{-1}) are within one order of magnitude of net primary production (~ 300 g m^{-2} yr^{-1}).[18–20] We are impacting larger and larger domains: oceans, not just coastal waters; the stratosphere, not just urban air; deep groundwater aquifers, not just surface waters.

Despite intensive research, we only partially understand how chemical pollutants move between atmosphere, land, and water and what transformations they undergo during transport. Some 1000–1500 new chemicals are manufactured each year with perhaps 60,000 chemicals in daily use.[21] Most of these chemicals are organic chemicals. A list of some priority pollutants (organic chemicals) is presented in Table 1.2[22] with their various transformation reactions. We have made progress during the past ten years at predicting rates of reaction and partitioning. From knowledge of only four parameters (octanol/water partition coefficient, Henry's law constant, dissociation constant, and sorption spectrum) one can estimate rate constants for many compounds via strict chemical hydrolysis, photodegradation, volatilization, sorption, and bioconcentration. However, less progress has been made on predicting perhaps the most important pathway, biotransformation. Quantitative structure–activity relationships for biological transformations are discussed in Chapter 7, but the quest is in its infancy.

Heavy metal pollutants are pervasive and, perhaps, are a greater problem than organic chemicals based on their persistence (total heavy metal is conserved). Often, aquatic concentrations are on the order of water quality criteria. Water quality criteria or standards can be violated under natural conditions in a few cases, such as ^{226}Ra and Ba in drinking waters and copper in cold-water fisheries. But generally it is human activities (water pollution and air pollution at the air–water interface) that cause elevated concentrations of metals such as mercury, lead, cadmium, copper, and zinc in surface waters (Figure 1.5).[23] Arsenic and selenium are leached from agricultural soils in the arid West, and concentrations in agricultural return water (irrigation) often exceed water quality standards and are toxic to biota. Phosphate, nitrate, and ammonium are nonpoint source problems from agricultural runoff that

Periodic Table

KEY
- Atomic number
- Species in fresh waters
- **Symbol**
- pConc. in US Rivers (-log M)
- Atomic Mass

IA	IIA	IIIB	IVB	VB	VIB	VIIB	VIIIB			IB	IIB	IIIA	IVA	VA	VIA	VIIA	O
1 H^+ **H** 7.8 1.0079																	2 **He** 4.0026
3 Li^+ **Li** 6.941	4 $BeOH^+$ **Be** 9.012											5 **B** 10.81	6 HCO_3^- **C** 2.7 12.011	7 NO_3^-, NH_4^+ **N** 4.5 14.007	8 O_2 **O** 3.5 15.9994	9 F^- **F** 5.3 18.9984	10 **Ne** 20.179
11 Na^+ **Na** 3.1 22.990	12 Mg^{2+} **Mg** 3.3 24.31											13 $Al(OH)_3$ **Al** 6.0 26.98	14 H_4SiO_4 **Si** 4.5 28.09	15 HPO_4^{2-} **P** 5.4 30.974	16 SO_4^{2-} **S** 3.4 32.064	17 Cl^- **Cl** 3.4 35.453	18 **Ar** 39.948
19 K^+ **K** 4.1 39.102	20 Ca^{2+} **Ca** 3.0 40.08	21 **Sc** 44.96	22 **Ti** 47.90	23 **V** 50.94	24 $Cr^{6+,3+}$ **Cr** 6.7 52.00	25 $Mn^{4+,2+}$ **Mn** 6.4 54.94	26 $Fe^{3+,2+}$ **Fe** 6.0 55.85	27 Co^{2+} **Co** 58.93	28 Ni^{2+} **Ni** 7.3 58.71	29 $Cu^{2+,+}$ **Cu** 7.0 63.546	30 Zn^{2+} **Zn** 6.6 65.38	31 **Ga** 69.72	32 **Ge** 72.59	33 $HAsO_4^{2-}$ **As** 7.9 74.92	34 SeO_3^{2-} **Se** 8.6 78.96	35 Br^- **Br** 5.9 79.904	36 **Kr** 83.80
37 **Rb** 85.47	38 Sr^{2+} **Sr** 87.62	39 **Y** 88.91	40 **Zr** 91.22	41 **Nb** 92.91	42 **Mo** 95.94	43 **Tc** 98.91	44 **Ru** 101.07	45 **Rh** 102.91	46 **Pd** 106.4	47 **Ag** 107.868	48 Cd^{2+} **Cd** 8.1 112.4	49 **In** 114.82	50 Sn^{2+} **Sn** 118.69	51 **Sb** 121.75	52 **Te** 127.60	53 I^-, IO_3^- **I** 126.90	54 **Xe** 131.30
55 Cs^+ **Cs** 132.91	56 Ba^{2+} **Ba** 6.0 137.34	57 **La** 138.91	72 **Hf** 178.49	73 **Ta** 180.95	74 **W** 183.85	75 **Re** 186.2	76 **Os** 190.2	77 **Ir** 192.2	78 **Pt** 195.09	79 **Au** 196.97	80 $Hg(OH)_2$, + **Hg** 8.0 200.59	81 **Tl** 204.37	82 Pb^{2+}, Pb^+ **Pb** 7.7 207.20	83 **Bi** 206.96	84 **Po** (209)	85 **At** (210)	86 **Rn** (222)
87 **Fr** (223)	88 **Ra** 226.0	89 **Ac** (227)	104 **Rf**	105 **Ha**	106												

←—— Transition Elements ——→ (IIIB to IIB)

Figure 1.5 Periodic table and average freshwater concentrations.[23]

continue to cause the eutrophication of surface waters, oxygen depletion of sediments, habitat alteration, and ecological changes in the structure and function of the ecosystem that are often difficult to detect, quantify, and prevent.

The ability of a trace element to pose an environmental hazard depends not only on its enrichment in the atmosphere or hydrosphere but also on its chemical speciation (form of occurrence) and the details of its biochemical cycling. Bioavailability and toxicity depend strongly on the chemical species. For algae and lower organisms, the free metal aquo ions often determine the physiological and ecological response.[24–26]

Particles are scavengers for reactive chemical species in transport from land to rivers and from continents to the ocean floor. Hydrous oxide surfaces, as well as organically coated surfaces, contain functional surface groups that act as coordinating sites for reactive elements. Metals and adsorption to hydrous oxides are discussed in Chapters 4 and 8.

At present the open ocean and many lakes are more affected by pollution impacts through tropospheric transport than through riverine transport. Elements are termed atmophile when their mass transport to the sea is greater from the atmosphere than from transport by streams. This is the case for Cd, Hg, As, Se, Cu, Zn, Sn, and Pb. Atmophile elements are either volatile, or their oxides or other compounds have low boiling points. The elements Hg, As, Se, Sn, and perhaps Pb can also become methylated and are released in gaseous form into the atmosphere. The elements Al, Ti, Mn, Co, Cr, V, and Ni are termed lithophiles because their mass transport to the ocean occurs primarily by streams.

Soft Lewis acids, metals such as Cu^+, Ag^+, Cd^{2+}, Zn^{2+}, Hg^{2+}, and Pb^{2+}, and the transition metal cations (Mn^{2+}, Fe^{2+}, Ni^{2+}, Cu^{2+}) are of environmental concern, both from a point of view of anthropogenic emissions as well as hazard to ecosystems and human health (chemical reactivity with biomolecules).[27] Of special concern are organometallic compounds such as organotin compounds.[28,29]

Considering the schematic reaction,

$$\text{Igneous rock} + \text{Volatile substances} = \text{Air} + \text{Seawater} + \text{Sediments}$$

Sillen[30] has postulated that volatiles such as H_2O, CO_2, HCl, and SO_2 (that have been emitted from volcanos after being leaked from the interior of the earth) have reacted in a gigantic acid–base reaction with the bases of the rocks (silicates, carbonates, oxides). Similarly, he calculated from a model system the quantities of reduction and oxidation components that have participated in a redox titration. On the global average, the environment with regard to a proton and electron balance is in a stationary situation, which reflects the present-day atmosphere (20.9% O_2, 0.03% CO_2, 79.1% N_2), an ocean pH of ~8, and a redox potential corresponding to a partial pressure of O_2 equal to 0.21 atm.

The weathering cycle is affected markedly at least locally and regionally by our civilization (Chapter 10). In local environments H^+ and e^- balances may become upset and significant variations in pH and pε occur. In the present competition between anthropogenic, geochemical, and biochemical processes, redox conditions in the atmosphere are disturbed by enhanced rates of artificial weathering of fossil fuel.[31] The combustion of these fuels leads to a disturbance in the electron (reduction–oxidation) balance. The reactions of the oxidation of C, S, and N exceed reduction reactions in these elemental cycles. A net production of hydrogen ions (acids) in atmospheric precipitation is a necessary consequence.[32] Furthermore, many more potential atmospheric pollutants (photooxidants, polycyclic aromatic hydrocarbons, smog particles, etc.) are formed under the influence of photochemically induced interactions with OH radicals, H_2O_2, ozone, and hydrocarbons with fossil fuel combustion products. The disturbance is transferred to the terrestrial, aquatic (mostly freshwater), estuarine, or coastal marine environment. Atmospheric acid deposition creates an additional input of hydrogen ions and sulfate and nitrate (sulfuric and nitric acid) to terrestrial and aquatic ecosystems.

The atmosphere has become an important conveyor belt for many potential aquatic pollutants. Many persistent pollutants are present in a vapor phase during transport from land to fresh water and from continent to ocean. Even substances with vapor pressures as low as 10^{-10} atm are released into the atmosphere. These substances include many pesticides, such as DDT, more volatile metals (Hg), metalloids (As, Se), or their compounds. At present the open ocean is probably more affected by metal pollution inputs through tropospheric transport than through river transport (Pb). The hydrogeochemical cycles couple land, water, and air and make these reservoirs interdependent (Figure 1.6).

In Figure 1.6 the sizes of the various reservoirs, measured in number of molecules or atoms, are compared. The mean residence time of the molecules in these reservoirs is also indicated. The smaller the relative reservoir size and residence time, the more sensitive the reservoir toward perturbation. Obviously, the atmosphere, living biomass (mostly forests), and ground and surface fresh waters are most sensitive to perturbation. The anthropogenic exploitation of the larger sedimentary organic carbon reservoir (fossil fuels and by-products of their combustion such as oxides and heavy metals and the synthetic chemicals derived from organic carbon) can above all affect the small reservoirs. Over the past years we have started to recognize that biosphere processes play an important role in coupling the cycles of essential elements and in regulating the chemistry and physics of our environment. The living biomass (Figure 1.6) is a relatively small reservoir and thus subject to human interference; each species forming the biosphere requires specific environmental conditions for sustenance and survival.[33]

Transport of pollutants from air to water and from land to water have become increasingly important pathways for the occurrence of water pollution (Figure 1.7). Degradation of groundwater from soil pollution is a major environmental problem (e.g., infiltration of pesticides from agricultural applications or leachate from hazardous waste landfills). Also, impacts of acid deposition on surface waters and oxidants on forests and soils illustrate the importance of transport through the air–water interface. We need to know about the aquatic chemistry of these pollutants to estimate their speciation and fate in the environment, and we need to know how to

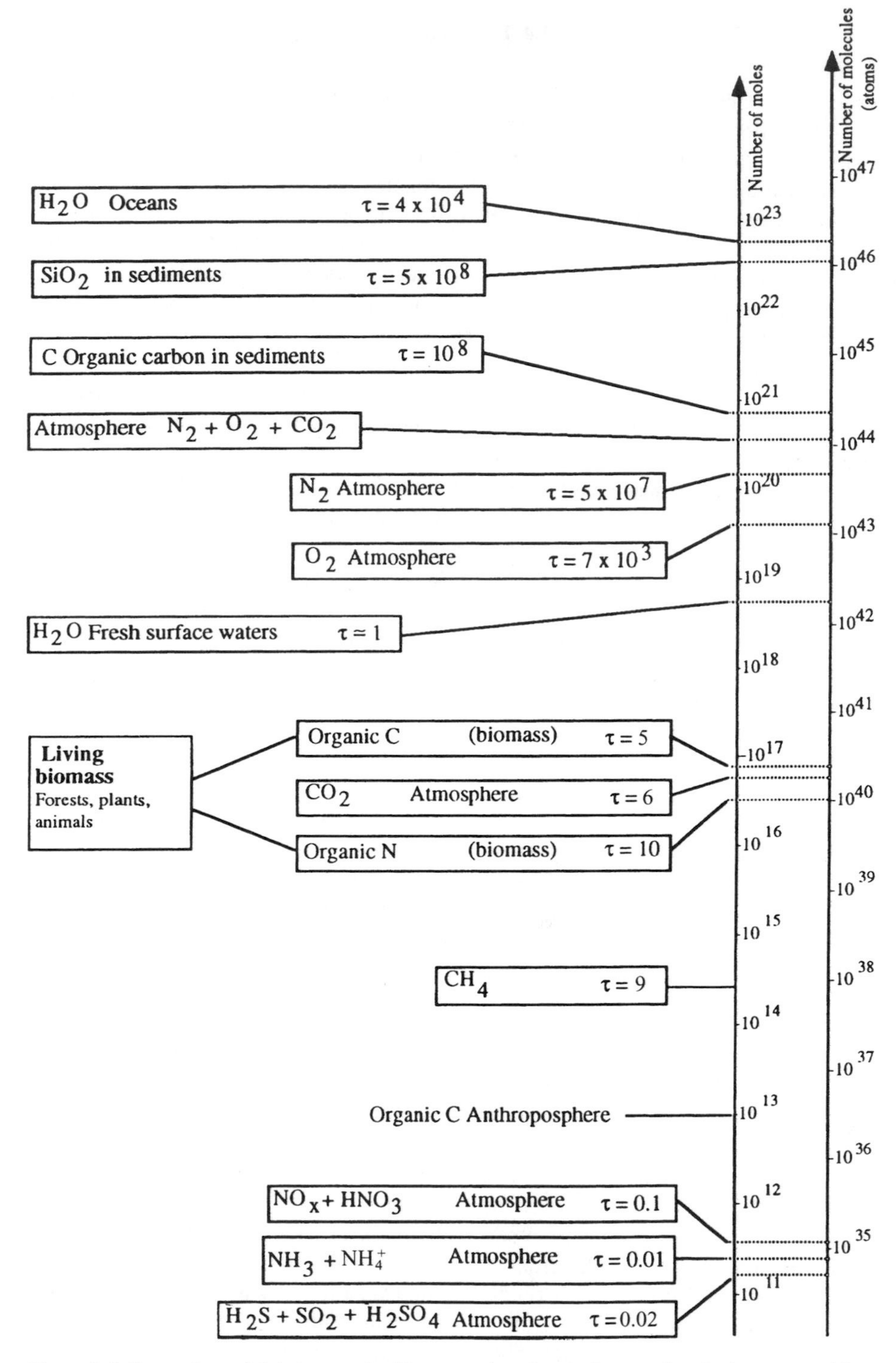

Figure 1.6 Comparison of global reservoirs. The reservoirs of atmosphere, surface fresh waters, and living biomass are significantly smaller than the reservoirs of sediment and marine waters. The total groundwater reservoir may be twice that of fresh water. However, groundwater is much less accessible. (τ = respective residence time [years] of molecule [atoms].)[35]

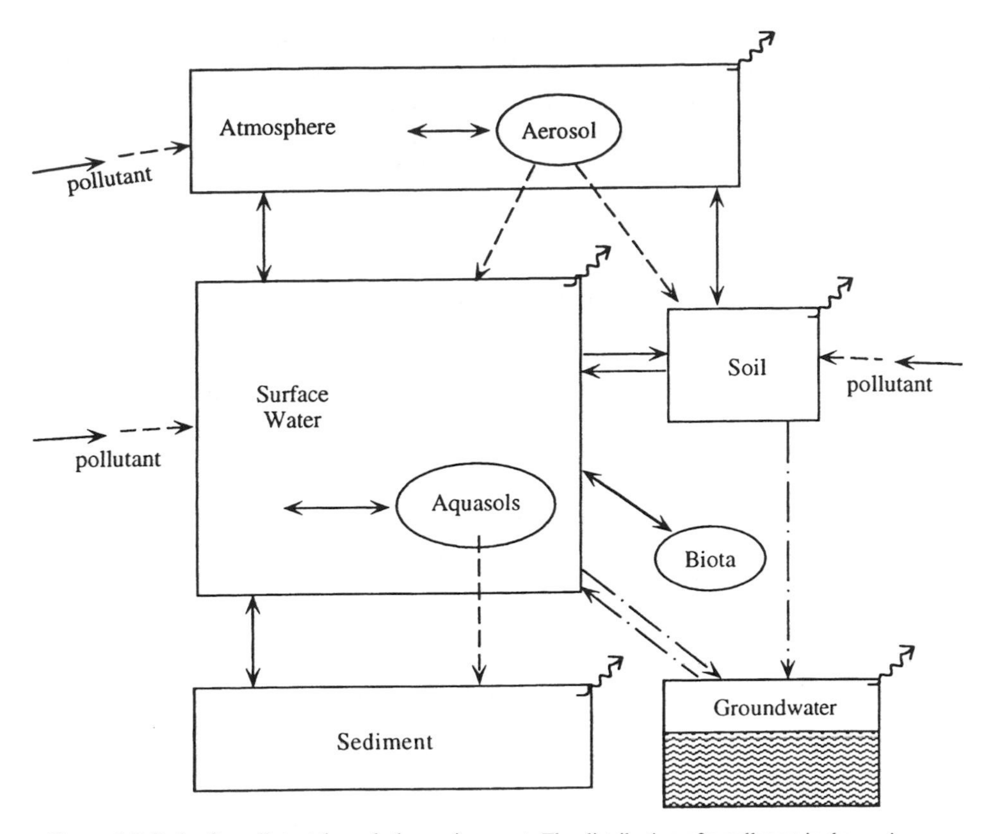

Figure 1.7 Path of a pollutant through the environment. The distribution of a pollutant in the environment is dependent on its specific properties. Of particular ecological relevance is fat solubility or, in other words, lipophilicity, as lipophilic substances accumulate in organisms and the food chain. Biodegradation and chemical or photochemical decomposition (indicated by ∿∿→), on the other hand, decrease residence time and residual concentrations. From Sigg and Stumm.[35]

construct and solve mathematical models (mass balance equations) to calculate simultaneously their transport and transformations.

Figure 1.8 provides a synopsis of the perspectives of aquatic ecotoxicology. Let us follow the various steps from the source to the potential ecological effects of a pollutant released into an aquatic ecosystem. The emission is measured as a flow or input load (capacity factor; mass per unit time or mass per unit time and volume or area). The resulting concentration is a consequence of the dilution, transport, and transformation of this chemical. At any point, the water condition is characterized by the interacting physical, chemical, and biological factors.

The concentrations of these variables determine the doses (concentration × time) an organism or an ecosystem will receive and, in turn, the type of community of organisms present in the ecosystem.[34] Thus, for an assessment of exposure and an evaluation of the ecological and human health risk, we need to know the residual concentrations of pollutants. An estimate of the ambient concentration of a pollutant based on a prediction of its relevant fate and residence time (considering relevant pathways, exchange processes, biological and chemical conversions), or on the basis of analytical determinations in the receiving waters, is essential to any risk assessment.

The ecological damage of a substance depends on its interaction with organisms or with communities of organisms (Figure 1.8). The intensity of this interaction depends on the specific structure and activity of the substance under consideration, but other factors such as temperature, turbulence, and the presence of other substances are also important. A thorough knowledge of the basic mechanisms by which toxic agents exert their injurious effects (pharmacokinetics, uptake, conversion, degradation and separation, disturbance of regulatory mechanisms in the biological community) is fundamental for an understanding of biological responses. Comparative toxicological research is necessary in order to extrapolate effects ob-

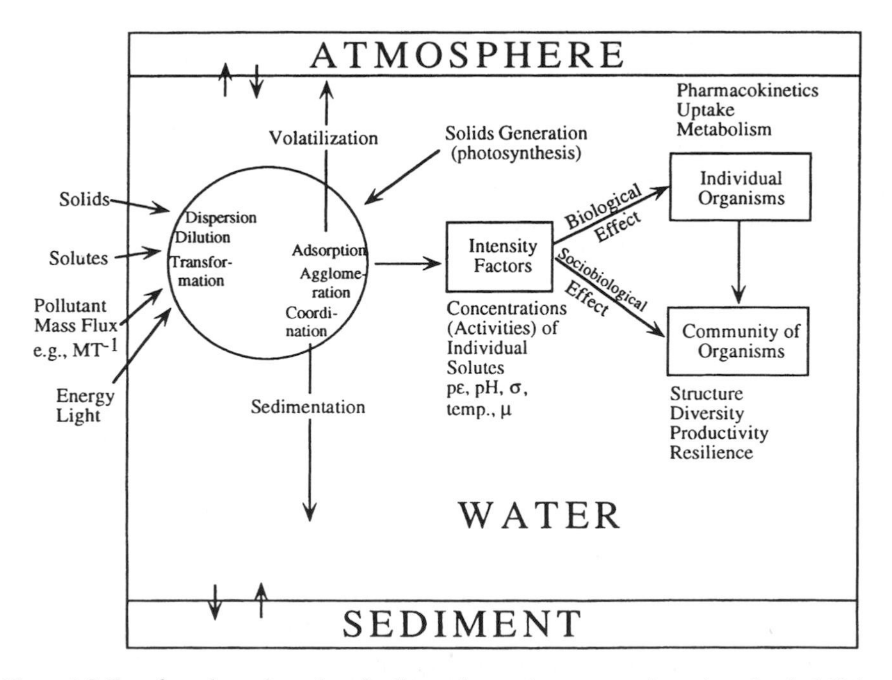

Figure 1.8 Transfer and transformation of pollutants in aquatic ecosystems by various chemical, biological, and physical processes and pathways. A substance introduced into the system becomes dispersed or diluted. It can become eliminated from the water by adsorption on particles or by volatilization. It may also be transformed chemically or biologically. From Stumm et al.[34]

served in laboratory experiments to organisms in nature, from one organism to another or to humans.

The natural distribution of organisms depends primarily on their ability to compete under given conditions and not merely on their ability to survive the physical and chemical environment. A population will be eliminated when its competitive power is reduced to such an extent that it can be replaced by another species. The competitive abilities of an organism are based on its reproductive rate (which is related to food and physiological potential), and the mortality rate from all causes, including predation and imposed toxicity. There are many ways in which an organism can die, but there is only a very narrow range of ways in which it can survive and leave offspring. Thus, in an ecosystem, a population may be eliminated by the presence of pollutants even at apparently trivial toxicity levels if its competitive ability is marginal, or if it is the most sensitive of the competitors.

Often, contaminants at very low concentrations cause changes in the structure of the population by interfering through chemotaxis with interorganismic communication. For example, the survival of a fish population may be rendered impossible by a pollutant (even if it exhibits neither acute nor chronic toxicity to the particular species of fish) if it impairs the food source (zooplankton) or disturbs chemotactical stimuli or mimics wrong signals (and thus, for example, interferes with food finding).

As a consequence of the many microhabitats (niches) that are typically present in a "healthy" water, many species can survive. Because of interspecies competition, most species are present in a low population density. Pollution destroys microhabitats, diminishes the chance of survival for some of the species, and thus in turn reduces the competition; the more tolerant species become more numerous. This shift in the frequency distribution of the species toward a lower diversity of the ecosystem is a general consequence of the chemical impact on waters by substances not indigenous to nature.

An understanding of the interaction of chemical compounds in the natural system hinges on the recognition of the compositional complexity of the environment. This requires an adequate analytical methodology, especially the ability to predict individual components (chemical species) selectively, to measure them accurately and with sensitivity, and to forecast their fate with environmental models. Table 1.3 lists water quality criteria toxicity thresholds, carcinogenicity, and maximum contaminant levels (M.C.L.) for many toxic chemicals discharged to natural waters. It was updated by EPA in 1994. Water quality criteria are the best scientific information from toxicological studies of the maximum concentration allowable that will not cause an observable biological effect. Water quality standards are enforceable by law; they include the water quality criteria, a designated use for the water body, and a nondegradation clause. As ecotoxicology becomes more sophisticated as a science, the list of chemicals will grow and species specific criteria will be promulgated under various environmental conditions.

Table 1.3 Water Quality Criteria and Acute and Chronic Threshold Levels for Various Toxicants[13]

	WQ Criteria, Concentrations in µg L^{-1}						Human Health Criteria, units per liter				
	Priority Pollutant	Carci nogen	Fresh Acute Criteria	Fresh Chronic Criteria	Marine Acute Criteria	Marine Chronic Criteria	Water and Organisms	Organisms Only	Drinking Water MCL	Date/ Reference[a]	Number of States with Aquatic Life Standard
Acenapthene	Y	N	1,700.[b]	520.[b]	970.[b]	710.[b]				1980/FR	1
Acrolein	Y	N	68.[b]	21.	55.[b]		320. µg[c]	780. µg		1980/FR	1
Acrylonitrile	Y	Y	7,550.[b]	2,600.[b]			0.058 µg[b]	0.65 µg		1980/FR	
Aldrin	Y	Y	3.0		1.3		0.074 ng[c]	0.079 ng[c]		1980/FR	16
Alkalinity	N	N		20,000.						1976/RB	
Ammonia	N	N								1985/FR	24
Antimony	Y	N	88.[b]	30.[b]	1,500	500	146. µg	45,000. µg		1980/FR	1
Arsenic	Y	Y					2.2 ng[c]	17.5 ng[c]	0.05 mg	1980/FR	21
Arsenic (pent)	Y	Y	850.[b]		2,319.[b]					1985/FR	21
Arsenic (tri)	Y	Y	360.	190.	69.	36.				1985/FR	21
Asbestos	Y	Y					30 kf L^{-1}[c]		7 mf/L	1980/FR	
Bacteria	N	N							<1/100ml	1986/FR	56
Barium	N	N					1. mg		2.0 mg	1976/RB	8
Benzene	Y	Y	5,300.[b]		5,100.[b]	700.	0.66 µg[c]	40. µg[c]	5 µg[d]	1980/FR	1
Benzidine	Y	Y	2,500.[b]				0.12 ng[c]	0.53 ng[c]		1980/FR	6
Beryllium	Y	Y	130.[b]	5.3[b]			3.7 ng[c]	64. ng[c]	4 µg	1980/FR	8
BHC	Y	N	100.[b]		0.34[b]					1980/FR	
Cadmium	Y	N	3.9[f]	1.1[f]	43.	9.3	10. µg		0.005 mg	1985/FR	21
Carbon tetrachloride	Y	Y	35,200.[b]		50,000.[b]		0.4 µg[c]	6.94 µg[c]	5 µg[d]	1980/FR	1
Chlordane	Y	Y	2.4	0.0043	0.09	0.004	0.46 ng[c]	0.48 ng[c]	2 µg[d]	1980/FR	12
Chlorinated benzenes	Y	Y	250.[b]	50.[b]	160.[b]	129.	488. µg[b]		75-100 µg[d]	1980/FR	1
Chlorinated naphthalenes	Y	N	1,600.[b]		7.5[b]					1980/FR	1
Chlorine	N	N	19.	11.	13.	7.5				1985/FR	21

Chloroalkyl ethers	Y	N	238,000.[b]							1980/FR	1
Chloroethyl ether (bis-2)	Y	Y					0.03 μg[c]	1.36 μg[c]		1980/FR	
Chloroform	Y	Y	28,900.[b]	1,240.[b]			0.19 μg[c]	15.7 μg[c]	100 μg[d]	1980/FR	1
Chlorophenol, 2-	Y	N	4,380.[b]							1980/FR	1
Chlorophenol, 4-	N	N			29,700[b]					1980/FR	
Chlorophenoxy herbicides (2,4,5,-TP)	N	N					10. μg		50 μg[d]	1980/FR	
Chlorophenoxy herbicides (2,4-D)	N	N					100. μg		70 μg[d]	1976/RB	7
Chloropyrifos	N	N	0.083	0.041	0.011	0.0056				1986/FR	
Chloro-4 methyl-3 phenol	N	N	30.[b]							1980/FR	
Chromium (hex)	Y	N	16.	11.	1,100.	50.	50. μg		0.10 mg	1985/FR	24
Chromium (tri)	Y	N	1,700.[f]	210.[f]	10,300.[b]		170. mg	3,433. mg	0.10 mg	1985/FR	24
Color	N	N								1976/RB	
Copper	Y	N	18.[f]	12.[f]	2.9				1.3 mg	1985/FR	20
Cyanide	Y	N	22.	5.2	1.		200. μg		0.2 mg	1985/FR	23
DDT	Y	Y	1.1	0.001	0.13	0.001	0.024 ng[c]	0.024 ng[c]		1980/FR	16
DDT metabolite (DDE)	Y	Y	1,050.[b]		14.[b]					1980/FR	
DDT metabolite (TDE)	Y	Y	0.6[b]		3.6[b]					1980/FR	
Demeton	Y	N		0.1		0.1				1976/RB	
Dibutylphthalate	Y	N					3.5. mg	154. mg		1980/FR	
Dichlorobenzenes	Y	N	1,120.[b]	763.[b]	1,970.[b]		400. μg	2.6 mg	0.075–0.6 mg[d]	1980/FR	1
Dichlorobenzidine	Y	Y					0.01 μg[c]	0.020 μg[c]		1980/FR	1
Dichloroethane 1,2	Y	Y	118,000.[b]	20,000.[b]	113,000.[b]		0.94 μg[c]	243. μg[c]	5 μg[d]	1980/FR	1
Dichloroethylenes	Y	Y	11,600.[b]		224,000.[b]		0.033 μg[c]	1.85 μg[c]	7-100 μg[d]	1980/FR	1
Dichlorophenol 2,4	Y	N	2,020.[b]	365[b]			3.09 mg			1980/FR	1
Dichloropropane	N	N	23,000.[b]	5,700.[b]	10,300.[b]	3,040.[b]			5 μg[d]	1980/FR	1
Dichloropropene	Y	N	6,060.[b]	244.[b]	790.[b]		87. μg	14.1 mg		1980/FR	1
Dieldrin	Y	Y	2.5	0.0019	0.71	.0019	0.071 ng[c]	0.76 ng[c]		1980/FR	16

(continued)

Table 1.3 *(Contimued)*

			WQ Criteria, Concentrations in $\mu g\ L^{-1}$				Human Health Criteria, units per liter				
	Priority Pollutant	Carci nogen	Fresh Acute Criteria	Fresh Chronic Criteria	Marine Acute Criteria	Marine Chronic Criteria	Water and Organisms	Organisms Only	Drinking Water MCL	Date/ Reference[a]	Number of States with Aquatic Life Standard
Diethylphthalate	Y	N					350. mg	1.8g		1980/FR	
Dimethylphenol 2,4	Y	N	2,120.[b]							1980/FR	
Dimethylphthalate	Y	N					313. mg	2.9g		1980/FR	
Dinitrophenol	Y	N					70. μg	14.3 mg		1980/FR	
Dinitrotoluene 2,4	N	Y	330.[b]	230.[b]	590.[b]	370.[b]	0.11 μg	9.1 μg[c]		1980/FR	1
Dinitro-*o*-cresol 2,4	Y	N					13.4 μg	765. μg		1980/FR	
Dioxin (2,3,7,8-TCDD)	Y	Y	<0.01[b]	<0.00001[b]			0.000013 ng[c]	0.000014 ng[c]		1984/FR	1
Diphenylhydrazine 1,2	Y	Y	270.[b]				42 ng	0.56 μg		1980/FR	
Di-2-ethylhexylphthalate	Y	Y	2,000.	160.			15. mg	50. mg		1980/FR	
Endosulfan	N	N	0.22	0.056	0.034	0.0087	74. μg	159. μg		1980/FR	10
Endrin	Y	N	0.18	0.0023	0.037	0.0023	1. μg		2.0 μg[d]	1980/FR	18
Ethylbenzene	Y	N	32,000.[b]		430.[b]		1.4 mg	3.28 mg	0.7 mg[d]	1980/FR	
Fluoranthene	Y	N	3,980[b]		40.[b]	16.	42.[b] μg	54. μg		1980/FR	1
Gases, total dissolved	N	N								1976/RB	
Guthion	N	N		0.01		0.01				1976/RB	8
Haloethers	N	N	360.[b]	122.[b]						1980/FR	
Halomethanes	N	Y	11,000.[b]		12,000[b]	6,400.	0.19 μg[bc]	15.7 μg[c]		1980/FR	
Heptachlor	Y	Y	0.52	0.0038	0.053	0.0036	0.28 ng[c]	0.29 ng[c]	0.4 μg[d]	1980/FR	12
Hexachloroethane	N	Y	980.[b]	540.[b]	940.[b]		1.9 μg[bc]	8.74 μg		1980/FR	1
Hexachlorobenzene	Y	Y	6.0	3.68			0.72 ng[c]	0.74 ng[c]	1.0 μg	1980/FR	
Hexachlorobutadiene	Y	Y	90.[b]	9.3[b]	32.[b]		0.45 μg[c]	50. μg[c]		1980/FR	2
Hexachlorocyclohexane) (lindane)	Y	Y	2.0	0.08	0.16		18.6 ng	62.5 ng	0.2 μg	1980/FR	12

Hexachlorocyclohexane-alpha	Y	Y					9.2 ng[c]	31. ng[c]		1980/FR	
Hexachlorocyclohexane-beta	Y	Y					16.3 ng[c]	54.7 ng[c]		1980/FR	
Hexachlorocyclohexane-gamma	Y	Y	2.0	0.08	0.16		18.6 ng[c]	62.5 ng[c]		1980/FR	
Hexachlorocyclohexane-technical	N	Y					12.3 ng[c]	41.4 ng[c]		1980/FR	
Hexachlorocyclopentadiene	Y	N	7.[b]	5.2[b]	7.[b]		206. µg			1980/FR	3
Iron	N	N		1,000.			0.3 mg			1976/RB	15
Isophorone	Y	N	117,000.[b]		12,900.[b]		5.2 mg	520. mg		1980/FR	
Lead	Y	N	82.[f]	3.2[f]	220.	8.5	50. µg		0.05 mg	1985/FR	20
Malathion	N	N		0.1		0.1				1976/RB	7
Manganese	N	N					50. µg	100. µg		1976/RB	7
Mercury	Y	N	2.4	0.012	2.1	0.025	144. ng	146. ng	0.002 mg	1985/FR	17
Methoxychlor	N	N		0.03		0.03	100. µg		40 µg[d]	1976/RB	12
Mirex	N	N		0.001		0.001				1976/RB	7
Naphthalene	Y	N	2,300.[b]	620.[b]	2,350.[b]					1980/FR	1
Nickel	Y	N	1,400.[f]	160[f]	75	8.3	13.4 µg	100. µg		1986/FR	10
Nitrates	N	N					10. mg		45. mg	1976/RB	5
Nitrobenzene	Y	N	27,000.[b]		6,680.[b]		19.8 mg			1980/FR	1
Nitrophenols	Y	N	230.[b]	150.[b]	4,850.[b]					1980/FR	1
Nitrosamines	Y	Y	5,850.[b]		3,300,000[b]		0.8 ng[c]	1240. ng[c]		1980/FR	1
Nitrosodibutylamine	N	Y					6.4 ng[c]	587. ng[c]		1980/FR	
Nitrosodiethylamine	N	Y					0.0008 ng[c]	1.2. ng[c]		1980/FR	
Nitrosodimethylamine	Y	Y					1.4 ng[c]	16,000. ng[c]		1980/FR	
Nitrosodiphenylamine	Y	Y					4,900. ng[c]	16,100. ng[c]		1980/FR	
Nitrosopyrolidine	N	Y					16. ng[c]	91,900. ng[c]		1980/FR	

(continued)

Table 1.3 *(Contimued)*

			WQ Criteria, Concentrations in $\mu g\ L^{-1}$				Human Health Criteria, units per liter				
	Priority Pollutant	Carci nogen	Fresh Acute Criteria	Fresh Chronic Criteria	Marine Acute Criteria	Marine Chronic Criteria	Water and Organisms	Organisms Only	Drinking Water MCL	Date/ Reference[a]	Number of States with Aquatic Life Standard
Oil and grease	N	N								1976/RB	56
Oxygen dissolved	N	N		5,000	4,000					1986/FR	56
Parathion	N	N	0.065	0.013						1986/FR	8
PCBs	Y	Y	2.0	0.014	10.	0.03	0.079 ng[c]	0.079 ng[c]	0.5 μg[d]	1980/FR	16
Pentachlorinated ethane	N	N	7,240.[b]	1,100.[b]	390.[b]	281.[b]				1980/FR	1
Pentachlorobenzene	N	N					74. μg	85. μg		1980/FR	
Pentachlorophenol	Y	N	20.[g]	13.[g]	13.	7.9[b]	1.01 mg		1.0 μg[d]	1986/FR	2
pH	N	N		6.5-9		6.5–8.5	5–9			1976/RB	56
Phenol	Y	N	10,200.[b]	2,560.[b]	5,800.[b]		3.5 mg			1980/FR	23
Phosphorus elemental	N	N				0.1				1976/RB	
Polynuclear aromatic hydrocarbons	Y	Y			300.[b]		2.8 ng[c]	31.1 ng[c]		1980/FR	1
Selenium	Y	N	20.	5.0	300.	71.	10. μg		50. μg	1980/FR	15
Silver	Y	N	4.1[f]		0.12	2.3	0.92 μg	50. μg		1980/FR	14
Solids dissolved and salinity	N	N					250. mg			1976/RB	56
Solids suspended and turbidity	N	N								1976/RB	44

Sulfide–hydrogen sulfide	N	N		2.		2.				1976/RB	
Temperature	N	N	SPECIES-DEPENDENT CRITERIA							1976/RB	56
Tetrachlorobenzene 1,2,4,5	N	N					38. µg	48. µg		1980/FR	
Tetrachloroethane 1,1,2,2	Y	Y		2,400.[b]	9,020.[b]		0.17 µg[c]	10.7 µg[c]		1980/FR	1
Tetrachloroethanes	Y	N	9,320.[b]							1980/FR	1
Tetrachloroethylene	Y	Y	5,280.[b]	840.[b]	10,200.[b]	450.	0.8 µg[bc]	8.85 µg[c]	5 µg[d]	1980/FR	1
Tetrachlorophenol 2,3,5,6	N	N				440.[b]				1980/FR	
Thallium	Y	N	1,400.[b]	40.[b]	2,130.[b]		13. µg	48. µg	2 µg	1980/FR	2
Toluene	Y	N	17,500.[b]		6,300.[b]	5,000.[b]	14.3 mg	424. mg	1.0 mg[d]	1980/FR	1
Toxaphene	Y	Y	0.73	0.0002	0.21	0.0002	0.71 ng[c]	0.73 ng[c]	5 µg[d]	1986/FR	17
Trichlorinated ethanes	Y	Y	18,000.[b]							1980/FR	
Trichloroethane 1,1,1	Y	N			31,200.[b]		18.4 mg	1.03g	0.2 mg[d]	1980/FR	1
Trichloroethane 1,1,2	Y	Y		9,400.[b]			0.6 µg[c]	41.8 µg[c]	5 µg	1980/FR	1
Trichloroethylene	Y	Y	45,000.[b]	21,900.[b]	2,000.[b]		2.7 µg[c]	80.7 µg[c]	5 µg[d]	1980/FR	1
Trichlorophenol 2,4,5	N	N	100.	63.	240.	11.	2,600. µg			1980/FR	
Trichlorophenol 2,4,6	Y	Y		970.[b]			1.2 µg[c]	3.6 µg[c]		1980/FR	
Vinyl chloride	Y	Y					2. µg[c]	525. µg[c]	2 µg[d]	1980/FR	
Zinc	Y	N	120.[f]	110[f]	95	86				1987/FR	19

g, grams; mg, miligrams; µg, micrograms; ng, nanograms; f, fibers; Y, Yes; N, No; MCL, Maximum contaminant level.

[a] FR, *Federal Register*; RB, Quality Criteria for Water, 1976 (Redbook).

[b] Insufficient data to develop criteria. Value presented is the LOEL—Lowest observed effect level.

[c] Human health criteria for carcinogens reported for three risk levels. Value presented is the 10^{-6} risk level.

[d] MCL established by 1992.

[e] Classified as carcinogen, 1994.

[f] Hardness-dependent criteria (100 mg L^{-1} used).

[g] pH-dependent criteria (7.8 pH used).

1.5 REFERENCES

1. Streeter, H.W., and Phelps, E.B., A Study of the Pollution and Natural Purification of the Ohio River, III, Factors Concerned in the Phenomena of Oxidation and Reaeration, U.S. Public Health Service, Public Health Bulletin No. 146 (1925).
2. O'Connor, D.J., and Dobbins, W.E., *Trans. Am. Soc. Civ. Eng.,* 123, 655 (1958).
3. O'Connor, D.J., and Mueller, J.A., *J. Sanit. Eng. Div., Am. Soc. Civ. Eng.,* 96, 955 (1970).
4. Vollenweider, R.A., *Schweiz. Z. Hydrol.* 37, 53 (1975).
5. Chapra, S.C., *J. Environ. Eng. Div. Am. Soc. Civ. Eng.,* 103, 147 (1977).
6. Burns, L.A., and Cline, D.M., Exposure Analysis Modeling System Reference Manual for EXAMS II, EPA-600/3-85-038, U.S. Environmental Protection Agency, Athens, GA (1985).
7. Thomann, R.V., and Connolly, J.P., *Environ. Sci. Technol.,* 18:2, 65 (1984).
8. DiToro, D.M., and Connolly, J.P., Mathematical Models of Water Quality in Large Lakes, Part 2: Lake Erie, EPA-600/3-80-065, U.S. Environmental Protection Agency, Duluth, MN (1980).
9. Mackay, D., *Environ. Sci. Technol.,* 16:12, 654A (1982).
10. Morel, F.M.M., Westall, J.C., O'Melia, C.R., and Morgan, J.J., Environ. Sci. Technol., 9, 757 (1975).
11. Felmy, A.R., MINTEQ—A Computer Program for Calculating Aqueous Geochemical Equilibria, EPA-600/3-84-032, U.S. Environmental Protection Agency, Athens, GA (1984).
12. U.S. Environmental Protection Agency, Notice of Final Ambient Water Quality Criteria Documents. *Fed. Reg.,* Vol. 50, No. 145, July 29, 1985.
13. Office of Water Regulations and Standards, Quality Criteria for Water, EPA-440/5-86-001, U.S. Environmental Protection Agency, Washington, DC (1986).
14. Carlson, A.R., Hammermeister, D., and Nelson, H., *Environ. Toxicol. Chem.,* 5, 997 (1986).
15. Hedtke, S.F., and Arthur, J.W., in *Aquatic Toxicology and Hazard Assessment: Seventh Symposium,* ASTM STP 854, R.D. Cardwell, R. Purdy and R.C. Bahner, Eds., American Society for Testing and Materials, Philadelphia, PA (1985).
16. Winter, T.C., *Limnol. Oceanogr.,* 26, 925 (1981).
17. Winter, T.C., in *Modeling of Total Acid Precipitation Impacts,* Vol. 9, J.L. Schnoor, Ed., Butterworth Publishers, Boston, MA, pp. 89–119 (1984).
18. Stumm, W., *Ambio,* 15, 201 (1986).
19. Korte, F., *Oekologische Chemie,* Thieme Publishers, Stuttgart, Germany (1980).
20. Stumm, W., and Morgan, J.J., *Aquatic Chemistry,* Wiley-Interscience, New York (1981).
21. Organization for Economic Cooperation and Development, *Guidelines for Testing of Chemicals,* OECD, Paris (1981).
22. Schnoor, J.L., Sato, C., McKechnie, D., and Sahoo, D., Processes, Coefficients, and Models for Simulating Toxic Organics and Heavy Metals in Surface Waters, EPA-0600/4-87, U.S. Environmental Protection Agency, Athens, GA, (1987).
23. Smith, R.A., Alexander, R.B., and Wolman, M.G., *Science,* 235, 1607 (1987).
24. Zoller, W.H. in *Changing Metal Cycles and Human Health,* J.O. Nriagu, Ed., Springer, Berlin (1985).

25. Sigg, L., Stumm, W., and Zinder, B., in *Complexation of Trace Metals in Natural Waters,* C.J.M. Kramer and J.C. Duinkard, Eds., M. Nijhoff–W. Junk Publishing, The Hague (1984).
26. Sunda, W.G., and Huntsman, S.A., *Limnol. Oceanogr.,* 28, 924 (1983).
27. Sposito, G., in *Metal Ion in Biological Systems,* Vol. 20, H. Sigel, Ed., Dekker, New York, (1985).
28. Andreae, M.O., et al., in *Changing Metal Cycles and Human Health,* J.O. Nriagu, Ed., Springer, Berlin (1985).
29. Laughlin, R.B. Jr., and Linden, O., *Ambio,* 14, 88 (1985).
30. Sillen, L.G., in *Oceanography,* M. Sears, Ed., AAAS, Washington, DC, (1961).
31. Paces, T., *Ambio,* 14, 354 (1985).
32. Schnoor, J.L.,, and Stumm, W., in *Chemical Processes in Lakes,* W. Stumm, Ed., Wiley-Interscience, New York (1985).
33. Fyfe, W.S., in *Global Change,* T.F. Malone and J.G. Roederer, Eds., Cambridge University Press, Cambridge (1984).
34. Stumm, W., Schwarzenbach, R., and Sigg, L., *Angew. Chem.,* 22, 380 (1983).
35. Sigg, L., and Stumm, W., *Aquatische Chemie,* VDF Publishers, Zurich (1989).

1.6 PROBLEMS

1. The Fibonacci number sequence is a famous geometrical series that often occurs in nature (0, 1, 1, 2, 3, 5, 8, 13, 21, 34, 55, 89, etc.). It has been used to describe rabbit population growth, and tree branching among other things. Joe Navel has proposed a mathematical model for predicting belly-button heights using the modified Fibonacci number sequence:

$$\left(\ldots, \frac{3}{5}, \frac{5}{8}, \frac{8}{13}, \frac{13}{21}, \frac{21}{34}, \text{etc.}\right) \quad \text{which converges to } 0.618$$

The model is $B = 0.618 \times H$, where B is the predicted belly-button height and H is the subject's overall height. Joe gathered data from the Environmental Engineering and Science Program staff, which appears below. Add your own personal datum to that listed below and plot the predicted vs. the measured belly button heights. In your opinion, how well does the model work? Use some statistical framework to decide. What is a model? (Be brief, 1–2 sentences.)

Subject Height, in.	Measured Navel Height, in.	Subject Height, in.	Measured Navel Height, in.
71.5	43.6	65	40.5
66.5	43.6	75	44.0
72.0	43.5	70	43.0

(*continued*)

Subject Height, in.	Measured Navel Height, in.	Subject Height, in.	Measured Navel Height, in.
64.0	40.0	73	44.2
67.0	41.5	69	42.7
68.0	42.0	69	41.5
71.0	43.5	71.5	44.0

2. The inflow to a small reservoir during a storm may be described by the equation

$$Q = 1000\,(1 - e^{-kt})$$

where Q = flow in cfs
k = runoff rate = 1.0 per hour
t = time in hours

The outflow through the dam gates is held constant at 800 cfs. The initial volume of the reservoir is 10×10^6 cubic feet. Write the differential equation describing the rate of change in reservoir volume with time. Plot the volume of the reservoir versus time during the first 8 hours of the storm. At what time does the minimum volume occur? Does a maximum volume occur?

3. Von Foerster et al. (1960) published a paper on population growth. They proposed the following mathematical model, where N is the world population and t is time in years from 1800–2026 A.D. Answer the following questions. Model equation:

$$N = \frac{1.79 \times 10^{11}}{(2026.87 - t)^{0.99}}$$

a. What is the significance of 2026.87, the empirical constant?

b. The model fits data prior to 1960 (when it was published) very well, and it is reasonably accurate between 1960 and 1990 (a 30-year post audit). Is it a good model? (You must look up world population data for comparison to model.)

c. Propose an alternate model for population growth from 1800 to 2020.

4. Animals that emerge in periodic intervals are often found in nature. Periodical cicadas with 7-, 13-, or 17-year intervals between generations are extreme examples, but fish, birds, and other animals reproduce in discrete time intervals also. In mathematical models for such systems, time is a discrete rather than a continuous variable, and one has difference equations rather than differential equations: the population at generation $t + 1$ is N_{t+1} and is related to that at time t, which was N_t. The equations are of the form

$$N_{t+1} = f(N_t)$$

Robert May (1975) has proposed a mathematical model for insect emergence from natural waters:

$$N_{t+1} = aN_t - bN_t^2$$

where a = the generation rate or birth rate
b = the death rate = 0.1
N_0 = initial insect density = 1.0

Note that in this model the population grows at low densities but declines very quickly when N is large. The biological reasons for such behavior are many: when fish populations grow large, they are often quickly decreased by disease or fungal infections. Plot the population versus discrete time (0, 1, 2, 3, . . . , 20) for a = 2.0, 3.5, and 4.5. You will make three plots altogether. Comment on the model behavior.

5. The following toxic chemical concentrations were measured in various water bodies. Which of them, if any, could cause water quality and/or ecotoxicology problems at these levels? Do you know why?

Dieldrin in the Iowa River	0.2 ppb
PCBs in Lake Superior	1.0 ng L^{-1}
Arsenic at Whitewood Creek, SD	60 $\mu g\ L^{-1}$
Mirex in the James River, Virginia	2.0 ng L^{-1}
Trichloroethylene in groundwater	10 ppb
Lead in rainwater	50 $\mu g\ L^{-1}$
Chromate in groundwater	10 $\mu g\ L^{-1}$

6. Review of differential equations. Solve the following differential equations, where C = concentration, t = time, and x = distance.

(a) $\dfrac{dC}{dt} = -kC$ if $C = C_0$ at $t = 0$ and k = constant

(b) $\dfrac{dC}{dt} + k_1 C = k_2$ if $C = C_0$ at $t = 0$ and k_1 and k_2 are constant

(c) The initial-value problem

$$\frac{d^2C}{dx^2} + \frac{dC}{dx} - 2C = 0; \quad C(0) = 3 \text{ and } C'(0) = 0$$

2

TRANSPORT PHENOMENA

. . . they program various combinations of known facts about the present world and read out from computers' analyses the social and ecological consequences of various courses of actions. But, in fact, the real future is likely to be very different from any of the predictable futures.

—René Dubos, *A God Within*

2.1 INTRODUCTION

The reactions that a chemical may undergo are an important aspect of a chemical's fate in the environment, but an equally important process has to do with the rate of a chemical's transport in the aquatic environment. In this chapter, we shall discuss three processes of mass transport in aquatic ecosystems: transport by the current of the water (advection), transport due to mixing within the water body (dispersion), and transport of sediment particles within the water column and between the water and the bed.

Toxic chemicals, at low concentrations in natural waters, exist in a dissolved phase and a sorbed phase. Dissolved substances are transported by water movement with little or no "slip" relative to the water. They are entirely entrained in the current and move at the water velocity. Likewise, chemicals that are sorbed to colloidal material or fine suspended solids are essentially entrained in the current, but they may undergo additional transport processes such as sedimentation and deposition or scour and resuspension. These processes may serve to retard the movement of the sorbed substances relative to the water movement. Thus in order to determine the fate of toxic organic substances, we must know both the water movement and suspended sediment movement.

The transport of toxic chemicals in water principally depends on two phenomena: advection and dispersion. *Advection* refers to movement of dissolved or very fine particulate material at the current velocity in any of three directions (longitudinal, lateral or transverse, and vertical). *Dispersion* refers to the process by which these substances are mixed within the water column. Dispersion can also occur in three directions. A schematic for advection, turbulent diffusion, and dispersion in a stream is given in Figure 2.1. Three processes contribute to mixing (dispersion):

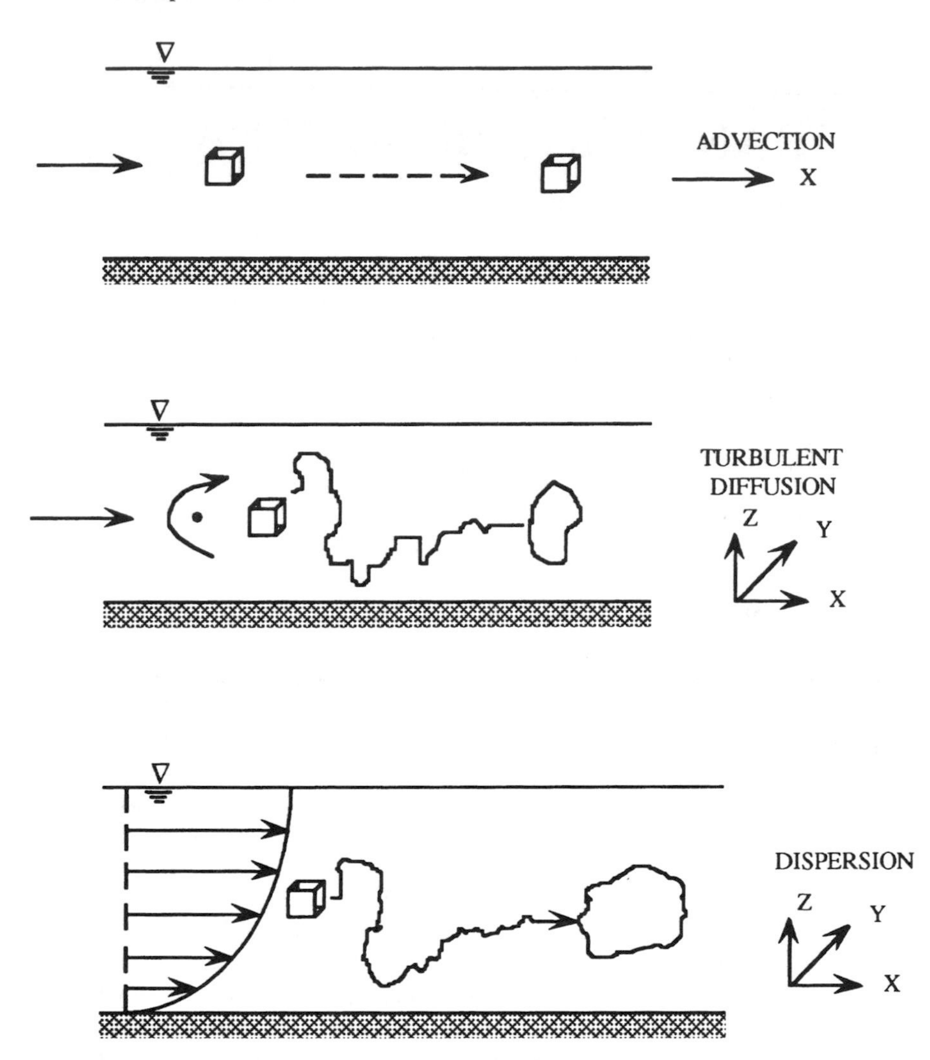

Figure 2.1 Schematic of transport processes: (1) advection, movement of chemical entrained in current velocity; (2) turbulent diffusion, spread of chemical due to eddy fluctuations; and (3) dispersion, spread of chemical due to eddy fluctuations in a macroscopic velocity gradient field.

1. *Molecular diffusion.* Molecular diffusion is the mixing of dissolved chemicals due to the random walk of molecules within the fluid. It is caused by kinetic energies of molecular vibrational, rotational, and translational motion. In essence, molecular diffusion corresponds to an increase in entropy whereby dissolved substances move from regions of high concentration to regions of low concentration according to Fick's laws of diffusion. It is an exceedingly slow phenomenon, such that it would take on the order of 10 days for 1 mg L^{-1} of dissolved substance to diffuse through a 10-cm water column from a concentration of 10 mg L^{-1}. It is generally not an important process in the

transport of dissolved substances in natural waters except relating to transport through thin and stagnant films at the air–water interface or transport through sediment pore water.

2. *Turbulent diffusion.* Turbulent or eddy diffusion refers to mixing of dissolved and fine particulate substances caused by microscale turbulence. It is an advective process at the microscale level caused by eddy fluctuations in turbulent shear flow. Shear forces within the body of water are sufficient to cause this form of mixing. It is several orders of magnitude larger than molecular diffusion and is a contributing factor to dispersion. Turbulent diffusion can occur in all three directions but is usually anisotropic (i.e., there exist preferential directions for turbulent mixing due to the direction and magnitude of shear stresses).
3. *Dispersion.* The interaction of turbulent diffusion with velocity gradients caused by shear forces in the water body causes a still greater degree of mixing known as *dispersion.* Transport of toxic substances in streams and rivers is predominantly by advection, but transport in lakes and estuaries is often dispersion-controlled. Velocity gradients are caused by shear forces at the boundaries of the water body, such as vertical profiles due to wind shear at the air–water interface, and vertical and lateral profiles due to shear stresses at the sediment–water and bank–water interfaces (Figure 2.2). Also, velocity gradients can develop within the water body due to channel morphology and sinuosity or meandering of streams. Secondary currents develop in stream and river channels that account for a great deal of mixing. Figure 2.3 shows the helical current that forms as a result of morphology in stream channels.

Thermal or density stratification in lakes and estuaries serves to decrease dispersion by stabilizing water into layers of equal density. Morphological causes of dispersive mixing in rivers also include dead spots, side channels, and pools where back-mixing occurs. When turbulent diffusion causes a parcel of fluid containing dissolved substances to change position, that parcel of fluid becomes entrained in the water body at a new velocity, either faster or slower. This causes the parcel of fluid and the toxic substance to mix forward or backward relative to its neighbors. In estuaries, tidal oscillations create a large degree of mixing. The mixing process is called dispersion and results in a mass flux of toxic substances from areas of high concentration to areas of low concentration. The process is analogous to molecular diffusion but occurs at a much more rapid rate.

In Chapter 1, the concept of a mass balance taken around a control volume was introduced. The accumulation of mass in the control volume is always equal to the inputs of mass, minus the outflows of mass, plus or minus the reactions occurring in the control volume. In this chapter, we will quantify the first two terms as transport terms.

$$\text{Accumulation} = \underbrace{\text{Inputs} - \text{Outflows}}_{\text{Transport}} \pm \text{Reactions} \tag{1}$$

where "Transport" can be by (1) advection (flow) and/or (2) diffusion/dispersion (mixing). In Chapters 3–6, we will quantify the "Reactions" term, so a complete mass balance equation may be written thereafter.

2.2 ADVECTION

Advective transport is the movement of mass entrained in a current and traveling from one point to another. For a chemical traveling in a stream or a river, advective transport is the product of the volumetric flowrate and the mean concentration [equation (2)]. Figure 2.4 illustrates the movement of mass from point a to point b by advection.

$$J = \bar{u}AC = QC \tag{2}$$

where J is the mass discharge rate in units of MT^{-1}, $\bar{u}$ is the mean current velocity in units of LT^{-1}, C is concentration in units of ML^{-3}, and Q is the volumetric flowrate in units of L^3T^{-1}.

During steady flow conditions ($\partial Q/\partial t = 0$) and steady state ($\partial C/\partial t = 0$), the mass discharge rate is constant with respect to time. If either the flowrate or the concentration becomes time-variable, then mass discharge rate (advective mass transport) becomes variable with time. Mass inside the control volume (Figure 2.4), at any instant, may be written as volume times concentration ($V \cdot C$), where V is the volume in units of L^3 and C is the concentration, ML^{-3}. The change in mass with respect to time due to advection may be written as a difference equation:

$$\Delta(VC) = (Q_aC_a - Q_bC_b)\,\Delta t \tag{3}$$

$$\Delta\text{mass} = (\text{Mass inflow rate} - \text{Mass outflow rate})\,\Delta t$$

where C_a is the concentration entering the incremental control volume and C_b is the concentration leaving the control volume. Dividing through by Δt gives equation (4),

$$\frac{\Delta(VC)}{\Delta t} = Q_aC_a - Q_bC_b \tag{4}$$

and further division by the incremental volume $V = A\,\Delta x$ yields a difference equation (equation 5a), which taken to the limit as $\Delta x \to 0$ is a partial differential equation describing advection of mass under time-varying conditions

$$\frac{\Delta C}{\Delta t} = \frac{-\Delta(QC)}{A\,\Delta x} \tag{5a}$$

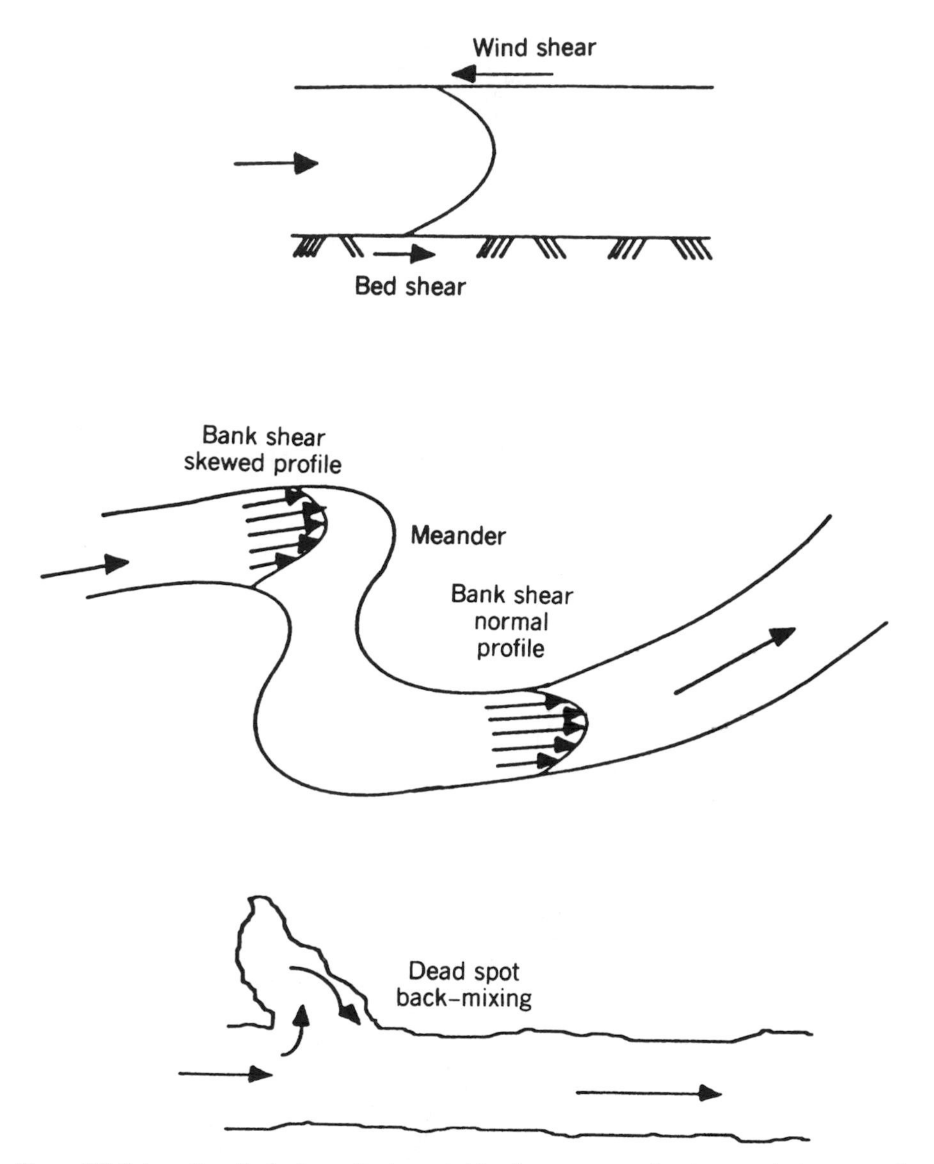

Figure 2.2 Schematics of velocity gradients created by shear stresses at the air–water, bed–water, and bank–water interfaces.

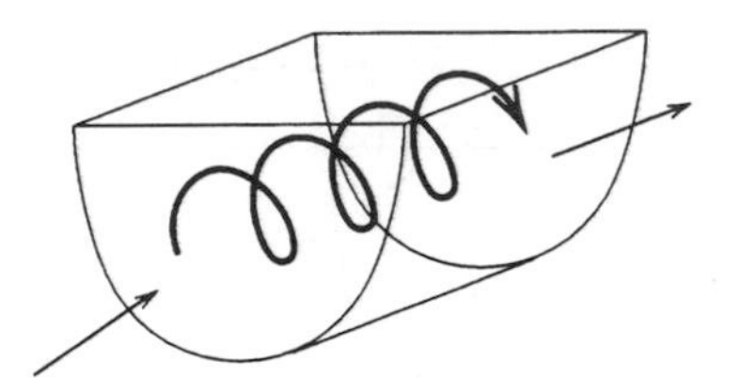

Figure 2.3 Secondary current in a stream responsible for lateral and longitudinal dispersion.

$$\frac{\partial C}{\partial t} = -\frac{1}{A}\frac{\partial(QC)}{\partial x} = -\bar{u}\frac{\partial C}{\partial x} \tag{5b}$$

where Δx is the incremental distance of the control volume and x is longitudinal distance. The last term of equation (5b) is valid under steady flow conditions such that

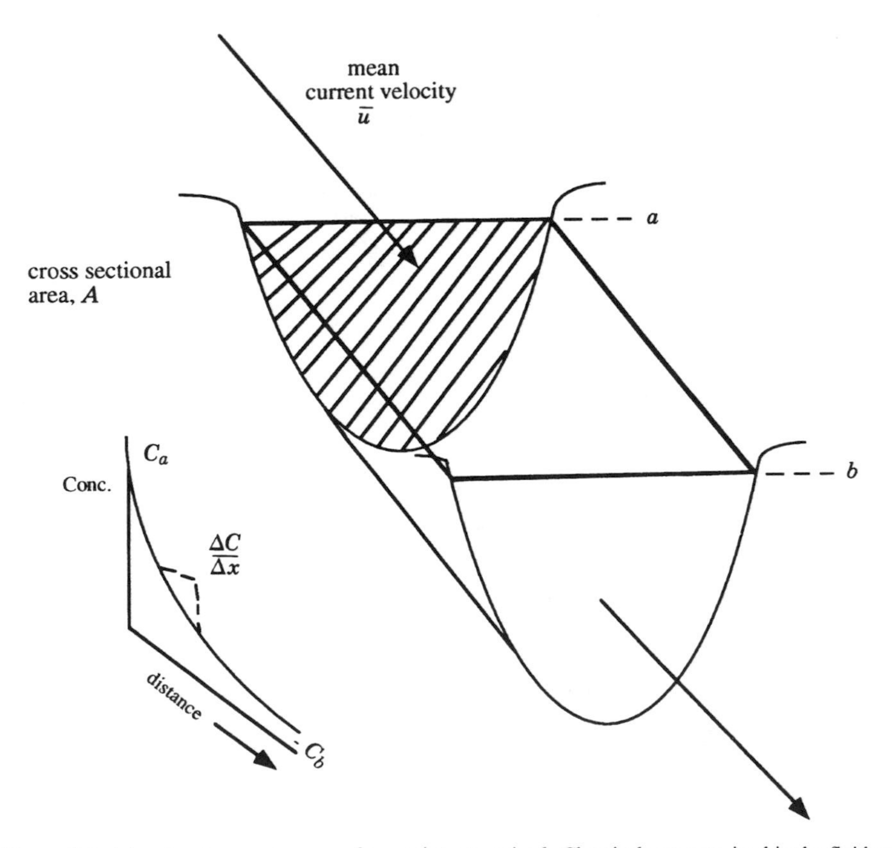

Figure 2.4 Advective transport process from point a to point b. Chemicals are entrained in the fluid at the mean current velocity, u. A hypothetical concentration gradient is shown below as concentration versus distance, x. Note that the slope of the line, $\Delta C/\Delta x$, is negative when mass is transported from point a into the incremental element of volume $V = A\Delta x$.

$\bar{u} = Q/A$, a constant mean velocity. The negative sign in equations (4) and (5) is necessary to reflect an increasing concentration within the control volume if mass inflow is greater than mass outflow. (Note Figure 2.4, in which a negative slope or concentration gradient, $\Delta C/\Delta x$, produces a positive change in concentration with respect to time, $\Delta C/\Delta t$.) The negative sign in equations (4) and (5) converts a negative slope (concentration gradient) into a positive mass flux into the incremental control volume, as needed.

Equation (5) is a mathematical description of advection when flowrate and/or concentration are changing. It is a *time-variable* equation (note that time is the derivative on the left-hand side of the equation), as opposed to the *steady-state* equation (2).

Under time-variable conditions, it is sometimes necessary to estimate the total mass that has passed a point in a given amount of time. This can be accomplished by integrating the mass discharge rate over time

$$M = \int_0^{t_1} Q(t) \cdot C(t)\, dt \tag{6}$$

where M is the total mass and t gives the prescribed time interval of interest ($0 \rightarrow t_1$). If steady flow conditions prevail (Q is constant with time), then equation (7) is a special case:

$$M = Q \int_0^{t_1} C(t)\, dt \tag{7}$$

If the change in concentration with respect to time can be described by a mathematical equation, then it may be possible to integrate the equation directly. The process is equivalent to estimating the area under the concentration versus time curve and multiplying by the flowrate to get the total mass.

Example 2.1 Advective Transport of Pesticide in a River

Calculate the average mass flux (in kg d^{-1}) of the pesticide alachlor passing a point in a river draining a large agricultural basin. The mean concentration of pesticide is 1.0 $\mu g\ L^{-1}$, and the mean flow is 50. $m^3\ s^{-1}$. Is this an accurate estimate of the total mass passing this point in a year, considering high runoff events?

Solution

$$J = QC$$

$$J = \frac{50\ \text{m}^3}{\text{s}}\ \frac{1\ \mu\text{g}}{\text{L}}\ \frac{1000\ \text{L}}{\text{m}^3}\ \frac{\text{kg}}{10^9\ \mu\text{g}}\ \frac{86{,}400\ \text{s}}{\text{d}} = 4.3\ \text{kg d}^{-1}$$

Using the annual average concentration and annual average flowrate does not give the total annual mass discharge passing a point because flowrate and concentration are positively correlated. High flows create high pesticide concentrations due to agricultural runoff. The mass discharge rate calculated above underestimates the total average mass discharge rate ($\overline{QC}$) for the year. Use equation (6) to estimate the total mass passing a point during the year.

$$\overline{QC} \neq \overline{Q} \cdot \overline{C}$$

$$\begin{aligned}\overline{QC} &= \overline{(\overline{Q} + Q')(\overline{C} + C')} \\ &= \overline{\overline{Q}\,\overline{C}} + \underbrace{\overline{\overline{Q}C'} + \overline{Q'\overline{C}}}_{0} + \overline{Q'C'}\end{aligned}$$

$$M = \int_0^{1\ \mathrm{yr}} Q(t) \cdot C(t)\, dt$$

Thus the average mass discharge, $\overline{QC}$, is equal to the average flowrate times the average concentration, $\overline{Q} \cdot \overline{C}$, plus the average mass fluctuation, $\overline{Q'C'}$. The deviation from the mean is denoted by the prime (Q' and C') and the average mass fluctuation depends on the correlation between flowrate and concentration.

2.3 DIFFUSION/DISPERSION

In 1855, Fick published his first law of diffusion based on movement of chemicals through fluids under quiescent conditions. He recognized the analogy to Fourier's law of heat conduction. Molecular diffusion results from the translational, vibrational, and rotational movement of molecules through a fluid, in this case water. Energetically it is a spontaneous reaction, and it results in an increase in entropy (tendency toward the random state). Fick determined that mass transfer by diffusion was proportional to the cross-sectional area of the apparatus and the steepness of the concentration gradient.

$$J_m \propto A \frac{dC}{dx} \tag{8}$$

where J_m is the mass flux rate due to molecular diffusions, MT^{-1}; A is the cross-sectional area, L^2; and dC/dx is the concentration gradient, $ML^{-3}L^{-1}$. In Figure 2.5, a doubling of the tubular cross-sectional area would result in twice the mass rate of flux, and a doubling in the concentration gradient (driving force) would do likewise.

A proportionality constant was needed to change the proportionality [equation (8)] into an equation:

$$J_m = -DA \frac{dC}{dx} \tag{9}$$

Or Fick's first law of diffusion can be written on an areal basis,

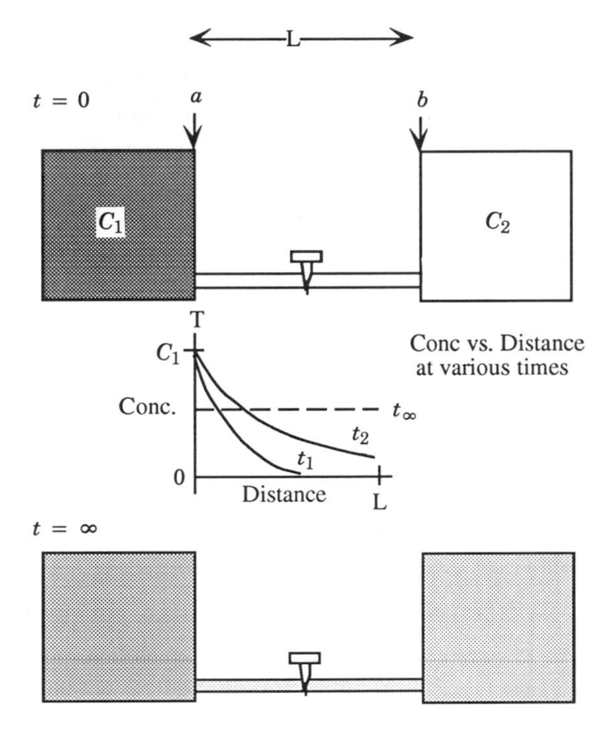

Figure 2.5 Diffusive transport from point a to point b. At the beginning of the experiment ($t = 0$) all of the chemical is dissolved in the beaker on the left-hand side. When the experiment begins, mass moves from areas of high concentration to areas of low concentration according to Fick's laws of diffusion until equilibrium is established (see inset of concentration versus distance at various times t_1, t_2, t_∞).

$$F_m = -\frac{D\,dC}{dx} \tag{10}$$

where D is the molecular diffusion coefficient, L^2T^{-1}, and F is the areal mass flux rate, $ML^{-2}T^{-1}$. The negative signs on the right-hand side of equations (9) and (10) are necessary to convert a negative concentration gradient into a positive flux in the x-direction according to mathematical convention. Note that proportionality constant D, the molecular diffusion coefficient, has units of L^2T^{-1}, which is required to obtain a mass flux rate in units of mass per time. The molecular diffusion coefficient is a fundamental property of the chemical and the solvent (water). Molecular diffusion coefficients are tabulated for many chemicals in water in *Handbooks of Chemistry and Physics*, and they may be estimated from chemical and thermodynamic properties (see Lyman et al.[1]). On the order of 10^{-5} cm^2 s^{-1}, molecular diffusion coefficients indicate a very slow movement of mass. Molecular diffusion occurs in nature as chemical transport through thin, laminar boundary layers, such as might occur at interfaces (air–water, sediment–water, particle–water) or in quiescent sediment pore waters.

Example 2.2 Molecular Diffusion of a Chemical in Water

Calculate the mass flux rate in mg d^{-1} for a chemical diffusing between two beakers, such as shown in Figure 2.5. Assume the chemical is diffusing through a 10-cm distance with a concentration gradient of -1 mg L^{-1} cm^{-1}.

$$D = 10^{-5}\ \text{cm}^2\ \text{s}^{-1}$$

$$A = 3.14\ \text{cm}^2$$

Solution:

$$J = -DA\frac{dC}{dx}$$

$$J = -\frac{10^{-5}\ \text{cm}^2}{\text{s}}\ 3.14\ \text{cm}^2\ \frac{-1\ \text{mg}}{\text{L cm}}\ \frac{\text{L}}{1000\ \text{cm}^3}\ \frac{86{,}400\ \text{s}}{\text{d}} =$$

$$J = 0.00271\ \text{mg d}^{-1}$$

This is an incredibly slow rate of mass transfer considering that it would take 1 year to transport 1 mg of chemical if the concentration gradient was held constant over time. (The experiment in Figure 2.5 actually shows a nonsteady state condition.)

Example 2.3 Molecular Diffusion Through a Thin Film

The molecular diffusivity of caffeine (C_9H_8O) in water is 0.63×10^{-5} cm^2 s^{-1}. For a 1.0 mg L^{-1} solution, calculate the mass flux in mg s^{-1} through an intestinal membrane (0.1-m^2 area) with a liquid film approximately 60 μm thick. How long would it take 1 mg of caffeine to move through 0.1 m^2 of intestine, assuming the above flux rate? (*Note*: We assume transport through the film is the rate-limiting step in transport and metabolism.)

Solution:

$$J = -DA\frac{dC}{dx}$$

$$\Delta C = (0 - 1.0)\ \text{mg L}^{-1} \quad \text{(assume zero caffeine inside intestine)}$$

$$\Delta x = 60\ \mu\text{m}$$

$$J = \frac{0.63 \times 10^{-5}\ \text{cm}^2}{\text{s}}\ 0.1\ \text{m}^2\ \frac{-1.0\ \text{mg}}{\text{L}}\ \frac{1}{60 \times 10^{-6}\ \text{m}}\ \frac{10^4\ \text{cm}^2}{1\ \text{m}^2}\ \frac{\text{L}}{1000\ \text{cm}^3}\ \frac{\text{m}}{100\ \text{cm}}$$

$$J = 0.00105\ \text{mg s}^{-1}$$

$$t = \text{mass/flux rate}$$

$$t = 1\ \text{mg}\ \frac{\text{s}}{0.00105\ \text{mg}}\ \frac{\text{min}}{60\ \text{s}} = 15.9\ \text{min}$$

2.3.1. Analogies Between Mass, Momentum, and Heat Transfer

In 1877, Boussinesq first proposed that turbulent momentum transport is analogous to viscous momentum transfer in laminar flow. It was also suggested by Reynolds in 1894, following his famous experiment in 1883, in which he showed clearly the critical dimensionless number (Re = 2300) necessary to change from laminar flow to turbulent flow in pipes.[2–4]

$$\text{Re} = \frac{\bar{u}d}{\nu} \tag{11}$$

where Re is the Reynolds number; $\bar{u}$ is the mean velocity, LT^{-1}; d is the pipe diameter, L; and ν is the kinematic viscosity, L^2T^{-1}. Turbulence transfers momentum as a viscous force per unit area (the shear stress) in proportion to the vertical velocity gradient, just as viscous shear stress does in laminar flow, but with a proportionality constant that is much greater than that for laminar flow. In Table 2.1, $\varepsilon_v \gg \nu$.

Figure 2.6 demonstrates that mass, heat, and momentum transport can occur simultaneously and they are all analogous. The flux rate per unit area is a gradient driving force ($\partial/\partial z$) times a proportionality constant for the turbulent flow field (ε). As in Fick's first law of diffusion for mass, the negative sign on the flux is a mathemat-

Table 2.1 Mass, Momentum, and Heat Transfer Coefficients

	Molecular/Laminar Regime	Turbulent Regime
Mass	D	ε_m
Momentum	$\nu = \mu/\rho$	ε_v
Thermal (heat)	$\alpha = \dfrac{k}{\rho C_p}$	ε_t

where D = molecular diffusion coefficient, L^2T^{-1}
ε_m = eddy mass diffusivity, L^2T^{-1}
ν = kinematic viscosity, L^2T^{-1}
ε_v = eddy viscosity, L^2T^{-1}
α = thermal diffusivity, L^2T^{-1}
ε_t = eddy thermal diffusivity, L^2T^{-1}
μ = viscosity, $ML^{-1}T^{-1}$
ρ = mass density, ML^{-3}
k = coefficient of thermal conductivity, $HL^{-1}\theta^{-1}T^{-1}$
C_p = heat capacity, $HM^{-1}\theta^{-1}$

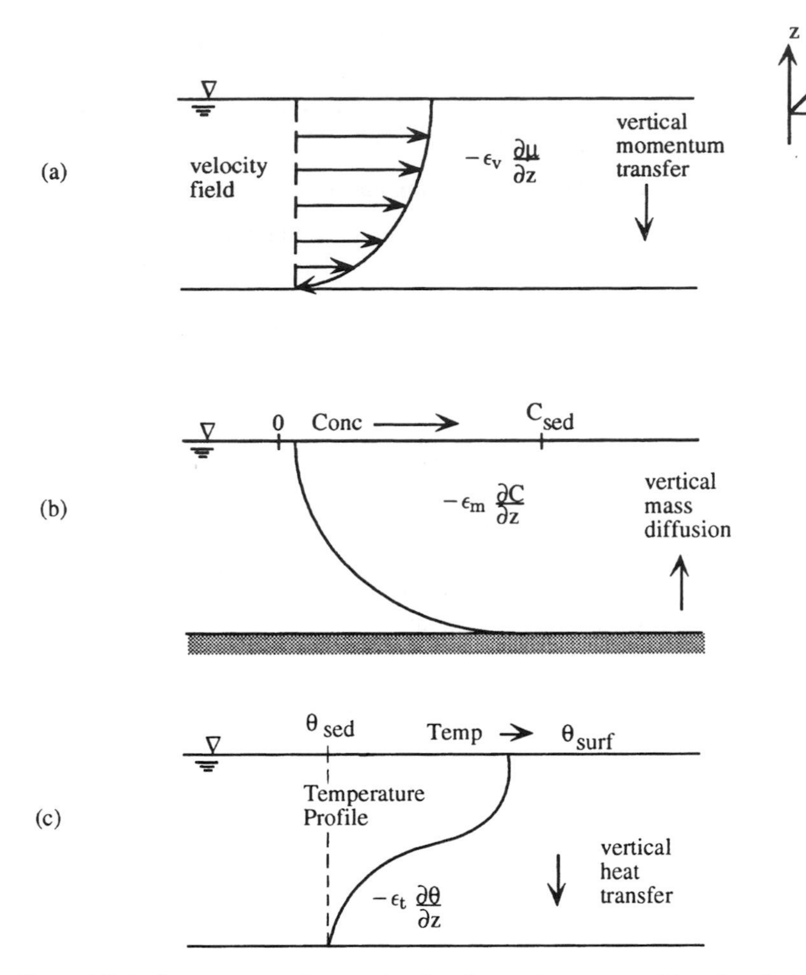

Figure 2.6 Analogous and simultaneous transfer of momentum, mass, and heat transfer in a turbulent river. (a) Shear forces at the sediment–water interface create a vertical velocity field that transfers momentum downward [viscous force per unit area is the shear stress $\tau(z) = \varepsilon_v \, \partial u/\partial z$]. (b) A contaminated sediment transfers mass to the water column by vertical turbulent diffusion [mass flux per unit area is defined as $F(z) = -\varepsilon_m \, \partial C/\partial z$]. (c) Summer stratification of the river creates a vertical heat flux downward (heat flux per unit area is $q = -\varepsilon_t \, \partial\theta/\partial z$).

ical convention to show that when the gradient increases away from the $z = 0$ location (the sediment in Figure 2.6), the flux is toward the sediment.

Table 2.1 gives the definition for the "proportionality" constants among these three analogous transport processes under laminar or turbulent flow conditions. All the proportionality constants have the same units L^2T^{-1}. In each case the turbulent transport "proportionality constant" is much greater than the corresponding one for laminar flow:

$$\varepsilon_m \gg D, \quad \varepsilon_v \gg \nu, \quad \varepsilon_t \gg \alpha \tag{12}$$

Analogies between mass, momentum, and heat transfer are qualitatively very useful, but sometimes they can introduce errors into aquatic chemical models. The dimensionless ratio between thermal diffusivity and mass diffusion coefficient is termed the Lewis number. Its value determines the extent of analogy between heat and mass transfer. Particularly under turbulent conditions, the Lewis number can vary from 1.0. When a heat budget model is used to calibrate mixing in a mass balance model, errors may be introduced but it is, in general, appropriate practice. Dimensionless numbers relating kinematic viscosity to mass diffusion coefficients (the Schmidt number) and kinematic viscosity to thermal diffusivity (Prandtl number) may also vary from 1.0 in simultaneous heat mass, and momentum transfer.

$$\mathrm{Le} = \alpha/D, \quad \mathrm{Sc} = \nu/D, \quad \mathrm{Pr} = \nu/\alpha \tag{13}$$

The final analogy of importance is between turbulent diffusion and dispersion. These processes are physically very different, but both are mixing processes and, under certain circumstances, they may take the same form. They are mixing processes that may be written similar to equation (9) for molecular diffusion, but the proportionality constants are much larger.

$$J_t = -\varepsilon_m A \frac{dC}{dx} \tag{14a}$$

$$J_d = -EA \frac{dC}{dx} \tag{14b}$$

where J_t is the mass flux rate due to turbulent diffusion, MT^{-1}, em is the turbulent diffusion coefficient (or turbulent diffusivity), L^2T^{-1}; J_d is mass flux rate due to dispersion, MT^{-1}; and E is the dispersion coefficient, L^2T^{-1}. These mixing processes have orders of magnitude difference among the proportionality constants for dispersion, turbulent diffusion, and molecular diffusion, but they all are expressed in the same units and are used in a "Ficks law" type of equation. The driving force in each case is the concentration gradient, dC/dx. Dispersion coefficients are much greater than eddy diffusivities which are, in turn, much larger than molecular diffusion coefficients.

$$E \gg \varepsilon_m \gg D \tag{15}$$

Note that the molecular diffusion coefficient D depends on the fluid and chemical properties, but the turbulent mass diffusivity ε_m and the dispersion coefficient E depend only on the flow regime.

At the molecular or turbulent scale, it is the local concentration that we are modeling, not a spatial average. As soon as one considers dispersion due to the interaction between diffusion and differential advection, it is the dispersion of some spatially averaged measure of the contaminant mass. For example, when one talks about Fischer's one-dimensional (1-D) longitudinal mixing coefficient for rivers (Section 2.3.5), or Taylor's 1-D longitudinal mixing coefficient for pipes, one is necessarily dealing with the mixing of the cross-sectional average concentration. The concepts are meaningless for any other measure; for example, they are meaningless for the local concentration. The coefficients increase in magnitude as the scales increase, but the different coefficients are relevant for quite different spatial measures of contaminant mass. Elder's work (Section 2.3.6) dealt with depth-averaged concentration, Fischer's mostly with cross-sectional averages. Taylor's diffusive work dealt with the local, but turbulence time-averaged, concentration.

There is an initial period of mixing, especially in medium to large rivers, during which the Fickian analogy for dispersion is not valid. This is due to the scale of turbulent diffusion and differential advection filling a fully developed flow regime. After the initial release of a slug of tracer, it takes some time for conditions to develop that can be described by the Fickian analogy.

2.3.2 Fick's Second Law

Fick's second law of diffusion follows from the first law of diffusion under non-steady state. The second law is necessary to predict the concentration with respect to time at any location, such as the curves presented for the two-beaker experiment shown in Figure 2.5.

Beginning with Fick's first law of diffusion, we may write it as a difference equation [equation (16)] and then divide by the incremental volume, $V = A\ \Delta x$ [equations (17 and 18)]:

$$J = -DA\frac{\Delta C}{\Delta x} \tag{16}$$

$$V\frac{\Delta C}{\Delta t} = -DA\frac{\Delta C}{\Delta x} \tag{17}$$

$$\frac{\Delta C}{\Delta t} = -D\frac{\Delta C}{\Delta x\ \Delta x} \tag{18}$$

$$\lim \Delta t \to 0 \qquad \frac{\partial C}{\partial t} = D\frac{\partial^2 C}{\partial x^2} \tag{19}$$

The negative sign in equation (18) switches to a positive sign when the second derivative is considered. Equation (19) is the mathematical expression for time-variable diffusion—it is a partial differential equation accounting for concentration differences in space (1-D) and time. Fick's first law of diffusion is applicable at any point in space and time, but the driving force for movement of mass (the concentration gradient) is always changing. Nevertheless, mass always diffuses from areas of high concentration to areas of low concentration until equilibrium is achieved, where the concentration gradient is zero everywhere (the concentration is constant in space and time).

Equation (19) is a second-order partial differential equation so it requires two boundary conditions (one for each order) and one initial condition in order to solve. Solutions to equation (19) are many and varied—there is a different solution for each set of boundary and initial conditions that may be posed. Integrating equation (19) may be accomplished by La Place Transformation or by trial and error methods, depending on the boundary conditions posed. As with the steady state equations [equations (9), (13), and (14)], Fick's second law has analogies with other mixing processes for turbulent diffusion and dispersion, so ε_m and E may be substituted for D in equation (19) under certain circumstances. The reader is referred to Fischer et al.[4] for a full discussion of the limitations and theory on the development of coefficients for turbulent diffusion and dispersion.

Example 2.4 Fick's Second Law—Diffusion from a Contaminated Sediment

Solve Fick's second law of diffusion. The problem is for vertical eddy diffusion from a planar source (a contaminated lake sediment) to the overlying water column. The initial condition (IC) is

$$C(x) = 0 \quad \text{at } t = 0 \qquad \text{(IC)}$$

and the boundary conditions (BC1 and BC2) are for a constant chemical mass M diffusing into a semi-infinite water column (early stages of diffusion).

$$C(+\infty) = 0 \quad \text{for all } t \qquad \text{(BC1)}$$

$$M = \int_{-\infty}^{\infty} C\,dx \qquad \text{(BC2)}$$

Hint: Trial-and-error solution for an instantaneous planar source. Try the solution below for the partial differential equation.

$$\frac{\partial C}{\partial t} = E\,\frac{\partial^2 C}{\partial x^2}$$

$$C = \frac{A}{t^{1/2}} \exp\left(\frac{-x^2}{4Et}\right)$$

where A is an arbitrary constant.

Solution: We must have

$$\frac{\partial C}{\partial t} = E\,\frac{\partial^2 C}{\partial x^2}$$

Take the partial derivative with respect to time and the second partial derivative w.r.t. x. Use integral tables for "derivative of a product" and e^u:

$$\frac{d}{dx}(uv) = u\frac{dv}{dx} + v\frac{du}{dx}$$

$$\frac{d}{dx}(e^u) = (e^u)\frac{du}{dx}$$

The partial derivative w.r.t. time is:

$$\frac{\partial C}{\partial t} = \left[\left(\frac{A}{t^{1/2}}\right)\left(\frac{x^2}{4Et^2}\right)e^{-x^2/4Et} - \tfrac{1}{2}At^{-3/2}e^{-x^2/4Et}\right]$$

$$\frac{\partial C}{\partial t} = \left(\frac{Ax^2}{4Et^{5/2}} - \frac{A}{2t^{3/2}}\right)e^{-x^2/4Et}$$

The second derivative with respect to vertical distance, x, is

$$E\frac{\partial^2 C}{\partial x^2} = E\frac{\partial C}{\partial x}\left[\left(\frac{A}{t^{1/2}}\right)\left(\frac{-2x}{4ET}\right)e^{-x^2/4Et}\right]$$

$$E\frac{\partial^2 C}{\partial x^2} = \left(\frac{Ax^2}{4Et^{5/2}} - \frac{A}{2t^{3/2}}\right)e^{-x^2/4Et}$$

This is the same result as for $\partial C/\partial t$ above, which proves that the solution will work. Now it is necessary to solve for the arbitrary constant, A, in terms of the mass diffusing, M:

$$M = \int_{-\infty}^{\infty} C\,dx = \int_{-\infty}^{\infty} \frac{A}{t^{1/2}} e^{-x^2/4Et}\,dx$$

Using the definite integral identity below, the exponent is of the form

$$\int_0^{\infty} e^{-a^2x^2}\,dx = \frac{1}{2a}\sqrt{\pi}$$

where

$$a = \sqrt{\frac{1}{4Et}} = \frac{1}{2\sqrt{Et}}$$

By the principle of reflection at the $x = 0$ plane: $M \int_{-\infty}^{\infty} C\,dx$; $\tfrac{1}{2}M = \int_0^{\infty} C\,dx$

$$M = 2\int_0^\infty \frac{A}{t^{1/2}} e^{-x^2/4Et}\, dx = \frac{2A}{t^{1/2}} \int_0^\infty e^{-x^2/4Et}\, dx$$

$$M = \frac{2A}{t^{1/2}} \frac{2\sqrt{Et}}{2} \sqrt{\pi} = 2A\sqrt{\pi E}$$

$$\therefore \quad A = \frac{M}{2\sqrt{\pi E}}$$

and

$$C = \frac{M}{2\sqrt{\pi E t}} e^{-x^2/4Et} \qquad \text{Q.E.D.}$$

Under the influence of larger and larger mixing scales, the eddys that influence a given solute molecule increase the proportionality of mixing. Prandtl in 1925 introduced the concept of mixing length, a measure of the average distance a fluid element would stray from the mean streamline. As a chemical solute travels, it is entrained in a large variety of eddies. In fully developed turbulent flow, all scales of eddies are present, from the largest ones that can fit within the spatial confines of the physical environment to the smallest scale allowed by dissipative structures. In the ocean, a large current such as the Gulf Stream, traveling thousands of miles, can be viewed as an eddy down to microscopic turbulent vortices on the orders of micrometers.

The probability of a chemical solute being entrained by ever-larger eddies increases with time and the scale of the problem. It was proposed by Richardson that horizontal eddy diffusivities in the ocean increase with the length of the solute plume to the four-thirds power.

$$\varepsilon_m = 0.01\, L^{4/3} \tag{20}$$

where ε_m is the mass eddy diffusivity in $cm^2\ s^{-1}$, and L is the length scale of the plume in cm, and 0.01 is a proportionality constant for units in the cgs system. Under these conditions the diffusion equation becomes

$$\frac{\partial C}{\partial t} = \frac{1}{A} \frac{\partial}{\partial x}\left(\varepsilon_m(x) A \frac{\partial C}{\partial x}\right) \tag{21}$$

Figure 2.7 shows how the vertical eddy diffusivity can vary with depth in a stratified reservoir.

Earlier in the chapter, equation (1) gave the basic mass balance equation around a control volume; equation (5) was the transport due to advection; and equation (19) gave the transport due to diffusion (and by analogy turbulent diffusion and disper-

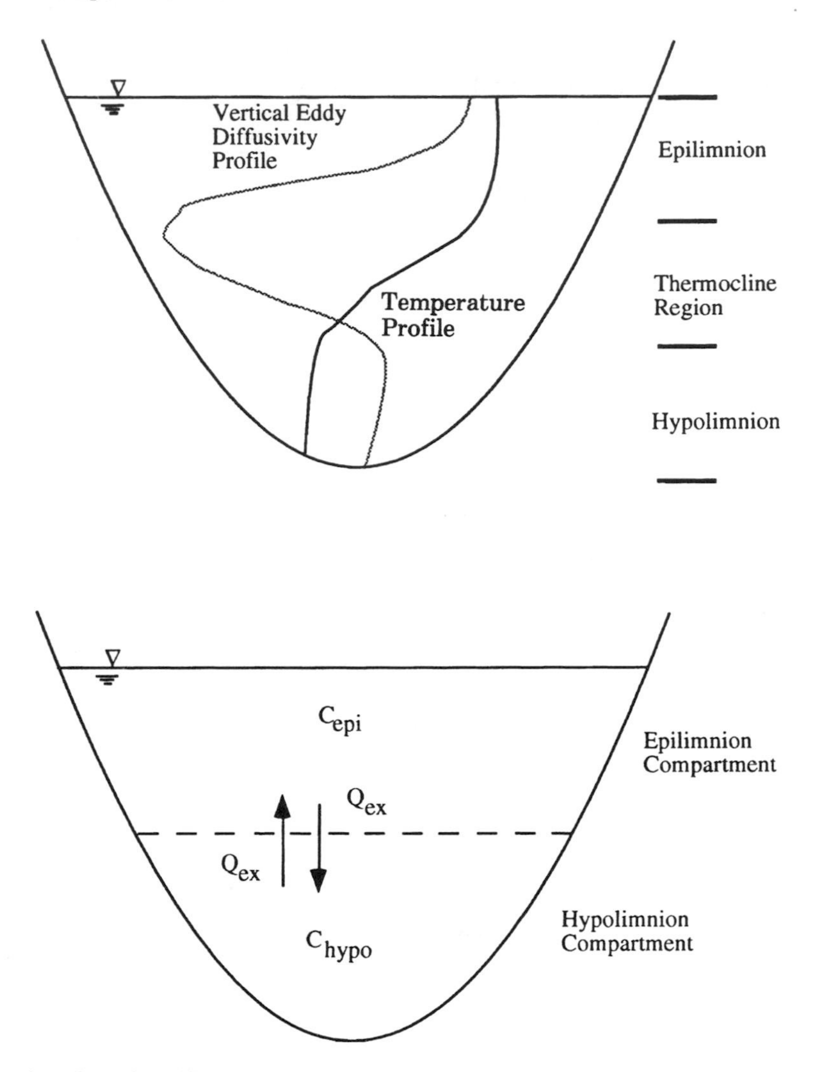

Figure 2.7 Thermal stratification in a lake and the assumption of mixing between two compartments.

sion). In the next section, we shall learn more about the mathematics of the advection–dispersion equation.

2.3.3 Advection-Dispersion Equation

The basic equation describing advection and dispersion of dissolved matter is based on the principle of conservation of mass and Fick's law. For a conservative substance, the principle of conservation of mass can be stated:

Rate of change of mass in control volume	=	Rate of change of mass in control volume due to advection	+	Rate of change of mass in control volume due to diffusion	–	Transformation reaction rates (degradation)

$$\frac{\partial C}{\partial t} = -u_i \frac{\partial C}{\partial x_i} + \frac{\partial}{\partial x_i} E_i \frac{\partial C}{\partial x_i} - R \tag{22}$$

where C = concentration, ML^{-3}
t = time, T
u_i = average velocity in the ith direction, LT^{-1}
x_i = distance in the ith direction, L
R = reaction transformation rate, $ML^{-3}T^{-1}$

E_i is the diffusion coefficient in the ith direction. In Fickian theory, it is assumed that dispersion resulting from turbulent open-channel flow is analogous to molecular diffusion. The dispersion coefficients in the x, y, and z directions are assumed to be constants, given by E_x, E_y, and E_z. The resulting equation, expressed in Cartesian coordinates, is

$$\frac{\partial C}{\partial t} + u_x \frac{\partial C}{\partial x} + u_y \frac{\partial C}{\partial y} + u_z \frac{\partial C}{\partial z} = E_x \frac{\partial^2 C}{\partial x^2} + E_y \frac{\partial^2 C}{\partial y^2} + E_z \frac{\partial^2 C}{\partial z^2} - R \tag{23}$$

The solution of equation (23) depends on the values of E_x, E_y, and E_z and on the initial conditions and boundary conditions. Various authors have arrived at equations to approximate the values of the dispersion coefficients (E) in the longitudinal (x), lateral (y), and vertical (z) directions.

Under nonsteady flow conditions, the velocity in the longitudinal direction can vary in space and time. For a one-dimensional river,

$$\frac{\partial (AC)}{\partial t} = -\frac{\partial (QC)}{\partial x} + \frac{\partial}{\partial x}\left(EA \frac{\partial C}{\partial x}\right) - AR \tag{24}$$

where Q = volumetric flowrate, L^3T^{-1}
A = cross-sectional area, L^2

To solve equation (24) analytically would require exact (and simple) functional relationships for A, Q, and E, but in practice the nonsteady transport equation is solved numerically, and it is coupled to numerical solutions of open-channel flow such as the St. Venant equations:

$$\frac{\partial z}{\partial t} = -\frac{1}{b}\frac{\partial Q}{\partial x} + \frac{1}{b} q_i \tag{25}$$

$$\frac{\partial Q}{\partial t} = \left(\frac{Q^2 b}{A^2} - gA\right)\frac{\partial z}{\partial x} - \frac{2Q}{A}\frac{\partial Q}{\partial x} + \frac{Q^2}{A^2}\frac{\partial A}{\partial x} - gAS_f \tag{26}$$

where $S_f = \frac{f}{8g}\frac{P}{A}\frac{Q^2}{A^2}$ (27)

b = width at water level, L
f = Darcy–Weisbach friction factor, dimensionless
g = gravitational acceleration constant, L^2T^{-1}
P = wetted perimeter, L
q_i = lateral inflow per unit length of river, L^2T^{-1}
Q = flowrate (discharge), L^3T^{-1}
z = absolute elevation of water level above datum, L

If the velocity and cross-sectional area of the river are nearly constant with respect to time (steady flow) but increasing with longitudinal distance, equation (24) can be simplified to

$$\frac{\partial C}{\partial t} = -\frac{1}{A}\frac{\partial (QC)}{\partial x} + \frac{1}{A}\frac{\partial}{\partial x}\left(EA\frac{\partial C}{\partial x}\right) - R \tag{28}$$

Note that the cross-sectional area (A) and flowrate (Q) are permitted to vary with distance in equation (28), but the cross-sectional area and velocity are no longer functions of time as the left hand side of equation (24) implies.

The simplest form of the advection–dispersion equation for one-dimensional rivers is given by equation (29) when A, Q, and E are all constant with respect to time and distance.

$$\frac{\partial C}{\partial t} = -u_x\frac{\partial C}{\partial x} + E_x\frac{\partial^2 C}{\partial x^2} - R \tag{29}$$

Equation (29) may not be exact for many model applications where river velocity and dispersion coefficient vary with longitudinal distance, but it can be useful if applied in segments of the river in which u_x and E_x are constant. The river can be segmented into pieces of relatively constant flow and morphometry, and a new segment can be initiated at each point source discharge of the chemical of interest.

2.3.4 Solutions to Advection-Dispersion Equation and Tracer Experiments

Dyes (e.g., Rhodamine WT) are often used to determine the transport characteristics of a natural water body. Releases of salts and stable isotopes are also used to analyze transport. The advection and dispersion of a dye (tracer) in a three-dimensional (3-D) velocity field under anisotropic conditions (variable dispersion in each direction) has been solved. The solution to equation (23) without reaction in an unbounded domain is

$$C = \frac{M}{2\pi^{3/2}(2E_x t)^{1/2}(2E_y t)^{1/2}(2E_z t)^{1/2}} \times$$
$$\times \exp\left[-\frac{1}{2}\left\{\frac{(x-u_x t)^2}{2E_x t} + \frac{(y-u_y t)^2}{2E_y t} + \frac{(z-u_z t)^2}{2E_z t}\right\}\right] \quad (30)$$

If the dye has been well mixed with depth, the 3-D equation can often be released to a two-dimensional problem with solution as given by equation (31):

$$C = \frac{M}{2\pi\,(2E_x t)^{1/2}(2E_y t)^{1/2}} \exp\left[-\frac{1}{2}\left\{\frac{(x-u_x t)^2}{2E_x t} + \frac{(y-u_y t)^2}{2E_y t}\right\}\right] \quad (31)$$

For large rivers and estuaries, the tracer can be assumed to be well mixed with depth and across the channel (lateral or transverse direction) if the release is engineered with manifolds or if one waits until after the initial mixing period.

2.3.5 Longitudinal Dispersion Coefficient in Rivers

Liu[5,6] used the work of Fischer[7–10] to develop an expression for the longitudinal dispersion coefficient in rivers and streams (E_x, which has units of length squared per time):

$$E_x = \beta \frac{u_x^2 B^3}{U_* A} = \beta \frac{Q_B^2}{U_* D^3} \quad (32)$$

where Liu[6] defined the following:

$\beta = 0.5\,(U_*/u_x)^2$
D = mean depth, L
B = mean width, L
U_* = bed shear velocity, LT^{-1}
u_x = mean stream velocity, LT^{-1}
A = cross-sectional area, L^2
Q_B = river discharge, L^3T^{-1}.

β does not depend on stream morphometry but on the dimensionless bottom roughness. Based on existing data for E_x in streams, the value of E_x can be predicted to within a factor of 6 by equation (32). The bed shear velocity is related empirically to the bed friction factor and mean stream velocity:

$$U_* = \sqrt{\frac{\tau_0}{\rho}} = \sqrt{\frac{f}{8}\,u_x^2} \quad (33)$$

where τ_0 = bed shear stress, $ML^{-1}T^{-2}$
f = Darcy–Weisbach friction factor ≈ 0.02 for natural, fully turbulent flow
ρ = density of water, ML^{-3}

2.3.6 Lateral Dispersion Coefficient in Rivers

Elder[11] proposed an equation for predicting the lateral dispersion coefficient, K_y:

$$E_y = \phi\, DU_* \tag{34}$$

where ϕ is equal to 0.23. The value of $\phi = 0.23$ was obtained by experiment in long, wide laboratory flumes.

Many authors have since investigated the value of ϕ in both laboratory flumes and natural streams. Sayre[12] and Sayre and Chang[13] reported $\phi = 0.17$ in a straight laboratory flume. Yotsukura and Cobb[14] and Yotsukura and Sayre[15] report values of ϕ for natural streams and irrigation canals varying from 0.22 to 0.65, with most values being near 0.3. Other reported values of ϕ range from 0.17 to 0.72. The higher values for ϕ are all for very fast rivers and bends. The conclusions drawn are: (1) that the form of equation (34) is correct in predicting E_y, but ϕ may vary; and (2) that application of Fickian theory to lateral dispersion is correct as long as there are no appreciable lateral currents in the stream.

Okoye[16] refined the determination of ϕ somewhat by use of the aspect ratio, $\lambda = D/B$, the ratio of the stream depth to stream width. He found that ϕ decreased from 0.24 to 0.093 as λ increased from 0.015 to 0.200.

The effect of bends in the channel on E_y is significant. Yotsukura and Sayre[15] reported that ϕ varies from 0.1 to 0.2 for straight channels, ranging in size from laboratory flumes to medium size irrigation channels; from 0.6 to 10 in the Missouri River; and from 0.5 to 2.5 in curved laboratory flumes. Fischer[8] reports that higher values of ϕ are also found near the banks of rivers.

2.3.7 Vertical Dispersion Coefficient in Rivers

Very little experimental work has been done on the vertical dispersion coefficient, K_z. Jobson and Sayre[17] reported a value for marked fluid particles of

$$E_z = \kappa U_* z\left(1 - \frac{z}{D}\right) \tag{35}$$

for a logarithmic vertical velocity distribution, where z is the vertical depth dimension. κ is the von Karman coefficient, which is shown experimentally to be approximately = 0.4.[18] Equation (35) agrees with experimental data fairly closely.

2.3.8 Vertical Eddy Diffusivity in Lakes

Vertical mixing in lakes is not mechanistically the same as that in rivers. The term "eddy diffusivity" is often used to describe the turbulent diffusion coefficient for dissolved substances in lakes. Chemical and thermal stratification serve to limit vertical mixing in lakes, and the eddy diffusivity is usually observed to be a minimum at the thermocline.

Many authors have correlated the vertical eddy diffusivity in stratified lakes to the mean depth, the hypolimnion depth, and the stability frequency. Mortimer[19] first correlated the vertical diffusion coefficient with the mean depth of the lake. He found the following relationship.

$$E_z = 0.0142\ Z^{1.49} \tag{36}$$

where E_z = vertical eddy diffusivity, $m^2\ d^{-1}$
Z = mean depth, m

Vertical eddy diffusivities can be calculated from temperature data by solving the vertical heat balance or by the simplified estimations of Edinger and Geyer.[20] Schnoor and Fruh[21] demonstrated that the mineralization and release of dissolved substances from anaerobic sediment can be used to calculate average hypolimnetic eddy diffusivities. This approach avoids the problem of assuming that heat (temperature) and mass (dissolved substances) will mix with the same rate constant, that is, that the eddy diffusivity must equal the eddy conductivity. A summary of dispersion coefficients and their order of magnitude appears below.

Condition	Dispersion Coefficient, $cm^2\ s^{-1}$
Molecular diffusion	10^{-5}
Compacted sediment	10^{-7}–10^{-5}
Bioturbated sediment	10^{-5}–10^{-4}
Lakes—vertically	10^{-2}–10^{1}
Large rivers—lateral	10^{2}–10^{3}
Large rivers—longitudinal	10^{4}–10^{6}
Estuaries—longitudinal	10^{6}–10^{7}

A literature summary of longitudinal dispersion coefficients for streams and rivers is reported in Table 2.2. The wide range of values reflects the site-specific nature of longitudinal dispersion coefficients and the many hydrologic and morphologic properties that affect mixing processes. An excellent reference for mixing processes in natural waters is that of Fischer et al.[4]

Vertical dispersion coefficients in lakes (eddy diffusivities) have most commonly been determined by the heat budget method[20–22] or by McEwen's method[23]. Radiochemical methods also have been used with success.[23–28] Table 2.3 gives some liter-

Table 2.2 Summary of Dispersion Measurements in Streams

Reach	Depth, m	Width, m	U_* cm s^{-1}	Slope	Velocity, m s^{-1} (Flow, m^3 s^{-1})	Longitudinal Dispersion Coefficient, m^2 s^{-1}	Reference
Chicago Ship Canal	8.07	48.8	1.91			3	10
Sacramento River	4.00		5.1			15	10
River Derwent, Australia	0.25		14			4.6	10
South Platte River, NB	0.46		6.9			16.2	10
Yuma Mesa Canal	3.45		3.45			0.76	10
Green-Duwamish River WA	1.10	20	4.9			6.5–8.5	10
Copper Creek, VA	0.49	16	8			20	10
	0.85	18	10			21	10
	0.49	16	8			9.5	10
	0.40	19	11.6			9.9	10
Clinch River, TN	0.85	47	6.7			14	10
	2.10	60	10.4			54	10
	2.10	53	10.7			47	10
Powell River, TN	0.85	34	5.5			9.5	10
Clinch River, VA	0.58	36	4.9			8.1	10
Coachella Canal, CA	1.56	24	4.3			9.6	10
Monocacy River, MD		35.1		0.0006	0.11 (2.41)	4.6	30
		36.6			0.21 (5.21)	13.9	30
		47.6			0.38 (18.41)	37.2	30
Antietam Creek, MD		15.9		0.0001	0.20 (1.98)	9.3	30
		19.8			0.27 (4.36)	16.3	30
		24.4			0.42 (8.92)	25.6	30
Missouri River, NB–IA		182.9		0.0002	0.91 (379.50)	464.7	30
		201.2			1.24 (911.92)	836.4	30
		196.6			1.48 (934.58)	1,487.0	30
Clinch River, TN		47.3		0.0006	0.21 (9.20)	13.9	30
		53.4			0.44 (50.98)	46.5	30
		59.5			0.65 (84.96)	55.8	30
Bayou Anacoco, LA		19.8		0.0005	0.21 (2.44)	13.9	30
		25.9			0.34 (8.21)	32.5	30
		36.6			0.40 (13.45)	39.5	30
Nooksack River, WA		64.0			0.68 (32.57)	34.9	30
		86.0		0.0098	1.3 (303.03)	153.3	30
Wind/Bighorn Rivers, WY		67.1		0.0013	0.89 (59.33)	41.8	30
		68.6			1.56 (230.81)	162.6	30
Elkhorn River, NB		32.6		0.00073	0.34 (4.25)	9.3	30
John Day River, OR		25.0		0.00355	(14.16)	13.9	30
		34.1		0.00135	(69.10)	65.1	30
Comite River, LA		12.5		0.00078	0.23 (0.99)	7.0	30
		15.9			0.35 (2.41)	13.9	30
Amite River, LA		36.6		0.00061	0.24 (8.64)	23.2	30

Table 2.2 *(Continued)*

Reach	Depth, m	Width, m	U_* cm s^{-1}	Slope	Velocity m s^{-1} (Flow, m^3 s^{-1})	Longitudinal Dispersion Coefficient, m^2 s^{-1}	Reference
					0.36 (14.16)	30.2	30
Sabine River, LA		42.4		0.00015	0.57 (118.95)	316.0	30
		127.4			0.65 (389.41)	669.1	30
Yadkin River, NC		70.1		0.00044	0.44 (70.80)	213.8	30
Muddy Creek, NC		13.4		0.00083	0.30 (3.96)	13.9	30
		19.5			0.38 (10.62)	32.5	30
Sabine River, TX		35.1		0.00018	0.18 (7.36)	39.5	30
White River, IN		67.1		0.00036	0.30 (12.74)	30.2	30
Chattahoochee River, GA		65.5		0.0052	0.34	32.5	30
Susquehanna River, PA		202.7		0.00032	0.33	92.9	30
Miljacka River MI,	0.285	11.28	5.5		0.342/1.02	2.22	2
	0.3	8.6	6.6		0.368/1.02		2
	0.29	10.53	6.2		0.35/1.02	5.66	2
	0.295	12.0	49		0.332/1.02	0.07	2
Uvas Creek, CA		(0.3)*			(0.0125)	0.12	31
		(0.45)*			(0.0125)	0.48	31
		(0.30)*			(0.0125)	0.12	31
		(0.42)*			(0.0125)	0.15	31
		(0.72)*			(0.0133)	0.24	31
		(0.82)*			(0.0136)	0.31	31
		(2.08)*			(0.0140)	0.40	31

*Cross-sectional area, m^2.

ature values for the vertical dispersion coefficient at the thermocline (minimum value), and Table 2.4 reports the mean vertical dispersion coefficient for the entire water column. Vertical dispersion is a function of the depth and morphometry of the lake, fetch-to-wind direction relationship, solar insulation and light penetration, and other factors. Example calculations for lake dispersion coefficients are presented at the end of this chapter; data for these calculations were taken from actual field measurements.

Modeling hydrophobic chemical contaminants (e.g., DDT, PCB, kepone, dioxin, dieldrin) that are strongly sorbed to sediments requires knowledge of the diffusion and release rates from contaminated sediments into overlying waters. Radiotracers that occur naturally and from bomb-testing have been used with success in analyzing sediment pore-water diffusion rates. Table 2.5 reports some values found in the literature. Most pore-water diffusion coefficients are on the order of molecular diffusion coefficients (~10^{-5} cm^2 s–1) or smaller. Bioturbation by benthic fauna or fish may significantly increase pore-water transfer to overlying water.

Table 2.3 Vertical Dispersion Coefficient for Stratified Lakes Across the Thermocline

Site	Month	Vertical Dispersion, $cm^2\ s^{-1}$	Thermocline Depth, m	Data from	Reference
Lake Zurich, Switzerland	April	0.71	5	Temperature	32
	May	0.14	10	Temperature	32
	June	0.064	10	Temperature	32
	July	0.039	12.5	Temperature	32
	Aug	0.026	10	Temperature	32
	Sept	0.020	10–12.5	Temperature	32
	Oct	0.074	20	Temperature	32
Lake Greifensee, Switzerland		0.25	10	PO_4^{3-}	26
Lake Baldeggersee, Switzerland (limno-corral)	May	0.0021	9	Temperature	27
	June	0.08	8	Temperature	27
	July	0.003	7–9	Temperature	27
	Aug	0.08	7–9	Temperature	27
	Sept	0.0013	9	Temperature	27
	June	0.08	7	^{222}Rn	27
	Aug	0.09	6	Rn	27
	Sept	0.05	9.5	Rn	27
	Oct	0.05	9.5	Rn	27
Lake Onondaga Lake, NY	May	0.04	11.5	Temperature	33
	June	0.09	10.5	Temperature	33
	July	0.03	11.5	Temperature	33
	Aug	0.005	11.5	Temperature	33
	Sept	0.008	12.5	Temperature	33
	Oct	0.015	—	Temperature	33
Lake Baikal, Russia		2.5–7.4		Temperature	34
Lake Tahoe, NV		0.178		Temperature	34
Lake Ontario		0.125, 0.063		Temperature	34
Lake Cayuga, NY		0.178, 0.25		Temperature	34
Lake Luzern, Switzerland		0.10		Temperature	34
Lake Zurich, Switzerland		0.03		Temperature	34
Lake Washington, WA		0.03		Temperature	34
Lake Tiberias, Israel		0.063		Temperature	34
Lake Sammamish, WA		0.03		Temperature	34
Lake ELA 305, Ontario		0.01		Temperature	34
Lake Mendota, WI		0.025		Temperature	34
Linsley Pond, CT		0.003		Temperature	34
ELA 240, Ontario		0.004		Temperature	34
ELA 227, Ontario		0.003		Temperature	34
Cayuga Lake, NY		0.253	21	Temperature	35
Castle Lake, CA	July	0.011	6	Temperature	36
	July	0.0091	6	Temperature	36
	July	0.0069	7	Temperature	36
	July	0.0068	7	Temperature	36
	July	0.0068	7	Temperature	36

Table 2.3 *(continued)*

Site	Month	Vertical Dispersion, $cm^2\ s^{-1}$	Thermocline Depth, m	Data from	Reference
	July	0.0042	7	Temperature	36
	Aug	0.0004	9	Temperature	36
	Aug	0.0062	9	Temperature	36
	Aug	0.0041	9	Temperature	36
	Aug	0.0061	9	Temperature	36
	Aug	0.0036	9	Temperature	36
	Aug	0.0076	9	Temperature	36
	Sept	0.0077	9	Temperature	36
ELA 227, Ontario		0.0017	8	Tritium	23
ELA 224, Ontario		0.018	18	Tritium	23
Lake Valencia, Venezuela		0.114	20	Temperature	37
Lake Erie		0.21	16	$\delta\ ^3He$	29

Table 2.4 Whole Lake Average Vertical Dispersion Coefficient

Site	Month	Vertical Dispersion, $cm^2\ s^{-1}$	Data from	Reference
Lake Erie		0.58	δ^3 He	29
Lake Huron		1.16	δ^3 He	29
Lake Ontario		3.47	δ^3 He	29
Wellington Reservoir, Australia		1.00	Temperature	38
White Lake, MI		0.4	PO_4^{3-}	39
Lake LBJ, TX	Feb–Apr	0.18	Temperature	22
	May–June	0.12	Temperature	22
	July–Jan	0.01	Temperature	22
Lake Erie	Average	15	Temperature	40
Lake Huron	(No stratification)	1.16	Temperature	41
Lake Erie	Unstratified	102	Temperature	42,43
	Stratified	0.05–0.25	Temperature	42,43
Cayuga Lake		2.31	Temperature	3
Lake Greifensee	April	0.2	^{222}Rn	8
	May–Aug	0.15	^{222}Rn	28
	Sept–Nov	0.05	^{222}Rn	28

Table 2.5 Interstitial Sediment Pore-Water Diffusion Coefficients

Site	Vertical Dispersion, $cm^2\ s^{-1}$	Data from	Reference
White Lake, MI	2×10^{-6}		39
Lake Erie	4×10^{-6}	^{90}Sr	44
	2×10^{-5}	^{137}Cs	44
Lake Ontario	2×10^{-5}	^{90}Sr	44
	2×10^{-6}	^{90}Sr	44
	2×10^{-5}	^{137}Cs	44
Green Bay, Lake Michigan	1.3×10^{-7}	^{210}Pb	45
	6.3×10^{-9}	^{210}Pb	45
Lake Greifensee	10^{-10}	^{230}Th	26
	10^{-9}	^{226}Ra	26
	0.8×10^{-5}	^{222}Rn	26

2.4 COMPARTMENTALIZATION

2.4.1 Choosing a Transport Model

It is possible to estimate the relative importance of advection compared to dispersion with the Peclet number:

$$\mathrm{Pe} = uL/E \tag{37}$$

where Pe = Peclet number, dimensionless
u = mean velocity, $L\ T^{-1}$
L = segment length, L
E = dispersion coefficient, $L^2\ T^{-1}$.

If the Peclet number is significantly greater than 1.0, advection predominates; if it is much less than 1.0, dispersion predominates in the transport of dissolved, conservative substances.

If there is a significant transformation rate, the reaction number can be helpful:

$$\text{Rxn No.} = \frac{kE}{u^2} \tag{38}$$

where k is the first-order reaction rate constant, T^{-1}. If the reaction number is less than 0.1, then advection predominates and a model approaching plug flow is appropriate. If the reaction number is greater than 10, then dispersion controls the transport and the system is essentially completely mixed. Otherwise a plug flow with dispersion model or a number of compartments in series will best simulate the prototype water body.

2.4.2 Compartmentalization and Box Models

Compartmentalization refers to the segmentation of model ecosystems into various "completely mixed" boxes of known volume and interchange. Interchange between compartments is simulated via bulk dispersion or equal counterflows between compartments. Compartmentalization is a popular assumption in pollutant fate modeling because the assumption of complete mixing reduces the set of partial differential equations (in time and space) to one of ordinary differential equations (in time only). Nevertheless, it is possible to recover some coarse spatial information by introducing a number of interconnected compartments.

A completely mixed flow-through (CMF) compartment contains an ideal mixing of fluid in which turbulence is so large that no concentration gradients can exist within the compartment. This corresponds to the assumption that $E_{x,y,z} = \infty$. The 3-D partial differential equation (23) becomes much simpler, and it takes the form of an ordinary differential equation with exchange counterflows between compartments.

Accumulation of mass within Compartment j	=	Mass inflows to j	+	Dispersive inflows to j	−	Mass outflows from j	−	Dispersive outflows from j	−	Transformation reactions within j

$$V_j \frac{dC_j}{dt} = \sum_{k=1}^{n} Q_{j,k}C_k + \sum_{k=1}^{n} Q'_{j,k}C_k - \sum_{k=1}^{n} Q_{k,j}C_j - \sum_{k=1}^{n} Q'_{k,j}C_j - kC_jV_j \tag{39}$$

where V_j = volume of j compartment, L^3
C_j = concentration within j compartment, $M\ L^{-3}$
t = time, T
n = number of adjacent compartments to j
$Q_{j,k}$ = inflow from compartment k to compartment j, $L^3\ T^{-1}$
C_k = concentration in compartment k, $M\ L^{-3}$
Q = dispersive (interchange) flow from k to j, $L^3\ T^{-1}$
$Q_{k,j}$ = outflow from j to k, $L^3\ T^{-1}$
$Q'_{k,j}$ = dispersive (interchange) flow from j to k, $L^3\ T^{-1}$
k = pseudo-first-order rate constant for transformation, T^{-1}
$Q'_{j,k} = Q'_{k,j}$ a symmetric matrix with zero diagonal

Equation (39) can be rewritten in terms of bulk dispersion coefficients:

$$V_j \frac{dC_j}{dt} = \sum_{k=1}^{n} Q_{j,k}C_k - \sum_{k=1}^{n} Q_{k,j}C_j + \sum_{k=1}^{n} E'_{j,k}A_{j,k}(C_k - C_j)/\ell_{j,k} - kC_jV_j \tag{40}$$

where E' = bulk dispersion coefficient, $L^2\ T^{-1}$
$A_{j,k}$ = interfacial area between compartments j and k, L^2

$\ell_{j,k}$ = distance between midpoints of compartments, L

There is one mass balance equation [e.g., equation (40)] for each of the j compartments. This set of ordinary differential equations is solved simultaneously by numerical computer methods.

Bulk dispersion coefficients between compartments are dependent on the scale chosen for the compartments. They are not equivalent to measured dispersion coefficients from dye studies, which are usually derived from the continuous partial differential equations. The very nature of the compartmentalized system introduces considerable mixing into the model. Such mixing or numerical dispersion is in addition to the bulk dispersion specified by the bulk dispersion coefficient.

Streams and swift-flowing rivers may approach a 1-D plug-flow system (i.e., the water is completely mixed in the lateral and vertical dimensions, but there is no mixing in the longitudinal dimension). In an ideal plug-flow system, the longitudinal dispersion coefficient is equal to zero because no forward or backward mixing occurs. For this case, an infinite number of compartments (of infinitesimal length in the longitudinal direction) would be required to produce zero longitudinal mixing. Because it is impossible to specify an infinite number of compartments, one chooses a finite number of compartments and accepts the artificial dispersion that accompanies that choice. One method of estimating the artificial or numerical dispersion of a compartmentalized model for an ideal, plug-flow system is given by equation (41) for an explicit upwind differencing method.

$$E_x = \frac{u\,\Delta x}{2}\left(1 - \frac{u\,\Delta t}{\Delta x}\right) \tag{41}$$

where E_x = artificial numerical dispersion coefficient, $L^2\,T^{-1}$
u = mean longitudinal velocity, $L\,T^{-1}$
Δx = longitudinal length of equally spaced compartments, L
Δt = time step for numerical computation, T

One approach would be to set the artificial dispersion coefficient equal to the measured or estimated dispersion coefficient from equation (32). With this approach, it is not necessary to use bulk dispersion coefficients; rather, one allows the artificial dispersion of the model to account for the actual dispersion of the prototype.

A better approach is to adjust the time step to minimize E_x while preserving stability:

$$\Delta t = \min_i \left(\frac{\Delta x_i}{u_i}\right)$$

where i refers to the physical compartments.

In general, most river simulations require many compartments due to their nearly plug flow nature, as indicated by their large Peclet number [equation (37)]. The greater the number of compartments, the greater the tendency toward plug-flow conditions. It is poor practice to simulate a riverine environment with one completely mixed compartment.

Lakes, reservoirs, and embayments may require a number of compartments if one desires some spatial detail, such as concentration profiles. These compartments should be chosen to relate to the physical and chemical realities of the prototype. For example, a logical choice for a stratified lake is to have two compartments: an epilimnion and a hypolimnion (Figure 2.7). Mixing between compartments can be accomplished by interchanging flows:

$$J = Q_{ex}C_{epi} - Q_{ex}C_{hypo} \tag{42}$$

or

$$J = Q_{ex}\,(C_{epi} - C_{hypo})$$

where J = net mass flux from epilimnion to hypolimnion due to vertical mixing, $M\,T^{-1}$

Q_{ex} = exchange flow, $L^3\,T^{-1}$

C = concentration of organic, $M\,L^{-3}$

The magnitude of the interchange flow, Q_{ex}, can be determined from tracer studies or from temperature profiles and simulations. Bulk dispersion coefficients can then be calculated based on the interchange flow as $E' = Q_{ex}\ell_{epi/hypo}/A$, where $\ell_{epi/hypo}$ is the distance between centroids of two adjacent compartments.

Sometimes only coarse information is required for a given use of a model. The literature offers many examples of modeling efforts based on very simple transport models. The Great Lakes have often been simulated as single-compartment, completely mixed lakes in series.[46–48] Toxic chemical screening methodologies are usually based on organic chemical properties that are known only within an order of magnitude. In such cases, it may not be necessary to simulate transport with great accuracy. A distinct trade-off exists between errors in transport formulations and errors in reaction rate constants as shown in Table 2.6. If the sum of the pseudo-first-order reaction rate constant is accurately determined in the field or laboratory, then an accurate model simulation will require a realistic transport formulation. If the reaction rate constant and/or the detention time are low ($k\tau = 0.01$), the choice of the number of compartments is not very critical. Errors in outflow concentration of greater than 10% will occur, however, if the dimensionless number $k\tau$ becomes greater than 1.0.

For example, consider a hypothetical lake whose steady-state outlet concentration of a toxic chemical is determined to be 0.01 times the inflow concentration. Suppose the hydraulic detention time, t, of the lake is 10 days, and the transformation reaction rate constant is determined to be 1.0/day ($k\tau = 10$). The lake is behaving like three compartments in series according to Table 2.6. The model calibration,

Table 2.6 Outflow Concentration Divided by Inflow Concentration at Steady State as a Function of Number of Completely Mixed Compartments and $k\tau$

	C/C_{in} Values				
	Rate Constant × Detention Time				
Model	$k\tau = 0.01$	$k\tau = 0.1$	$k\tau = 1$	$k\tau = 10$	$k\tau = 100$
CMF[a]					
One-compartment	0.99	0.91	0.50	0.09	0.01
Three-compartment[b]	0.99	0.91	0.42	0.01	2×10^{-5}
Ten-compartment[b]	0.99	0.91	0.39	1×10^{-3}	4×10^{-11}
Plug flow[c] (∞ compartments)	0.99	0.90	0.37	5×10^{-5}	4×10^{-44}

[a] $C/C_{in} = 1/(k\tau + 1)$

[b] $C/C_{in} = 1/[(k\tau/n) + 1]^n$

[c] $C/C_{in} = \exp(-k\tau)$

where τ = total hydraulic detention time
k = first-order reaction rate constant
n = number of compartments
C_{in} = input concentration

however, would have required a reaction rate constant of 10/day in order to obtain the observed result of $C/C_0 = 0.01$ if only the completely mixed compartment had been assumed.

If better than order-of-magnitude accuracy is required in the model, one should estimate the dispersion coefficients from dye studies, temperature simulations, or equations (32) through (36). This allows the proper compartmental configuration to be selected, including consideration of numerical dispersion based on equation (41).

2.5 SEDIMENT TRANSPORT

2.5.1 Partitioning

A chemical is partitioned into a dissolved and particulate adsorbed phase based on its sediment-to-water partition coefficient, K_p.[49] The dimensionless ratio of the dissolved to the particulate concentration is the product of the partition coefficient and the concentration of suspended solids, assuming local equilibrium:

$$C_p/C = K_p M \tag{43}$$

where C_p = particulate chemical concentration, $\mu g\ L^{-1}$
C = dissolved chemical concentration, $\mu g\ L^{-1}$
K_p = sediment/water partition coefficient, $L\ kg^{-1}$
M = suspended solids concentration, $kg\ L^{-1}$

The particulate and dissolved concentrations can be calculated from knowledge of the total concentration, C_T, as stated in equations (44) and (45):

$$C_p = \frac{K_p M}{1 + K_p M} C_T \tag{44}$$

$$C = \frac{1}{1 + K_p M} C_T \tag{45}$$

These concentrations can be calculated for the water column or the bed sediment, by using the concentration of suspended solids in the water (M) or in the bed (M_b), where $M_b = M/n$, the bed sediment concentration in kg L^{-1} of pore water, and n = the porosity of the bed sediment.

2.5.2 Suspended Load

The suspended load of solids in a river or stream is defined as a flowrate times the concentration of suspended solids (e.g., kg d^{-1} or tons d^{-1}); the mean load is greatly affected by peak flows. Peak flows cause large inputs of allochthonous material from erosion and runoff as well as increases in scour and resuspension of bed and bank sediment.

The average suspended load is not equal to the average flow times the average concentration, as stated in equation (46),

$$\overline{Q} \times \overline{C} \neq \overline{QC} \tag{46}$$

but it is calculated in equation (47) as per Example 2.1:

$$\overline{QC} = (\overline{Q} \times \overline{C}) + \overline{Q'C'} \tag{47}$$

The mean fluctuation of mass, $\overline{Q'C'}$, is usually greater than the first term of equation (47) and contributes greatly to the average suspended load. These equations hold true for the mass of suspended solids as well as the mass of adsorbed chemical.

2.5.3 Bed Load

Several formulas have been reported to calculate the rate of sediment movement very near the bottom. These equations were developed for rivers and noncohesive sediments, that is, fine-to-coarse sands and gravel. It is important to note that it is not sands, but rather silts and clays, to which most chemicals sorb. Therefore these equations are of limited predictive value in environmental chemical modeling. Generally, bed load transport is a small fraction of total sediment transport (suspended load plus bed load). In estuaries, however, bed load transport of fine silts and clays may be an important contributor to the fate of chemical contaminants. Bed load consists of those particles that creep, flow, or saltate very near to the bottom (within a few particle diameters). Figure 2.8 is a schematic of bed load and suspended load in a stream or river.

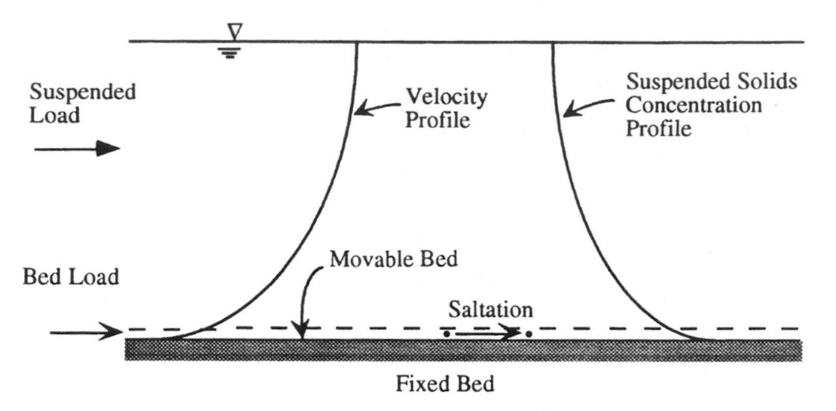

Figure 2.8 Suspended load and bed load. Bed load is operationally defined as whatever the bed load sampler can measure. Bed load occurs within a few millimeters of the fixed bed.

2.5.4 Sedimentation

Suspended sediment particles and adsorbed chemicals are transported downstream at nearly the mean current velocity. In addition, they are transported verticially downward by their mean sedimentation velocity. Generally, silt and clay-size particles settle according to Stokes' law,[50] in proportion to the square of the particle diameter and the difference between sediment and water densitities:

$$W = 8.64 \left(\frac{g}{18\mu} \right) (\rho_s - \rho_w) d_s^2 \tag{48}$$

where W = particle fall velocity, ft s^{-1}
ρ_s = density of sediment particle, 2–2.7 g cm^{-3}
ρ_w = density of water, 1 g cm^{-3}
g = gravitational constant, 981 cm s^{-2}
d_s = sediment particle diameter, mm
μ = absolute viscosity of water, 0.01 poise (g cm^{-1} s^{-1}) at 20 °C

Generally, it is the washload (fine silt and clay-size particles) that carries most of the mass of adsorbed chemical. These materials have very small fall velocities, on the order of 0.3–1.0 m d^{-1} for clays of 2–4 μm nominal diameter and 3–30 m d^{-1} for silts of 10–20 μm nominal diameter.

Once a particle reaches the bed, a certain probability exists that it can be scoured from the bed sediment and resuspended. The difference between sedimentation and resuspension represents net sedimentation. Often it is possible to utilize a *net* sedimentation rate constant in a pollutant fate model to account for both processes. In many ecosystems where the bed is aggrading, sedimentation is much larger than resuspension.[51] The net sedimentation rate constant can be calculated as follows:

$$k_s = \frac{W}{H} - k_u \simeq \frac{W}{H} \tag{49}$$

where k_s = net sedimentation rate constant, T^{-1}
W = mean particle fall velocity, L T^{-1}
H = mean depth, L
k_u = scour/resuspension rate constant, T^{-1}

2.5.5 Scour and Resuspension

Quantitative relationships to predict scour and resuspension of cohesive sediments are difficult to develop due to the number of variables involved. Sayre and Chang[13] reported on the vertical scour and dispersion of silt particles in flumes. DiToro et al.[43] recommended a resuspension velocity (W_{rs}) of about 1–30 mm yr^{-1} based on model calibration studies. The turbulent vertical eddy diffusivity for sediment (ε_s) is also related to the scour coefficient and/or resuspension velocity.

Under steady-state conditions, the sedimentation of suspended sediment must equal the scour and resuspension of sediment.

$$w\overline{C} + \varepsilon_s \frac{\partial \overline{C}}{\partial z} = 0 \tag{50}$$

where w = sedimentation velocity, L T^{-1}
ε_s = suspended sediment vertical eddy diffusivity, L^2 T^{-1}
C = concentration of suspended sediment, M L^{-3}.

Under time-varying conditions, however, the boundary condition at the bed–water interface is more complex. According to Onishi and Wise,[52] the following equation applies, based on the work of Krone[53] and Partheniades[54]:

$$pw\overline{C} + \varepsilon_s \frac{\partial \overline{C}}{\partial z} = S_D - S_R \tag{51}$$

where p = probability that descending particle will "stick" to the bed
$S_D = \frac{2w\overline{C}}{h}(1 - \tau_0/\tau_{cD})$ = rate of bed deposition, M L^{-2} T^{-1}
$S_R = M_j\left(\frac{\tau_0}{\tau_{cR}} - 1\right)$ = rate of bed scour, M L^{-2} T^{-1}
M_j = erodibility coefficient, M L^{-2} T^{-1}
τ_{cR} = critical bed shear required for resuspension, M $L^{-1}T^{-2}$
τ_{cD} = critical bed shear stress that prevents deposition, M $L^{-1}T^{-2}$
h = ratio of depth of water to depth of active bed layer

Equation (51) shows that the bed can either be aggrading or degrading at any time or location depending on the relationship between S_D and S_R.

2.5.6 Desorption/Diffusion

In addition to sedimentation and scour/resuspension, an adsorbed chemical can desorb from the bed sediment. Likewise, dissolved chemical can adsorb from the water to the bed. Both pathways can be presented by a diffusion coefficient and a concentration gradient or difference between pore-water and overlying dissolved chemical concentrations.

Sediment mass balances must include terms for advection, sedimentation, scour/resuspension, and possibly vertical dispersion. At the bottom, bed load movement may be included. Processes that affect the fate of dissolved substances include desorption from the bed (or adsorption from the water column), advection, dispersion, and transformation reactions. Adsorbed particulate chemical is removed from the water column by sedimentation and returned to the water column by scour. Models used to evaluate transport and transformation should include these processes.

Often, it is possible to neglect the kinetics of adsorption and desorption in favor of a local equilibrium assumption. Over the time scales of interest, this may be a good assumption. Bed load is sometimes small relative to wash load movement of absorbed chemicals and can be neglected. Under steady-state conditions, net sedimentation rates are often used to simplify the transport of sedimentation and scour.

2.6 LAKE DISPERSION CALCULATIONS

The steps for calculating vertical dispersion across the thermocline in a lake from temperature data are presented below. These methods were derived from the heat dispersivity equation, assuming that E does not vary much with depth over the region of interest, particularly the thermocline:

$$\frac{\partial \theta}{\partial t} = E \frac{\partial^2 \theta}{\partial z^2} \tag{52}$$

where θ is temperature, t is time, E is thermal dispersivity, and z is distance. It is assumed that no heat has entered the lower part of the water column by any mechanism other than vertical turbulent transport, E. The assumption is made that mass transfer through dispersion occurs at the same rate as heat transfer. The analogy is applied by substituting concentration or mass (C) into the equation, thus

$$\frac{\partial C}{\partial t} = E \frac{\partial^2 C}{\partial z^2} \tag{53}$$

2.6.1 McEwen's Method

This method of computing lake dispersion is based on fitting an exponential curve to the mean temperature data in the thermocline and hypolimnion. If the data are of a linear or otherwise nonexponential shape, this method is inappropriate. The reader is referred to Hutchinson.[55,56]

2.6.2 Heat Budget Method

The vertical thermal dispersivity also can be estimated from the total heat entering and leaving the lake. A number of field measurements are necessary (pyroheliometer data, air temperature) as well as temperature profiles throughout the lake. A heat budget results in vertical dispersion coefficients that are a function both of depth and time, $E = f(z, t)$.[22]

The basic equation is based on heat transfer and is formulated similar to a mass balance:

$$\begin{aligned}\frac{d(V_j\theta_j)}{dt} &= (Q_{ij}\theta_{ij} - Q_{oj}\theta_{oj}) + (Q_{vj}\theta_{j-1} - Q_{vj+1}\theta_j) \\ &- (E_j a_j/\Delta z)\,(\theta_j - \theta_{j-1}) + (E_{j+1}a_{j+1}/\Delta z) \\ &(\theta_{j+1} - \theta_j) + V_j h_{\text{net}}/(\rho)c\,\Delta z\end{aligned} \tag{54}$$

where V_j is the volume of the jth slice (m^3); θ_j is the mean temperature in the jth slice (°C), Q_i and Q_o are inflows and outflows to slice j, respectively, as θ_o and θ_i are the temperature associated with those flows (Q in m^3 s^{-1} and θ in °C), Q_{vj} is the vertical flowrate at the bottom of the jth element, where upward flow is positive (m^3 s^{-1}), E_j is the dispersion across the bottom of slice j (cm^2 s^{-}), a_j is the bottom surface area of slice j (m^2), t is time (s), h_{net} is the net heat flux (cal s^{-1} cm^{-3}), ρ is density (g cm^{-3}), c is specific heat (cal g^{-1} °C^{-1}), and Δz is the thickness (m) of a slice, which must be the same for all slices.

The heat flux at the surface, h', equals $h_{\text{net}}/\Delta z$, and

$$h_{\text{net}} = \beta h_s + h_a - h_b - h_e + h_c \tag{55}$$

where β is the fraction of short-wave radiation absorbed at the surface, h_s is short-wave radiative flux, h_a is long-wave or atmospheric radiative flux, h_b is back radiation, h_e is evaporative energy flux, and h_c is convective energy flux.

The net heat flux to the jth slice is

$$h_{\text{net}} = h_j'\,\Delta z \tag{56}$$

Below the surface, only short-wave radiation is absorbed. In deeper slices, h_j' is an exponential function of depth. The quantity of solar radiation absorbed by the jth element is expressed as

$$h_j' = \frac{\Phi_{j+1}\, a_{j+1} - \Phi_j\, a_j}{(a_j + a_{j+1})/2\Delta z} \tag{57}$$

where

$$\Phi(z) = (1 - \beta)\, h_s \exp\,[-h\,(z_n - z)]$$

and β and h_s are defined above. The short-wave radiative flux can be estimated from pyroheliometer data (where the field data are given in cal cm^{-2} s^{-1}), and β is 0.4. h is the exponential decay constant for the absorption of solar radiation with depth, z_n is the elevation of the bottom of the surface element, and z is the elevation of interest.

The long-wave radiative flux is

$$h_a = 1.17 \times 10^{-18}\,(\theta_a + 273)^6 C_L \tag{58}$$

where θ_a is the air temperature 2 m above the water surface (°C), $C_L = 1 + 0.17(\text{fraction cloudy})^2$. The back radiation is

$$h_b = 1.192 \times 10^{-14}\,(\theta_s + 273)^4 \tag{59}$$

where θ_s is the surface water temperature (°C). The evaporative heat flux, h_e, at the surface is

$$h_e = 2.23 \times 10^{-5}\, w(e_s - e_a) \tag{60}$$

where w is the wind speed (kph), e_s is the saturation vapor pressure at the water surface (mm Hg), and e_a is the water vapor pressure (mm Hg). The convective heat flux, h_c, at the surface is

$$h_c = 1.89 \times 10^{-4}\, h_e(\theta_a - \theta_s)P_a/(e_s - e_a) \tag{61}$$

where h_e, θ_a, θ_s, e_s and e_a are defined above, and P_a is the atmospheric pressure (mm Hg).

Once the net heat flux, h_{net}, is determined for each elemental slice of the lake in the vertical dimension, it is possible to solve the system of equations represented in equation (54). One must specify the slices (elevations) where inflows and outflow occur ($Q_{oj}\theta_{oj}$ and $Q_{ij}\theta_{ij}$), and the vertical flows for each slice (Q_{vj}) based on continuity. There are n equations (one for each slice) and n unknowns (E_j) when we solve the inverse problem. The temperature profile is known and we solve the system of equations for $E_j(z)$, the vertical thermal diffusivity values. Alternatively, a numerical model can be utilized to solve the system of equations assuming various E_j profiles until the vertical temperature data are accurately simulated.

2.7 SIMPLE TRANSPORT MODELS

In this Section, analytical solution techniques are described for simple transport models of idealized systems. It should be recognized, however, that the models described here represent only the simplest, most ideal mixing conditions. In spite of this fact, they may be quite useful in checking more complicated numerical results and in gaining insight to the dynamics of chemical movement in the environment.

2.7.1 Completely Mixed Systems

An ideal completely mixed system is illustrated, using a lake as an example, in Figure 2.9. The major assumptions involved in this model are that the concentration of chemicals in the lake is uniform (completely mixed) and the lake outlet has a concentration, C, and the concentration is the same everywhere within the lake. The mass balance yields

Change in mass in the lake	=	Mass in inflow	−	Mass in outflow	±	Mass reacting in the lake

This can be expressed mathematically as

$$\frac{\Delta(VC)}{\Delta t} = Q_{\text{in}}\, C_{\text{in}} - Q_{\text{out}}\, C \pm rV$$

where C_{in} = chemical concentration in inflow, ML^{-3}
C = chemical concentration in the lake and in outflow, ML^{-3}
Q_{in} = volumetric inflow rate, L^3T^{-1}
Q_{out} = volumetric outflow rate, L^3T^{-1}
V = volume of the lake, L^3
r = reaction rate, $ML^{-3}T^{-1}$; positive and negative signs indicate formation and decay reactions, respectively
t = time, T

The limit as Δt approaches zero gives the ordinary differential equation below.

$$\frac{d(VC)}{dt} = Q_{\text{in}}\, C_{\text{in}} - Q_{\text{out}}\, C \pm rV \tag{62}$$

The volume of the lake V, flows Q_{in} and Q_{out}, and inflow concentration C_{in} can be time-dependent variables. In addition to the completely mixed assumption, assumptions also may be made to simplify the equation:

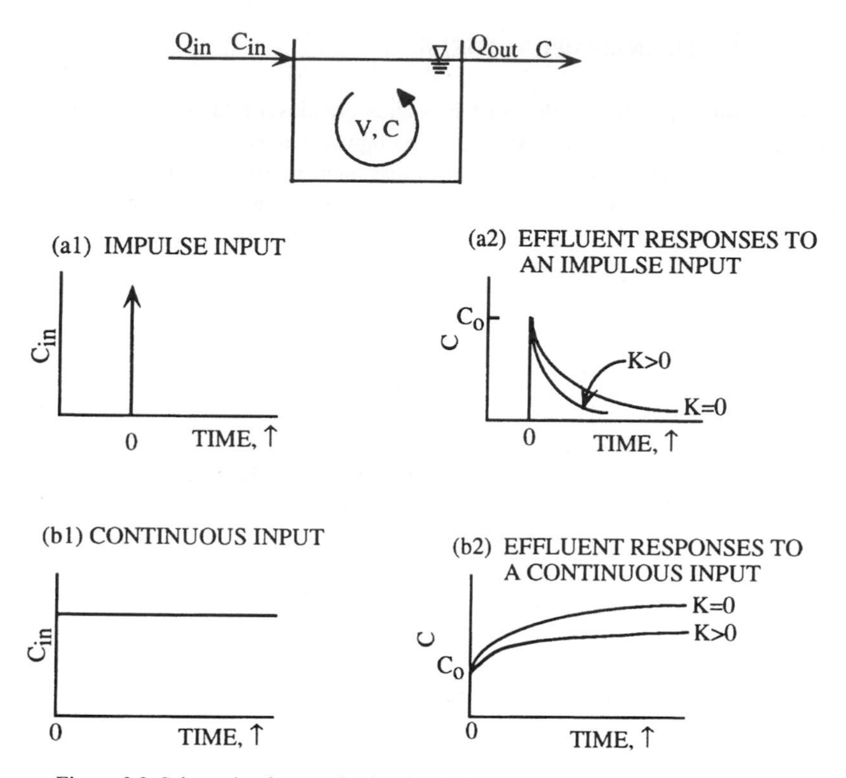

Figure 2.9 Schematic of a completely mixed lake, with inputs and effluent responses.

1. The inflow concentration C_{in} is constant.
2. The volumetric flowrate into and out of the lake is constant ($Q_{in} = Q_{out} = Q =$ constant), and the volume of the lake V is constant ($dV/dt = 0$).
3. The rate of change in the concentration C that is occurring within the lake is governed by a first-order reaction ($r = -kC$; note that the negative sign refers to a decay reaction).

Incorporating these assumptions, equation (62) can be written as

$$V\frac{dC}{dt} = QC_{in} - QC - kCV \tag{63}$$

Equation (63) is the general first-order decay equation for a completely mixed system.

A response to an accidental spill of chemical into a lake, for example, can be formulated using the impulse (or delta) function if the discharge of chemical occurred for a relatively short time period. For a simple case, that conservative tracer is instantaneously injected into the lake as in impulse input, equation (63) may be reduced to

$$V \frac{dC}{dt} = -QC \tag{64}$$

Dividing by V yields

$$\frac{dC}{dt} = -\frac{Q}{V} C = -\frac{C}{\tau} \tag{65}$$

where $\tau = V/Q$ = mean hydraulic detention time (T). With an initial condition of $C = C_0$ at $t = 0$, equation (65) can be integrated as

$$\int_{C_0}^{C} \frac{1}{C}\, dC = -\frac{1}{\tau} \int_0^t dt \tag{66}$$

Integrating equation (66) for the time interval 0 to t yields

$$C = C_0 \exp(-t/\tau) \tag{67}$$

Equation (67) is the analytical solution to an impulse input for a conservative tracer.

In the event that a reactive chemical was spilled into the lake, equation (63) may be reduced to

$$V \frac{dC}{dt} = -QC - kCV \tag{68}$$

which can be solved similarly:

$$C = C_0 \exp[-(k + 1/\tau)t] \tag{69}$$

Equation (69) is the analytical solution to an impulse input for reactive substances. A graphical sketch of the responses to an impulse input for reactive ($k > 0$) and nonreactive ($k = 0$) chemicals is shown in Figure 2.9.

Response to a continuous load, such as a waste discharge from a municipality or an industry to a lake, is also represented by equation (63), which can be rewritten as

$$\frac{dC}{dt} + \left(\frac{1}{\tau} + k\right)C = \frac{C_{\text{in}}}{\tau} \tag{70}$$

Equation (70) has the form of a first-order, nonhomogeneous linear differential equation. If only the steady-state concentration is desired, then solution of equation (70) can be obtained noting that the change in concentration is zero ($dC/dt = 0$). The steady-state solution of equation (70) is given as:

$$C_{ss} = \frac{QC_{in}}{Q + kV} = \frac{C_{in}}{1 + k\tau} \tag{71}$$

where C_{ss} is the steady-state concentration (ML^{-3}). Note that there is no change in concentration with respect to time, t.

If one desires to see the change in concentration with time, the nonsteady-state solution can be obtained for equation (70), a first-order linear differential equation that has a general form:

$$y' + p(x)y = q(x) \tag{72}$$

with the general solution

$$y = y_0 \exp[-P(t)] + \exp[-P(t)] \int_0^t \exp[+P(t)]\, q(t)\, dt \tag{73}$$

where $P(t) = \int p(t)\, dt$

This solution technique is a form of the integrating factor method. The solution of equation (70) can be obtained as the integral equation:

$$C = C_0\, e^{-(k + 1/\tau)t} + e^{-(k + 1/\tau)t} \int_0^t e^{(k + 1/\tau)t} \frac{C_{in}}{\tau}\, dt \tag{74}$$

Integrating this equation over the interval 0 to t yields

$$C = C_0\, e^{-(k + 1/\tau)t} + \frac{C_{in}}{k\tau + 1}\left(1 - e^{-(k + 1/\tau)t}\right) \tag{75}$$

Note that the solution is composed of two concentration changes; the first term on the right-hand side of the equal sign represents "die-away" of the initial concentration, and the second term represents "buildup" of concentration due to continuous input. When t approaches infinity, equation (75) reduces to equation (71), the steady-state equation.

If a number of lakes are present in the series, these water bodies can be analyzed collectively. Figure 2.10 shows a series of lakes that consists of n equal-volume, completely mixed lakes. As was done above for a single lake, the approach is based on a mass balance around each lake of the series. Before deriving time-variable solutions, the steady-state solution will be developed.

The mass balance for the first lake is given as

$$V \frac{dC_1}{dt} = QC_{in} - QC_1 - kC_1V$$

and solved for

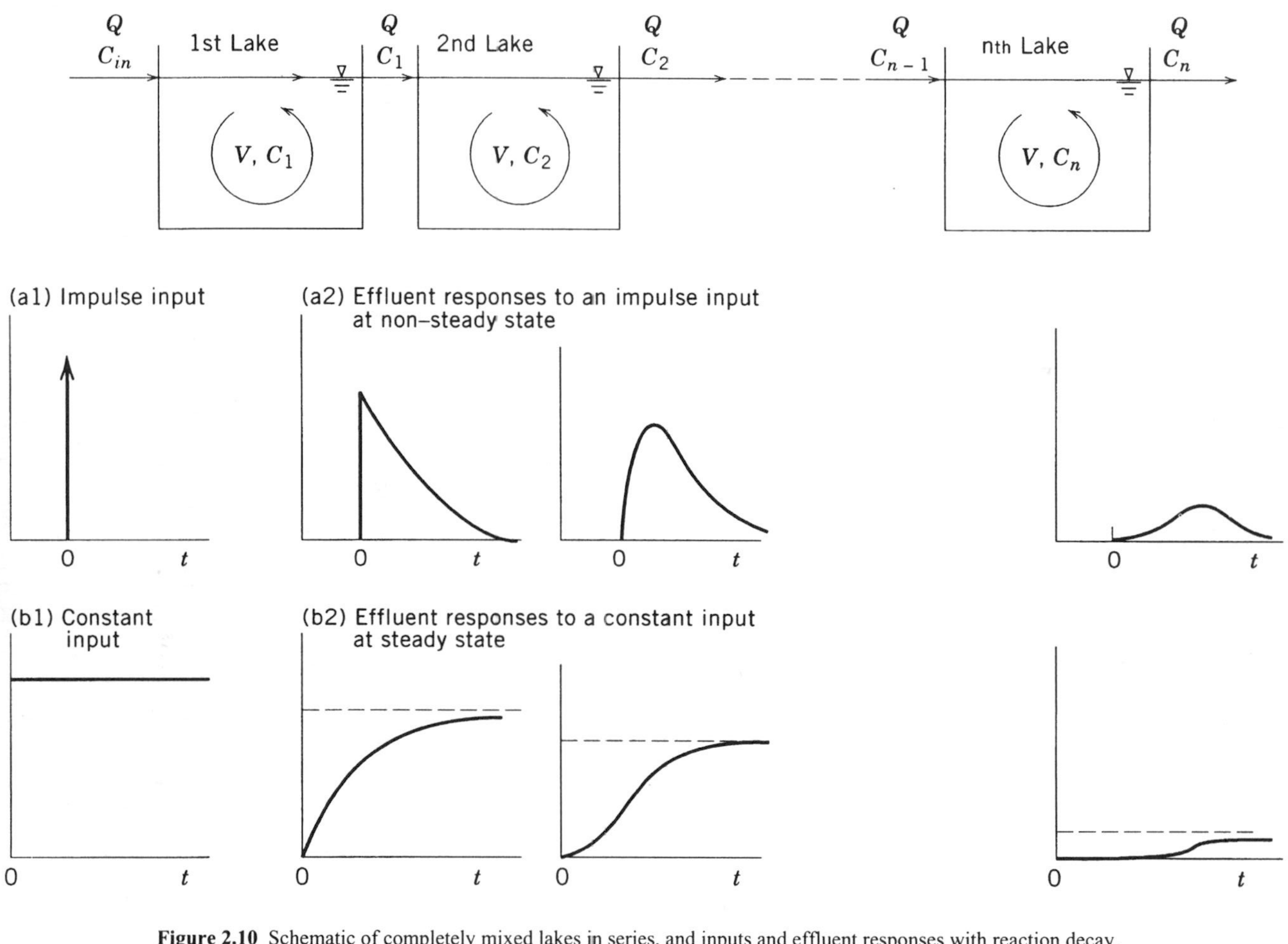

Figure 2.10 Schematic of completely mixed lakes in series, and inputs and effluent responses with reaction decay.

$$C_1 = \frac{C_{\text{in}}}{1 + k\tau} \tag{76}$$

For the second lake,

$$V \frac{dC_2}{dt} = QC_1 - QC_2 - kC_2V$$

and solved for

$$C_2 = \frac{C_1}{1 + k\tau} \tag{77}$$

Substitution of equation (76) into equation (77) yields

$$C_2 = \frac{C_{\text{in}}}{(1 + k\tau)^2} \tag{78}$$

where τ is the detention time of each individual lake, not the overall detention time.
The mass balance for the nth lake is given as

$$V \frac{dC_n}{dt} = QC_{n-1} - QC_n - kC_nV$$

and solved for

$$C_n = \frac{C_{n-1}}{(1 + k\tau)} \tag{79}$$

where n is the number of lakes in question and $n - 1$ designates the upstream lake. Therefore the analytical solution for the nth lake is given as

$$C_n = \frac{C_{\text{in}}}{(1 + k\tau)^n} \tag{80}$$

The time-variable solution can be obtained for an impulse input of conservative tracer. The mass balance for the first lake may be given as

$$V \frac{dC_1}{dt} = -QC_1 \tag{81}$$

Integrating equation (81) for the time interval 0 to t with an initial condition of $C_1 = C_{1(0)}$ at $t = 0$, yields

$$C_1 = C_{1(0)} \exp(-t/\tau) \tag{82}$$

The mass balance for the second lake gives

$$\frac{dC_2}{dt} = \frac{C_1}{\tau} - \frac{C_2}{\tau} \tag{83}$$

Substituting equation (82) into equation (83) and rearranging yields

$$\frac{dC_2}{dt} + \frac{C_2}{\tau} = \frac{C_{1(0)} \exp(-t/\tau)}{\tau} \tag{84}$$

Equation (84) can be solved using the integrating factor method for

$$C_2 = \frac{C_{1(0)}t}{\tau} \exp(-t/\tau) \tag{85}$$

For the third lake, the mass balance yields

$$\frac{dC_3}{dt} = \frac{C_2}{\tau} - \frac{C_3}{\tau} \tag{86}$$

Substituting equation (85) into equation (86) and solving using the integrating factor yields

$$C_3 = \frac{C_{1(0)}t^2}{2\tau^2} \exp(-t/\tau) \tag{87}$$

Thus the general formula for n lakes in series that receive an impulse input of conservative tracer is given as

$$C_n = \frac{C_{1(0)}}{(n-1)!} \left(\frac{t}{\tau}\right)^{n-1} \exp(-t/\tau) \tag{88}$$

where τ is the detention time of an individual lake, V/Q.

In the case that a lake or reactor vessel is segmented into n compartments as shown in Figure 2.11, the effluent response to an impulse input of nonreactive chemical may be given by

$$C_n = \frac{C_0 n^n}{(n-1)!} \left(\frac{t}{\tau'}\right)^{n-1} \exp(-nt/\tau') \tag{89}$$

where τ' represents the detention time of the entire vessel (V_{total}/Q) and C_0 is the initial concentration if the impulse input were delivered to the entire vessel (M/V_{total}). The effluent responses with respect to the number of compartments are illustrated in Figure 2.11. The greater the number of compartments, the greater the tendency toward plug-flow conditions.

Equation (89) is powerful because it provides effluent responses that are intermediate between the ideal plug-flow model and the ideal completely mixed model ($n = \infty$ and $n = 1$ in Figure 2.11). For lakes and reservoirs that have, in reality, plug-flow and dispersion characteristics, equation (89) can be used with a hypothetical number of compartments (n) to obtain the best fit to an impulse injection of tracer, and thus one can obtain the mixing characteristics of the system for modeling of other pollutants.

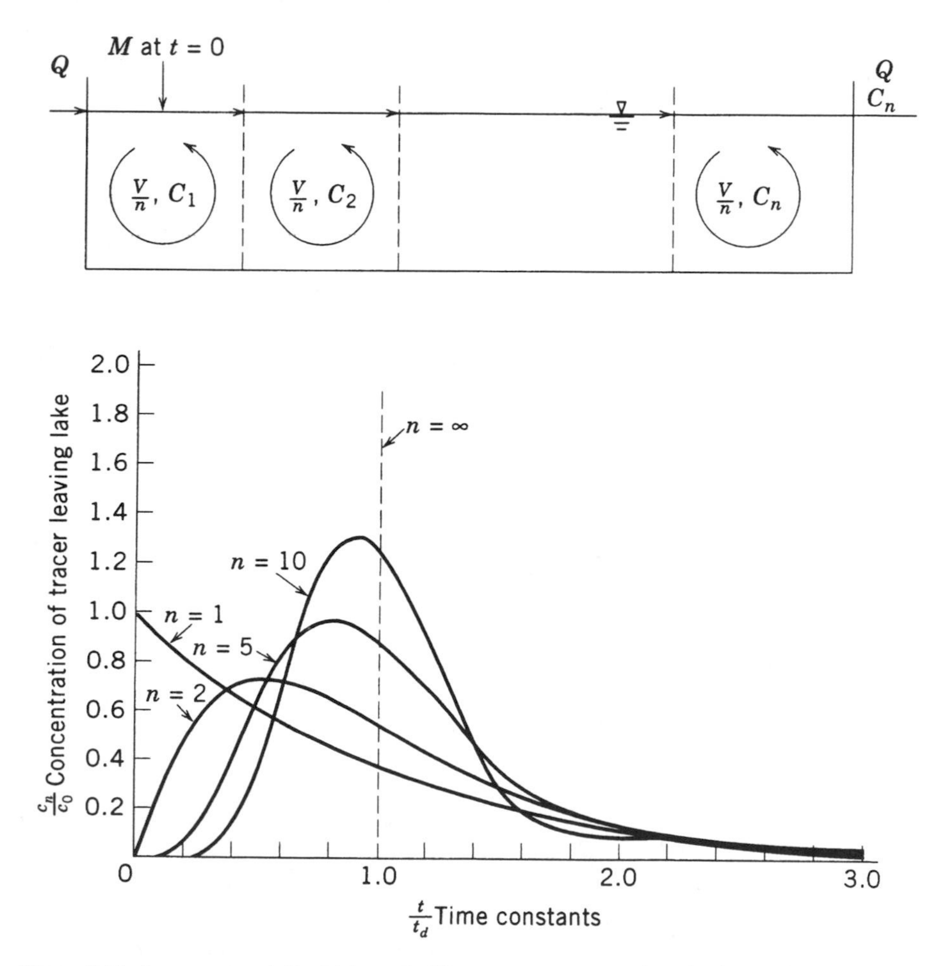

Figure 2.11 A compartmentalized lake and effluent responses to an impulse input of conservative tracer.

2.7.2 Plug-Flow Systems

An ideal plug-flow system is illustrated, using a river as an example, in Figure 2.12. The major assumptions involved in this model are that the bulk of water flows downstream with no longitudinal mixing (as a plug) and that instantaneous mixing occurs in the lateral and vertical directions. It is a one-dimensional (1-D) model. The mass balance is developed around an incremental volume V and is given as

$$\frac{\Delta(VC)}{\Delta t} = QC - Q(C + \Delta C) - kCV \tag{90}$$

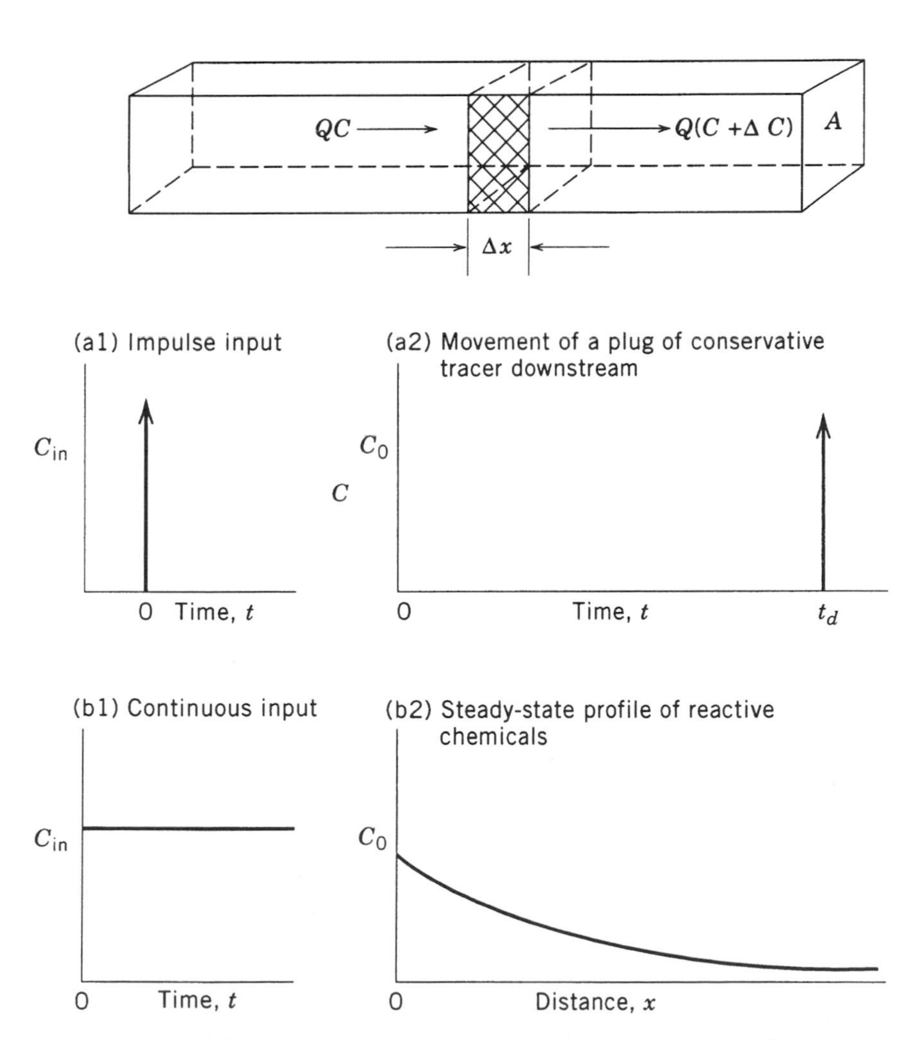

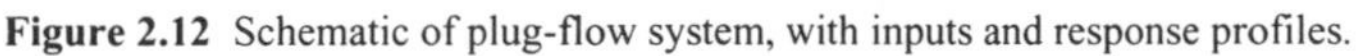
Figure 2.12 Schematic of plug-flow system, with inputs and response profiles.

where $V = A\,\Delta x$, L^2L

A = cross-sectional area, L^2

Δx = an increment of finite thickness of the stream, L

Δt = a time interval, T

K = a first-order decay rate, T^{-1}

Dividing equation (90) by V and simplifying yields

$$\frac{\Delta C}{\Delta t} = -\frac{Q\,\Delta C}{A\,\Delta x} - kC \tag{91}$$

The limit of equation (91) as $\Delta t \to 0$ is:

$$\frac{\partial C}{\partial t} = -\frac{Q}{A}\frac{\partial C}{\partial x} - kC = -u\frac{\partial C}{\partial x} - kC \tag{92}$$

where $u = Q/A$ = mean velocity. This is the general equation for a plug-flow system. Note that the concentration, C, is a function of both time, t, and distance, x.

At steady state ($\partial C/\partial t = 0$), equation (92) reduces to

$$\frac{\partial C}{\partial x} = -\frac{k}{u}C \tag{93}$$

With a boundary condition of $C = C_0$ at $x = 0$, equation (93) can be integrated by separation of variables to yield

$$C = C_0 \exp(-kx/u) \tag{94}$$

This is the steady-state solution to the plug-flow equation. Effluent responses to an impulse and constant inputs are illustrated in Figure 2.12.

2.7.3 Advective-Dispersive Systems (Plug Flow with Dispersion)

An ideal plug-flow system is illustrated, using an estuary as an example, in Figure 2.13. As was done in the plug-flow model, the mass balance is written around an incremental control volume of small but finite volume.

Accumulation		Advective transport inputs		Dispersive transport inputs		
	=		+			
	−	Advective transport outputs	−	Dispersive transport outputs	±	Reactions

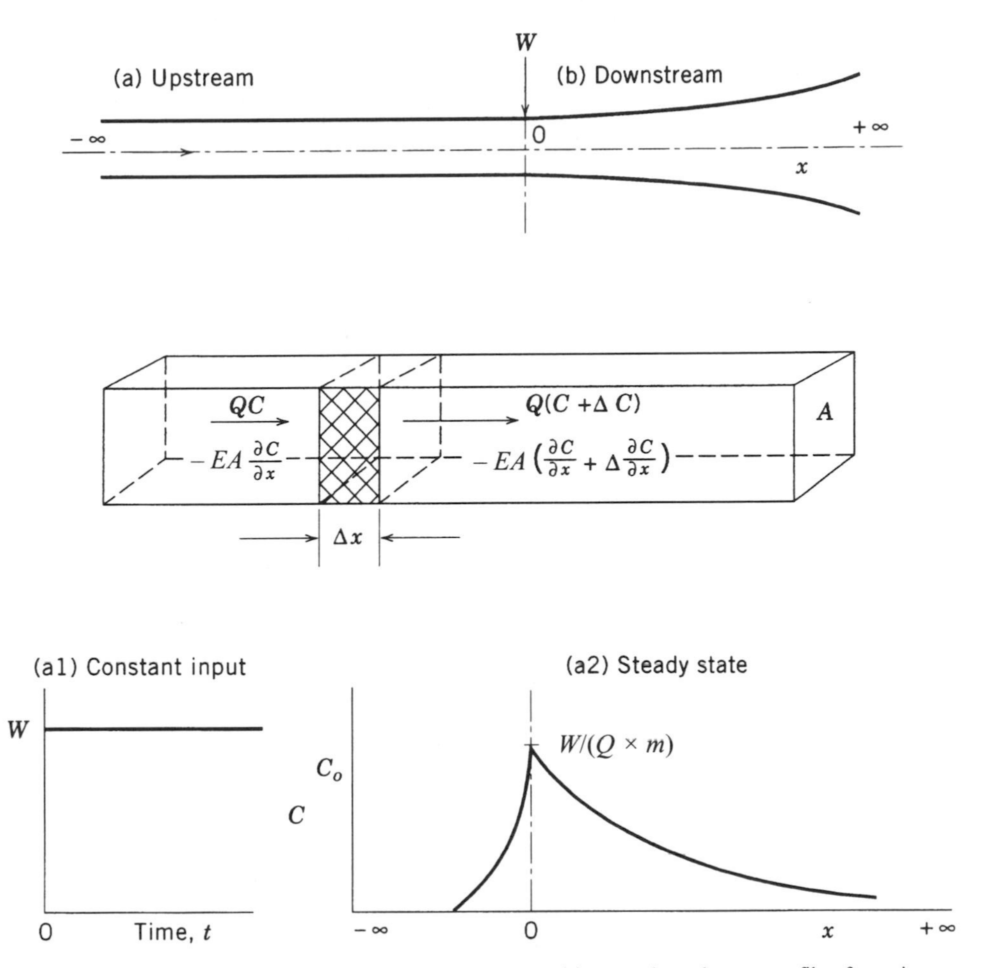

Figure 2.13 Schematic of advective–dispersive system, and input and steady-state profile of reactive chemicals.

A mass balance over an infinitesimal time interval can written for the differential volume, $A\ \Delta$x, as

$$V\frac{\partial C}{\partial t}=\left[QC+\left(-EA\frac{\partial C}{\partial x}\right)\right]-\left\{Q(C+\Delta C)+\left[-EA\left(\frac{\partial C}{\partial x}+\Delta\frac{\partial C}{\partial x}\right)\right]\right\}-kCV \tag{95}$$

where $V = A\ \Delta$x, L^2L
$E =$ dispersion coefficient, L^2T^{-1}
$k =$ first-order reaction coefficient, T^{-1}

Simplifying equation (95) yields

$$V \frac{\partial C}{\partial t} = -Q \frac{\partial C}{\partial x} \Delta x + EA \frac{\partial}{\partial x} \left(\frac{\partial C}{\partial x} \right) \Delta x - kCV \tag{96}$$

Dividing by $V = A\ \Delta x$, we find

$$\frac{\partial C}{\partial t} = -\frac{Q}{A} \frac{\partial C}{\partial x} + E \frac{\partial}{\partial x} \left(\frac{\partial C}{\partial x} \right) - kC$$

or

$$\frac{\partial C}{\partial t} = E \frac{\partial^2 C}{\partial x^2} - u \frac{\partial C}{\partial x} - kC \tag{97}$$

Equation (97) is the time-variable equation for advective–dispersive systems with constant coefficients, Q, A, E, and k.

The steady-state equation for an estuary system may be obtained by setting the left-hand side of equation (97) equal to zero, $\partial C/\partial t = 0$.

$$0 = E \frac{d^2 C}{dx^2} - u \frac{dC}{dx} - kC \tag{98}$$

Equation (98) is a second-order, ordinary, homogeneous, linear differential equation, which has the general form

$$0 = ay'' + by' + cy$$

where $y = f(x)$, and the general form of the solution is given as

$$y = B \exp(gx) + D \exp(jx)$$

where

$$g \cdot j = \frac{-b \pm \sqrt{b^2 - 4ac}}{2a}$$

The roots of the quadratic equation of coefficients gives g and j ($g \cdot j$); and B and D are integration constants obtained from boundary conditions. Note that g and j refer to positive and negative rotos, respectively. Accordingly, the solution of equation (98) is obtained as

$$C = B \exp(gx) + D \exp(jx) \tag{99}$$

where

$$g \cdot j = \frac{u \pm \sqrt{u^2 + 4Ek}}{2E} = \frac{u}{2E}(1 \pm m)$$

$$m = \sqrt{1 + \frac{4Ek}{u^2}}$$

In order to solve equation (99), boundary conditions must be used. To establish boundary conditions, the estuary system in question may be divided into upstream and downstream segments at the point of chemical discharge (Figure 2.13).

In the upstream segment, we can set two boundary conditions (BC 1 and BC 2) as follows.

BC 1: At the upstream segment, far from the discharge point, the concentration approaches zero, that is, $C = 0$ at $x = -\infty$. Under this condition, we obtain

$$D = 0 \quad \text{and} \quad C = B\exp(gx) \tag{100}$$

BC 2: The concentration at the point of discharge is C_0, that is, $C = C_0$ at $x = 0$. Under this condition, we obtain

$$B = C_0$$

Therefore the concentration in the upstream segment is given as

$$C_a = C_0 \exp(gx) = C_0 \exp\left(\frac{ux}{2E}(1 + m)\right) \tag{101}$$

In the downstream segment, two additional boundary conditions can be set.

BC 1: The concentration approaches zero downstream, far from the discharge point, that is, $C = 0$ at $x = +\infty$. Under this condition, we obtain

$$B = 0 \quad \text{and} \quad C = D\exp(jx) \tag{102}$$

BC 2: At the point of discharge, the concentration is C_0; that is, $C = C_0$. Under this condition, we obtain

$$D = C_0$$

Therefore the concentration in the downstream segment is given as

$$C_b = C_0 \exp(jx) = C_0 \exp\left(\frac{ux}{2E}(1 - m)\right) \tag{103}$$

The boundary concentration (C_0) at the discharge point can be obtained by making a mass balance at $x = 0$ (Figure 2.13).

Mass in = Mass out

$$QC_0 - EA\left.\frac{dC_a}{dx}\right|_{x=0} + W = QC_0 - EA\left.\frac{dC_b}{dx}\right|_{x=0} \tag{104}$$

The reaction is negligible because Δx is infinitesimally small. From equation (100),

$$\left.\frac{dC_a}{dx}\right|_{x=0} = gB \tag{105}$$

and from equation (102),

$$\left.\frac{dC_b}{dx}\right|_{x=0} = jD \tag{106}$$

Substituting equations (105) and (101) into (104) yields

$$-EAgB + W = -EAjD$$

Since $B = D = C_0$, we obtain

$$C_0 = \frac{W}{EA(g - j)} \tag{107}$$

Substituting $g \cdot j$ of equation (99) into equation (107) and simplifying yields

$$C_0 = \frac{W}{mQ} \tag{108}$$

The final solution is summarized as follows:

$$C = C_0 \exp(gx) \quad \text{at } x \leq 0$$

$$C = C_0 \exp(jx) \quad \text{at } x \geq 0 \tag{109}$$

where $C_0 = W/mQ$ and

$$g \cdot j = \frac{u}{2E}(1 \pm m)$$

$$m = \sqrt{1 + \frac{4kE}{u^2}}$$

The response to a constant input under steady-state conditions is presented in Figure 2.13.

2.7.4 Closure

The simplified transport models for completely mixed systems, reactors in series, plug-flow systems, and plug flow with dispersion (1-D) will prove to be quite useful in preliminary assessments of environmental pollutants. In Chapter 5, we will see that a completely mixed flow-through system is a useful approximation for a lake and eutrophication analyses. The plug-flow model is the ideal case for a swiftly flowing river or stream, and it is the basis for the classical Streeter–Phelps model of dissolved oxygen (Chapter 6). In Chapters 7 and 9, we will make use of the plug flow with dispersion equations for modeling toxic organics in estuaries and groundwater systems.

2.8 REFERENCES

1. Lyman, W.J., et al., Eds., *Handbook of Chemical Property Estimation Methods,* McGraw-Hill, New York (1982).
2. Bajraktarevic-Dobran, H., Dispersion in Mountainous Natural Streams, *J. Env. Eng. Div., ASCE,* 108:3, 502–514 (1982).
3. Bedford, K.W., and Babajimopoulos, C. Vertical Diffusivities in Areally Averaged Models, *J. Env. Eng. Div., ASCE,* 103:1, 113–125 (1977).
4. Fischer, H.B. et al., Mixing in Inland and Coastal Waters, Academic Press, New York (1979).
5. Liu, H., Predicting Dispersion Coefficient of Streams, *J. Env. Eng. Div., ASCE,* 103:1, 59–69 (1977).
6. Liu, H., Discussion of, "Predicting Dispersion Coefficient of Streams," *J. Env. Eng. Div., ASCE,* 104:4, 825–828 (1978).
7. Fischer, H.B., The Mechanics of Dispersion in Natural Streams, *J. Hydraul. Div., ASCE,* 93:6, 187–216 (1967).
8. Fischer, H.B., Dispersion Predictions in Natural Streams, *J. Sanit. Eng. Div., ASCE,* 94:5, 927–943 (1968).
9. Fischer, H.B., Methods for Predicting Dispersion Coefficients in Natural Streams, with Applications to Lower Reaches of the Green and Duwamish Rivers, Washington, U.S. Geological Survey Professional Paper 582-A (1968).

10. Fischer, H.B., Longitudinal Dispersion and Turbulent Mixing in Open-channel Flow, *Ann. Rev. Fluid Mech.*, 5:59–78 (1973).
11. Elder, J.W., The Dispersion of Marked Fluid in Turbulent Shear Flow, *J. Fluid Mech.*, 5:4, 544–560 (1959).
12. Sayre, W.W., Natural Mixing Processes in Rivers, In: Environmental Impact on Rivers, H.W. Shen, Ed., H.W. Shen Publishers, Fort Collins, CO (1973).
13. Sayre, W.W., and Chang, F.M. A Laboratory Investigation of Open-Channel Dispersion Processes for Dissolved, Suspended and Floating Dispersants, U.S. Geological Survey professional Paper 433–E (1968).
14. Yotsukura, N., and Cobb, E.D., Transverse Diffusion of Solutes in Natural Streams, U.S. Geological Survey Professional Paper 582-C (1972).
15. Yotsukura, N., and Sayre, W.W., Transverse Mixing in Natural Channels, *Water Resour. Res.*, 12:695–704 (1976).
16. Okoye, J.D., Characteristics of Transverse Mixing in Open-Channel Flows, Report No. KH-R-23, W.M. Kech Laboratory of Hydraulics and Water Resources, California Institute of Technology, Pasadena, CA (1970).
17. Jobson, H.E., and Sayre, W.W., Vertical Transfer in Open Channel Flow, *J. Hydraul. Div., ASCE,* 96:3, 703–724 (1970).
18. Tennekes, H., and Lumley, J.L., *A First Course in Turbulence,* MIT Press, Cambridge, MA (1972).
19. Mortimer, C.H., The Exchange of Dissolved Substances Between Mud and Water in Lakes, *J. Ecol.,* 29:280–329 (1941).
20. Edinger, J.E., and Geyer, J.C., Heat Exchange in the Environment, Cooling Water Studies for the Edison Electric Institute, Johns Hopkins University, Baltimore, MD (1965).
21. Schnoor, J.L., and Fruh, E.G., Dissolved Oxygen Model of a Short Detention Time Reservoir with Anaerobic Hypolimnion, *Water Resour. Bull.*, 15:2, 506–518 (1979).
22. Park, G.G., and Schmidt, P.S., Numerical Modeling of Thermal Stratification in a Reservoir with Large Discharge-to-Volume Ratio, *Water Resour. Bull.*, 9:5, 932-941 (1973).
23. McEwen, G.F., A Mathematical Theory of the Vertical Distribution of Temperature and Salinity in Water Under the Action of Radiation, Conduction, Evaporation and Mixing Due to the Resulting Convection, *Bull. Scripps Inst. Oceanogr., Tech. Ser.,* 2, 197–306 (1929).
24. Quay, P.D., et al., Vertical Diffusion Rates Determined by Tritium Tracer Experiments in the Thermocline and Hypolimnion of Two Lakes, *Limnol. Oceanogr.,* 25:2, 201–218 (1980).
25. Imboden, D.M., and Emerson, S., Study of Transport through the Sediment–water Interface in Lakes Using Natural Radon-222, In: *Interaction Between Sediments and Water,* H.L. Gotterman, Ed., xxx–xxx (1976).
26. Imboden, D.M., and Emerson, S., Natural Radon and Phosphorus as Limnologic Tracers: Horizontal and Vertical Eddy Diffusion in Greifensee, *Limnol. Oceanogr.,* 23:1, 77–90 (1978).
27. Imboden, D.M., et al., MELIMEX, an Experimental Heavy Metal Pollution Study: Vertical Mixing in a Large Limno-Corral, Schweiz. *Z. Hydrol.,* 41:2, 177–189 (1979).
28. Imboden, D.M., Natural Radon as a Limnological Tracer for the Study of Vertical and Horizontal Eddy Diffusion, Isotopes in Lake Studies, International Atomic Energy Agency, Vienna, pp. 213–218 (1979).

29. Torgersen, T., et al., A New Method for Physical Limnology–Tritium–Helium-3 Ages—Results for Lakes Erie, Huron and Ontario, *Limnol. Oceanogr.,* 22:2, 181–193 (1977).
30. McQuivey R.S., and Keefer, T.N., Simple Method for Predicting Dispersion in Streams, *J. Env. Eng. Div., ASCE,* 100:4, 997–1011 (1974).
31. Bencala, K.E., and Walters, R.A., Simulation of Solute Transport in a Mountain Pool-and-Riffle Stream: A Transient Storage Model, *Water Resour. Res.,* 19:3, 718–724 (1983).
32. Li, Y.H., Vertical Eddy Diffusion Coefficient in Lake Zurich, *Schweiz.* Z. Hydrol., 35:1–7 (1973).
33. Wodka, M.C., et al., Diffusivity-Based Flux of Phosphorus in Onondaga Lake, *J. Env. Eng. Div., ASCE,* 109:6, 1403–1415 (1983).
34. Snodgrass, W.J., and O'Melia, C.R., Predictive Model for Phosphorus in Lakes, *Environ. Sci. Technol.,* 9:10, 937–944 (1975).
35. Powell T., and Jassby, A., The Estimation of Vertical Eddy Diffusivities Below the Thermocline, *Water Resour. Res.,* 10:2, 191–198 (1974).
36. Jassby, A., and Powell, T., Vertical Patterns of Eddy Diffusivity During Stratification in Castle Lake, California, *Limnol. Oceanogr.,* 20:4, 530–543 (1975).
37. Lewis, W.M. Jr, Temperature, Heat, and Mixing in Lake Valencia, Venezuela, *Limnol. Oceanogr.,* 28(2, 273–286 (1983).
38. Imberger, J., et al., Dynamics of Reservoir of Medium Size, *J. Hydraul. Div., ASCE,* 104:5, 725–743 (1978).
39. Lung, W.S., and Canale, R.P., Projections of Phosphorus Levels in White Lake, *J. Env. Eng. Div., ASCE,* 103:4, 663–676 (1977).
40. Heinrich, J., Lick, W., and Paul, J., Temperatures and Currents in a Stratified Lake: A Two-Dimensional Analysis, *J. Great Lakes Res.,* 7:3, 264–275 (1981).
41. DiToro, D.M., and Matystik, W.F., Mathematical Models of Water Quality in Large Lakes, Part 1: Lake Huron and Saginaw Bay, EPA-600/3-80-056, U.S. Environmental Protection Agency, Duluth, MN (1979).
42. DiToro, D.M., and Connolly, J.P., Mathematical Models of Water Quality in Large Lakes, Part 2: Lake Erie, EPA-600/3-80-065, U.S. Environmental Protection Agency, Duluth, MN (1980).
43. DiToro, D.M., et al., Simplified Model of the Fate of Partitioning Chemicals in Lakes and Streams, In: *Modeling the Fate of Chemicals in the Aquatic Environment,* K.L. Dickson, A.W. Maki, and J. Cairns, Jr., Eds., Ann Arbor Science Publishers, Ann Arbor, MI, pp. 165–190 (1982).
44. Lerman A., and Lietzke, T.A., Uptake and Migration of Tracers in lake Sediments, *Limnol. Oceanogr.,* 20:4, 497–510 (1975).
45. Christensen, E.R., A Model for Radionuclides in Sediments Influenced by Mixing and Compaction, *J. Geophys. Res.,* 87:C_1, 566–572 (1982).
46. O'Connor, D.J., and Mueller, J.A., A Water Quality Model of Chlorides in Great Lakes, *J. San. Eng. Div., ASCE,* 96:4, 955–975 (1970).
47. Chapra, S.C., Total Phosphorus Model for the Great Lakes, *J. Env. Eng. Div., ASCE,* 103:147–161 (1977).
48. Schnoor, J.L., and O'Connor, D.J., A Steady State Eutrophication Model for Lakes, *Water Res.,* 14:1651–1665 (1980).

49. Karickhoff, S.W., Brown, D.S., and Scott, T.A., Sorption of Hydrophobic Pollutants on Natural Sediments, *Water Res.*, 13:421–428 (1979).
50. Rubey, W.W., Equilibrium-Conditions in Debris-Laden Streams, *Trans. Am. Geophys. Union* (1933).
51. Schnoor, J.L., and McAvoy, D.C., A Pesticide Transport and Bioconcentration Model, *J. Env. Eng. Div., ASCE,* 107:6, 1229–1246 (1981).
52. Onishi, Y., and Wise, S.W., Mathematical Model, SERATRA, for Sediment and Pesticide Transport in Rivers and Its Application to Pesticide Transport in Four Mile and Wolf Creeks in Iowa, EPA-600/3-82-045, U.S. Environmental Protection Agency, Athens, GA (1982).
53. Krone, R.B., Flume Studies of the Transport of Sediment in Estuarial Shoaling Processes, Hydr. Eng. Lab., University of California, Berkeley (1962).
54. Partheniades, E., Erosion and Desposition of Cohesive Soils, *Proc. Am. Soc. Eng.* (1961).
55. Hutchinson, G.E., Limnological Studies in Connecticut, IV, The Mechanism of Intermediary Metabolism in Stratified Lakes, *Ecol. Monogr.*, 11:21–60 (1941).
56. Hutchinson, G.E., A Treatise on Limnology, Vol. 1, Wiley, New York, pp. 466–468 (1957).

2.9 PROBLEMS

1. Estimate the longitudinal, lateral, and vertical dispersion coefficients for the Mississippi River at Keokuk Iowa, for the following conditions.

$D = 10$ m
$f = 0.02$
$u_x = 0.37$ m s^{-1}
$Q = 550$ m^3 s^{-1}

Which one is largest? Why? Which one is smallest? Why?

2. Estimate the summertime vertical eddy diffusivity for Lake Superior and Coralville Reservoir using Mortimer's equation, given the following information.

Superior mean depth = 145 m
Coralville Reservoir mean depth = 2.1 m

How accurate do you think this estimate is? Consider mixing scale.

3. The James Estuary in Virginia from Hopewell to Newport has the following physiographic and hydrologic data:

Freshwater flow = 10,000 ft^3 s^{-1}

Cross-sectional area = 75,000 ft^2
Longitudinal dispersion coefficient = 15 $mi^2\ d^{-1}$
Longitudinal distance = 100 mi

Estimate the net nontidal velocity $\overline{u}_x$. Calculate the Peclet number. Which phenomenon predominates—advection or dispersion? How do you know?

4. Develop the mass balance equation found in Table 2.6 for: (1) the steady-state, completely mixed model with reaction decay and (2) the three-compartment steady state model with reaction decay. Start from the mass balance equation assuming a first-order decaying substance and steady input concentration, C_{in}. Solve for C/C_{in}. Assuming $k\tau = 1.0$, why does the three-compartment model result in a lower outflow concentration from an equal volume lake?

5. Write the mass balance equation for a completely mixed lake with an impulse input of mass, M, added at time $t = 0$. Assume the lake is of constant volume, V, with continuous outflow, Q, and a conservative substance. Solve for the outlet concentration in terms of the initial concentration C_0, the mean hydraulic detention time (τ), and time t. Plot the concentration C/C_0 versus t/τ. How long does it take for 95% of the impulse input of material to be depleted from the system?

6. Write the mass balance equation for an ideal 1-D plug-flow stream under steady state conditions with a first-order decaying substance (e.g., BOD). Solve for the concentration as a function of the concentration at $x = 0$, the rate constant k_d, longitudinal distance x, and mean velocity u. Assume the rate constant for BOD decay is 0.2 d^{-1}, the mean velocity of the stream is 0.5 m s^{-1}, and the initial concentration is 100 mg L^{-1} at $x = 0$. Plot the resulting concentration with downstream distance x in km.

7. Equation (75) gives the time-variable solution to a step function change in inflow concentration for an ideal, completely mixed lake. If the initial concentration of toxic chemical in the lake is 10 $\mu g\ L^{-1}$, the first-order decay rate constant is 0.005 per day, the mean hydraulic detention time of the lake is 100 days, and the inflow concentration is 50 $\mu g\ L^{-1}$, plot the concentration versus time (in days) of the toxic chemical in the lake outflow.

8. Define and compare the molecular diffusion coefficient (D), the turbulent diffusion coefficient (ε), and the longitudinal dispersion coefficient (E) in rivers.

a. Which is largest? Which is smallest?
b. What are their units?
c. Give an example of when each applies and write the diffusion/dispersion equation for that application.

9. The concentration profile given below is observed in a river with advection and dispersion under steady state conditions. Is mass transport greater due to advection or due to dispersion at mile point 5? Calculate the transport due to advection and dispersion in mg d^{-1}.

x, mi	C, mg L^{-1}
0	10
2.5	7.5
5.0	5.0
7.5	2.5
10.0	0.0

$E = 2$ mi^2 d^{-1}
$A = 10{,}000$ ft^2
$u = 0.5$ ft s^{-1}

3

CHEMICAL REACTION KINETICS

One must wait until the evening to see how splendid the day has been.
—Sophocles

Chemical kinetics refers to the rates and mechanisms by which chemical reactions take place. In this chapter, we define the concept broadly to include the rates and mechanisms by which chemical, physical, and ecological reactions are affected.

3.1 LAW OF MASS ACTION

Guldberg and Waage proposed the law of mass action in 1867.[1] In the law of mass action, the rate of a reaction is proportional to the product of the concentration of each substance participating in the reaction raised to the power of its stoichiometric coefficients. For the general chemical reaction equation, the following rate expressions can be written:

$$aA + bB \longleftrightarrow cC + dD$$

$$\text{Rate of the forward reaction} = k_f [A]^a [B]^b \tag{1}$$

$$\text{Rate of the reverse reaction} = k_r [C]^c [D]^d \tag{2}$$

$$\text{Overall rate of Rxn} = k_f [A]^a [B]^b - k_r [C]^c [D]^d \tag{3}$$

where the brackets [ ] denote chemical concentration (activity) in solution. If the reaction proceeds to chemical equilibrium, the rate of the forward reaction becomes equal to the reverse reaction:

$$k_f [A]^a [B]^b = k_r [C]^c [D]^d \tag{4}$$

The equilibrium constant is defined as the ratio of the forward rate constant divided by the reverse rate constant of chemical equilibrium:

$$K_{eq} = \frac{k_f}{k_r} = \frac{[C]^c[D]^d}{[A]^a[B]^b} \tag{5}$$

The equilibrium expression is valid regardless of the actual mechanism by which the reaction takes place, but the rates of the forward and reverse reactions depend on the actual mechanism. Equilibria chemical modeling is the subject of Chapter 4 in this book.

Elementary reactions are those that occur in a single step. For these reactions, the law of mass action holds. For example, in the case of a simple unimolecular reaction where 1 mole of chemical A decomposes to form 1 mole of B irreversibly, we may write

$$A \rightarrow B$$

$$\text{Rxn rate} = \frac{d[A]}{dt} = -k\,[A] \tag{6}$$

Also, bimolecular elementary reactions are common in the environment, such as

$$2A \rightarrow B \quad \text{or} \quad A + B \rightarrow C$$

$$\frac{d[A]}{dt} = -k\,[A]^2 \quad \text{or} \quad \frac{d[A]}{dt} = -k\,[A][B] \tag{7}$$

Trimolecular elementary reactions are less common (it is less probable that three molecules will collide simultaneously to effect a reaction), and more complicated stoichiometric equations than trimolecular do not occur. Thus all stoichiometric chemical reaction equations involving the collision of three or more reactants almost certainly are not elementary reactants, and one must consult the literature or perform experiments to determine the exact mechanism and the rate expression. For the purposes of modeling simple reactions, we will begin with the law of mass action.

3.2 RATE CONSTANTS AND TEMPERATURE

The reaction rate constant k is simply a proportionality constant between the rate of the reaction and the reactant concentration mass law expression. The rate constant carries its own units necessary to convert the mass law expression into a reaction rate. For example, for the first-order decay reaction [equation (6)], the units on k are inverse time:

$$\frac{d[A]}{dt} = -\mathrm{k}\,[A]$$

where k has units of T^{-1}. But in the second-order reaction (equation 7), the units on k are $L^3M^{-1}T^{-1}$ (such as L mol^{-1} s^{-1} or L mg^{-1} d^{-1}). Each type of reaction mechanism (reaction order) yields different units for its rate constant. Units should always be carried along with values reported for rate constants and equilibrium constants for this reason.

In Eyring's transition state theory, a reaction must overcome an activation energy before it can proceed. Figure 3.1 shows a schematic of how the reaction proceeds. At point A + BC, the reactant mixture has a certain energy content (internal energy) derived from its chemical potential at a given temperature and pressure. If the reaction occurs, the system proceeds through a peak in energy, a metastable transition state that may involve an activated complex ($ABC^{\ddagger}$).

$$A + BC \rightarrow ABC^{\ddagger}$$

$$ABC^{\ddagger} \rightarrow AB + C$$

An activation energy ($E_{act} = \Delta G^{\ddagger}$) is required to lift the reactant mix over its initial inertial energy requirement. The transition state complex may then decompose to form products (AB and C), and the resulting mixture of components must have less energy content than the reactants alone. The change in internal energy ΔG is an extensive property of the reaction system, and a negative ΔG is required for the reaction to occur spontaneously and to form a stable product mixture. Entropy increases in the process and/or heat is released.

Reaction rates increase with increasing temperature. Svante Arrhenius determined the relationship between the reaction rate constant and temperature

$$k = Ae^{-E_{act}/RT} \tag{8}$$

$$\ln \frac{k}{A} = -\frac{E_{act}}{RT} \tag{9}$$

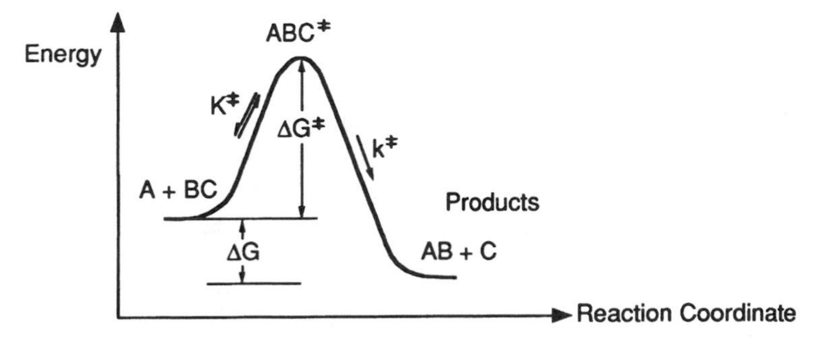

Figure 3.1 Diagram for transition reaction A + BC → AB + C. The free activation energy $\Delta G^{\ddagger}$ is necessary to form the activated complex $ABC^{\ddagger}$, which is in equilibrium with the reactants. The products AB + C are formed from the dissociation of $ABC^{\ddagger}$.

where A is a constant that is characteristic of the reaction, E_{act} is the activation energy (J mol^{-1} or cal mol^{-1}), T is the absolute temperature in K, and R is the universal gas constant (8.314 J mol^{-1} K^{-1} or 1.987 cal mol^{-1} K^{-1}). Thus a plot of ln k versus $1/T$ from equation (9) reveals the E_{act} from the slope of the straight line (Figure 3.2). Experiments in the laboratory or the field, where rates of reaction are measured at different temperatures, can provide considerable information about the reaction (E_{act}) and clues about its mechanism.

Occasionally, it is desirable to convert a rate constant from a known or reference temperature to a second temperature. In this case, equation (8) can be used for two different temperatures, the equations for k equated, and the constant A drops out.

$$\frac{k_{T_1}}{k_{T_2}} = \exp\left(\frac{E_{act}}{RT_2} - \frac{E_{act}}{RT_1}\right) \tag{10}$$

or

$$k_{T_1} = k_{T_2} \exp\left(\frac{E_{act}}{RT_1T_2}(T_1 - T_2)\right) \tag{11}$$

In modeling environmental systems, we generally are interested in the temperature range from 0 to 35 °C, and under these conditions equation (11) simplifies to equation (12) because E_{act}/RT_1T_2 is approximately constant:

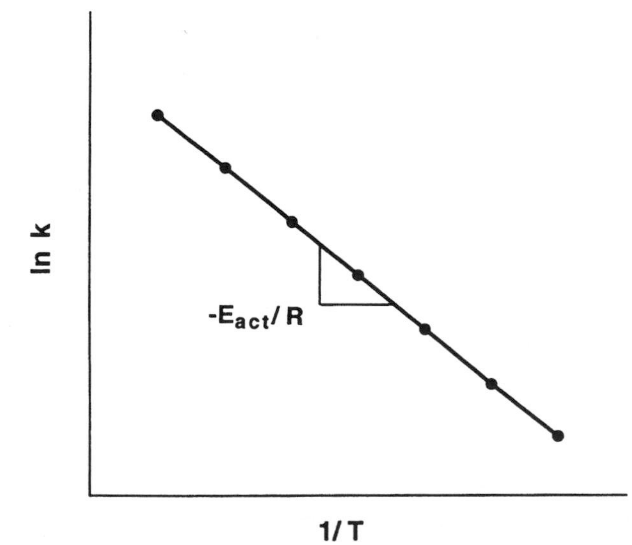

Figure 3.2 Arrhenius plot of reaction rate constant at any temperature. Activation energies for the reaction can be obtained from the slope of the line.

$$k_{T_1} = k_{T_2}\, \theta^{(T_1 - T_2)} \tag{12}$$

or

$$k_T = k_{20}\, \theta^{T-20} \tag{13}$$

where θ is a constant temperature coefficient greater than 1.0 and usually within the range 1.0–1.10, and k_{20} is the rate constant at the reference temperature 20 °C. Either of the two equations [equation (11) or (13)] is acceptable and each results in an exponential increase in reaction rate with temperature (Figure 3.3).

Biological reactions work on the basis of enzyme kinetics and have a surprisingly small range of activation energies in natural systems. Activation energies are within the range of 30–60 kJ mol^{-1} or 7–14 kcal mol^{-1} (1 cal = 4.184 J).[2]

Example 3.1 Effect of Temperature on Reaction Rate Constants

The Q_{10} rule in biology states that for a 10 °C increase in temperature, the rate of the reaction will approximately double. Solve for the activation energy and θ value necessary for a doubling of the reaction rate constant from 20 → 30 °C.

Solution: From equations (11) and (13),

$$\frac{k_{20}}{k_{30}} = 0.5 = \exp\{[E_{act}/(8.314)\,(293)\,(303)]\,(-10)\}$$

$$\ln 0.5 = -0.693 = [E_{act}/(8.314)\,(293)\,(303)]\,(-10)$$

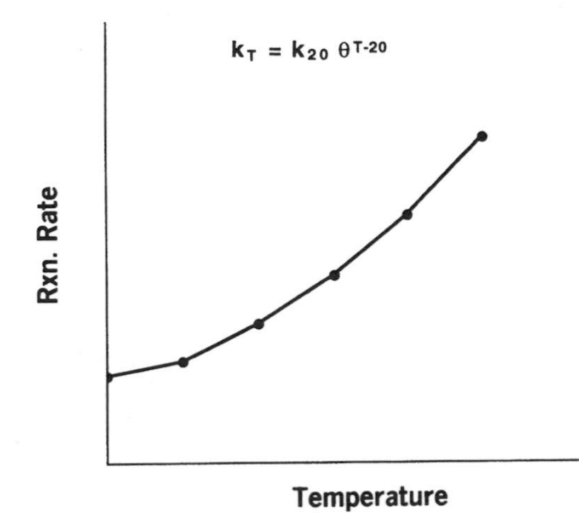

Figure 3.3 Effect of temperature on reaction rate.

solving we find

$$E_{act} = 51{,}200 \text{ J mol}^{-1}$$

Solving for θ, we find

$$\frac{k_{30}}{k_{20}} = 2.0 = \theta^{(30-20)}$$

$$\ln 2.0 = 0.693 = 10 \ln \theta$$

$$\theta = 1.072$$

Table 3.1 shows some typical values of q that are used in *Environmental Modeling*. The values should be used only within the range of temperatures where the reaction occurs. Enzyme reactions may decrease abruptly at temperatures higher and lower than the temperature range to which the organism is adapted.

Enzymes are catalysts that speed the rate of reaction but are not consumed in the reaction. For example,

$$\text{S} + \text{E} \longleftrightarrow \text{SE} \longrightarrow \text{P} + \text{E}$$

where S is the substrate, E is the enzyme, SE is the substrate–enzyme complex, and P is the product(s). The role of the enzyme is to lower the activation energy of the reaction in Figure 3.1, resulting in a greater probability that reactants will interact successfully to form products.

In environmental modeling, important catalysts include both homogeneous and heterogeneous catalysts. Homogeneous catalysts are dissolved in the aqueous phase together with the reactants. Heterogeneous catalysts are usually solid surfaces, and surface coordination reactions are one of the steps in the overall reaction. The surfaces bind a soluble reactant and create an activated complex.

Table 3.1 Effect of Temperature on Reaction Rate Constants [equation (13)]

Process	θ
Biochemical oxygen demand[2]	1.047
Reaeration[2]	1.024–1.037
Algae photosynthesis[3]	1.065
Nitrification[4]	1.04
Chemical weathering	1.02
Bacterial respiration	1.03
Zooplankton grazing	1.07
Zooplankton respiration	1.06

Table 3.2 Catalysts in Selected Aquatic Chemical Reactions

Reaction	Catalyst
Homogeneous Fe oxidation	Cu
Heterogeneous Fe oxidation	MnO_2, $FeOOH_{(s)}$
Sulfite oxidation (homogeneous)	Co
Bacterial + algal synthesis	Enzymes
Organic hydrolysis (homogeneous)	Bases + acids
Organic photooxidations (homogeneous)	NO_3^-, NO_2^-
Organic photooxidations (heterogeneous)	TiO_2, $FeOOH_{(s)}$

These catalysts can reduce the activation energy by an order of magnitude and greatly speed the rate of reactions.

3.3 REACTION ORDER AND TESTING REACTION RATE EXPRESSIONS

In the arbitrary reaction between species A, B, and C, the overall *reaction order* is defined as the sum of the exponents in the rate expression ($a + b + c$). For a reaction rate that can be written as an elementary reaction:

$$a\text{A} + b\text{B} + c\text{C} + \cdots \rightarrow \text{Products}$$

$$\text{Rxn rate} = k\,[\text{A}]^a\,[\text{B}]^b\,[\text{C}]^c \cdots$$

$$\text{Rxn order} = a + b + c + \cdots$$

The overall reaction would be said to be of order $a + b + c$, but the reaction rate could also be said to be a order in reactant A, b order in reactant B, and c order in reactant C. Most elementary reactions are either zero, first, or second order. Fractional order reactions are also observed because they occur in a series of steps. One step may be the rate-determining step that defines the reaction rate expression.

3.3.1 Zero-Order Reactions

An example of a zero-order reaction would be the irreversible degradation of a reactant in natural waters that did not depend on the concentration of the reactant in solution.

$$\frac{d[\text{A}]}{dt} = -k_0 \tag{14}$$

where k is the rate constant of the zero-order reaction. Examples of zero-order reactions in natural waters include heterogeneous complex reactions that occur in many steps, and a physical constraint such as surface area actually limits the rate of reaction. Methane production and release of hydrolysis products (e.g., NH_3, PO_4^{3-}) from anaerobic sediments are examples of zero-order reactions.[4]

Modelers need to determine rate constants and to test the rate expressions that are used in mathematical models. It is best to test the rate expression in field experiments under similar conditions to those used in the model. But if field experimentation is not feasible, it may be possible to perform an analogous experiment in the laboratory or to find a similar experiment in the literature. Batch experiments are frequently used because of the lack of transport in and out of the reactor, which simplifies the mathematics.

For a zero-order reaction, integration of the rate expression results in a straight line, and the rate constant k_0 can be determined as the slope of the line. Zero order reactions may be either a product formation (+ Rxn) or decay reaction (– Rxn) in our mass balance equation from Chapter 2. The results are plotted in such a way as to linearize the data. From the results of the batch experiment, the modeler can determine two important facts about the reaction.

- The proposed rate expression is correct if the line is straight (the measurements fall on a straight line to within some acceptable statistical limit).
- The rate constant can be obtained from the slope of the line.

In the case of the zero-order reaction, the observed data should fall on a straight line (concentration versus time) with a slope of k_0 if the reaction is, indeed, zero order (Figure 3.4). Units on k_0 are $ML^{-3}T^{-1}$.

3.3.2 First-Order Reactions

First-order reactions are the most commonly employed in modeling environmental chemistry. In the absence of any greater knowledge or experimental evidence, modelers will often assume that a reaction is first order. While it is a logical assumption, and it results in linear models that are more easily solved, it is not always correct mechanistically and it may give erroneous results. Our world is so nonlinear!

A first-order reaction is one in which the reaction rate is proportional to the concentration of the reactant to the first power:

$$A \xrightarrow{k_1} B$$

$$\frac{d[A]}{dt} = -k_1\,[A]^{1.0} \qquad (15)$$

$$\frac{d[B]}{dt} = +k_1\,[A]^{1.0} \qquad (16)$$

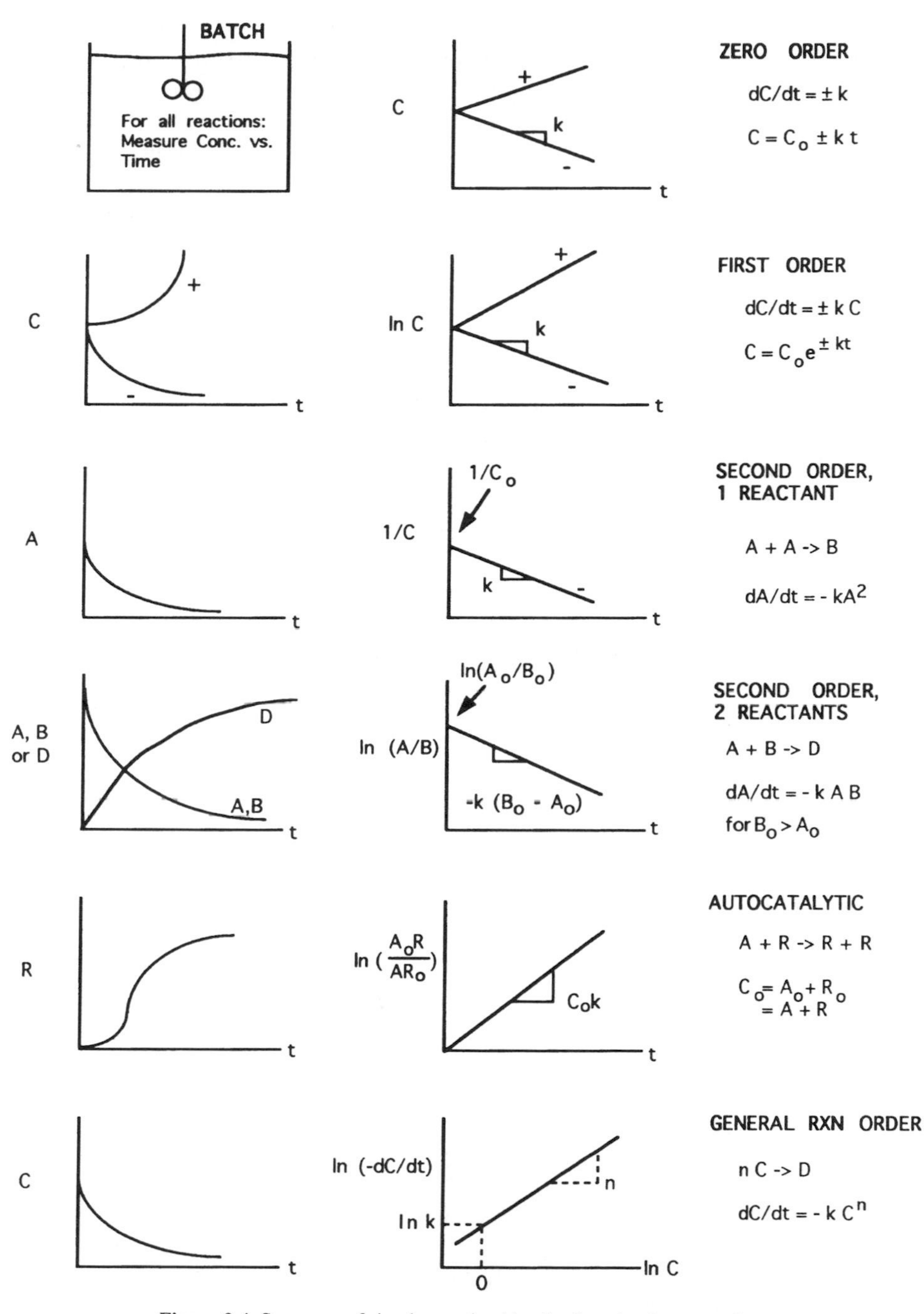

Figure 3.4 Summary of simple reaction kinetics from batch reactor data.

where k_1 is the first-order rate constant with units of T^{-1}. Batch kinetic data for equation (15) yield a plot with an exponentially declining concentration versus time, and for equation (16) the concentration will increase but then level off and approach a maximum value after a long time. Both reaction equations (15) and (16) are considered first order in reactant A; equation (15) is a decay reaction and equation (16) is a production reaction.

Equation (15) can be solved by separation of variables and integrated. It is a linear, ordinary, first-order differential equation.

$$\int_{A_0}^{A} \frac{dA}{A} = -k_1 \int_0^t dt$$

$$\ln \frac{A}{A_0} = -k_1 t \qquad (17)$$

$$A = A_0 e^{-k_1 t} \qquad (18)$$

Equation (17) is the linearized form of the equation that can be plotted graphically to determine whether the proposed reaction mechanism is correct and to obtain the rate constant (Figure 3.4). Equation (18) is the final solution giving the concentration of species A as a function of time and its initial concentration A_0 at $t = 0$.

Likewise, equation (16) can be integrated, but it is one ordinary differential equation with two unknowns (A and B). We must substitute for A from equation (18) into equation (16) and solve.

$$\frac{dB}{dt} = k_1 A_0 e^{-k_1 t} \qquad (19)$$

Equation (19) is a linear, ordinary, nonhomogeneous differential equation that is solvable by first-order methods.

$$\int_0^B dB = k_1 A_0 \int_0^t e^{-k_1 t} dt$$

$$B = A_0 (1 - e^{-k_1 t}) \qquad (20)$$

It is difficult to linearize equation (20) except by substituting equation (18):

$$B = A_0 - A_0 e^{-k_1 t}$$

$$A_0 - B = A_0 e^{-k_1 t} \qquad (21)$$

$$\ln(A_0 - B) = -k_1 t + \ln A_0 \qquad (22)$$

A plot of $\ln(A_0 - B)$ versus t should yield a straight line with a slope of $-k_1$.

A final type of first-order reaction results in exponential (logarithmic) growth. Cell A may synthesize new cellular mass, divide, and result in two cells of A.

$$\mathrm{A} \rightarrow 2\mathrm{A}$$

$$\frac{d[\mathrm{A}]}{dt} = +\,\mathrm{k}_1[\mathrm{A}] \tag{23}$$

$$\int_{\mathrm{A}_0}^{\mathrm{A}} \frac{d\mathrm{A}}{\mathrm{A}} = k_1 \int_0^t t$$

$$\ln \frac{\mathrm{A}}{\mathrm{A}_0} = k_1 t \tag{24}$$

$$\mathrm{A} = \mathrm{A}_0\, e^{+k_1 t} \tag{25}$$

Equation (24) is the linearized version to test the reaction rate expression and to derive the rate constant. Equation (25) shows the exponential form that grows without bound.

Examples of first-order reactions include:

- Radioisotope decay.
- Biochemical oxygen demand in a stream.
- Sedimentation of noncoagulating solids.
- Death and respiration rates for bacteria and algae.
- Reaeration and gas transfer.
- Log growth phase of algae and bacteria (production reaction).

Probably the only one that is "truly" and "exactly" first order is radioisotope decay. We set our atomic clocks by this reaction! But the other reactions may be sufficiently close to first-order reactions that we may assume the reaction mechanism as an approximation. The only production reaction listed above is the log growth phase of algae and bacteria. Of course, log growth phases cannot last forever because the concentration increases exponentially with time (it explodes!). Tanner[5] analyzed 72 animal populations for the log growth equation (25), and only one of them showed a positive correlation between growth rate and population over the entire period of record—it was humans! Even the growth rate of humans is gradually declining. In 1970, the rate was 2.0% per year, and in 1995, it is approximately 1.7% per year. But demographers project a population of 10–12 billion by the end of the 21st century.

3.3.3 Second-Order Reactions

There are several different types of second-order reactions that are common in aquatic chemistry including the second-order reaction with one reactant, with two reactants, and the autocatalytic second-order reaction.[6]

$$\text{One reactant} \qquad A + A \xrightarrow{k_2'} B \tag{26}$$

$$\text{Two reactants} \qquad A + B \rightarrow D \tag{27}$$

$$\text{Autocatalytic} \qquad A + R \rightarrow 2R \tag{28}$$

Using the law of mass action, all three reactions result in a second-order rate expression. We will integrate, linearize, and solve several of these reaction rate expressions.

For the second-order reaction with one reactant, the reaction rate expression is derived from reaction (26).

$$\frac{d[A]}{dt} = -2k_2'[A]^2 = -k_2[A]^2 \tag{29}$$

Note that it is a *nonlinear* ordinary differential equation because the dependency on concentration of A is other than zero or first order. We solve the differential equation by separating variables:

$$\int_{A_0}^{A} \frac{dA}{A^2} = -k_2 \int_0^t dt \tag{30}$$

The linearized equation is obtained after integration:

$$\frac{1}{A} - \frac{1}{A_0} = k_2 t \tag{31}$$

A plot of inverse concentration of A with time (1/A versus time) will yield a straight line with a slope of k_2 if the reaction is indeed second order in A (Figure 3.4). Equation (27) represents the second-order reaction with two reactants. The rate expression involves only one equation with two unknowns:

$$\frac{d[A]}{dt} = -k_2[A][B] \tag{32}$$

In order to solve equation (32), we must utilize the mass balance stoichiometry that relates species A to species B. One mole of A reacts with 1 mole of B. Thus

$$B = B_0 - B_{\text{reacted}} = B_0 - (A_0 - A) \tag{33}$$

Substituting equation (33) into equation (32), we obtain

$$\frac{d[A]}{dt} = -k_2 A (B_0 - A_0 + A) \tag{34}$$

Table 3.3 Second Order Reactions

$A + A \rightarrow B$	Zooplankton death rates Atmospheric gas phase reactions Second-order chemical reactions
$A + B \rightarrow D$	Microbial kinetics (substrate + cells) Sorption reactions (adsorbent + adsorbate) Predator–prey (phyto + zooplankton) Aquatic redox Rxns (oxidant + reductant)
$A + R \rightarrow 2R$	Autocatalytic reaction Microbial kinetics (substrate + cells) Some photo redox Rxns Nucleation/crystal growth

$$\frac{d[\mathrm{A}]}{dt} = -k_2\,(\mathrm{AB_0} - \mathrm{AA_0} + \mathrm{A}^2) \tag{35}$$

Equation (35) now has only one unknown (A), and it relies on knowledge of the initial concentrations of A and B, namely, A_0 and B_0. One can use integration tables and separation of variables to solve equation (35), and for the condition where $B_0 > A_0$, one obtains

$$\ln\frac{\mathrm{A}}{\mathrm{B}} - \ln\frac{\mathrm{A_0}}{\mathrm{B_0}} = -k_2\,(\mathrm{B_0} - \mathrm{A_0})t \tag{36}$$

A plot of ln(A/B) versus time should yield a straight line with the slope of $-k_2\,(B_0 - A_0)$.

A similar treatment of the autocatalytic equation will yield a straight line when one plots ln (A_0R/AR_0) versus time, where A_0 is the initial reactant concentration (e.g., substrate) and R_0 is the initial autocatalyst concentration (e.g., microbial biomass). In an autocatalytic reaction, one needs to know the stoichiometry (or yield coefficient) such that

$$\mathrm{C_0} = \mathrm{A_0} + \mathrm{R_0} \quad \text{and} \quad \mathrm{A} + \mathrm{R} = \mathrm{C_0}$$

at all times. The autocatalytic reaction is commonly used in environmental modeling when a little of the product is necessary to accelerate the rate of the forward reaction (nucleation-crystal growth, bacterial kinetics, etc.)

3.3.4 Other Reaction Orders

Many times a reaction is not elementary (single-step), and the rate expression is not simply zero, first, or second order. In this case the change in reaction rate can be plotted against the concentration (ln versus ln) to get an estimate of the order of the reaction.

$$\frac{dC}{dt} = -kC^n \tag{37}$$

$$\ln\left(-\frac{dC}{dt}\right) = n \ln C + \ln k \tag{38}$$

It may be fractional order ($0 < n < 1$) or some other noninteger order because the reaction mechanism is not elementary, and one reaction may not be fully rate-determining in the series of steps that occur.

Fractional order kinetics occur in precipitation and dissolution reactions. For example, in the dissolution of oxides and aluminosilicate minerals during chemical weathering, the reaction is surface-controlled by the slow detachment of the central metal ion (activated complex) into solution[7,8]:

$$\exists\!\!-OH + H^+ \xrightarrow{\text{fast}} \exists\!\!-OH_2^+$$

$$\exists\!\!-OH_2{}^+ \xrightarrow{\text{slow}} M(H_2O)_x^{z+} + \exists\!\!-$$

where $\exists\!\!-OH$ = hydrous oxide or aluminosilicate mineral
z = charge on the central metal ion and the number of protons bound to the central metal atom
M = central metal ion of valence $z+$
$\exists\!\!-$ = renewed surface

The rate of the reaction is proportional to $(\exists\!\!-OH_2^+)^z$, that is, the degree of protonation of the surface to the z power, which is the mass law for the precursor of the activated complex (the reactant of the rate-determining slow step). But the degree of protonation of the surface site is, in turn, proportional to a fractional order Freundlich isotherm, where n = the slope of the ln–ln adsorption isotherm. Therefore the final rate of dissolution is fractional order. Fractional order kinetics in this case resulted from a series of steps involving two phases, solid and aqueous.

$$\text{Rate} \propto \{\exists\!\!-OH_2^+\}^z$$

$$\{\exists\!\!-OH_2^+\} \propto \{H^+\}^n$$

$$\therefore \quad \text{Rate} \propto \{H^+\}^m$$

where $0 < m < 1$ and $m = n \times z$.

3.3.5 Michaelis–Menton Enzyme Kinetics

Enzyme kinetics often result in rather complicated rate expressions that are not elementary reactions and do not yield a simple rate expression or reaction order. The classic case of Michaelis–Menton enzyme kinetics follows a two-step reaction mechanism.

$$E + S \underset{k_2}{\overset{k_1}{\rightleftarrows}} [ES] \underset{\text{slow}}{\overset{k_3}{\longrightarrow}} E + P$$

where E is the enzyme, S is the substrate, ES is the enzyme–substrate complex, and P is the product of the reaction. Note that the enzyme is a catalyst that speeds the rate of the reaction (lowers the activation energy) but is not consumed in the reaction.

The rate of formation of ES complex involves all three reactions with rate constants k_1, k_2, and k_3.

$$\frac{d[ES]}{dt} = k_1\,[E]\,[S] - k_2\,[ES] - k_3\,[ES] \tag{39}$$

And the rate of formation of products is first order in the ES complex.

$$\frac{d[P]}{dt} = +\,k_3\,[ES] \tag{40}$$

If one assumes steady state for the formation of ES complex, $d[ES]/dt = 0$ in equation (39), and recognizing that k_3 is small compared to k_2 and is the rate determining step, then equation (41) results:

$$(k_2 + \overset{\text{small}}{\cancel{k_3}})\,[ES] = k_1[E]\,[S]$$

$$\frac{k_2}{k_1} = K_M = \frac{[E]\,[S]}{[ES]} \tag{41}$$

If we consider that the total enzyme in the system is (E + ES),

$$E = E_T - ES \tag{42}$$

then we substitute and rearrange for E in equation (41):

$$[ES] = \frac{[E_T]\,[S]}{(K_M + [S])} \tag{43}$$

Substituting equation (43) into equation (40) for ES, we derive

$$\frac{dP}{dt} = k_3 \frac{[E_T]\,[S]}{K_M + [S]} \tag{44}$$

The total enzyme E_T is seen to increase the rate of the reaction (catalyze it), but it is not consumed in the reaction. The formation rate of product increases with increasing E_T concentration. If the product P is cellular synthesis (cell biomass), then $k_3[E_T]$ represents the maximum growth rate of the product, and we obtain the final expression for Michaelis–Menton kinetics:

$$\frac{dP}{dt} = \frac{\mu_{max}\,[P]\,[S]}{K_M + [S]} \tag{45}$$

where μ_{max} is the maximum growth rate of product (cells).

The reaction rate expression in equation (40) is not second order overall, nor is it first order. It is intermediate between the two cases. At low substrate concentrations $S \ll K_M$, it is second order overall,

$$\frac{dP}{dt} = \frac{\mu_{max}}{K_M}\,[P]\,[S] \tag{46}$$

and at high substrate concentrations $S \gg K_M$, it is first order overall and represents a log-growth phase.

$$\frac{dP}{dt} = \mu_{max}\,[P] \tag{47}$$

The growth rate is a maximum when $S \gg K_M$ (the substrate concentration is very large), and it is first order with respect to substrate concentration for small substrate concentrations. Figure 3.5 is a plot of the growth rate μ as a function of substrate concentration.

$$\mu = \frac{dP/dt}{[P]} = \frac{\mu_{max}\,[S]}{K_M + [S]} \tag{48}$$

Because the rate expression is intermediate and variable between first and second order, it is difficult to linearize equation (45) to test the reaction model and to derive the parameters μ_{max} and K_M. Lineweaver and Burk used a double-reciprocal plot of equation (48) to linearize the experimental data:

$$\frac{1}{\mu} = \frac{K_M}{\mu_{max}}\,\frac{1}{[S]} + \frac{1}{\mu_{max}} \tag{49}$$

Figure 3.6 is a schematic of the plot in which batch experimental data for substrate concentration and growth rate can be plotted to obtain the parameters K_M and μ_{max} for the model [equation (45)].

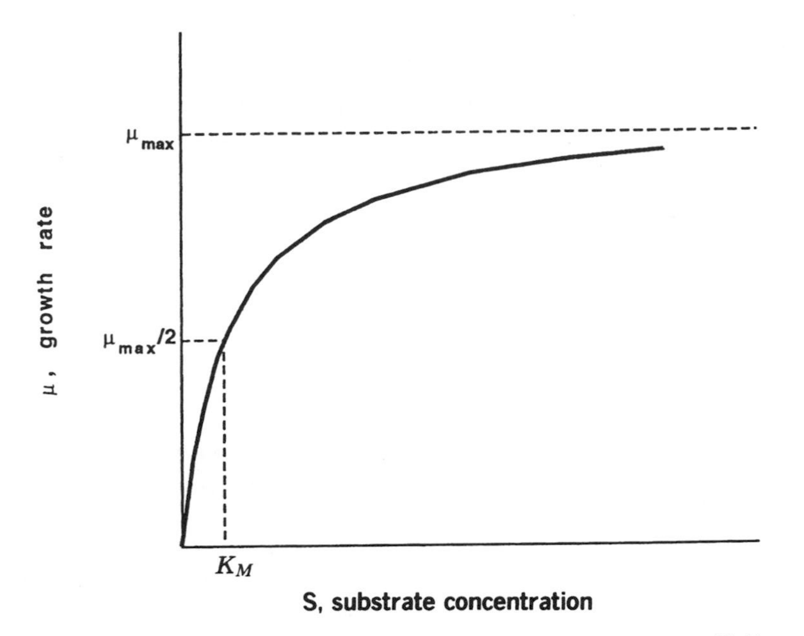

Figure 3.5 Michaelis–Menton enzyme kinetics showing maximum growth rate μ_{max} and half-saturation constant (Michaelis constant) K_M.

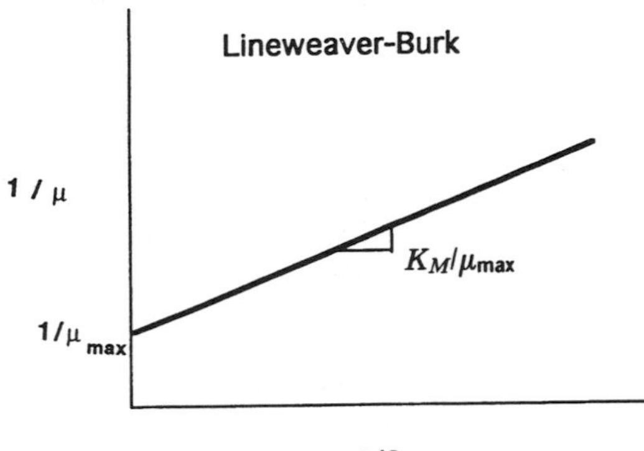

Figure 3.6 Lineweaver–Burk plot to linearize data using Michaelis–Menton enzyme kinetics to obtain the parameters μ_{max} and K_M. It is a double-reciprocal plot of growth rate and substrate concentration.

3.4 CONSECUTIVE REACTIONS

Consecutive reactions occur commonly in nature and they are usually irreversible. The reaction proceeds essentially to completion without a significant back reaction. Examples of consecutive reactions are

$$A \xrightarrow{k_1} B \xrightarrow{k_2} C \quad (50)$$

$$NH_3\text{-N} \rightarrow NO_2^-\text{-N} \rightarrow NO_3^-\text{-N} \quad (51)$$

$$\text{CBOD} \rightarrow D \rightarrow 0 \quad (52)$$

Nitrification and carbonaceous biochemical oxygen demand (CBOD) in a stream are examples, where D is the dissolved oxygen deficit that is created when CBOD exerts itself.

In reaction equation (51), ammonia-nitrogen is oxidized to nitrite-nitrogen, which is, in turn, oxidized to nitrate-nitrogen. Because each species is expressed in terms of *nitrogen,* the stoichiometric coefficients are unity (1 mole of NH_3-N yields 1 mole of NO_2-N, etc.) Bacteria catalyze the reactions in equations (51) and (52). For consecutive nitrification reactions, *Nitrosomonas* spp. mediate the first reaction and *Nitrobacter* spp. mediate the second reaction. The overall balanced chemical reaction for nitrification is

$$NH_3 + \tfrac{3}{2}\,O_2 \rightarrow NO_2^- + H^+ + H_2O$$
$$NO_2^- + \tfrac{1}{2}\,O_2 \rightarrow NO_3^-$$
$$\overline{NH_3 + 2\,O_2 \rightarrow NO_3^- + H^+ + H_2O}$$

Oxygen concentration does not enter into the rate expression except at very low oxygen concentrations. Thus we may assume that the first oxidation reaction for $NH_3 \rightarrow NO_2$ is a first order irreversible reaction in NH_3-N concentration, and that the second oxidation reaction is a first-order irreversible reaction in the concentration of NO_2-N.

On a molar basis, 1 mole of ammonia combines with 2 moles of oxygen to form 1 mole of nitrate, overall. On a mass basis, 1.0 gram of ammonia-nitrogen consumes 4.57 grams of oxygen to form 1.0 gram of nitrate-nitrogen. However, the rate of the reactions are determined by the reaction kinetics:

$$\frac{dA}{dt} = -k_1\,A \quad (53)$$

$$\frac{dB}{dt} = +k_1\,A - k_2\,B \quad (54)$$

$$\frac{dC}{dt} = + k_2 B \tag{55}$$

where A is the ammonia-nitrogen concentration, B is the nitrite-nitrogen concentration, and C is the nitrate-nitrogen concentration. Equations (53–55) represent a set of three ordinary differential equations that must be solved simultaneously, three equations with three unknowns.

Reaction equation (52) represents a common deoxygenation problem in streams, rivers, and estuaries that receive a waste discharge. The concentration of biodegradable organic material can be measured using a biochemical oxygen demand test. It measures the concentration of dissolved oxygen that is consumed via microbial oxidation of the organics. This process results in a dissolved oxygen deficit (D) in equation (52). The deficit, in turn, reaerates away due to the absorption of oxygen from the atmosphere to the stream. Instead of forming a product, the deficit goes to zero as atmospheric reaeration proceeds to chemical equilibrium (saturation).

$$D = C_{sat} - D.O.$$

where C_{sat} is the saturated concentration of dissolved oxygen in equilibrium with the atmosphere, and D.O. is the dissolved oxygen concentration. C_{sat} depends on temperature and salinity of the water body. Each reaction in equation (52) is expressed in terms of dissolved oxygen, so the stoichiometric coefficients are unity. The rate equations (53) and (54) are directly applicable, but equation (55) is not necessary because no product is formed.

Using batch reactor data from the laboratory (which is analogous to a Lagrangian approach or "floating along" in a plug-flow stream while measuring the reaction rates), we may derive the rate constants k_1 and k_2 after solving the set of ordinary differential equations (53–55). Equation (53) is one equation with one unknown, and it can be "decoupled" from the set of equations and integrated to determine the concentration of A as a function of time.

$$A = A_0 \, e^{-k_1 t} \tag{56}$$

We may substitute for the concentration of A into the rate expression for B, equation (54)

$$\frac{dB}{dt} = k_1 \, A_0 \, e^{-k_1 t} - k_2 \, B \tag{57}$$

Equation (57) is a first-order, linear, ordinary differential equation that is solvable by the integration factor method. For the general case,

$$\frac{dy}{dt} + p(t)\, y = q(t) \tag{58}$$

where $p(t)$ is the integrating factor that can be a function of time, and the nonhomogeneous forcing function is $q(t)$. The solution to equation (58) when the concentration varies from y_0 to y, and the time interval is from $0 \rightarrow t$ is equation (59):

$$y = y_0 \, e^{-P(t)} + e^{-P(t)} \int_0^t e^{+P(t)} \, q(t) \, dt \tag{59}$$

where $P(t) = \int p(t) \, dt$

$P(t)$ is the indefinite integral of the integrating factor. Making the proper substitutions from equation (57) into (58), we have equation (60):

$$p(t) = k_2$$

$$P(t) = k_2 t$$

$$y = \mathrm{B}$$

$$q(t) = k_1 \, \mathrm{A}_0 e^{-k_1 t}$$

$$\mathrm{B} = \mathrm{B}_0 \, e^{-k_2 t} + e^{-k_2 t} \int_0^t e^{k_2 t} \, (k_1 \, \mathrm{A}_0 \, e^{-k_1 t}) \, dt \tag{60}$$

Integrating over the time interval from $0 \rightarrow t$, one can derive the solution:

$$\mathrm{B} = \mathrm{B}_0 \, e^{-k_2 t} + \frac{\mathrm{A}_0 \, k_1}{k_2 - k_1} \, (e^{-k_1 t} - e^{-k_2 t}) \tag{61}$$

The solution for the intermediate concentration B in the series of consecutive irreversible reactions is given by equation (61). It gives the characteristic "humped-shape with a long tail" seen in Figure 3.7 for consecutive reactions. It is typical of the buildup of nitrite concentrations in a stream before further oxidation reduces them to nitrate. It is also characteristic of the buildup of dissolved oxygen deficit in a stream below a waste discharge prior to reaeration and self-purification of the stream. Equation (61) is the classic D.O. sag curve of Streeter–Phelps if we make the substitution

$$\mathrm{D.O.} = C_{\mathrm{sat}} - \mathrm{B}$$

where B is now the dissolved oxygen deficit concentration, and D.O. is the dissolved oxygen concentration.

The solution of the third species (e.g., nitrate) requires the substitution of equation (61) into equation (55) or the mass balance principle that

$$N_T = \mathrm{A} + \mathrm{B} + \mathrm{C} = \mathrm{A}_0 + \mathrm{B}_0 + \mathrm{C}_0$$

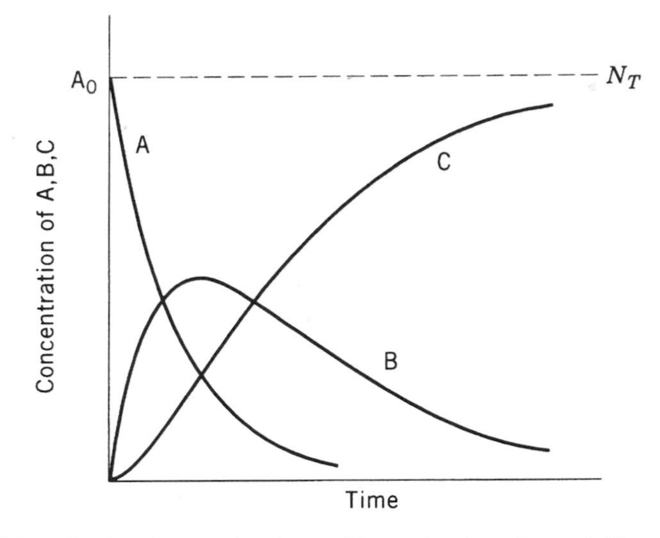

Figure 3.7 Schematic plot of consecutive, irreversible reaction A → B → C [Rxn rate equations (53)–(55) and resulting equations (56), (61), (62)].

where N_T is the total moles of species A, B, and C or the sum of their initial concentrations. In either technique, the solution for the third species is

$$C = N_T - A_0 e^{-k_1 t} - B_0 e^{-k_2 t} - \frac{A_0 k_1}{k_2 - k_1} (e^{-k_1 t} - e^{-k_2 t}) \tag{62}$$

Equation (62) gives the characteristic sigmoidal shape curve for the buildup of the final product, species concentration C, shown in Figure 3.7.

It is possible to evaluate the rate constants in a sequential procedure. k_1 can be obtained from linearization of equation (56) and a semilogarithmic plot, ln A versus t. Next, k_2 can be obtained from a nonlinear least-squares curve fitting of equation (60) and/or (61) to the data. k_2 can also be estimated from a plot of B versus time (Figure 3.7). At the point in time where B is a maximum, k_2 can be estimated from the following relationship, equation (63):

$$\frac{dB}{dt} = k_1 A - k_2 B = 0$$

$$k_2 = \frac{k_1 A}{B_{max}} \tag{63}$$

The rate constant k_2 can be estimated by equation (63), but it is better to use all the experimental data (not just the maximum point of B) in a nonlinear least-squares fit.

3.5 REVERSIBLE REACTIONS

Reversible reactions are characterized by a mixture of products and reactants remaining at chemical equilibrium. Instead of the forward reaction going to essential completion, the reaction only proceeds spontaneously to a mixture that results in the largest change in free energy ΔE. At chemical equilibrium, the rate of the forward reaction is equal to the rate of the reverse reaction. Most acid–base and complexation reactions in natural waters are reversible reactions. Also, many physical chemical reactions that occur in nature are results of forward and reverse reactions coming into a chemical equilibrium. To summarize, some examples of reversible reactions are:

- Acid–base reactions.
- Gas transfer.
- Adsorption–desorption.
- Bioconcentration–depuration.

The simplest case of a reversible reaction is when 1 mole of chemical A reacts to form 1 mole of chemical B and vice versa:

$$\mathrm{A} \underset{k_2}{\overset{k_1}{\rightleftharpoons}} \mathrm{B}$$

The total concentration of chemical in the system (e.g., batch reactor) is constant throughout time.

$$C_T = \mathrm{A} + \mathrm{B} \tag{64}$$

We may write the rate expressions as differential equations that describe the reaction of A and the formation of B as a function of time:

$$\frac{d\mathrm{A}}{dt} = -\,k_1\mathrm{A} + k_2\mathrm{B} \tag{65}$$

$$\frac{d\mathrm{B}}{dt} = -\,k_2\mathrm{B} + k_1\mathrm{A} \tag{66}$$

The simplest reversible reaction results in a set of two linear ordinary differential equations with two unknowns, A and B, with time. We can substitute equation (64) into (65) to obtain one equation with one unknown and solve for A:

$$\frac{d\mathrm{A}}{dt} = -\,k_1\mathrm{A} + k_2\,(C_T - \mathrm{A}) \tag{67}$$

Rearranging, we arrive at equation (68):

$$\frac{dA}{dt} + (k_1 + k_2)\,A = k_2\,C_T \tag{68}$$

It is solvable by the integration factor method that was used previously [equations (58) and (59)] or by the use of integration tables and separation of variables. In this case, for the integration factor method:

$$p(t) = k_1 + k_2$$

$$P(t) = (k_1 + k_2)\,t$$

$$y = A$$

$$q(t) = k_2\,C_T$$

The solution is

$$A = A_0\,e^{-(k_1 + k_2)t} + \frac{k_2 C_T}{k_1 + k_2}\,(1 - e^{-(k_1 + k_2)t}) \tag{69}$$

and

$$B = C_T - A \tag{70}$$

The first term in equation (70) represents the die-away of initial conditions, and the second term is the "buildup" of reactant A over time due to the reverse reaction.

The equilibrium mixture is observed at steady state when the concentrations of A and B no longer change with respect to time (Figure 3.8). The ratio of product B to reactant A depends on the ratio of the rate constants. At equilibrium $dA/dt = 0$, so equation (65) can be rearranged to

$$k_1 A = k_2 B \tag{71}$$

$$\frac{\hat{B}}{\hat{A}} = \frac{k_1}{k_2} = K_{eq} \tag{72}$$

where $\hat{B}$ and $\hat{A}$ are steady-state concentrations and K_{eq} is the equilibrium constant. If we substitute for $\hat{B}$ in equation (72), we can obtain the steady-state solution for reactant A at $t = \infty$.

$$\hat{B} = C_T - \hat{A} \tag{73}$$

$$\hat{A} = \frac{k_2 C_T}{k_1 + k_2} \tag{74}$$

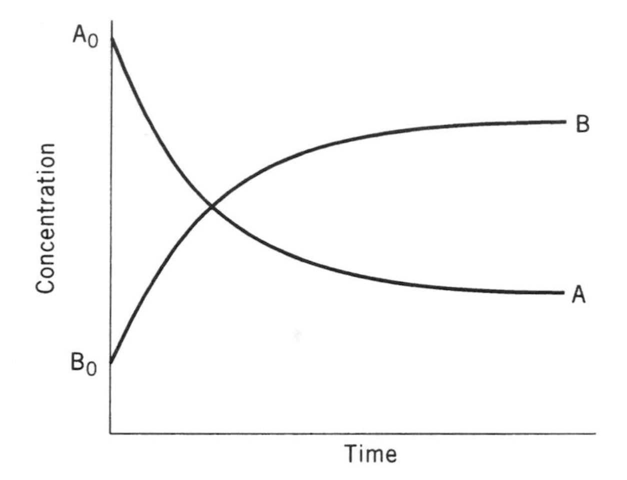

Figure 3.8 Reversible reaction showing the mixture of products and reactants at chemical equilibrium ($t = \infty$).

The final steady-state concentration of reactant A depends on the total concentration of A and B added (i.e., on C_T) and the rate constants.

3.6 PARALLEL REACTIONS, CYCLES, AND FOOD WEBS

Parallel reactions can be reversible or irreversible, but they include different pathways that a given reactant may take. For example, in the nitrification of ammonia that was discussed previously [equation (51)], parallel reactions might include the uptake of ammonia by algae and the stripping of ammonia from the water body to the atmosphere at high pH.

$$
\begin{aligned}
& \nearrow \xrightarrow{k_3} \quad 0 \text{ (atmosphere)} \\
NH_3 & \xrightarrow{k_1} \quad NO_2^- \xrightarrow{k_2} NO_3^- \\
& \searrow \xrightarrow{k_4} \quad \text{algae-N}
\end{aligned}
$$

The pathway that ammonia disappears from the environment, then, depends on the relative magnitude of the rate constants k_1, k_3, and k_4. A rate expression must include all three reactions, and the parallel paths for ammonia would affect the nitrite and nitrate concentrations as well.

$$\frac{d[NH_3\text{-N}]}{dt} = -(k_1 + k_3 + k_4)\,[NH_3\text{-N}] \tag{75}$$

$$\frac{d[\mathrm{NO_2\text{-}N}]}{dt} = + k_1\,[\mathrm{NH_3\text{-}N}] - k_2\,[\mathrm{NO_2\text{-}N}] \tag{76}$$

$$\frac{d[\mathrm{NO_3\text{-}N}]}{dt} = + k_2\,[\mathrm{NO_2\text{-}N}] \tag{77}$$

The resulting system of equations (75), (76), and (77) could be solved simultaneously to yield a plot of concentration versus time for ammonia algae-N, nitrite, and nitrate. Total nitrogen concentration C_T would no longer be constant with time because NH_3-N escapes the system via stripping. Parallel reactions are common in natural systems.

Elemental cycles are common in nature where a reactant is eventually regenerated by a product from a series of consecutive reactions. An example is shown in Figure 3.9 for the sulfur cycle. Each reaction in the cycle is represented by an arrow extending between two circles (state variables). To model the system, we would write a rate expression (dC/dt) for each state variable—there are five state variables shown and eight reactions. The rate expression for the concentration of sulfate with respect to time would be

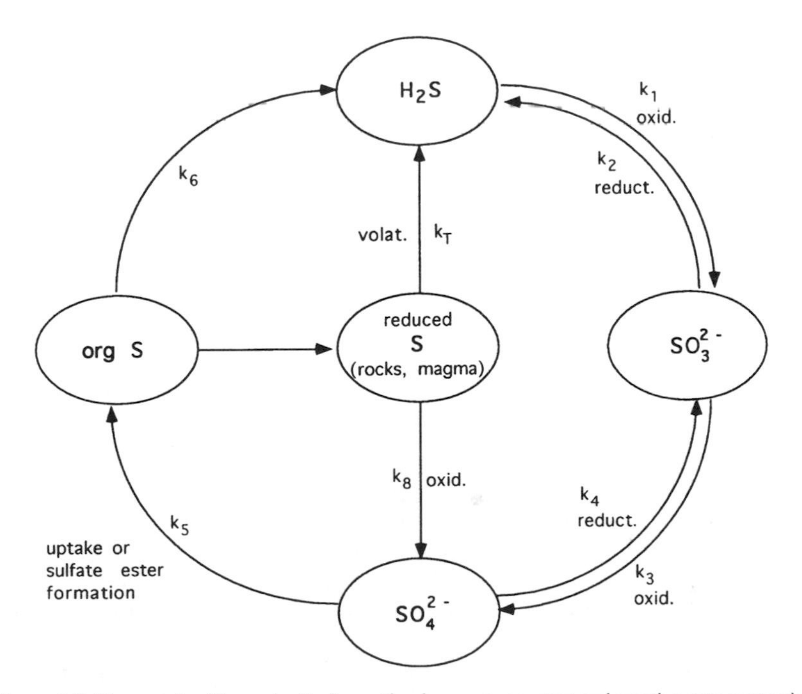

Figure 3.9 Elemental sulfur cycle. Each reaction has a rate constant and reaction rate expression.

$$\frac{d[SO_4]}{dt} = k_8\,[S] + k_3\,[SO_3] - (k_4 + k_5)\,[SO_4] \tag{78}$$

if we assume linear first-order reaction kinetics. A similar procedure could be used to write a set of five ordinary differential reaction equations that could be solved simultaneously. Heavy metals, nitrogen, carbon, sulfur, and phosphorus are elemental cycles that can be modeled at the microscale, mesoscale, or even global scale using chemical reaction kinetics.

Food webs are similar to elemental cycles for carbon or biomass. In a lake, one might be interested in modeling the transport and transformation of a contaminant (e.g., polychlorinated biphenyls or PCBs) as they move through the aquatic food web. The entire system is driven by primary production involving photosynthesis (the sun's energy) and the uptake of carbon dioxide by algae and rooted plants. Photosynthesis could not occur without the macro- and micronutrients required for algal growth including nitrogen, phosphorus, and silica. Food webs work because organisms die, settle to the bottom of the lake, mineralize by the action of decomposing bacteria and fungi, and are eventually returned to the pool of nutrients that sustain primary production. Ecosystem models have been constructed that examine the relationships between the various trophic levels and human activities such as waste inputs. In Figure 3.10, every line between each box represents a reaction in which a reaction rate expression (differential equation) could be written to solve the food web dynamics. However, this assumes that the proper kinetic formulation and rate constant are known for each reaction. Each reaction could probably be (and has been) the subject of at least one Ph.D. dissertation!

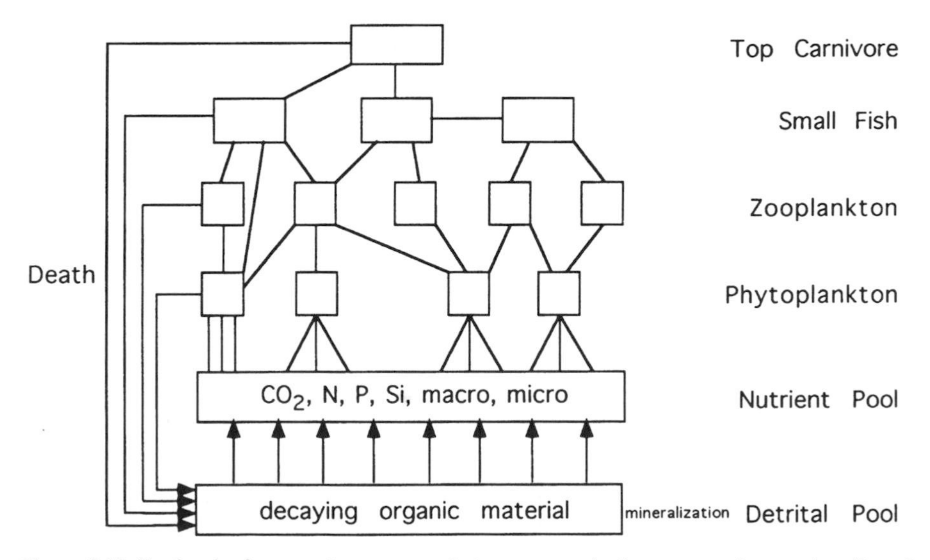

Figure 3.10 Food web of an aquatic ecosystem. It demonstrates the interconnectedness and cycling of elements in natural waters.

With this introduction to chemical reaction kinetics, we are ready to study equilibrium reactions in more detail and to formulate the reaction term (± Rxn) in the mass balance equations for environmental modeling.

3.7 TRANSITION STATE THEORY

In Figure 3.1, we showed schematically that an activation energy has to be overcome for a reaction to occur. Transition state theory considers the free-energy requirements of a chemical reaction. Rate expressions based on transition state theory provide an important bridge between thermodynamics (energetics and equilibrium reactions of Chapter 4) and rates of reactions (kinetics in Chapter 3).

Let's consider the reaction

$$\mathrm{A + BC \xrightarrow{k} AB + C} \tag{79}$$

The reaction may proceed by first forming an activated complex $ABC^‡$, a transition state:

$$\mathrm{A + BC \rightleftharpoons ABC^‡} \tag{80}$$

which then dissociates into products, irreversibly.

$$\mathrm{ABC^‡ \rightarrow AB + C} \tag{81}$$

The higher the activation energy (standard free energy of activation, $\Delta G^‡$, the less is the probability that the reaction occurs, and the smaller is the rate of reaction.

$$\frac{d\{\mathrm{AB}\}}{dt} = \frac{d\{\mathrm{C}\}}{dt} = k^‡\,\{\mathrm{ABC}^‡\} \tag{82}$$

$k^‡$, the rate constant, can be expressed as

$$k^‡ = k_\mathrm{B}\,T/h \tag{83}$$

where k_B is Boltzmann's constant (1.38×10^{-23} J K^{-1}) and h is Planck's constant (6.63×10^{-34} J s^{-1}), and T is the absolute temperature in degrees Kelvin.

The activated complex, $ABC^‡$, is in equilibrium with the reactants

$$K^‡ = \frac{\{\mathrm{ABC}^‡\}}{\{\mathrm{A}\}\,\{\mathrm{BC}\}} \tag{84}$$

Thus the rate of reaction of equation (82) can be expressed as

$$\frac{dB}{dt} = k^{\ddagger} K^{\ddagger} \{A\} \{BC\} \tag{85}$$

The standard free energy of activation is defined as

$$\Delta G^{\ddagger} = -RT \ln K^{\ddagger} \tag{86}$$

The rate constant for equation (79) follows from equation (85):

$$k = k^{\ddagger} K^{\ddagger} = k^{\ddagger} \exp(-\Delta G^{\ddagger}/RT) \tag{87}$$

Equation (87) illustrates that there is a relation between thermodynamics and kinetics.

Using thermodynamics, we can replace $\Delta G^{\ddagger}$, the activation energy in equation (87), by

$$\Delta G^{\ddagger} = \Delta H^{\ddagger} - T \Delta S^{\ddagger} \tag{88}$$

where $\Delta H^{\ddagger}$ is the standard enthalpy of activation, and $\Delta S^{\ddagger}$ is the standard entropy of activation. Equation (87) is similar to the Arrhenius equation (8) presented earlier in the chapter to examine temperature effects on reaction rate constants, where E_{act} is approximately equal to the standard free energy of activation $\Delta G^{\ddagger}$.

$$k = k^{\ddagger} \exp(-\Delta G^{\ddagger}/RT) = k^{\ddagger} \exp(\Delta S^{\ddagger}/R) \exp(-\Delta H^{\ddagger}/RT) \tag{89}$$

Slow reactions can be accelerated by a catalyst, a substance that alters the velocity of the overall reaction and is both a reactant and a product of the reaction. In the framework of transition state theory, a catalyst provides an alternative pathway that lowers $\Delta G^{\ddagger}$ or the activation energy. Lowering the activation energy from 100 to 30 kJ mol^{-1} increases the reaction rate by approximately a factor of 10!

3.8 LINEAR FREE-ENERGY RELATIONSHIPS

When we know the rate-determining step in a reaction, it is often possible to establish quantitative relationships for a series of *related* processes, empirical equations between the rate constants of the reaction and the equilibrium constants. For two related reactions, the following relationship can be established:

$$\ln k_2 - \ln k_1 = \alpha (\ln K_2 - \ln K_1) \tag{90}$$

where k_1 and k_2 are the rate constants of two reactions, and K_1 and K_2 are the equilibrium constants for the same reactions. Alternatively, we can write the linear free-energy relation in terms of thermodynamics:

$$-\left(\frac{\Delta G_2^{\ddagger} - \Delta G_1^{\ddagger}}{RT}\right) = \left(\frac{-\Delta G_2^{o} + \Delta G_1^{o}}{RT}\right) \tag{91}$$

where $\Delta G_2^{\ddagger}$ and $\Delta G_1^{\ddagger}$ are the free activation energies, and ΔG_2^{o} and ΔG_1^{o} are the free energies of the related reactions.

For a series of i reactants, we have the final linear free energy relationships

$$\ln k_i = \alpha \ln K_i + \ln \beta \tag{92a}$$

or

$$\Delta G_i^{\ddagger} = \alpha \, \Delta G_i^{o} + \beta \tag{92b}$$

where α is the slope of the linear plot and β is the intercept. From the results of experimental data, we can plot the logarithms of k_i versus K_i and determine α and β empirically.

An example is presented in Figure 3.11 from Wehrli[9] where the rate of oxygenation of three Fe(II) species (Fe^{2+}, $FeOH^+$, and $Fe(OH)_{2(aq)}$) are plotted versus the equilibrium constant for the interaction of the species with $O_{2(aq)}$. The equilibrium reactions are presented in equations (93a–c), and the rate expressions follow the law of mass action as the product of the Fe(II) species times the molar oxygen concentration in solution.

$$Fe^{2+} + O_2 + H_2O \overset{k}{\rightleftharpoons} FeOH^{2+} + O_2^- + H^+ \tag{93a}$$

$$FeOH^+ + O_2 + H_2O \rightleftharpoons Fe(OH)_2^+ + O_2^- + H^+ \tag{93b}$$

$$Fe(OH)_2^0 + O_2 + H_2O \rightleftharpoons Fe(OH)_3^0 + O_2^- + H^+ \tag{93c}$$

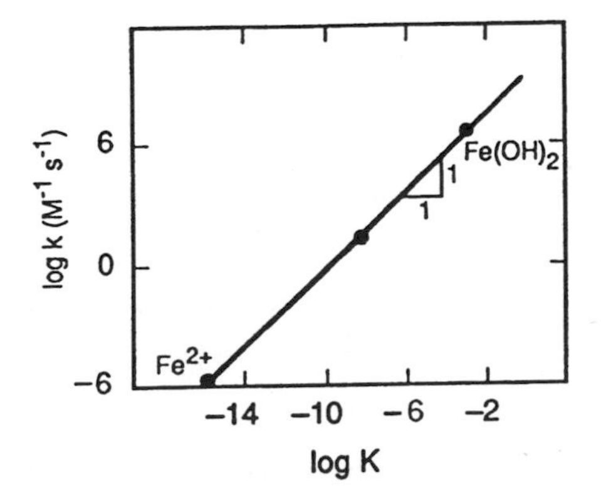

Figure 3.11 Linear free-energy relationship for the oxidation of various Fe(II) species (Fe^{2+}, $FeOH^+$, and $Fe(OH)_2^0$) with $O_{2(aq)}$ and the equilibrium constant for the reaction. (Modified from Wehrli.[9].)

The Fe(II) species in equations (93a–c) are oxidized by a one-electron transfer with dissolved oxygen to form the superoxide anion radical $O_2^{\bar{\cdot}}$. Three points representing the rate constant versus the equilibrium constant for the reactions are shown in Figure 3.11. The rate constant in each case is defined by

$$\frac{d[\mathrm{Fe(II)}]}{dt} = -k[\mathrm{Fe(II)}]\,[O_2] \tag{94}$$

where [Fe(II)] represents concentrations of the three species Fe^{2+}, $FeOH^{+}$, and $Fe(OH)_2^0$. It is much easier to oxidize $Fe(OH)_{2(aq)}$ than the other species because the complexing ligands (OH^- groups) have pulled electrons and deformed the electron cloud of the *d* orbital in the Fe transition element. Thus both the rate constant and the equilibrium constant are much greater (Figure 3.11). A word of caution: the relationship between kinetics and equilibrium reactions is not always so clear-cut. Reaction mechanisms and rate determining steps must be similar for hopes of developing structure–activity relationships.

Another structure–activity relationship is shown in Figure 3.12 from the work of Alan Stone[10] for the oxidation of variously substituted phenols by manganese dioxide. Manganese is reduced from Mn(IV) to Mn(II) and dissolves into solution while the substituted phenols are oxidized. The rate of oxidation of the organics is related

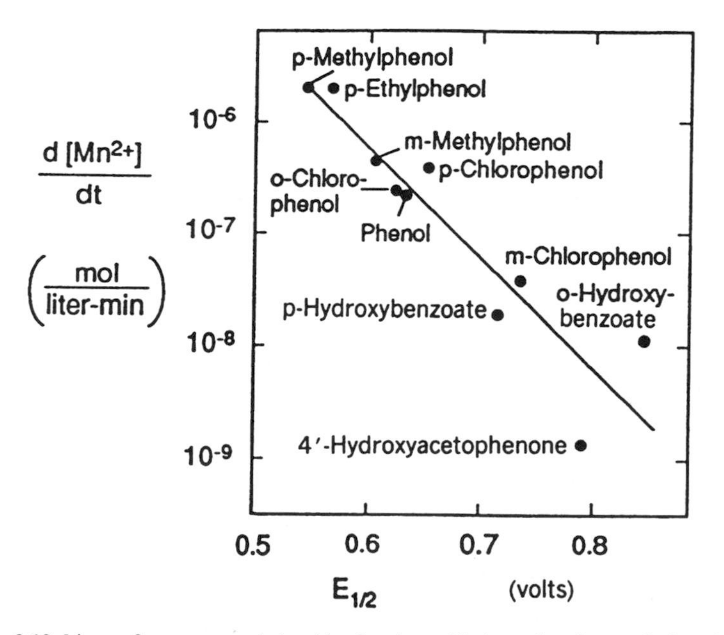

Figure 3.12 Linear free-energy relationship for the oxidation of various substituted phenols by $MnO_{2(s)}$.[10] The rate of oxidation is given by $d[Mn^{2+}]/dt$, which is related to the thermodynamic tendency of an electrode to oxidize these phenols. This tendency is measured by the half-wave potential, $E_{1/2}$, similar to the electrode potential.

to the tendency of an electrode to oxidize the phenolic groups. This tendency is measured electrochemically by the half wave potential in volts (a thermodynamic measure). The more reducing is the half-wave potential, the greater is the rate of the reaction as measured by $d[Mn^{2+}]/dt$ (manganese reduction and dissolution).

3.9 REFERENCES

1. Eyring, H., and Eyring, E.M., *Modern Chemical Kinetics,* Reinhold Publishing, New York (1963).
2. Tchobanoglous, G., and Schroeder, E.D., *Water Quality,* Addison-Wesley Publishing, Reading, MA (1985).
3. Parsons, T., and Takahashi, M., *Biological Oceanographic Processes,* Pergamon Press, New York (1973).
4. Schnoor, J.L., and Fruh, E.G., *Water Resour. Bull.,* 15, 506 (1979).
5. Tanner, J.T., *Ecology,* 47, 733 (1966).
6. Levenspiel, O., *Chemical Reaction Engineering,* 2nd ed., Wiley, New York (1970).
7. Furrer, G., and Stumm, W., *Chimia,* 37, 338 (1983).
8. Schnoor, J.L., in *Aquatic Chemical Kinetics,* W. Stumm, Ed., Wiley-Interscience, New York (1990).
9. Wehrli, B., in *Aquatic Chemical Kinetics,* W. Stumm, Ed., Wiley-Interscience, New York (1990).
10. Stone, A., *Geochim. Cosmochim. Acta,* 51, 919 (1987).

3.10 PROBLEMS

1. The reaction below proceeds reversibly.

a. Write the rate equation, $d[SO_3^{2-}]/dt$, according to the law of mass action.

b. What are the units on k_1 and k_2?

c. Draw a rough sketch (graph) of the concentrations of sulfite and sulfate versus time, assuming the presence of excess oxygen. Would the curves appear differently in the absence of oxygen?

$$SO_3^{-2} + \tfrac{1}{2}O_2 \underset{k_2}{\overset{k_1}{\rightleftharpoons}} SO_4^{-2}$$

2. What are the units on k_1, k_2, k_3, k_4, and k_5 below? Write the rate equations for dA/dt in each case.

a. $A \underset{k_2}{\overset{k_1}{\rightleftharpoons}} B$

b. $A + B \xrightarrow{k_3} C$

c. $A + R \xrightarrow{k_4} 2R$

d. $A \xrightarrow{k_5} B$

3. A pilot plant experiment was conducted on the oxygen transfer efficiency of a mechanical aerator. The temperature was maintained by 27 °C. The contents of the vessel were deaerated by the addition of sodium sulfite. The dissolved oxygen concentration at the start of the test was not properly measured but the data at subsequent intervals are satisfactory and are tabulated. Estimate the initial concentration of dissolved oxygen and the oxygen transfer coefficient, k_{La}. Dissolved oxygen deficit ($D = C_{sat} - C$) decreases in a first-order reaction according to the notation: $D \xrightarrow{k_{La}} 0$.

Time, min	D.O., mg L^{-1}
0	—
5	2.6
10	3.2
15	5.3
20	5.8
25	6.6
30	6.9

4. Write the kinetic reaction rate expression (dC/dt) for the following reactions. Define your variables and constants where necessary.

a. Carbonaceous biochemical oxygen demand (CBOD) exertion in a bottle.
b. Monod kinetics for bacteria, B, and their growth on substrate, S.
c. Oxygen transfer or aeration, C = D.O. concentration.
d. Photodegradation of chemical C (first-order decay).
e. Radioactive decay of ^{226}Ra in water.
f. Adsorption of chemical C onto solids M.

$$C + M \rightleftharpoons CM$$

5. Determine the order of the reaction of substance A from the data given below. The reaction was studied in a constant-temperature bath. The initial concentration of material A was 10.0 mg L^{-1}. The amount of chemical A for various intervals of time was measured and recorded as indicated below. Plot the data as if it were a zero-, first-, or second-order reaction and determine which model best fits the data. Report the reaction order, the rate expression, and the reaction rate constant. Show your plots of model fit and data in each case.

Time		Concentration of A, $mg\ L^{-1}$
Hours	Minutes	
0	0	10.0
0	42	8.9
2	00	7.5
3	36	6.3
5	05	5.4
7	11	4.5

6. Below are four kinetic expressions (mathematical models), which have been proposed to simulate coliform die-away during disinfection of a wastewater. Given the batch laboratory data that follow, which of the four models best describes the coliform die-away? Linearize the equations and make four graphical presentations of the linearized equation and the laboratory data.

a. $\frac{dC}{dt} = \text{Rxn rate} = -k_1 C$

b. $\frac{dC}{dt} = \text{Rxn rate} = -k_2 C^2$

c. $\frac{dC}{dt} = \text{Rxn rate} = -k_3 CX$

where X is the concentration of chlorine disinfectant

$X = X_0 e^{-rt} = (10\ mg\ L^{-1})\ e^{-(0.05/min)t)}$

d. $\frac{dC}{dt} = \text{Rxn rate} = -k_4 Ct$

in which C is the concentration of coliform bacteria (cfu mL^{-1}) and t is time (min).

For the best model, report the rate constant that you obtained from your plot.

Bacteria Concentration, cfu mL^{-1}	Time, min
1×10^8	0
1×10^3	2.5
3.3×10^2	5
2.5×10^2	10
1.7×10^2	15
1.1×10^2	20
1.0×10^2	25
8.3×10^1	30

7. Nitrite NO_2^- is toxic to fish and other aquatic microorganisms.

a. Solve the differential rate equations for the concentration of nitrite with respect to time given the following consecutive reactions:

$$(NH_3\text{-}N) \xrightarrow{k_1} (NO_2^-\text{-}N) \xrightarrow{k_2} (NO_3^-\text{-}N)$$

b. Now solve for nitrite concentration as a function of time if, in a parallel reaction, ammonia is taken up by algae:

$$(NH_3\text{-}N) \xrightarrow{k_1} (NO_2^-\text{-}N) \xrightarrow{k_2} (NO_3^-\text{-}N)$$

$$k_3 \;\llcorner\!\!\longrightarrow \text{algae}$$

8. Bacteria often grow autocatalytically; that is, it takes an initial bacterial inoculum to produce more bacteria R from substrate A. We may express this reaction as the following:

$$A + R \xrightarrow{k} R + R$$

a. Write the rate expression for the formation of R.
b. Solve for the concentration of R as a function of time and C_T, where $C_T = A_0 + R_0$.
c. Can you determine how you might plot the data to linearize it for determination of k?

9. The following data show the uptake and depuration of oysters (*Crassostrea* sp.) exposed to 0.39 μg L^{-1} of the pesticide kepone continuously. Depuration refers to the clearance of pesticide after it is removed from the aquaria water at day 28. The residue is the concentration of kepone in the oysters in units of μg kepone per gram of biomass (wet).

a. Develop a model for bioconcentration of kepone.

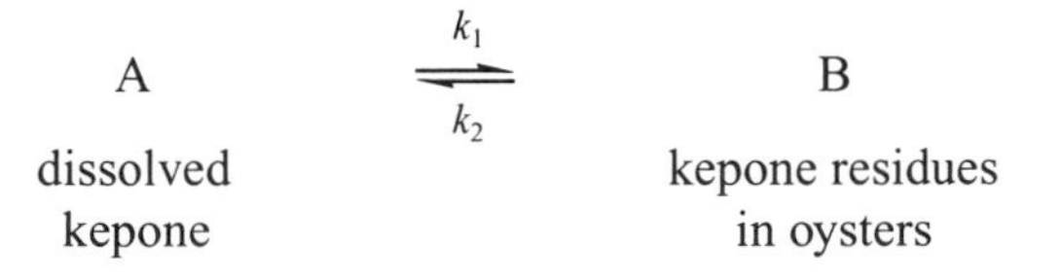

b. Solve for the rate constants using the data below. What does the ratio k_1/k_2 signify?
c. Plot your model results as a continuous line and the lab data (the points below) on a graph to show the model fit.

Time, days	Residue, $\mu g\ g^{-1}$	Kepone Concentration, $\mu g\ L^{-1}$
0.6	0.032	0.39
0.7	0.186	0.39
0.9	0.513	0.39
1.8	0.933	0.39
4.7	1.35	0.39
8.9	2.19	0.39
13.1	2.51	0.39
16.0	2.63	0.39
19.8	3.55	0.39
22.0	3.63	0.39
25.8	2.82	0.39
29.1	2.24	0
30.0	1.45	0
32.9	0.18	0
36.0	0.17	0
40.0	0.051	0
42.9	0.036	0
50.0	0.035	0

10. Investigators have determined that the rate equation for bacterial kill in batch chlorine contact chambers is

$$\text{Rxn rate} = \frac{dC}{dt} = -kCt$$

where C is the concentration of bacteria, k is the rate constant, and t is time. Integrate the rate equation and solve for the concentration as a function of time. What would you plot in order to get a straight-line graph of the data? How do you find k from the plot?

4

EQUILIBRIUM CHEMICAL MODELING

Everything that can be invented has been invented.
—Charles Duell, U.S. Patent Officer, 1899

4.1 INTRODUCTION

Reaction kinetics are sometimes fast relative to transport processes. In these cases, it is valid to assume that the system is at chemical equilibrium and to calculate the concentrations of all species in solution. Furthermore, even when reactions are slow, sometimes we may assume that chemical equilibrium applies as a limiting case (i.e., the system is tending toward equilibrium). Acid–base and complexation reactions are usually fast relative to transport reactions; dissolution and precipitation reactions are variable (sometimes they are quite slow); and redox reactions are generally slow. Even in the case of slow redox reactions, we gain insight by computing the chemical equilibrium speciation and achieve an understanding of where the system is tending. In this chapter, we will learn how to calculate chemical concentrations under the assumption of chemical equilibrium.

In modeling environmental systems, one makes use of mass balance equations such as those in Chapters 1 and 2 for a clearly defined control volume and chemical system.

$$\text{Accumulation of mass} = \text{Mass inputs} - \text{Mass outflow} \pm \text{Reactions} \qquad (1)$$

The "Reactions" term refers to chemical, physical, or biological kinetics that take place within the control volume, representing a reaction rate times the volume. For a given point in time and space, equation (1) is solved to yield the total chemical concentration, which is then used as input to a chemical equilibrium speciation model. Figure 4.1 is a flowchart of fate and transport models with chemical equilibrium speciation.

Sillen[1] and Garrels and Thompson[2] were the first to use simple chemical equilibrium models in the early 1960s to describe seawater chemistry. The first computer solutions came from the RAND Corporation using an energy minimization approach.[3,4] Numerical methods and computer codes followed in the early 1970s

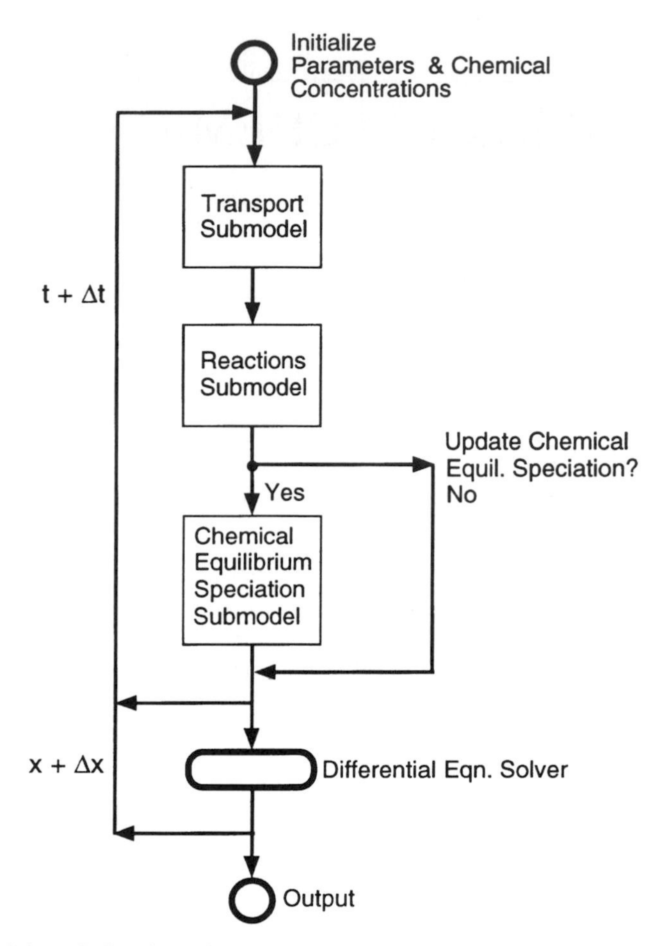

Figure 4.1 Schematic flowchart of a chemical equilibrium submodel within a fate and transport mass balance model.

based on solution of mass law expressions: REDEQL by Morel and Morgan[5] and WATEQ by Truesdell and Jones.[6] Most current geochemical equilibrium models emanate from these two early models. REDEQL was followed by Westall's[7] MINEQL and MICROQL,[8] and WATEQ evolved through a number of generations to WATEQIVF.[9] These two lineages of chemical equilibrium models had different principal objectives: REDEQL and its progeny were designed to estimate equilibrium chemical composition for natural waters and wastewaters; and WATEQ and its offspring were used primarily to determine if natural water samples were at chemical equilibrium with respect to mineral phases. Thus WATEQ emphasized an extensive thermodynamic database, while REDEQL was designed to be numerically efficient and easy-to-use. The two lineages merged in 1984 with the publication of MINTEQ ("MIN" came from MINEQL and "TEQ" came from WATEQ).[10] Now most geochemical models operate on similar principles with Newton–Raphson root-finding solution techniques to mass law equilibrium equations. Models differ in their databases, input/output formats, methods of handling heterogeneous equilibria, programming codes, and computer compatibility, but the basic structure of the programs are similar.[11,12]

Equilibrium models allow speciation of chemicals, which is important for determining not only fate and transport but also toxicity of the chemical. Site-specific water quality criteria require estimation of the toxic species under different environmental conditions (ligands, pH, redox, dissolved organic matter, etc.).

4.2 EQUILIBRIUM PRINCIPLES

4.2.1 Ionization of Water

Water undergoes self-ionization to form hydrated protons and hydroxide ions.

$$H_2O + H_2O \rightleftharpoons H_3O^+ + OH^- \tag{2}$$

According to the mass law, we can express the ion product of water at chemical equilibrium with an equilibrium constant, K_w:

$$K_w = \{H_3O^+\}\ \{OH^-\} \tag{3}$$

Its value at different temperatures is given in Table 4.1. For computer programming, a continuous algebraic function for K_w as a function of temperature can be written as equation (4).

$$\log K_w = -(4470.99/T) + 6.0875 - 0.01706T \tag{4}$$

where T is absolute temperature in K.[14]

A combination pH electrode measures the pH relative to the junction potential of $\{H_3O^+\}$. We do not need to use activity coefficients and concentrations when using pH data because it measures the activity of the H^+ ion directly.

Table 4.1 Ion Product of Water

Temperature, °C	K_w	pK_w
0	0.12×10^{-14}	14.93
15	0.45×10^{-14}	14.35
20	0.68×10^{-14}	14.17
25	1.01×10^{-14}	14.00
30	1.47×10^{-14}	13.83
50	5.48×10^{-14}	13.26

Note: For seawater (34.82% salinity, 25 °C),[13] $\log\,[H^+]\,[OH^-] = -13.199$.

Water can act as either an acid or a base. It can exhibit both properties depending on the reactants.

4.2.2 Acids and Bases

An acid can react with water or a base to form an acid and its conjugate base, and a base can react with water or an acid to form its conjugate acid and a base. Examples of acids are presented in equation (5). Brønsted defined an acid as any substance that can donate a proton to another substance, and a base as any substance that can accept a proton from another substance.

$$HCl + H_2O \rightleftharpoons H_3O^+ + Cl^- \tag{5a}$$

$$H_2CO_3 + H_2O \rightleftharpoons H_3O^+ + HCO_3^- \tag{5b}$$

$$NH_4^+ + H_2O \rightleftharpoons H_3O^+ + NH_3 \tag{5c}$$

$$HAc + H_2O \rightleftharpoons H_3O^+ + Ac^- \tag{5d}$$

$$\text{Acid} + \text{Base} \rightleftharpoons \text{Acid} + \text{Base}$$

where HAc represents acetic acid (CH_3COOH), a weak acid. HCl is a strong acid so its equilibrium reaction goes to virtual completion [almost completely to products on the right-hand side of Equation (5a)]. Examples of bases are given by equations (6a)–(6c).

$$NH_3 + H_2O \rightleftharpoons NH_4^+ + OH^- \tag{6a}$$

$$CO_3^{2-} + H_2O \rightleftharpoons HCO_3^- + OH^- \tag{6b}$$

$$Ac^- + H_2O \rightleftharpoons HAc + OH^- \tag{6c}$$

$$\text{Base} + \text{Acid} \rightleftharpoons \text{Acid} + \text{Base}$$

We can define corresponding acidity constants K_a and basicity constants K_b using equilibrium equations:

$$K_a = \frac{\{H^+\}\,\{A^-\}}{\{HA\}}; \qquad K_b = \frac{\{B^+\}\,\{OH^-\}}{\{B\}} \tag{7}$$

where $\{HA\}$ is the acid and $\{B\}$ represents the base. Curly brackets { } define activities in solution, and square brackets [ ] will be used to denote concentrations. Mixtures of acids and bases have multiple equilibrium expressions, and as the problem gets more complicated, we use chemical equilibrium models to solve it. Table 4.2 gives a compilation of some common acidity and basicity constants at 25 °C.

Table 4.2 Acidity and Basicity Constants of Acids and Bases in Aqueous Solutions at 25 °C

Acid[a]		–Log Acidity Constant, p*K* (approximate)	Base[b]	–Log Basicity Constant, p*K* (approximate)
$HClO_4$	Perchloric acid	–7	ClO_4^-	21
HCl	Hydrogen chloride	~ –3	Cl^-	17
H_2SO_4	Sulfuric acid	~ –3	HSO_4^-	17
HNO_3	Nitric acid	–1	NO_3^-	15
H_3O^+	Hydronium ion	0	H_2O	14
HSO_4^-	Bisulfate	1.9	SO_4^{2-}	12.1
H_3PO_4	Phosphoric acid	2.1	$H_2PO_4^-$	11.9
$[Fe(H_2O)_6]^{3+}$	Aquo ferric ion	2.2	$[Fe(H_2O)_5(OH)]^{2+}$	11.8
CH_3COOH	Acetic acid	4.7	CH_3COO^-	9.3
$[Al(H_2O)_6]^{3+}$	Aquo aluminum ion	4.9	$[Al(H_2O)_5(OH)]^{2+}$	9.1
$H_2CO_3^*$	Carbon dioxide[c]	6.3	HCO_3^-	7.7
H_2S	Hydrogen sulfide	7.1	HS^-	6.9
$H_2PO_4^-$	Dihydrogen phosphate	7.2	HPO_4^{2-}	6.8
$HOCl$	Hypochlorous acid	7.6	OCl^-	6.4
HCN	Hydrogen cyanide	9.2	CN^-	4.8
H_3BO_3	Boric acid	9.3	$B(OH)_4^-$	4.7
NH_4^+	Ammonium ion	9.3	NH_3	4.7
$Si(OH)_4$	*o*-Silicic acid	9.5	$SiO(OH)_3^-$	4.5
HCO_3^-	Bicarbonate	10.3	CO_3^{2-}	3.7
H_2O_2	Hydrogen peroxide	—	HO_2^-	2.3
$SiO(OH)_3^-$	Silicate	12.6	$SiO_2(OH)_2^{2-}$	1.4
HS^-	Bisulfide[d]	14	S^{2-}	0
H_2O	Water	14	OH^-	0
NH_3	Ammonia	~ 23	NH_2^-	–9
OH^-	Hydroxide ion	~ 24	O^{2-}	~ –10
CH_4	Methane	~ 34	CH_3^-	~ –20

[a]In order of decreasing acid strength.
[b]In order of increasing base strength.
[c]Total un-ionized CO_2 in water, that is, the acidity constant is a composite constant for the analytical sum

$$[H_2CO_3^*] = [CO_2\ aq] + [H_2CO_3].$$

[d]Considerable uncertainty on bisulfide constant. Recent evidence suggests pK = 17–19.
Source: Modified from Stumm and Morgan.[14]

The maximum coordination number of aluminum is 6 (2 × valence), and it coordinates six solvent molecules (H_2O) around it in solution. It is an acid because of its tendency to donate a proton into solution (pK = 4.9 in Table 4.2).

$$Al(H_2O)_6^{3+} + H_2O \rightleftharpoons H_3O^+ + Al(OH)(H_2O)_5^{2+} \quad (8)$$

$$K = \frac{\{H_3O^+\}\ \{Al(OH)(H_2O)_5^{2+}\}}{\{Al(H_2O)_6^{3+}\}} \quad (9)$$

Because the activity of water in dilute solution is constant (mole fraction is unity, 55.4 M H_2O at 25 °C), the hydration of the proton in equation (8) can be neglected. The activity of water $\{H_2O\}$ is 1.0 as a solvent. Thermodynamic convention sets the equilibrium constant for equation (10) equal to $K_b = 1$, which corresponds to a free-energy change $\Delta G = 0$:

$$H_2O + H^+ \rightleftharpoons H_3O^+ \tag{10}$$

Thus it is permissible to write equation (8) in the following manner, keeping in mind that aluminum and protons coordinate H_2O molecules.

$$Al^{3+} + H_2O \rightleftharpoons Al(OH)^{2+} + H^+ \tag{11}$$

The equilibrium constant for equation (11) is equal to that for equation (8). They are identical reactions.

4.2.3 Equilibrium Constants from Thermodynamic Data

There is a thermodynamic basis, Gibbs' free energy, for calculation of equilibrium constants. By definition, the energy content of a system is equal to the heat content minus its state of randomness:

$$G = H - TS \tag{12}$$

where G is Gibb's free energy in kJ mol^{-1}, H is its enthalpy (heat content) in kJ mol^{-1}, T is absolute temperature (K), and S is the entropy in kcal K^{-1}.

For a chemical reaction, the change in free energy of the system (ΔG) is related to the change in enthalpy of the reaction (ΔH) and the change in entropy ($T\,\Delta S$) at constant temperature and pressure:

$$\Delta G = \Delta H - T\,\Delta S \tag{13}$$

where ΔG is the free-energy change of the reaction (kJ mol^{-1}), ΔH is the change in enthalpy of reaction (kJ mol^{-1}), T is absolute temperature (K), and ΔS is the change in the entropy of reaction (kJ K^{-1}). If ΔG is positive, the reaction will not occur as written; if ΔG is negative, the reaction proceeds spontaneously to the right; and if $\Delta G = 0$, the reaction is at chemical equilibrium.

Suppose the reaction has the hypothetical stoichiometry given by equation (14):

$$aA + bB \rightleftharpoons cC + dD \tag{14}$$

Then the change in free energy can be calculated from its free energy of formation of the products and reactants and the mass law equation.

$$\Delta G = \Delta G^{\circ} + RT \ln Q \tag{15}$$

and

$$Q = \frac{\{C\}^c \{D\}^d}{\{A\}^a \{B\}^b} \tag{16}$$

where Q is the reaction quotient at any time in the reaction, and ΔG° is the Gibbs standard free energy of the reaction products minus reactants.

$$\Delta G^\circ = \sum_i \nu_i (\Delta G_f^\circ)_{\text{products}} - \sum_j \nu_j (\Delta G_f^\circ)_{\text{reactants}} \tag{17}$$

where ΔG_f° is the standard free energy of each reactant and product in units of kJ mol^{-1}, and the ν_i are the stoichiometric coefficients of the reactants.

At chemical equilibrium, ΔG must be equal to zero, and equation (15) can be reduced to

$$\Delta G^\circ = -RT \ln K \tag{18}$$

where K is the equilibrium constant. Q, the reaction quotient, is equal to K at chemical equilibrium. For reactions *not* at chemical equilibrium, we may write

$$\Delta G = RT \ln \frac{Q}{K} \tag{19}$$

so if $Q/K > 1$ (ΔG is positive), the reaction does not proceed as written; if $Q/K < 1$ (ΔG is negative), the reaction proceeds spontaneously from left to right; and if $Q/K = 1$ ($\Delta G = 0$), the reaction is at chemical equilibrium.

There are a few thermodynamic conventions for use of equations (15–18).

- Aqueous species are expressed as activities in solution, usually mol L^{-1} or mol kg^{-1}.
- Pure solids and the solvent (H_2O) have activities equal to 1.
- Gas components are expressed in units of partial pressure.

Example 4.1 Calculation of* K *from Standard Free Energies of Formation

Find $\log_{10} K$ from ΔG_f° at 25 °C for the reaction

$$H_2CO_3^* \text{ (aq)} \rightleftharpoons H^+ + HCO_3^-$$

Solution: In chemistry texts, you will find a thermodynamic database that gives values of ΔG_f° for all products and reactants for the dissociation of carbonic acid. Here, we interpret the problem to be for $H_2CO_3^*$, including "true" molecular $H_2CO_{3(aq)}$ and $CO_2 \cdot H_2O$. $RT = (0.008314 \text{ kJ mol}^{-1} \text{ K}^{-1})(298.15 \text{ K})$ at 25 °C.

$$\begin{array}{lccc} & H_2CO^*_{3(aq)} \rightleftharpoons & H^+ & + HCO_3^- \\ \Delta G_f^\circ & -623.2 & 0 & -586.8 \quad \text{kJ mol}^{-1} \end{array}$$

$$\Delta G^\circ = \sum_i (\nu_i \, \Delta G_f^\circ)_{\text{products}} - \sum_j (\nu_j \, \Delta G_f^\circ)_{\text{reactants}}$$

$$\Delta G^\circ = -586.8 + 0 - (-623.2) = +36.50 \text{ kJ mol}^{-1}$$

$$\Delta G^\circ = -RT \ln K \Rightarrow \ln K = \frac{-\Delta G^\circ}{RT} \quad \text{or} \quad \log_{10} K = \frac{-\Delta G^\circ}{RT\,(2.303)}$$

$$\log_{10} K = \frac{-36.50 \text{ kJ mol}^{-1}}{(0.008314)(298.15)(2.303)} = -6.39$$

$$K = 4.04 \times 10^{-7} = \frac{\{H^+\}\,\{HCO_3^-\}}{\{H_2CO_3^*\}}$$

This is quite close to the measured acidity constant given in Table 4.2 for $H_2CO^*_{3(aq)}$.

4.2.4 Temperature and Pressure Correction for Equilibrium Constants

Equilibrium constants must be adjusted for temperatures other than 25 °C. Chemical equilibrium models use the van't Hoff approximation to adjust for temperature change over small ranges. We begin with Gibbs' equation that relates equilibrium constants to thermodynamics.

$$\ln K = \frac{-\Delta G^\circ}{RT} \tag{20}$$

Taking the derivative of both sides, we have

$$\frac{d(\ln K)}{dT} = \frac{\Delta G^\circ}{RT^2} \simeq \frac{\Delta H^\circ}{RT^2} \tag{21}$$

Assuming that for small changes in temperature $\Delta S \approx 0$, we can approximate $\Delta G^\circ \simeq \Delta H^\circ$, the standard enthalpy. Integrating the equation, we have equation (22):

$$\int_{K_1}^{K_2} d(\ln K) = \int_{T_1}^{T_2} \frac{\Delta H^\circ}{RT^2}\, dT \tag{22}$$

This results in the van't Hoff equation after rearranging.

$$\ln \frac{K_1}{K_2} = \frac{\Delta H^\circ}{R}\left(\frac{1}{T_2} - \frac{1}{T_1}\right) \tag{23}$$

where K_1 is the equilibrium constant at the new temperature, K_2 is the equilibrium constant at 25°C (reference), ΔH° is the standard enthalpy (kJ mol^{-1}) at 25 °C, T_1 is the new temperature (K), and T_2 is the reference temperature (25 °C = 298.15 K). For temperature changes more than ±30 °C, errors will increase using the van't Hoff equation, and a more rigorous analysis including entropy effects on ΔG° in equation (20) is required.

Example 4.2 Calculation of K *at Different Temperatures*

Find the change in K acidity constant of $H_2CO^*_{3(aq)}$ from Example 4.1 at 10 °C, rather than 25 °C.

Solution: Find the standard enthalpies for the products and the reactants from a table of thermodynamic data such as Stumm and Morgan.[14]

$$H_2CO^*_{3(aq)} \rightleftharpoons H^+ + HCO_3^-$$

$$\Delta H_f^\circ \quad -699.6 \qquad 0 \quad -692.0 \quad \text{kJ mol}^{-1}$$

$$\Delta H_f^\circ = \sum_i (\nu_i\, \Delta H_f^\circ)_{\text{products}} - \sum_j (\nu_j\, \Delta H_f^\circ)_{\text{reactants}}$$

$$\Delta H_f^\circ = -692.0 + 0 - (-699.6) = +7.60 \text{ kJ mol}^{-1}$$

Use equation (23) to solve for the new equilibrium constant at 10 °C.

$$\ln \frac{K_1}{K_2} = \frac{6.60}{0.008314}\left(\frac{1}{298.15} - \frac{1}{283.15}\right) = -0.16$$

$$\frac{K_1}{K_2} = 0.85$$

$$\therefore \quad K(10\ ^\circ\text{C}) = 0.85\,(4.04 \times 10^{-7}) = 3.43 \times 10^{-7}$$

At the lower temperature, the acidity constant is smaller, 0.85 times the K value at 25 °C. This is because the reaction is endothermic and ΔH_f° is positive; higher temperatures result in a greater acidity constant (stronger acid) and the reaction proceeds farther to the right.

The influence of pressure on equilibrium constants can be neglected in most cases of natural waters, except where pressures reach 10–1000 atm in seawater. Under these circumstances, the standard molar volumes of the products and reactants can be used to calculate K as a function of pressure according to equation (24) at constant temperature:

$$\frac{d(\ln K)}{dP} = -\frac{\Delta V^{\circ}}{RT} \tag{24}$$

where ΔV° is the change in partial molar volume of the reaction at standard state in $cm^3\ mol^{-1}$ (P = 1 atm), K is the equilibrium constant, and P is the pressure in atmospheres. Assuming that the molar volume ΔV° is independent of pressure and integrating equation (24) yields

$$\ln \frac{K_2}{K_1} = -\frac{\Delta V^{\circ}(P_2 - P_1)}{RT} \tag{25}$$

where K_1 is the equilibrium constant at the reference pressure, K_2 is the new equilibrium constant, P_2 is the new pressure, and P_1 is the reference pressure (P_1= 1 atm).

Stumm and Morgan[14] give a detailed treatment on the effect of temperature on equilibrium constants, particularly regarding calcite dissolution in the deep ocean. At 1000 atm pressure (10,000 m of seawater) the solubility product K_{s0} for calcite is 8.1 times greater than at standard state.

4.2.5 Activity Corrections

The theory of ideal solutions implies that there is no interaction between individual species. In real solutions, particularly solutions of ionic species in water, these conditions are not met. There are electrostatic interactions between charged ions, and the ions are generally surrounded by regions in which the water molecules are ordered in a structure somewhat different from that of pure water.

The ratio of the activity of a species to its concentration is called the activity coefficient. The activity coefficient f_A of species A is

$$f_A = \frac{\{A\}}{[A]} \tag{26}$$

In general, activity coefficients of uncharged species are near unity in dilute solutions and rise above unity in concentrated solutions, largely because much of the water in concentrated solutions is involved in the hydration shells of ions and less water is available to solvate uncharged species. The activity coefficient of an ion in electrolyte solution (natural waters also) is usually smaller than one.

From a purely thermodynamic point of view there is no way in which the activity coefficient of a single ion can be measured without introducing nonthermodynamic (and nonverifiable) assumptions. For dilute solutions, however, it is convenient to use single-ion activity coefficients even though they are not well-defined thermodynamically.

The Debye–Hückel theory is a model that allows activity coefficients to be calculated on the basis of the effect ionic interactions should have on the free energy.

Different equations have been proposed for the estimation of individual activity coefficients (Table 4.3).

In dilute solutions ($I < 10^{-2}$ M), that is, in fresh waters, our calculations are usually based on the infinite dilution activity convention and corresponding thermodynamic constants. In these dilute electrolyte mixtures, deviations from ideal behavior are primarily caused by long-range electrostatic interactions. The Debye–Hückel equation or one of its extended forms (see Table 4.3) is assumed to give an adequate description of these interactions and to define the properties of the ions. Correspondingly, individual ion activities are estimated by means of individual ion activity coefficients calculated with the help of the Güntelberg or Davies equations [(3) and (4) of Table 4.3]; or sometimes it is more convenient to calculate, with these activity coefficients, a concentration equilibrium constant valid at a given I:

$$K = \frac{\{AB\}}{\{A\}\,\{B\}} \tag{27}$$

$$K = \frac{[AB]}{[A]\,[B]}\,\frac{f_{AB}}{f_A f_B} \tag{28}$$

$$K = {}^{c}K\,\frac{f_{AB}}{f_A f_B} \tag{29}$$

Table 4.3 Activity Coefficients (f) of Individual Ions

Approximation	Equation[a]	Applicability (Ionic Strength, M)
(1) Debye–Hückel (simplified)	$\log f = -AZ^2\sqrt{I}$	$< 10^{-2.3}$
(2) Debye–Hückel (extended)	$\log f = -AZ^2\,\frac{\sqrt{I}}{1 + Ba\sqrt{I}}$	$< 10^{-1}$
(3) Güntelberg	$\log f = -AZ^2\,\frac{\sqrt{I}}{1 + \sqrt{I}}$	$< 10^{-1}$
(4) Davies	$\log f = -AZ^2\left(\frac{\sqrt{I}}{1 + \sqrt{I}} - 0.24\right)$	< 0.5

[a] I (ionic strength) $= \frac{1}{2}\sum C_i Z_i^2$;
$A = 1.82 \times 10^6\,(\varepsilon T)^{-3/2}$ (where ε = dielectric constant);
$A \approx 0.5$ for water at 25 °C;
Z = charge of the ion;
$B = 50.3\,(\varepsilon T)^{-1/2}$;
$B \approx 0.33$ water at 25 °C;
a = adjustable parameter (unit Ångström dependent on the size of the ion; see Table 4.4).
Source: Modified from Stumm and Morgan.[14]

Table 4.4 Parameter *a* and Activity Coefficient of Individual Ions[14, 15]

Parameter for Ion Diameter, *a*, (Ångström = 10^{-10} m)	Ion	Activity Coefficient Calculated According to Equation (2) of Table 4.3 for Ionic Strength, *I*				
		$I = 10^{-4}$	10^{-3}	10^{-2}	0.05	10^{-1} M
9	H^+	0.99	0.97	0.91	0.86	0.83
	Al^{3+}, Fe^{3+}, La^{3+}, Ce^{3+}	0.90	0.74	0.44	0.24	0.18
8	Mg^{2+}, Be^{2+}	0.96	0.87	0.69	0.52	0.45
6	Ca^{2+}, Zn^{2+}, Cu^{2+}, Sn^{2+}, Mn^{2+}, Fe^{2+}	0.96	0.87	0.68	0.48	0.40
5	Ba^{2+}; Sr^{2+}, Pb^{2+}, CO_3^{2-}	0.96	0.87	0.67	0.46	0.38
4	Na^+, HCO_3^-, $H_2PO_4^-$, CH_3COO^-	0.99	0.96	0.90	0.81	0.77
	SO_4^{2-}, HPO_4^{2-}	0.96	0.87	0.66	0.44	0.36
	PO_4^{3-}	0.90	0.72	0.40	0.16	0.10
3	K^+, Ag^+, NH_4^+, OH^-, Cl^-, ClO_4^-, NO_3^-, I^-, HS^-	0.99	0.96	0.90	0.80	0.76

In this text, we will assume that $\{H^+\}$ is the activity used in chemical equilibrium expressions for acids and bases. It can be replaced by an activity coefficient times a concentration $f_{H^+}[H^+]$. When making an H^+ measurement with a combination pH electrode and calibrating with National Bureau of Standards buffer solutions, the measurement is closest to an activity measurement, not concentration. Only if one calibrates with a strong acid such as 10^{-3} M H_2SO_4 does one measure the H^+ concentration with a pH meter.

Example 4.3 Activity Corrections for the Carbonate System

In exact calculations with the carbonate system in a groundwater ($I = 5 \times 10^{-3}$ M) we need activity corrections. pH is measured operationally with standard buffers, that is, as p^{c}H (infinite dilution). It is convenient to retain concentration units for the carbonate species (conservative quantities like total inorganic carbon or alkalinity have to be expressed as concentrations). Thus we apply our corrections to the equilibrium constants, K', and then make our calculations numerically or graphically as before.

Solution:

$$K_1 = \frac{\{H^+\}\,\{HCO_3^-\}}{\{H_2CO_3^*\}} = \frac{\{H^+\}\,[HCO_3^-]}{[H_2CO_3^*]}\,\frac{f_{HCO_3^-}}{f_{H_2CO_3^*}} = K_1'\,\frac{f_{HCO_3^-}}{f_{H_2CO_3^*}} \tag{i}$$

$$K_2 = \frac{\{H^+\}\,\{CO_3^{2-}\}}{\{HCO_3^-\}} = \frac{\{H^+\}\,[CO_3^{2-}]}{[HCO_3^-]}\,\frac{f_{CO_3^{2-}}}{f_{HCO_3^-}} = K_2'\,\frac{f_{CO_3^{2-}}}{f_{HCO_3^-}} \tag{ii}$$

Using the Güntelberg approximation [equation (3), Table 4.3], we obtain

$$pK_1' = pK_1 - \frac{0.5\sqrt{I}}{1+\sqrt{I}} \quad \text{(iii)}$$

and for K_2, the divalent carbonate ion has a valence $Z = +2$.

$$pK_2' = pK_2 - \frac{0.5\,(4-1)\sqrt{I}}{1+\sqrt{I}} \quad \text{(iv)}$$

For $I = 5 \times 10^{-3}$ the corrections are

$$pK_1' = pK_1 - 0.03 = 6.3 - 0.03 = 6.27$$

$$pK_2' = pK_2 - 0.1 = 10.3 - 0.1 = 10.2$$

In this example, the corrections on pK values for activity coefficients are not very large. We did not use an activity correction for $\{H^+\}$ assuming that pH measurements correspond most closely to an activity measurement.

Example 4.4 Activity Coefficient of Nonelectrolytes

Find the activity coefficient of dissolved oxygen $O_{2(aq)}$ in seawater ($I \simeq 0.7$ M).

Solution: Morel and Hering[16] give the general empirical equation for neutral molecules:

$$\log_{10} f = 0.1\, I$$

where f is the activity coefficient of the neutral molecule and I is the ionic strength, M. At ionic strengths less than 0.1 M, the activity correction is less than 2%. In seawater it becomes significant. For molecular oxygen in seawater, the activity coefficient is $f = 1.17$ according to the equation above.

Actual studies of dissolved oxygen in seawater in the laboratory have yielded the following (more accurate) equation for the activity coefficient.

$$\log_{10} f_{O_2} = 0.132\, I$$

which yields an activity coefficient of $f = 1.24$. Neutral molecules have activity coefficients greater than 1.0 in electrolyte solutions and this is known as the "salting out" effect.

4.2.6 Solving Chemical Equilibrium Problems

A simple example can be used to illustrate how a chemical equilibrium problem is solved. It is assumed that the reader has had a general chemistry course and is knowledgeable in solving such problems. In the next section, we will set up a chemical equilibrium model in matrix form.

Suppose that we have a weak acid in solution—acetic acid, CH_3COOH, 0.01 M at 25 °C. (We will use the notation HAc to designate acetic acid.) The relevant equilibrium equation is defined by the acidity constant K_a, given in Table 4.2 as $pK_a = 4.7$.

The system is defined at the zero level as HAc and H_2O.

$$HAc \rightleftharpoons H^+ + Ac^- \tag{30}$$

$$K_a = \frac{\{H^+\}\ \{Ac^-\}}{\{HAc\}} = 10^{-4.7} = 2.00 \times 10^{-5} \tag{31}$$

Other relevant equations are the mass law equation for the ionization of water, the mass balance on total acetate in solution, the charge balance, and the proton condition.

$$\text{H}_2\text{O ionization} \qquad K_w = \{OH^-\}\ \{H^+\} = 10^{-14} \tag{32}$$

$$\text{Ac mass balance} \qquad C_T = 0.01\text{ M} = [Ac^-] + [HAc] \tag{33}$$

$$\text{Charge balance} \qquad [H^+] = [Ac^-] + [OH^-] \tag{34}$$

$$\text{Proton condition} \qquad [H^+] = [OH^-] + [Ac^-] \tag{35}$$

Equations (31)–(35) represent everything that we know about the system. We have four unknown concentrations or activities H^+, OH^-, Ac^-, and HAc, so we only need four equations. Note that the proton condition [equation (35)] and the charge balance [equation (34)] gave identical equations in this case. Let us use equations (31)–(34) to solve for all the concentrations and the pH of solution.

There is one small problem. Equilibrium expressions [equations (31) and (32)] make use of activities { } rather than concentrations [ ] so we must replace the activities with their activity coefficients times the concentration of each species. We have learned from Example 4.4 that the activity coefficient of a neutral molecule at such low ionic strength is approximately 1.0, therefore $\{HAc\} \simeq [HAc]$.

$$K_a = \frac{\{H^+\}\ \{Ac^-\}}{\{HAc\}} = \frac{f_{H^+}\,[H^+]\,[Ac^-]\,f_{Ac^-}}{[HAc]} \tag{36}$$

This problem sets up an iterative solution because the activity coefficients are dependent on the ionic strength I in equation (3) of Table 4.3, and we cannot calculate I until we know all the concentrations in solution (or unless the ionic strength of the solution is given initially). In numerical chemical equilibrium models, you must specify the ionic strength or correct the equilibrium constants for ionic strength effects iteratively.

An approximate solution is possible for equations (31)–(35) neglecting activity coefficients for the moment. HAc is a weak acid, so it is likely that $[H^+] \gg [OH^-]$. Therefore equation (34) can be rearranged and $[OH^-]$ neglected.

$$[H^+] - [OH] = [Ac^-] \tag{37}$$

$$\therefore \quad [H^+] \simeq [Ac^-] \tag{38}$$

Let $x = [H^+] = [Ac^-]$; then $[HAc] = 0.01 - x$, and we can substitute into the equilibrium expression [equation (31)]:

$$2.00 \times 10^{-5} = \frac{x^2}{0.1 - x} \tag{39}$$

We have the quadratic equation

$$x^2 + (2 \times 10^{-5})\,x - (2 \times 10^{-7}) = 0 \tag{40}$$

which is of the form $ax^2 + bx + c = 0$ with the general quadratic equation solution:

$$x = \frac{-b \pm \sqrt{b^2 - 4\,ac}}{2a} = 4.37 \times 10^{-4}\ \text{M} \tag{41}$$

Only the positive root in equation (41) is used in this case. The approximate solution is pH 3.36 and $[H^+] = [Ac^-] = 4.37 \times 10^{-4}$ M. The approximate solution seems to be quite accurate because we can check the error using equation (37) and the ionization of water.

$$[H^+] - [OH^-] = [Ac^-] \tag{42}$$

$$4.37 \times 10^{-4} - 2.29 \times 10^{-11} = 4.37 \times 10^{-4}$$

$$\%\ \text{error} = \left(1 - \frac{\text{left-hand side}}{\text{right-hand side}}\right) 100\% = 5.2 \times 10^{-6} \tag{43}$$

The error is very small. Even in numerical solutions, there will always be error. The modeler must determine what is the acceptable error and tolerance allowed for closure. In most environmental problems, a known background electrolyte is present and/or the ionic strength is measured, so Table 4.3 can be used to estimate activity coefficients and to correct K values.

For the sake of completeness, we can return to equations (31)–(34) now and solve them with activity coefficients. We will use the estimated concentrations from equation (41) and the approximate solution. Using the Güntelberg activity coefficient formula [equation (3) of Table 4.3], we find

$$I = \tfrac{1}{2}\Sigma\, C_i Z_i^2 = \tfrac{1}{2}[0.000437(1) + 0.000437(1)] \quad (44)$$

$$I = 4.37 \times 10^{-4}\ \text{M}$$

and

$$\log f_{\text{Ac}^-} = -0.5(1)^2 \frac{\sqrt{I}}{1+\sqrt{I}} = \log f_{\text{H}^+} \quad (45)$$

$$f_{\text{Ac}^-} = f_{\text{H}^+} = 0.9995$$

The activity coefficients are very near to 1.0 (ideal dilute solution) in this case (a weak acid dissolved in water with no other electrolytes present). The equilibrium constant corrected for activity coefficients is given by equation (46). (We will assume that the hydrogen ion activity coefficient is needed as well as that of the acetate ion, in this case.)

$$K_a' = \frac{K_a}{f_{\text{H}^+} f_{\text{Ac}^-}} = \frac{[\text{H}^+][\text{Ac}^-]}{[\text{HAc}]} = 2.002 \times 10^{-5} \quad (46)$$

Now let's substitute equation (32) into equation (34). K_w' is 1.001×10^{-14}, the corrected ionization constant.

$$[\text{H}^+] = [\text{Ac}^-] + \frac{K_w'}{[\text{H}^+]} \quad (47)$$

Rearrange and substitute equation (47) into the mass balance equation (33).

$$[\text{HAc}] = 0.01 - \left([\text{H}^+] - \frac{K_w'}{[\text{H}^+]}\right) \quad (48)$$

Lastly, substitute equation (47) for $[\text{Ac}^-]$ and equation (48) into the equilibrium expression, equation (46). We now have just one equation with one unknown.

$$K_a' = \frac{[\text{H}^+]\left[\text{H}^+ - \dfrac{K_w'}{[\text{H}^+]}\right]}{0.01 - \left[\text{H}^+ - \dfrac{K_w'}{[\text{H}^+]}\right]} \quad (49)$$

$$[\text{H}^+]^3 + K_a'[\text{H}^+]^2 - (0.01\,K_a' + K_w')[\text{H}^+] - K_a' K_w' = 0 \quad (50)$$

Equation (50) is a third-degree polynomial. The easiest way to solve it algebraically is by trial and error. In this case, the answer is almost exactly that given by equation (41). The Newton–Raphson method can solve this iteratively using a computer.

The preceding example was very simple. As one increases the number of solutes and the number of equations, the polynomial becomes very large, an n-degree polynomial. Trial-and-error solutions become impractical, and we must use a numerical solution technique.

4.3 NUMERICAL SOLUTION TECHNIQUE

Consider our example of 0.01 M acetic acid from the previous section. The first thing one needs to determine in setting up a numerical solution is the number of *species* present in solution. In this case, we have four.

- Species: HAc, Ac^-, OH^-, and H^+

Second, we need to determine a minimum number of species necessary to solve the system of equilibrium equations. These are the independent variables, which we will call *components*. We have two equilibrium equations, the acid dissociation of HAc with equilibrium constant K_a, and the ionization of water with equilibrium constant K_w.

$$K_a = \frac{\{H^+\}\ \{Ac^-\}}{\{HAc\}} = 10^{-4.73} \tag{51}$$

$$K_w = \{H^+\}\ \{OH^-\} = 10^{-14.0} \tag{52}$$

These two equilibrium expressions contain the four chemical species (two equations and four unknowns), but we also know the mass balance for total acetate in the system [equation (33)] and the charge balance [equation (34)], which can be substituted into equations (51) and (52) to leave us with two equations and two unknowns. Thus we really only need two components (independent variables) to specify the system. Let us choose HAc and H^+ as our components.

- Components: HAc and H^+

(Note that we could have chosen HAc and OH^-, or Ac^- and H^+, or Ac^- and OH^-. The choice is arbitrary as long as you have two independent components.) An independent equation can be written for every species in terms of the components.

Next, we need to write chemical equations for the four species in terms of the components, including the formation of the components themselves.[17]

- Formation of HAc: $HAc \rightleftharpoons HAc$ $\qquad \log K_1 = 0$

- Formation of Ac^-: $HAc \rightleftharpoons Ac^- + H^+$ $\log K_2 = -4.73$
- Formation of OH^-: $H_2O \rightleftharpoons OH^- + H^+$ $\log K_3 = -14$
- Formation of H^+: $H^+ \rightleftharpoons H^+$ $\log K_4 = 0$

The chemical equations are, in reality, mass action expressions that form a set of algebraic equations for the activity of each species in terms of the components. We will assume that all activity coefficients are 1.0; therefore all equilibrium constants must be corrected for activity coefficients (conditional stability constants) as we did in Example 4.3. For our example of 0.01 M HAc in H_2O, the activity corrections are negligible.

$$[HAc] = K_1\,[HAc]^1 \tag{53}$$

$$[Ac^-] = K_2\,[HAc]^1\,[H^+]^{-1} \tag{54}$$

$$[OH^-] = K_3\,[H^+]^{-1} \tag{55}$$

$$[H^+] = K_4\,[H^+]^1 \tag{56}$$

These equations define the system completely, together with the mass balance equation for acetate [equation (33)] and the charge balance [equation (34)]. Taking the logarithm of both sides of equations (53)–(56), we find

$$\log HAc = 1.0 \log HAc + \log K_1 \tag{57}$$

$$\log Ac^- = 1.0 \log HAc - 1.0 \log H^+ + \log K_2 \tag{58}$$

$$\log OH^- = -1.0 \log H^+ + \log K_3 \tag{59}$$

$$\log H^+ = 1.0 \log H^+ + \log K_4 \tag{60}$$

Equations (57)–(60) are a linear set of equations that can be solved using matrix algebra. The species comprise the rows of the matrix and the components are the columns.

	Components		
Species	HAc	H^+	$\{\log K\}$
HAc	1	0	0
Ac^-	1	−1	−4.73
OH^-	0	−1	−14.0
H^+	0	1	0

There is one final step before solving the matrix equation. We must provide an initial estimate of the $[H^+]$ to get the program started, and we must provide the mass balance equation. The initial estimate of $[H^+]$ is in lieu of the charge balance. The manner in which we provide this information in MacμQL, MINEQL$^+$, and MINTEQ is to specify the *mode* of each component, that is, whether the concentration specified is the *total* or *free* concentration. The total concentration of acetate is 10^{-2} M and we specify it at the bottom of the column, and the initial guess for $[H^+]$ is 10^{-7} M.

	HAc total	H^+ free	{log K}
HAc	1	0	0
Ac^-	1	−1	-4.73
OH^-	0	−1	−14.0
H^+	0	1	0
	10^{-2}	10^{-7}	

The programs will recognize the concentrations specified to be the mass balance $HAc + Ac^- = 0.01$ M and pH = 7 (initial guess). We will calculate the distribution of species as a function of pH.

In matrix notation, we will use { } to designate one-dimensional arrays (column vectors) and square brackets [ ] for two-dimensional matrices. The stoichiometric coefficients, which came from the exponents of the mass action expressions [equations (53)–(56)], comprise the matrix, and all the species are expressed in terms of concentration, mol L^{-1}, or M.

In matrix notation, we can rewrite equations (57)–(60) in a general form. This discussion follows the work of Westall.[18]

$$\{C^*\} = [A] \cdot \{X^*\} + \{K^*\} \tag{61}$$

where $\{C^*\}$ = the column vector of log species concentrations, n dimension
$[A]$ = the matrix of stoichiometric coefficients, $n \times m$
$\{X^*\}$ = the column vector of log component concentrations, m dimension
$\{K^*\}$ = the column vector of log equilibrium constants, n dimension
n = the number of species
m = the number of components

In our case the number of species is 4 and the number of components is 2, so the matrix of stoichiometric coefficients $[A]$ has dimensions of $m \times n$ (4 rows and 2 columns).

The material balance equations are determined in the computer program from the modes that were specified and the concentrations given at the bottom. There is exactly one material balance equation for each component.

$$0.01 = C_{TOT} = 1\ HAc + 1\ Ac^- \tag{62}$$

$$10^{-7} = C_{\text{free}} = 1\ \text{H}^+ \tag{63}$$

The material balance equations can also be written, in general, in matrix notation. It is the transpose of the $[A]$ matrix of stoichiometric coefficients times the concentration vector for the species, minus the total concentration of each component. This is *not* a logarithmic equation as is equation (61). The equation is cast in terms of a differential error term that will be calculated by the Newton–Raphson numerical solution technique until it is within an acceptable closure tolerance.

$${}^{T}[A] \cdot \{C\} - \{C_{\text{TOT}}\} = \{Y\} \tag{64}$$

where ${}^{T}[A]$ = the transposed stoichiometric coefficient matrix, $m \times n$
$\{C\}$ = the column vector of species concentration, n dimension
$\{C_{\text{TOT}}\}$ = the column vector of component total concentrations, m dimension
$\{Y\}$ = the column vector of errors or remainders, in the material balance equation for each component, m dimension

The equilibrium problem is solved when $\{Y\} = 0$, or when $\{Y\} < \varepsilon$, the acceptable closure tolerance.

In the numerical scheme, the log concentration of each species is computed from equation (61) based on an initial guess for the concentration of the components. Then the error, or remainder, is estimated from equation (64). An iterative technique is used to find improved values of $\{X\}$, the component concentrations, such that the value of $\{Y\}$ is reduced, that is, the Newton–Raphson method. Improved values for $\{X\}$, the component concentration array, are found from the matrix equation

$$[Z] \cdot \{\Delta X\} = \{Y\} \tag{65}$$

where Z = the square matrix ($m \times m$) that is the Jacobian of Y with respect to X
= $\partial Y / \partial X$
ΔX = the column vector for the improvement in component concentrations, m dimension, $X_{\text{original}} - X_{\text{improved}}$
Y = the column vector, which is the remaining error in the material balance equations, m dimension

The Jacobian operator can be written in terms of the stoichiometric coefficients a_{ij}.[18]

$$Z_{jk} = \frac{\partial Y_j}{\partial X_k} = \sum_{i=1}^{n} a_{ij}\, a_{ik}\, C_i / X_k \tag{66}$$

for all species ($i = 1, \ldots, n$)
for all components ($j = 1, \ldots, m$)
and ($k = 1, \ldots, m$) and $j \neq k$

$[Z]$ is a square matrix, $m \times m$, with dimensions of the number of components. Equation (65) can be solved for ΔX by inversion of the Z matrix.

$$\{\Delta X\} = [Z]^{-1} \cdot \{Y\} \tag{67}$$

$$\therefore \quad \{X\}_{\text{improved}} = \{X\}_{\text{original}} - [Z]^{-1} \cdot \{Y\} \tag{68}$$

Since $\{Y\}$ is an array of error terms that may vary widely (orders of magnitude), the convergence criterion is chosen to reflect the magnitude of Y_j relative to the maximum of terms of which Y_j is the sum. A criterion for convergence is

$$\frac{|Y_j|}{\max(Y_j)} < \varepsilon \quad \text{for all components } (j = 1, \ldots, m) \tag{69}$$

When the error term is within closure limits, the program can exit and the equilibrium problem is "solved." A flowchart is presented in Figure 4.2 of a typical computer program, and Table 4.5 is a summary of the mathematical solution technique. In Appendix E is a complete FORTRAN program for the Newton–Raphson solution to an example problem of aluminum speciation in a natural water at pH ~ 5 in the presence of gibbsite and with fluoride, hydroxide, and sulfate complexation.

4.4 SURFACE COMPLEXATION AND ADSORPTION

So far we have discussed modeling homogeneous equilibrium reactions involving solutes in water. But particles and surfaces of many types are present in natural waters including inorganic clays and oxides, organic polysaccharides and proteins, and algae and bacteria. Solutes can be adsorbed to the surface of these particles either as outer-sphere or inner sphere complexes (Figure 4.3). In inner-sphere complexation, a covalent bond is formed, such that the solute is held very closely to the surface, as in Figure 4.3a (the black atom). (Sometimes inner-sphere complexation is termed *chemisorption.*) Suppose the adsorbent is a metal ion. The chemical bond with the surface is strong enough that the ion loses its coordinated H_2O molecules as it draws near to the surface and forms a bond.

In outer-sphere complexation, the solute draws close enough to form a coordinated arrangement, but not close enough to lose all its coordinated water molecules, as shown in Figure 4.3a. This is analogous to ion pairs in homogeneous solution, for example, $(CaSO_4)^0_{(aq)}$. Some solutes may not approach closely enough to the surface to be a part of the inner-sphere or outer-sphere coordinated atoms; these molecules are part of the bulk solution.

The chemistry of the surface affects its coordinated ions. Ligands that form strong complexes in solution tend to form strong complexes at the surface also.[19] For example, the stability constant for $FePO_{4(aq)}$ is quite large—it is a strong complex in solution. This indicates that PO_4^{3-} ^{3}will form strong inner-sphere complexes with hydrous ferric oxide, FeOOH, for example.

Figure 4.3b shows an oxide such as Al_2O_3. In aqueous solution, ligands form inner-sphere surface complexes with the central metal ion, aluminum. Phosphate anions will adsorb; fluoride anions and hydroxy complexes are common. Oxide sur-

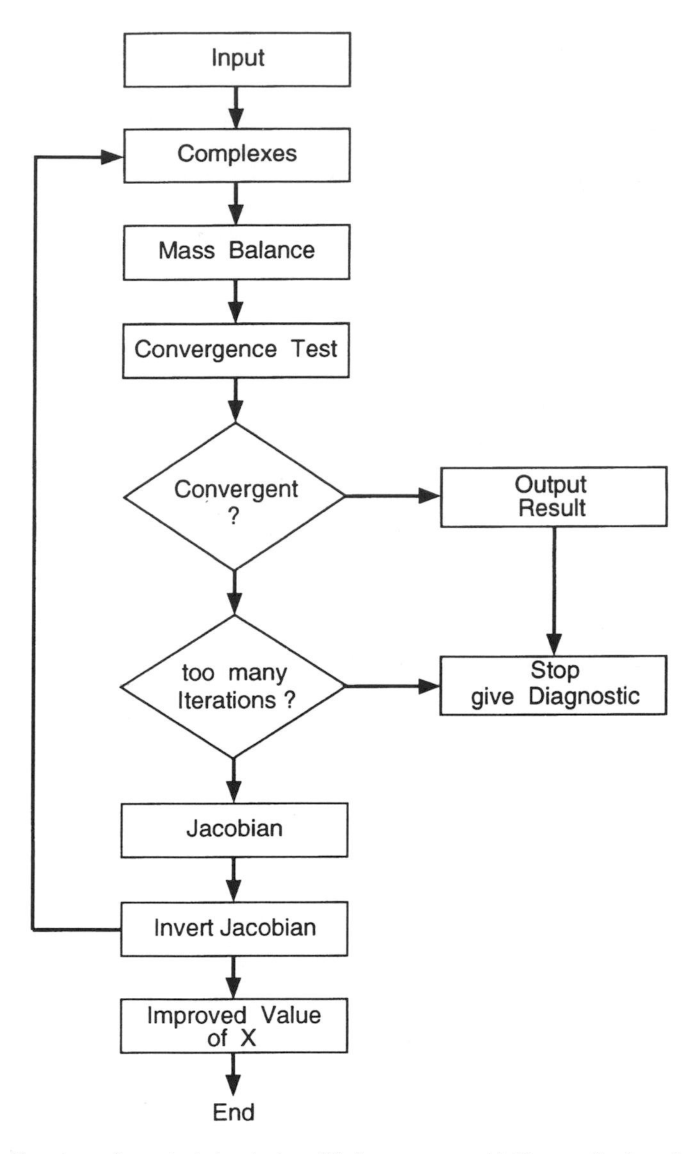

Figure 4.2 Flowchart of a typical chemical equilibrium program with Newton–Raphson iterative solution technique.

faces in water become polyhydroxy surfaces because of coordination with water, in which OH^- ions complex the central metal ion and H^+ ions align with oxygen atoms in the surface layer of the oxides. As shown in Figure 4.3b, Cu^{2+} ions or H^+ ions can adsorb to the oxygen donor atom at the surface. The surface has amphoteric properties, both donating and accepting protons. Outer-sphere complexes of Na^+ and Cl^-

Table 4.5 Mathematical Description of Chemical Equilibrium Problem

Description	Matrix or Vector
Stoichiometric coefficient of component j in species i	A
Free concentration of species i (log C^*)	C (C^*)
Stability constant of species i (log K^*)	K (K^*)
Total analytic concentration of component j	C_{TOT}
Free concentration of component j (log X^*)	X (X^*)
Residual in material balance equation for component j	Y
Jacobian of Y with respect to X ($\partial Y_j / \partial X_k$)	Z

Mass Action Equations

$$\log C_i = \log K_i + \Sigma\, a_{ij} \log X_j$$

$$C^* = K^* + A \cdot X^*$$

Mass Balance Law Equation

$$Y_j = \Sigma\, a_{ij}\, C_i - C_{TOTj}$$

$$Y = {}^{T}\!A \cdot C - C_{TOT}$$

Iteration Procedure (Newton–Raphson)

$$Z \cdot \Delta X = Y$$

$$\Delta X = X_{original} - X_{improved}$$

Source: Modified from Westall.[18]

are shown with their coordinated H_2O molecules surrounding them. Water molecules are lost as ions form bonds close to the surface, inner-sphere complexes.

Clay particles usually carry a negative charge in solution at their surface because of isomorphic substitution of ions in the interlayers of the clay. This is known as *permanent charge*. In addition, charge at the surface of the particle can be altered by lattice imperfections and surface complexation of ions. Other particles such as clays, aluminosilicates, oxides, and calcite can undergo similar charge-altering surface complexation in aqueous solution. Figure 4.4 (top) shows how the charge on the surface of particles (measured in coulombs m^{-2}) is a function of pH because protons are adsorbed on the surface. At pH 7, SiO_2, montmorillonite, kaolinite, and α-MnO_2 are negatively charged; and calcite, γ-Al_2O_3, and MgO are positively charged. Biological particles, detritus, and organic phases such as oil emulsions are negatively charged at pH 7, and they show significant electrophoretic mobility when current is passed through a solution (Figure 4.4, bottom). Algae and bacteria have negative surface charge at neutral pH because of the loss of protons from sur-

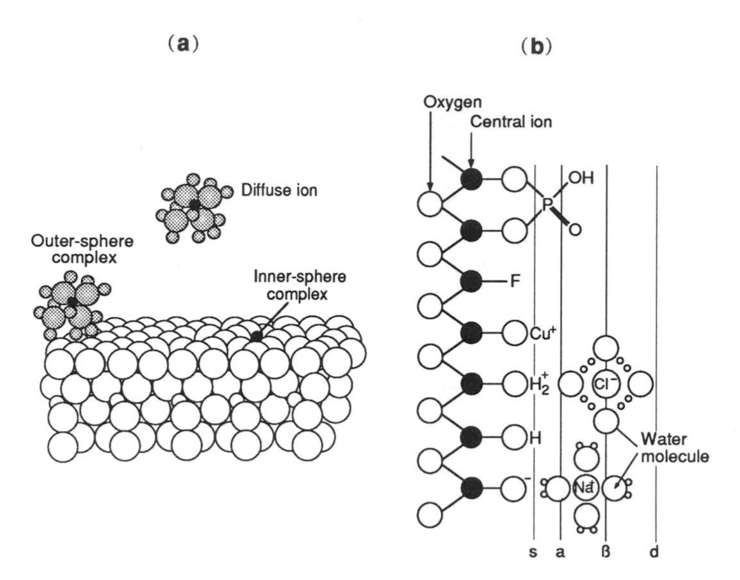

Figure 4.3 (a) Surface complex formation of a cation on the hydrous oxide surface. The ion may form an inner-sphere complex (covalent bond), an outer-sphere complex with hydrated waters, or be in the diffuse layer. (b) Hydrous oxide surface with central metal ions (dark circles), oxygen atoms (light circles), and inner-sphere coordinated ions PO_4^{3-}, F^-, Cu^{2+}, and H^+. Na^+ and Cl^- are outer-sphere complexes. The surface of the solid is defined by plane *s*. The beginning of the diffuse layer is defined by plane *d*. Outer-sphere complexes lie between planes *a* and *b*. (From Stumm.[20])

face functional groups such as carboxylic acids, quinones, and amines. Solutes adsorb to all types of particles in natural waters.

4.4.1 K_d Adsorption Model

An empirical expression of equilibrium adsorption is frequently used, the distribution coefficient between the solid phase and aqueous phase, K_d. K_d is defined as the ratio of the concentration of adsorbate bound to solids to the total dissolved concentration at equilibrium.

$$K_d = \frac{[\text{S-M}]}{[\text{M}]_T} \tag{70}$$

where S represents the solid surface and M represents the adsorbate, $[M_T]$ is the total dissolved aqueous concentration ($\mu g\ L^{-1}$), [S-M] is the adsorbed concentration ($\mu g\ kg^{-1}$), and K_d is the distribution coefficient ($L\ kg^{-1}$). K_d is an equilibrium constant that is too empirical (conditional) for some purposes, but it can be quite useful for comparisons of adsorption in complex natural waters.

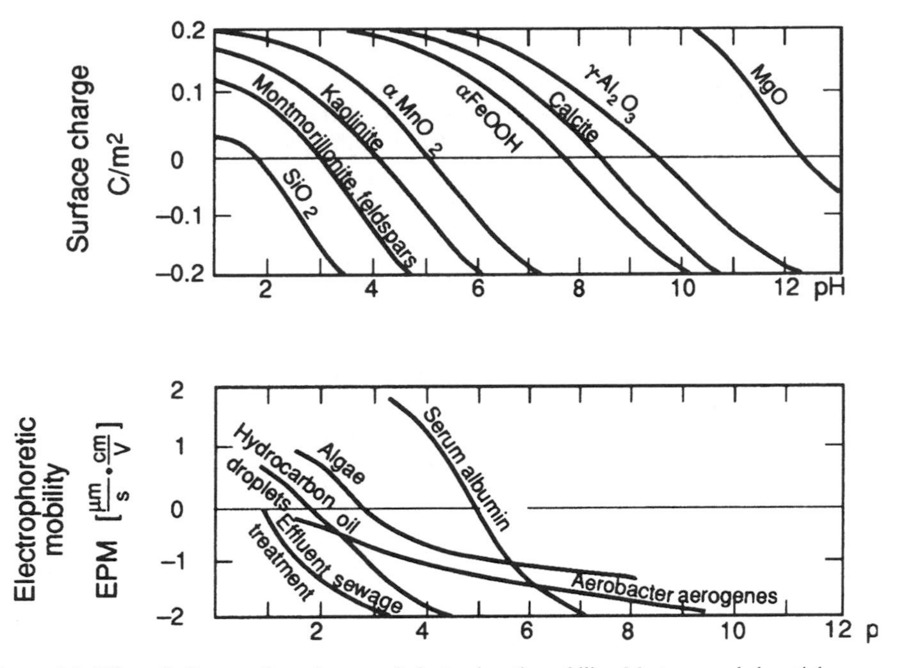

Figure 4.4 Effect of pH on surface charge and electrophoretic mobility. Most suspended particles are negatively charged at neutral pH. Curves depend on exact solution chemistry and should be considered operational. (Modified from Stumm and Morgan.[14])

Chemical equilibrium models such as MINTEQA2/PRODEFA2[21] have improvised the concept by defining an "activity K_d" adsorption model such that

$$K_d^{\text{act}} = \frac{\{\text{S-M}\}}{\{\text{M}\}} \tag{71}$$

where {M} is the activity of free dissolved species in solution and {S-M} represents the activity of the adsorbed species. A unit activity coefficient is assumed for the adsorbent phase and the activity coefficient of the free adsorbate in solution f_{M} is estimated from the Güntelberg approximation or related equation.[21]

$$K_d^{\text{act}} = \frac{[\text{S-M}]}{f_{\text{M}}[\text{M}]} \tag{72}$$

If [S-M] is expressed in units of moles adsorbed per liter of solution, then K_d^{act} *is* dimensionless. K_d^{act} can be calculated from the empirical K_d (L kg^{-1}) times the suspended solids concentration (kg L^{-1}). Normally, equation (72) is used when adsorbate concentrations are relatively low, and linear adsorption is expected.

4.4.2 Langmuir Adsorption Model

The Langmuir adsorption model is defined by a limiting (maximum) adsorption capacity that is related to a monolayer coverage of surface sites. We may think of the chemical reaction equation below with K_L^{act} as the equilibrium constant.

$$S + M \rightleftharpoons S\text{-}M$$

$$K_L^{act} = \frac{\{S\text{-}M\}}{\{M\}\ \{S\}} = \frac{\{S\text{-}M\}}{f_M\ [M]\ [S]} \tag{73}$$

where [S-M] is the moles adsorbed per liter of solution, [M] is the concentration of free species in solution in mol L^{-1}, and [S] is the concentration of available sites (in mol L^{-1}). Considering the mass balance equation (74), it is possible to combine equations (73) and (74) to derive the Langmuir adsorption isotherm equation with activity corrections [equation (75)]:

$$\text{Mass balance} \qquad [S_T] = [S\text{-}M] + [S] \tag{74}$$

$$\text{Langmuir} \qquad [S\text{-}M] = \frac{K_L^{act}\ [S_T] f_M\ [M]}{1 + K_L^{act} f_M\ [M]} \tag{75}$$

where $[S_T]$ is the total concentration of sites (in mol L^{-1}) and the other quantities are as previously defined.

One advantage of the Langmuir activity model is that it can be used to model competitive adsorption.[21] The total mass balance on sites is then equation (76), and an equilibrium constant K_L^{act} must be specified for each adsorbate.

$$[S_T] = [S\text{-}M_1] + [S\text{-}M_2] + \cdots + [S\text{-}M_n] + [S] \tag{76}$$

where S-M, $S\text{-}M_2$, and $S\text{-}M_n$ are competing adsorbates for the total concentration of sites $[S_T]$.

The Langmuir adsorption model is one of the most useful options in chemical equilibrium models, and the assumption of limited (monolayer) coverage is representative of a wide range of equilibrium sorption isotherm data in the literature for both metal and organic adsorbates on particles in natural waters.

Example 4.5 Langmuir Adsorption Model

The following experimental data are for the adsorption of a neutral organic chemical onto 5.98 grams of powdered activated carbon in 1.0 L of water (from Schecher and McAvoy[22]). Derive the Langmuir equilibrium constant, K_L^{act}, and the total concentration of sites for use in a chemical equilibrium model.

Dissolved Concentration [M], mol L^{-1}	Adsorbed Concentration [S-M], mol L^{-1}
4.57×10^{-7}	5.43×10^{-7}
6.28×10^{-5}	4.34×10^{-5}
0.000148	6.37×10^{-5}
0.000244	7.24×10^{-5}
0.000345	7.71×10^{-5}
0.000450	7.69×10^{-5}
0.000655	8.16×10^{-5}
0.000759	8.35×10^{-5}

Solution: We can linearize equation (75) by taking the reciprocal of both sides and rearranging:

$$\frac{[\text{M}]}{[\text{S-M}]} = \frac{1}{K_L^{\text{act}} f_{\text{M}} [\text{S}_T]} + \frac{[\text{M}]}{[\text{S}_T]}$$

A plot of the ratio of dissolved/adsorbed concentrations versus the dissolved concentration should yield a straight line with a slope $1/[\text{S}_T]$ and an intercept $1/K_L^{\text{act}} f_{\text{M}}[\text{S}_T]$. In this case, the dissolved concentrations are $< 10^{-3}$ M, and the activity coefficient $f_{\text{M}} = 1.0$ can be assumed (otherwise we would use the activity coefficient formula for nonelectrolytes from Example 4.4).

Figure 4.5 is a plot of the original isotherm data. It shows the classical "maximum adsorption capacity" at high dissolved adsorbate concentrations of $\sim 9 \times 10^{-5}$ mol L^{-1}. The shape is characteristic of a Langmuir-type model with monolayer coverage.

A plot of the linearized equation is given in Figure 4.6. The model fits very well with a coefficient of determination $r^2 = 0.999$. The slope is 1.1039×10^4 L/mol which is equal to $1/[\text{S}_T]$. Therefore, the total number of sites is $[\text{S}_T] = 9.06 \times 10^{-5}$ mol L^{-1}. The intercept is $1/K_L^{\text{act}}[\text{S}_T] = 0.74943$, and thus the equilibrium constant is $K_L^{\text{act}} = 1.47 \times 10^4$. Q.E.D.

4.4.3 Freundlich Adsorption Model

The Freundlich adsorption model is used when more than monolayer coverage of the surface is expected and the sites are heterogeneous, that is, they have different binding energies. An unlimited supply of unreacted sites is assumed to be available, and the activity coefficient of the surface species is assumed equal to 1.0.

$$K_F^{\text{act}} = \frac{[\text{S-M}]}{\{\text{M}\}^{1/n}} = \frac{[\text{S-M}]}{(f_{\text{M}}[\text{M}])^{1/n}} \tag{77}$$

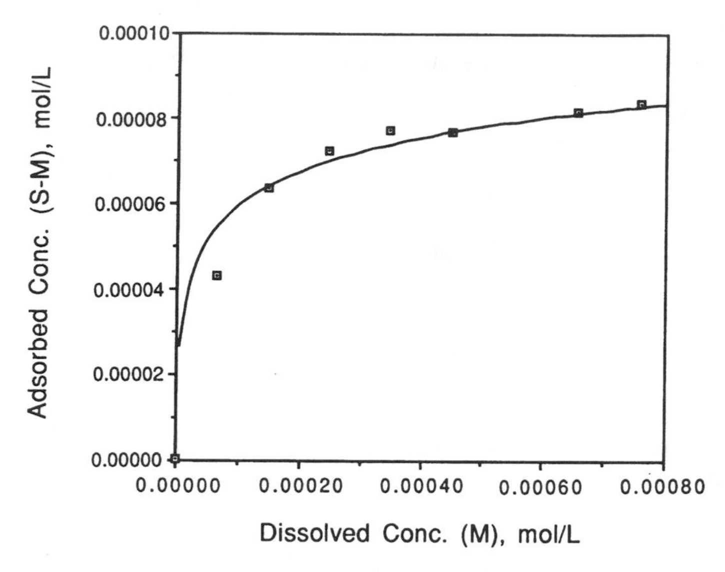

Figure 4.5 Langmuir adsorption isotherm for an organic chemical on activated carbon. Maximum adsorption capacity (monolayer coverage) is at ~ 0.00009 mol L^{-1}.

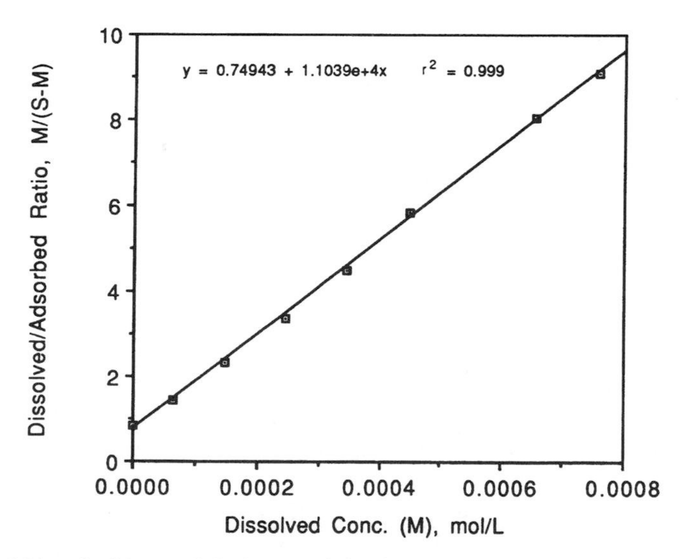

Figure 4.6 Linearized Langmuir isotherm to derive the maximum adsorption (S_T, total sites available) and the equilibrium constant, K_L.

The Freundlich isotherm is often used to describe organic chemical sorption onto activated carbon at relatively high concentrations in water and wastewater. It is less frequently used for modeling adsorption in natural waters. The exponent $1/n$ is usually less than 1.0 ($n < 1$) because sites with the highest binding energies are utilized first, followed by weaker sites, and so on. If $n = 1$, the Freundlich model reduces to the linear adsorption K_d model.

The conventional way to express the Freundlich isotherm is given by equations (78) and (79). Some chemical equilibrium codes require activity corrections [equation (77)][21] and some do not.[22]

$$[\text{S-M}] = K_F[\text{M}]^{1/n} \tag{78}$$

$$\log\,[\text{S-M}] = (1/n)\log\,[\text{M}] + \log K_F \tag{79}$$

where [S-M] is the concentration of adsorbed chemical (in mol L^{-1}), [M] is the dissolved chemical concentration (mol L^{-1}), $1/n$ is the slope on a log–log plot of adsorbed versus dissolved concentrations, and $\log K_F$ is the intercept of the log-log plot. K_F can be used to estimate the activity Freundlich coefficient according to the following equation:

$$K_F = K_F^{\text{act}}(f_{\text{M}})^{1/n} \tag{80}$$

Units other than mol L^{-1} can be used for adsorption isotherms (see Chapter 7, Toxic Organic Chemicals), but most chemical equilibrium codes use molar concentrations, so that is what is presented in this chapter.

4.4.4 Ion Exchange Model

Ion exchange is available in some chemical equilibrium codes. We may think of ion exchange as the competition of ions for exchange sites at the surface of the suspended particle. It is theoretically difficult to express the activity of the species on the exchange site. For ions of equal valence:

$$(\text{S-M}_1) + \text{M}_2 \rightleftharpoons (\text{S-M}_2) + \text{M}_1 \tag{81}$$

$$K_{\text{ie}} = \frac{[\text{S-M}_2]\,\{\text{M}_1\}}{[\text{S-M}_1]\,\{\text{M}_2\}} \tag{82}$$

where $[\text{S-M}_1]$ is the mole fraction of sites with ion M_1, $[\text{S-M}_2]$ is the mole fraction of sites with ion M_2, and M_1 and M_2 are the free ions of equal charge in solution. K_{ie} is the selectivity coefficient, an equilibrium constant.

When ions of different charge are considered, equivalent fractions should be used. One can write selectivity coefficients for any pair of ions that compete for exchange sites, even if they have different charge, but each selectivity expression must be independent of one another or else the system is overspecified.

$$(S_2\text{–}A) + 2\,B^+ \rightleftharpoons 2(S\text{-}B) + A^{2+} \tag{83}$$

$$K_{ie} = \frac{[S\text{-}B]^2\,\{A^{2+}\}}{[S\text{-}A]\,\{B^+\}^2} \tag{84}$$

$$[S_T] = [S] + [S\text{-}A] + [S\text{-}B] \tag{85}$$

where $\{A^{2+}\}$ is the divalent ion activity in solution (in mol L^{-1}), $\{B^+\}$ is the monovalent cation activity (in mol L^{-1}), [S-A] is the equiv L^{-1} of species A^{2+} on exchange sites, [S-B] is the equiv L^{-1} of species B^+ on exchange sites, [S] is the equivalents of unoccupied sites, and $[S_T]$ is the total equiv L^{-1} of sites. Alternatively, [S], [S-A], and [S-B] can be expressed as equivalent fractions of total sites and the mass balance is then

$$1 = [S] + [S\text{-}A] + [S\text{-}B] \tag{86}$$

Once the selectivity coefficient is formulated as an equilibrium constant, it can be used in chemical equilibrium programs.

4.4.5 Diffuse Double-Layer Model: Surface Complexation Approach

Previous adsorption models in this section have resulted from empirical observations (K_d model) or mass law equations (Langmuir model). In the late 1960s and early 1970s, Schindler et al. [23,24] and Stumm et al.[25] developed the surface complexation approach for adsorption of ions in natural waters. They recognized the critical importance of surface sites (surface functional groups) that can interact with H^+, OH^-, metal ions, or ligands. They also accounted for the influence of surface charge resulting from these reactions using Gouy–Chapman theory and electric double-layer (EDL) theory.[26] They realized that a coulombic term, fixed entirely by EDL theory, was necessary to model the continuous change in apparent equilibrium constants, for example, acidity constants of hydrous oxides at different pH values (see Figure 4.7).

In Figure 4.7, a representation of the diffuse double layer is given. A surface layer containing an excess of negatively charged functional groups (imparting a fixed negative charge to the surface) is balanced by a diffuse layer of counter-ions (and co-ions). Thus the two layers in the double-layer model are (1) the negatively charged surface layer including inner-sphere complexes, and (2) the diffuse layer of ions as one moves away from the surface into the bulk solution. The bottom panel of Figure 4.7 shows the net excess charge with distance x away from the surface. When the negative charge of the surface has been offset by the net positive charge of the outer-sphere ions and the diffuse swarm of ions, then the net excess charge of the solution is zero. Outside the diffuse layer, the bulk solution has an equal number of positively and negatively charged ions.

Figure 4.8 is a schematic of the fixed charge attached to the surface of a hydrous

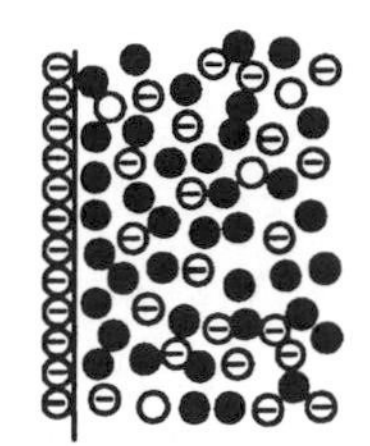

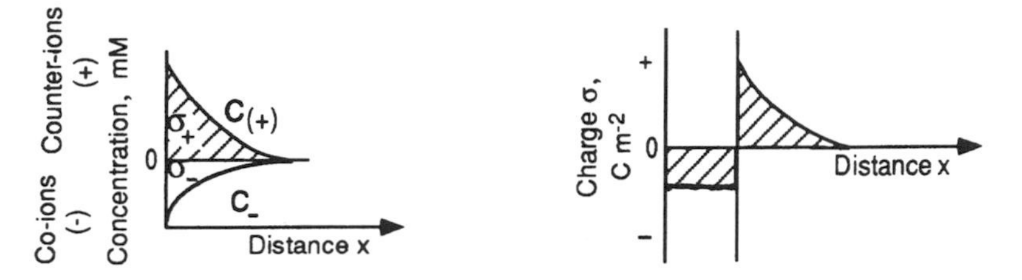

Figure 4.7 The diffuse double layer. Two layers are shown in the top panel: (1) a negatively charged surface and (2) the diffuse layer, that is, the swarm of counter-ions and co-ions in solution. Middle panel shows the variation in charge distribution with distance out from the surface. Right panel shows the two layers and the charge in coulombs per square meter (C m^{-2}).

oxide. The polyhydroxy surface is altered by the interaction with H^+, OH^-, metal ions, anions, and ligands. The net surface charge density of the hydrous oxide is given by

$$\sigma_P = \frac{F}{AS}\,[\{\text{S-OH}_2^+\} + \{\text{S-OM}^+\} - \{\text{S-O}^-\} - \{\text{S-A}^-\}] \tag{87}$$

where σ_P is net surface charge in coulombs per square meter (C m^{-2}), F is the Faraday constant (96,485 C mol^{-1}); A is the specific surface area (m^2 g^{-1}); S is the solids concentration (g L^{-1}) and $\{\text{S-OH}_2^+\}$, $\{\text{S-OM}^+\}$, $\{\text{S-O}^-\}$, and $\{\text{S-A}^-\}$ are the sorption densities of the specifically sorbed (i.e., coordinatively bound) inner-sphere cations and anions, respectively, in units of mol L^{-1}. More generally, we can write

$$\sigma_P = F[(\Gamma_H - \Gamma_{OH}) + \Sigma(Z_M\Gamma_M) + \Sigma(Z_A\Gamma_A)] \tag{88}$$

where σ_P is the net surface charge in C m^{-2}; F is the Faraday constant; Z is the valency of the sorbing ion; Γ_H, Γ_{OH}, Γ_M, and Γ_A, respectively, are the sorption densities (mol m^{-2}) of H^+, OH^-, metal ions, and anions. The net surface charge (σ_P) is related to the electric surface potential (ψ) by the Gouy–Chapman equation.[20,26]

$$\sigma_P = 0.1174\, c^{1/2} \sinh(Z\psi \times 19.46) \tag{89a}$$

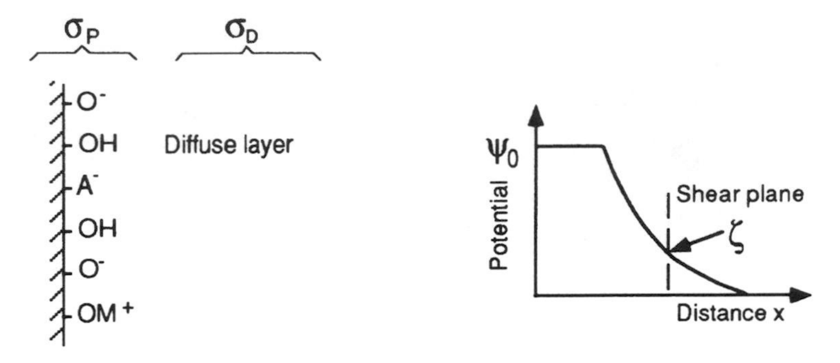

Figure 4.8 Surface complexation model. Left panel shows the substituted, negatively charged surface of a hydrous oxide with surface charge σ_P. O^-, OH, A^-, and OM^+ are considered inner-sphere complexes; they are a part of the surface. The surface potential ψ_0 is a constant at the surface, but it decreases exponentially in the diffuse layer. At the shear plane, ions lose their coordinated water molecules and a zeta potential can be established with the help of electrophoretic mobility measurements.

or

$$\sigma_P \simeq 2.5\, I^{1/2}\, \psi \tag{89b}$$

where σ_P is the net surface charge in C m^{-2}, c is the molar electrolyte concentration in mol L^{-1}, I is the ionic strength in mol L^{-1}, and ψ is the surface potential in volts. As demonstrated in Figure 4.8, the surface potential decreases exponentially with distance from the surface. We cannot measure ψ (zeta potential measurements are overestimates), so equation (89) allows us to estimate ψ from σ_P and a charge balance at the surface [equation (87)].

An ion that approaches the surface must overcome the electrostatic potential associated with the charged surface (Figure 4.8). It takes electrostatic work to transport ions through the interfacial potential gradient.

$$\Delta G^{\circ}_{\text{coul}} = \Delta Z\, F\psi \tag{90}$$

where $\Delta G^{\circ}_{\text{coul}}$ is the free energy term due to electrostatics. The total free energy includes both coulombic and chemical components. The chemical component is accounted for in the mass law expression with a coulombic correction.

$$\Delta G_{\text{tot}} = -RT \ln K \tag{91}$$

Combining equations (90) and (91) yields the coulombic correction term.

$$\ln K = -\frac{\Delta Z\, F\psi}{RT} \tag{92}$$

where K is the equilibrium constant, ΔZ is the charge of the ion, F is Faraday's constant (96,485 C mol^{-1}), ψ is the surface potential in volts, R is the universal gas law constant (8.314 V C mol^{-1} K^{-1}), and T is absolute temperature (K).

When we titrate hydrous oxide surfaces with acid or base, we can obtain an apparent equilibrium constant.

$$SOH + H^+ \rightleftharpoons SOH_2^+ \tag{93}$$

$$K^{app} = \frac{\{SOH_2^+\}}{\{SOH\}\ \{H^+\}} \tag{94}$$

But this equilibrium constant varies with pH and the degree of proton coverage of surface sites (surface charge); it is not constant. We can write the equilibrium constants in terms of their intrinsic constants, which correspond to sorption onto an uncharged surface and with a coulombic correction factor from equation (92). As before, we assume unity activity coefficients for the adsorbent, SOH.

$$K^{int} = \frac{\{SOH_2^+\}}{\{SOH\}\ \{H^+\}\ [e^{-\Delta Z\, F\psi/RT}]} \tag{95}$$

$$K^{app} = K^{int}\ [e^{-\Delta Z\, F\psi/RT}] = K^{int}\ P^{\Delta Z} \tag{96}$$

where P is the coulombic correction factor. We input the apparent equilibrium constant, but it is corrected within the computer program for the coulombic interaction. It can be interpreted as a surface activity coefficient necessary for bringing a proton close to the surface (equation 95). Westall[27] made the innovation to treat the coulombic correction factor, exp $(-\Delta Z\, F\psi/RT)$, as a "dummy chemical component" so that equation (95) fits neatly within the mathematical framework of most chemical equilibrium models.[27]

We always write the reaction assuming SOH is the neutral component serving as the reactant. The acidity constant is then

$$SOH \rightleftharpoons SO^- + H^+ \tag{97}$$

$$K^{int} = \frac{\{SO^-\}\ \{H^+\}\ [e^{-\Delta Z\, F\psi/RT}]}{\{SOH\}} \tag{98}$$

where $K^{int} = K_a^{app}$ exp $(-\Delta Z\ F\ \psi/RT)$. For a divalent cation adsorbing to the surface, the charge is $\Delta Z = +2$ (more work must be done bringing it to the surface).

$$SOH + M^{2+} \rightleftharpoons SOM^+ + H^+ \tag{99}$$

$$K^{int} = \frac{\{SOM^+\}\ \{H^+\}\ [e^{-F\psi/RT}]}{\{SOH\}\ \{M^{2+}\}\ [e^{-2F\psi/RT}]} \tag{100}$$

or

$$K^{\text{int}} = \frac{\{\text{SOM}^+\}\ \{\text{H}^+\}}{\{\text{SOH}\}\ \{\text{M}^{2+}\}\ [e^{-F\psi/RT}]} \qquad (101)$$

Once again, we treat the coulombic correction term as a dummy component in the chemical equilibrium model.

Stumm and co-workers applied these concepts originally to hydrous ferric oxides, FeOOH, and other oxides that are important in natural waters and groundwater. Dzombak and Morel[26] have begun the process of standardizing acidity, basicity, and adsorption constants for hydrous ferric oxides (HFO). They give complete tables of the coulombic correction factor P as a function of pH and ionic strength for HFO in 1:1 electrolytes (such as NaCl in Figure 4.9) and 2:2 electrolytes. Although the concepts have been developed most rigorously for hydrous oxides, chemical equilibrium models have used them for a number of sorption reactions, especially cations and anions on particles of all types in natural waters. Of course, the reactivity of these surfaces vary with origin, age, and chemical and biological character, but many apparent equilibrium constants have been used successfully in surface complexation models. In environmental research, it is now the state-of-the-art to use surface complexation models for metals adsorption to hydrous oxides.

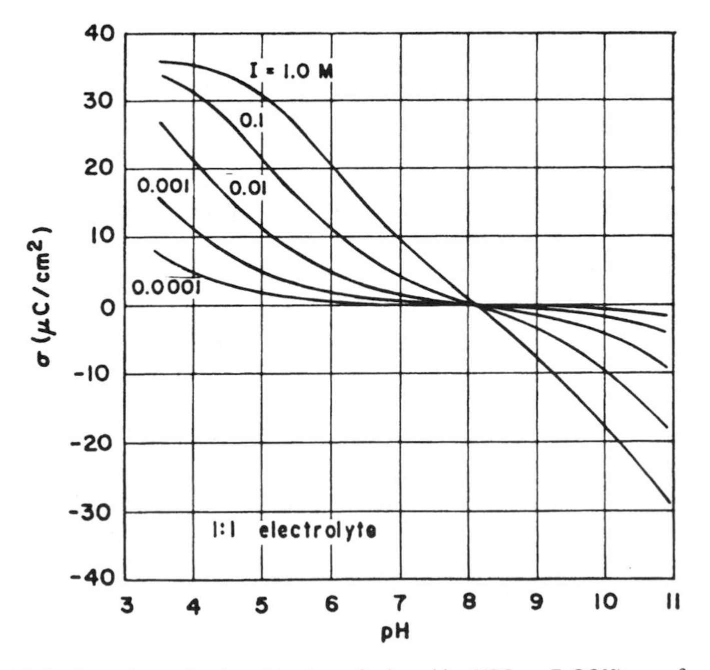

Figure 4.9 Surface charge density of hydrous ferric oxide (HFO or FeOOH) as a function of ionic strength and pH in a 1:1 electrolyte solution such as NaCl. (From Dzombak and Morel.[26])

Example 4.6 Coulombic Correction Factor for Surface Complexation Diffuse Double-Layer Model

For the acidity constant of hydrous ferric oxide (represented in equation 97 as SOH), find the coulombic correction factor at pH 7 and $I = 0.1$ M.

$$P^{\Delta Z} = \exp(-\Delta Z\, F\psi/RT)$$

As suggested in Dzombak and Morel,[26] we first estimate the surface charge density from titration data.

Solution: Use Figure 4.9 to find σ, the surface charge density, at $I = 0.1$ M and pH 7.

$$\sigma_P = 4.5\ \mu\text{C cm}^{-2} = 0.045\ \text{C m}^{-2}$$

Substitute σ_P into equation (89a) and solve for ψ, the surface potential, in volts.

$$\sigma_P = 0.045\ \text{C m}^{-2} = 0.1174\, c^{1/2} \sinh(Z\psi \times 19.46)$$

$$\sigma_P = 0.1174\,(0.1)^{1/2} \sinh(1 \times \psi \times 19.46) = 0.045$$

$$\sinh(\psi \times 19.46) = 1.212$$

$$\psi = 0.053\ \text{V}$$

Then solve for P:

$$P = \exp(-\Delta Z\, F\psi/RT) = \exp\left[\frac{-(96{,}485\ \text{C mol}^{-1})\,(0.053\ \text{V})}{(8.314\ \text{V C mol}^{-1}\ \text{K}^{-1})\,(298.15\ \text{K})}\right]$$

$$P = e^{-2.063} = 0.12708$$

$$\log P = -0.895$$

Under these conditions, the coulombic correction factor is quite important, analogous to a surface activity coefficient of 0.127 for H^+ ions approaching the surface of HFO at pH 7 and $I = 0.1$ M. Correction factors are particularly important for pH values less than 7 and greater than 9 in solutions with $I > 0.01$ M on HFO.

4.4.6 Constant Capacitance (Double-Layer) Model

Schindler et al.[23,24] used the constant capacitance model. It is the same as the diffuse double-layer model from the previous section except that the potential in the diffuse layer decreases linearly rather than exponentially with distance from the surface (see Figure 4.8). It is especially applicable when surface potentials are small (a small number of sites or weakly binding ions at near neutral pH) or when the double layer is compressed (high ionic strength solutions). The surface potential is directly proportional to the surface charge

$$\psi = \frac{\sigma_P}{C} \tag{102}$$

where C is the internal capacitance (farads m^{-2}), σ_P is the surface charge (C m^{-2}), and ψ is the surface potential (volts). One farad = one coulomb/volt.

In chemical equilibrium models, the procedure is the same as for the diffuse double-layer model except the coulombic correction factor takes on a different form. For the example given by equations (93)–(96), the constant capacitance coulombic correction would be the following:

$$K^{app} = K^{int}\,[e^{-\Delta Z\, F\sigma_P/RTC}] = K^{int}\,P^{\Delta Z} \tag{103}$$

The difference in the coulombic correction term between the diffuse double-layer model and the constant capacitance model is in most cases quite small. They both fit the experimental data equally well.[28] Schindler and Stumm[29] discuss the surface chemistry of oxides in terms of the constant capacitance model.

4.4.7 Triple-Layer Model

The triple-layer model was developed by Yates et al.[30] and Davis and Leckie.[31] It includes an additional surface plane for sorption. Returning to Figure 4.3, the triple-layer model could be envisioned with H^+ and OH^- ions binding to the innermost surface of the hydrous oxide mineral (plane s in Figure 4.3), while other coordinated ligands bind as partially hydrated ions, that is, as outer-sphere complexes (within the β plane). The third layer in the triple-layer model is the diffuse layer of counter-ions and co-ions as described for the diffuse double-layer model. Background electrolytes are allowed to adsorb.

Model input parameters are similar to those for the diffuse double-layer model except that two capacitance terms and *three* electrostatic components are required. Relating to Figure 4.3, capacitances C_1 and C_2 are specified at the s and β planes, while electrostatic potentials are required at the s, β, and d planes. For the sorption of a monovalent cation,

$$SOH + M^+ \rightleftharpoons SOM + H^+ \tag{104}$$

$$K^{int} = \frac{\{SOM\}\,\{H^+\}\,[e^{-\psi_s F/RT}]}{\{SOM\}\,\{M^+\}\,[e^{-\psi_\beta F/RT}]} \tag{105}$$

where two "dummy components" are shown as coulombic correction factors for the H^+ and M^+ ions, respectively, in the numerator and the denominator.

Westall and Hohl[28] and Dzombak and Morel[26] found that one cannot distinguish which model gives a better fit on the basis of the available data. The triple-layer model is the most detailed and contains the most parameters, but it may not always be necessary to explain surface complexation data. Most computer equilibrium models include both the diffuse double-layer and triple-layer model options.

4.5 PRECIPITATION AND DISSOLUTION IN EQUILIBRIUM MODELS

Because of the vast number of solid phase precipitates that are possible from aqueous solutions, chemical equilibrium models are quite structured in the manner that one designates precipitation and dissolution of species. Several models[21,22] use a format first developed for REDEQL.[5] Each chemical species is assigned a "Type" upon data entry.

- Type I species – all the aqueous chemical species designated as *components* according to our discussion in Section 4.3.
- Type II species—all other aqueous species not designated as Type 1.
- Type III species—all fixed species including solids designated to be present at chemical equilibrium and not subject to complete dissolution (an infinite supply of solid); all gases whose partial pressures are constrained to a fixed and constant value; fixed pH (H^+ ion) or pε (e^-); and ratios of species such as redox couples (SO_4^{2-}/S^{2-}) that may be designated as one species with fixed activity.
- Type IV species—all finite solids that are presumed to be present initially and have the potential to be completely dissolved if the solution becomes undersaturated.
- Type V species—all solids that may precipitate from solution but are presently undersaturated (possible solids).
- Type VI species—all species that are excluded from mass balance calculations; for example, electrons, all gases, and redox couples unless explicitly designated as Type III species, all solid phases unless explicitly designated as Types III, IV, or V, all electrostatic components according to discussion in Section 4.4.5, and all solid phases that are not expected to precipitate for kinetic reasons but would precipitate according to thermodynamics.

The number of degrees of freedom of the chemical equilibrium problem is the number of independent variables. These would ordinarily include temperature and pressure and we would use Gibbs' phase rule ($F = C + 2 - P$). Because temperature and pressure are specified by the user or the chemical equilibrium program, the phase rule for computer modeling purposes is

$$F = C - P \tag{106}$$

where F = the number of degrees of freedom
C = the number of chemical components
P = Type III + Type IV species

By specifying the activity of Type III and Type IV species, we decrease the number of degrees of freedom. As long as F is positive, the computer calculation will proceed. If F is zero or less, a diagnostic will prompt the user to restructure the problem.

To understand why Type III and Type IV species decrease the number of degrees of freedom, consider the case of $CaCO_{3(s)}$ dissolution as a Type IV solid.

$$CaCO_{3(s)} \rightleftharpoons Ca^{2+} + CO_3^{2-} \qquad \log K_{s0} = -8.42 \tag{107}$$

Because the $CaCO_{3(s)}$ is present in infinite supply, the product of calcium ions and carbonate ions is fixed; they are not both independent variables.

$$\{Ca^{2+}\} = 10^{-8.42}/\{CO_3^{2-}\} \tag{108}$$

If carbonate anion increases, calcium ion must decrease.

4.6 REDOX REACTIONS IN EQUILIBRIUM MODELS

Redox reactions can be handled just like acid–base reactions except one needs to specify the electron, $p\varepsilon = -\log\{e^-\}$, as a component. Half-reactions are included in the thermodynamic database, and it is possible to specify either E_H or $p\varepsilon$ with most programs.[21,22] One can perform fixed $p\varepsilon$ calculations and vary $p\varepsilon$ to make a pC versus $p\varepsilon$ plot or even a stability diagram ($p\varepsilon$ versus pH). It is possible to control $p\varepsilon$ with dissolved O_2 from atmospheric P_{O_2}, and one can calculate and control the $p\varepsilon$ by fixing the ratio of two species at chemical equilibrium, such as the Fe^{2+}/Fe^{3+} or the HS^-/SO_4^{2-} couples.

Remember that redox reactions can be catalyzed (mediated) by enzymes and microbes in natural waters, but other redox reactions can be very slow. Chemical equilibrium may not apply. In particular, E_H measurements in situ (redox potential measurements) can give a misleading picture of the redox status of the environment because the electrode does not respond very well to some redox reactions. A single platinum electrode does not respond to nonelectroactive couples such as NO_3^--NO_2^-, SO_4^{2-}-H_2S, or CH_4-CO_2. It can respond, theoretically, to $Fe^{2+} \leftrightarrow Fe^{3+}$ when concentrations are $>10^{-6}$ M and the reaction is at thermodynamic equilibrium. Mixed potentials also can cause problems because the electrode is receiving exchange current from a number of reactions, but the specialist doesn't know which ones. Stumm and Morgan[14] have proposed a partial remedy for this problem. They suggest analytical chemical determinations of key redox couples such as pH-O_2, Fe^{2+}-FeOOH, Mn^{2+}-MnO_2, HS^--SO_4^{2-}, and/or CH_4-CO_2, which gives a conceptually defined $p\varepsilon$ (or E_H)

provided the system is in chemical equilibrium. A singular $p\varepsilon$ value can be ascribed to a system if equal values of $p\varepsilon$ (or E_H) are obtained from each of the redox couples. Determinations of E_H to within a factor of 5–20 mV and $p\varepsilon$ values to within 0.1–0.3 units are possible under ideal circumstances.[14]

Stumm and Morgan[14] give the following examples of how to estimate $p\varepsilon$ (and E_H) from analytical determinations. Equilibrium constants are needed from a thermodynamic table. If $\log K$ equilibrium constants are given, one can convert to $p\varepsilon^o$ or E^o values using equations below ($p\varepsilon^o = 16.9\, E_H^o$).

$$p\varepsilon = p\varepsilon^o - \frac{1}{n}\log K \tag{109}$$

$$\log K = 16.9\, nE_H^o \tag{110}$$

where n is the number of electrons transferred in the reaction, K is the equilibrium constant, and $E_H{}^o$ is the standard electrode potential at 25 °C in volts.

A. Water from a hypolimnion of a lake has a dissolved O_2 concentration of 0.032 mg L^{-1} (very low) and a pH of 7. The redox reaction for oxygen in water in the thermodynamic table is

$$\tfrac{1}{4}O_{2(g)} + H^+ + e^- \rightleftharpoons \tfrac{1}{2}H_2O \qquad p\varepsilon^o \equiv \log K = 20.75 \tag{111}$$

We use equation (109) to find the $p\varepsilon$. The oxygen concentration is 10^{-6} M $O_{2(aq)}$, which is equivalent to ~6×10^{-4} atm $O_{2(g)}$ using Henry's law.

$$\begin{aligned} p\varepsilon &= 20.75 - \frac{1}{1}\log\frac{1}{[6\times10^{-4}]^{1/4}\,[1\times10^{-7}]} \\ &= 20.75 - 7.81 = 12.94 \end{aligned} \tag{112}$$

The $p\varepsilon$ is positive and surprisingly large ($E_H = +0.77$ V). Redox potentials in waters with dissolved oxygen remain large until very small oxygen concentrations are present (theoretically at chemical equilibrium).

B. An anaerobic digester has 65% CH_4 and 35% CO_2 gas in contact with water at pH 7. The $p\varepsilon$ and E_H are quite negative under these conditions.

$$\tfrac{1}{8}CO_{2(g)} + H^+ + e^- \rightleftharpoons \tfrac{1}{8}CH_{4(g)} + \tfrac{1}{4}H_2O \qquad p\varepsilon^o \equiv \log K = +2.87 \tag{113}$$

$$p\varepsilon = 2.87 - \frac{1}{1}\log\frac{(0.65)^{1/8}}{(0.35)^{1/8}\,(1\times10^{-7})} \tag{114}$$

$$p\varepsilon = 2.87 - 7.034 = -4.16$$

The electrode potential would be $E_H = -0.25$ V. This is quite reducing. In natural waters, we can expect a range of +0.75 V in aerobic waters with $O_{2(aq)}$ to – 0.25 V in anaerobic sediments.

The above examples may help to constrain input data when working redox problems using chemical equilibrium computer models.

Example 4.7 Acid–Base, Diprotic, Closed System

Find the concentration of all species for the system of 10^{-3} M $NaHCO_3$ in water at chemical equilibrium. Use a chemical equilibrium model of your choice and plot the results as log concentration versus pH. Show how the concentrations of species in solution vary with pH. At this low concentration, one can neglect activity corrections, but computer programs will automatically calculate and assign activity coefficients to each ion.

Solution: Use thermodynamic data or the acidity constants in Table 4.2 to write the mass law expressions. Designate all the species in solution and the components—remember, mass law equations must be written in terms of the components only (independent variables).

Species: $H_2CO_3^*$, HCO_3^-, CO_3^{2-}, Na^+, OH^-, H^+ (six species)
Components: HCO_3^-, Na^+, H^+ (three components)

The components are Type I species and all others are Type II species. After some thinking, it is clear that equations for each species can be written in terms of these three components. We could leave out Na^+ as a species and a component because we know that $NaHCO_3$ completely ionizes in water and that the concentration of Na^+ ions is 10^{-3} M, but for the sake of completeness, we will include it. The component species are independent variables, so mass law expressions are not necessary (they would be redundant). Identity relationships are used for components. All the species are written on the left-hand side of the following equations (dependent variables). They are defined in terms of the components (independent variables).

(i) $$H_2CO_3^* = HCO_3^- + H^+ \qquad K_{a1} = 10^{-6.3}$$

$$[H_2CO_3^*] = [HCO_3^-][H^+] \times 10^{6.3}$$

(ii) $$[HCO_3^-] = [HCO_3^-]$$

(iii) $$CO_3^{2-} = HCO_3^- - H^+ \qquad K_{a2} = 10^{-10.2}$$

$$[CO_3^{2-}] = [HCO_3^-][H^+]^{-1} \times 10^{-10.2}$$

(iv) $$[Na^+] = [Na^+]$$

(v) $$OH^- = -H^+ + H_2O \qquad K_w = 10^{-14.0}$$

$$[OH^-] = [H^+]^{-1} \times 10^{-14}$$

(vi) $$[H^+] = [H^+]$$

Now the matrix can be constructed based on equations (i)–(vi). We enter the exponent of each component on the right-hand side of the equation into the matrix as shown.

		Components			
	Species	HCO_3^-	Na^+	H^+	log K
(i)	$H_2CO_3^*$	1	0	1	6.3
(ii)	HCO_3^-	1	0	0	0
(iii)	CO_3^{2-}	1	0	–1	–10.2
(iv)	Na^+	0	1	0	0
(v)	OH^-	0	0	–1	–14
(vi)	H^+	0	0	1	0
	C_T	10^{-3}	10^{-3}		

The mass balance equations are readily seen from the vertical columns.

TOT Carbonate = $C_T = H_2CO_3^* + HCO_3^- + CO_3^{2-} = 10^{-3}$ M
TOT Na = $Na^+ = 10^{-3}$ M
TOT H = $H_2CO_3^* - CO_3^{2-} - OH^- + H^+ = 0$

The last mass balance for hydrogen is equal to zero because H^+ ions are not added to the system. This is a "fixed pH problem" where the H^+ ion concentration will be fixed, and then the programs will repeat the calculation at incremental pH values until the entire pH value to initiate the numerical scheme. The mass balance for TOT H is the proton balance, also derivable from the charge balance of a $NaHCO_3$ + H_2O solution.

Computer results will be stored as a column of numbers at each pH value after iteration. Depending on the computer program, these may be used as input to a graphics package and a plot like Figure 4.10 is generated. The proton condition valid for a 10^{-3} M $NaHCO_3$ solution is fulfilled at pH = 8.2, where $[CO_3^{2-}] \simeq [H_2CO_3^*]$. Thus, the approximate equilibrium composition is $[H_2CO_3^*] = [CO_3^{2-}] \simeq 10^{-5}$ M, pH = 8.2. The "exactly" computed result is $[H_2CO_3^*] = 1.2 \times 10^{-5}$ M, $[CO_3^{2-}] = 1.02 \times 10^{-5}$ M, $[HCO_3^-] = 9.8 \times 10^{-4}$ M, pH = 8.22 from the computer output.

Example 4.8 Solubility of Amorphous Fe(OH)$_{3(s)}$ as f(pH)

Iron hydrous oxides are good adsorbents of other cations and anions in natural waters, as we have discussed in this chapter. They are relatively insoluble at neutral pH

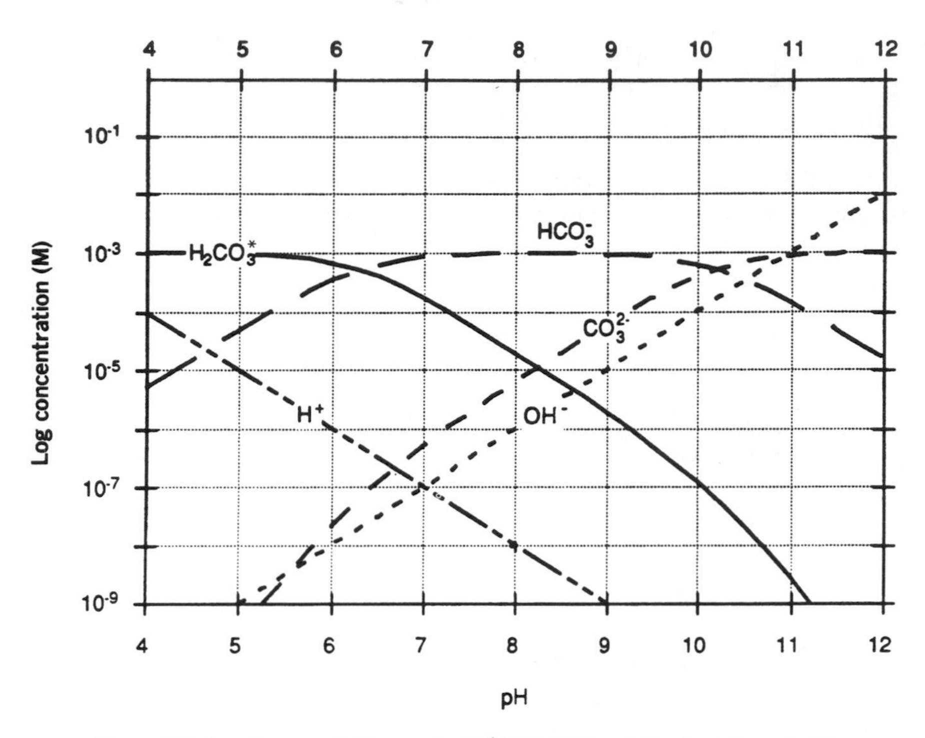

Figure 4.10 Log C versus pH diagram for 10^{-3} M $NaHCO_3$ solution from Example 4.7.

values. Calculate the solubility of $Fe(III)_{aq}$ species in pure water using a chemical equilibrium program. The K_{s0} and K_{eq} values for hydrolysis of Fe(III) are given below. (Your computer program will automatically estimate activity coefficients for you using the Güntelberg or modified Davies approximations.)

Solution: Use a thermodynamic database or your chemical equilibrium program, or a *Handbook of Chemistry and Physics* to obtain the equilibrium constants for mass law equations. Decide on the number of species in solution and the components that you will designate.

Usually solubility products are given in terms of the constituent species, that is,

$$Fe(OH)_{3(s)} = Fe^{3+} + 3\ OH^-$$

$$K_{s0} = [Fe^{3+}]\ [OH^-]^3 = 10^{-38.7}$$

but if OH^- is a constituent of a solid phase, its solubility can also be expressed conveniently in terms of $[H^+]$; that is,

$$Fe(OH)_{3(s)} = Fe^{3+} + 3\ OH^- ; \quad \log K_{s0} = -38.7$$
$$3\ OH^- + 3\ H^+ = 3\ H_2O; \quad \log K_w^{-3} = 42$$
$$Fe(OH)_3 + 3\ H^+ = Fe^{3+} + 3\ H_2O; \quad \log {}^*K_s = 3.3$$

or as in the first row $[Fe^{3+}] = [H^+]^3 \times 10^{3.3}$ of the matrix below.

Species: Fe^{3+}, $FeOH^{2+}$, $Fe(OH)_2^+$, $Fe(OH)_{3(aq)}$, OH^-, H^+
Components: $Fe(OH)_{3(s)}$, H^+

In choosing components, we could have chosen Fe^{3+} or any other aqueous Fe(III) species. It was easiest in this case to write the mass law equations in terms of $Fe(OH)_{3(s)}$, so it was chosen. Choose solid phases as components if you want to calculate dissolution or precipitation amounts. Hydrogen ions should always be chosen as a component in problems of fixed pH, where the chemical equilibrium program will fix the pH at each increment (say, pH increments of 0.2 units) in order to plot p*C* versus pH. (In MacμQL,[17] it is required to list H^+ as the last component of the matrix.)

The mass action expressions are rearranged below with all species on the left-hand side of the equation and all components on the right-hand side. We did not need an equation for $Fe(OH)_{3(s)}$ because it makes no sense to calculate its activity; it is a Type III solid with unit activity and infinite supply, and it is also designated as a component.

(i) $$[Fe^{3+}] = [H^+]^3 \times 10^{3.3}$$

(ii) $$[FeOH^{2+}] = [H^+]^2 \times 10^{0.5}$$

(iii) $$[Fe(OH)_2^+] = [H^+] \times 10^{-2.6}$$

(iv) $$[Fe(OH)_3] = 10^{-9.8}$$

(v) $$[Fe(OH)_4^-] = [H^+]^{-1} \times 10^{-18.5}$$

(vi) $$[OH^-] = 10^{-14.0}/[H^+]$$

(vii) $$[H^+] = [H^+]$$

The matrix is shown below; no mass balances are known. The program will ask for an initial value for pH.

	Species	Components: $Fe(OH)_{3(s)}$	Components: H^+	log K
(i)	Fe^{3+}	1	3	3.3
(ii)	$FeOH^{2+}$	1	2	0.5
(iii)	$Fe(OH)_2^+$	1	1	–2.6
(iv)	$Fe(OH)_{3(aq)}$	1	0	–9.8
(v)	$Fe(OH)_4^-$	1	–1	–18.5
(vi)	OH^-	0	–1	–14.0
(vii)	H^+	0	1	0

Results are shown in Figure 4.11. It is soluble in acidic solutions.

Example 4.9 Surface Complexation of Pb(II) to α-FeOOH

Solve for all concentrations in solution as a function of pH for 10^{-6}M Pb(II) sorption to geothite (α-FeOOH). Equilibrium constants are given below.[17,30,31] Hydrous ferric oxides are important particles in natural waters forming relatively strong surface complexes with Cd, Cu, Zn, Pb, and Hg among other ions. Use a chemical

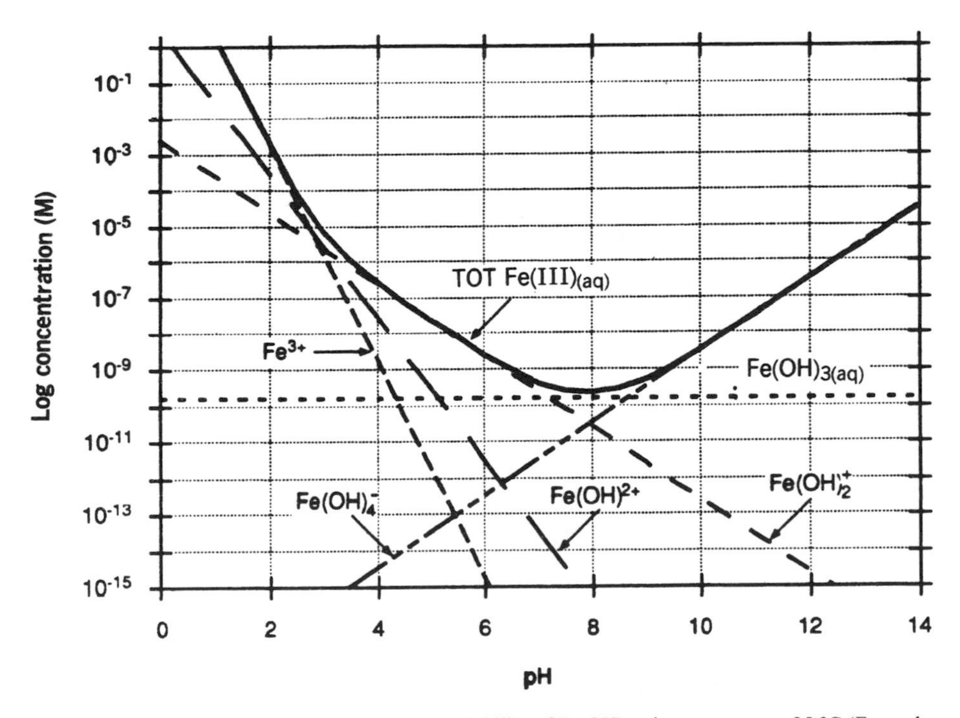

Figure 4.11 Log C versus pH diagram for the solubility of $Fe(OH)_{3(s)}$ in pure water at 25 °C (Example 4.8).

equilibrium model to solve and plot the results on a log C versus pH diagram. Choose the constant capacitance model with an internal capacitance of 2.9 F m^{-2}, assuming a specific surface area of 14.7 m^2 g^{-1}, concentration of active sites SOH = 0.000135 mol g^{-1}, concentration of suspended solids 0.1 g L^{-1}, and an ionic strength of 0.01 mol L^{-1}.

Solution: The species at equilibrium are:

Species: $FeOOH_{(s)}$, $> FeOO^-$, $> FeOOH_2^+$, $> FeOOPb^+$,
$> (FeOO)_2Pb$, Pb^{2+}, H^+ (seven species)

The character ">" designates the surface species of FeOOH. The first five (including FeOOH, the adsorbent) are surface species. $FeOOH_{(s)}$ is amphoteric, so it can act as an acid or a base. Pb(II) ions can bind to one oxygen donor atom $>FeOOPb^+$ or two oxygen donor atoms $>(FeOO)_2Pb$ to form inner-sphere complexes. We will ignore other aqueous phase complexes of Pb^{2+} because no information was given on other ligands in solution.

Let us choose the solid phase adsorbent FeOOH, Pb^{2+}, and H^+ as components. It is always convenient to choose the adsorbent, H^+ ion, and free metal ion as components in problems of this type. Equations for all seven species can be written in terms of these three components. Because the surface charge varies with pH according to the constant capacitance model, we must also include surface charge as a component.

Components: $FeOOH_{(s)}$, surface charge, Pb^{2+}, H^+ (four components)

The surface charge equation is written assuming FeOOH has a neutral charge.

$$\text{Surface charge} = FeOOH_2^+ + FeOOPb^+ - FeOO^-$$

Equilibrium constants are given below.

$$FeOOH \rightleftharpoons FeOO^- + H^+ \qquad \frac{[FeOO^-]\,[H^+]}{[FeOOH]} = 10^{-9}$$

$$FeOOH + H^+ \rightleftharpoons FeOOH_2^+ \qquad \frac{[FeOOH_2^+]}{[FeOOH]\,[H^+]} = 10^{6.7}$$

$$FeOOH + Pb^{2+} \rightleftharpoons FeOOPb^+ + H^+ \qquad \frac{[FeOOPb^+]\,[H^+]}{[FeOOH]\,[Pb^{2+}]} = 10^{-0.52}$$

$$2FeOOH + Pb^{2+} \rightleftharpoons (FeOO)_2Pb + 2H^+ \qquad \frac{[(FeOO)_2Pb]\,[H^+]^2}{[FeOOH]^2\,[Pb^{2+}]} = 10^{-6.27}$$

Rearranging the equilibrium constants, we can write an equation for each species in terms of the components that we have chosen.

(i) $[FeOOH] = 10^0 \cdot [FeOOH]$

(ii) $[FeOO^-] = 10^{-9} \cdot [FeOOH][H^+]^{-1}$

(iii) $[FeOOH_2^+] = 10^{6.7} \cdot [FeOOH][H^+]$

(iv) $[FeOOPb^+] = 10^{-0.52} \cdot [FeOOH][Pb^{2+}][H^+]^{-1}$

(v) $[(FeOO)_2Pb] = 10^{-6.27} \cdot [FeOOH]^2 [Pb^{2+}][H^+]^{-2}$

(vi) $[Pb^{2+}] = 10^0 \cdot [Pb^{2+}]$

(vii) $[H^+] = 10^0 [H^+]$

The input data are given below in the form of a spreadsheet from the chemical equilibrium program MacμQL by Beat Müller[17] and originally developed by Westall as MICROQL.[18]

	A	B	C	D	E	F
1		FeOOH	surfCharge	Pb^{+2}	H^+	{log *K*}
2		adsorbent	charge 1	total	free	
3	FeOOH	1	0	0	0	0
4	$FeOO^-$	1	–1	0	–1	–9
5	$FeOOH_2^+$	1	1	0	1	6.7
6	$FeOOPb^+$	1	1	1	–1	–0.52
7	$(FeOO)_2Pb$	2	0	1	–2	–6.27
8	Pb^{+2}	0	0	1	0	0
9	H^+	0	0	0	1	0
10		1.00	0.00	1.00E-06	1.00E-07	
11	constantCap	{*model: "constantCap," "diffuseLayer" or "tripleLayer"*}				
12	14.7	{*specific surface area in [m²]*}				
13	1.35E-04	{*no. of active surface sites in [mol/g]*}				
14	0.1	{*conc. of adsorbent in [g/l]*}				
15	0.01	{*ionic strength in [mol/l]*}				
16	2.9	{*capacitance of inner layer in [F m⁻²]*}				
17						

The FeOOH component is the adsorbent solid with an activity of 1.0 in row 10, column B. The concentration of sites is 0.1×0.000135 in mol L^{-1}.

For the sake of clarity only three of the seven species are plotted as results in Figure 4.12. Below pH 6.3, most of the lead is present as Pb^{2+} in solution, but there is a sharp "sorption edge" beginning at pH ~ 6. At pH values greater than 7.0, most of the lead in the system is adsorbed (in the form of a surface complex > $FeOOPb^+$). The concentrations of the five surface species (in mol L^{-1}) are given in tabular form for comparison below.

pH	>FeOOH	>$FeOO^-$	>$FeOOH_2^+$	>$FeOOPb^+$	>$(FeOO)_2Pb$
4	6.47×10^{-6}	3.00×10^{-8}	7.00×10^{-6}	4.22×10^{-11}	2.25×10^{-15}
5	8.35×10^{-6}	6.89×10^{-8}	5.07×10^{-6}	3.05×10^{-9}	3.73×10^{-13}
6	1.00×10^{-5}	1.60×10^{-7}	3.15×10^{-6}	1.59×10^{-7}	4.54×10^{-11}
7	1.09×10^{-5}	4.72×10^{-7}	1.26×10^{-6}	8.83×10^{-7}	7.42×10^{-10}
8	1.08×10^{-5}	1.27×10^{-6}	4.59×10^{-7}	9.94×10^{-7}	2.24×10^{-9}
9	9.67×10^{-6}	2.64×10^{-6}	1.77×10^{-7}	9.95×10^{-7}	4.68×10^{-9}
10	8.04×10^{-6}	4.37×10^{-6}	7.42×10^{-8}	9.92×10^{-7}	7.72×10^{-9}

Concentration of the binuclear surface complex >$(FeOO)_2Pb$ is always small. FeOOH sites and >$FeOOH_2^+$ predominate at low pH, while FeOOH sites, >$FeOO^-$, and > $FeOOPb^+$ predominate at high pH.

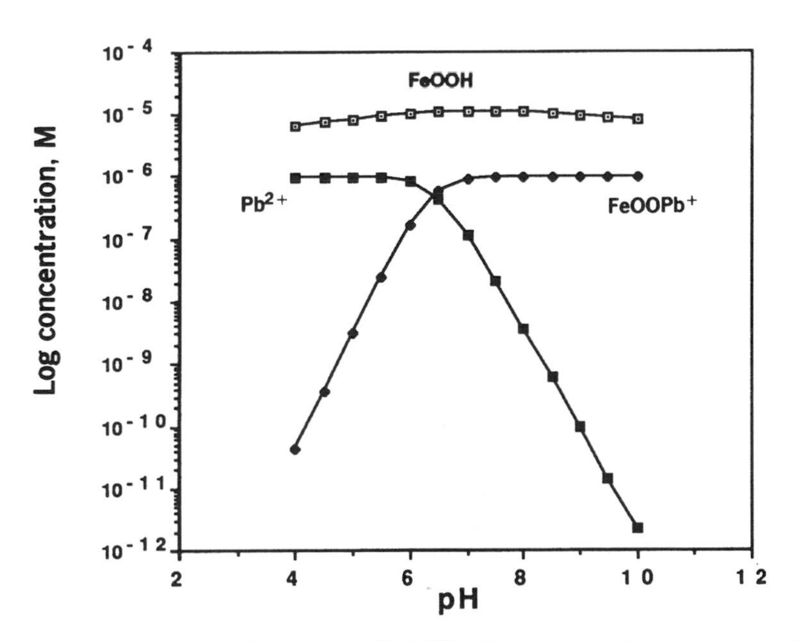

Figure 4.12 Log C versus pH diagram for 10^{-6} M Pb^{2+} with surface complexation onto geothite (α-FeOOH) from Example 4.9.

Example 4.10 Redox Reactions of NO_3^- and NH_4^+ in H_2O

Ammonia ions can be oxidized to nitrate in aerobic, natural waters. Bacteria (enzymes) mediate and catalyze the reaction, but the reaction proceeds spontaneously (thermodynamically) as well. Given 5×10^{-4} mol L^{-1} of total nitrogen, solve for the concentration of ammonium, ammonia, and nitrate as a function of pH. Assume slightly oxidizing conditions, $p\varepsilon = 5.52$. The example is from Müller.[17]

Solution: The nitrogen species are obvious, but in a redox reaction we also must consider the redox reactions of H_2O to form H_2 and O_2. This is most easily accomplished by designating $H_{2(g)}$ and $O_{2(g)}$ as species.

Species: NO_3^-, $NH_{3(aq)}$, NH_4^+, $O_{2(g)}$, $H_{2(g)}$, e^-, H^+ (seven species)
Components: NO_3^-, e^-, H^+ (three components)

In redox reactions, it is best to use $\{e^-\}$ as a component. For fixed pH problems, we also must choose H^+ as a component. We have a choice of which nitrogen species to designate as a component and, in this example, nitrate ion was chosen. Because nitrate is a component, we must rearrange the redox reaction equations so that all the species are expressed in terms of NO_3^- and the other components. This requires the user to algebraically combine equilibrium reactions in strange ways, sometimes. Half reactions are given in the thermodynamic database of MINTEQ and MINEQL+.[21,22]

$$NO_3^- + 9H^+ + 8e^- \rightleftharpoons NH_3 + 3H_2O \qquad \frac{[NH_3]}{[NO_3^-][H^+]^9[e^-]^8} = 10^{109.9}$$

$$NO_3^- + 10H^+ + 8e^- \rightleftharpoons NH_4^+ + 3H_2O \qquad \frac{[NH_4^+]}{[NO_3^-][H^+]^{10}[e^-]^8} = 10^{119.2}$$

$$2H_2O \rightleftharpoons O_{2(g)} + 4H^+ + 4e^- \qquad [O_{2(g)}][e^-]^4[H^+]^4 = 10^{-83}$$

$$2H^+ + 2e^- \rightleftharpoons H_{2(g)} \qquad \frac{[H_{2(g)}]}{[e^-]^2[H^+]^2} = 10^0$$

(The first two equilibrium reactions shown above include the acidity constant for ammonium $NH_4^+ \rightleftharpoons NH_3 + H^+$, but we couldn't use it as an equilibrium equation because it does not involve the component NO_3^-.)
Seven equations for the species are:

(i) $$[NO_3^0] = 10^0 \cdot [NO_3^-]$$

(ii) $$[NH_3] = 10^{109.9} \cdot [NO_3^-][e^-]^8[H^+]^9$$

(iii) $$[NH_4^+] = 10^{119.2} \cdot [NO_3^-][e^-]^8[H^+]^{10}$$

(iv) $$[O_{2(g)}] = 10^{-83}[e^-]^{-4}[H^+]^{-4}$$

(v) $$[H_{2(g)}] = 10^0 \cdot [e^-]^2[H^+]^2$$

(vi) $$[e^-] = 10^0 \cdot [e^-]$$

(vii) $$[H^+] = 10^0 \cdot [H^+]$$

Mass balance equations are given by summing each column in the table below

TOT N = $NO_3^- + NH_3 + NH_4^+ = 5 \times 10^{-4}$ M
TOT $e^- = 10^{-5.52} = 3.00 \times 10^{-6}$ M
TOT $H^+ = 3.0 \times 10^{-8}$ (initial value)

	A	B	C	D	E
1		NO_3^-	e^-	H^+	{log K}
2		total	free	free	
3	NO_3^-	1	0	0	0
4	NH_3	1	8	9	109.9
5	NH_4^+	1	8	10	119.2
6	$O_{2(g)}$	0	–4	–4	–83
7	$H_{2(g)}$	0	2	2	0
8	e^-	0	1	0	0
9	H^+	0	0	1	0
10		5.00E-04	3.00E-06	3.00E-08	(concentrations}

Key species are plotted in Figure 4.13. At acidic pH values, NH_4^+ predominates, and in basic solutions, nitrate should predominate at chemical equilibrium. As the pH increases from pH 5, $NH_{3(aq)}$ increases due to more basic conditions, but it is oxidized to nitrate ion at high pH. The pε was fixed in this calculation at pε = +5.52, and the amounts of $O_{2(g)}$ and $H_{2(g)}$ in equilibrium were very small (10^{-45} M at pH 4 and 10^{-21} M at pH 10 for oxygen; 10^{-19} at pH 4 and 10^{-31} M at pH 10 for hydrogen gas). It is essentially anoxic.

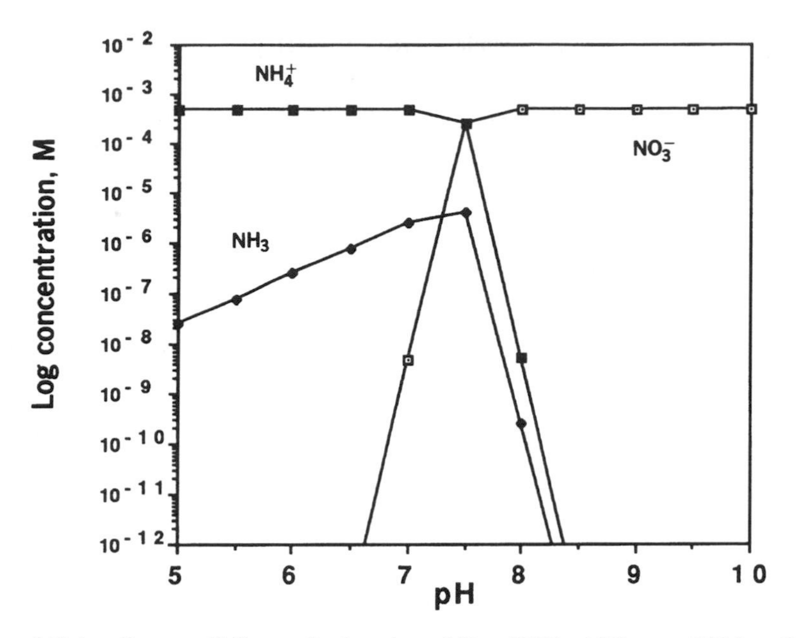

Figure 4.13 Log C versus pH diagram for the redox stability of NH_4^+ and NO_3^- at pε 5.52 from Example 4.10.

4.7 COMPUTER MODELS

There are many computer models available for chemical equilibrium calculations. Melchior and Bassett[11] reviewed them. They run on similar principles. Here, we will briefly discuss three of them that are freely available, are well documented, and have found wide use among environmental professionals. Other good models have been excluded for the sake of brevity.

4.7.1 MacμQL

MacμQL was developed by Müller[17] for Macintosh computers. Müller took Westall's MICROQL program[18] and added a user-friendly interface and made it interactive. As the name implies, it is the smallest of the chemical equilibrium models that we will discuss, but it runs on the same principles as the others (mass law equations plus mass balance equations, solved iteratively using a Newton–Raphson numerical technique). The code is small because it does not include a thermodynamic database (which takes considerable storage capacity and slows execution when it is being searched). MacμQL does not include activity or temperature corrections; the user must specify these by specifying K' conditional equilibrium constants if the ionic strength and temperature of the solution are known. There is not a large editor or output data manager, but it is possible to create the database in EXCEL for import

Table 4.6 Comparison of Some Chemical Equilibrium Models

	MacμQL	MINEQL$^+$	MINTEQA2
Thermodynamic database	No	Yes, 2300 species	Yes, 1000 species
Temperature corrections	No	van't Hoff	van't Hoff
Activity corrections	No	Yes	Yes
Adsorption models	Constant capacitance Diffuse layer Triple layer	Langmuir Freundlich Constant capacitance Triple layer model Two-layer model	K_d model Langmuir Freundlich Ion exchange Constant capacitance Diffuse layer Triple layer
Computer	Macintosh	IBM compatible DOS Windows	IBM compatible DOS or VAX system
Interactive	Yes	Yes	Yes, PRODEFA2
Output	Tables for export to spreadsheets	Graphics output or tables for spreadsheets	Tables for export to spreadsheets
Create your own database?	Yes, mandatory	Yes, optional	Yes, optional
Reference	17	22	21

to MacμQL, and to export the output data tables to Cricket Graph, EXCEL, or Kaleida Graph, for example. MacμQL is a good system for students to learn first because it requires a thorough knowledge of the mass action equations, thermodynamic data, and mass balances to set up a run.

4.7.2 MINEQL$^+$ 3.0

MINEQL$^+$ 3.0 was developed by Schecher and McAvoy[22] beginning with the MINEQL model of Westall and co-workers.[7] It runs on DOS operating systems with Windows® and includes a database manager for graphical output. The strength of the program is its interactive input and output, making it relatively simple to plot field data and/or laboratory results with model output for comparison.

MINEQL$^+$ includes the two-layer surface complexation model of Dzombak and Morel,[26] together with important thermodynamic data on hydrous ferric oxides (HFO). The two-layer model is the diffuse double-layer model with the addition of surface precipitation or coprecipitation at high adsorption densities by Farley et al.[32] and Dzombak and Morel.[26]

The thermodynamic database is the most extensive in MINEQL$^+$. It includes all of the data from MINTEQA1 that came from the WATEQ[9] model of the United States Geological Survey, and it includes the MINEQL[7] database plus some other.[26]

The database can be modified and you can create your own database; it is quite user-friendly.

4.7.3 MINTEQA2

MINTEQA2 has been supported by the U.S. Environmental Protection Agency Environmental Research Laboratory, Athens, Georgia, and it was originally developed by Battelle Pacific Northwest Laboratory.[10] MINTEQ took the best database that was available from WATEQ,[9] the USGS model, and wed it to the numerical engine of MINEQL.[7] The most recent development has been the addition of a user-friendly interactive program, PRODEF2, that helps the user assemble a coherent input database without making obvious mistakes. Like MINEQL$^+$ 3.0, it is IBM compatible, or is available for VAX mainframe computers. There are no graphics capabilities within PRODEFA2, but files can be exported to other software programs for plotting. Many users like the wide array of choices that are available for adsorption in MINTEQA2, including the simple equilibrium distribution coefficient K_d. All three models are well-documented and suitable for chemical equilibrium problems in natural waters (acid–base, precipitation–dissolution, complexation, surface complexation, and/or redox). Table 4.6 is a summary of the three selected models in this section.

(The previous discussion does not imply endorsement of any commercial products, computers, or software.)

4.8 REFERENCES

1. Sillen, L.G., in *Oceanography,* M. Sears, Ed., American Association for the Advancement of Science, Washington, DC (1961).
2. Garrels, R.M., and Thompson, M.E., *Am. J. Sci.,* 260, 57 (1962).
3. Clasen, R.J., *The Numerical Solution of the Chemical Equilibrium Problem,* RAND RM-4345-PR, New York (1965).
4. Deland, E.C., *Chemist—The Rand Chemical Equilibrium Program,* RAND RM-5404-PR, New York (1967).
5. Morel, F.M.M., and Morgan, J.J., *Environ. Sci. Technol.,* 6, 58 (1972).
6. Truesdell, D.H., and Jones, B.F., *WATEQ: A Computer Program for Calculating Chemical Equilibria of Natural Waters,* National Technical Information Service, NTIS Technical Report, PB2-20464, Washington, DC (1973).
7. Westall, J.C., Zachary, J.L., and Morel, F.M.M., *MINEQL, A Computer Program for the Calculation of Chemical Equilibrium Composition of Aqueous Systems,* MIT Technical Note 18, Cambridge, MA (1976).
8. Westall, J.C., *MICROQL, A Chemical Equilibrium Program in BASIC,* Report 86-02 Oregon State University, Corvallis, OR (1986).
9. Ball, J.W., Nordstrom, D.K., and Zachman, D.W., *WATEQIVF,* U.S. Geological Survey Open File Report 87-50 (1987).

10. Felmy, A.R., Girvin, D., and Jenne, E.A., *MINTEQ: A Computer Program for Calculating Aqueous Geochemical Equilibria,* U.S. Environmental Protection Agency, Environmental Research Laboratory, Athens, GA (1984).
11. Melchior, DC, and Bassett, R.L., Eds. *Chemical Modeling of Aqueous Systems II,* American Chemical Society, Washington, DC (1990).
12. Jenne, E.A., Ed. *Chemical Modeling of Aqueous Systems,* American Chemical Society, Washington, DC (1979).
13. Culberson, C., and Pytkowicz, R.M. *Mar. Chem.,* 1, 309 (1973).
14. Stumm, W., and Morgan, J.J., *Aquatic Chemistry,* 2nd ed., Wiley, New York (1981).
15. Kielland, J., *J. Am. Chem. Soc.,* 59, 1675 (1937).
16. Morel, F.M.M., and Hering, J.G., *Principles and Applications of Aquatic Chemistry,* Wiley, New York (1993).
17. Müller, B., *MacμQL v1.1—A Program to Calculate Chemical Speciation and Adsorption,* EAWAG/ETH, Limnological Research Center, Kastaneinbaum, Switzerland (1993).
18. Westall, J. MICROQL–A Chemical Equilibrium Program in BASIC, EAWAG/ETH Swiss Federal Institute for Environmental Science and Technology, Dübendorf, Switzerland (1979).
19. Sigg, L., and Stumm, W., *Colloids Surf.,* 2, 101 (1981).
20. Stumm, W., *Chemistry of the Solid–Water Interface,* Wiley, New York (1992).
21. Allison, J.D., Brown, D.S., and Novo-Gradac, K.J., *MINTEQA2/PRODEFA2—A Geochemical Assessment Model for Environmental Systems,* EPA/600/3-91/021, U.S. Environmental Protection Agency, Environmental Research Laboratory, Athens, GA (1991).
22. Schecher, W.D., and McAvoy, DC, *MINEQL*$^+$: A Chemical Equilibrium Program for Personal Computers, Version 3.0, Environmental Research Software, Hallowell, ME (1994).
23. Schindler, P.W., and Gamajager, H., *Kolloid Z. Z. Polym.,* 250, 759 (1972).
24. Schindler, P.W., and Kamber, H.R., *Helv. Chim. Acta,* 51, 1781 (1968).
25. Stumm, W., Huang, C.P., and Genkins, S.R., *Croat. Chem. Acta,* 42, 223 (1970).
26. Dzombak, D.A., and Morel, F.M.M., *Surface Complexation Modeling,* Wiley, New York (1990).
27. Westall, J.C., in *Particulates in Water,* M.C. Kavanaugh and J.O. Leckie, Eds., ACS Advances in Chemistry Series 189, American Chemical Society, Washington, DC (1980).
28. Westall, J.C., and Hohl, H., *Adv. Colloid Interface Sci.,* 12, 265 (1980).
29. Schindler, P.W., and Stumm, W., in *Aquatic Surface Chemistry,* W. Stumm, Ed., John Wiley, New York (1987).
30. Yates, D.E., Levine, S., and Healy, T.W., *J. Chem. Soc. Faraday Trans.,* 1, 70, 1807 (1974).
31. Davis, J.A., and Leckie, J.O., in *Chemical Modeling of Aqueous Systems,* E.A. Jenne, Ed., ACS Symposium Series 93, American Chemical Society, Washington, DC (1979).
32. Smith, R.M., and Martell, A.E., *Critical Stability Constants, Inorganic Complexes,* Vol. 4, Plenum Press, New York (1976).
33. Benjamin, M.M., and Leckie, J.O., *J. Colloid Interface Sci.,* 79, 209 (1981).
34. Farley, K.J., Dzombak, D.A., and Morel, F.M.M., *J. Colloid Interface Sci.,* 106, 226 (1985).

4.9 PROBLEMS

1. The solubility product log K_{s0} for the dissolution of calcite is –8.42 at 25 °C. Find the solubility of calcite (log K_{s0}) at 15 °C. Is it more or less soluble at the cooler temperature? Why?

$$CaCO_{3(s)} \rightleftharpoons Ca^{2+} + CO_3^{2-}$$

2. What happens if one adds table salt, NaCl, to beer? At chemical equilibrium, what is the activity coefficient of $CO_{2(aq)}$ if one adds enough salt to increase the ionic strength to $I = 0.1$ M? What is the effect on the acidity constant pK_1 for $H_2CO_3^*$?

3. a. Find the pK value for the following reaction at 25 °C in aqueous solution.

$$NH_4^+ \rightleftharpoons H^+ + NH_{3(aq)}$$

   b. Adjust the pK value from part (a) to 5 °C.
   c. Find the pH of a solution for 0.01 M NH_4Cl at 25 °C. Include activity corrections.

4. For the adsorption of an organic molecule by 5.98 g of activated carbon, plot the linearized versions of the Langmuir and Freundlich adsorption models. Which one fits the data more accurately?

Dissolved Concentrations, mol L^{-1}	5.89×10^{-8}	1.13×10^{-4}	2.01×10^{-4}	3.96×10^{-4}	5.85×10^{-4}	7.29×10^{-4}	8.77×10^{-4}
Adsorbed Concentrations, mol L^{-1}	9.41×10^{-7}	4.60×10^{-5}	6.30×10^{-5}	7.8×10^{-5}	1.00×10^{-4}	1.13×10^{-4}	1.23×10^{-4}

5. Find the coulombic correction factor P for complexation of Cd^{2+} onto hydrous ferric oxide at pH 4 and I = 0.01 M (analogous to freshwater acid mine drainage).

6. Determine the composition and the pH of an open system in equilibrium with a gas phase that contains $P_{NH3} = 10^{-5}$ atm. Find the mass law relationships, write the matrix, and solve using a chemical equilibrium program. [*Hint*: $NH_{3(aq)}$ is a Type III species of fixed activity determined by Henry's law; $NH_{3(g)}$ is a component.]

7. Find the pH of a 10^{-5} M Na_2CO_3 solution at 15 °C and 25 °C.

8. Find the pH of a solution of 10^{-3} M $NaHCO_3$ and 2×10^{-3} M $NH_{3(aq)}$ at an ionic strength of $I = 0.01$ M and 25 °C. Use activity coefficients.

9. To a solution we add 10^{-4} mol L^{-1} of Sr^{2+} together with 10^{-4} mol L^{-1} of SO_4^{2-}. Does $SrSO_{4(s)}$ precipitate at 25 °C and, if so, how much?

$$pK_{s0} = 7.8 \text{ at } 25\ °C$$

10. The species involved in the reactions of interest are NH_4^+, NH_3, NO_3^-, and NO_2^-.

a. Construct a pε–pH predominance diagram using a chemical equilibrium model for these species. The concentration of total nitrogen species is 10^{-4}. Neglect ionic strength effects.

b. What species should predominate in the surface water at chemical equilibrium (pH = 7, pε = 13)?

c. What species should predominate in the surface water at chemical equilibrium (pH = 6, pε = 3)?

d. Are natural waters always at chemical equilibrium? Why or why not?

Reactions	log K
(1) $NO_3^- + 2H^+ + 2e^- \leftrightarrow NO_2^- + H_2O$	28.4
(2) $NO_3^- + 10H^+ + 8e^- \leftrightarrow NH_4^+ + 3H_2O$	119.2
(3) $NO_2^- + 8H^+ + 6e^- \leftrightarrow NH_4^+ + 2H_2O$	90.0
(4) $NH_4^+ \leftrightarrow H^+ + NH_3$	–9.3

11. Dissolved aluminum species in water [especially $Al(OH)^{2+}$] is toxic to fish. It occurs under low pH conditions in acidified natural waters from the dissolution of gibbsite, $Al(OH)_3$, and the hydrolysis of Al^{3+}. Organic complexes in solution are known to decrease the Al toxicity to fish. Thus it is important to know the concentration of all species in natural waters. Use a chemical equilibrium program to determine the concentration of total dissolved monomeric aluminum in natural waters under the conditions listed below:

a. Sketch the pC versus pH solubility diagram and the aluminum species concentrations.

b. At which pH will the aluminum $Al(OH)^{2+}$ concentration be maximum?

c. What is the effect of organic complexation at pH 5?

	Reactions	log K_{eq} at 10 °C
Organic acid	$HA \leftrightarrow H^+ + A^-$	–4.5
Organic–Al complex	$Al^{3+} + A^- \leftrightarrow AlA^{2+}$	+5.0
Al solubility	$Al(OH)_{3(s)} + 3H^+ \leftrightarrow Al^{3+} + 3\ H_2O$	+8.77
Al hydrolysis	$Al^{3+} + H_2O \leftrightarrow AlOH^{2+} + H^+$	–4.99
	$Al^{3+} + 2\ H_2O \leftrightarrow Al(OH)_2^+ + 2\ H^+$	–10.1
	$Al^{3+} + 3\ H_2O \leftrightarrow Al(OH)_{3(aq)} + 3H^+$	–16.0
K_w, water	$H_2O \leftrightarrow H^+ + OH^-$	–14.0

Conditions

pH = 3, 5, 7, 9

TOT A = 10^{-3} M = $A^- + HA + AlA^{2+}$

12. Determine all species in solution and surface species for the adsorption of 10^{-7} M Zn (total zinc) in a system of 90 mg L^{-1} of hydrous ferric oxides (HFO = 10^{-3} M) with an ionic strength of 0.1 M and pH 8 and 6. At pH 8, the surface charge is equal to zero; at pH 6 it is +11.5 μC cm^{-2}.

HFO $pK_{a1}^{int} = 7.29$
HFO $pK_{a2}^{int} = 8.93$
$\log K_{SOZn^+}^{int} = 0.99$
HFO Surface area = 600 m^2 g^{-1}
HFO Site density = 0.005 mol mol^{-1}

Use a chemical equilibrium program of your choice with the constant capacitance model and a capacitance of the inner layer of 2.0 F m^{-2}.

ANSWERS TO SELECTED PROBLEMS—CHAPTER 4

1. $\Delta H_f^o = -12.53$ kJ mol^{-1}; ln K_{s0} (15 °C) = −19.21; $\log_{10} K_{s0} = -8.34$.

3. pK = 9.25 at 25 °C; pK = 9.9 at 5 °C; $f_{NH_3} \simeq 1.0$ so activity coefficients cancel out; pH = 5.62.

4. The Freundlich plot.

5. $\sigma = 21$ μC cm^{-2}; $\psi = 0.092$ V; $\log P = -3.11$. The electrostatic correction is very important in modeling this problem.

6. Matrix for a NH_3–Water System; $p_{NH_3} = 10^{-5}$ atm

		Components		
	Species	$NH_{3(g)}$	H^+	log K
(i)	NH_4^+	1	1	11.05
(ii)	$NH_{3(aq)}$	1	0	1.75
(iii)	$NH_{3(g)}$	1	0	0
(iv)	OH^-	0	−1	−14
(v)	H^+	0	1	0

$[NH_{3(aq)}] = 5.6 \times 10^{-4}$ M; $[NH_4^+] = 1.0 \times 10^{-4}$M; and pH = 10.0.

7. $pK_{a2} = 10.43$, $pK_w = 14.35$ at 15 °C; pH = 9.32.

9. 8.5×10^{-5} mol L^{-1} of $SrSO_4$ should precipitate.

11. $[Al^{3+}] = 0.59$ M, 5.89×10^{-7} M, 5.89×10^{-13} M, and 5.89×10^{-19} M at pH 3, 5, 7, and 9, respectively.

5

EUTROPHICATION OF LAKES

Like winds and sunsets, wild things were taken for granted until progress began to do away with them.

—Aldo Leopold, *A Sand County Almanac*

5.1 INTRODUCTION

Eutrophication refers to the excessive rate of addition of nutrients, usually in reference to anthropogenic activities and the addition of phosphorus and nitrogen to natural waters. Nutrient additions result in the excessive growth of plants including phytoplankton (free-floating algae), periphyton (attached or benthic algae), and macrophytes (rooted, vascular aquatic plants).

Eutrophication is a natural process taking place over geologic time that is greatly accelerated by human activities. For example, due to soil erosion and biological production, lakes normally fill with sediments over thousands of years. As the lakes fill up, the mean depth and detention time of the water body decreases. Sediments are in greater contact with overlying water, and nutrients recycle under anaerobic conditions via diffusion. With addition of anthropogenic nutrients, algae growth accelerates even further. Eventually a process that would have occurred over geologic time scales is accelerated to decades, and the lake becomes overly productive biologically. The process has some undesirable effects on water quality.

- Excessive plant growth (green color, decreased transparency, excessive weeds).
- Hypolimnetic loss of dissolved oxygen (anoxic conditions).
- Loss of species diversity (loss of fishery).
- Taste and odor problems.

Not all eutrophic lakes will exhibit all of these water quality problems, but they are likely to have one or more of them. Eutrophication, then, is the *excessive* rate of addition of nutrients. Finally, water quality may become so degraded that the lake's original uses are lost. It may no longer be swimmable or fishable. A schematic of a eutrophic lake is shown in Figure 5.1.

Of course, the degree of eutrophication is a continuum, and researchers have at-

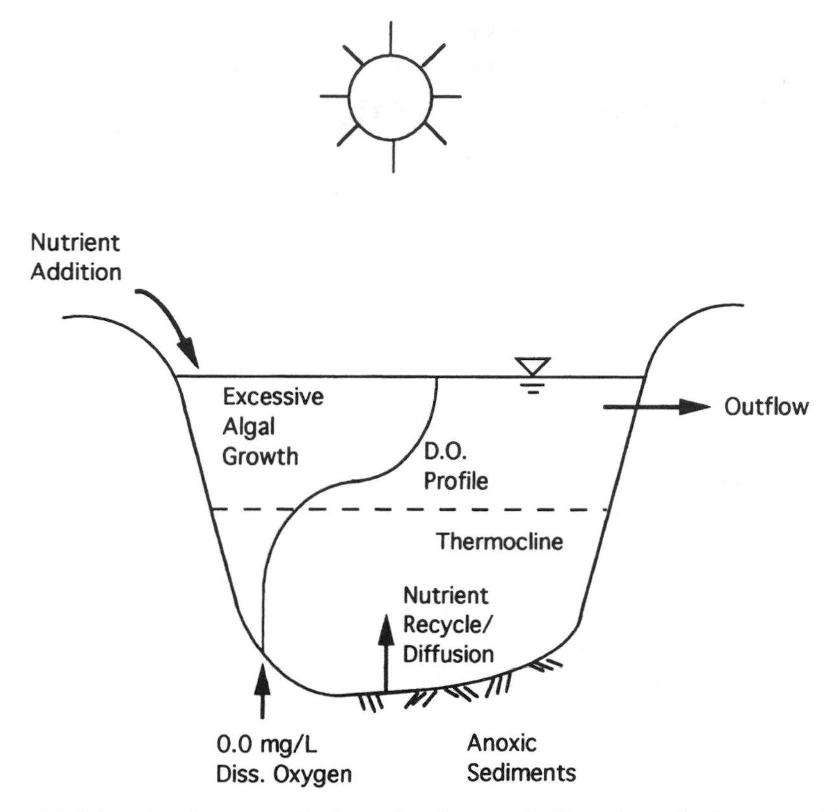

Figure 5.1 Schematic of lake eutrophication and nutrient recycle. Thermal stratification causes a high concentration of oxygen in near-surface waters, but the dissolved oxygen cannot mix vertically, and deep water eventually becomes anoxic. Anoxic conditions in the bottom waters and sediments cause anaerobic decomposition and the release of nutrients (phosphate, ammonia, dissolved iron).

tempted to classify lakes according to their relative extent of eutrophication since 1939.[1] It is referred to as the "trophic status" of the water body:

- Oligotrophic.
- Mesotrophic.
- Eutrophic.

"Eutrophic" comes from the Greek term meaning well-nourished. Oligotrophic systems are "undernourished," that is, biological production is limited by nutrient additions. Eutrophic systems are overfertilized by anthropogenic nutrient additions with concomitant water quality problems. Mesotrophic waters lie somewhere in between. Investigators have based these trophic classification schemes on the nutrient concentrations available for algal blooms in the spring and on the areal nutrient loading rate in g m^{-2} y^{-1} (based on the lake surface area).

Nutrients from human activities may emanate from point sources or from nonpoint sources. Because point sources are more easily controlled by tertiary treatment of wastewater, nonpoint source reduction is the subject of current research. Phosphate is often the limiting nutrient for algal growth, and point sources are reduced by precipitation of phosphate in wastewater with iron chloride or iron sulfate.

$$FeCl_{3(aq)} + PO_4^{3-} \rightleftharpoons FePO_{4(s)} + 3Cl^- \qquad (1)$$

For nitrogen nutrients, point source reductions sometimes involve ammonia stripping at high pH (pH > 9), NH_4^+ ion exchange at lower pH, or biological denitrification. Filtration of wastewater effluents can remove particulate and adsorbed nitrogen and phosphorus nutrients.

Nonpoint sources are caused by agricultural runoff, stormwater runoff, or combined-sewer overflow. After point source reductions are accomplished, nonpoint sources may become the majority of nutrient loadings to the water body. Both particulate and dissolved forms of nutrient additions are important because even particulate phases may become available to algae after desorption or transformation in the natural ecosystem. *Bioavailability* of nutrients is the subject of much concern and debate as to the relative contribution of particulate and adsorbed nutrients to the eutrophication of lakes.

5.2 STOICHIOMETRY

Primary production in natural waters is the photosynthetic process whereby carbon dioxide and nutrients are converted to plant protoplasm. Respiration is the reverse process in which protoplasm undergoes endogenous decay and/or lysis and oxidation[2,3]:

$$\begin{aligned} &106\,CO_2 + 16\,NO_3^- + HPO_4^{2-} + 122\,H_2O + 18\,H^+ \\ &\quad + \text{Trace elements (Fe, Mo, V, . . .)} \underset{R}{\overset{\text{light, P}}{\rightleftharpoons}} C_{106}H_{263}O_{110}N_{16}P_1 + 138\,O_2 \qquad (2) \end{aligned}$$

The chemical formula for algal protoplasm $C_{106}H_{263}O_{110}N_{16}P_1$ is not a strict chemical entity (a mole of algae), but it represents the ratios of elements in algal cells. Algal respiration occurs during both dark and daylight hours, but photosynthetic primary production can occur only in the presence of sunlight. Algae settle and decay, and the nutrients recycle from the sediment back to the overlying water (Figure 5.2).

Stoichiometry helps us to determine the ratios of elements assimilated during primary production for mathematical modeling (Example 5.1). It also helps us to understand the regulation of the chemical composition of natural waters. Redfield[4] in 1934 was the first to show that the chemical composition of the oceans is somewhat regulated by algal stoichiometry.[4,5] The molar ratio of nitrogen to phosphorus

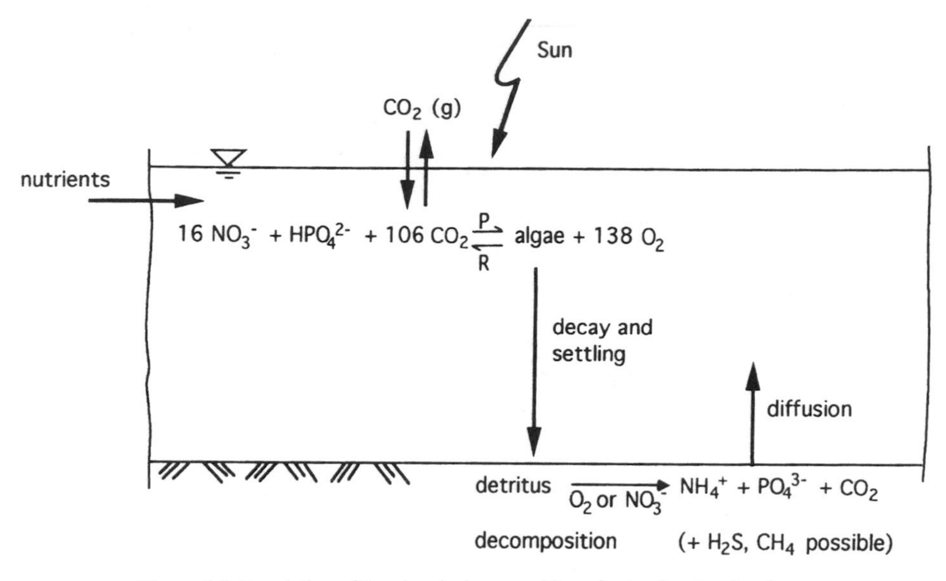

Figure 5.2 Regulation of the chemical composition of natural waters by algae.

is 16:1, as indicated by the stoichiometric equations. Algae assimilate nutrients in these ratios and when algal cells lyse and decompose (Figure 5.3), they release the N and P in the same molar ratios, thus regulating, to some extent, the chemical composition of natural waters. Sigg and colleagues[6,7] have shown that, due to active uptake by algae, even trace metal constituents may occur in lakes in the ratios found in algae.

Martin and co-workers[8–10] have hypothesized that the relatively high concentrations of nitrate and phosphate in Antarctic seas is due to the extremely small amounts of dissolved iron in solution, which limits phytoplankton growth. Theoretically, large amounts of CO_2 could be sequestered from the atmosphere (thus decreasing the global CO_2 greenhouse gas problem) if phytoplankton could be stimulated by addition of dissolved and colloidal iron. However, respiration of CO_2 by grazing zooplankton, mixing processes, and multiple limiting nutrients limit the amount of net CO_2 that could be assimilated.

Example 5.1 Nutrient Assimilation Rates Based on Algal Uptake Stoichiometry

Assuming the stoichiometric relationship for algal protoplasm, estimate the nutrient uptake rate for nitrate and CO_2 in Lake Ontario. It was observed that 5 μg per liter of phosphate was removed from the euphotic zone during the month of May in Lake Ontario. What is the rate of phytoplankton production in biomass, dry weight?

Solution: Use stoichiometric ratios to convert from phosphorus to biomass.

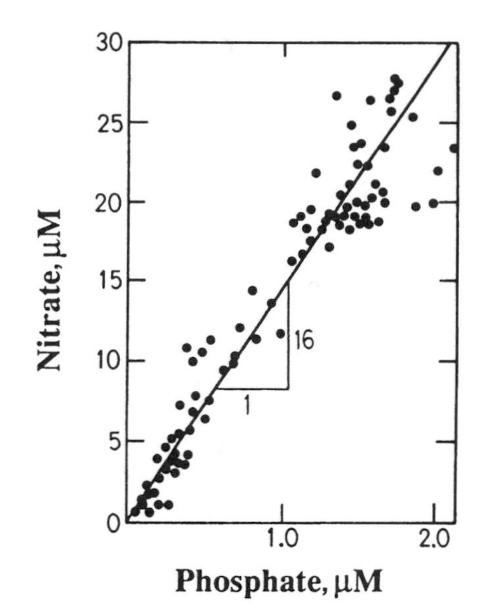

Figure 5.3 Ratio of nitrate to phosphate in surface ocean waters. (Modified from Redfield et al.[5])

$$\frac{5\ \mu\text{g P}}{\text{L}}\left|\frac{}{31\ \text{days}}\right. = 0.16\ \mu\text{g L}^{-1}\ \text{d}^{-1} \qquad \text{P}$$

$$\frac{0.16\ \mu\text{g}}{\text{L d}}\left|\frac{96\ \mu\text{g HPO}_4^{2-}}{30.97\ \mu\text{g P}}\right. = 0.50\ \mu\text{g L}^{-1}\ \text{d}^{-1} \qquad \text{HPO}_4^{2-}$$

$$\frac{0.16\ \mu\text{g}}{\text{L d}}\left|\frac{\mu\text{mol}}{30.97\ \mu\text{g P}}\right. = 0.0052\ \mu\text{mol L}^{-1}\ \text{d}^{-1} \qquad \text{P}$$

$$\frac{0.0052\ \mu\text{mol}}{\text{L d}}\left|\frac{16\ \text{mol NO}_3^-}{1\ \text{mol P}}\right. \begin{array}{l} = 0.083\ \mu\text{mol L}^{-1}\ \text{d}^{-1} \qquad \text{NO}_3^- \\ = 5.2\ \mu\text{g L}^{-1}\ \text{d}^{-1} \qquad \text{NO}_3^- \end{array}$$

$$\frac{0.0052\ \mu\text{mol}}{\text{L d}}\left|\frac{106\ \text{mol CO}_2}{1\ \text{mol P}}\right. \begin{array}{l} = 0.55\ \mu\text{mol L}^{-1}\ \text{d}^{-1} \qquad \text{CO}_2 \\ = 24\ \mu\text{g L}^{-1}\ \text{d}^{-1} \qquad \text{CO}_2 \end{array}$$

$$\frac{0.0052\ \mu\text{mol}}{\text{L d}}\left|\frac{1\ \text{mol algae}}{1\ \text{mol P}}\right. = 0.0052\ \mu\text{mol L}^{-1}\ \text{d}^{-1} \qquad \text{algae}$$

$$\frac{0.0052\ \mu\text{mol}}{\text{L d}} \bigg| \frac{3553.26\ \mu\text{g algae}}{\mu\text{mol}} = 18.5\ \mu\text{g L}^{-1}\ \text{d}^{-1} \quad \text{algae}$$

$$(31\ \text{days})\ (18.5\ \mu\text{g L}^{-1}\ \text{d}^{-1}) = 573\ \mu\text{g L}^{-1} \quad \text{algae}$$

0.57 mg L^{-1} of algae grew during the month of May in Lake Ontario

5.3 PHOSPHORUS AS A LIMITING NUTRIENT

It is possible that one of many nutrients could limit algal growth. Parsons and Takahashi[11] list the following micronutrients as being important for photosynthesis.

Fe	Zn	Mo	Co
Mn	B	Cl	
Cu	Na	V	

Macronutrients include CO_2 (the carbon source), phosphorus, nitrogen (ammonia or nitrate), Mg, K, Ca, and dissolved silica for diatom frustule formation. Algae and rooted aquatic plants are photoautotrophs—they use sunlight as their energy source and CO_2 for their carbon source. The terminal electron acceptor is usually dissolved oxygen in the electron transfer of photosynthesis. Suitable temperature is needed for algal growth also.

Any of these nutrients could, theoretically, become limiting for growth, as per Liebig's law of the minimum from the 19th century:

> Growth of a plant depends on the amount of food-stuff that is presented to it in limiting quantity."

Liebig envisioned a saturated response growth curve for each nutrient, similar to that shown for Monod[12] kinetics (Figure 5.4).

$$\mu = \frac{\mu_{\max} S}{K_s + S} \tag{3}$$

where $\mu_{\max}$ = maximum growth rate, T^{-1}
S = substrate or nutrient concentration, ML^{-3}
K_s = half-saturation constant, ML^{-3}

Of course, it is possible that more than one nutrient will become limiting for growth at the same time. DiToro and co-workers[13] have discussed this possibility and evaluated the expressions for an electrical resistance analogue in parallel:

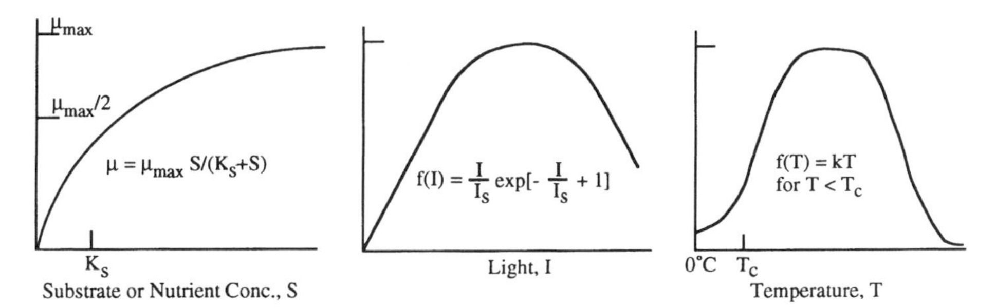

Figure 5.4 Response curves of algal growth rates to limiting nutrients, light, and temperature.

$$\frac{1}{\mu} = \frac{1}{\mu_1} + \frac{1}{\mu_2} + \cdots + \frac{1}{\mu_i} \tag{4}$$

where μ = overall limited growth rate, T^{-1}

μ_i = limited growth rate due to the ith nutrient, T^{-1}

and a multiplicative analogue:

$$\mu = \mu_{\max}\left(\frac{S_1}{K_{S_1} + S_1}\right)\left(\frac{S_2}{K_{S_2} + S_2}\right)\cdots\left(\frac{S_i}{K_{S_i} + S_i}\right) \tag{5}$$

The multiplicative analogue is much more limiting to growth (it usually results in a lower growth rate), but it is the one that is recommended for use with phytoplankton populations in the Great Lakes.[14]

Example 5.2 Limiting Nutrient Concentrations and Overall Growth Rates

Estimate the resulting growth rate for diatom phytoplankton in the Great Lakes by three different methods for the following data if the maximum growth rate is 1.0 per day.

a. Liebig's law of the minimum.
b. Electrical resistance analogy.
c. Multiplicative algorithm.

	NH_4^+	PO_4-P	Si
S, mg L^{-1}	0.02	0.005	1.6
K_s, mg L^{-1}	0.02	0.020	1.0

Solution:

a. Calculation with the Monod kinetic expression yields the following for

Liebig's law of the minimum.

$$\mu = \frac{(1.0)\,(0.02)}{0.02 + 0.02} = 0.5 \text{ day}^{-1} \quad \text{for } NH_4^+$$

$$\mu = \frac{(1.0)\,(0.005)}{0.02 + 0.005} = 0.2 \text{ day}^{-1} \quad \text{for } PO_4\text{-P}$$

$$\mu = \frac{(1.0)\,(1.6)}{1.0 + 1.6} = 0.62 \text{ day}^{-1} \quad \text{for } S_i$$

According to Liebig, the minimum growth rate would be the appropriate choice for the resulting overall growth rate. The answer is 0.2 d^{-1}, and phosphate limits the growth rate of the diatoms under these conditions.

b. Electrical resistance analogy:

$$\frac{1}{\mu} = \frac{1}{0.5} + \frac{1}{0.2} + \frac{1}{0.62} = 8.63$$

$$\mu = 0.12 \text{ d}^{-1}$$

All three nutrients contribute to an overall growth rate that is even lower than Liebig's law of the minimum predicted from part (a).

c. Multiplicative algorithm:

$$\mu = (1.0)\,(0.5)\,(0.2)\,(0.62) = 0.062 \text{ day}^{-1}$$

The multiplicative law also includes limitation due to all three nutrients, and it results in the lowest predicted growth rate of all three methods.

Experimental studies with all three nutrients in combination would be needed to confirm which model is most accurate.

Nitrogen requirements of algae are supplied by inorganic solutes (ammonium, nitrate, and nitrite) or fixed from the atmosphere by cyanobacteria (blue-green algae). Generally, algae prefer ammonium if it is present because nitrate and nitrite must be reduced before being utilized by the cell. Nitrogen is often the limiting nutrient in ocean and coastal waters.

Carbon requirements can be supplied by dissolved CO_2, bicarbonate, or carbonate (any source of total inorganic carbon). Negative correlations between total inorganic carbon and chlorophyll *a* (a measure of phytoplankton standing crop) have been observed during bloom conditions.[15]

Phosphorus was the limiting nutrient in most of the inland waters of the U.S. National Eutrophication Survey.[16] Approximately 90% of the lakes studied showed phosphorus limitation as defined by an inorganic nitrogen to dissolved phosphorus ratio of greater than 14:1 on a molar basis. Observations of algal bioassays supported the presumption of phosphorus as the limiting nutrient; phosphate additions stimulated algal growth to the greatest extent. Because phosphorus is the limiting nutrient in the Great Lakes, the first strategy to reduce algal blooms in Lakes Erie and Ontario was based on removal of phosphate from detergent formulations and tertiary treatment of domestic treated wastewater to remove phosphate. This strategy has been largely successful, and it has resulted in the need for nonpoint source pollution control if further improvements are to be realized.

5.4 MASS BALANCE ON TOTAL PHOSPHORUS IN LAKES

A simple mass balance can be developed on the limiting nutrient in a lake, for example, total phosphorus. In this application, total phosphorus refers to the unfiltered (whole water) concentration of inorganic, organic, dissolved, and particulate forms of phosphorus in the water column of the lake. For the condition of steady flow (inflows = outflows) and constant volume, we assume the lake can be approximated as a completely mixed flow-through system ($P_{out} = P_{lake}$). The average lake concentration is equal to the concentration of total phosphorus in the outflow.

$$V\frac{dP}{dt} = Q_{in}P_{in} - QP - k_s PV \tag{6}$$

Accumulation = Inputs – Outputs – Sedimentation

where V = lake volume, L^3
P = total phosphorus concentration in the lake (whole water), ML^{-3}
t = time, T
Q_{in} = inflow rate, L^3T^{-1}
P_{in} = inflow total phosphorus concentration, ML^{-3}
k_s = first-order sedimentation coefficient, T^{-1}
Q = outflow rate, L^3T^{-1}

The sedimentation coefficient is surrogate parameter for the mean settling velocity, reciprocal mean depth, and an alpha factor for the ratio of particulate phosphorus to total phosphorus.

$$k_s = \alpha\, v_s/H \quad \text{and} \quad \alpha = P_{part}/P$$

where α = ratio of particulate P to total P
v_s = mean particle settling velocity, LT^{-1}
H = mean depth of the lake, L

Under conditions of steady state (or as a crude estimate of annual average phosphorus concentration) the equation can be simplified:

$$V \frac{dP}{dt} = 0 = Q_{in} P_{in} - QP - k_s PV \tag{7}$$

If evaporation can be neglected, the inflow rate is approximately equal to the outflow rate ($Q_{in} = Q$). Also, we can divide through by the lake volume to express the right-hand side of the equation in terms of the hydraulic detention time ($\tau = V/Q$).

$$0 = \frac{P_{in}}{\tau} - \frac{P}{\tau} - k_s P \tag{8}$$

where τ = hydraulic detention time, T

After rearranging, the concentration of total phosphorus in the lake is

$$P = \frac{P_{in}}{k_s \tau + 1} \tag{9}$$

The total phosphorus concentration is directly related to the total P concentration in the inflow (P_{in}), and it is inversely related to the hydraulic detention time and the sedimentation rate constant (the main removal mechanism of total phosphorus from the water column). Thus the fate of total phosphorus in lakes is determined by an important dimensionless number ($k_s \tau$). There is a trade-off between the detention time of the lake and the sedimentation rate constant—it is the product of these two parameters that determines the ratio of total phosphorus in the lake to the inflowing total P:

$$\frac{P}{P_{in}} = \frac{1}{k_s \tau + 1} \tag{10}$$

The fraction of total phosphorus that is trapped or removed from the water column may be of interest.

$$R = \text{fraction removed} = 1 - \frac{P}{P_{in}} = \frac{k_s \tau}{k_s \tau + 1} \tag{11}$$

A simple mass balance such as that developed in this section has served as the basis for a number of research papers on the eutrophication of lakes dating back to the work of Vollenweider in 1969.[17]

Example 5.3 Mass Balance on Total Phosphorus in Lake Lyndon B. Johnson

Lake Lyndon B. Johnson is a flood control and recreational reservoir along a chain of reservoirs in central Texas along the Colorado River. It has an average hydraulic detention time of 80 days, a volume of 1.71×10^8 m^3, and a mean depth of 6.7 m. The ratio of particulate to total phosphorus concentration in the lake is 0.7, and the mean particle settling velocity is 0.1 m d^{-1}. If the flow-weighted average inflow concentration of total phosphorus to Lake LBJ is 72 μg L^{-1}, estimate the average annual total phosphorus concentration in the lake.

Solution:

$$k_s = \alpha v_s / H$$

$$k_s = (0.7)\,(0.1 \text{ m d}^{-1})/(6.7 \text{ m}) = 0.0104 \text{ day}^{-1}$$

$$k_s\tau = (0.0104)\,(80) = 0.832$$

$$\mathrm{P} = \frac{\mathrm{P_{in}}}{k_s\tau + 1} = \frac{72}{(0.0104)\,(80) + 1}$$

$$\mathrm{P} = 39 \ \mu\text{g L}^{-1}$$

The answer is a total phosphorus concentration of 39 μg L^{-1} in the lake, 70% of which is particulate material and the remaining 30% is dissolved (most of the dissolved P is PO_4^{3-}, available for phytoplankton uptake). The mass flux of total P to the bottom sediment is $k_s PV$, a flux rate equal to 69 kg d^{-1}.

A plot of the fraction of total P remaining in any lake as a function of the dimensionless number $k_s\tau$ is given by Figure 5.5. The fraction of total P removed to the lake sediments is equal to 0.454 $(1 - \mathrm{P/P_{in}})$; the mass of total phosphorus entering the lake is 152 kg d^{-1}, and the mass outflow is 83 kg d^{-1} (Figure 5.6).

5.5 NUTRIENT LOADING CRITERIA

Clair Sawyer was one of the first investigators to consider adoption of a criterion for the classification of eutrophic lakes.[1,18,19] In 1947, Sawyer noted that lakes in New England would exhibit algal blooms if total phosphorus concentrations in the early spring were greater than 30 μg L^{-1}.

Vollenweider argued that it is not the nutrient concentration that mattered as much as the nutrient supply rate (the areal nutrient loading rate expressed in g m^{-2} yr^{-1}).[17,20] He published "permissible" and "dangerous" loading rates for lakes based on a log–log plot of annual phosphorus loading versus mean depth for about 100 lakes in Europe and North America. These loading rates became widely adopted as nutrient loading criteria by many countries throughout the world. They comprised a criteria to classify lakes as oligotrophic, mesotrophic, and eutrophic. Vollenweider's nutrient loading rates are given in Table 5.1. Trophic classifications were based on observations of algal blooms and nuisance conditions (an empirical classification criteria).

Actually, both approaches to classifying the trophic status of lakes are valid and,

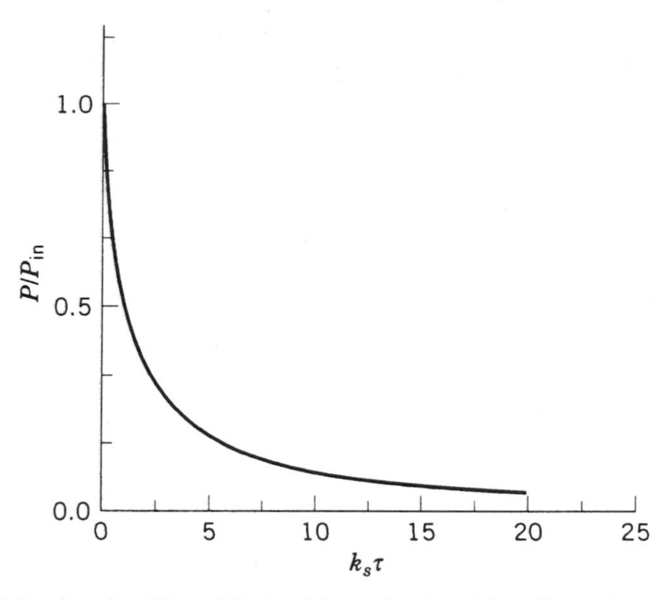

Figure 5.5 Fraction of total P remaining in a lake as a function of the sedimentation coefficient k_s times the detention time τ.

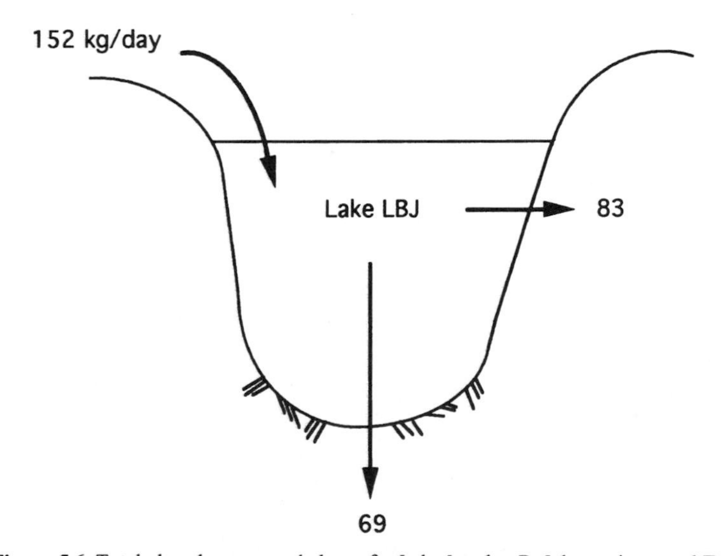

Figure 5.6 Total phosphorus mass balance for Lake Lyndon B. Johnson in central Texas.

Table 5.1 Allowable Nitrogen and Phosphorus Loading Levels in Lakes[20]

	Permissible Loading (g/m^2 of surface area/yr)		Dangerous Loading (g/m^2 of surface area/yr)	
Mean Depth (m)	N	P	N	P
5	1.0	0.07	2.0	0.13
10	1.5	0.10	3.0	0.20
50	4.0	0.25	8.0	0.50
100	6.0	0.40	12.0	0.80
150	7.5	0.50	15.0	1.00
200	9.0	0.60	18.0	1.20

in fact, they are interrelated. Dillon and Rigler[21] modified the Vollenweider approach to account for lakes of varying hydraulic detention times and phosphorus retention (fraction removed). This modification not only improved the empirical fit of the observed lake conditions and their trophic classification, but it also firmly grounded the classification criteria on mass balance principles. It is now possible to determine the mean total phosphorus concentrations that defined the classification scheme (Figure 5.7). Oligotrophic lakes are defined as those with less than 10 $\mu g\ L^{-1}$ total phosphorus on an annual average basis, and eutrophic lakes are those with greater than 20 $\mu g\ L^{-1}$ total phosphorus concentrations. Mesotrophic lakes are intermediate with total P concentrations between 10 and 20 $\mu g\ L^{-1}$. Nutrient supply is very important in determining the trophic status of lakes as Vollenweider suggested, but the nutrient supply also helps to define the total nutrient concentration in the lake that ultimately controls algal growth.

It is straightforward to derive the equation for Figure 5.7 from a mass balance similar to that shown in Section 5.4. If we express the areal nutrient (phosphorus) loading rate as L, the mass balance becomes

$$V\frac{d\mathrm{P}}{dt} = LA_{\mathrm{surf}} - Q\mathrm{P} - k_s \mathrm{P} V \tag{12}$$

Accumulation = Inputs − Outflows − Sedimentation

where L = total P loading rate, g m^{-2} yr^{-1}
A_{surf} = surface area of the lake, m^2

Dividing both sides of the equation through by the volume of the lake ($V = A_{\mathrm{surf}} H$) and substituting for k_s:

$$\frac{d\mathrm{P}}{dt} = \frac{L}{H} - \frac{\mathrm{P}}{\tau} - \frac{v_s}{H}\mathrm{P} \tag{13}$$

where v_s = the mean apparent settling velocity of total phosphorus.

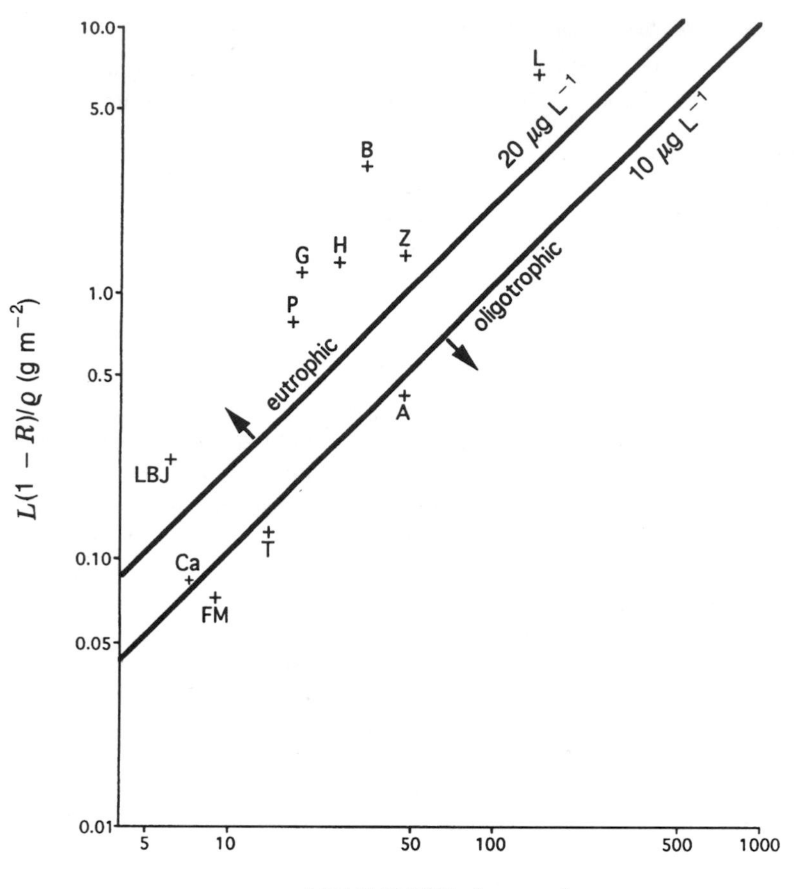

Figure 5.7 Nutrient loading criteria and classification. (Modified from Dillon and Rigler.[21,22])

At steady state ($d\text{P}/dt = 0$) the equation is simplified to

$$\text{P}\left(\frac{1}{\tau} + \frac{v_s}{H}\right) = \frac{L}{H} \tag{14}$$

$$0 = \frac{L}{H}(1 - R) - \frac{\text{P}}{\tau} \tag{15}$$

where R is the the fraction of phosphorus removed or retained due to sedimentation. The final equation describing Figure 5.7 is

$$\text{P} = \frac{L\tau\,(1 - R)}{H} = \frac{L(1 - R)}{H\rho} \tag{16}$$

where ρ is the hydraulic flushing rate of the lake ($1/\tau$). A log–log plot of nutrient loading remaining, $L(1 - R)/\rho$, versus mean depth (H) will have a slope of 1.0 and an intercept where $H = 1$ m of log P.

$$\log\left(\frac{L(1-R)}{\rho}\right) = 1.0 \log H + \log \mathrm{P} \tag{17}$$

By empirical observation of lake characteristics, the lines that divide oligotrophic lakes and eutrophic lakes are at 10 and 20 μg L^{-1} of total phosphorus on an annual average basis (Figure 5.7).[22]

Similar simple mass balance models for predicting total phosphorus concentrations in lakes have been developed by Lorenzen, Larsen, and others.[23–26] They rely on the concepts of total phosphorus loading on an areal basis and an apparent settling velocity at steady state.

$$0 = \frac{L}{H} - \frac{\mathrm{P}}{\tau} - \frac{v_s}{H}\mathrm{P} \tag{18}$$

Multiplying both sides of the equation by the mean depth H, we find

$$0 = L - \frac{Q\mathrm{P}}{A_{\text{surf}}} - v_s\mathrm{P} \tag{19}$$

$$\mathrm{P}\,(q_s + v_s) = L \tag{20}$$

$$\mathrm{P} = \frac{L}{q_s + v_s} \tag{21}$$

where $q_s = Q/A_{\text{surf}}$ (the surface overflow rate for the lake). All of these approaches are equivalent because they are founded on the mass balance principle. In terms of classifying the trophic status of lakes, they are nonetheless empirical because the observation that lakes less than 10 μg L^{-1} total P are oligotrophic and that lakes greater than 20 μg L^{-1} are eutrophic is empirical.

5.6 RELATIONSHIP TO STANDING CROP

The nuisance conditions of a eutrophic lake are not caused directly by nutrient concentrations, rather it is excessive algal blooms, decreased transparency, and decaying algae in the sediment that consumes oxygen, ruins aquatic habitats, and causes taste and odor problems. The previous mass balance models[21–26] that predict steady-state or annual average total phosphorus concentrations in lakes do not predict biomass or chlorophyll *a* (phytoplankton pigments as a measure of standing crop). Other investigators have shown that it is possible to correlate the summer concentration of total phosphorus to the chlorophyll *a* concentration, as shown in Figure 5.8.[27,28]

Phytoplankton primarily use ortho-phosphate (PO_4^{3-}) measured as molybdate-reactive phosphorus. Figure 5.8 is valid because total phosphorus is correlated with ortho-phosphorus, which is bioavailable to phytoplankton. The ratio of ortho-phosphate to total phosphorus may be relatively constant for groups of lakes, and the cyclic mineralization of organic particulate phosphorus yields ortho-phosphate to sustain algal growth.[29,30]

Other limnological measures can be correlated with chlorophyll *a* or trophic status of lakes. Total phosphorus is one of the better correlates, but Secchi disk depth, oxygen consumption rates of bottom sediments, and carbon dioxide fixation rates (primary productivity) have also been utilized. Greater than 50 mg C m^{-3} h^{-1} of primary production is generally considered to be eutrophic, measured by ^{14}C radiocarbon light and dark bottles, incubated for 4 hours on floating racks in situ.

Other indicators of eutrophic conditions are:

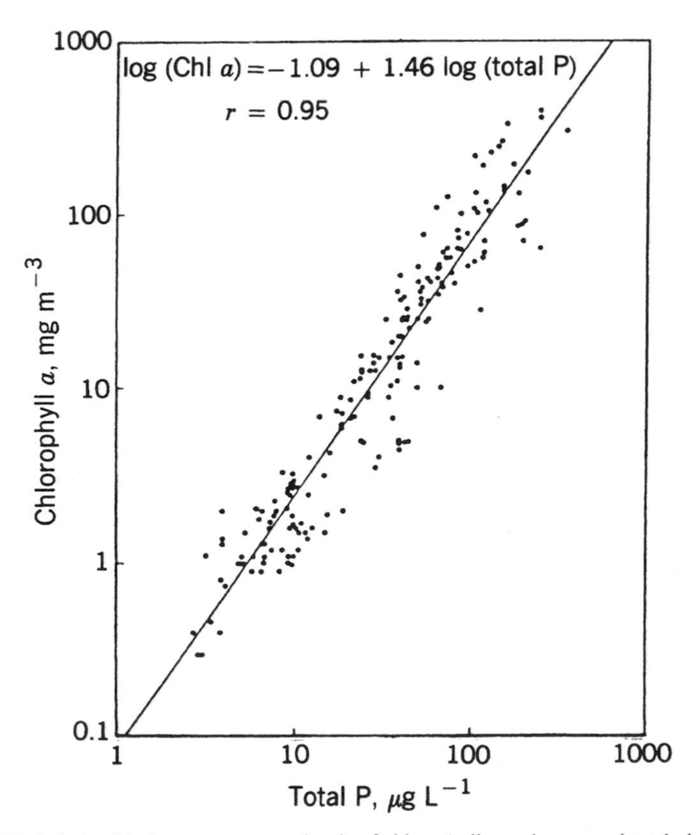

Figure 5.8 Relationship between summer levels of chlorophyll *a* and measured total phosphorus concentration for 143 lakes. (From Sakamoto.[27])

- Cyanobacteria (blue-green algae) blooms.
- Loss of benthic invertebrates such as mayfly larvae (*Hexagenia* spp.).
- Secchi disk depths less than 2.0 meters.
- Loss of fishery and presence of "rough" fish.
- Taste and odor problems.
- Aquatic weeds in littoral zones.
- Chlorophyll *a* concentrations greater than 10 $\mu g\ L^{-1}$.

Some or all of these indicators may be exhibited by any given eutrophic lake. Sediment oxygen depletion rates have been another good indicator of the trophic status of lakes (Table 5.2).

Lake Erie was often cited as "dead" in the 1960s and 1970s, but it was not really dead. It was overfertilized by anthropogenic wastes. It was eutrophic. In 1974, the average oxygen depletion rate in the Central Basin was -12 mmol $O_2\ m^{-2}\ d^{-1}$, which classified it as eutrophic (Table 5.2). A massive reduction of total phosphorus loadings was undertaken to decrease the inputs from 32,000 metric tonnes per year to 22,000 by requiring effluent limitations on phosphorus discharges from domestic wastewater treatment plants in the watershed and by rudimentary nonpoint source controls. Chlorophyll, total P, and anoxia in the Central Basin of Lake Erie have been much reduced by these actions, and the recycle of phosphate from anoxic sediments has also been reduced. The recovery of Lake Erie is an environmental success story, but further improvement depends on nonpoint source controls. Now, Lake Erie suffers most from toxic organic chemical problems (Chapter 7) and introduced species (zebra mussels).

5.7 LAND USE AND BIOAVAILABILITY

Nonpoint source runoff, particularly from intensive agriculture, urban stormwater runoff, and combined-sewer overflow are major inputs to streams and lakes that present a difficult challenge for water quality management and control. Runoff concentrations of total phosphorus and total nitrogen are shown in Tables 5.3 and 5.4 by land use and region of the United States. In many cases, these are the predominant

Table 5.2 Sediment Oxygen Depletion Rates as a Measure of the Trophic State of Lakes (mmol $O_2\ m^{-2}\ d^{-1}$)

Classification	Hutchinson Criteria	Mortimer Criteria
Oligotrophic lakes	< -5.3	< -7.7
Mesotrophic lakes	-5.3 to -10.3	-7.7 to -17.2
Eutrophic lakes	> -10.3	> -17.2

Table 5.3 Mean Total Phosphorus Concentration (mg L^{-1}) in Surface Waters by Land Use Type and Region of the United States[28]

	North and Northeast	Corn Belt + Dairy Region	East and Central Farming + Forest	Piedmont + Coastal Plain Farming + Forest
Forest	0.015	—	0.013	0.045
Mostly forest	0.028	0.055	0.031	0.044
Mixed	0.033	0.054	0.039	0.033
Mostly urban	0.070	0.077	0.088	0.022
Mostly agriculture	0.054	0.092	0.058	0.060
Agriculture	0.081	0.142	0.077	0.086

sources of nutrients to inland lakes and waterways, and they are very expensive to control.

Bioavailability of the nutrients becomes an issue when much of the runoff from nonpoint sources is in organic or particulate forms that algae cannot use directly.[30] It is only after hydrolysis and mineralization that some nutrient chemical species become bioavailable (PO_4^{3-}, NO_3^-, and NH_4^+). Also, the oxygen content of bottom waters has an effect on bioavailability. Anoxic or anaerobic conditions liberate dissolved, available nutrient species from particulate organic debris that has settled-out.

Several empirical models have been reviewed by Chapra[31] and Dillon[32] for estimating nonpoint source nutrient loadings to natural waters and the net phosphorus retention of lakes. It is too simplistic to assume that all of the phosphorus that settles to the bottom sediments of a lake remains there. Some is recycled, especially if anoxic conditions prevail.

When costly decisions are necessary for water quality management, it is desirable to use a more detailed and mechanistic approach to the eutrophication problem, one that includes nonlinear interactions between nutrients, plankton, and dissolved oxygen. Although nutrient-loading mass balance models have been extremely useful to decision-makers, they suffer from the following limitations and assumptions:

Table 5.4 Mean Total Nitrogen Concentration (mg L^{-1}) in Surface Waters by Land Use Type and Region of the United States[28]

	North and Northeast	Corn Belt + Dairy Region	East and Central Farming + Forest	Piedmont + Coastal Plain Farming + Forest
Forest	0.80	—	0.7	0.7
Mostly forest	0.95	1.3	0.8	0.9
Mixed	1.65	1.4	1.1	1.4
Mostly urban	1.5	1.6	1.9	1.9
Mostly agriculture	1.7	2.1	1.7	1.7
Agriculture	2.2	4.4	2.3	5.2

- Steady state, completely mixed assumptions.
- One limiting nutrient assumption.
- No recycle of nutrients from sediments.
- Lack of a dissolved oxygen mass balance that triggers sediment release of nutrients under anoxic conditions.

In the next section, we will see how various improvements in environmental models seek to address these shortcomings.

5.8 DYNAMIC ECOSYSTEM MODELS FOR EUTROPHICATION ASSESSMENTS

In many cases, a better methodology is needed rather than a steady-state, total phosphorus model for assessing the eutrophication of lakes. Chapra[33] was one of the first researchers to extend the concept of the Vollenweider[34] approach to a dynamic (time-variable) lake model for total phosphorus in the Great Lakes with sedimentation as the major loss mechanism (Figure 5.9). It was used to show the advantage of point source controls on phosphate at wastewater treatment plants within the Great Lakes Basin.

At the same time (late 1970s), a variety of dynamic ecosystem models were developed by DiToro, Thomann, and O'Connor that included nitrogen, silica, and phosphorus limitation, as well as zooplankton grazing and loss terms.[35–40]

A general schematic of a dynamic ecosystem model is shown in Figure 5.10.[37] Water quality includes nutrients, chlorophyll *a*, transparency, and dissolved oxygen. Kinetic formulations may vary somewhat for each major limiting nutrient species (phosphate, nitrate, ammonium, and silica). A typical flowchart for phosphorus in the Lake Erie model[36] is presented in Figure 5.11. Note that the nutrient is partitioned into an available compartment and an unavailable compartment to try to address the issue of bioavailability. Of course, this scheme requires independent measurement of each water quality constituent in each box in order to calibrate and verify the model. Available phosphorus is commonly taken to be ortho-phosphate (PO_4^{3-}) that passes a 0.45-μm membrane filter. Unavailable particulate phosphorus is everything else that is measured in a whole water sample for total phosphorus.

It is important to consider that more than one taxa of phytoplankton contribute to the aggregate measurement of chlorophyll *a* in lakes. Often, chlorophyll *a* is the easiest estimate of standing crop (or biomass) that a researcher can measure. Water quality standards or goals for lake restoration are sometimes written in terms of chlorophyll *a*, so it is a logical water quality parameter to simulate. In northern temperate lakes, a spring bloom of diatoms (cold-water phytoplankton with silica frustules) is often followed by a summer bloom of green algae or a late summer–early fall bloom of cyanobacteria (blue-green algae). Figure 5.12 illustrates a double phytoplankton bloom in Lake Ontario.[41]

Blue-green algae especially can cause water quality problems because they float

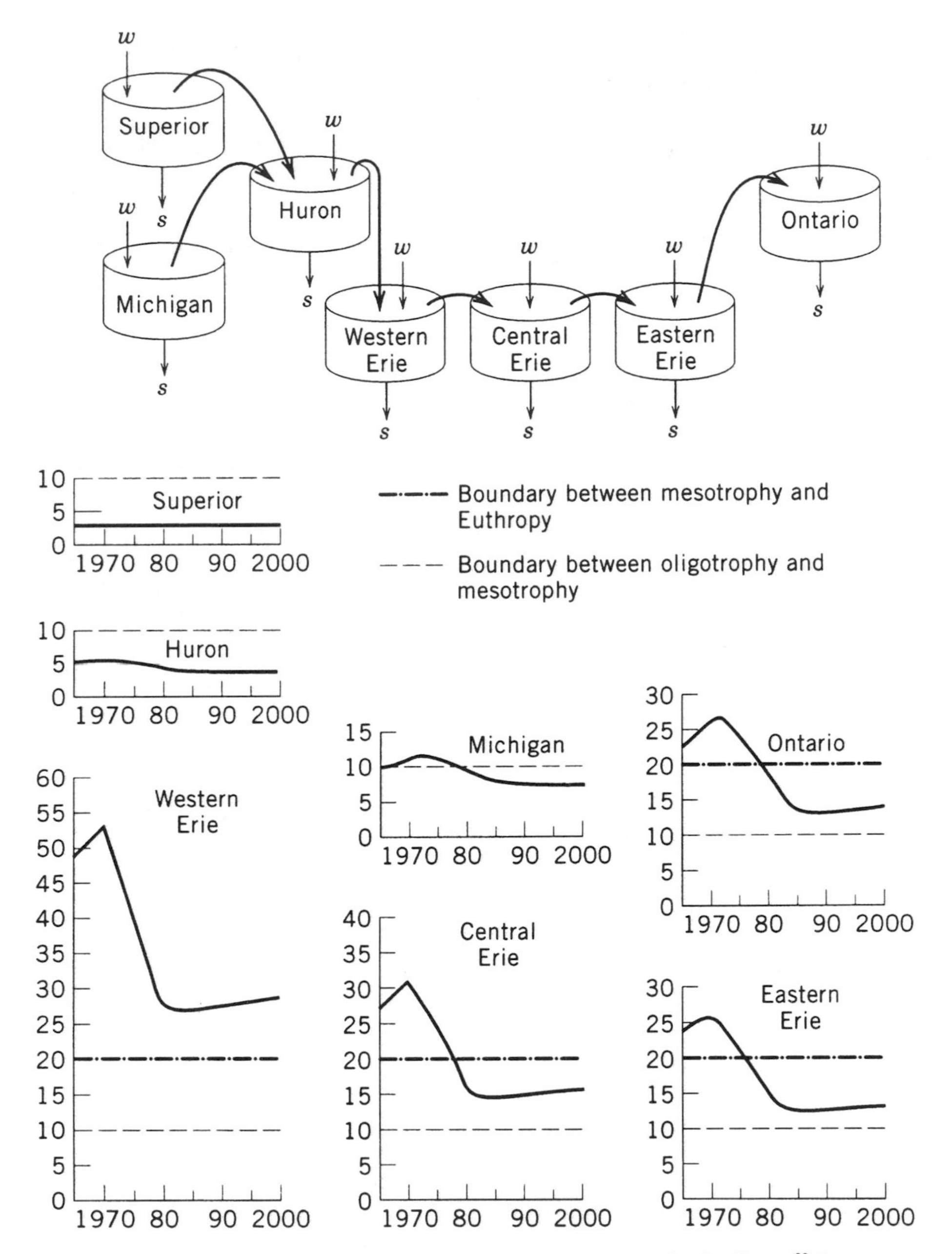

Figure 5.9 Dynamic model simulation of total phosphorus in the Great Lakes by Chapra.[33] Concentrations are for total phosphorus (in $\mu g\ L^{-1}$) in response to a linear decrease in point source loadings from 1970 values to 1 mg L^{-1} effluent-P in 1980. Reprinted with permission from *Journal Environ. Eng. Division.* Copyright (1977). American Society of Civil Engineers.

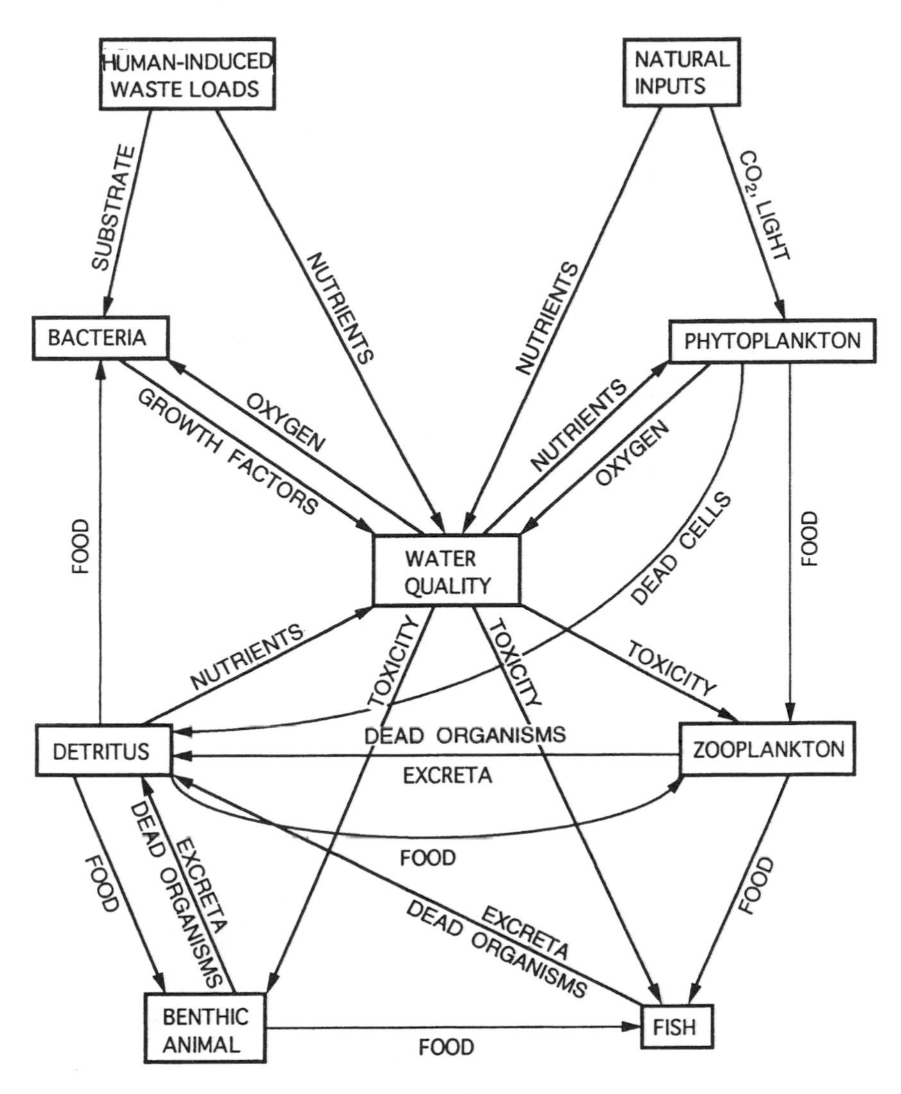

Figure 5.10 Schematic of an aquatic ecosystem.

and are known to be associated with taste and odor problems and toxins during their decay following the bloom. Blue-green algal blooms are sometimes so thick that the floating algal mat may blow onto the shoreline and create a stench due to decay and decomposition. The ability to float is caused by gas vacuoles in the cell that are formed during nitrogen-fixation of some blue-green algae with heterocysts. As a nuisance organism, blue-green algae are difficult to control because they do not need many nutrients and they have very low loss rates (they do not sink).[42] If they are nitrogen-fixing, they may require zero concentrations of nitrate and ammonium

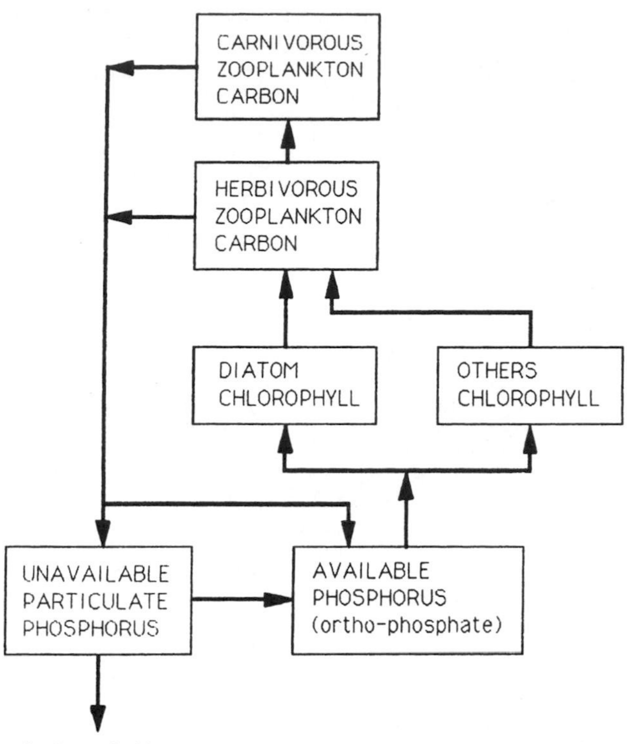

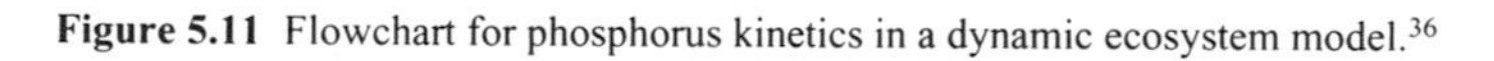

Figure 5.11 Flowchart for phosphorus kinetics in a dynamic ecosystem model.[36]

and still grow actively. Various phytoplankton taxa are given below in rough order of their seasonal succession.

Diatoms	Dissolved silica limiting nutrient (frustules) Cold water, P- or Si-limited
Green algae	Usually P-limited, summer
Dinoflagellates	Flagellated, motile, occasional toxins (red tides)
Blue-green algae	Low N requirements (N-fixing) Low sinking velocities (gas vacuoles) Warm water, late summer or early fall

Dinoflagellates can swim to the photic zone during the day and dive to the nutrient-rich thermocline during the night, demonstrating that luxury uptake of phosphorus or other nutrients may be possible. Growth rates sometimes depend on the nutrient content of the cell rather than the nutrient concentration of the surrounding water. "Phytoplankton," which translates from Greek roughly as "free-floating plants," is a misnomer. Sometimes phytoplankton can float or exhibit limited motility.

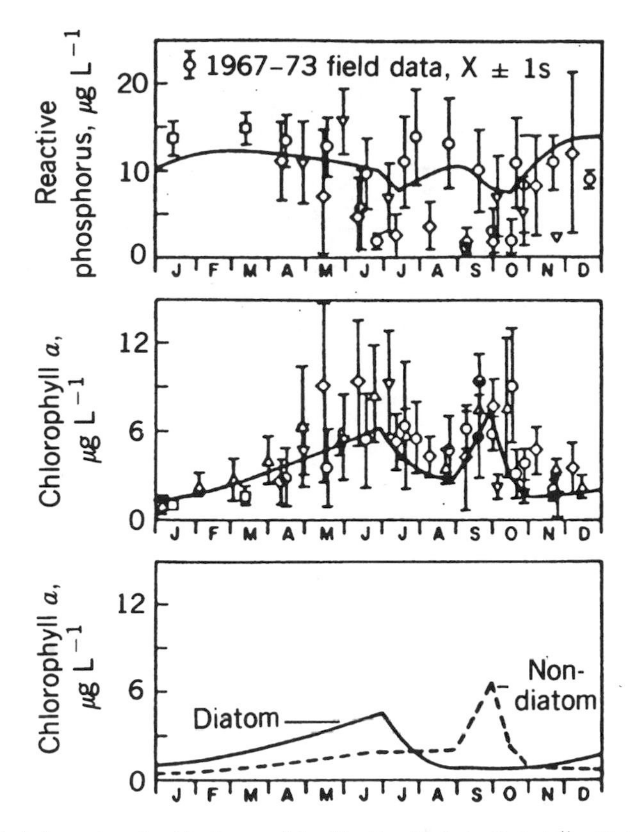

Figure 5.12 A dynamic eutrophication model calibration for Lake Ontario[41] with two phytoplankton species (diatoms and nondiatoms).

Phytoplankton must be modeled within the context of their transport regime. Within a lake, there are various ecological niches that are important to consider.

- Littoral—shallow water, near shore.
- Pelagic—open water, deep.
- Euphotic zone—light penetration to 1–10% of near-surface light.
- Epilimnion—above the thermocline, well-mixed.
- Hypolimnion—below the thermocline, well-mixed.
- Sediment–water interface—bioturbation, diffusion of pore water, variable redox conditions.

Figure 5.13 illustrates a transport regime for a large lake ecosystem. It is necessary to compartmentalize the lake (a box model) into a number of compartments to account for transport and unique ecological features of each zone. As a hypothetical example, Figure 5.13 shows a six-compartment model with three surface layers

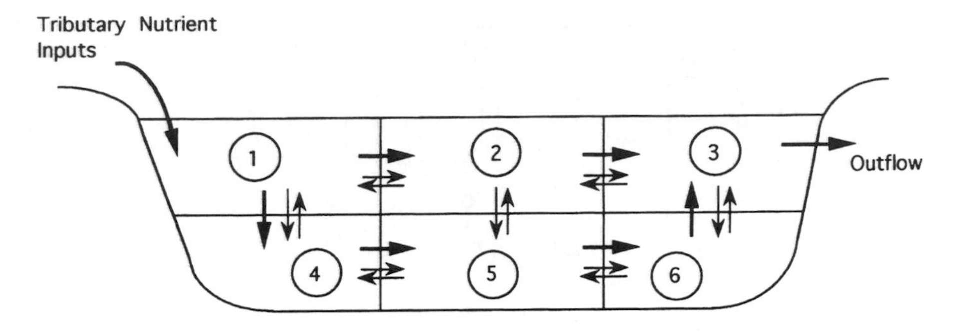

Figure 5.13 Transport regime for a compartmentalized lake ecosystem model. Large arrows represent current flow (advection) and double; small arrows represent mixing (dispersion) between compartments.

(epilimnetic compartments) and three deep layers (hypolimnetic compartments). The dark arrows represent current flow between compartments and the light double arrows represent exchange (turbulent eddy diffusion and dispersive mixing) between compartments.

The first step in developing a dynamic ecosystem water quality model is to determine the transport regime quantitatively. This is usually accomplished by simulating momentum, heat, or mass transport of a constituent that is well known and independent of water quality. The method of determining the correct transport regime is one of the following:

- Velocity field (momentum).
- Temperature distribution (heat).
- Conservative tracer (chloride or total dissolved solids).

A thermal heat budget is often used to set up the transport model, especially for 1-D vertical transport in thermally stratified lakes (see Chapter 2). A hydrodynamic model may be used to determine the momentum balance for a water body, which yields the velocity (current) vectors. Movement of water is required to quantify the advective flowrates depicted in Figure 5.13 as well as the dispersion coefficients. For a conservative tracer such as chloride in large 3-D lakes, we find for the jth compartment

$$\frac{d(V_jC_j)}{dt} = \sum_{i=1}^{n} -Q_{ij}C_j - \sum_{i=1}^{n} \frac{E_{ij}A_{ij}(C_j - C_i)}{\ell_{ij}} + \sum_{i=1}^{n} Q_{ji}C_i \qquad (22)$$

where V_j = volume of the jth compartment, L^3
C_j = concentration of a conservative tracer in the jth compartment, ML^{-3}
t = time, T
Q_{ij} = flowrate from the jth to the ith adjacent compartment, L^3T^{-1}

E_{ij} = bulk dispersion coefficient between the jth and the ith adjacent compartments, L^2T^{-1}

A_{ij} = interfacial area between the jth and the ith adjacent compartments, L^2

C_i = concentration of a conservative tracer in the ith compartment, ML^{-3}

ℓ_{ij} = half-distance (connecting distance between the middle of the two adjacent compartments), L

Note that the bulk dispersion coefficient is scale dependent as an aggregate formulation of all mixing processes. For example, it will not be equal to point measurements of dye dispersion in large lakes, and it cannot be determined explicitly. Bulk dispersion coefficients may be thought of as a bulk exchange flow going each way between compartments in Figure 5.13.

$$Q'_{ij} = \frac{E_{ij}A_{ij}}{\ell_{ij}} \tag{23}$$

where Q'_{ij} is the bulk exchange flow with units of L^3T^{-1}. The reader is referred to the section on compartmentalization in Chapter 2.

Procedures for model calibration and verification are:

1. Calibrate the model using field data for a conservative substance (e.g., chloride) or heat to determine Q_{ij} and E_{ij} values.
2. Using the same advective flows and bulk dispersion coefficients between compartments, calibrate the model for all water quality constituents for the same set of field data to determine all other adjustable parameters and coefficients (reaction rate constants, stoichiometric coefficients, etc.).
3. Verify the model with an independent set of field data, keeping all transport and reaction coefficients the same. Determine the "goodness of fit" between the model results and field data.

A suitable criterion for acceptance of the model calibration and verification should be determined *a priori,* depending on the use of the model in research or water quality management. A median relative error of 10–20%, model results/field data, is typical in water quality models.[43]

Ecosystem models consist of a series of mass balance equations within each compartment. The number of compartments will vary depending on the spatial extent of water quality data that is available and the resolution that is needed from the model. A schematic of a typical ecosystem model is shown in Figure 5.14. There are nine state variables shown, but there could be more or less depending on the design of the model. Each of the nine state variables will undergo advection and dispersion as shown in the previous equation; however, fish and zooplankton would not disperse at the same rate as other water quality constituents because they are motile. We assume four limiting nutrients and a multiplicative expression here. Biomass, rather than chlorophyll *a* or carbon, is simulated.

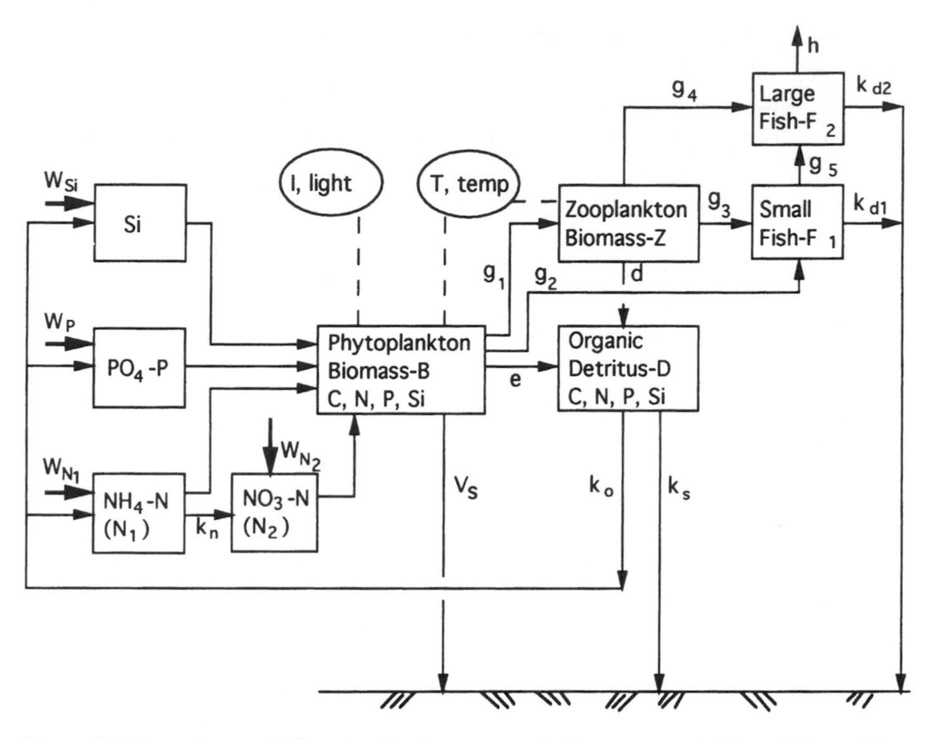

Figure 5.14 Ecosystem model for eutrophication assessments. Boxes represent state variables; solid arrows are mass fluxes; and dotted lines represent external forcing functions.

The model employs a multiplicative Michaelis–Menton expression for phytoplankton growth rate and linear kinetics (first-order decay) on hydrolysis and death rates of phytoplankton, zooplankton, and fish. Nonlinear second-order kinetic expressions are utilized for grazing rates of zooplankton and fish on their prey. The model assumes that phytoplankton fulfill their nitrogen requirement by taking up NH_4-N and NO_3-N in proportion to their fractional concentration. (This is not strictly the case—it is generally accepted that NH_4-N is preferred due to its higher energy content.)

Temperature affects every rate constant, increasing the rate of the reaction.

$$\frac{d[\mathrm{Si}]}{dt} = -\mu B a_{\mathrm{Si}} + \frac{W_{\mathrm{Si}}(t)}{V_i} + k_0(\theta_0^{T-20})\, D f_{\mathrm{Si}} \tag{24}$$

$$\frac{d[\mathrm{P}]}{dt} = -\mu B a_{\mathrm{P}} + \frac{W_{\mathrm{P}}(t)}{V_i} + k_0(\theta_0^{T-20})\, D f_{\mathrm{P}} \tag{25}$$

$$\frac{d[\mathrm{N_1}]}{dt} = -\mu B a_{\mathrm{N}} \left(\frac{\mathrm{N_1}}{\mathrm{N}_T}\right) + \frac{W_{\mathrm{N_1}}(t)}{V_i} + k_0(\theta_0^{T-20})\, D f_{\mathrm{N}} - k_n(\theta_n^{T-20})\, \mathrm{N_1} \tag{26}$$

Table 5.5 Rate Constants and Stoichiometric Coefficients for Dynamic Ecosystem Models[38–43]

Parameter	Symbol	Typical Value at 20 °C	Range at 20 °C
Mineralization			
Organic C	k_0	0.1 d^{-1}	0.001–0.2 d^{-1}
Organic N	$k_0 f_N$	0.03	0.001–0.2
Organic P	$k_0 f_P$	0.03	0.001–0.2
Organic Si	$k_0 f_{Si}$	0.1	0.001–0.2
Settling velocity			
Chlorophyll *a*	$V_S a_{Chl}$	0.1 m d^{-1}	0.005–0.8 m d^{-1}
Organic detritus	$k_S H$	0.1	0.01–0.3
Phytoplankton	V_S	0.2	0.00–2.0
Apparent sedimentation, total P	k_s	0.1 m d^{-1}	0.005–0.4 m d^{-1}
Phyto ratios			
N:Chlorophyll *a*		10	5–20
P:Chlorophyll *a*		1.0	0.5–2.0
C:Chlorophyll *a*		65	35–100
Si:Chlorophyll *a* (diatoms)		40	30–50
C:dry wt		0.4	0.33–0.43
N:dry wt	f_N	0.05	0.04–0.09
P:dry wt	f_P	0.01	0.006–0.03
Phytoplankton			
Death rates	e	0.1 d^{-1}	0.05–0.25 d^{-1}
Maximum growth rates	μ_{max}	1.5	1.0–2.0
Half-saturation constants			
PO_4-P	K_P	10 μg L^{-1}	6–25 μg L^{-1}
NH_4-N	K_N	10	1–20
NO_3-N	K_N	15	1–30
Si	K_{Si}	75	50–100
Zooplankton			
Grazing rates	g_1	0.4 L mg^{-1} d^{-1}	0.2–2.0 L mg^{-1} d^{-1}
Death + respiration	d	0.07 d^{-1}	0.02–0.3 d^{-1}

$$\frac{d[N_2]}{dt} = -\mu B a_N \left(\frac{N_2}{N_T}\right) + \frac{W_{N_2}(t)}{V_i} + k_n(\theta_n^{T-20})\, N_1 \tag{27}$$

$$\frac{d[B]}{dt} = +\mu B - eB\theta_e^{T-20} - g_1 BZ\theta_g^{T-20} - g_2 BF_1\theta_{f1}^{T-20} - V_s(T)B/H \tag{28}$$

$$\frac{d[D]}{dt} = -eB\theta_e^{T-20} - k_0\theta_0^{T-20} D - k_s(T)\, D + dZ\theta_d^{T-20} \tag{29}$$

$$\frac{d[Z]}{dt} = g_1 BZ\theta_z^{T-20} - dZ\theta_z^{T-20} - g_3 ZF_1\theta_{f1}^{T-20} - g_4 ZF_2\theta_{f2}^{T-20} \quad (30)$$

$$\frac{d[F_1]}{dt} = g_2 BF_1\theta_{f1}^{T-20} + g_3 ZF_1\theta_{f1}^{T-20} - k_{d1}F_1 - g_5 F_1F_2\theta_{f2}^{T-20} \quad (31)$$

$$\frac{d[F_2]}{dt} = g_4 ZF_2\theta_{f2}^{T-20} + g_5 F_1F_2\theta_{f2}^{T-20} - k_{d2}F_2 - hF_2 \quad (32)$$

where:

$$\mu = \mu_{\max}\left[\frac{I}{I_s}\exp\left(-\frac{I}{I_s}+1\right)\right]\theta_b^{T-20}\left(\frac{\text{Si}}{K_{\text{Si}}+\text{Si}}\right)\left(\frac{\text{P}}{K_{\text{P}}+\text{P}}\right)\left(\frac{\text{N}_T}{K_{\text{N}}+\text{N}_T}\right) \quad (33)$$

and

$$N_T = N_1 + N_2 \quad (34)$$

Equations (24)–(34) describe quantitatively the interactions of Figure 5.14. Table 5.5 gives a range and a typical value for the parameters, coefficients, and rate constants used in the model, and Table 5.6 gives the definitions of the variables in the model equations.

Eutrophication remains one of the world's greatest water quality problems. In developed countries the issue has evolved from controlling point sources to one of watershed control and nonpoint source runoff. This is difficult and expensive, but further improvements in water quality will require nonpoint source controls. When state-of-the-art modeling is needed, a comprehensive ecological model of the type presented in this section should be considered. Sediment dynamics, oxygen balances at the sediment–water interface, and diffusive flux modeling will be crucial to obtaining long-term reliable estimates of eutrophication in lakes and estuaries.

Table 5.6 Definition of Variables in Ecological Model

State Variables

- Si = silica concentration, ML^{-3}
- P = PO_4-P concentration, ML^{-3}
- N_1 = NH_4-N concentration, ML^{-3}
- N_2 = NO_3-N concentration, ML^{-3}
- B = biomass of phytoplankton, ML^{-3} (dry weight)
- D = organic detritus concentration, ML^{-3} (dry weight)
- Z = zooplankton biomass, ML^{-3} (dry weight)
- F_1 = small fish biomass, ML^{-3} (dry weight)
- F_2 = large fish biomass, ML^{-3} (dry weight)

(continued)

Table 5.6 *(Continued)*

Independent Variable

t = time, T

Inputs (Forcing Functions)

I = average light intensity in control volume, foot candles
I_s = saturated-growth rate light intensity of phytoplankton, 2000 foot candles
T = temperature, °C
$W_{Si}(t)$ = input rate of Si to the control volume, MT^{-1}
$W_P(t)$ = input rate of P to the control volume, MT^{-1}
$W_{N_1}(t)$ = input rate of NH_3-N to the control volume, MT^{-1}
$W_{N_2}(t)$ = input rate of NO_3-N to the control volume, MT^{-1}

Physical Constants and Parameters

μ = growth rate of phytoplankton, T^{-1}
μ_{max} = maximum growth rate of phytoplankton, T^{-1}
a_{Si} = ratio of Si:biomass (mass fraction)
a_P = ratio of P:biomass
a_N = ratio of N:biomass
V_i = volume of the ith control volume, L^3
k_0 = hydrolysis or mineralization rate constant, T^{-1}
T = temperature, °C
θ_j = temperature factor for each rate constant, $j = 1 \rightarrow n$ processes
f_{Si} = mass fraction of Si in detritus
f_P = mass fraction of P in detritus
f_N = mass fraction of N in detritus
k_N = nitrification rate constant, T^{-1}
e = endogenous decay and death rate constant for phytoplankton, T^{-1}
V_S = settling velocity of phytoplankton as a function of viscosity and water temperature, LT^{-1}
H = mean depth of compartment, L
g_1 = grazing rate of zooplankton on phytoplankton, $L^3M^{-1}T^{-1}$
g_2 = grazing rate of small fish on phytoplankton, $L^3M^{-1}T^{-1}$
g_3 = grazing rate of small fish on zooplankton, $L^3M^{-1}T^{-1}$
g_4 = grazing rate of large fish on zooplankton, $L^3M^{-1}T^{-1}$
g_5 = feeding rate of large fish on small fish, $L^3M^{-1}T^{-1}$
k_s = apparent sedimentation rate constant of organic detritus, T^{-1}
d = death and excretion rate of zooplankton, T^{-1}
k_{d1} = death and excretion rate of small fish, T^{-1}
k_{d2} = death and excretion rate of large fish, T^{-1}
h = harvesting rate constant, T^{-1}
K_{Si} = half-saturation constant for Si (diatoms), ML^{-3}
K_P = half-saturation constant for P, ML^{-3}
K_N = half-saturation constant for N, ML^{-3}
N_T = total inorganic nitrogen concentration, ML^{-3}

We have seen how powerful simple mass balance models of phosphorus (as the limiting nutrient) can be in assessing the eutrophication status of lakes. In some cases, the costs of eutrophication are sufficient to cause governments to reduce point sources and even some nonpoint source discharges of nutrients to waterways. Dynamic ecosystem modeling is a valuable tool to assess the efficacy of nutrient control measures and their benefits.

5.9 REFERENCES

1. Sawyer, C.N. *J. N. Engl. Water Works Assoc.*, 61, 109 (1947).
2. Odum, E.P., *Fundamentals of Ecology,* Saunders, Philadelphia (1961).
3. Stumm, W., and Morgan, J.J., *Aquatic Chemistry,* Wiley-Interscience, New York (1981).
4. Redfield, A.C., James Johnstone Memorial Volume, Liverpool (1934).
5. Redfield, A.C., Ketchum, B.H., and Richards, F.A. in *The Sea,* Vol. 2, M.N. Hill, Ed., Wiley-Interscience, New York (1963).
6. Sigg, L., and Stumm, W., *Aquatische Chemie,* VDF Publishers, Zürich, Switzerland (1989).
7. Stumm, W., in *Chemistry of the Solid–Water Interface* (Chapter 11), Wiley-Interscience, New York (1992).
8. Martin, J.H., and Fitzwater, S.E., *Nature,* 331, 341 (1988).
9. Martin, J.H., and Gordon, R.M., *Deep Sea Res.*, 35, 177 (1988).
10. Martin, J.H., Gordon, R.M., Fitzwater, S.E., and Broenkow, W.W., *Deep Sea Res.*, 36, 649 (1989).
11. Parsons, T.R., and Takahashi, M., *Biological Oceanographic Processes,* Pergamon Press, Oxford (1973).
12. Monod, J., The Growth of Bacterial Cultures, *Annu. Rev. Microbiol.*, III (1949).
13. DiToro, D.M., and Matystik, W.F., Mathematical Models of Water Quality in Large Lakes, EPA-600/3-80-056, U.S. Environmental Protection Agency, Washington, DC (1980).
14. Thomann, R.V., and Mueller, J.A., *Principles of Surface Water Quality Modeling and Control,* Harper & Row, New York (1987).
15. Gordon, L.I., Park, P.K., Hager, S.W., and Parsons, T.R., *J. Oceanogr. Soc. Jpn.*, 37, 81 (1971).
16. U.S. Environmental Protection Agency. A Compendium of Lake and Reservoir Data Collected by the National Eutrophication Survey in the Northeast and North-Central United States, Working Paper No. 474, Corvallis, OR (1975).
17. Vollenweider, R.A., *Arch. Hydrobiol.*, 66, 1 (1969).
18. Sawyer, C.N., *Sewage Indust. Wastes,* 26, 317 (1954).
19. Sawyer, C.N., *Sewage Indust. Wastes,* 24, 768 (1952).
20. Vollenweider, R.A., in *Scientific Fundamentals of the Eutrophication of Lakes and Flowing Waters with Particular Reference to Nitrogen and Phosphorus as Factors in Eutrophication,* Organization for Economic Cooperation and Development, Paris (1970).

21. Dillon, P.J., and Rigler, F.H., *J. Fish. Res. Bd. Can.,* 31, 1771 (1974).
22. Dillon, P.J., and Rigler, F.H. *Limnol. Oceanogr.,* 19, 767 (1974).
23. Lorenzen, M.W., in *Modeling the Eutrophication Process,* E.J. Middlebrooks, Ed., Ann Arbor Science, Ann Arbor, MI, pp. 205–210 (1974).
24. Lorenzen, M.W., Smith, D.J., and Kimmel, L.V., in *Modeling Biochemical Processes in Aquatic Ecosystems,* R.P. Canale, Ed., Ann Arbor Science, Ann Arbor, MI, pp. 75–91 (1976).
25. Larsen, D.P., Maleug, K.W., Schults, D.W., and Brice, R.M., *Verh. Int. Ver. Limnol.,* 19, 884 (1975).
26. Larsen, D.P., Van Sickle, J., Maleug, K.W., and Smith, P.D., *Water Res.,* 13, 1259 (1979).
27. Sakamoto, C., *J. Water Pollu. Control Fed.,* 48, 2177 (1979).
28. U.S. EPA, The Relationships of Phosphorus and Nitrogen to the Trophic State of Northeast and North-Central Lakes and Reservoirs, Working Paper No. 23, Corvallis, OR (1974).
29. Lung, W.S., Canale, R.P., and Freedman, P.L., *Water Res.,* 10, 1101 (1976).
30. Verhoff, F.H., and Heffner, M.R., *Environ. Sci. Technol.,* 13, 844 (1979).
31. Chapra, S.C., *Water Resour. Res.,* 11, 1033 (1975).
32. Dillon, P.J., and Kirchner, W.B., *Water Resour. Res.,* 11, 1035 (1975).
33. Chapra, S.C., *J. Environ. Eng. Div., Am. Soc. Civ. Eng.,* 103, 147 (1977).
34. Vollenweider, R.A., *Schweiz. Z. Hydrol.,* 37, 53 (1975).
35. Thomann, R.V., DiToro, D.M., Winfield, R.P., and O'Connor, D.J., Mathematical Modeling of Phytoplankton in Lake Ontario—Model Development and Verification, EPA-660/3-75-005, Washington, DC (1975).
36. DiToro, D.M., and Connolly, J.P., Mathematical Models of Water Quality in Large Lakes, Part 2: Lake Erie. EPA, Washington, DC (1979).
37. Chen, C.W., and Orlob, G.T., Ecologic Simulation for Aquatic Ecosystems, Office of Water Resources Research OWRRC-2044, U.S. Dept. of Interior, Washington, DC (1972).
38. DiToro, D.M., O'Connor, D.J., and Thomann, R.V., in *Nonequilibrium Systems in Natural Water Chemistry, Advances in Chemistry Series,* Vol 106, American Chemical Society, Washington, DC, pp. 132–180 (1971).
39. Thomann, R.V., DiToro, D.M., and O'Connor, D.J., *J. Environ. Eng. Div., ASCE,* 100 (SA3), 699 (1974).
40. DiToro, D.M., O'Connor, D.J., Thomann, R.V., and Mancini, J.L., in *Systems Analysis and Simulation in Ecology,* Vol. III, Academic Press, New York, pp. 423–474 (1975).
41. Schnoor, J.L., and O'Connor, D.J., Water Res., 14, 1651 (1980).
42. Schnoor, J.L., and DiToro, D.J., *Ecol. Modeling,* 9, 233 (1980).
43. Thomann, R.V., *J. Environ. Eng. Div., ASCE,* 108, 923 (1982).

5.10 PROBLEMS

1. The following water quality parameters were measured for the nutrient concentrations in Green Lake near Seattle, Washington, in the spring time.

Ortho-phosphate $[PO_4^{3-}] = 10\ \mu g\ L^{-1}$
Total phosphorus (dissolved) $= 15\ \mu g\ L^{-1}$
$[NO_3^- \text{-N}] = 15\ \mu g\ L^{-1}$
$[NH_4^+ \text{-N}] \leq 10\ \mu g\ L^{-1}$

a. What is likely to be the limiting nutrient for algal growth in Green Lake considering the 14:1 stoichiometric guideline for inland lakes?
b. How much algae in mg L^{-1} biomass would be produced based on the stoichiometry of equation (2)?

2. Estimate the resulting growth rate for phytoplankton in Lake Erie from the following data. The maximum growth rate under ideal conditions of light, temperature, and nutrients is 1.3/day. Use three approaches: (a) phosphorus as the limiting nutrient, (b) the multiplicative Michaelis–Menton expression, and (c) the electrical resistance analogue.

	$[NH_4^+ + NO_3^-]$ as N	$[PO_4^{3-} - P]$
Concentration, $\mu g\ L^{-1}$	50	5
K_s, $\mu g\ L^{-1}$	25	5

Based on (1) growth rate and (2) stoichiometry, which nutrient is likely to be most limiting to phytoplankton growth?

3. The light extinction coefficient in Lake Erie in 1974 was 1.2 m^{-1}. In 1995, it was approximately 0.2 m^{-1}.

a. Plot the light intensity in depth for Lake Erie in 1974 and 1995. Surface intensity = 400 cal $cm^{-2}\ d^{-1}$ (langleys d^{-1}).
b. What do you think are the causes of improvement in light penetration in Lake Erie? (Note: In 1995 there was also a proliferation of zebra mussels, an introduced species, in Lake Erie.)
c. If the saturated light intensity that is optimal for growth is 350 cal $cm^{-2}\ d^{-1}$, what depth was optimal for phytoplankton growth in 1974 versus 1995?

$$I(z) = I_0 \exp(-k_e z)$$

where $I(z)$ = light intensity with depth
I_0 = surface light intensity
k_e = extinction coefficient, L^{-1}
z = depth, L

4. Plot the light intensities versus depth for the following water bodies.

Area	Light Extinction Coefficient k_e, m^{-1}
Lake Erie—1974	1.2
Lake Ontario—1974	0.22
Long Island Sound	0.5
Ocean (open)	0.05
James River Estuary	5.0
Lake Tahoe	0.08
Lake Lyndon B. Johnson, Texas	0.30

If the Secchi disk depth measures the point of ~10% light penetration, list the Secchi disk depth transparencies (in meters) for each of the water bodies. Give a formula for the light extinction coefficient as a function of Secchi disk depth, d. [*Note:* $(k_e = f(d)$.]

5. Plot the relationship of μ/μ_{max} versus light intensity (I) due to light limitation if the following relationship applies:

$$\frac{\mu}{\mu_{max}} = \frac{I}{I_s} \exp\left(1 - \frac{I}{I_s}\right)$$

where μ = growth rate, d^{-1}
μ_{max} = maximum growth rate, d^{-1}
I = light intensity, langleys d^{-1}
I_s = saturated light intensity, langleys d^{-1}

6. a. Given a lake with the following characteristics, classify it according to its trophic status (eutrophic, mesotrophic, or oligotrophic).

$H = 10$ m	Mean depth
$L = 1.0$ g P m^{-2} yr^{-1}	P-loading rate
$\tau = 1.0$ yr	Retention time
$R = 0.4$	Fraction of P sedimented

What is the total phosphorus concentration in the lake?

b. What would be the trophic classification of a nearby lake with L, τ, and R identical except that it has a mean depth of 50 m?

c. What would be the trophic classification of the same lake as in part a, except that it has a mean hydraulic retention time of 0.1 yr?

7. Classify Lake Tahoe according to eutrophication criteria. What is the mean total phosphorus concentration?

$H = 120$ m
$L = 0.01$ g P m^{-2} yr^{-1}

$\tau = 170$ yr

$R = 0.8$

What is the approximate sediment oxygen demand (Table 5.2)? What is the approximate chlorophyll *a* concentration (Table 5.5 or Figure 5.8)?

8. For a step function decrease in total phosphorus loading, how long would it take to reach 95% of steady state (improved concentration) for Lake Tahoe (Problem 7) versus the lake in problem 6a? [*Hint:* Solve for the time to 95% of steady state using equation (75) in Chapter 2.]

9. Model the step function response to a 50% decrease in total phosphorus loading to Lake Erie assuming a completely mixed lake using the mass balance model, equation (13).

$H = 18.6$ m

$L = 1.27 \text{ g P m}^{-2} \text{ yr}^{-1} \rightarrow 0.635$

$\tau = 2.6$ yr

$k_s = 0.005 \text{ d}^{-1}$

$P_o = 25 \text{ } \mu\text{g L}^{-1} \text{ P}$

$V = 470 \text{ km}^3$

10. The following simple model was used by Schnoor and O'Connor[41] to model phytoplankton concentrations in the epilimnion of Lake Ontario. List the assumptions in this simplified approach. For the input parameters given below, solve for the steady-state concentrations of inorganic phosphorus, phytoplankton-phosphorus, and organic phosphorus in Lake Ontario. The model assumes first order kinetics for all reactions except phytoplankton uptake, which is modeled as second order.

$$\frac{dN_i}{dt} = \frac{W_1}{V} + k_0 N_0 - k_g N_i P - \frac{N_i}{\tau}$$

$$\frac{dP}{dt} = k_g N_i P - k_1 P - k_s P - \frac{P}{\tau}$$

$$\frac{dN_0}{dt} = k_1 P + \frac{W_2}{V} - k_s N_0 - k_0 N_0 - \frac{N_0}{\tau}$$

where $\tau = 534$ d

$k_1 = 0.25 \text{ d}^{-1}$

$k_g = 0.0375 \text{ L } \mu\text{g}^{-1} \text{ d}^{-1}$

$k_s = 0.024 \text{ d}^{-1}$

$k_0 = 0.14 \text{ d}^{-1}$

V = epilimnion lake volume
N_0 = organic P conc., μg/L
N_i = inorganic P conc., μg/L
P = phytoplankton P conc., μg/L
t = time, days
W_1 = inorganic waste load
W_2 = organic waste load
τ = deletion time, days

$$\frac{W_1}{V} = \frac{14.3\ \mu\text{g L}^{-1}}{534\ \text{d}} = 0.0268\ \mu\text{g L}^{-1}\ \text{d}^{-1}$$

$$\frac{W_2}{V} = \frac{45.7\ \mu\text{g L}^{-1}}{534\ \text{d}} = 0.0856\ \mu\text{g L}^{-1}\ \text{d}^{-1}$$

11. Write a dynamic simulation program that solves the set of ordinary differential equations (24) through (33) simultaneously. Runge Kutta or predictor-corrector numerical methods are suggested.

ANSWERS TO SELECTED PROBLEMS—CHAPTER 5

1. Nitrogen is limiting; 0.32 mg L^{-1} biomass produced.

4. Secchi Depth = $2.3/k_e$

6. a) 0.060 mg L^{-1} total phosphorus
b) 0.012
c) 0.006

8. Time to 95% of steady state = $3.0/(k_s + 1/\tau)$
Lake Tahoe = 100 years
Lake in Problem 6a = 1.8 years

6

CONVENTIONAL POLLUTANTS IN RIVERS

If we could first know where we are and whither we are tending, we could better judge what to do, and how to do it.

—Abraham Lincoln

6.1 INTRODUCTION

The use of aerobic biological treatment for domestic wastewater dates back to the late 19th century, and it has been a standard method in the United States since the 1930s.[1–3] Wastewater treatment has been directed primarily toward removal of "conventional pollutants" since that time. Conventional pollutants are biochemical oxygen demand (BOD), ammonia-nitrogen, total suspended solids (TSS), and total and fecal coliform bacteria. BOD is an important pollutant because it is a measure of biodegradable organics in water and the oxygen that will be consumed in the process of microbial degradation.

$$\underset{\substack{\text{"organics"} \\ \text{BOD}}}{CH_2O} + \underset{\substack{\text{dissolved} \\ \text{oxygen}}}{O_2} \xrightarrow{\text{bacteria}} \underset{\substack{\text{carbon} \\ \text{dioxide}}}{CO_2} + \underset{\text{water}}{H_2O} \qquad \text{(i)}$$

Dissolved oxygen is probably the single most important chemical parameter that is required to ensure the ecological health of a receiving water. Fish kills, noxious taste and odors, and low biological diversity are often indications of low dissolved oxygen in a receiving water. Due to the importance of oxygen in natural waters, states have set water quality standards that require dissolved oxygen (D.O.) concentration of equal to or greater than 5.0 mg L^{-1} most of the time, and 4.0 mg L^{-1} at all times.

It is important to control ammonia because it is toxic, even at low concentrations, to fish and other aquatic biota (Table 1.1), and it is an additional oxygen-demanding chemical in aquatic systems via nitrification.

$$NH_3 + 2\,O_2 \xrightarrow{\text{bacteria}} NO_3^- + H_2O + H^+ \qquad \text{(ii)}$$

Ammonia is in chemical equilibrium with ammonium ions in natural waters. Ammonium is a weak acid with a pK_a of about 9.2 at 25 °C, so over the pH range of natural waters, 6.0–9.0, ammonium ion predominates [equation (iii) goes rather far to the right]:

$$NH_3 + H^+ \leftrightarrow NH_4^+ \qquad \text{(iii)}$$

Water quality standards have been promulgated for total ammonia-nitrogen (NH_3 – N + NH_4^+ – N). A total allowable ammonia-nitrogen of 2.0 mg L^{-1} is often set as the water quality standard. Alternatively, standards may be set based on the toxic species, un-ionized ammonia (usually 0.02 mg N L^{-1} allowable). Un-ionized ammonia increases with pH and temperature (Table 6.1).

Total suspended solids are the third conventional pollutant because they exacerbate a dissolved oxygen problem by sedimentation and forming an oxygen-demanding sludge deposit; they cause turbidity in the receiving water and may alter the habitat of aquatic biota; and, perhaps most importantly, they can harbor pathogens (disease-causing microorganisms). TSS are controlled in aerobic wastewater treatment plants by sedimentation and thickening processes, and they are the ultimate by-product of what amounts to a bioconversion process of transforming soluble organic material into settleable (and some not so settleable) solids. Coliform-bacteria (both fecal and total) are used as indicators of pathogens in rivers and streams, and they are sometimes included in consideration of conventional pollutants. They can be efficiently killed by chlorination of the effluent at wastewater treatment plants, but at the risk of formation of toxic chlorinated organic compounds. The natural die-away of coliform bacteria can be modeled as a first-order decay reaction. It is much faster in marine waters than fresh water.

Table 6.1 Percent Un-ionized Ammonia [$NH_{3(aq)}$/($NH_{3(aq)}$ + $NH_{4(aq)}^+$)] as a Function of Water Temperature and pH

	Un-ionized Ammonia Values, %						
Temperature, °C	pH 6	6.5	7	7.5	8	8.5	9
0	0.010	0.027	0.083	0.27	0.82	2.6	7.7
5	0.013	0.040	0.12	0.40	1.2	3.6	10.5
10	0.019	0.058	0.19	0.58	1.8	5.4	15.
15	0.028	0.086	0.25	0.82	2.5	7.5	20.
20	0.035	0.12	0.39	1.1	3.3	10.0	26.
25	0.052	0.18	0.50	1.6	5.0	13.	32.
30	0.080	0.23	0.72	2.1	7.0	18.	40.
35	0.10	0.31	0.92	2.8	9.0	21.	47.

Source: Modified from Mills et al.[3]

Sawyer et al.[4] have demonstrated that the energetics of biological wastewater treatment can be estimated by dividing the overall stoichiometric equation into three parts: (1) a portion of the energy that goes into cell synthesis, (2) a portion of the energy that goes into respiration, and (3) the source of energy from oxidation of soluble organics.

$$\text{Cell synthesis:} \quad \tfrac{1}{5}CO_2 + \tfrac{1}{20}HCO_3^- + \tfrac{1}{20}NH_4^+ + H^+ + e^- = \tfrac{1}{20}C_5H_7NO_2 + \tfrac{9}{20}H_2O \quad \text{(iv)}$$

$$\text{Respiration:} \quad \tfrac{1}{4}O_2 + H^+ + e^- = \tfrac{1}{2}H_2O \quad \text{(v)}$$

$$\text{Organics oxidation:} \quad \tfrac{1}{50}C_{10}H_{19}O_3N + \tfrac{9}{25}H_2O = \tfrac{9}{50}CO_2 + \tfrac{1}{50}NH_4^+ + \tfrac{1}{50}HCO_3^- + H^+ + e^- \quad \text{(vi)}$$

If one assumes that approximately 28% of the energy that is made available from soluble organics is channeled into cell synthesis and about 72% is for respiration, then the overall stoichiometric equation can be written, as below, by multiplying equation (iv) by 0.28, multiplying equation (v) by 0.72, and summing the equations.

$$0.056\ CO_2 + 0.014\ HCO_3^- + 0.014\ NH_4^+ + 0.28\ H^+ + 0.28\ e^- = 0.014\ C_5H_7NO_2 + 0.126\ H_2O \quad \text{(vii)}$$

$$0.180\ O_2 + 0.72\ H^+ + 0.72\ e^- = 0.36\ H_2O \quad \text{(viii)}$$

$$0.36\ H_2O + 0.02\ C_{10}H_{19}O_3N = 0.180\ CO_2 + 0.02\ NH_4^+ + 0.02\ HCO_3^- + H^+ + e^- \quad \text{(ix)}$$

$$0.02\ C_{10}H_{19}O_3N + 0.18\ O_2 = 0.124\ CO_2 + 0.006\ HCO_3^- + 0.006\ NH_4^+ + 0.014\ C_5H_7NO_2 + 0.126\ H_2O \quad \text{(x)}$$

Equation (x) can be restated in terms of its reactants and products, without stoichiometry, as

$$\text{Soluble BOD} + \text{D.O. } (O_2) = CO_2 + H_2O + \text{Alkalinity} + \text{Ammonium} + \text{Cells} \quad \text{(xi)}$$

Equation (xi) is the net effect of aerobic wastewater treatment. Key to the process is to make the reaction go as far as possible to the right and to be able to separate the cells from the effluent by sedimentation. If the reaction goes to the right, one is able to minimize BOD in the effluent and to settle out the cells (TSS). The ammonium can be oxidized in the wastewater treatment plant, but unless the process is designed for efficient nitrification, a portion of the ammonium will be discharged, as will some BOD and TSS. Usually rivers are the receiving waters.

Effluent permits on municipal wastewater discharges of BOD and TSS are 30 mg L^{-1} in effluent-limited water quality segments. Discharges of ammonia and ammonium are not regulated except in cases of ammonia toxicity (> 2 mg N L^{-1}) or in water-quality-limited segments, where ammonia-nitrogen or D.O. concentrations may be expected to violate water quality standards in the receiving stream. However, states with critical water quality problems due to development have begun to enact increasingly stringent effluent guidelines for municipal wastewater discharges and even individual septic tank discharges. In Florida in 1994, the state passed an effluent requirement of 5 mg L^{-1} BOD, 5 mg L^{-1} TSS, 3 mg L^{-1} total nitrogen, and 1 mg L^{-1} phosphate-P on new septic tank discharges in the Keys. This is especially stringent, and it demonstrates the continued attention that conventional pollutants must receive.

Situations in which the ammonia-nitrogen or dissolved oxygen water quality standards are expected to be violated require a waste load allocation, that is, a mathematical model of the stream segment with important wastewater discharges included as inputs. By decreasing the mass of BOD and NH_4-N as input from each plant and running a new simulation, the model can be used to estimate allowable loadings for each discharge that will keep the stream in compliance with water quality standards.

Waste load allocations are performed as a special designation for stream segments that are not in compliance with water quality standards. Mathematical models are necessary to determine allowable loadings for each plant as a part of their permit.

Water quality modeling of rivers goes back to 1925 and the Streeter–Phelps equation for dissolved oxygen concentrations in the Ohio River. There have been many developments since that time that make it possible to model other settings in greater detail and with more realistic processes. New research questions include the following:

- Coagulation of particles in natural waters. We know that assumptions of discrete particle settling are not very accurate. Coagulation affects the rate of chemical sorption and TSS sedimentation to bottom sediments.[5]
- In-place pollutants and sediment oxygen demand (SOD). Many pollutants at the sediment–water interface were deposited historically but continue to affect water quality. How to model the sediment oxygen demand, as it develops through time, is a particularly vexing problem considering complex, coupled redox reactions and transport limitations.
- Sorption of toxic organics and metals by suspended solids and macromolecules. It is important to know the fate of suspended solids in natural waters that sorb organic pollutants and create oxygen demand.
- Organics (BOD) degradation. Fate of individual trace organic compounds in rivers and natural waters is needed. One can define the bulk of organic material (BOD) as a primary substrate and trace organics (Chapter 7) as a secondary substrate. We know that photochemistry, strong oxidants (H_2O_2), and transients (singlet oxygen, superoxide, hydrated electrons, hydroxyl radicals, peroxyl radicals) can play a role in the degradation of organics and BOD.

To model the aquatic chemistry of BOD, TSS, NH_3-N, and dissolved oxygen in natural waters requires mathematical formulations for mass transport in a stream or a river and reaction kinetics. We begin with the basic mass balance equation.

6.2 MASS BALANCE EQUATION: PLUG-FLOW SYSTEM

Conventional pollutants are discharged to rivers or streams because the current removes them from the source swiftly, and eutrophication is not as great a problem in flowing waters as compared to lakes. We begin by writing a mass balance equation around a control volume, an incremental element (slice) of stream volume (Figure 6.1). If the current is sufficiently fast, we may model the stream as a plug-flow system. We may check this assumption via the Peclet and Reaction numbers presented in Chapter 2.

$$\text{Accumulation} = \text{Inputs} - \text{Outputs} \pm \text{Rxns} \tag{1}$$

$$\frac{d(VC)}{dt} = QC - Q(C + \Delta C) \pm rV \tag{2}$$

where V is the incremental control volume, C is the concentration, Q is the flowrate, and r is the reaction rate. Equation (2) is written for steady flow ($dQ/dt = 0$) conditions. Let us also assume a constant cross-sectional area ($dA/dx = 0$) that also requires a constant velocity and incremental volume. With a constant incremental volume, we may divide through by the control volume ($V = A\ \Delta x$).

$$\frac{dC}{dt} = \frac{QC}{A\ \Delta x} - \frac{QC}{A\ \Delta x} - \frac{Q\ \Delta C}{A\ \Delta x} \pm r \tag{3}$$

If we take the limit as $\Delta x \rightarrow 0$, we obtain a partial differential equation in space and time.

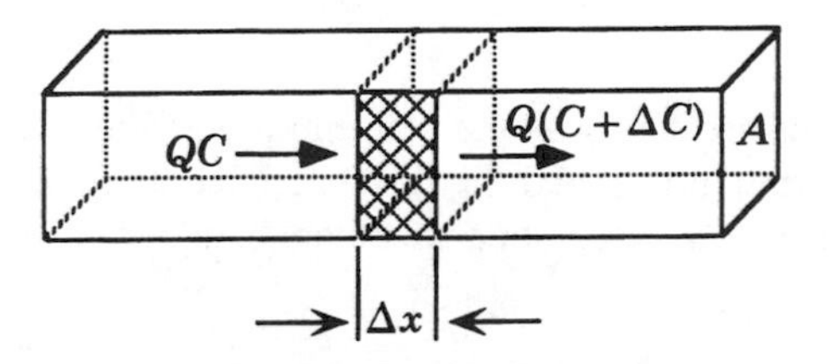

Figure 6.1. Schematic of flow through a stream and incremental control volume. Cross-sectional area (A) multiplied times the incremental length (Δx) gives the incremental volume $V = A\ \Delta x$. Mass transport into the upstream face of the control volume is taken to be QC, and the mass leaving the control volume is $Q(C + \Delta C)$. The concentration change in the control volume is due to reactions under steady flow conditions and a constant cross-sectional area.

$$\frac{\partial C}{\partial t} = -\frac{Q}{A}\frac{\partial C}{\partial x} \pm r \tag{4}$$

Because Q/A is equal to the mean stream velocity ($\overline{u}$), we may also write equation (5) as the general plug-flow equation with reactions.

$$\frac{\partial C}{\partial t} = -\overline{u}\frac{\partial C}{\partial x} \pm r \tag{5}$$

At steady state, the change in concentration with respect to time is zero, $\partial C/\partial t = 0$, and equation (5) results in the steady-state ordinary differential equation for a stream with reactions,

$$\frac{dC}{dx} = \pm\, r/\overline{u} \tag{6}$$

or written as an integral equation,

$$\int dC = \pm\frac{1}{\overline{u}}\int r\,dx \tag{7}$$

Suppose the reaction in equation (6) was a first-order decay reaction such that $r = -kC$, where k is the first-order decay rate constant (T^{-1}).

$$\frac{dC}{dx} = -\frac{k}{\overline{u}}C \tag{8}$$

We may solve equation (8) (an ordinary differential equation) by the method of separation of variables. We move the dependent variable (C) to the left-hand side of the equation and the independent variable to the right-hand side, and then we may integrate to solve for the concentration as a function of stream distance.

$$\int_{C_0}^{C}\frac{dC}{C} = -\frac{k}{\overline{u}}\int_0^x dx \tag{9}$$

The definite integrals are set up as the concentration from C at $x = 0$ (C_0) to any downstream concentration (C), and the distance interval from $x = 0$ (arbitrary) to any downstream distance x. Integrating we obtain equations (10) and (11):

$$\ln C\,\Big|_{C_0}^{C} = \frac{-kx}{\overline{u}}\Big|_0^x \tag{10}$$

$$\ln\frac{C}{C_0} = \frac{-kx}{\overline{u}} \tag{11}$$

Taking the exponent of both sides, we find

$$C = C_0 \exp(-kx/\overline{u}) \tag{12}$$

in which C_0 is the concentration at the source ($x = 0$). The solution to equation (12) is a first-order decaying exponential, similar to radioisotope decay except that the independent variable here is distance (x) or travel time ($x/\overline{u}$) rather than actual time.

Example 6.1 Steady State Stream with a First-Order Decaying Substance, BOD

Calculate the relative concentration remaining (C/C_0) after one, two, and three "time constants" of downstream distance for a stream with a first-order decaying substance under steady-state conditions. The time constant is the dimensionless number in the exponent of equation (12), $kx/\overline{u}$. Also show how to linearize field data to obtain the rate constant for the decaying substance using equation (11).

Solution: Use equation (12) to calculate C/C_0. Answers are given below:

$kx/\overline{u}$	C/C_0
0	1.0
1	0.368
2	0.135
3	0.050

Note that the reaction decay of the chemical is exponential. After three time constants, only 5% of the original material remains.

Using equation (11), we can linearize field data collected from the stream to obtain the rate constant for the decay reaction. A plot of ln C versus x or ln (C/C_0) versus $x/\overline{u}$ can be used to estimate the rate constant. If the stream has a mean velocity of 0.4 m s^{-1} and the concentration profile is given below from field measurements, estimate the in situ rate constant for BOD degradation below a wastewater discharge at $x = 0$. Ultimate BOD concentrations are given at the km point where the sample was collected.

x, km	C, mg L^{-1}
0	10.0
5	9.4
10	9.1
20	8.4
30	7.6
50	6.4
100	4.2
200	1.9

Plot ln (C/C_0) versus $x/\bar{u}$ (travel time) to obtain the rate constant for the degradation of BOD in the stream. Data and results are plotted in Figure 6.2.

x, km	C/C_0	ln C/C_0	$x/\bar{u}$, days
0	1.0	0.0	0
5	0.94	-0.062	0.145
10	0.91	-0.094	0.289
20	0.84	-0.174	0.579
30	0.76	-0.274	0.868
50	0.64	-0.446	1.447
100	0.42	-0.868	2.894
200	0.19	-1.661	5.787

6.2.1 Steady Flow Mass Balance in a Stream: Variable Coefficients

Equations (6) and (7) are the general forms of the stream equation assuming steady flow ($dQ/dt = 0$), constant velocity, and general reaction kinetics. It is frequently the case that flow is not steady or the parameters and coefficients (Q, A, $\bar{u}$, k) are vari-

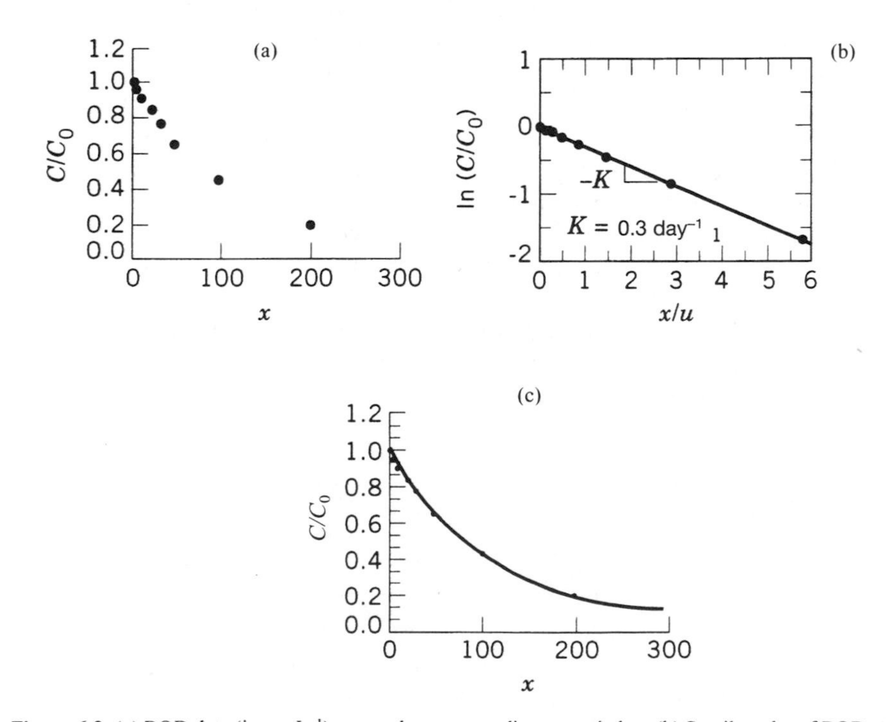

Figure 6.2 (a) BOD data (in mg L^{-1}) versus downstream distance; x in km. (b) Semilog plot of BOD versus x/u to solve for deoxygenation rate constant. (c) Model results and field data for BOD versus distance.

able with distance or time. As an example, suppose that the flowrate and cross-sectional area increase over a stretch of stream in the following manner:

$$Q = Q_0 \exp(qx) \tag{13a}$$

$$A = A_0 \exp(ax) \tag{13b}$$

in which Q_0 is the flowrate at $x = 0$, q is the exponential rate coefficient with distance (L^{-1}), and x is stream longitudinal distance. Then the mass balance around the incremental control volume (shown in Figure 6.1) is now

$$\frac{\partial(VC)}{\partial t} = QC - (Q + \Delta Q)(C + \Delta C) - rV \tag{14}$$

where the second term on the right-hand side is the outflow mass discharge, a product of the increased outflow from the elemental control volume ($Q + \Delta Q$) and the modified concentration out ($C + \Delta C$). Multiplying through we obtain equation (15):

$$C \overset{0}{\cancel{\frac{\partial V}{\partial t}}} + V \frac{\partial C}{\partial t} = -(\Delta Q)C - Q(\Delta C) - (\Delta Q)(\Delta C) - rV \tag{15}$$

For constant-volume systems (steady flow), the first term on the left-hand side of equation (15) is zero ($\partial V/\partial t = 0$), and dividing through by the incremental control volume ($V = A\,\Delta x$) we obtain

$$\frac{\partial C}{\partial t} = -\frac{(\Delta Q)C}{A(\Delta x)} - \frac{Q}{A}\frac{\Delta C}{\Delta x} - \frac{(\Delta Q)(\Delta C)}{A(\Delta x)} - r \tag{16}$$

$$\lim \Delta x \to 0 \quad \frac{\partial C}{\partial t} = -\frac{1}{A} C \frac{\partial Q}{\partial x} - \frac{Q}{A}\frac{\partial C}{\partial x} - \overset{\text{neg.}}{\cancel{\frac{(\partial Q)(\partial C)}{A\,\partial x}}} - r \tag{17}$$

The third term on the right-hand side of equation (17) is assumed to be negligible because the product of two incremental changes should be small relative to other terms in the equation.

A final result may be expressed as:

$$\frac{\partial C}{\partial t} = -\frac{1}{A}\frac{\partial (QC)}{\partial x} - r \tag{18}$$

Equation (18) is the general equation for one-dimensional transport in a stream or river (neglecting dispersion). All coefficients (Q, A, r) may be functions of distance and time, and the resulting equation may be solved analytically, in simple cases, or numerically by the method of characteristics.

$$\frac{\partial C}{\partial t} = -\frac{1}{A(x,t)} \frac{\partial [Q(x,t)\,C(x,t)]}{\partial x} - r(C,x,t) \tag{19}$$

Under steady-state conditions, the partial differential equation reduces to an ordinary differential equation (equation 20):

$$0 = -\frac{1}{A} \frac{d(QC)}{dx} - r \tag{20}$$

Example 6.2 Exponentially Increasing Flow and Area in a Stream at Steady State

Develop the mass balance differential equation and solve (integrate) for the concentration as a function of time for the following cases:

a. Steady state, increasing flow and cross-sectional area with distance (x) and first-order reaction decay.
b. Steady state, exponentially decreasing rate constant as a function of distance in the stream (easiest-to-degrade material is most rapidly degraded near the point of discharge leaving recalcitrant compounds)

$$Q = Q_0 \exp(qx)$$
$$A = A_0 \exp(ax)$$
$$k = k_0 \exp(-rx)$$

in which q, a, and r are exponential coefficients for flowrate, area, and degradation rate constant with distance.

Solution:

a. Use equation (20) for steady-state conditions.

$$0 = -\frac{1}{A(x)} \frac{d[Q(x)\,C(x)]}{dx} - kC$$

$$0 = -\frac{1}{A_0 e^{ax}} \frac{d[Q_0 e^{qx}\,C(x)]}{dx} - kC$$

$$0 = -\frac{Q_0}{A_0 e^{ax}} \left(e^{qx} \frac{dC}{dx} + C\,q e^{qx} \right) - kC$$

$$0 = -\frac{Q_0}{A_0} e^{(q-a)x} \frac{dC}{dx} - \left(\frac{qQ_0}{A_0} e^{(q-a)x} \right) C - kC$$

$$\frac{Q_0}{A_0} e^{(q-a)x} \frac{dC}{dx} = -\left(\frac{qQ_0}{A_0} e^{(q-a)x} + k \right) C$$

$$\int_{C_0}^{C} \frac{dC}{C} = \int_0^x -\left(q + \frac{kA_0 e^{(a-q)x}}{Q_0} \right) dx$$

$$\ln \frac{C}{C_0} = \left[-qx - \frac{kA_0}{(a-q)Q_0} e^{(a-q)x} \right|_0^x$$

$$\ln \frac{C}{C_0} = \left[-qx - \frac{kA_0}{(a-q)Q_0} e^{(a-q)x} + \frac{kA_0}{(a-q)Q_0} \right]$$

$$C = C_0 \exp \left[-qx - \frac{kA_0}{(a-q)Q_0} e^{(a-q)x} + \frac{kA_0}{(a-q)Q_0} \right]$$

The solution to the problem of first-order reaction decay (with exponentially increasing flowrate and area) is an exponentially decreasing function with distance (x), but the exact shape of the concentration versus x plot depends on Q_0/A_0 and $(a - q)$.

b. For solution of this case we have

$$0 = -\frac{Q}{A} \frac{dC}{dx} - (k_0 e^{-rx}) C$$

$$\overline{u} = Q/A$$

$$\overline{u} \frac{dC}{dx} = -(k_0 e^{-rx}) C$$

$$\int_{C_0}^{C} \frac{dC}{C} = -\frac{k_0}{\overline{u}} \int_0^x e^{-rx} \, dx$$

$$\ln \frac{C}{C_0} = -\frac{k_0}{\overline{u}} \left| \frac{-1}{r} e^{-rx} \right|_0^x = \frac{k_0}{\overline{u}r} \left| e^{-rx} \right|_0^x$$

$$\ln \frac{C}{C_0} = \frac{k_0}{\overline{u}r} (e^{-rx} - 1)$$

$$C = C_0 \exp \left[-k_0 (1 - e^{-rx}) / \overline{u}r \right]$$

The solution to the problem of a first-order reaction decay (with exponentially decreasing rate constant) is a decreasing concentration with x, but the rate of decrease is slowed due to the declining reaction rate constant with distance.

6.3 STREETER–PHELPS EQUATION

In 1925, Streeter and Phelps[6] published a seminal work on the dissolved oxygen "sag curve" in the Ohio River. They were able to demonstrate that the dissolved oxygen decreased with downstream distance due to degradation of soluble organic biochemical oxygen demand (BOD), and they proposed a mathematical equation to describe the phenomenon, which has since become widely known as the Streeter–Phelps equation.

The oxidation of carbonaceous biochemical oxygen demand is generally written as a first-order reaction, although there have been some studies indicating a dependency on oxygen concentration as well as BOD concentration.[7] For a constant-velocity stream and steady-state condition, equation (20) is applicable and is rewritten below for a first-order degradation reaction.

$$\bar{u}\frac{dL}{dx} = -k_d L \tag{21}$$

where L = ultimate BOD concentration, ML^{-3}
$\bar{u}$ = mean velocity, LT^{-1}
k_d = first-order deoxygenation rate constant, T^{-1}

For dissolved oxygen, it is possible to write a mass balance equation directly as in equation (20).

$$\bar{u}\frac{dC}{dx} = \underbrace{-k_d L}_{\text{deoxygenation}} + \underbrace{k_a(C_s - C)}_{\text{reaeration}} \tag{22}$$

where C = dissolved oxygen concentration, ML^{-3}
L = ultimate BOD concentration, ML^{-3}
C_s = saturated dissolved oxygen concentration, ML^{-3}
k_a = first-order reaeration rate constant, T^{-1}

The two terms on the right-hand side of equation (22) represent opposing processes in streams, the rate of deoxygenation due to carbonaceous BOD versus the rate of reaeration. $C_s - C$ is the concentration driving force serving to reoxygenate the water from atmospheric oxygen at the air–water interface. It is preferable to write equation (22) as a D.O. deficit equation (D).

$$\bar{u}\frac{dD}{dx} = \underset{\text{deoxygenation}}{+k_d L} - \underset{\text{reaeration}}{k_a D} \tag{23}$$

Note that the sign changed on the deoxygenation and reaeration terms in going from equation (22) to equation (23) because D.O. deficit (D) is equal to the driving force of oxygen reaeration (which is opposite in sign to the D.O. concentration).

$$D = C_s - C \tag{24}$$

We need a simultaneous solution of equations (21) and (23) in order to reproduce the Streeter–Phelps equation. It is a set of ordinary differential equations, but they are uncoupled since equation (21) can be solved directly for the BOD concentration (L) with distance, and the equation for L can then be substituted into equation (23). The solution of equation (21) for the BOD concentration is given by equations (25) and (26) below:

$$\int_{L_0}^{L} \frac{dL}{L} = -\frac{k_d}{\bar{u}} \int_0^x dx \tag{25}$$

$$L = L_0\, e^{-k_d x/\bar{u}} \tag{26}$$

Substituting equation (26) into equation (23), we may solve for the D.O. deficit (D) by separation of variables or the integration factor method:

$$\bar{u}\frac{dD}{dx} = +k_d L_0\, e^{-k_d x/\bar{u}} - k_a D \tag{27}$$

Equation (27) can be rearranged so that all of the terms involving the dependent variable (D) are on the left-hand side of the equation, and the forcing function is on the right-hand side:

$$\bar{u}\frac{dD}{dx} + k_a D = k_d L_0\, e^{-k_d x/\bar{u}} \tag{28}$$

Using the integration factor method, equation (28) is of the form

$$\frac{dy}{dt} + p(t)y = q(t) \tag{29}$$

with the solution over the interval from 0 to t:

$$y = y_0\, e^{-\mu(t)} + e^{-\mu(t)} \int_0^t e^{\mu(t)} q(t)\, dt \tag{30}$$

where $\mu(t) = \int p(t)\, dt$, the integrating factor; $q(t)$ is the forcing function; y is the dependent variable; and t is the independent variable. A comparison of equations (28) and (29) allows for the following definitions:

$$y = D$$

$$t = x$$

$$p(t) = k_a / \overline{u}$$

$$q(t) = \frac{k_d L_0}{\overline{u}}\, e^{-k_d x/\overline{u}}$$

The solution to equation (28) may be determined from the general solution given as equation (30).

$$D = D_0\, e^{-k_a x/\overline{u}} + e^{-k_a x/\overline{u}} \int_0^x e^{-k_a x/\overline{u}} \left(\frac{k_d L_0}{\overline{u}}\, e^{-k_d x/\overline{u}} \right) dx \tag{31}$$

$$D = D_0\, e^{-k_a x/\overline{u}} + \frac{k_d L_0}{\overline{u}}\, e^{-k_a x/\overline{u}} \left| \frac{\overline{u}}{k_a - k_d}\, e^{(k_a - k_d)x/\overline{u}} \right|_0^x \tag{32}$$

$$D = D_0\, e^{-k_a x/\overline{u}} + \frac{k_d L_0}{k_a - k_d}\, e^{-k_a x/\overline{u}} \left(e^{(k_a - k_d)x/\overline{u}} - 1 \right) \tag{33}$$

$$D = D_0\, e^{-k_a x/\overline{u}} + \frac{k_d L_0}{k_a - k_d} \left(e^{-k_d x/\overline{u}} - e^{-k_a x/\overline{u}} \right) \tag{34}$$

Equation (34) is the final solution for dissolved oxygen deficit versus distance after a point source discharge of BOD in a one-dimensional, steady-state, plug-flow system. We could have also solved for the dissolved oxygen concentration [equation (22)] rather than the deficit. The solution is given by equation (35) below by substituting $(C_s - C)$ for D in equation (34).

$$C = C_s - (C_s - C_0)e^{-k_a x/\overline{u}} - \frac{k_d L_0}{k_a - k_d} \left(e^{-k_d x/\overline{u}} - e^{-k_a x/\overline{u}} \right) \tag{35}$$

Equations (26) and (34) are the solutions to differential equations (21) and (23). Hypothetical solutions to the differential equations are illustrated in Figure 6.3. Ultimate BOD is expected to decrease in an exponential fashion with downstream distance [as in equation (26)]. Dissolved oxygen deficit concentration increases to a maximum critical deficit (D_c) at the critical distance (x_c), and then it decreases to near zero as $x \to \infty$ [equation (34) and Figure 6.3]. Dissolved oxygen concentration decreases to a minimum and then increases in the classic "D.O. sag curve" represented by Figure 6.3 and given by equation (35). Eventually the stream purifies itself by reaeration after the point source input of BOD at $x = 0$.

As shown in Figure 6.3, the rate of deoxygenation ($k_d L$, mg L^{-1} d^{-1}) can be compared to the rate of reaeration ($k_a D$, mg L^{-1} d^{-1}) at any point in the stream. Between the discharge point ($x = 0$) and the critical distance ($x = x_c$), the rate of deoxygenation exceeds the rate of reaeration ($k_d L > k_a D$) because the BOD concentration is relatively large and the D.O. deficit (D) has not yet reached its maximum point over that distance. Beyond the critical distance ($x > x_c$), the rate of reaeration exceeds deoxygenation and the dissolved oxygen concentration approaches the saturated concentration (C_s). Exactly at the critical distance ($x = x_c$), the rate of reaeration is exactly equal to the rate of deoxygenation, and we have a maximum where the rate of

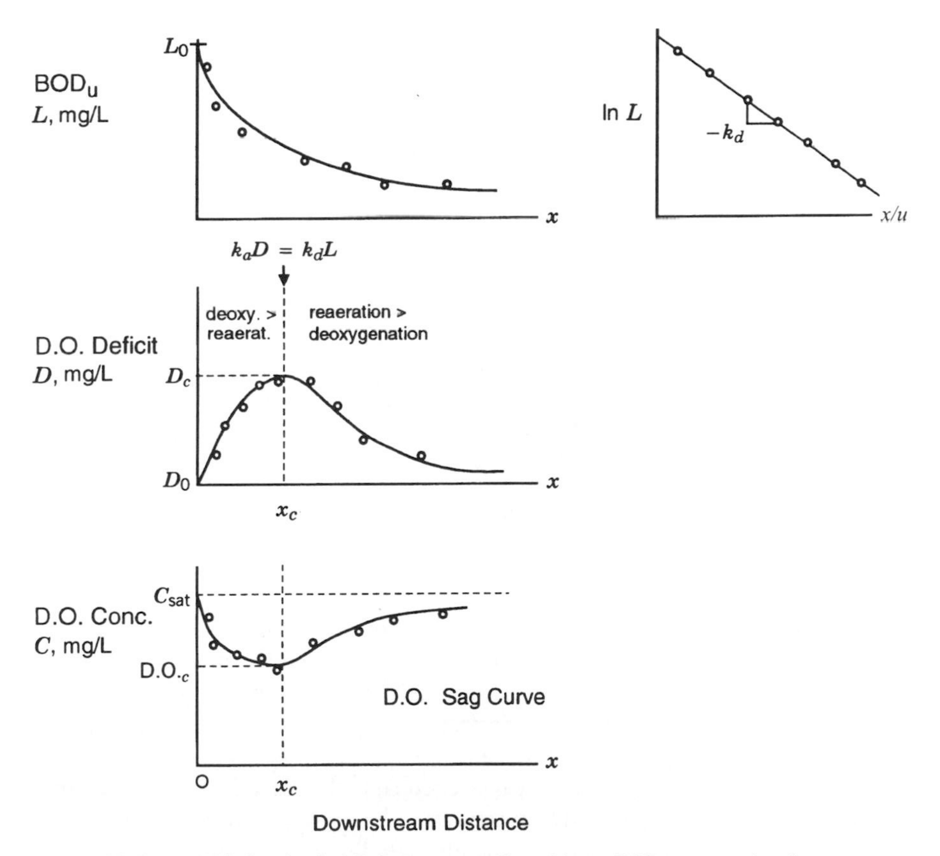

Figure 6.3 Streeter–Phelps classical D.O. sag curve. *Top:* ultimate BOD concentration decreases exponentially with distance. *Middle:* D.O. deficit reaches a maximum point where the deoxygenation rate in the stream equals the reaeration rate. *Bottom:* D.O. sag curve is critical at distance x_c.

change of oxygen concentration (dC/dx) or deficit (dD/dx) is equal to zero. This fact can be demonstrated from equation (23) when $dD/dx = 0$:

$$\bar{u}\overbrace{\frac{dD}{dx}}^{0} = k_d L - k_a D \tag{36}$$

$$k_d L = k_a D \quad \text{at } x = x_c \tag{37}$$

6.3.1 Critical Deficit and Distance (D_c, x_c)

The critical deficit and the downstream distances under steady-state conditions can be solved explicitly using equation (28).

$$\bar{u}\overbrace{\frac{dD}{dx}}^{0 \text{ at } x_c} + k_a D = k_d L_0 e^{-k_d x/\bar{u}} \tag{38}$$

Solving for the critical deficit (D_c), one arrives at equation (39):

$$D_c = \frac{k_d L_0}{k_a} e^{-k_d x/\bar{u}} \tag{39}$$

Substituting equation (39) into the equation for dissolved oxygen deficit [equation (34)] allows one to solve for the critical distance (x_c).

$$x_c = \frac{\bar{u}}{k_a - k_d} \ln \frac{k_a}{k_d}\left(1 - \frac{(k_a - k_d)}{k_d} \frac{D_0}{L_0}\right) \tag{40}$$

If the initial dissolved oxygen deficit is zero at $x = 0$ (D.O. is saturated, C_s), then equation (40) simplifies to

$$x_c = \frac{\bar{u}}{k_a - k_d} \ln \frac{k_a}{k_d} \tag{41}$$

Various inputs and parameters have an effect on the D.O. sag curve and D.O. deficit curve shown schematically in Figure 6.3. The self-purification ratio was defined by Fair and Geyer[8] to be k_a/k_d, the ratio of the reaeration rate constant to the deoxygenation rate constant. It is an important dimensionless number that strongly affects both D_c and x_c, as predicted by equations (39) and (40). The initial concentration of ultimate BOD (L_0) is directly proportional to the critical deficit concentration; the greater is the initial BOD concentration, the greater will be the dissolved oxygen deficit concentration at x_c. L_0, the ultimate BOD concentration at $x = 0$, is equal to the mass discharge rate of ultimate BOD diluted by the flowrate of the river. If only BOD_5 (5-day BOD) concentrations are available, a suitable conversion factor should be used to convert to ultimate BOD concentrations ($BOD_u \simeq 1.47 \times BOD_5$ for municipal wastewater).

$$L_0 = W/Q \tag{42}$$

where W = wastewater mass discharge rate, MT^{-1}
and Q = flowrate of the river, L^3T^{-1}

Here, instantaneous mixing of the wastewater discharge in both the lateral dimension (across the width of the stream) and the vertical dimension with depth have been assumed. If the flowrate of the wastewater discharge cannot be neglected relative to the stream discharge rate and/or if the ultimate BOD concentration of the upstream segment cannot be neglected relative to the BOD concentration below the discharge, then a flow-weighted average concentration should be calculated for L_0, according to equation (43):

$$L_0 = \frac{Q_w C_w + Q_s C_s}{Q_w + Q_s} \tag{43}$$

where the subscript w refers to the waste discharge and s refers to the stream. The effect of the wastewater discharge as measured by W or L_0 is to increase the D.O. deficit curve throughout all x; the critical distance (x_c) does not change, but the critical deficit concentration (D_c) changes linearly with W or L_0. See Figure 6.4.

Temperature plays a role in D.O. reaeration kinetics, D.O. solubility (C_s), and deoxygenation (microbial degradation of BOD). The three effects are in opposition.

1. D.O. reaeration rate constant *increases* with increasing temperature.
2. D.O. solubility *decreases* with increasing temperature so the driving force (D) for reaeration decreases somewhat. See Appendix A.
3. Deoxygenation rate constant *increases* with increasing temperature.

Arrhenius developed the equation for the dependence of rate constants on absolute temperature in 1889. If activation energy is constant over a small temperature range (not always the case), the Arrhenius equation can be simplified

$$k = A_e^{-E_a/RT} \tag{44}$$

where E_a is the activation energy, R is the universal gas constant, T is absolute temperature, and A is the pre-exponential constant. Equation (44) is often simplified still further to a logarithmic relationship for changing temperature with respect to a reference value. For deoxygenation and reaeration rate constants:

$$k_d = k_{d,20}\, \theta^{(T-20)} \tag{45}$$

$$k_a = k_{a,20}\, \theta^{(T-20)} \tag{46}$$

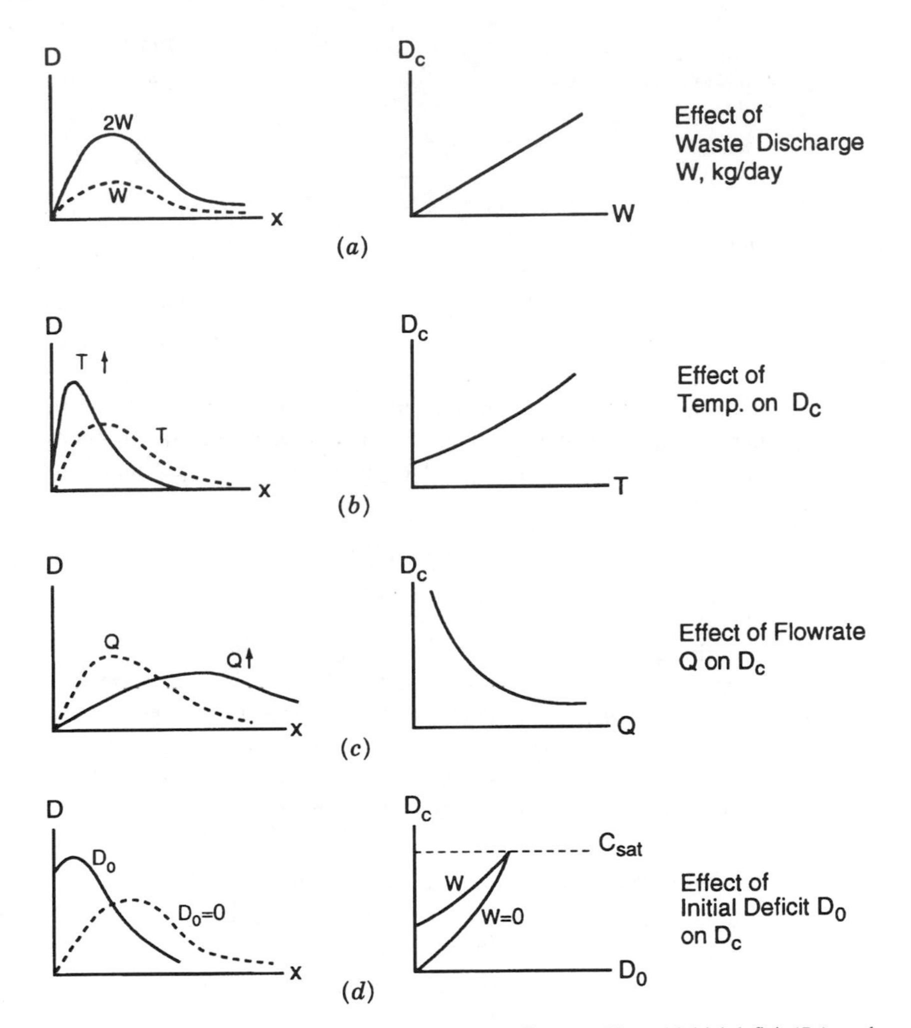

Figure 6.4 Effect of waste discharge (W), temperature (T), flowrate (Q), and initial deficit (D_0) on the critical dissolved oxygen deficit concentration (D_c).

where $k_{d,20}$ = rate constant for deoxygenation at a reference temperature 20 °C, day^{-1}

$k_{a,20}$ = rate constant for reaeration at a reference temperature of 20 °C, day^{-1}

k_d = deoxygenation rate constant at temperature T, day^{-1}

k_a = rate constant for reaeration at temperature T, day^{-1}

θ = number raised to the power $(T - 20)$

T = any temperature within the range, usually 0–30 °C

Theta (θ) for deoxygenation is normally taken to be 1.048 and for reaeration it is

1.024. Thus the effect of increasing temperature is greater on deoxygenation compared to reaeration, and this causes the critical deficit (D_c) to increase and to move upstream as depicted in Figure 6.4b. Increasing temperature also results in a marked decrease in dissolved oxygen saturation concentration, which decreases the dissolved oxygen concentration (C) by limiting the driving force for reaeration ($C_s - C$).

Increasing stream flow has two effects on the dissolved oxygen deficit profile. First, the most important effect is on initial BOD concentration (L_0), which decreases proportionately with flowrate: $L_0 = W/Q$. Second, the flowrate influences reaeration rate constants in complex fashion. Reaeration rate constants are often expressed as a power function of mean velocity and an inverse power function of mean depth.

$$k_a \propto \frac{\bar{u}^m}{H^n} \tag{47}$$

in which $\bar{u}$ is the mean stream velocity, H is the mean depth, m is an exponent usually less than 1.0, and n is an exponent usually greater than 1.0. Increasing the flowrate increases both the mean velocity and the mean depth according to the friction factor energy slope and backwater profile. Generally, the effect of increasing flowrate on mean depth (to the n power) is larger than the effect on mean velocity (to the m power), so the reaeration rate constant *decreases* with increasing stream flow. This is a somewhat counterintuitive result that is not true in the tailwaters of dam releases and rapids. At any rate, the overriding effect of increased flowrate is on the initial ultimate BOD concentration, L_0 at $x = 0$, and the net effect is to "push" the D.O. deficit curve further downstream and to decrease the critical deficit (D_c) as shown in Figure 6.4c. Effects of flowrate on stream physical parameters can be summarized by equations (48)–(51):

$$\text{Mean depth} \qquad H \propto Q^a \tag{48}$$

$$\text{Mean velocity} \qquad \bar{u} \propto Q^b \tag{49}$$

$$\text{Mean width} \qquad B \propto Q^c \tag{50}$$

$$a + b + c = 1 \tag{51}$$

where Q is the flowrate and $a = 0.4$–0.7, $b = 0.3$–0.5, $c = 0$–0.25. The cross-sectional area of the stream varies with flowrate to the power $(1 - b)$ by virtue of continuity with equation (49).

6.3.2 Reaeration Rate Constants

O'Connor and Dobbins[9] developed the first model equation for reaeration rate constants in streams in 1958. It was based on Danckwert's surface renewal theory, which can be approximated by

$$k_L = \sqrt{D_m r} \coth \sqrt{\frac{r\ell^2}{D_m}} \tag{52}$$

where k_L = controlling liquid-film mass transfer coefficient, LT^{-1}
D_m = molecular diffusion coefficient for oxygen in water, L^2T^{-1}
r = surface renewal rate, T^{-1}
ℓ = Prandtl mixing length

The hyperbolic cotangent portion of equation (52) was approximately equal to 1.0, and it was dropped. Rate of surface renewal is proportional to shear stress ($\propto d\bar{u}/dy$) at the surface because shear at the surface causes vertical mixing (and renewal of oxygen-saturated water down into the stream). A definition of surface renewal can be stated as

$$r = \frac{\bar{u}_y'}{\ell} \tag{53}$$

where $\bar{u}_y'$ = average vertical velocity fluctuation due to turbulent shear, LT^{-1}
ℓ = Prandtl's mixing length (the average distance that the vertical velocity fluctuation travels)

O'Connor and Dobbins[9] argued that, for flume and field studies of the Mississippi River, the average vertical velocity fluctuation was approximately one-tenth of the average longitudinal velocity, and that the vertical mixing length was about one-tenth of the mean depth of the stream. Therefore, for isotropic turbulence,

$$r = \frac{\bar{u}}{H} \tag{54}$$

where $\bar{u}$ is the mean longitudinal stream velocity and H is the mean depth. Substitution of equation (54) into equation (52) yields the O'Connor–Dobbins reaeration formula, which is theoretically based except for the assumptions on vertical velocity fluctuations and mixing length.

$$k_L = \sqrt{D_m \bar{u}/H} \tag{55}$$

For uniform dissolved oxygen concentration with depth, the reaeration rate constant is simply the mass transfer coefficient divided by the mean depth.

$$k_a = k_L/H \tag{56}$$

In English units, the final reaeration rate constant is given by equation (57):

$$k_a = \frac{12.9\overline{u}^{0.5}}{H^{1.5}} \tag{57}$$

where $\overline{u}$ is the mean longitudinal stream velocity (in ft s^{-1}), H is the mean depth (in ft), and 12.9 is a units conversion factor (and contains $\sqrt{D_m}$) to determine the reaeration rate constant k_a (base e) in units of day^{-1}.

For a smooth quiescent surface, the mass transfer coefficient of atmospheric oxygen into streams is approximately 0.7 m d^{-1}. Thus a stream of 2 m mean depth would have a reaeration rate constant equal to 0.35 day^{-1} under quiescent conditions. The presence of ripples or a broken surface would increase the river reaeration rate constant markedly.

There have been numerous other reaeration formulas developed, pertaining to different stream velocities, usually empirical power-function relationships of the form

$$k_a = \frac{c\overline{u}^m}{H^n} \tag{58}$$

Of these formulas, the Churchill–Elmore–Buckingham has been the most applied, especially for large rivers (Table 6.2).

Example 6.3 Streeter–Phelps Equation

A plug-flow stream has a velocity of 0.3048 m s^{-1}, a mean depth of 1.056 m, and a deoxygenation rate constant of 0.6 day^{-1}. The initial ultimate BOD concentration at $x = 0$ is 10.0 mg L^{-1}, and the initial deficit is 0.0 mg L^{-1}. Estimate the reaeration rate constant using the O'Connor–Dobbins reaeration formula and plot: (a) the BOD concentration versus downstream distance, (b) the D.O. deficit versus downstream distance, and (c) the reaeration rate (k_aD in mg L^{-1} d^{-1}) and the deoxygenation rate (k_dL in mg L^{-1} d^{-1}) versus distance. Where is the reaeration rate equal to the deoxygenation rate? What is the critical deficit and distance?

Solution: Using the O'Connor–Dobbins formula, the reaeration rate constant is 2.0 day^{-1}.

$$k_a = \frac{12.9 \text{ m}^{1/2}}{H^{3/2}} = \frac{12.9(1 \text{ ft s}^{-1})^{1/2}}{(3.465 \text{ ft})^{3/2}} = 2.0 \text{ day}^{-1}$$

The BOD equation and the Streeter–Phelps D.O. deficit equations are given below, where $k_a = 2.0$ day^{-1}, $k_d = 0.6$ day^{-1}, $L_0 = 10$ mg L^{-1}, $D_0 = 0$ mg L^{-1}, and $u = 16.4$ mi d^{-1}.

$$L = L_0 e^{-k_d x/u}$$

Table 6.2 Stream Reaeration Formulas

O'Connor–Dobbins[9]	$k_a = \dfrac{12.9u^{0.5}}{H^{1.5}}$		
Owens–Edwards–Gibbs[10]	$k_a = \dfrac{23u^{0.73}}{H^{1.75}}$	for	$H = 1–2.5$ $\overline{u} = 0.1–0.5$ $Q = 4–36$
Churchill–Elmore–Buckingham (TVA)[11]	$k_a = \dfrac{11u}{H^{1.67}}$	for	$H = 2–11$ $\overline{u} = 2–5$ $Q = 1000–17{,}000$
USGS	$k_a = \dfrac{7.6u}{H^{1.33}}$		
Tsivoglou	$k_a = \dfrac{0.048\ \Delta S}{t}$	for	$Q = 5–3000$

where k_a = reaeration rate constant (base e), day^{-1}
$\overline{u}$ = mean stream velocity, ft sec^{-1}
H = mean stream depth, ft
ΔS = water surface elevation change, ft
Q = flowrate, ft^3 s^{-1}
t = travel time, days

$$D = D_0 e^{-k_a x/u} + \frac{k_d L_0}{k_a - k_d}\left(e^{-k_d x/u} - e^{-k_a x/u}\right)$$

Answer:

x, mi	L, mg L^{-1}	k_dL, mg L^{-1}d^{-1}	D, mg L^{-1}	k_aD, mg L^{-1}d^{-1}
0	10.0	6.00	0	0
5	8.33	5.00	1.24	2.48
10	6.94	4.16	1.71	3.42
15	5.78	3.47	1.79	3.58
20	4.81	2.89	1.69	3.38
30	3.33	2.00	1.32	2.64
40	2.31	1.39	0.96	1.92
50	1.60	0.96	0.68	1.36
75	0.64	0.38	0.27	0.54

The critical deficit and distance are given by the following equations:

$$x_c = \frac{u}{k_a - k_d} \ln\left(\frac{k_a}{k_d}\right) = 14.1 \text{ mi}$$

$$D_c = \frac{k_d L_0}{k_a} \exp\left(\frac{-k_d x_c}{u}\right) = 1.8 \text{ mg L}^{-1}$$

Note that the rate of reaeration, $k_a D$, is exactly equal to the deoxygenation rate, $k_d L$, where the deficit is a maximum ($x_c = 14.1$ mi).

6.4 MODIFICATIONS TO STREETER–PHELPS EQUATION

When Streeter and Phelps (1925) developed their equation for dissolved oxygen concentrations in the Ohio River, the principal dissolved oxygen (D.O.) sink was carbonaceous biochemical oxygen demand (CBOD), and reaeration was the primary source of D.O. But several other sources and sinks of dissolved oxygen are common in streams and rivers:

D.O. Sinks	D.O. Sources
CBOD	Reaeration
NBOD	Primary production – respiration
Sediment oxygen demand	
Initial D.O. deficit	
Background B.O.D. (nonpoint source runoff)	

Nitrogenous biochemical oxygen demand (NBOD) is analogous to CBOD, except that it emanates from reduced nitrogen species, especially ammonium. Often, ultimate NBOD is assumed to be the stoichiometric equivalent of the ammonium concentration. We can estimate the mass equivalency as 64/14 = 4.57 mg NBOD per mg NH_4^+-N on a stoichiometric basis:

$$NH_4^+ + 2\,O_2 \xrightarrow{\text{bacteria}} NO_3^- + H_2O + 2H^+ \tag{59}$$

$$14 \text{ mg N} + 64 \text{ mg } O_2$$

Thus, if the ammonium concentration in the stream is 2.0 mg NH_4^+ as N per liter, the concentration of NBOD would be equal to 9.14 mg L^{-1} NBOD.

Nitrifying bacteria are ubiquitous in flowing waters; they are often periphytic, attached to rocks, or they can be planktonic. Nitrifiers are usually autotrophs, utilizing CO_2 as a carbon source, but heterotrophic nitrifying organisms have been reported also as a part of the microbial consortia. Nitrifying organisms mediate reaction (59) and display a strong temperature dependency between 10 and 30 °C. Below 4 °C, autotrophic nitrifiers are not sufficiently active to accomplish nitrification. Both the CBOD and NBOD deoxygenation rate constants are typically in the range of 0.05–0.5 day^{-1}.

Sediment oxygen demand is a major sink for dissolved oxygen in heavily polluted streams and rivers. A consequence of particulate BOD and algae sedimenting to the stream bed, the organic "ooze" can create a major oxygen demand on the overlying water. In areas of intense pollution, the sediment oxygen demand (SOD) is in the range of 5–10 g m^{-2} d^{-1}. For natural streams and rivers with small wastewater discharges, the SOD is typically 0.1–1.0 g m^{-2} d^{-1}. Sediment oxygen demand is measured with a benthic respirometer, a box that is lowered onto the sediment with a flange for a tight fit, a D.O. probe, and a small stirrer. Dissolved oxygen concentrations are recorded continuously on shipboard to determine the sediment oxygen demand (decrease in dissolved oxygen concentration within the respirometer) per time and unit area of the box. In highly polluted waters that have since had controls placed on wastewater discharges, sediment oxygen demand may be the largest sink of dissolved oxygen in the system.

In Chapter 5, we learned that eutrophication results in the sedimentation of algae to the bed–water interface, which creates a subsequent oxygen demand. It is the prediction of this SOD after wastewater and nutrient discharges have been mitigated that poses a particularly vexing problem for modelers. DiToro[12] recommends that the redox potential of the sediment be modeled in order to estimate the flux (recycle) of nutrients from the sediment to overlying water. Ammonium (NH_4^+) and organic acids (CBOD) are released from anoxic sediments to overlying water, creating an oxygen demand. Furthermore, reduced iron Fe(II), manganese Mn(II), and hydrogen sulfide (HS^-) can cause large sediment oxygen demands from overlying water.

Net primary production is defined as gross primary production minus respiration. Biomass of algae predominates in most streams and rivers compared to fish, higher plants, and macroinvertebrates. Therefore dissolved oxygen models include a term for net primary production by algae and phytoplankton ($P - R$). Usually the dissolved oxygen model is applied under steady-state conditions, and the net primary production ($P - R$) is considered as a daily average value. At peak photosynthesis (~ 12:00 noon and plenty of nutrients) gross primary production values in North American streams and rivers is in the range of 10–40 mg L^{-1} d^{-1} of dissolved oxygen produced. Photosynthesis can be considered as varying sinusoidally during the photoperiod, while respiration occurs continuously. Peak concentrations of oxygen in a stream are expected to occur at midday, but if one wants to measure the worst case conditions, they would occur just before sunrise when algal blooms have consumed significant concentrations of dissolved oxygen during night respiration. Respiration amounts to an average of 2.5% of gross photosynthesis, so the net effect of phytoplankton is to supply dissolved oxygen to the stream prior to death and sedimentation.

Background BOD is indicative of nonpoint source pollution to streams and rivers. A continuous input of CBOD occurs due to runoff from agricultural land, stormwater discharges, and highway runoff. These background concentrations of BOD cause a general background D.O. deficit under steady-state conditions. For a plug-flow stream, one can write the mass balance equation for D.O. deficit, D:

$$u \frac{dD}{dx} = -k_a D + k_d L \tag{60}$$

where D = dissolved oxygen deficit concentration, mg L^{-1}
x = longitudinal distance, km
u = mean velocity, km d^{-1}
k_a = reaeration rate constant, day^{-1}
k_d = deoxygenation rate constant, day^{-1}
L = carbonaceous ultimate BOD, mg L^{-1}

Under steady-state conditions and a continuous background concentration of CBOD, the background D.O. deficit concentration is a constant.

$$D_b = \frac{k_d L_b}{k_a} \tag{61}$$

in which D_b is the background concentration of D.O. deficit and L_b is the background concentration of CBOD due to nonpoint sources. Background deficit concentrations typically are in the range of 0.5–2 mg L^{-1}. One may think of D_b as a kind of safety factor that water quality engineers apply in waste load allocations to assure a conservative estimate of the stream's assimilative capacity. In addition, background D.O. deficits of some magnitude are almost always present in streams as evidenced by the lack of saturated D.O. concentrations far from waste discharges.

Table 6.3 gives ranges of model parameters that can be used in dissolved oxygen models for streams and rivers. It includes sediment oxygen demand, net primary production, and background D.O. deficit values that are significant improvements over the Streeter–Phelps equation of Section 6.3. It also includes the first-order die-away rate constants for decay of viruses and coliform bacteria. Coliform bacteria (both fecal and total) are considered as conventional pollutants, and their downstream concentrations are estimated by the following simple exponential equation:

$$C = C_0 e^{-kx/u} \tag{62}$$

where C is the coliform concentration in colony forming units, cfu mL^{-1} or most probable number, mpn mL^{-1}; C_0 is the initial concentration of coliform bacteria at x = 0; and k is the first-order die-away rate constant, day^{-1}.

All model parameters in Table 6.3 must be corrected for temperature, in equations (45) and (46). Theta values (θ) given in Table 6.3 are those that are most typical of riverine and wastewater discharge conditions. We are now ready to make some specific modifications and improvements to the Streeter–Phelps equation.

Table 6.3 Method of Determination and Range of Dissolved Oxygen Model Rate Constants at 20 °C with Typical Temperature Correction Factors for Rivers

Parameter	Value	θ, Temperature Correction[a]	Method of Determination
CBOD deoxygenation, k_d	0.05–0.5 day^{-1}	1.048	Figure 6.5 or lab
CBOD deoxygenation plus sedimentation, k_r	0.5–5 day^{-1}	1.04	Figure 6.5
NBOD deoxygenation, k_n	0.05–0.5 day^{-1}	1.08	Semilog N versus $x/\overline{u}$
Reaeration, k_a			
Slow, deep rivers	0.1–0.4 day^{-1}	1.024	Calibration, field isotopes, or Table 6.2
Typical conditions	0.4–1.5 day^{-1}	1.024	Calibration field isotopes or Table 6.2
Swift, deep rivers	1.5–4.0 day^{-1}	1.024	Calibration field isotopes or Table 6.2
Swift, shallow rivers	4.0–10 day^{-1}	1.024	Calibration field isotopes or Table 6.2
Sediment oxygen demand, S			
Natural to low pollution	0.1–1.0 g m^{-2} d^{-1}	1.065	Calibration or in situ respirometer
Moderate to heavy pollution	5–10 g m^{-2} d^{-1}	1.065	Calibration or in situ respirometer
Net primary production, $(P - R)$			
Daily average value $(P - R)$	0.5–10 mg L^{-1} d^{-1}	1.066	In situ light and dark bottles or calibration
P_{max}, maximum daily production	2–20 mg L^{-1} d^{-1}		In situ light and dark bottles or calibration
R, respiration only	1–10 mg L^{-1} d^{-1}		In situ light and dark bottles or calibration
Background D.O. Deficit, D_b	0.5–2 mg L^{-1}	NA	Model calibration
Coliform bacteria die-away, k			
Freshwater	0.5–5 day^{-1}	1.07	Lab or field semilog plot
Saltwater	2–40 day^{-1}	1.10	Lab or field semilog plot
Virus particles in marine waters	0.03–0.16 day^{-1}	1.10	Lab or field semilog plot

[a] $k_T = k_{20}\,\theta^{(T-20)}$ from Thomann and Mueller.[13]

6.4.1 Sedimentation of CBOD

Streeter–Phelps equations (26) and (34) imply that CBOD is entirely soluble and that loss of CBOD in the stream is only by deoxygenation (oxidation of soluble organics). However, particulate CBOD is discharged at wastewater treatment plants as well. Domestic treated wastewater is permitted to contain 30 mg L^{-1} of CBOD and 30 mg L^{-1} of total suspended solids (TSS). The total suspended solids contain organic matter that may be exerted in a BOD test. If CBOD is in particulate form, it

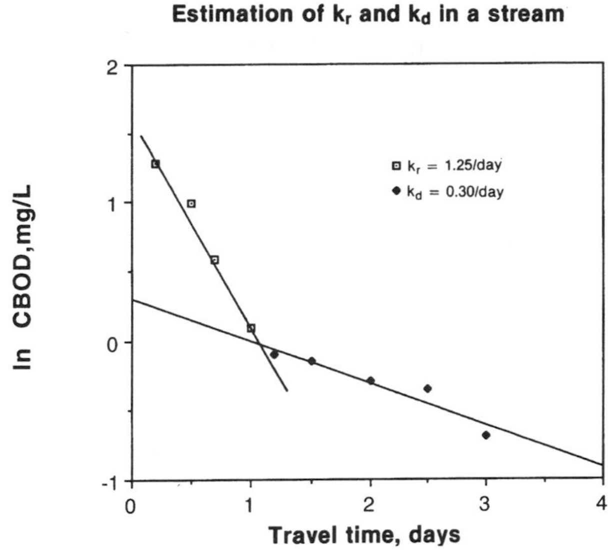

Estimation of k_r and k_d in a stream

Figure 6.5 Semilog plot of CBOD concentration in a stream versus travel time in days. Two slopes are evident: k_r includes sedimentation of particulate matter and deoxygenation; k_d is for deoxygenation alone.

will settle out of the water column. (Perhaps it will form a sludge deposit below the discharge point and require the specification of a sediment oxygen demand.) We may modify the Streeter–Phelps equations slightly to include the possibility of CBOD sedimentation.

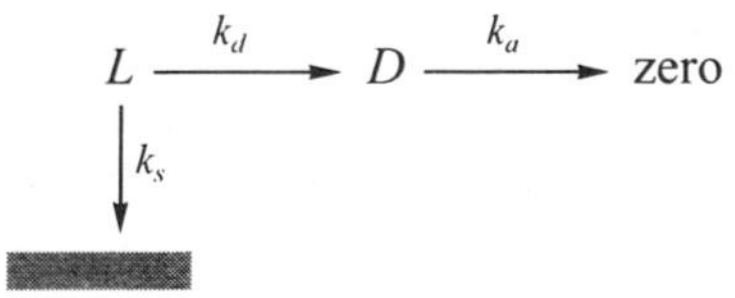

$$L \xrightarrow{k_d} D \xrightarrow{k_a} \text{zero}, \qquad L \xrightarrow{k_s} \text{(sediment)}$$

$$u\frac{dL}{dx} = -k_r L \tag{63a}$$

$$L = L_0 \exp(-k_r x/u) \tag{63b}$$

$$u\frac{dD}{dx} = k_d L - k_a D \tag{64a}$$

$$u\frac{dD}{dx} = k_d L_0 \exp(-k_r x/u) - k_a D \tag{64b}$$

where u = mean velocity, LT^{-1}

L = carbonaceous BOD, ML^{-3}

k_r = total CBOD loss rate constant, T^{-1}

= $k_s + k_d$

L_0 = initial CBOD concentration at $x = 0$, ML^{-3}

x = longitudinal distance, L

D = D.O. deficit, ML^{-3}

k_d = CBOD deoxygenation rate constant, T^{-1}

k_a = reaeration rate constant, T^{-1}

Equation (63a) is the mass balance equation for a plug-flow stream, and its solution is given by equation (63b).

By substituting for L in equation (64a), we can solve equation (64b) by the integration factor method.

$$D = D_0 \exp\left(\frac{-k_a x}{u}\right) + \frac{k_d L_0}{k_a - k_r}\left[\exp\left(\frac{-k_r x}{u}\right) - \exp\left(\frac{-k_a x}{u}\right)\right] \quad (65)$$

It is similar to the Streeter–Phelps equation (34) except that k_r replaces k_d in two places. The shape of the D.O. deficit curve (the D.O. "sag" curve) is similar to Streeter–Phelps , but the CBOD concentration in the stream decreases more quickly near the discharge point due to sedimentation. Figure 6.5 shows how to determine k_r and k_d from a semilog plot of carbonaceous BOD (L) versus travel time (x/u). Two distinctly different slopes can be estimated that define the rate constants.

6.4.2 Principle of Superposition

Modifications to the Streeter–Phelps steady-state dissolved oxygen model include consideration of NBOD, sediment oxygen demand, net primary production, and background D.O. deficit (nonpoint source runoff of BOD). These additional sources and sinks of dissolved oxygen can be solved individually, and their solutions can be summed to yield the overall solution. Because the dissolved oxygen mass balance equation is a linear differential equation, we can invoke the principle of superposition to obtain the solution (i.e., summation of the various solutions for each source and sink).

The schematic of the D.O. model is shown in Figure 6.6. There we can see how each source and sink term contributes to the D.O. deficit concentration (D). If respiration is greater than photosynthesis (nighttime conditions), then net photosynthesis is negative, and the term ($R - P$) contributes to the D.O. deficit. During daylight hours, primary production of oxygen is greater than respiration and dissolved oxygen is created. Using a steady-state approach implies that a daily average ($P - R$) value is used in the model. Of course, critical (worst case) conditions occur before sunrise when $R > P$.

The overall mass balance equation for D.O. deficit in a plug-flow river can be developed from a control volume approach, as in Section 6.2.

$$\frac{\partial D}{\partial t} = -u\frac{\partial D}{\partial x} - k_a D + k_d L + k_n N + \frac{S}{H} + (R - P)$$

Accumulation = Transport − Sinks + Sources of deficit (66)

Parameters and units are defined in Table 6.2, and H is the mean depth of the river. Assuming steady-state, we can rearrange the equation into the form of a linear ordinary differential equation with forcing functions on the right-hand side of the equation (nonhomogeneous). After substituting for L and N in equation (66) we have the final equation:

$$u\frac{dD}{dx} + k_a D = k_d L_0\, e^{-k_r x/u} + k_n N_0 e^{-k_n x/u} + \frac{S}{H} + (R - P) \tag{67}$$

Equation (67) is a general nonhomogeneous ODE (ordinary differential equation) that is solvable by the integration factor method. The overall solution is given by equation (68):

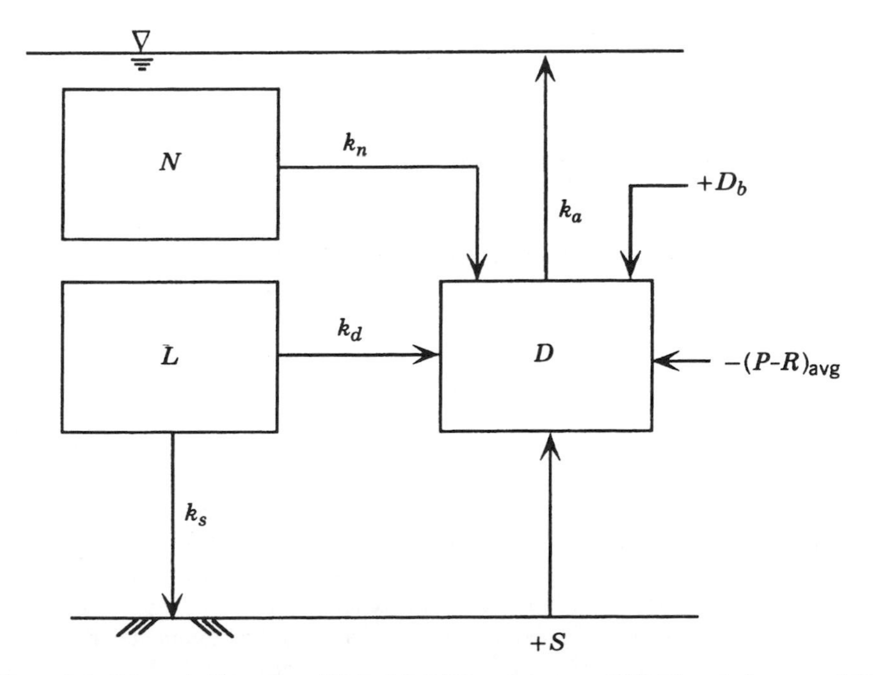

Figure 6.6 Schematic illustration of D.O. deficit (D), carbonaceous BOD (L), and nitrogenous BOD (N) with nitrogenous deoxygenation (k_n), carbonaceous deoxygenation (k_d), reaeration (k_a), sedimentation (k_s) of CBOD, net photosynthesis ($P - R$), and sediment oxygen demand (S).

$$D = D_0 e^{-k_a x/u} + \frac{k_d L_0}{k_a - k_r}(e^{-k_r x/u} - e^{-k_a x/u})$$

$$+ \frac{k_n N_0}{k_a - k_n}(e^{-k_n x/u} - e^{-k_a x/u}) + \frac{S}{k_a H}(1 - e^{-k_a x/u})$$

$$+ \frac{R-P}{k_a}(1 - e^{-k_a x/u}) + \frac{k_d L_b}{k_a} \quad (68)$$

The first term on the right-hand side of the equation is the die-away of initial deficit due to reaeration; the second term is due to initial carbonaceous BOD (L_0); the third term is due to initial nitrogenous BOD (N_0); the fourth term is due to sediment oxygen demand; the fifth term is due to $(R - P)$; and the last term is due to background BOD (L_b) caused by nonpoint sources. Table 6.4 shows the individual solutions for each source and sink term that can be summed to give equation (68). Figure 6.7 shows the graphical solution for each source and sink term that can be summed to give the overall solution.

The solutions for carbonaceous BOD and nitrogenous BOD in the river are given by equations (69) and (70):

$$L = L_0 e^{-k_r x/u} \quad (69)$$

$$N = N_0 e^{-k_n x/u} \quad (70)$$

Other sources and sinks such as sediment oxygen demand, net primary production, and background BOD are assumed to be constant within the stretch of the river that is being modeled. If these sources/sinks are variable, the river must be segmented into stretches of constant coefficients, and equation (68) is used to simulate each stretch as discussed in the next section on waste load allocations.

6.5 WASTE LOAD ALLOCATIONS

When water quality standards are expected to be violated even under conditions that all dischargers meet their effluent permits, a waste load allocation must be performed using water quality models. Waste load allocations are performed for critical conditions (worst case); usually this occurs under low flow (7-day, 1-in-10-year recurrence) in the summer at night (no primary production). Sometimes winter low-flow conditions with ice cover (no reaeration) are most critical.

The QUAL-2E model is supported by the U.S. EPA Environmental Research Laboratory at Athens, Georgia,[3] for waste load allocations. QUAL-2EU simulates dissolved oxygen with all the terms included in Table 6.4, but it also includes eutrophication and nutrients discussed in Chapter 5, including organic nitrogen, ammonia, nitrite, nitrate, organic P, phosphate-P, phytoplankton biomass and chloro-

Table 6.4 Source and Sink Contributions to the Dissolved Oxygen Deficit Equation (67) and Solution Equation (68) in a Plug-Flow River

Sources of Deficit	Mass Balance Equation	Solution
Initial deficit, D_0	$u\frac{dD}{dx} = -k_a D$	$D = D_0 e^{-k_a x/u}$
CBOD, L	$u\frac{dD}{dx} = -k_a D + k_d L$	$D = \frac{k_d L_0}{k_a - k_r}(e^{-k_r x/u} - e^{-k_a x/u})$
NBOD, N	$u\frac{dD}{dx} = -k_a D + k_n N$	$D = \frac{k_n N_0}{k_a - k_n}(e^{-k_n x/u} - e^{-k_a x/u})$
Sediment oxygen demand	$u\frac{dD}{dx} = -k_a D + \frac{S}{H}$	$D = \frac{S}{k_a H}(1 - e^{-k_a x/u})$
Net (respiration – production)	$u\frac{dD}{dx} = -k_a D + (R - P)_{avg}$	$D = \frac{R - P}{k_a}(1 - e^{-k_a x/u})$
Background deficit, D_b	$u\frac{dD}{dx} = -k_a D_b + k_d L_b$ (with $\frac{dD}{dx} \to 0$)	$D_b = \frac{k_d L_b}{k_a}$
Summation of sources and sinks	Overall equation (67)	Solution equation (68)

phyll a, coliform bacteria, and temperature (heat balance). It allows for dispersion in large rivers and includes a Monte Carlo uncertainty analysis option. It can be run under steady-state or dynamic water quality conditions [e.g., equation (66)], but the flow regime is steady.

6.5.1 River Segmentation

A new river segment is necessary when initial conditions or parameters change in a dissolved oxygen model that is to be used for a waste load allocation. This process of dividing the river reach into segments of constant coefficients is termed "segmentation." For example, in equation (68), changes in the initial concentrations D_0, L_0, and N_0 would require a new stream segment. Initial concentrations may change due to:

- Tributary input or confluence.
- Wastewater discharge.
- Dam or rapids (rapid reaeration).

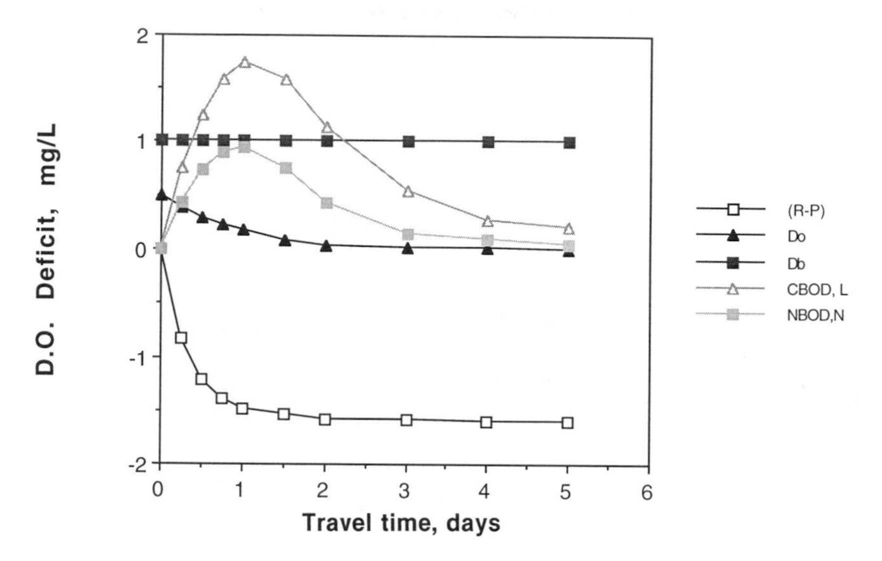

Figure 6.7 Sources and sinks of D.O. deficit in a stream.

When a wastewater discharge enters the river, a flow-weighted average mass balance of the streams is computed for the new initial conditions, assuming instantaneous complete mixing at $x = 0$.

$$L_0 = \frac{W + L_s Q}{Q + Q_w} \tag{71}$$

where L_0 = initial CBOD concentration at $x = 0$, ML^{-3}
W = wastewater discharge, MT^{-1}
L_s = upstream CBOD concentration entering $x = 0$, ML^{-3}
Q = river flowrate, L^3T^{-1}
Q_w = wastewater flowrate, L^3T^{-1}

If the wastewater mass of CBOD is large compared to upstream sources and if the wastewater flowrate is small relative to the river, equation (71) reduces to the following:

$$L_0 = \frac{W}{Q} \tag{72}$$

Similar mass balances are performed for NBOD to obtain N_0 at $x = 0$.

The initial D.O. deficit concentration is somewhat more complicated. The dissolved oxygen concentrations and temperatures should be averaged first, then the new D.O. deficit (D_0) is calculated based on the flow-weighted average temperature.

$$C_0 = \frac{Q_1 C_1 + Q_2 C_2}{Q_1 + Q_2} \tag{73}$$

$$T = \frac{Q_1 T_1 + Q_2 T_2}{Q_1 + Q_2} \tag{74}$$

$$D_0 = (C_{sat} - C_0) \quad \text{at } T_0 \tag{75}$$

where C_0 = initial mixed D.O. concentration, ML^{-3}
Q_1 = main river flowrate, L^3T^{-1}
C_1 = main river D.O. concentration entering $x = 0$, ML^{-3}
Q_2 = tributary flowrate, L^3T^{-1}
C_2 = tributary D.O. concentration entering $x = 0$, ML^{-3}
T_0 = initial mixed temperature, °C
T_1 = main river temperature, °C
T_2 = tributary temperature, °C
D_0 = initial mixed D.O. deficit, ML^{-3}
C_{sat} = saturation concentration of D.O. at T_0, ML^{-3}

Changes in other parameters that are assumed constant in equation (68) would require a new segment, such as:

- Velocity, u (changes in Q or area).
- Sediment oxygen demand, S.
- Net primary production, $P - R$.
- Background CBOD due to nonpoint source pollution, L_b.
- Reaeration rate constant (changes in velocity or depth).

Equation (68) includes the assumption that u, S, L_b, and $(P - R)$ are constant throughout the entire segment that is being modeled.

Figure 6.8 shows how a river has been divided into four segments of constant coefficients and the subsequent model simulation. Segment 1 was created due to a new waste discharge at $x = 0$, which affected D_0, L_0, and N_0. Segment 2 was necessary because a small dam at $x = 20$ km caused reaeration and a step-function change in D_0 (the distance variable, x, is reset to zero at $x = 20$ and the model equation is applied serially). Segment 3 was created due to another waste discharge, which affected D_0, L_0, and N_0 at $x = 35$ km (distance variable is reset to zero). Segment 4 was created due to the confluence of the river with a large tributary, which changed D_0, L_0, N_0, u, and k_a. At the beginning of Segment 4, $x = 45$ km, the x variable was once again reset to $x = 0$ and model equation (68) was reapplied.

Figure 6.8 is an example in which a waste load allocation and model simulation are necessary to achieve the water quality standard of 5 mg L^{-1} dissolved oxygen. Waste discharges W_1 and W_2 would be reduced until the D.O. concentration in the stream was greater than or equal to 5 mg L^{-1} everywhere. Once the allowable levels of W_1 and W_2 are determined, a new discharge permit would be written for the dischargers. They would need to minimize waste, prevent pollution within the plant where possible, and/or investigate further treatment alternatives to meet the new, more stringent permit values.

Example 6.4

A tributary merges with a river. Calculate the initial concentrations of dissolved oxygen, D.O. deficit, CBOD, and temperature at the confluence for use in a D.O. model.

River	Tributary
$T = 26.3$ °C	$T = 24.9$ °C
$C_{\text{sat},r} = 8.4$ mg L^{-1}	$C_{\text{sat},t} = 8.6$ mg L^{-1}
$C_r = 7.3$ mg L^{-1} (D.O.)	$C_t = 6.8$ mg L^{-1} (D.O.)
$L_r = 3.0$ mg L^{-1} (CBOD)	$L_t = 6.0$ mg L^{-1} (CBOD)
$Q_r = 2000$ cfs	$Q_t = 500$ cfs

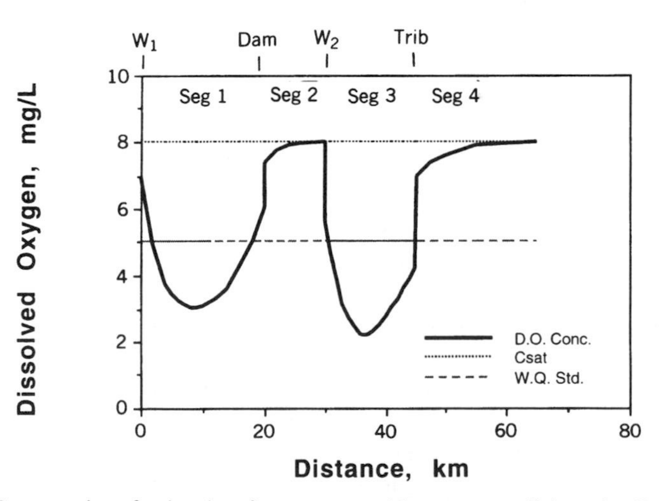

Figure 6.8 Segmentation of a river into four segments with constant coefficients for dissolved oxygen modeling.

Solution: We use flow-weighted average concentrations and assume complete mixing of the two streams. The following concentrations and temperature can be used as the initial conditions at $x = 0$ of the new segment beginning at the confluence. Subscript r refers to the river and t to the tributary.

$$L_0 = \frac{Q_r L_r + Q_t L_t}{Q_r + Q_t} = \frac{(2000)(3.0) + (500)(6.0)}{2000 + 500}$$

$$L_0 = 3.6 \text{ mg L}^{-1} \text{ (CBOD)}$$

$$C_0 = \frac{Q_r C_r + Q_t C_t}{Q_r + Q_t} = \frac{(2000)(7.3) + (500)(6.8)}{2000 + 500}$$

$$C_0 = 7.2 \text{ mg L}^{-1} \text{ (D.O. concentration)}$$

To determine the D.O. deficit concentration, one must first mix the temperature of the two streams and then look up the saturation concentration at that temperature.

$$T_0 = \frac{Q_r T_r + Q_t T_t}{Q_r + Q_t} = \frac{(2000)(26.3) + (500)(24.9)}{2000 + 500}$$

$$T_0 = 26.02 \text{ °C}$$

At 26 °C, the saturation concentration for dissolved oxygen is approximately 8.5 mg L^{-1} (see Appendix A). Therefore the initial D.O. deficit is 1.3 mg L^{-1}.

$$D_0 = C_{sat} - C_0 = 8.5 - 7.2 = 1.3 \text{ mg L}^{-1}$$

6.5.2 Diurnal Variations

Waste load allocations for small rivers are often complicated by algal productivity and diurnal swings in dissolved oxygen concentrations during critical conditions (7-day, 1-in-10-year low flow). In these cases, it is necessary to solve the time-variable mass balance differential equation. Time-variable (or dynamic) solutions are necessary because water quality standards are written in terms of a daily average minimum allowable D.O. concentration (5 mg L^{-1}) and a minimum concentration that cannot be violated at any time (4 mg L^{-1}). Thus it is necessary to know the minimum D.O. concentration in the diurnal cycle at the critical point in the stream. Normally it occurs at sunrise after an entire night of respiration without photosynthesis. Daily average D.O. concentrations can then be computed from the mean of the daily fluctuations.

The mass balance equations are written in terms of D.O. concentration (rather than D.O. deficit) because the standards are in terms of concentrations necessary to protect fish and aquatic biota.

$$\frac{\partial C}{\partial t} = -\frac{Q}{A}\frac{\partial C}{\partial x} + k_a(C_s - C) - k_d L(x) - k_n N(x) + P(x, t) - R(x) - \frac{S(x)}{H} \qquad (76)$$

where C = D.O. concentration, ML^{-3}
t = time, T
Q = flowrate, L^3T^{-1}
A = cross-sectional area, L^2
x = longitudinal distance, L
k_a = reaeration rate constant, T^{-1}
C_s = saturation D.O. concentration, ML^{-3}
k_d = CBOD deoxygenation rate constant, T^{-1}
k_n = NBOD deoxygenation rate constant, T^{-1}
$L(x)$ = CBOD concentration with distance, ML^{-3}
$N(x)$ = NBOD concentration with distance, ML^{-3}
$P(x, t)$ = primary production as a function of x and t, $ML^{-3}T^{-1}$
$R(x)$ = plant respiration with distance, $ML^{-3}T^{-1}$
$S(x)$ = sediment oxygen demand with distance, $ML^{-2}T^{-1}$
H = mean depth, L

Equation (76) can be solved numerically using the method of characteristics. In essence, the method follows a parcel of fluid down the river at constant velocity, computing variations in time. We obtain the concentration profile in space and time by recovering the results at each spatial increment (Δx) and time step (Δt).

In order to get started, one needs a concentration profile with distance at $t = 0$, and a diurnal cycle of concentrations at the head end of the river segment ($x = 0$). The diurnal cycle is repeated at $x = 0$ until a periodic solution is obtained throughout the segment. The solution is representative of worst case (critical) conditions. This model is valid under steady flow conditions, constant temperature, and constant wastewater discharges.

The model should be calibrated and verified with independent sets of field data that are roughly representative of low-flow conditions. Then the actual waste load allocation can be performed under mandated conditions (7-day, 1-in-10-year low flow) to obtain the allowable discharge(s) to the river that do not violate water quality standards.

Field survey data are very sensitive to cloud cover, which reduces primary production, lowering the average D.O. concentration throughout the day. Also, estimates of $P(x, t)$ and $R(x)$ are difficult to obtain. $R(x)$ includes only respiration due to plants. Yet, a light–dark bottle measurement of $R(x)$ includes deoxygenation due to

CBOD and NBOD. Furthermore, respiration by zooplankton may be stimulated or deterred within floating bottles. Primary production measurements in light bottles do not include rooted aquatic plants (macrophytes); they include only planktonic algae and are an underestimate of total primary production. For these reasons, it is desirable to estimate primary production, $P(x, t)$, and respiration, $R(x)$, by an analysis of the diurnal cycle obtained at each sampling station within the stream. This is equivalent to obtaining $P(x, t)$ and $R(x)$ from calibration of the model with field data.

Equation (76) requires a forcing function for primary production in each segment of the river. The rate of photosynthetic oxygen production, $P(t)$, may be represented as a half-cycle sine wave (O'Connor and DiToro[14]):

$$P(t) = P_m \sin\left(\frac{\pi}{f}(t - t_s)\right) \quad \text{when } t_s \leq t \leq (t_s + f) \tag{77a}$$

$$P(t) = 0 \quad \text{when } (t_s + f) \leq t \leq t_s \tag{77b}$$

where P_m = maximum rate of primary oxygen production, mg L^{-1} d^{-1}

t_s = time at which the source begins, days

f = fraction of the day over which the source is active (usually $\frac{1}{2}$ day)

In order to extend the periodic function of primary production for more than 1 day, a Fourier series is used:

$$P(t) = P_m\left(\frac{2f}{\pi} + \sum_{n=1}^{\infty} b_n \cos\left[2\pi n(t - t_s - f/2)\right]\right) \tag{78}$$

where $b_n = \cos(n\pi f) \, x \, \dfrac{4\pi/f}{(\pi/f)^2 - (2\pi n)^2}$

n = the number of cycles.

There is a diurnal variation in oxygen concentration at the head end of the segment, which can be handled in a manner similar to equation (78).

$$C_0(t) = \frac{A_0}{2} + \sum_{n=1}^{\infty}\left(A_n \cos\frac{n\pi}{T} + B_n \sin\frac{n\pi t}{T}\right) \tag{79}$$

where T = one-half period ($T = \frac{1}{2}$ day)

Dropping all but the first two terms of the Fourier series, we have

$$C_0(t) = \frac{A_0}{2} + A_1 \cos\left(\frac{\pi}{12}t\right) + B_1 \sin\left(\frac{\pi}{12}t\right) + A_2 \cos\left(\frac{\pi}{6}t\right) + B_2 \sin\left(\frac{\pi}{6}t\right) \quad (80)$$

where A_0, A_1, A_2, B_1, and B_2 are fitting parameters that determine the best representation of the diurnal D.O. cycles at the upstream boundary. Figure 6.9 shows data for the South Fork of the Shenandoah River in northwestern Virginia (Deb and Bowers[15]). Note that the dissolved oxygen concentrations are lowest at approximately 7–8 a.m. (hours 7 and 31) because respiration was active during the entire preceding night in the absence of primary production. Maximum D.O. concentrations occurred in the early evening around 7–8 p.m. (hours 19 and 43). This does not correspond to the time of maximum photosynthetic production, P_m, because of a time lag associated with accumulation of dissolved oxygen from primary production and reaeration. The maximum rate of photosynthetic oxygen production occurred from 12:00 noon to 4:00 p.m. The time lag between P_m and maximum D.O. concentrations is determined by the model parameters k_a, $R(x)$, and $P(x, t)$. The best fit to observed data in Figure 6.9 is given by Fourier series coefficients $A_0 = 16.16$, $A_1 = 0.575$, $A_2 = -0.1$, $B_1 = -1.025$, and $B_2 = 0.0$. The proper representation of upstream diurnal D.O. variations is very important in predicting diurnal D.O. variations at the downstream stations.[15]

Deb and Bowers[15] presented diurnal D.O. cycles for the South Fork of the Shenandoah River, Virginia. There was a large wastewater discharge of CBOD and NBOD at mile point $x = 0$ that necessitated a waste load allocation. They obtained the field data presented in Figure 6.10 at three downstream locations during summer low-flow conditions ($Q = 390$ cfs), but it was not at the 7-day, 1-in-10-year extreme ($Q = 160$ cfs). The diurnal variation reached a minimum at approximately 6:00 a.m. and a maximum at 6:00 p.m. (12-hour half-cycle). The areas under the diurnal curves were integrated according to the method of Erdmann[16] to determine P_m, $P(x, t)$, and $R(x)$ at each sampling station. Input parameters that resulted in a successful calibration are shown in Table 6.5. Model results fit the field data quite closely.

The model was verified with an independent set of field data from a different year, and a waste load allocation was performed subsequently. To determine allowable discharges of CBOD and NBOD, the flowrate was adjusted to the required 7-day, 1-in-10-year low flow of 160 cfs. Critical conditions occurred at 7:00 a.m., and results are given by Figure 6.11. The classic D.O. sag curve is characteristic of a large waste discharge, which occurs at $x = 0$. The critical point is at $x_c = 4.6$ mi downstream, and the allowable discharges to maintain dissolved oxygen concentrations (D.O. > 4 mg L^{-1}) were estimated to be 1740 kg d^{-1} CBOD and 644 kg d^{-1} NBOD. Interestingly, the average D.O. concentration over the day was nearly 7 mg L^{-1} and it did not come close to violating the water quality standard of 5 mg L^{-1}. But the diurnal fluctuations and the minimum D.O. concentration created by nighttime respiration were critically important. In such cases, time-variable solutions to the D.O. equations are necessary for waste load allocations.

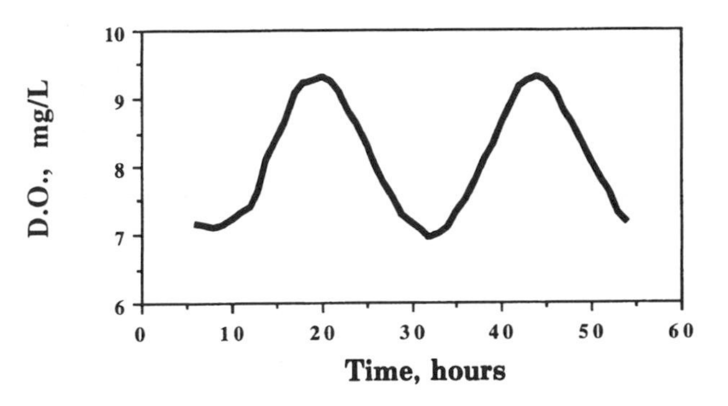

Figure 6.9 Diurnal dissolved oxygen concentration at $x = 0$, fit with Fourier series ($t = 0$ is at midnight).

6.6 UNCERTAINTY ANALYSIS

In waste load allocations and, for that matter, most environmental modeling, government regulators would like to know more than the average concentrations that one expects for a particular pollutant. They would also like to know how certain we are of the answer. In statistical terms, this means that the modeler should provide the decision-maker with the best estimate of the pollutant concentration (the mean) and the variation expected (the standard deviation). This allows the decision-maker more information to decide whether to act. Under random sampling, the expected value of a variable is its mean, and the measure of its variation is its variance (the standard deviation squared).

Because environmental models are simplified approximations of reality, they contain errors. In using models and field data, we embrace errors of several types. Errors propagate through models to cause uncertainty in the final result. This section is about "error analysis" or the equivalent term, uncertainty analysis. How certain are we of the answer (the best estimate)?

Models include the following types of error:

- Model error.
- Errors in the state variables (dependent variables and initial conditions).
- Errors in the input data used to drive the model.
- Parameter error (rate constants, coefficients, and independent variables).

The first type of error, model error, is the incorrect formulation of the model, that is, the difference between the environment and the differential equations in the model. Processes may have been omitted or included improperly; more state variables may be needed to describe the ecosystem; and unforeseen reactions may become important in different ranges of parameter space. In all cases, the only indica-

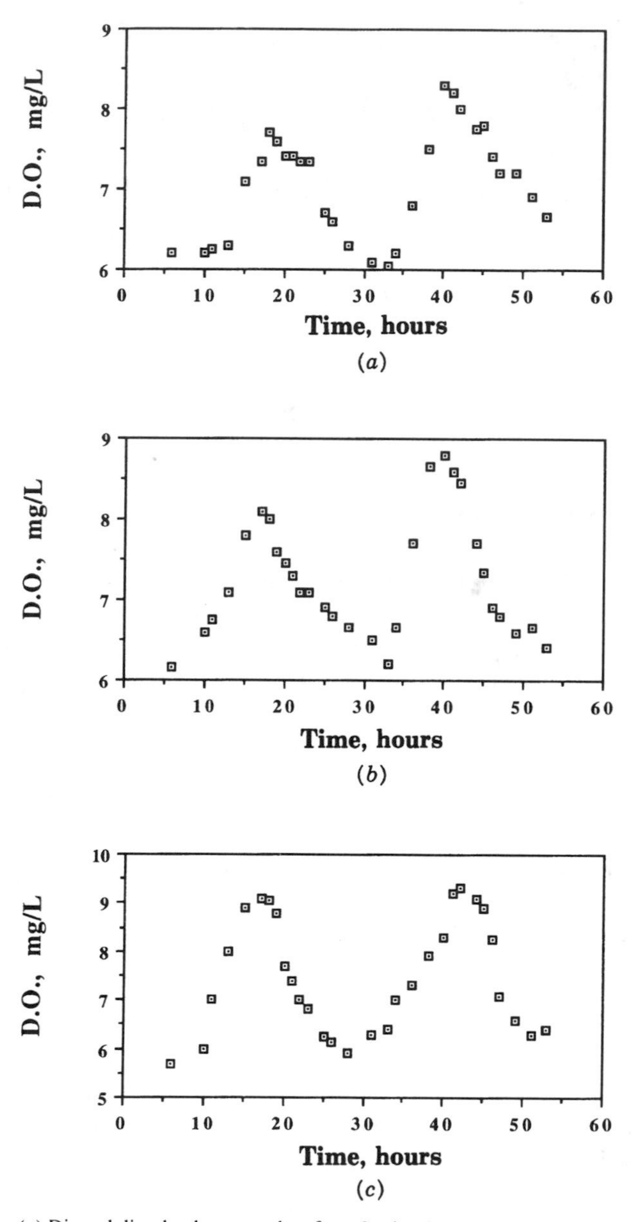

Figure 6.10 (a) Diurnal dissolved oxygen data from Station 2A at $x = 1.6$ mi.(b) Data from Station 2 at $x = 2.7$ mi. (c) Data from Station 3 at $x = 6.8$ mi.

Table 6.5 Model Input Data for Calibration, South Fork Shenandoah River[15]

	Model Reach Numbers								
Parameters	1	2	3^a	4^a	5	6^a	7	8	9
Sampling stations	1A	2A	2A–2	2–2B	2B–3	3–4	4-5	5-6	6-7
x, mi	0	0.02	1.6	2.7	4.6	6.8	8.2	8.8	9.3
Length of reach (mi)	0.02	1.58	1.1	1.9	2.2	1.4	0.6	0.5	4.2
Flow (cfs)	301.0	315.4	315.4	315.5	315.5	315.5	315.5	315.5	315.7
Velocity (mi d^{-1})	6.68	6.68	9.9	12.2	11.7	7.9	2.8	2.8	11.6
Travel time (day)	0.003	0.24	0.12	0.16	0.18	0.18	0.20	0.20	0.36
Temperature (°C)	22.8	22.8	23.0	23.6	23.6	22.8	22.5	22.8	22.2
Reaeration coefficient (k_a, day^{-1})	1.8	1.8	2.2	2.0	2.0	1.6	0.43	0.37	2.64
Deoxygenation coefficient (k_d, day^{-1})	0.20	0.40	0.40	0.40	0.40	0.40	0.40	0.40	0.40
Nitrification rate coefficient (k_n, day^{-1})	0.20	1.25	1.25	1.25	1.25	1.25	0.20	0.20	0.20
Benthic coefficient (S)	0	0	0	0	0	0	0	0	0
Maximum photosynthetic rate (P_m, mg L^{-1} d^{-1})	14.23	14.23	18.38	18.91	18.91	6.35	6.35	6.35	6.35
Respiration rate (R, mg L^{-1} d^{-1})	6.50	6.50	4.13	5.24	5.24	1.12	1.12	1.12	1.12

^{a}Segments with diurnal field data shown in Figure 6.10.

tion of model error is that the field data do not match model predictions. But we don't know why! Model error is the most difficult type to consider in uncertainty analyses, but one approach is to use several different models for the same application. Then one can obtain an estimate of the model error by virtue of the differences in predictions among the models under many different circumstances. By "flexing" the models in simulations of a wide variety of field conditions, it helps to accentu-

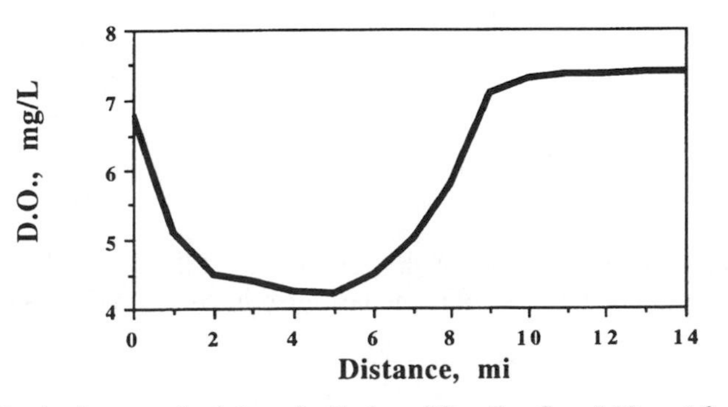

Figure 6.11 Dissolved oxygen simulation of critical conditions (low flow, 7:00 a.m.) for the South Fork of the Shenandoah River.[15]

ate the differences among the models. If a model performs better under certain circumstances, one can discover the cause of model error.

Other errors in state variables, input data, and parameters are best addressed through the following methods: sensitivity analysis, first-order analysis, or Monte Carlo simulations. Let's discuss each one and their relative advantages and disadvantages.

6.6.1 First-Order Analysis

First-order analysis is based on the *approximation* of variation in the model-dependent variable(s) with variation in the independent variable(s) (or parameters). We measure the variation in the parameters by their statistical moments (the variance). Here, we follow the discussions of Benjamin and Cornell,[17] Reckhow and Chapra,[18] and Beck and van Straten.[19]

Consider a function or a set of equations where y is the dependent variable and x is the independent variable:

$$y = f(x) \tag{81}$$

If the function is smooth and well behaved, and provided that the variance of x is not too large, the expected value of y is approximately equal to the value of the function with the expected value of x:

$$E(y) = E(f(x)) \simeq f(E(x)) \tag{82}$$

where E is the expectation operator. The expected value of x is the mean (average) value of x. We can write a Taylor series expansion of $f(x)$ to approximate the function:

$$f(x) = f(x_0) + (x - x_0) \left.\frac{\partial f(x)}{\partial x}\right|_{x_0} + \frac{(x - x_0)^2}{2} \left.\frac{\partial^2 f(x)}{\partial x^2}\right|_{x_0} + \cdots \tag{83}$$

where x_0 is any point that satisfies the function. If we "linearize" equation (83) by using only the first two terms of the equation, we can consider the expansion of the function about its mean, $\bar{x}$.

$$f(x) \approx f(\bar{x}) + (x - \bar{x}) \left.\frac{\partial f(x)}{\partial x}\right|_{\bar{x}} \tag{84}$$

We now write an equation for the variance of $f(x)$ directly. The variance about the mean $S^2(f(\bar{x}))$ is zero so only the last term in equation (84) is considered. The variance of $f(x)$ is

$$S^2(f(x)) \approx S^2(x) \left.\left(\frac{\partial f(x)}{\partial x}\right)^2\right|_{\bar{x}} \tag{85}$$

where S is the sample standard deviation and S^2 is the sample variance of the function about any point from the mean, $\bar{x}$. Equation (85) is valid for a bivariate relationship involving only one independent variable. For a multivariate relationship, we must consider the covariation among independent variables:

$$S_y^2 \sim \sum_{i=1}^{n} \left(\frac{\partial f}{\partial x_i}\right)^2 S_{x_i}^2 + 2\sum_{j=1}^{n-1}\sum_{i=j+1}^{n} \left(\frac{\partial f}{\partial x_i}\right)\left(\frac{\partial f}{\partial x_j}\right) S_{x_i} S_{x_j}\, \rho_{x_i\, x_j} \tag{86}$$

where $\rho_{x_i\, x_j}$ is the correlation coefficient in a linear least-squares regression between x_i and x_j variables (Reckhow and Chapra[18]).

Equation (86) provides the basis for first-order uncertainty analysis. It is called "first-order" because we dropped higher order terms in the Taylor series expansion [equation (83)]. The variance in the independent variable depends on the variance (uncertainty) of the individual variables or parameters S_{x_i}, the sensitivity of the independent variable compared to changes in each parameter df/dx_i, and the correlation between variables $\rho_{x_i\, x_j}$.

First-order uncertainty analysis has the disadvantages of being an approximation to the real solution—it may be erroneous for highly nonlinear equations due to the variance of higher order terms that were dropped in equation (83). Nevertheless, it has been shown to be quite an effective predictor of error in many environmental models,[19] and it preserves the covariance structure of the model parameters properly. It is a parametric statistical approach that requires the use of mean and standard deviations (normal probabilities).

Example 6.5 First-Order Analysis

The Streeter–Phelps equation for BOD in a stream contains parameters L_0 and k_d for initial BOD and deoxygenation, respectively. Suppose many measurements of L_0 and k_d are obtained for the stream that is being modeled and they have the following mean and standard deviations:

$$L_0 = 10.0 \pm 2.0 \text{ mg L}^{-1}$$

$$k_d = 0.6 \pm 0.1 \text{ day}^{-1}$$

The stream velocity is exactly 0.305 m s^{-1}.

$$L = L_0 e^{-k_d x/u}$$

The parameters are weakly correlated with each other. When L_0 is large, there is a tendency for faster deoxygenation $\rho_{L_0\, k_d} = +\,0.7$.

Solution: Use the first-order analysis, equation (86), to estimate the variance in the BOD concentration at downstream distances 0, 10, 20, 40, 60, and 100 km.

$$\frac{\partial L}{\partial k_d} = -\frac{L_0 x}{u} e^{-k_d x/u}$$

$$\frac{\partial L}{\partial L_0} = e^{-k_d x/u}$$

$$S_L^2 = \left(\frac{\partial L}{\partial k_d}\right)^2 S_{k_d}^2 + \left(\frac{\partial L}{\partial L_0}\right)^2 S_{L_0}^2$$

$$+ 2\left(\frac{\partial L}{\partial k_d}\right)\left(\frac{\partial L}{\partial L_0}\right)^2 S_{k_d} S_{L_0} \rho_{k_d L_0}$$

At $x = 20$ km, for example, we can solve the equation for the estimate of variance in the BOD concentration in the stream due to uncertainty in L_0 and k_d. The units must be converted and careful attention to the units will help to prevent mistakes.

$$S_L^2 = \left(\frac{7.2521 \text{ mg d}}{\text{L}}\right)^2 (0.1 \text{ d}^{-1})^2 + (0.6342)^2 (2.0 \text{ mg L}^{-1})^2$$

$$+ 2\left(\frac{7.2521 \text{ mg d}}{\text{L}}\right) (0.6342) (0.1 \text{ d}^{-1}) (2.0 \text{ mg L}^{-1})\ 0.7$$

$$S_L^2 = 3.4225 \text{ (mg L}^{-1})^2 \quad \text{variance estimate}$$

The mean expected value for L is given by the BOD equation assuming average parameter values.

$$L_{20} = L_0 e^{-k_d x/u} = (10 \text{ mg L}^{-1}) \exp (0.6 \text{ d}^{-1} \times 0.759 \text{ d})$$

$$L_{20} = 6.342 \text{ mg L}^{-1}$$

The mean and standard deviation of BOD at 20 km downstream from the discharge point is 6.342 ± 1.850 mg L^{-1}. There is a rather large uncertainty due to the variation in L_0 and the positive correlation between L_0 and k_d.

x, km	Travel Time, days	$\overline{L}$, mg L^{-1} (mean)	$S_{(L)}$, mg L^{-1} (standard deviation)
0	0	10.000	2.000
10	0.3795	7.9637	1.817
20	0.7590	6.3421	1.850
40	1.5179	4.0222	1.3067

x, km	Travel Time, days	$\bar{L}$, mg L^{-1} (mean)	$S_{(L)}$, mg L^{-1} (standard deviation)
60	2.2769	2.5509	1.0062
100	3.7948	1.0261	0.5528

The error term (standard deviation) is well-behaved—it does not grow; but the coefficient of variation (std. dev./mean) increases from 20% to 53.9% from $x = 0$ to $x =$ 100 km.

6.6.2 Monte Carlo Analysis

Monte Carlo analysis is a "brute force" solution to the problem of uncertainty analysis in environmental models. It is conceptually very simple. One uses a random number generator to select parameter values from a known or suspected distribution (probability density function). After the proper number of parameters have been randomly selected (synthetic sampling), the model is run and the results of the simulation are saved. Then the process is repeated over and over until a statistically significant number of simulations (large number of realizations) have been recorded. From the distribution of results, a mean predicted value and a standard deviation can be calculated that reflects the combined uncertainty of all parameter errors.

The advantages of Monte Carlo simulation are that it does not require linearization of the model equations and it is, in theory, nonparametric. Error terms in parameters and independent variables can be represented as any probability density functions and sampled without the requirement of standard deviation and mean (Gaussian) statistics. However, Monte Carlo analysis is computationally expensive for large models, and it doesn't preserve the covariance structure among the parameters (unless one performs the synthetic sampling in a special manner to obtain a realistic set of outcomes).

Parameter distributions may be determined by multiple samples collected in the field, determined in microcosm experiments in the laboratory, or from expert judgment. Typical distributions of parameters are normal distributions, lognormal, "hat" functions (i.e., constant probabilities over the range of parameter values), and trapezoid functions (i.e., constant probabilities in a predominant range and decreasing linearly at both ends of the function). Normal distributions are thought to arise from "additive" processes in the environment.

$$f(x) = \frac{1}{\sigma\sqrt{2\pi}} \exp\left[-\frac{1}{2}\left(\frac{x-\mu}{\sigma}\right)^2\right] \tag{87}$$

where μ is the mean value of the distribution and σ is the standard deviation. Equation (87) gives the probability density function (pdf) for a normal distribution. It is a bell-shaped Gaussian distribution. The area under the curve at $x = \pm 1\sigma$ contains 68.3% of the values; the area under the curve at $x = \pm 2\sigma$ contains 95.5% of values.

Perhaps the most common distribution of parameters in the environment is the lognormal distribution. If you don't know anything about how a parameter varies, you should assume a lognormal distribution. It accounts for environmental measurements with skewness in their distribution, which is often the case. Lognormal distributions are thought to arise from multiplicative processes. They can be made into a Gaussian normal distribution simply by performing a log transformation on the data. Equation (88) is the pdf for a lognormal distribution:

$$f(y) = \frac{1}{y\sqrt{2\pi}\,\sigma_x} \exp\left[-\frac{1}{2}\left(\frac{\ln y - \mu_x}{\sigma_x}\right)^2\right] \tag{88}$$

where μ_x is the geometric mean of the parameter values, σ_x is the geometric deviation, and y are the parameter values themselves, which must be positive. The y values are transformed by taking their natural logarithm.

$$x = \ln y \tag{89}$$

where $y \geq 0$ and $-\infty \leq x \leq +\infty$. The area under a probability density function is normalized to 1.0, so the use of pdf's allows a ready comparison to be made in the shape of various parameters.

Monte Carlo analysis offers some advantages over first-order analysis. Due to the low unit cost of computing, it has become the favored method among environmental scientists and engineers for uncertainty analysis. Steps in the analysis and computer program for Monte Carlo include the following:

1. Determine the uncertainty distributions (pdf) for each parameter, input function, and variable that you want to analyze.
2. Sample each of the parameter distributions using a random number generator that takes into account the probability of each value. Most computer operating systems and some spreadsheets have a random number generator that you can call to synthetically sample your parameter distributions prior to each model simulation.
3. Use the set of parameter values selected as input for the first simulation. Run the model and save the output to tape or file.
4. Repeat steps 2 and 3 a large number of times, usually 100–200, or until the statistical output no longer changes by repeated realizations of the model.
5. Sort the stored output data and plot the output as the mean value plus or minus the probability range that is desired.

The following two examples illustrate how to call a random number generator to sample distributions of reaeration and deoxygenation rate constants for the Streeter–Phelps equation.

Example 6.6 Random Number Generation

From field sampling, it is known that the deoxygenation rate constant and reaeration rate constants follow a normal distribution with mean and standard deviations shown below:

$$k_a = 2.0 \pm 0.3 \text{ day}^{-1}$$

$$k_d = 0.6 \pm 0.1 \text{ day}^{-1}$$

Write a FORTRAN program that calls a random number generator to sample the distributions for k_a and k_d.

Solution: The following program was written that generates the parameter values, prints them out, and writes them to tape for input to the S-P simulation program. Fifty values are generated here. The random numbers are called from an external subroutine of the International Mathematics and Statistics Library (IMSL).

```
      Program simulation
      parameter (Numr=50)
      real X(Numr), Y(Numr)
      External RNSET, RNUN, RNNOR, SSCAL, SADD
c     open(2, file-'deoxy')
c     write(2, 111)
      Iseed=12345
      call RNSET (Iseed)
      avg=0.6
      avg1=2.0
      std=0.1
      std1=0.3
      DO I=1, 50
c         call RNUN (Numr, X)
          call RNNOR (Numr, X)
          call RNNOR (Numr, Y)
          call SSCAL (Numr, std, X, 1)
          call SSCAL (Numr, std1, Y, 1)
          call SADD (Numr, avg, X, 1)
          call SADD (Numr, avg1, Y, 1)
      ENDDO
      print 111
111   format (T10, '50 numbers generated for kd(0.6+0.1)')
      print 333, X
c     write (2, 333), X
333   format (T20, F6.2)
      print 222
```

```
222   format (//T10, '50 numbers generated for ka(2.0+0.3)')
      print 444, Y
444   format (T20, F6.2)
      STOP
      END
```

The values of k_d and k_a must be used in pairs to generate 50 realizations of the D.O. deficit (D) versus distance. The value of deficit concentration is saved at each km or 5-km distance for 50 simulations. Then, the deficit concentrations are sorted and the mean (or median) and quartile values can be plotted. Reporting the expected value and the uncertainty (± 25% quartile values) gives regulators more information on which to base their decisions for water quality management. Figure 6.12 shows the range of uncertainty in BOD and D.O. deficit (D) with downstream distance.

50 Numbers Generated for kd(0.6±0.1)	50 Numbers Generated for ka(2.0±0.3)
0.61	1.49
0.69	1.98
0.76	1.93
0.49	2.52
0.71	2.58
0.63	1.89
0.50	1.56
0.61	2.06
0.56	2.56
0.51	2.11
0.75	1.82
0.56	1.52
0.55	1.91
0.52	1.70
0.70	1.77
0.71	2.50
0.59	1.76
0.77	2.03
0.53	1.97
0.64	1.88
0.80	2.11
0.57	1.86
0.60	1.97
0.56	2.31
0.58	2.18
0.51	1.82
0.67	1.77
0.59	1.91
0.51	2.08

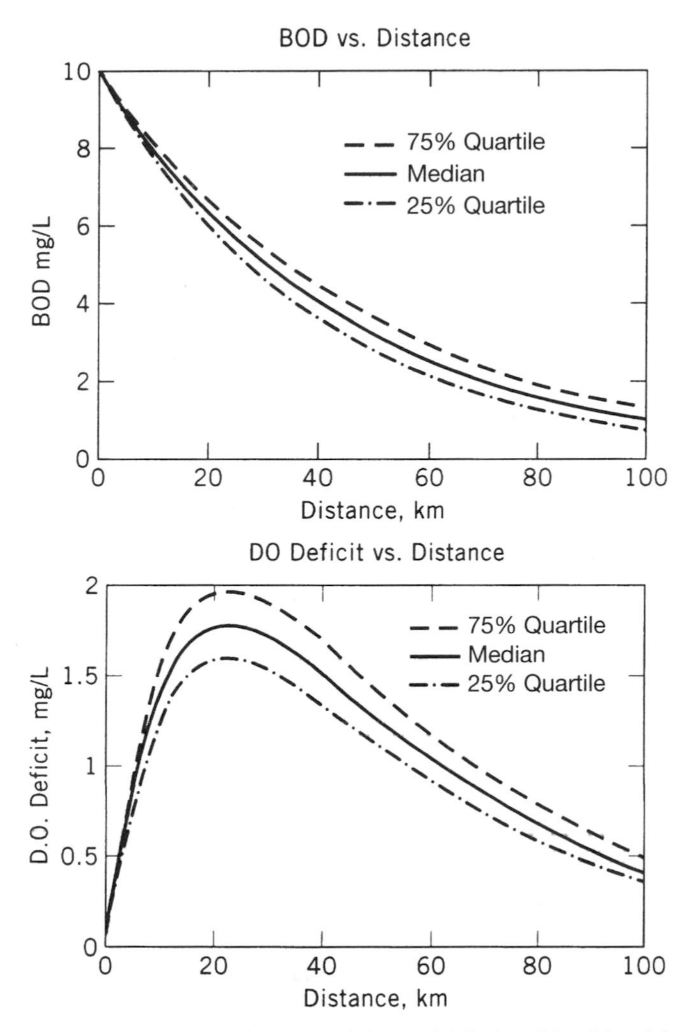

Figure 6.12 Monte Carlo simulation of Streeter–Phelps model for $k_a = 2.0 \pm 0.3$ and $k_d = 0.6 \pm 0.1$

50 Numbers Generated for kd(0.6±0.1)	50 Numbers Generated for ka(2.0±0.3)
0.39	1.94
0.59	2.58
0.54	2.15
0.72	1.47
0.52	2.08
0.51	2.02
0.56	2.19

50 Numbers Generated for kd(0.6±0.1)	50 Numbers Generated for ka(2.0±0.3)
0.78	1.80
0.69	1.81
0.66	1.88
0.69	1.67
0.54	2.01
0.64	2.45
0.44	2.36
0.62	1.25
0.64	2.42
0.72	1.98
0.61	2.15
0.48	1.68
0.80	2.35
0.67	1.76

The D.O. deficit at the critical point is shown to be 1.76 ± 0.15 mg L^{-1} within a 50% confidence level (two quartiles).

Increasingly, environmental scientists and engineers will be asked to provide uncertainty analyses in all modeling studies. Decision-makers need to know the expected value (best estimate) and the second moment or variance of the estimate. Better decisions should result from more information. In this section, two methods have been presented for uncertainty analysis, first-order and Monte Carlo simulation. Table 6.6 provides a summary of the advantages and disadvantages of each approach.

6.7 DISSOLVED OXYGEN IN LARGE RIVERS AND ESTUARIES

6.7.1 BOD–D.O. Deficit, Steady State

The control volume approach can be used to write mass balance equations for BOD and D.O. (or D.O. deficit) in a large river or estuary that has considerable longitudinal mixing. For BOD and D.O. deficit, we may consider the following reactions in the simplest case:

$$L \xrightarrow{k_d} D \xrightarrow{k_a} 0 \tag{89}$$

where L = BOD concentration, mg L^{-1}
D = D.O. deficit, mg L^{-1}
k_d = deoxygenation rate constant, day^{-1}
k_a = reaeration rate constant, day^{-1}

Table 6.6 A Comparison of Methods for Uncertainty Analysis

Method	Theory	Advantages	Disadvantages
Sensitivity analysis	Vary one parameter at a time by ±1% to determine the change in state variable.	Easy to use and conceptually simple.	Not truly uncertainty analysis. Parameter may not be independent of other parameters.
First order analysis	Solve the variance equation directly for the state variable as a function of the uncertainty in the parameter value. (It is a linearized approximation that is solved.)	Preserves the covariance relationships among the parameters. Uses less computer time for complex models.	It is an approximation to the real solution. For highly nonlinear equations, the method is not very accurate. It is parametric and requires the use of mean and standard deviations.
Monte Carlo analysis	Select parameters randomly from a known or hypothetical error distribution. Run the model many times to determine the expected value and the variance (uncertainty)	Exact solution. Does not depend on parametric statistics (mean and standard deviation). Less error at the extremes usually.	Uses lots of computer time to generate statistically significant number of realizations. May not preserve covariance structure among parameters and variables.

Transport within an infinitesimal slice of control volume is considered as discussed in Chapter 2 for a plug flow with dispersion model. Within the control volume

$$\text{Accumulation} = \text{Inputs} - \text{Outputs} \pm \text{Reactions} \tag{90}$$

Inputs and outputs are from advection and dispersion. The infinitesimal slice of the river/estuary is of volume $V = A\,\Delta x$, where A is the cross-sectional area and Δx is the incremental distance in the longitudinal direction. The mass balance equation for BOD in the river/estuary is represented by equation (91):

$$V\frac{\partial L}{\partial t} = \underbrace{QL}_{\text{advection in}} \underbrace{- Q\left[L + \frac{\partial L}{\partial x}\Delta x\right]}_{\text{advection out}}$$

$$\underbrace{-EA\frac{\partial L}{\partial x}}_{\text{dispersion in}} \underbrace{+ EA\left[\frac{\partial L}{\partial x} + \frac{\partial}{\partial x}\left(\frac{\partial L}{\partial x}\right)\Delta x\right]}_{\text{dispersion out}}$$

$$\underbrace{-\,k_d LV}_{\text{deoxygenation}} \tag{91}$$

where E = longitudinal dispersion coefficient, L^2T^{-1}
Q = volumetric flowrate, L^3T^{-1}
t = time

The "advection out" term accounts for the outflow of BOD from the control volume. At the downstream end of the control volume, the BOD concentration is different from that at the upstream end; it is modified by the change in the BOD concentration over the incremental distance, $(+\partial L/\partial x)\Delta x$. The term for transport into the control volume by dispersive mixing has a negative sign because mass is transferred in the positive x-direction only when the BOD gradient $(\partial L/\partial x)$ is negative. Conversely, the term for dispersion out of the control volume is positive because mass is transferred out when the gradient is negative. The term for dispersion out of the control volume contains an expression for the change in the BOD gradient (slope) over the incremental distance, $(\partial/\partial x)(\partial L/\partial x)\Delta x$. The reaction term accounts for BOD decay in the river/estuary and deoxygenation.

Equation (91) can be expanded and terms canceled as shown below:

$$V\frac{\partial L}{\partial t} = \cancel{QL} - \cancel{QL} - Q\left(\frac{\partial L}{\partial x}\right)\Delta x$$

$$-\cancel{EA\frac{\partial L}{\partial x}} + \cancel{EA\frac{\partial L}{\partial x}} + EA\frac{\partial}{\partial x}\left(\frac{\partial L}{\partial x}\right)\Delta x - k_d LV \tag{92}$$

Rearranging and dividing both sides of equation (91) by the volume of the control slice ($V = A\,\Delta x$), we obtain the proper equation for 1-D transport and decay of BOD in a large river or estuary.

$$\frac{\partial L}{\partial t} = -\frac{Q}{A}\frac{\partial L}{\partial x} + E\frac{\partial^2 L}{\partial x^2} - k_d L \tag{93}$$

Equation (93) applies for the case of time-variable BOD in a river or estuary with constant freshwater flowrate, cross-sectional area, and dispersion coefficient. It is a parabolic, partial differential mass balance equation. A more general form of the equation is given by equation (94):

$$\frac{\partial L}{\partial t} = -\frac{1}{A}\frac{\partial (QL)}{\partial x} + \frac{1}{A}\frac{\partial}{\partial x}\left(EA\frac{\partial C}{\partial x}\right) - k_d L \tag{94}$$

in which the cross-sectional area, flowrate, and dispersion coefficient can vary in time and space.

The D.O. deficit equation is developed in a similar manner to equations (93) and (94).

$$\frac{\partial D}{\partial t} = -\frac{Q}{A}\frac{\partial D}{\partial x} + E\frac{\partial^2 D}{\partial x^2} + k_d L - k_a D \tag{95}$$

$$\frac{\partial D}{\partial t} = -\frac{1}{A}\frac{\partial (QD)}{\partial x} + \frac{1}{A}\frac{\partial}{\partial x}\left(EA\frac{\partial D}{\partial x}\right) + k_d L - k_a D \tag{96}$$

Under steady-state conditions and constant coefficients, equations (93) and (95) can be simplified and solved for the BOD–D.O. deficit in a large river or estuary. They become a set of ordinary rather than partial differential equations:

$$+E\frac{d^2 L}{dx^2} - u\frac{dL}{dx} - k_d L = 0 \tag{97}$$

$$+E\frac{d^2 D}{dx^2} - u\frac{dD}{dx} - k_a D + k_d L = 0 \tag{98}$$

where u is the mean velocity (freshwater) in the river or estuary (LT^{-1}). The mean velocity is equal to the freshwater flowrate divided by the average cross-sectional area. Equations (97) and (98) are the steady-state formulation for BOD and D.O. deficit in plug flow systems with dispersion. They are analogous to the Streeter–Phelps equation for swift flowing streams and rivers in Section 9.3.

The analytical solution to equation (97) is that of a second-order, linear, ordinary differential equation with constant coefficients. It is a homogeneous equation (without forcing functions) of the general algebraic quadratic form:

$$ay'' + by' + cy = 0 \tag{99}$$

and, in this case, we have the following identities based on equation (97):

$$a = E \tag{100a}$$

$$b = -u \tag{100b}$$

$$c = -k_d \tag{100c}$$

The general solution is of the form

$$Y = Ae^{gx} + Be^{jx} \tag{101}$$

where A and B are integration constants determined from the boundary conditions; and g and j are the roots of the algebraic equation (the notation $g \cdot j$ denotes two roots, "g and j").

$$g \cdot j = \frac{-b \pm \sqrt{b^2 - 4ac}}{2a} = \frac{u \pm \sqrt{u^2 + 4k_d E}}{2E} \tag{102}$$

$$g \cdot j = \frac{u}{2E}\left[1 \pm \sqrt{1 + \frac{4k_d E}{u^2}}\right] \tag{103a}$$

or

$$g \cdot j = \frac{u}{2E}[1 \pm m] \tag{103b}$$

where

$$m = \sqrt{1 + \frac{4k_d E}{u^2}}$$

In order to determine the integration constants, A and B, for the general equation (101), we must use upstream and downstream boundary conditions. Figure 6.13 is a schematic of the problem. We will divide the analytical solution into two parts: (1) an upstream segment above the wastewater discharge, W; and (2) a downstream segment below the discharge stretching all the way to the ocean, for the case of a coastal plain estuary.[13] Very far upstream, the concentration of BOD is expected to be zero. At the discharge point itself, $x = 0$, the concentration of BOD will be a maximum (L_0). For the upstream solution to equation (101), where x is negative above the wastewater discharge, we have the following boundary conditions:

$$\text{BC 1:} \quad \text{at } x = -\infty, L = 0 \tag{104}$$

$$\text{BC 2:} \quad \text{at } x = 0, L = L_0 \tag{105}$$

Based on the first boundary condition—equation (104)—the integration constant B in equation (101) must be equal to zero because the root g is always positive and j is always negative. Otherwise, the second term in equation (101) would be infinite since both j and x are negative at $x = -\infty$. The first term is zero at $x = -\infty$ because $e^{-\infty} = 0$.

$$L = A\, e^{g(-\infty)} + B\, e^{j(-\infty)} = 0 \quad \text{at } x = -\infty \tag{106}$$

$$\therefore \quad B = 0$$

From the second boundary condition—equation (105)–we know that the integration constant A is equal to L_0.

$$L = Ae^0 + \cancel{B}^{\,0} e^0 = L_0 \quad \text{at } x = 0 \tag{107}$$

$$\therefore \quad A = L_0$$

The solution to the upstream segment for BOD is

$$L = L_0\, e^{gx} \quad \text{for } x \le 0 \tag{108}$$

where $g = \frac{u}{2E}[1 + m]$

$$m = \sqrt{1 + \frac{4k_d E}{u^2}}$$

Similarly, the solution to equation (101) for the downstream segment is based on the boundary conditions.

$$\text{BC 1:} \quad \text{at } x = +\infty, L = 0 \tag{109}$$

$$\therefore \quad A = 0$$

$$\text{BC 2:} \quad \text{at } x = 0, L = L_0 \tag{110}$$

$$\therefore \quad B = L_0$$

The solution is given by equation (111):

$$L = L_0 \, e^{jx} \quad \text{for } x \geq 0 \tag{111}$$

where $j = \frac{u}{2E}[1 - m]$

$$m = \sqrt{1 + \frac{4k_d E}{u^2}}$$

In equations (108) and (111), the absolute value of g is always greater than the absolute value of j. Thus, the exponential decline in BOD due to transport and decay is much greater in the upstream direction than in the downstream direction, as depicted in Figure 6.13. BOD concentration declines exponentially in both directions as a result of dispersive mixing and deoxygenation (decay) according to solution equations (108) and (111).

One problem remains: the concentration of BOD at the source of the wastewater discharge is not known. Unlike the case of a plug-flow stream, we cannot assume that the concentration is equal to the mass input rate of BOD divided by the flowrate (W/Q). Now we must consider a mass balance equation around an infinitesimal slice of the estuary at $x = 0$.

The mass of material entering the upstream elemental slice at the discharge point ($x = 0$) is QL_0, where the waste discharge (W) enters. Similarly, the mass flux entering the upstream element is given by Fick's first law analogy.

$$\left(QL_0 - EA\frac{dL_1}{dx}\bigg|_{x=0} + W\right) \rightarrow \underset{x=0}{\overset{\Delta x}{\|}} \rightarrow \left(QL_0 - EA\frac{dL_2}{dx}\bigg|_{x=0}\right) \tag{112}$$

For steady-state conditions in equation (112), mass inputs on the left-hand side must be equal to mass outflows on the right-hand side: where Q is the flowrate (L^3T^{-1}); L_0 is the BOD concentration at $x = 0$ (ML^{-3}); E is the longitudinal tidal-averaged dispersion coefficient (L^2T^{-1}); A is the average cross-sectional area (L^2); L_1 is the upstream solution [equation (108)] for the BOD concentration, and the derivative is evaluated at $x = 0$; x is the longitudinal distance (L); W is the mass discharge rate (MT^{-1}); and L_2 is the downstream solution [equation (111)] whose derivative is evaluated at the discharge point, $x = 0$.

Evaluating the derivatives dL_1/dx and dL_2/dx at the discharge point gives

$$\left.\frac{dL_1}{dx}\right|_{x=0} = gL_0 \tag{113a}$$

$$\left.\frac{dL_2}{dx}\right|_{x=0} = jL_0 \tag{113b}$$

Substituting equations (113a) and (113b) into equation (112), we have

$$-g\,EAL_0 + W = -j\,EAL_0 \tag{114}$$

Rearranging equation (114), we can substitute the definition of g and j from equations (108) and (111).

$$L_0 = \frac{W}{EA(g-j)} = \frac{W}{EA\left[\frac{u}{2E}(1+m) - \frac{u}{2E}(1-m)\right]} \tag{115}$$

Recognizing that the flowrate is equal to the mean freshwater velocity times the cross-sectional area, the final solution for the BOD concentration at the discharge point is given by equation (116).

$$L_0 = \frac{W}{Qm} \tag{116}$$

where $m = \sqrt{1 + \frac{4k_dE}{u^2}}$

We see that the BOD concentration at the discharge point is less than that in a plug-flow stream (W/Q) because m is a number always greater than 1.0. As depicted in Figure 6.13, the BOD concentration is mixed upstream due to longitudinal dispersion. In estuaries, tidal action mixes the waste upstream from the discharge point. This causes some depletion of dissolved oxygen upstream from the waste discharge.

Now we must solve equation (98) for the D.O. deficit concentration in the large river or estuary. The analytic solution technique is similar to that for BOD except that we have a second-order, linear, *nonhomogeneous* ordinary differential equation.

$$E \frac{d^2D}{dx^2} - u \frac{dD}{dx} - k_a D = -k_d L \quad (117)$$

The forcing function on the right-hand side of equation (117) involves the BOD concentration, so we must substitute equations (108) and (111) for the upstream and downstream solutions of the D.O. deficit equation. The solution(s) are of the general form

$$y = Ae^{gx} + B\, e^{jx} + y_p(x) \quad (118)$$

where $y_p(x)$ is particular solution to the nonhomogeneous equation. We can solve equation (118) by the method of undetermined coefficients using boundary conditions as we did before. By trial and error, the particular solution (y_p) is given by equation (119):

$$y_p = \frac{k_d W}{(k_a - k_d) Q m_1} \exp\left[\frac{ux}{2E}(1 \pm m_1)\right] \quad (119)$$

where $m_1 = \sqrt{1 + \frac{4k_d E}{u^2}}$

which is the same definition as that of m previously, but we now must distinguish the root of the particular solution from that of the homogeneous equation. Use the positive root for the upstream portion of the solution ($x \leq 0$) and the negative root for the downstream ($x \geq 0$).

We solve equation (117) with the general solution [equation (118)] and boundary conditions to determine A and B in the upstream and downstream directions from the discharge point.

$$\text{BC 1:} \quad \text{at } x = +\infty,\ D = 0 \quad (120a)$$

$$\text{BC 2:} \quad \text{at } x = 0,\ D = D_0 \quad (120b)$$

$$\text{BC 3:} \quad \text{at } x = -\infty,\ D = 0 \quad (120c)$$

Evaluation of D_0 is accomplished via a mass balance (inputs = outputs) around an infinitesimally thin slice at the discharge point. Assuming that the discharge is saturated with dissolved oxygen, we have inputs and outputs of D.O. deficit due to dispersion and advection. Reactions do not have enough time to affect the D.O. deficit balance over an infinitesimal slice.

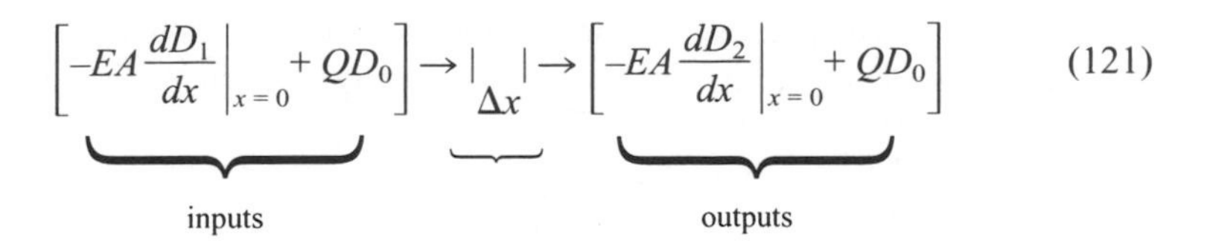

$$\left[-EA\frac{dD_1}{dx}\bigg|_{x=0} + QD_0\right] \rightarrow \big|_{\Delta x}\big| \rightarrow \left[-EA\frac{dD_2}{dx}\bigg|_{x=0} + QD_0\right] \qquad (121)$$

Because the advection terms on both sides of equation (121) cancel out, the equation tells us that the mass flux of D.O. deficit into the discharge point is equal to the mass flux out from the discharge point—it is a smooth D.O. deficit profile even though there is a cusp at $x = 0$ in the BOD profile (Figure 6.13).

The final analytic solution for the D.O. deficit concentration is given by equation (122).

$$D = \frac{k_d W}{(k_a - k_d)Q}\left[\frac{1}{m_1}\exp\left\{\frac{ux}{2E}(1 \pm m_1)\right\} - \frac{1}{m_2}\exp\left\{\frac{ux}{2E}(1 \pm m_2)\right\}\right] \qquad (122)$$

where $m_2 = \sqrt{1 + \frac{4k_a E}{u^2}}$

$m_1 = \sqrt{1 + \frac{4k_d E}{u^2}}$

In equation (122), one uses the + signs when $x \leq 0$ (upstream) and the – signs when $x \geq 0$ (downstream from the discharge). Equation (116) and (122) give the solution for the D.O. deficit in large rivers and estuaries as depicted by Figure 6.13.

The values for m_1 and m_2 contain important dimensional numbers that help to define the relative importance of dispersion to advection in rivers and estuaries with biochemical reactions.

$$\text{if } \frac{kE}{u^2} > 20, \text{ dispersion predominates} \qquad (123)$$

$$\text{if } \frac{kE}{u^2} < 0.05, \text{ advection predominates} \qquad (124)$$

For a typical coastal plain estuary in the United States, the BOD deoxygenation rate would be approximately 0.1 day^{-1}, the longitudinal dispersion coefficient due to tidal mixing is on the order of 25 mi^2 d^{-1}, and the freshwater (tidal-averaged) velocity is ~2 mi d^{-1}.

$$\frac{k_d E}{u^2} = \frac{(0.1)(25)}{(2)^2} = 0.63 \qquad (125)$$

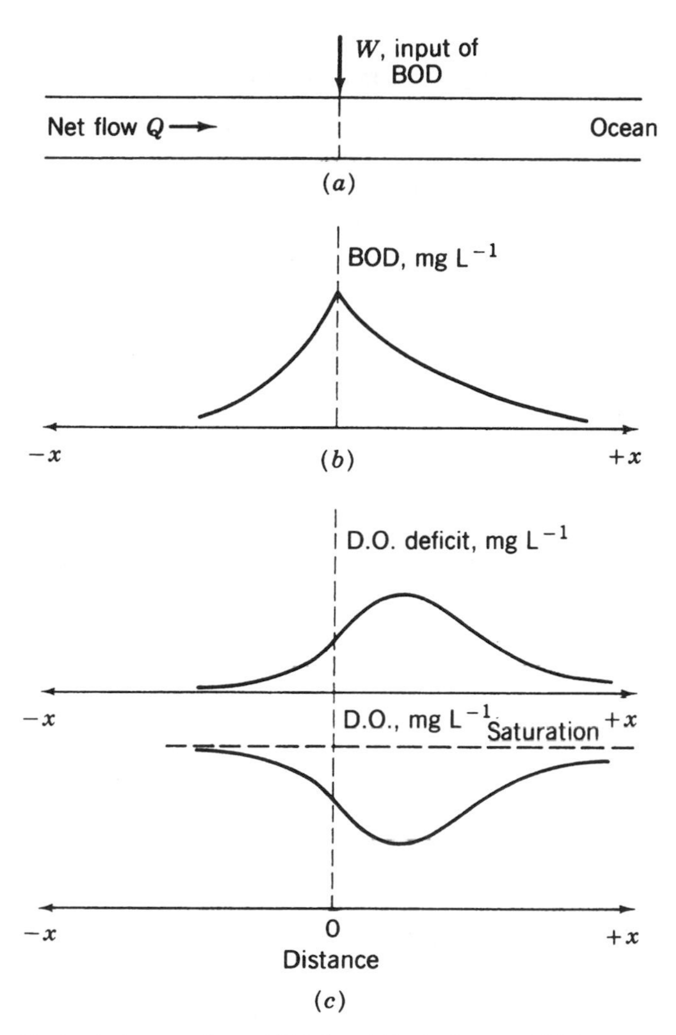

Figure 6.13 Biochemical oxygen demand (BOD) and dissolved oxygen (D.O.) concentration in an estuary, plut flow with dispersion model. (from Thomann and Mueller[13])

The above calculation indicates that both advection and dispersion are important in the transport of BOD in a coastal plain estuary. The next section describes how to estimate model parameters k_d, k_a, E, and u for field applications of the model.

Example 6.7 BOD–D.O. Deficit in an Estuary at Steady State

A continuous wastewater BOD discharge is released to a coastal plain estuary. Laboratory and field measurements were performed to estimate the parameters given below.

$E = 15\ \text{mi}^2\ \text{d}^{-1}$	Dispersion coefficient
$k_d = 0.15\ \text{day}^{-1}$	Deoxygenation rate constant
$u = 3.57\ \text{mi d}^{-1}$	Net (tidal-averaged) velocity
$k_a = 0.18\ \text{day}^{-1}$	Reaeration rate constant
$W/Q = 18.5\ \text{mg L}^{-1}$	Wastewater mass discharge rate ÷ flow rate

Solve for the steady-state BOD concentration and the D.O. deficit both upstream and downstream from the discharge point ($x = 0$).

Plot (a) the BOD versus x
(b) the D.O. deficit versus x
(c) k_dL and k_aD versus x
(d) the advective and dispersive fluxes:

$$uL,\ uD,\ -E\frac{dL}{dx},\ -E\frac{dD}{dx}$$

Solution: Solve for the BOD concentration and the D.O. deficit concentration using equations (108), (111), and (122).

$$L = L_0 \exp\left[\frac{ux}{2E}(1 \pm m_1)\right]$$

Use the + sign when $x \leq 0$ (upstream) and the – sign for $x \geq 0$ (downstream).

$$D = \frac{k_d W}{(k_a - k_d)Q}\left\{\frac{\exp\left[\frac{ux}{2E}(1 \pm m_1)\right]}{m_1} - \frac{\exp\left[\frac{ux}{2E}(1 \pm m_2)\right]}{m_2}\right\}$$

$$m_1 = \sqrt{1 + \frac{4k_d E}{u^2}} = 1.31$$

$$m_2 = \sqrt{1 + \frac{4k_a E}{u^2}} = 1.36$$

$$L_0 = \frac{W}{Qm_1} = 14.16\ \text{mg L}^{-1}$$

$$\therefore \quad L = 14.16\, e^{0.2744x} \quad \text{for } x \leq 0 \text{ (upstream)}$$

$$L = 14.16\, e^{-0.0364x} \quad \text{for } x \geq 0 \text{ (downstream)}$$

and

$$D = 92.5(0.763\ e^{0.2744x} - 0.735\ e^{0.2808x}) \quad \text{for } x \leq 0$$

$$D = 92.5(0.763\ e^{-0.369x} - 0.735\ e^{-0.0428x}) \quad \text{for } x \geq 0$$

The above equations can be solved by writing a simple program or a spreadsheet with calculations made at every spatial step (Δx = 1 mi). To obtain the fluxes, we take the derivatives of each equation.

$$\frac{dL}{dx} = 3.886\ e^{0.2744x} \quad \text{for } x < 0$$

$$\frac{dL}{dx} = -0.515\ e^{-0.0364x} \quad \text{for } x > 0$$

$$\frac{dD}{dx} = 92.5(0.209\ e^{0.2744x} - 0.206\ e^{0.2808x}) \quad \text{for } x \leq 0$$

$$\frac{dD}{dx} = 92.5(-0.028\ e^{-0.0369x} + 0.0315\ e^{-0.0428x}) \quad \text{for } x \geq 0$$

Note that the cusp at $x = 0$ for the BOD concentration causes the dispersive flux to be undefined at that point.

Results are given in Figure 6.14. The deoxygenation rate ($k_d L$) is equal to the reaeration rate ($k_a D$) at x = 25 mi. This is a little bit downstream from the point where the critical deficit D_c occurs at x_c = 20 mi due to dispersive mixing. In general, mass fluxes due to advection (uL and uD) are greater in magnitude than dispersive fluxes, but both transport processes are important. The units on the mass fluxes (mg mi L^{-1} d^{-1}) in Figure 6.14 are actually $ML^{-2}T^{-1}$, and they can be converted to mg ft^{-2} d^{-1} or mg m^{-2} d^{-1} quite easily.

6.7.2 Estimation of Parameters and Circulation in Estuaries

Circulation in estuaries is density-driven by saline intrusion of bottom waters. Neglecting Coriolis forces, we may simplify the problem to that of two-dimensional circulation as depicted in Figure 6.15. This discussion follows the seminal work of O'Connor and co-workers.[20,21]

Saline waters intrude below the plane of no-net motion, and there is a reversal in current direction with depth in the estuary (Figure 6.15). Continuity requires vertical mixing upward of bottom waters, which increases dispersion, as does the vertical shear velocity. Ebb and flood velocities at any time can be as much as five times the net (tidal-averaged) longitudinal velocity, u (Figure 6.16).

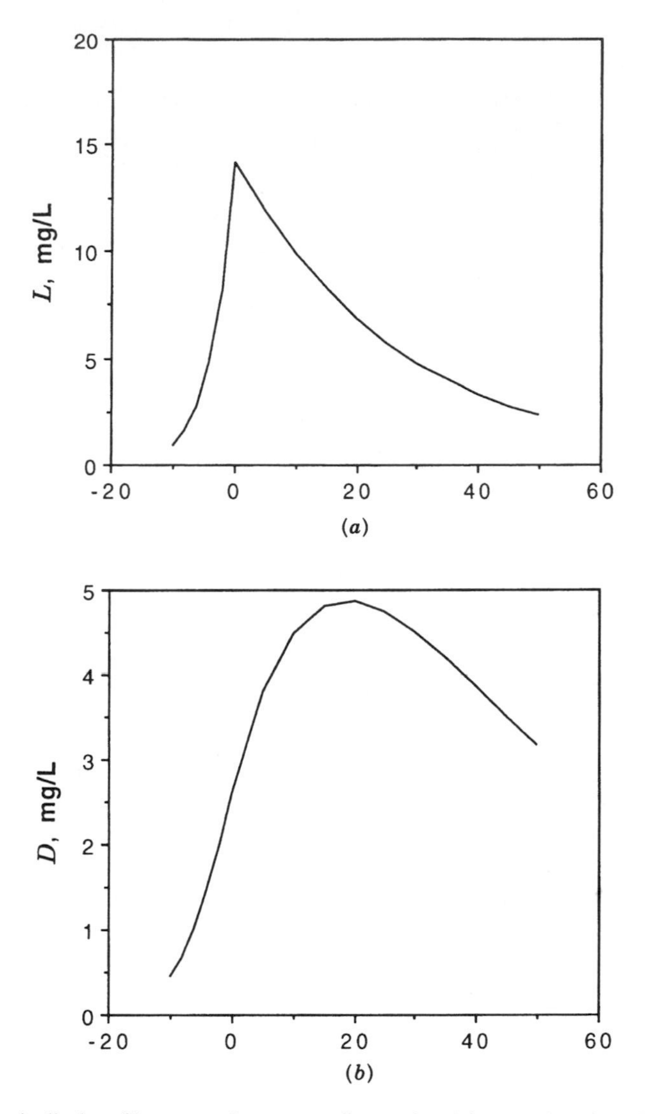

Figure 6.14 Longitudinal profiles versus downstream distance in mi for (a) BOD, (b) D.O. deficit, (c) reaction rates, (d) advective fluxes, and (e) dispersive fluxes for a waste discharge to an estuary at mile point $x = 0$ (Example 6.7).

$$u = \frac{Q_f}{A} \tag{126}$$

where u = net (tidal-averaged) velocity due to freshwater flow, LT^{-1}

Q_f = freshwater discharge, L^3T^{-1}

A = average cross-sectional area

The middle panel of Figure 6.16 shows the ebb and flood velocity during a tidal cy-

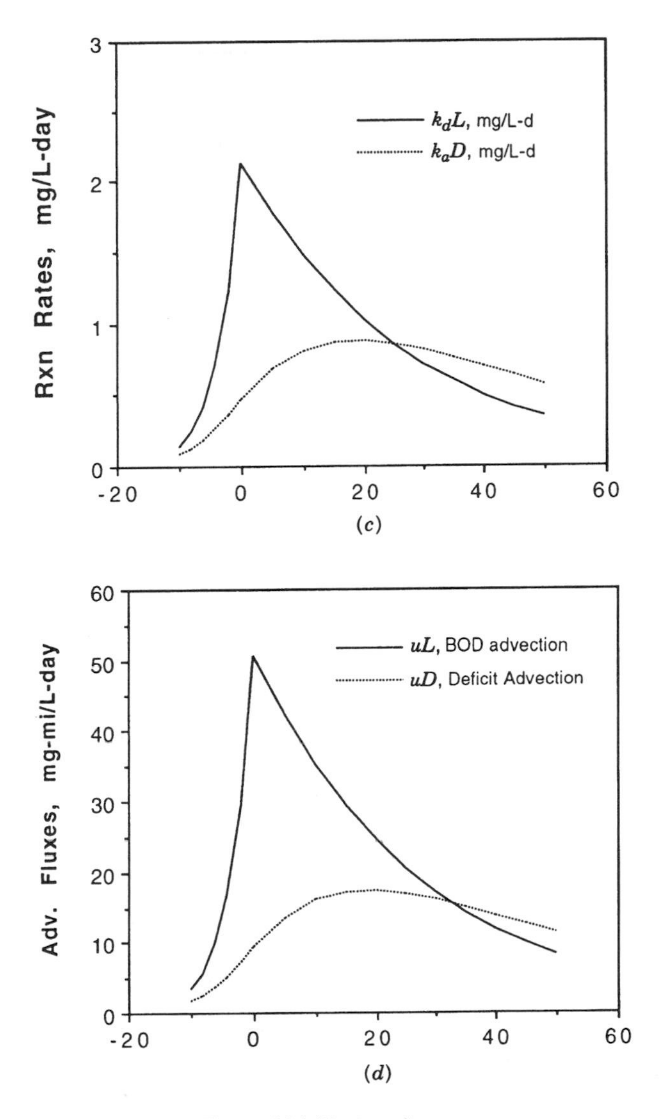

Figure 6.14 *(Continued).*

cle at a given point in the estuary. The average velocity ($u = Q_f/A$) is constant with time, and it is added to the ebb and flood velocity to give the total downstream velocity at any point in time.

In a coastal plain estuary, the vertically averaged salinity concentration is roughly synchronous with stage. But the stage lags the ebb and flood velocity time-series by approximately 90° (one-quarter of a tidal cycle in Figure 6.16). Tidal action, the ebb and flood cycles when current velocity actually reverses, causes a tremendous amount of longitudinal dispersion. The estuary runs backward for part of the tidal cycle!

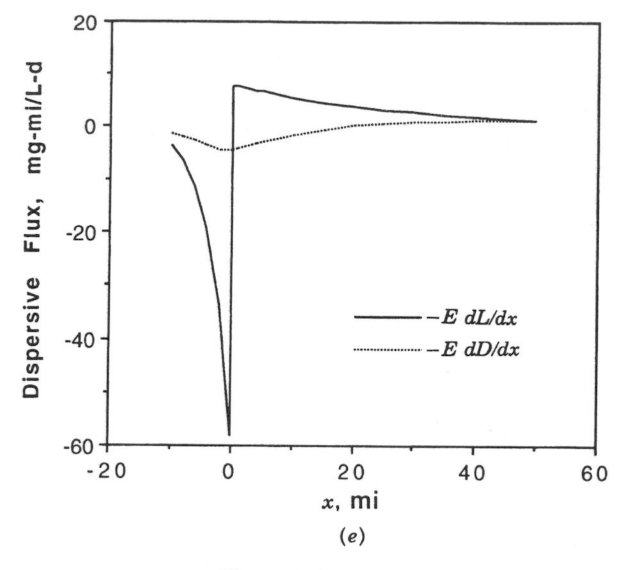

Figure 6.14 *(Continued).*

It is difficult to actually measure the net nontidal velocity due to freshwater discharge given the dynamics of an estuary. Usually, u is estimated in one of two ways: (1) using equation (126), the freshwater flowrate is estimated from the last gaging station on the river, and the average cross-sectional area is estimated from bathymetric maps of the channel; or (2) from release of a fluorescent dye (e.g., Rhodamine WT) that can be measured sensitively with a fluorometer. In dye studies, it is recommended to release the dye at high water slack tide (stage is a maximum in Figure 6.16) or at low water slack tide (stage is a minimum) and to make all measurements at high water or low water slack tide. By this methodology, accurate measurements of net nontidal velocity can be obtained. The net nontidal velocity is

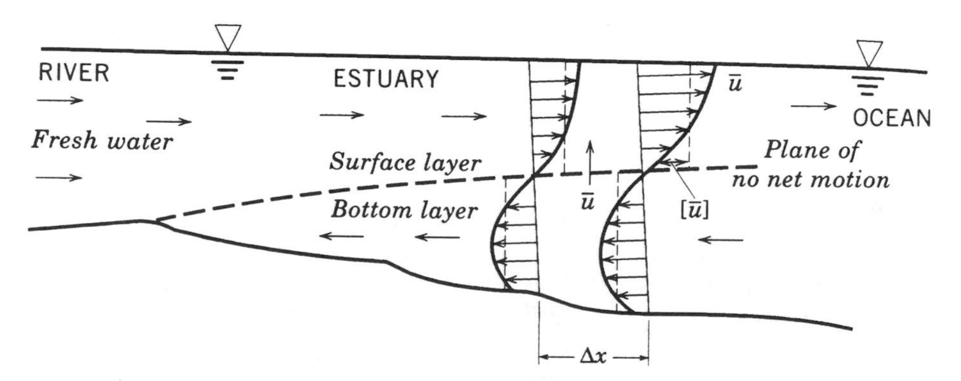

Figure 6.15 Schematic diagram of two-dimensional estuarine circulation. (Modified from Manhattan College Summer Institute Notes.[20]).

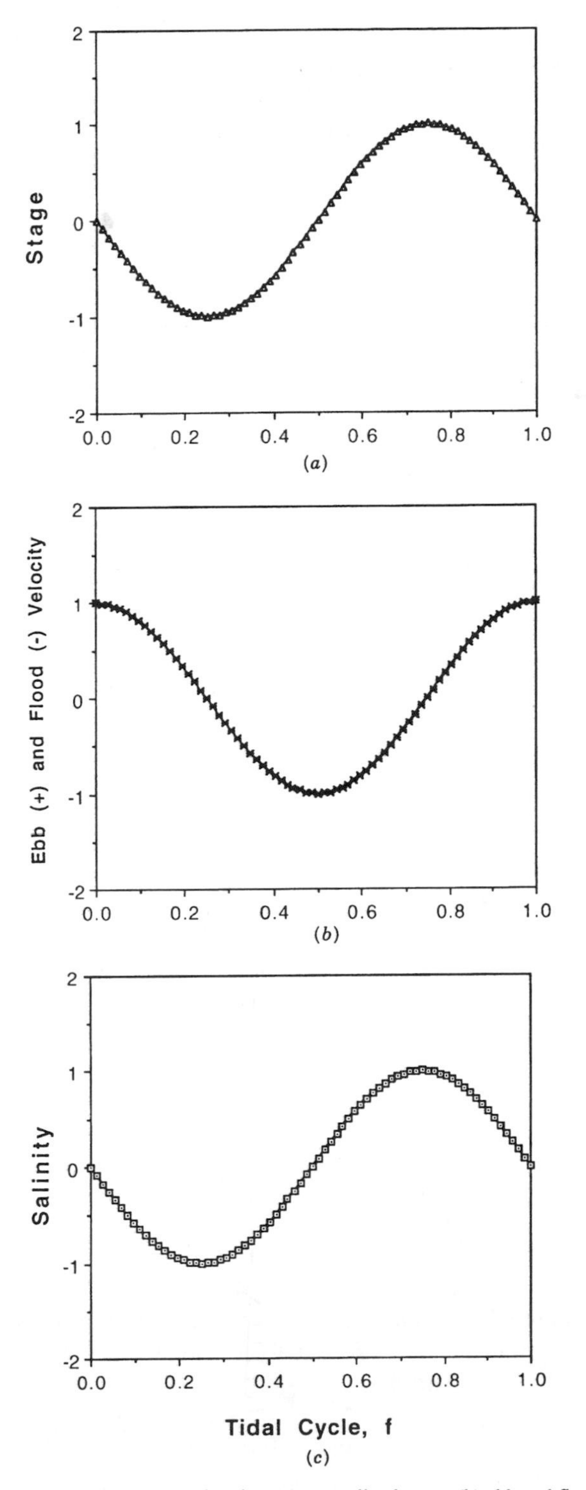

Figure 6.16 Tidal cycle of an estuary showing (a) normalized stage, (b) ebb and flood velocity, and (c) salinity concentrations.

based on the distance from where the dye was released to the maximum concentration at a later tidal cycle.

$$u = \frac{x_p}{t_n} \tag{127}$$

where x_p is the distance from the release point to the maximum concentration, and t_n is the time since release of the dye (n, number of tidal cycles). An average tidal cycle is 12.42 hours. If the mass of dye traveled 14 km in five complete tidal cycles, the average velocity, u, would be $14/5 \times 12.42$ or 0.22 km h^{-1}.

There are no simple formulas to estimate the longitudinal dispersion coefficient in an estuary. It can be obtained from:

Salinity data.
Spatial distribution of dye (tracer) experiments.
Temporal distribution of dye (tracer) experiments.

We begin with the use of salinity data to estimate E. Salinity is nature's own tracer of the longitudinal dispersion in an estuary. It decreases roughly exponentially from the source (the ocean) to the upstream portion of the salinity intrusion. Seawater consists of approximately 35,000 ppm of total dissolved solids with an ionic strength of 0.65 M and a concentration of NaCl $\simeq$ 0.5 M. Estuaries often empty into bays or straights behind barrier islands, and the source salinity is somewhat less than the open ocean (20,000–25,000 ppm salinity).

A mass balance differential equation for salinity, a conservative substance, at steady-state is

$$E \frac{d^2S}{dx^2} - u \frac{dS}{dx} = 0 \tag{128}$$

Two boundary conditions are necessary to solve the second-order ordinary differential equation.

$$\text{BC 1:} \quad S = S_0 \quad \text{at } x = 0 \tag{129a}$$

$$\text{BC 2:} \quad S = 0 \quad \text{at } x = -\infty \tag{129b}$$

where S is the salinity concentration and S_0 is the source concentration (the ocean) at $x = 0$. The solution to equation (128) is for negative values of distance, moving from the mouth to the upstream river.

$$S = S_0\, e^{ux/E} \quad \text{for } x \leq 0 \tag{130}$$

A semilog plot of salinity versus distance should have a slope of u/E and an intercept of $\ln S_0$.

$$\ln S = \frac{ux}{E} + \ln S_0 \tag{131}$$

From the slope of the semilog plot, we can solve for the average longitudinal dispersion coefficient, E.

Example 6.8 Estimation of Dispersion Coefficient from Salinity Data in an Estuary

Measurements have been made for the salinity concentration versus upstream distance from the mouth of the James River and Estuary, Virginia. Find the longitudinal dispersion coefficient, E, in $mi^2\ d^{-1}$ and $cm^2\ s^{-1}$. The freshwater velocity, u, has been estimated from dye studies to be 1.8 mi d^{-1}.

Miles from Mouth	x, mi	Salinity Concentration, ppm
0	0	20,000
5	–5	15,500
10	–10	12,000
15	–15	9,300
27	–27	5,000

Solution: Use the linearized form of the salinity equation (131) to find E. Plot ln S versus x—the slope is equal to u/E, then solve for E. There are only five data points and they plot as a straight line. The slope is given by the regression line in Figure 6.17.

$$\frac{u}{E} = 0.051\ \text{mi}^{-1}$$

Solving for E,

$$E = (1.8\ \text{mi d}^{-1})/(0.051\ \text{mi}^{-1}) = 35.3\ \text{mi}^2\ \text{d}^{-1}$$

This is a large dispersion coefficient indicating dominant tidal action.

A second method to obtain longitudinal dispersion coefficients is from dye (tracer) studies. Either an impulse input (slug) of dye or a continuous input can be released from a bridge at an upstream location. The dye is followed with boats and a fluorometer over several tidal cycles. The mass balance partial differential equation that describes an impulse discharge to a 1-D estuary is

$$\frac{\partial C}{\partial t} = \frac{1}{A}\frac{\partial}{\partial x}\left(EA\frac{\partial C}{\partial x}\right) - \frac{1}{A}\frac{\partial}{\partial x}(QC) - kC \tag{132}$$

where C is the concentration of dye tracer (ML^{-3}) and k is the first-order degrada-

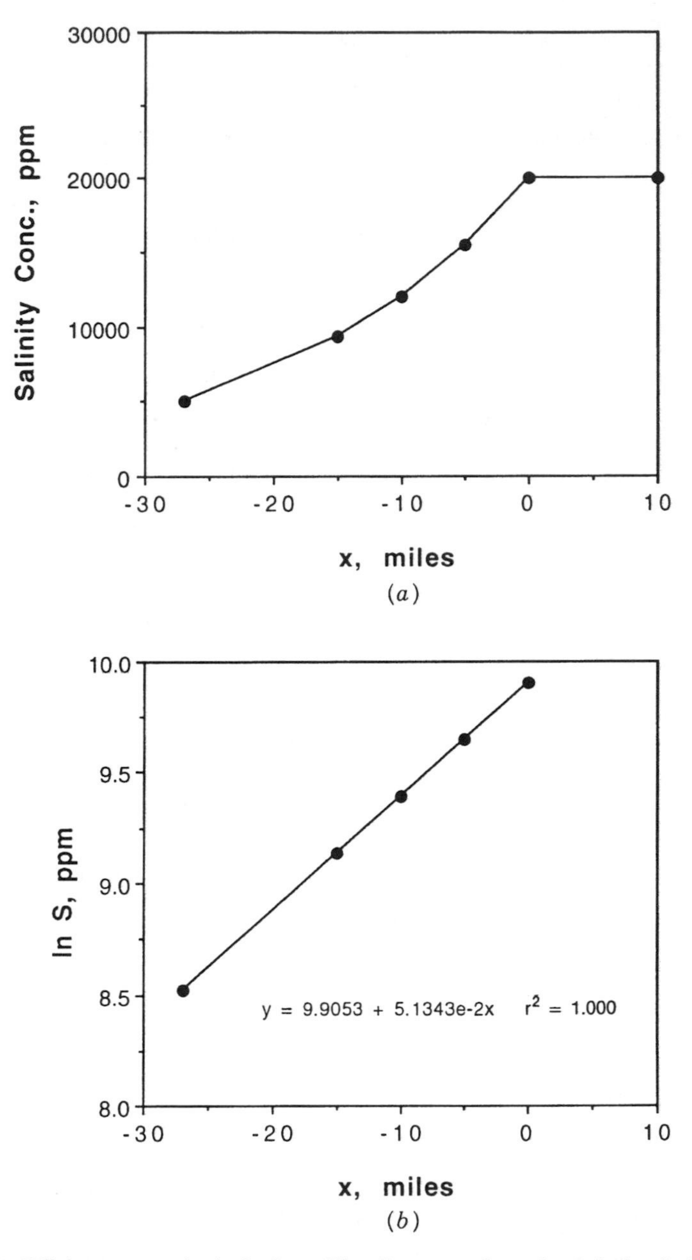

Figure 6.17 Salinity concentration in the James River Estuary used to estimate the longitudinal dispersion coefficient, E. (Ocean source of salinity is located at the mouth, $x = 0$; see Example 6.8.) (b) Semilog plot of salinity versus distance to determine E, the longitudinal dispersion coefficient (the slope is equal to u/E).

tion rate constant (T^{-1}) for the dye if it is not perfectly conservative. Organic dyes display small rates of degradation due to photolysis, adsorption to particles, and/or biotransformations. Under steady flow conditions (tidal-averaged conditions) and a relatively constant cross-sectional area, equation (132) can be simplified to a partial differential equation (PDE) with constant coefficients.

$$\frac{\partial C}{\partial t} = E\frac{\partial^2 C}{\partial x^2} - u\frac{\partial C}{\partial x} - kC \tag{133}$$

The solution to equation (133) in a 1-D estuary with impulse input of tracer is

$$C = \frac{M}{2A\sqrt{\pi E t}} \exp\left[\frac{-(x-ut)^2}{4Et} - kt\right] \tag{134}$$

where C = concentration of dye, ML^{-3}

M = mass of dye injected, M

A = average cross-sectional area of the estuary, L^2

E = longitudinal dispersion coefficient, L^2T^{-1}

t = time since release of the dye

x = longitudinal distance with $x = 0$ being the location where the dye was released and downstream distance is positive

u = net nontidal velocity, LT^{-1}

k = sum of the first-order reaction rate coefficients for disappearance of the dye, T^{-1}.

Dye should be released at slack tide and all measurements should be performed at slack tide also.

Theoretically, the spatial distribution of dye concentrations should form Gaussian (normal) profiles after an initial mixing period. Figure 6.18 shows the results of a tracer experiment performed in a physical model of the Delaware Estuary. The concentration profiles are roughly Gaussian. The standard deviation of the spatial distributions are related mathematically to the dispersion coefficient

$$\sigma_x = \sqrt{2Et} \tag{135}$$

where σ_x is the standard deviation of the distribution with units of length. It is equivalent to the physical distance across each bell-shaped concentration profile under which 68.3% of the mass is contained. The spread of the dye increases with the square root of time according to equation (135).

The degradation or disappearance of the dye tracer can be estimated by integrating the mass under each concentration profile [equation (136)] and by plotting ln M versus time [equation (138)]. The slope of the semilog plot yields the first-order degradation rate constant, k.

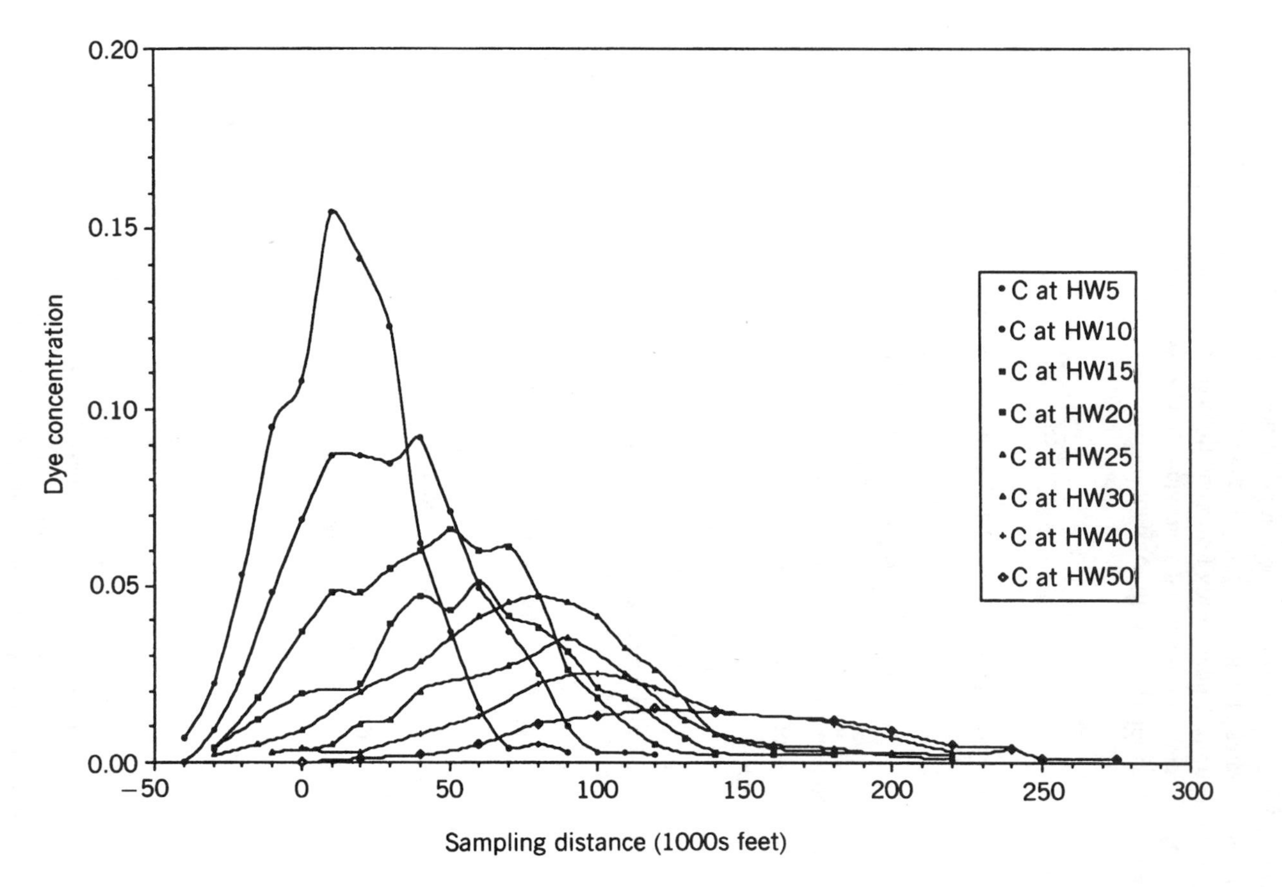

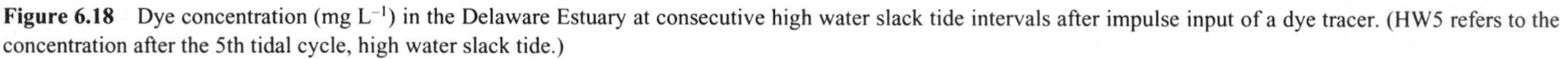

Figure 6.18 Dye concentration (mg L^{-1}) in the Delaware Estuary at consecutive high water slack tide intervals after impulse input of a dye tracer. (HW5 refers to the concentration after the 5th tidal cycle, high water slack tide.)

$$M_t = A \int_{-\infty}^{\infty} C\,dx \tag{136}$$

$$M_t = Me^{-kt} \tag{137}$$

$$\ln M_t = -kt + \ln M \tag{138}$$

where M_t is the integrated mass at any tidal cycle.

In Figure 6.18, the peak concentration (C_{px}) of dye for each tidal cycle provides a method to measure the net nontidal velocity [equation (127)]. Also, we can obtain an estimate of the longitudinal dispersion coefficient, E, by manipulating the following equations:

$$C_{px} = \frac{M}{2A\sqrt{\pi Et}}\,e^{-kt} \tag{139}$$

and

$$C = C_{px} \exp(-\phi^2/4Et) \tag{140}$$

where C_{px} = peak concentration in a spatial distribution of dye at a given tidal cycle, ML^{-3}

ϕ = distance of the dye measurement from the peak concentration, L

$$\phi = x - ut \tag{141}$$

Taking the natural logarithm of both sides of equation (140) yields

$$\ln C = \frac{-\phi^2}{4Et} + \ln\left(\frac{M}{2A\sqrt{\pi Et}}\right) - kt \tag{142}$$

A plot of ln C versus ϕ^2 from a spatial distribution of data at a given tidal cycle will yield a straight line with a slope of $-1/4Et$. Similarly, temporal data taken at a given location at successive tidal cycles can also be used to estimate E. A plot of $Ce^{kt}\sqrt{t}$ versus ϕ^2/t will linearize temporal data based on equation (134), and the straight line slope will be equal to $-1/4E$. Thus an impulse input of dye to an estuary can be used to estimate transport parameters for advection and dispersion in estuary models.

Example 6.9. Estimation of Dispersion Coefficient from Dye (Tracer) Data in an Estuary

A dye has been injected in a pulse to a model of the Delaware Estuary from Trenton to Cape May. Based on spatial dye measurements from the 30th tidal cycle and temporal data from x = 100,000 ft downstream, estimate the mean longitudinal dispersion coefficient. The dye was released at high water slack tide at $x = 0$, and measurements were recorded every five tidal cycles thereafter. The average tidal cycle was 12.42 hours.

Spatial Data, 30th Tidal Cycle

Dye Concentration, mg L^{-1}	Distance, ft
0.003	−10,000
0.005	10,000
0.011	20,000
0.012	40,000
0.020	50,000
0.027	80,000
0.035	100,000
0.025	120,000
0.012	140,000
0.005	160,000
0.004	180,000
0.002	200,000
0.001	220,000

Temporal Data at $x = 100{,}000$ ft

Number of Tidal Cycles	Dye Concentration, mg L^{-1}
10	0.003
15	0.018
20	0.031
25	0.041
30	0.035
40	0.025
50	0.013

Solution:

a. It is first necessary to determine if there is a decrease in the mass of dye in the estuary over time, due to reaction(s). The area from each tidal cycle was integrated and a semilog plot was constructed (Figure 6.19, top).

$$\text{slope} = -k$$

$$k = 0.0017226\ \text{h}^{-1} = 0.041\ \text{day}^{-1}$$

The dye was degrading during the course of the experiment and the first-order rate constant was 0.041 day^{-1}.

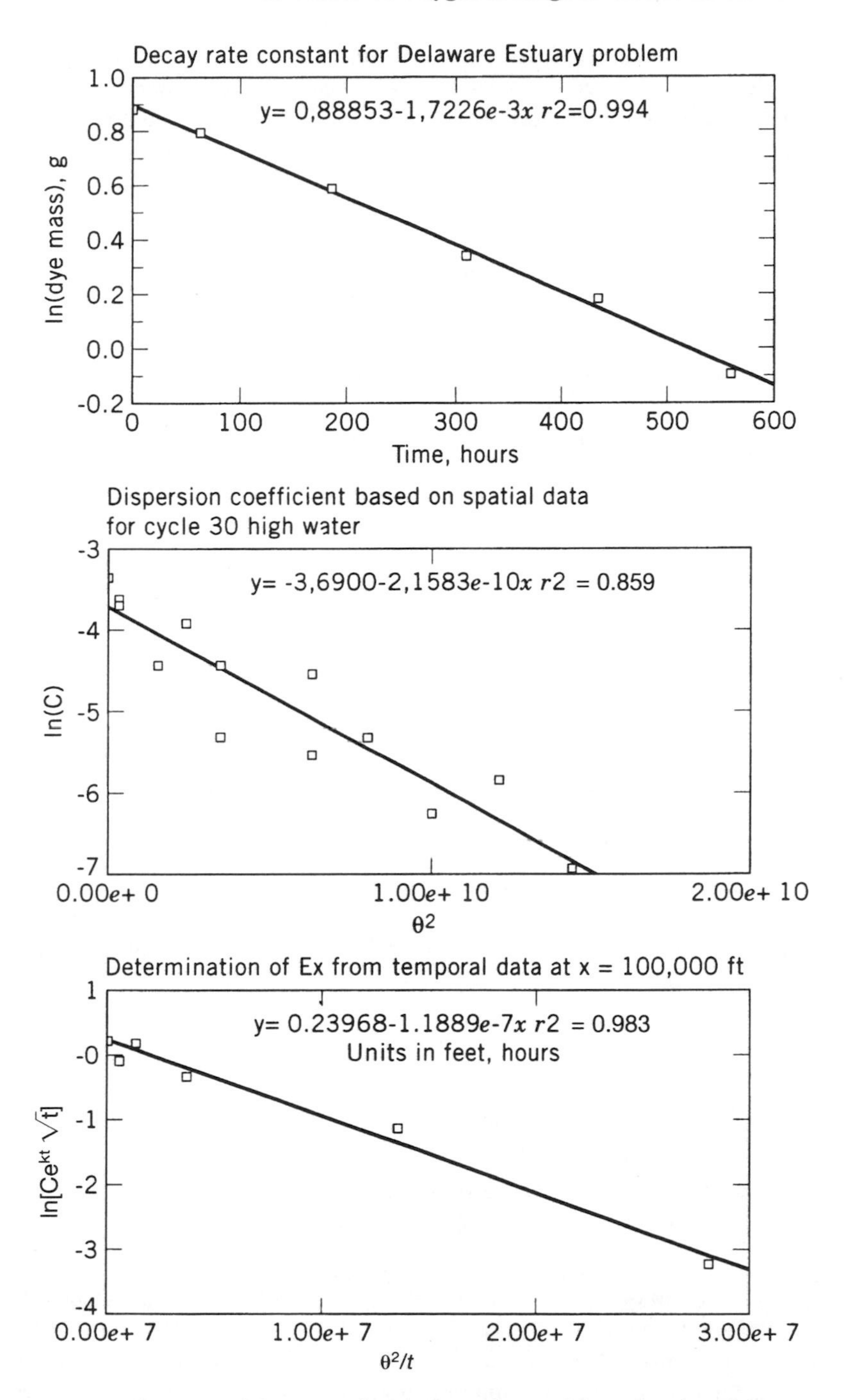

Figure 6.19 Semilog mass of dye versus time (*top*); semilog spatial concentration of dye versus ϕ^2 (*middle*); and semilog temporal concentration of dye versus ϕ^2/t for the Delaware Estuary model (Example 6.9).

b. Using the spatial dye distribution data from the 30th tidal cycle, a plot was made to determine E based on equation (142). The data were somewhat scattered with a linear least-squares regression coefficient of $r^2 = 0.859$ (Figure 6.19, middle).

$$\text{slope} = -1/4Et = -2.1583 \times 10^{-10}$$

$$E = 3.109 \times 10^6 \text{ ft}^2 \text{ h}^{-1} = 2.68 \text{ mi}^2 \text{ d}^{-1}$$

c. Using the temporal dye distribution data at $x = 100{,}000$ ft, a plot was made to determine $E(\ln Ce^{kt}\sqrt{t}$ versus $\phi^2/t)$.

$$\text{slope} = -1/4E = 1.1889 \times 10^{-7}$$

$$E = 2.103 \times 10^6 \text{ ft}^2 \text{ h}^{-1} = 1.81 \text{ mi}^2 \text{ d}^{-1}$$

Note that the estimates in parts (b) and (c) were significantly different. In this case, an average of the two values may be a prudent choice for the model.

We have tried to estimate transport characteristics for large rivers and estuaries to solve BOD–D.O. modeling problems. The reaeration constant is one of the most difficult reaction rate constants to estimate due to the dynamic nature of estuaries. Empirical and fundamental equations are not available. One method to estimate k_a, the reaeration rate constant, is by model calibration. BOD and D.O. data must be collected at slack tide throughout the estuary, and k_a can be estimated by model calibration or by nonlinear least-squares parameter estimation techniques. Reaeration in estuaries is enhanced by tidal action, and k_a can be approximated crudely by a mass transfer coefficient of 3 m d^{-1} divided by the mean depth of the estuary.

To summarize our estimates of transport and reaction parameters in estuaries, the following ranges apply:

$k_d = 0.05–0.5$ day^{-1}	Deoxygenation
$k_a = k_L/\text{depth} = (3 \text{ m d}^{-1})/H$	Reaeration
$E = 0.1 - 1.0$ mi^2 d^{-1}	Dispersion, large rivers
$E = 1–5$ mi^2 d^{-1}	Dispersion, tidal rivers
$E = 5–40$ mi^2 d^{-1}	Dispersion, estuaries
$u = Q_f/A = x_p/t_n$	Advection, estuaries

Large rivers have no tidal action and no salinity intrusion; tidal rivers display tidal action (changes in stage) but little saline intrusion; and estuaries have both tidal action and salinity profiles. Models similar to those in this section can be applied to other water quality problems in estuaries such as toxic metals and organic chemicals.

6.8 REFERENCES

1. Alleman, J.E., and Prakasam, T.B.S., *J. Water Pollut. Control Fed.*, 55, 436 (1983).
2. Metcalf and Eddy, Inc., *Wastewater Engineering*, 2nd ed., McGraw-Hill, New York (1978).
3. Mills, W.B., et al., Stream Sampling for Waste Load Allocation Applications, EPA/625/6-86/013, U.S. Environmental Protection Agency, Washington, DC (1986).
4. Sawyer, C.N., McCarty, P.L., and Parkin, G.F., *Chemistry for Environmental Engineers*, 3rd ed., McGraw-Hill, New York (1994).
5. Weilenmann, U., O'Melia, C.R., and Stumm, W., *Limnology and Oceanography*, 34, 1 (1989).
6. Streeter, H.W., and Phelps, E.B., U.S. Public Health Service, Bulletin No. 146 (1925).
7. Stones, T., *Effluent and Water Treatment J.*, 22, 75 (1982).
8. Fair, G.M., Geyer, J.C., and Okun, D.A., *Water and Wastewater Engineering*, Wiley, New York (1968).
9. O'Connor, D.J., and Dobbins, W.E., *Trans. Am. Soc. Civ. Eng.*, 123, 655 (1958).
10. Owens, M., Edwards, R.W., and Gibbs, J.W., *Air & Water Pollut. Int. J.* , 8, 469 (1974).
11. Churchill, M.A., Elmore, H.L., and Buckingham, R.A., *Air & Water Pollut. Int. J.* , 6, 467 (1962).
12. DiToro, D.M., *Sediment Flux Modeling*, Wiley-Interscience, New York (1996).
13. Thomann, R.V., and Mueller, J.A., *Principles of Surface Water Quality Modeling and Control*, Harper & Row, New York (1987).
14. O'Connor, D.J., and DiToro, D.M., *J. Sanit. Eng. Div., ASCE,* 96, 547 (1970).
15. Deb, A.K., and Bowers, D., *J. Water Pollut. Control Fed.* , 55, 1476 (1983).
16. Erdmann, J.B., *J. Water Pollut. Control Fed.* , 51, 7 (1979).
17. Benjamin, J.R., and Cornell, C.A., *Probability, Statistics, and Decision for Civil Engineers,* McGraw-Hill, New York (1970).
18. Reckhow, K.H., and Chapra, S.C., *Engineering Approaches for Lake Management*, Vol. 1, Butterworth, Boston (1983).
19. Beck, M.B., and van Straten, G., *Uncertainty and Forecasting of Water Quality,* Pergamon Press, New York (1982).
20. O'Connor, D.J., Fate of Toxic Substances in Natural Waters, Manhattan College Summer Institute Notes, Riverdale, NY, (1977).
21. O'Connor, D.J., Mueller, J.A., and Farley, K.J., J. Environ. Eng., *ASCE*, 109, 396 (1983).

6.9 PROBLEMS

1. Does an increase in the following factors serve to *increase, decrease*, or have *no effect* on the stream reaeration rate constant, k_a?

a. Stream velocity.
b. Mean depth.

c. Surface renewal rate..
d. Energy slope
e. Temperature.
f. Flowrate.
g. The molecular diffusion coefficient of O_2 in H_2O.
h. The film thickness.
i. The D.O. deficit.
j. The oxygen mass transfer coefficient, $K_{L.}$

2. Write the mass balance equation for a plug-flow stream with a second-order decaying substance, *C*. Solve for the steady-state concentration of *C* as a function of the second-order rate constant (k), longitudinal distance (x), mean velocity (u), and the concentration at $x = 0$ (C_0).

3. Write the time-variable mass balance differential equations for the growth of bacteria (X) utilizing a substrate (S) in an ideal plug-flow stream. Assume steady flow conditions and Monod kinetics with a maximum rate of substrate utilization (k) and a half-saturation constant (K_s).

4. The following data are from the Ohio River in the vicinity of Cincinnati (mile point 470) during low flow, September 1967.

a. If the mean velocity of the river was 0.3 m s^{-1}, calibrate the Streeter–Phelps model using the field data, and solve for the rate constants k_a and k_d. The saturation concentration for dissolved oxygen at the time of the survey was 8.3 mg L^{-1}, and the main wastewater discharge was at mile point 472.

Miles Below Pittsburgh	Ultimate BOD, mg L^{-1}	D.O. Concentration, mg L^{-1}
460	4.0	7.5
465	4.0	7.8
470	4.0	8.0
472	12	8.0
475	11	6.7
480	9.3	5.3
490	7.5	3.8
500	6.0	3.2
510	5.3	2.8
520	4.5	2.8
540	4.0	3.1
560	3.0	4.1
580	2.5	5.3

b. Waste load allocation. If the flow of the river was 5000 cfs, estimate the wastewater discharge in kg BOD d^{-1}. How much waste is allowed (in kg BOD d^{-1}) if the D.O. concentration must be greater than 5.0 mg L^{-1}?

5. Solve for the reaeration rate constant of the Missouri River (Iowa–Nebraska) using all five of the formulas given in Table 6.1. What are the mean and standard deviations? Which formula is most appropriate?

$S = 0.0002 \text{ m m}^{-1}$

$u = 1.24 \text{ m s}^{-1}$

$H = 3.7 \text{ m}$

$Q = 912 \text{ m}^3 \text{ s}^{-1}$

6. Write the mass balance equations for BOD (L), NBOD (N), and dissolved oxygen deficit (D) in a stream (plug flow) for a point source of CBOD (L_0) and NBOD (N_0) and benthic oxygen demand (S). Begin with an infinitesimal slice of the stream as depicted in Figure 6.1. Solve your differential equations for L, N, and D concentrations as a function of distance (x). Assume steady-state!

7. A wastewater discharge to a plug-flow stream includes some particulate BOD that settles out of the water column in addition to causing deoxygenation. Develop the mass balance equation and solve for the deficit concentration as a function of x, u, L_0, D_0, k_a, k_d, and k_s in the stream. L_0 and D_0 are the initial concentrations of BOD and D.O. deficit in the stream at $x = 0$.

$$L \xrightarrow{k_d} D \xrightarrow{k_a} 0$$
$$ \xrightarrow[k_s]{} 0$$

8. A plug-flow stream has nitrification and algal uptake of ammonium occurring as first-order irreversible reactions. For an input of NH_4-N of 5000 kg d^{-1} and the following conditions, solve the set of four differential equations with downstream distance assuming steady-state. You should use a Runge–Kutta numerical scheme or an equation solver. Plot your results for the four state variables versus distance.

$k_1 = 0.3 \text{ day}^{-1}$

$k_2 = 0.5 \text{ day}^{-1}$

$k_u = 0.5 \text{ day}^{-1}$ maximum uptake rate

$K_s = 0.1 \text{ mg L}^{-1}$ half saturation constant

$W_N = 5000 \text{ kg d}^{-1}$

$Q = 100 \text{ m}^3 \text{ s}^{-1}$

$A = 200 \text{ m}^2$

Initial conditions:

Algal-N = 0.1 mg L^{-1}

NO_2-N = NO_3-N = 0.0

NH_4-N = W_N / Q = 0.579 mg L^{-1} as N

$$NH_4\text{-N} \xrightarrow{k_1} NO_2\text{-N} \xrightarrow{k_2} NO_3\text{-N}$$
$$\llcorner \xrightarrow{k_u} \text{Algal-N}$$

9. Write the mass balance equation and analytic solution for D.O. deficit (D) in a plug-flow stream under steady-state conditions with the following reactions/processes occurring:

Reaeration, k_a
CBOD decay, k_d
$NH_4^+ \xrightarrow[k_n]{} NO_3^-$
Phytoplankton ($P - R$)
Sediment oxygen demand, S
Background deficit, D_b

Solve for D as a function of the rate constants and downstream distance, x.

10. The following ultimate CBOD data were measured in a stream.

L, mg L^{-1}	15	14.1	12.9	11.1	9.6	8.2	4.5
t, days	0	0.1	0.25	0.5	0.75	1.0	2.0

Find the deoxygenation rate constant, k_d.

11. Use the Streeter–Phelps model to calculate the following and plot:

a. The BOD (ultimate) versus downstream distance.
b. The D.O. deficit versus x.
c. The in-stream reaeration rate ($k_a D$) and deoxygenation rate ($k_d L$) versus x.

What is the minimum D.O. concentration and where does it occur? Use the Owens–Edwards–Gibbs reaeration rate constant (Table 6.2).

$k_d = 0.60$ day^{-1}
$L_0 = 10$ mg L^{-1}
$D_0 = 0$ mg L^{-1}
$H = 1.058$ m
$u = 0.305$ m s^{-1}

12. Determine the sensitivity due to rate constants for a river with a CBOD waste discharge ($L_0 = W/Q = 22$ mg L^{-1}) and an initial D.O. deficit ($D_0 = 1.6$ mg L^{-1}). Plot your results as D.O. deficit concentration (D) versus travel time (x/u). Give the values of D_c and x_c in each case.

a. $k_d = 0.69\ \text{day}^{-1}$
   $k_a = 0.92\ \text{day}^{-1}$
b. $k_d = 0.32\ \text{day}^{-1}$
   $k_a = 0.92\ \text{day}^{-1}$
c. $k_d = 0.62\ \text{day}^{-1}$
   $k_a = 0.69\ \text{day}^{-1}$
d. $k_d = k_a = 0.92\ \text{day}^{-1}$
e. $k_d = k_a = 0.62\ \text{day}^{-1}$

13. Use model equation (68) to simulate the dissolved oxygen concentration and deficit for a river with the following parameters and inputs.

$$D_0 = 0.5\ \text{mg L}^{-1}$$
$$L_0 = 10\ \text{mg L}^{-1}$$
$$N_0 = 5\ \text{mg L}^{-1}$$
$$k_a = 0.4\ \text{day}^{-1}$$
$$k_d = 0.3\ \text{day}^{-1}$$
$$k_n = 0.25\ \text{day}^{-1}$$
$$u = 0.4\ \text{m s}^{-1}$$
$$S = 5\ \text{g m}^{-2}\text{-d}^{-1}$$
$$H = 3\ \text{m}$$
$$P - R = 5\ \text{mg L}^{-1}\text{-d}^{-1}$$
$$L_b = 1.0\ \text{mg L}^{-1}$$

Solve for each of the six terms in equation (68) separately and then sum the terms to get the overall solution.

14. Perform a Monte Carlo uncertainty analysis on the Ohio River using the Streeter–Phelps model. The following parameters have been measured. In situ analyses of k_a and k_d have led to the following estimates of their mean and standard deviation values. You may consider that the parameter distributions are distributed normally.

$$k_a = 0.3 \pm 0.1\ \text{day}^{-1}$$
$$k_d = 0.4 \pm 0.05\ \text{day}^{-1}$$
$$u = 0.3\ \text{m s}^{-1}$$
$$D_0 = 0.0\ \text{mg L}^{-1}$$
$$L_0 = 12\ \text{mg L}^{-1}$$

Perform 100 simulations and plot the median BOD and D.O. deficit concentrations and their inner quartile values (median, 75% quartile, and 25% quartile values) at a distance of every 10 miles from the discharge point (0–100 miles total distance simulated).

15. A continuous wastewater discharge of BOD enters an estuary. Salinity measurements have been performed with distance, and the following estimates of model parameters are available. Assuming a one-dimensional, steady-state problem, find the longitudinal dispersion coefficient from salinity data, and plot the BOD concentration and the D.O. deficit concentration versus distance.

Salinity, ppm	20,000	6,000	1,850	171
Miles from mouth	0	5	10	20

$k_d = 0.1 \text{ day}^{-1}$
$k_a = 0.4 \text{ day}^{-1}$
$W/Q = 15 \text{ mg L}^{-1}$
$Q_f = 10{,}000 \text{ cfs}$
$A = 100{,}000 \text{ ft}^2$

The wastewater discharge is located 30 miles from the mouth of the estuary.

16. A dye (tracer) has been injected into the Delaware Estuary at high water slack tide. Spatial data are provided below from the 20th tidal cycle (high water slack tide). Find the dispersion coefficient, E.

$u = 1.3 \text{ mi d}^{-1}$
$k = 0.04 \text{ day}^{-1}$ (dye degradation)
$t_n = 20 \text{ cycles} \times 12.42 \text{ h}$

C, mg L^{-1}	0.032	0.042	0.049	0.051	0.048	0.029	0.015
x, ft Downstream Distance	40,000	50,000	60,000	70,000	80,000	100,000	125,000

17. Salinity was measured in the Delaware Estuary below.

a. Estimate the longitudinal dispersion coefficient. The mouth of the estuary is the beginning of the mid-Atlantic Bight (ocean) at $x = 0$. The net nontidal velocity is 0.06 ft s^{-1}.

Distance from mouth, km	Salinity, ‰, parts per thousand
0	30.0
20	26.5
40	20.0
60	14.0
80	8.0
100	4.0

b. There is a wastewater discharge 80 km upstream from the mouth. Plot the concentration of BOD and D.O. deficit versus distance (above and below the discharge).

$k_d = 0.05 \text{ day}^{-1}$
$k_a = 0.16 \text{ day}^{-1}$
$W/Q = 20 \text{ mg L}^{-1}$ BOD

c. Convert the D.O. deficit concentrations from part (b) into dissolved oxygen concentrations considering a temperature of 15 °C and the salinity concentrations given in part (a). Plot the D.O. concentration and the % oxygen saturation profiles with distance.

18. Solve the differential equation for salinity concentration in an estuary below. *Note*: You should set boundary conditions and solve by using the method of undetermined coefficients.

$$0 = E \frac{d^2S}{dx^2} - u \frac{dS}{dx}$$

19. The longitudinal dispersion coefficient is derived from salinity data for a coastal plain estuary. For the data given below, specify which model (advective, advective–dispersive, or dispersive only) is needed to solve the problem. Use an analysis of dimensionless numbers to determine if advection or dispersion predominates in this case.

$E = 15 \text{ mi}^2 \text{ d}^{-1}$
$k_d = 0.1 \text{ day}^{-1}$
$u = 0.2 \text{ ft s}^{-1}$

20. The general steady-state solution for a continuous waste discharge in an estuary is given below for a first-order reaction decay. Set the boundary conditions to evaluate the coefficients B_1, D_1, B_2, D_2, and C_0.

$C_1 = B_1 e^{gx} + D_1 e^{jx}$
$C_2 = B_2 e^{gx} + D_2 e^{jx}$

where

$$g = \frac{u}{2E} [1 + m]$$

$$j = \frac{u}{2E} [1 - m]$$

$$m = \sqrt{1 + \frac{4kE}{u^2}}$$

C is the concentration of the pollutant, C_1 is the upstream concentration, and C_2

is the downstream concentration. The wastewater discharge occurs at $x = 0$, and the problem is divided into two segments (an upstream and a downstream portion).

21. A tracer dye was released into an estuary. The rate of disappearance of the dye was 0.041 day^{-1}. Find the longitudinal dispersion coefficient based on the spatial and temporal dye concentrations given below.

Spatial Data, 15th tidal cycle (high water slack tide)

Point of Tracer Release: $x = 0$

Distance, ft	Concentration, mg L^{-1}
-30,000	0.004
-15,000	0.018
0	0.037
10,000	0.048
20,000	0.048
30,000	0.055
40,000	0.060
50,000	0.066
60,000	0.060
70,000	0.061
80,000	0.047
90,000	0.026
100,000	0.018
120,000	0.005
140,000	0.002

Temporal Data, $x = 50{,}000$ ft

1 Tidal Cycle = 12 h 25 min

Number Tidal Cycles	Concentration, mg L^{-1}
5	0.037
10	0.071
15	0.066
20	0.047
25	0.035
30	0.020
40	0.011
50	0.004

22. An estuary has a point source discharge W of chemical at $x = 0$. The chemical is known to degrade via a *zero-order* reaction.

a. Write the mass balance PDE for an estuary with constant coefficients and zero order decay.

b. Assuming steady-state, write the general solution for concentration in the upstream and downstream portion of the estuary. Give the roots $g \cdot j$ of the homogeneous second-order linear ODE.

c. Solve the nonhomogeneous equation using the general and particular solutions and the method of undetermined coefficients. Use boundary conditions in the upstream and downstream portion.

d. Sketch a picture of how the solution should look at steady-state (C versus x).

ANSWERS TO SELECTED PROBLEMS—CHAPTER 6

1. a. increase

c. increase

e. increase

g. increase (but the molecular diffusion coefficient is a physical chemistry constant at a given temperature)

i. no effect

4. a. (1) Plot ln BOD versus x to get an estimate of k_d ($k_d \approx 0.38$ day^{-1}) in the initial stretch of the river

(2) Then choose various k_a values to fit the D.O. deficit data ($k_a \approx 0.3$ day^{-1}).

(3) Adjust both rate constants to optimize the fit to the D.O. deficit data and the BOD data in the stream. (*Note*: You will not be able to get a perfect fit on both D and L. This is a real data set and calibrations are not perfect!)

b. (1) Estimate the waste discharge at mile point 472.

$L_0 = 12$ mg L^{-1} at mile point 472

$L_{w'} = W/Q = 12 - 4 = 8$ mg L^{-1} due to wastewater discharge

$\therefore\ W = 9.8 \times 10^4$ kg d^{-1}

(2) Find the dissolved oxygen concentration at the critical point.

$$\text{D.O.}_c = C_{\text{sat}} - D_c$$

Then calculate the decrease in D.O. deficit concentration necessary to achieve the water quality standard of 5 mg L^{-1} D.O. at the critical point.

$$-(\Delta D) = 5 - \text{D.O.}_c$$

(3) Conclusions: L_0 must decrease by approximately 37% to meet water quality standards at the critical point. But L_0 includes ultimate BOD due to upstream BOD sources:

$$L_0 = 12 \text{ mg L}^{-1} = L_{w'} + L_{\text{up}}$$

$$L_0 = 8 + 4$$

$$L_0 = 3.56 + 4 \text{ (under waste load allocation)}$$

Therefore the wastewater discharge must be decreased by more than 50% in order to meet water quality standards. (We are assuming that upstream BOD sources cannot be controlled and that the only wastewater discharge to be regulated occurs at mile point 472).

$$W_{\text{allowable}} = 4.4 \times 10^4 \text{ kg d}^{-1}$$

$$\text{Total} = W_{w'} + W_{\text{up}}$$

$$93{,}000 = 44{,}000 + 49{,}000$$

5. O'Connor or Churchill formula is most appropriate for large rivers.

   O'Connor $k_a = 0.615$ day^{-1}
   Churchill $k_a = 0.692$ day^{-1}

10. $k_d = 0.60$ day^{-1}

12.

Question	t (days)	D_c (mg L^{-1})
a	1.2	7.0
b	1.6	4.7
c	1.45	8.1
d	1.0	8.3
e	1.6	8.6

15. $E = 6.8$ mi^2 d^{-1}

16. $E = 1.7$ mi^2 d^{-1}

7

TOXIC ORGANIC CHEMICALS

If we live as if it matters and it doesn't matter, it doesn't matter.
If we live as if it doesn't matter, and it matters, then it matters.
—*The Precautionary Principle,* International Conference on an Agenda of Science for Environment and Development into the 21st Century, Vienna, Austria (1991)

There are 4 million organic chemicals with names given by the International Union of Pure and Applied Chemistry (IUPAC). Approximately 1000 new organic chemicals are synthesized and used commercially each year.[1] Only a fraction of these prove to be toxic or carcinogenic, and the vast majority of them break down in the environment. If they do not, if they are persistent as well as toxic, we may need to use mathematical models to determine if they pose an unreasonable risk to humans or the environment. We estimate their fate and transport in the environment, their exposure concentration to humans and wildlife, and perform waste load allocations to meet water quality standards.

Organic chemistry is the chemistry of compounds of *carbon.* Organic chemicals are obtained from material produced originally by living organisms (such as petroleum, coal, and plant residues) or they are synthesized from other organic compounds or inorganics (e.g., carbonates or cyanides).[2] Organic chemists are concerned with the synthesis of new organic chemicals to make new products; but environmental scientists are concerned with the degradation of these chemicals in the environment. We want them to be below toxicity thresholds and to be nonpersistent.

7.1 NOMENCLATURE

There are many different ways to name organic compounds including common names, IUPAC names, and trade names. In this book, we will use common names wherever possible. Figure 7.1 shows some classes of organic compounds that are widely used. The left-hand side of the figure gives some general classes of compounds and the right-hand side is a specific example of each. In the environment, alkanes are slowly oxidized to form an alcohol, beginning a chain of reactions. These reactions are usually microbially mediated (enzymes catalyze the reactions), but other abiotic processes such as photolysis, hydrolysis, chemical oxidation or reduction may also be important.

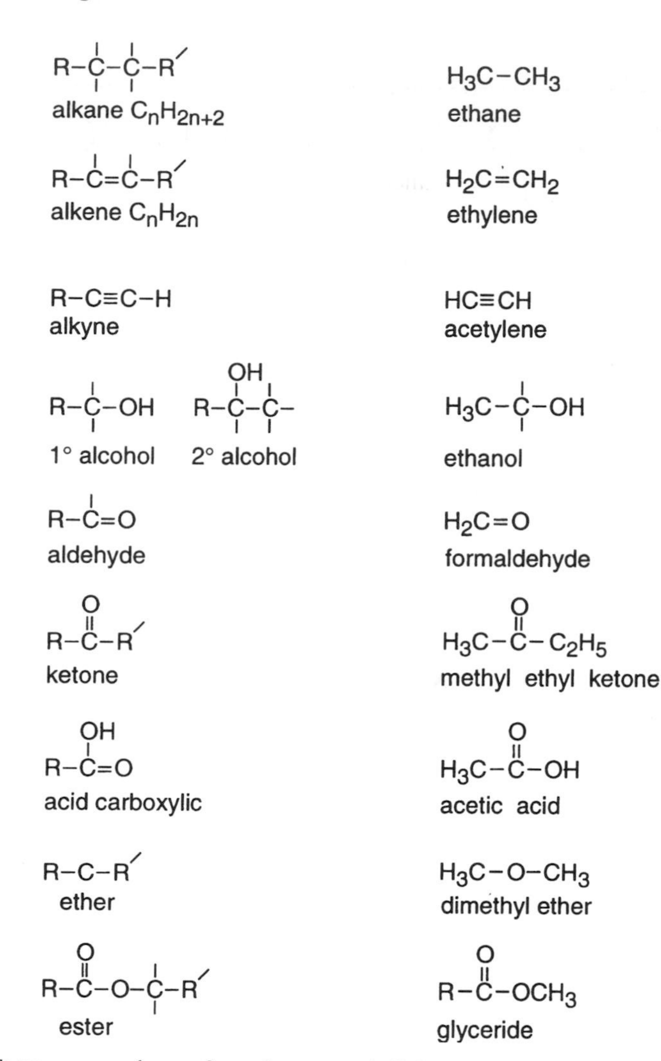

Figure 7.1 Some common classes of organic compounds (*left*) and examples (*right*). R and R′ indicate different alkyl groups.

$$\text{Alkanes} \xrightarrow[\text{oxidation}]{\Omega} \text{Alcohol} \rightarrow \text{Aldehyde/ketone} \rightarrow \text{Acid} \xrightarrow[\text{oxidation}]{\beta} CO_2 + H_2O$$

Microbial "infallibility" would state that all organic chemicals that are synthesized can be mineralized all the way to carbon dioxide and water as shown above. But many synthetic compounds have not been shown to degrade in the environment, or they might degrade extremely slowly, or only under special conditions. Microbes are not infallible, although given the proper conditions, enough time, and in concert with other physical and chemical reactions, they can often help to break down most organic chemicals. On the other hand, microbes and plants can sometimes synthesize chemicals in nature that are quite toxic and rather slow to degrade. Chlorinated organic chemicals are not purely man-made (xenobiotics), but now we know that some chlorinated organic chemicals are synthesized by plants and quite common in nature.

Figure 7.2 shows some examples of cyclic organic chemicals that are sometimes difficult to degrade in the environment.

$$C_6H_6 + \frac{15}{2} O_2 \xrightarrow{\text{microbes}} 6\ CO_2 + 3\ H_2O$$

To oxidize benzene to carbon dioxide and water requires that the very stable benzene ring must be cleaved. Under anaerobic conditions this can be a difficult task. There are many toxic organic chemicals that cause problems in the environment and comprise various "priority pollutant" lists. One of the most important lists is that for drinking water standards. Organic chemicals for which maximum allowable drinking water standards have been established are shown in Figure 7.3. They include a variety of widely used chemicals that are relatively persistent and toxic in drinking water.

7.2 ORGANICS REACTIONS

The mechanisms and kinetics discussed in Chapter 3 apply equally well to organic chemical reactions as to inorganics. The types of reactions can be classified and will be discussed in this section of the text. Biological transformations, chemical hydrolysis, oxidation/reduction, photodegradation, volatilization, sorption, and bioconcentration are among the important reactions that organic chemicals undergo in natural waters.

7.2.1 Biological Transformations

Biological transformations refer to the microbially mediated transformation of organic chemicals, often the predominant decay pathway in natural waters. It may occur under aerobic or anaerobic conditions, by bacteria, algae, or fungi, and by an array of mechanisms (dealkylation, ring cleavage, dehalogenation, etc.). It can be an intracellular or extracellular enzyme transformation.

The term "biodegradation" is used synonymously with "biotransformation," but some researchers reserve "biodegradation" only for oxidation reactions that break down the chemical. Reactions that go all the way to CO_2 and H_2O are referred to as "mineralization." In the broadest sense, "biotransformation" refers to any microbially mediated reaction that changes the organic chemical. It does not have to be an oxidation reaction, nor does it have to yield carbon or energy for microbial growth or maintenance. The term "secondary substrate utilization" refers to the utilization of organic chemicals at low concentrations (less than the concentration required for growth) in the presence of one or more primary substrates that are used as carbon and energy sources. "Co-metabolism" refers to the transformation of a substrate

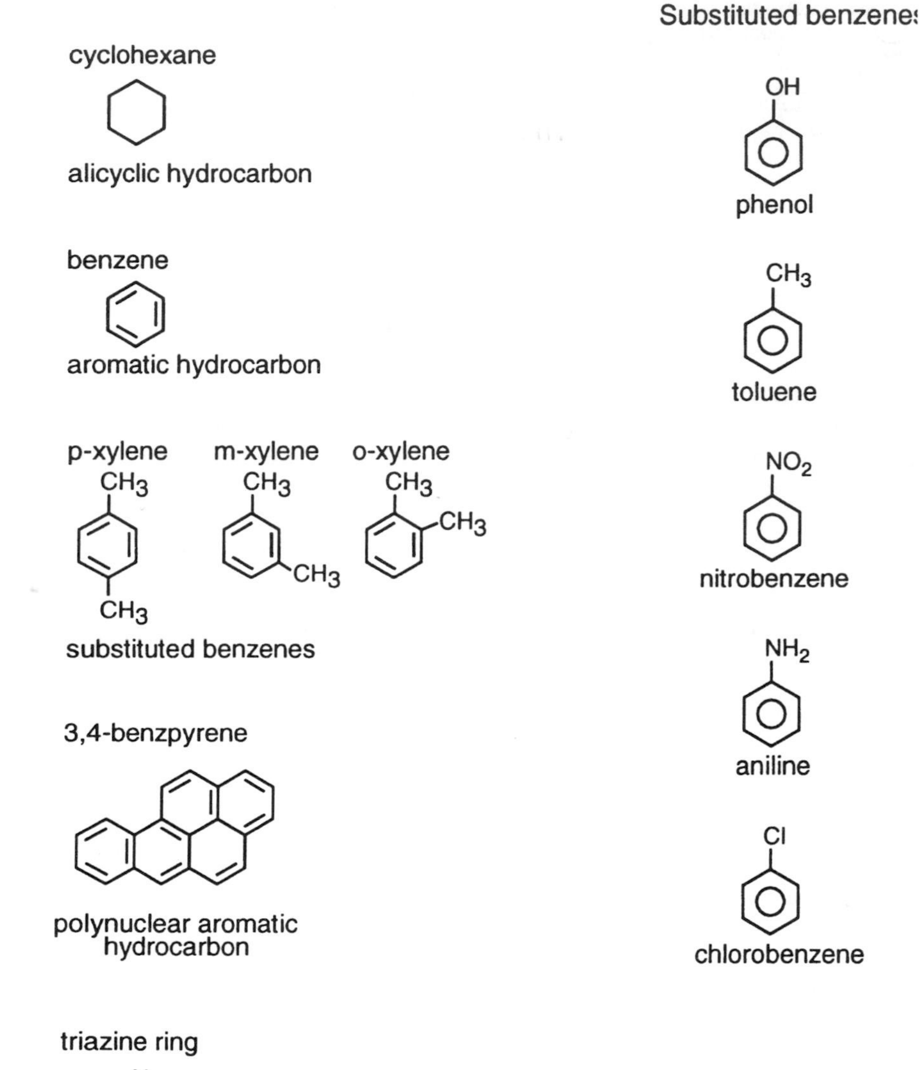

Figure 7.2 Examples of cyclic organic compounds (including alicyclic, aromatic, and heterocyclic compounds).

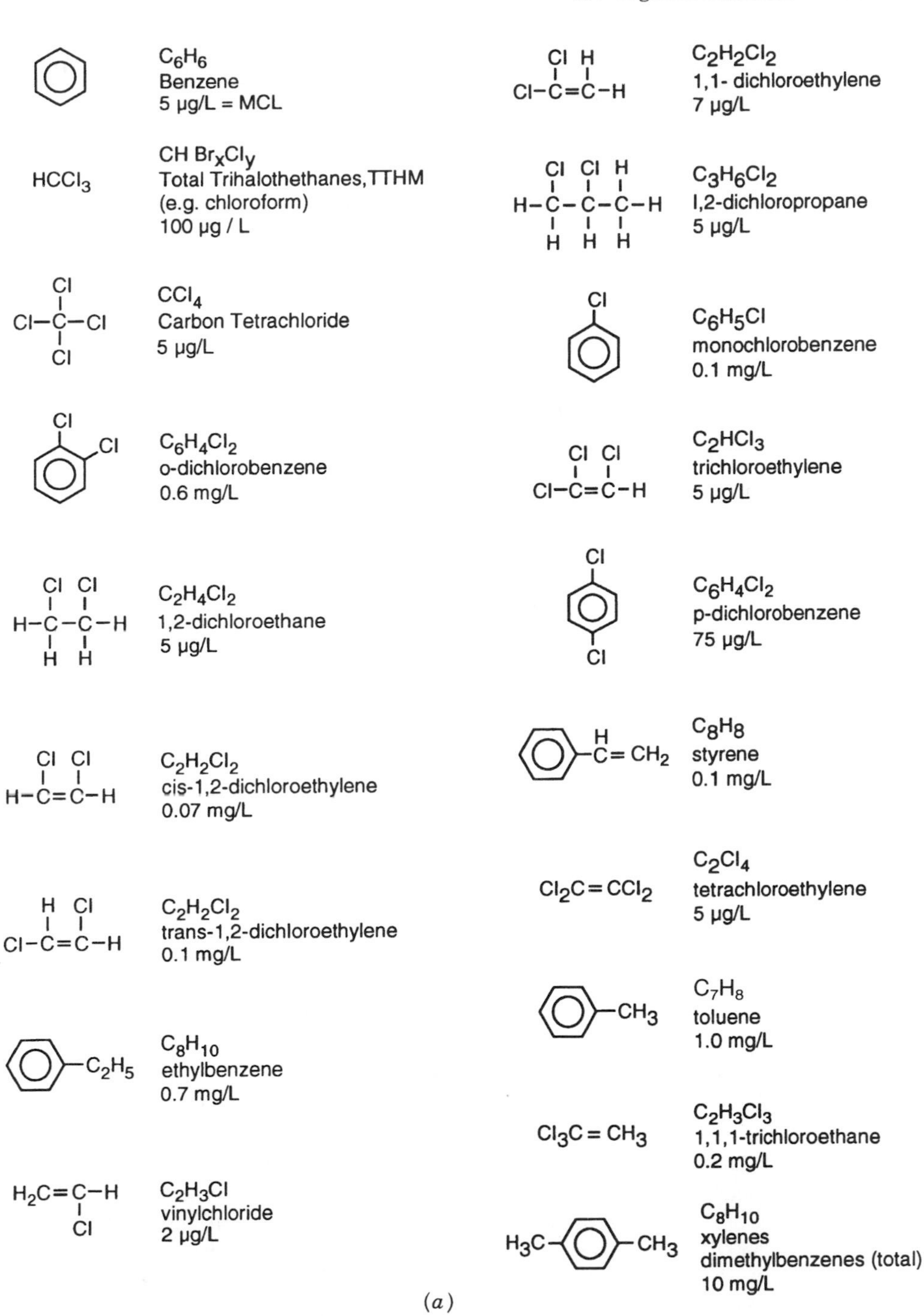

Figure 7.3 (a) Volatile organic compounds that have maximum contaminant level (MCL) drinking water standards. (b) Some synthetic organic chemicals for which maximum contaminant levels (MCLs) have been established. (*Continued on next page.*)

$C_{14}H_{20}ClNO_2$ alachlor MCL=2 μg/L	$C_{12}H_8OCl_6$ endrin 0.2 μg/L	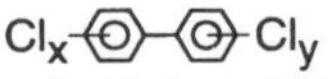 Cl_x-⟨◯⟩-⟨◯⟩-Cl_y polychlorinated biphenyls, PCB (as decachlorobiphenyl) 0.5 μg/L
$C_7H_{14}O_2N_2S$ aldicarb 3 μg/L	$C_2H_2Br_2$ ethylene dibromide 0.05 μg/L	C_6Cl_5OH pentachlorophenol 0.1 μg/L
$C_8H_{14}ClN_5$ atrazine 3 μg/L	$C_{10}H_7Cl_7$ heptachlor 0.4 μg/L	$C_{10}H_{10}C_8$ toxaphene 3 μg/L
$C_{12}H_{13}O_3N$ carbofuran 40 μg/L	$C_{10}H_5Cl_7O$ heptachlor epoxide 0.2 μg/L	$C_8H_6Cl_2O_3$ 2,4-D (2,4-dichlorophenoxy acetic acid) 70 μg/L
$C_{12}H_8OCl_6$ chlordane 2 μg/L	$C_6H_6Cl_6$ lindane 0.2 μg/L	$C_8H_5Cl_3O_3$ 2,4,5-T 50 μg/L
	$C_{16}H_{15}O_2Cl_3$ methoxychlor 40 μg/L	$C_3H_5Br_2Cl$ dibromochloropropane 0.2 μg/L

(*b*)

Figure 7.3 (*continued*)

that cannot be used as a sole carbon or energy source but can be degraded in the presence of other substrates. For example, no microbial species has been found that can grow on DDT as a sole carbon and energy source, but microbial consortia are known to degrade DDT in the presence of other substrates.

Many toxic organic reactions in natural waters are microbially mediated with both bacteria and fungi degrading a wide variety of pesticides. Dehalogenation, dealkylation, hydrolysis, oxidation, reduction, ring cleavage, and condensation reactions are all known to occur either metabolically or via co-metabolism (see Table 7.1). In co-metabolism, the microbe does not even derive carbon or net energy from the degradation; rather, the pesticide is "caught up" in the overall metabolic reactions as a detoxification or other enzymatic reaction.[3] On the other hand, several bacterial genera are known that are capable of utilizing certain organics as the sole carbon, energy, or nitrogen source. *Pseudomonas* (with 2,4-D and paraquat), *Nocardia* (with dalapon and propanil), and *Aspergillus* species (with trifluralin and picloram) are poignant examples.[4]

Table 7.1 Biological Transformations Common in the Aquatic/Terrestrial Environment[5]

Oxidation	$RCH_3 \rightarrow RCH_2OH$
Oxidative dealklylation	$ROCH_3 \rightarrow ROH + HCHO$
Decarboxylation	$RCOOH \rightarrow R\text{-}H + CO_2$
Aromatic hydroxylation	$Ar \rightarrow ArOH$
Ring cleavage	$Ar(OH)_2 \rightarrow CHOCHCHCHCOHCOOH$
β-Oxidation	$CH_3CH_2CH_3COOH \rightarrow CH_3COOH + CH_3COOH$
Epoxidation	RC═CR → RC—CR (with O bridging the two C atoms)
Oxidation of sulfur	$R_2S \rightarrow R_2SO$
Oxidation of amino acid	$RNH_2 \rightarrow RNO_2$
Hydrolytic dehalogenation	$RCHClCH_3 \rightarrow RCHOHCH_3 + Cl^-$
Reductive dehalogenation	$RCCl_2R \rightarrow RCHClR + Cl^-$
Dehydrohalogenation	$RCH_2CHClCH_3 \rightarrow RHC{=}CHCH_3$
Nitroreduction	$RNO_2 \rightarrow RNH_2$

It is convenient when possible to express rate expressions for organic transformations as pseudo-first-order-reactions, such as equation (1) below. That way, one can sum up all the various parallel reactions that may occur for a given chemical and compare the relative magnitudes of the rate constants to determine if any predominate. The reaction rate expression is then

$$\frac{dC}{dt} = -k_bC \tag{1}$$

where C is the toxic organic concentration in solution and k_b is the pseudo-first-order biotransformation rate constant. Table 7.2 is a summary of pseudo-first-order and second-order rate constants k_b for the disappearance of toxic organics from natural waters and groundwater via biotransformation.[6]

The actual microbial biotransformation rate follows the Monod or Michaelis–Menten enzyme kinetics expression, where

$$\frac{dC}{dt} = \frac{-\hat{\mu}XC}{Y(K_M + C)} = -k_bC \tag{2}$$

where k_b = pseudo-first-order biological transformation rate constant, T^{-1}
$\hat{\mu}$ = maximum growth rate, T^{-1}
X = viable microbial biomass concentration, $M\ L^{-3}$
Y = cell yield, microbial cell conc yield/organic conc utilized
K_M = Michaelis half saturation constant, $M\ L^{-3}$

Typical cell concentrations in surface waters would be 10^6–10^7 cells mL^{-1} and in groundwater somewhat less.

Table 7.2 Selected Biotransformation Rate Constants[6,7]

Chemical	Pseudo-First-Order Rate Constant Range, day^{-1}	Estimated Second-Order Rate Constants, $mL\ cell^{-1}\ h^{-1}$
Pesticides		
Carbofuran	0.03	1×10^{-8}
DDT	0.0–0.10	3×10^{-12}
Parathion	0.0–0.12	—
Dioxin TCDD	< 0.01	1×10^{-10}
Atrazine	0.02–0.03	1×10^{-8}
Alachlor	0.05–0.06	3×10^{-8}
PCBs		
Aroclor 1248	0.0–0.007	10^{-9}–10^{-12}
Halogenated aliphatic hydrocarbons		
Chloroform	0.09–0.10	1×10^{-10}
Carbon tetrachloride		
Halogenated ethers		
2-Chloroethyl vinyl ether	0.0–0.20	1×10^{-10}
Monocyclic aromatics		
2,4-Dimethylphenol	0.24–0.66	1×10^{-7}
Pentachlorophenol	0.00–33.6	3×10^{-8}
Benzene 14	0.006–0.01	1×10^{-7}
Toluene 14	0.01–0.019	1×10^{-7}
Phenol	—	3×10^{-6}
Phthalate esters		
Bis(2-ethylhexyl)phthalate	0.00–0.14	1×10^{-7}
Polycyclic aromatic hydrocarbons		
Anthracene	0.007–15.0	3×10^{-9}
Benzo[*a*]pyrene	0.0–0.075	3×10^{-12}
Nitrosamines and miscellaneous		
Benzidine	< 0.001	1×10^{-10}
Dimethyl nitrosamine	< 0.001	3×10^{-10}

Under typical environmental conditions, the concentration of dissolved organics ($C < 10\ \mu g\ L^{-1}$) is less than that of the Michaelis half-saturation constant ($K_M \simeq 0.1 - 10\ mg\ L^{-1}$). Therefore the equation becomes

$$\frac{dC}{dt} = \frac{-\hat{\mu}}{YK_M} XC = -k_b' XC \tag{3a}$$

where $k_b' = \hat{\mu}/YK_M$. This is essentially second-order biotransformation kinetics.[7] It is first order in bacterial biomass (X) and first order in chemical concentration (C). If the biomass can be assumed constant, then the reaction kinetics are pseudo-first-order, as given by k_bC in equation (1). Sometimes organic chemicals that are adsorbed to suspended particulate matter are biodegraded in addition to soluble chemicals. In this case, equation (3a) must be rewritten in terms of both dissolved and adsorbed chemical concentrations

$$\frac{dC_T}{dt} = -(k_b' XC + k_b'' XC_p) \tag{3b}$$

where C_T is the total whole water chemical concentration, C is the dissolved phase concentration, and C_p is the particulate adsorbed concentration.

If the substrate concentration C is very large such that $C \gg K_M$ (not likely in natural waters), then the microorganisms are growing exponentially, and the rate expression in equation (2) reduces to

$$\frac{dC}{dt} = -\frac{\hat{\mu}}{Y} X \tag{4}$$

which is a zero-order rate expression in C and first-order in X.

Biotransformation experiments are conducted by batch, column, and chemostat experimental methods. Other fate pathways (photolysis, hydrolysis, volatilization) must be accounted for in order to correctly evaluate the effects of biodegradation.

It is incumbent on the fate modeler to understand the range of breakdown products (metabolites) in biological transformation reactions. Metabolites can be as toxic (or more toxic) than the parent compound. For example, in the anaerobic transformation of trichloroethylene (TCE), vinyl chloride can be formed, which is much more toxic than TCE.

Following all the metabolites and pathways in the biological degradation of organic chemicals can be complicated. Polychlorinated biphenyls (PCBs) are mixtures of many isomers—the total number of different organic chemicals is 209 congeners. The structures are shown in Figure 7.3b, where x and y represent the combinations of chlorine atoms (one to five) at different positions on the biphenyl rings. Each congener has distinct properties that result in a different reactivity than the others. Both the *rate* of the biological transformation and the *pathway* can be different for each of the congeners. Although modelers sometimes sum all of the congener concentrations as "total PCBs," it is a simplification that results in a site-specific model that is only valid when it is applied under the same environmental conditions and with the same distribution of congeners as the prototype. It would be most accurate to simulate each congener and its metabolites, but it is not very practical and reactivity data are usually not available.

There are several basic types of biodegradation experiments. Natural water samples from lakes or rivers can have organic toxicant added to them in batch experiments. Disappearance of toxicant is monitored. Organic xenobiotic chemicals can be added to a water–sediment sample to simulate *in situ* conditions, or a contaminated sediment sample alone may be used with or without a spiked addition. Primary sewage, activated sludge, or digester sludge may be used as a seed to test degradability and measure xenobiotic disappearance. Radiolabeled organic chemicals can be used to estimate metabolic degradation (mineralization) by measuring $^{14}CO_2$ off-gas and synthesis into biomass. These experiments are called heterotrophic uptake experiments. The organic chemical may be added in minute concentrations to simulate exposure in natural conditions, or it may be the sole carbon source to the culture to determine whether transformation reactions are possible.

Biodegradation is affected by numerous factors that influence biological growth:

1. *Temperature.* Temperature effects on biodegradation of toxics are similar to those on biochemical oxygen demand (BOD) using an Arrhenius-type relationship.
2. *Nutrients.* Nutrients are necessary for growth and often limit growth rate. Other organic compounds may serve as a primary substrate so that the chemical of interest is utilized via co-metabolism or as a secondary substrate.
3. *Acclimation.* Adaptation is necessary for expressing repressed (induced) enzymes or fostering those organisms that can degrade the toxicant through gradual exposure to the toxicant over time. A shock load of toxicant may kill a culture that would otherwise adapt if gradually exposed. On the other hand, presence of some toxic organics (e.g., phenol or toluene) sometimes spurs on the induction or expression of an enzyme that helps to degrade a given organic chemical.
4. *Population density or biomass concentration.* Organisms must be present in large enough numbers to significantly degrade the toxicant (a lag often occurs if the organisms are too few).

7.2.2 Chemical Oxidation

Chemical oxidation takes place in the presence of dissolved oxygen in natural waters. Oxygen is reduced and the organic chemical is oxidized, but the reaction can be slow. Alternatively, chemical oxidation can be triggered by photochemical transients that may have considerable oxidizing power but low concentrations. Oxidants such as peroxyl radicals ROO·, alkoxy radicals RO·, hydrogen peroxide H_2O_2, hydroxyl radicals ·OH, singlet oxygen $^1O_2$, and solvated electrons are produced in low concentrations and react quickly in natural waters. Because of their large oxidizing power, they may react with a variety of trace organics in solution, but each transient reacts rather specifically with certain trace organic moieties. Thus it is not useful to consider a general second-order rate constant for all oxidants in a given water body. It is better to determine the relevant oxidant chemistry and to measure the oxidant concentration when possible. Since the transient chemical oxidants are often generated photochemically, light-absorbing chromophores, such as humic and fulvic acids and algal pigments, and sunlight intensity will influence oxidation rates.

Alkyl peroxyl radicals (~1×10^{-9} M in sunlit natural waters) react rapidly with phenols and amines in natural waters to form acids and aromatic radicals:

$$\mathrm{ROO\cdot + ArOH \rightarrow ROOH + ArO\cdot}$$

$$ROO\cdot + ArNH_2 \rightarrow ROOH + Ar\dot{N}H$$

Singlet oxygen reacts specifically with olefins:

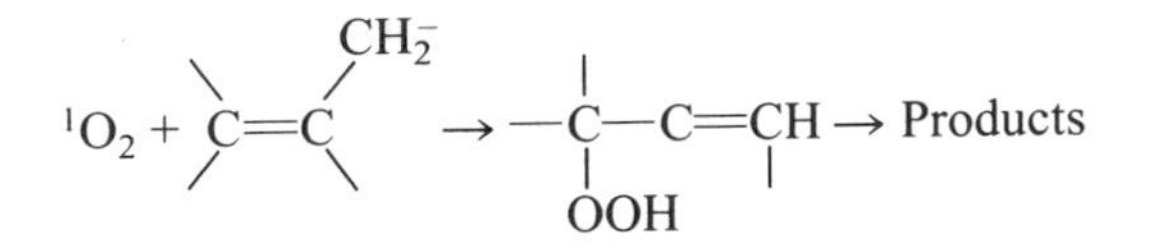

Singlet oxygen concentrations in sunlit natural waters are on the order of 1×10^{-12} M.[7] All of these oxidation reactions may be assumed to be second-order reactions

$$\frac{dC}{dt} = \text{Rxn rate} = -k_{ox}\,[\text{Ox}]\,[C] \tag{5}$$

where C is the organic concentration and Ox is the oxidant concentration. Table 7.3 gives the second-order rate constants for chemical oxidation of selected priority organic chemicals with singlet oxygen and alkyl peroxyl radicals.

Table 7.3 Second-Order Reaction Rate Constants for Chemical Oxidation: Summary Table of Oxidation Data with Singlet Oxygen $^1O_2$ and Alkyl Peroxyl Radicals ROO·

	Oxidation Range of Values ($M^{-1}\ h^{-1}$)	
Chemical	with $^1O_2$	with ROO·
Pesticides		
DDT	< 3600	3600
PCBs		
Aroclor 1248	< 360	< 1.0
Halogenated aliphatic hydrocarbons		
Chloroform	< 360	0.7
Halogenated ethers		
2-Chloroethyl vinyl ether	1×10^{10}	34
Monocyclic aromatics		
2,4-Dimethylphenol	$< 4 \times 10^6$	1.1×10^8
Pentachlorophenol	$< 7 \times 10^3$	1.0×10^5
Phthalate esters		
Bis(2-ethylhexyl)phthalate	≪360	7.2
Polycyclic aromatic hydrocarbons		
Anthracene	5×10^8	2.2×10^5
Benzo[*a*]pyrene	5×10^8	2.0×10^4
Nitrosamines and miscellaneous		
Benzidine	$< 4 \times 10^7$	1.1×10^8
Dimethyl nitrosamine	No Rxn	No Rxn

Source: Mabey et al.[7]

Free radical oxidation requires a chain or series of reactions involving an initiation step, propagation, and subsequent termination.[7] We will illustrate the free radical reaction using the alkyl peroxyl radical ROO· as an example.

Initiation:

$$A\text{-}B \xrightarrow[\text{or heat}]{\text{light}} A^{\cdot} + B^{\cdot} \qquad \text{(a)}$$

$$A^{\cdot} + RH \xrightarrow{\text{fast}} R^{\cdot} + AH \qquad \text{(b)}$$

$$RH + O_2 \xrightarrow{\text{slow}} R^{\cdot} + HOO^{\cdot} \qquad \text{(c)}$$

Propagation:

$$R^{\cdot} + O_2 \xrightarrow{k_1} ROO^{\cdot} \qquad \text{(d)}$$

$$ROO^{\cdot} + RH \xrightarrow{k_2} ROOH + R^{\cdot} \qquad \text{(e)}$$

Termination:

$$2\,ROO^{\cdot} \xrightarrow{k_1} \text{Termination products} \qquad \text{(f)}$$

$$ROO^{\cdot} + R^{\cdot} \longrightarrow \text{Termination products} \qquad \text{(g)}$$

$$2\,R^{\cdot} \longrightarrow \text{Termination products} \qquad \text{(h)}$$

The chemical is represented as an arbitrary organic, RH. A-B is the initiator, which is any free radical source including peroxides, H_2O_2, metal salts, and azo compounds. Investigators have utilized a commercially available azo initiator to estimate the reactivity of pesticides to ROO· in natural waters.[8] If no initiators are available in the water, then reaction (c) represents the probable oxidation pathway, a slow reaction with dissolved oxygen. Otherwise steps (a) and (b) lead to peroxide formation, step (d). Once the highly reactive peroxide radical is formed, it continues to react with the organic chemical, RH, and regenerates another free radical, $R^{\cdot}$, as given in reaction (e). This step may be repeated thousands of times for every photon of light absorbed. Chance collisions between free radicals can terminate the reaction, reactions (f), (g), and (h). At the low pollutant concentrations found in natural waters, reaction (f) is the most likely termination step. Hydrogen peroxide may also be formed, especially when natural dissolved organic matter (DOC) and humates are present. H_2O_2 is a powerful oxidant in natural waters.[9]

If the initiation step is rapid, then the rate-limiting step is the rate of oxidation of the organic in reaction (e):

$$r_{ox} = \frac{d(\text{RH})}{dt} = -k_2\,[\text{ROO}^\cdot]\,[\text{RH}] \tag{6}$$

Provided that reaction (d) is more rapid than reaction (e), the rate of peroxide formation is

$$\frac{d(\text{ROO}^\cdot)}{dt} = \underbrace{k_1[\text{R}^\cdot]\,[\text{O}_2]}_{\text{rate of formation, } r_f} - \underbrace{k_t\,[\text{ROO}^\cdot]^2}_{\text{rate of termination, } r_t} \tag{7}$$

and assuming steady state, the rate of radical initiation must be equal to the rate of termination:

$$k_1\,[\text{R}^\cdot]\,[\text{O}_2] = k_t\,[\text{ROO}^\cdot]^2 \tag{8}$$

so that

$$\text{ROO}^\cdot = \frac{(k_1\,[\text{R}^\cdot]\,[\text{O}_2])^{1/2}}{k_t^{1/2}} = \frac{r_f^{1/2}}{k_t^{1/2}} \tag{9}$$

Substituting equation (9) into equation (6), we find the final reaction rate for the oxidation of the organic chemical is

$$r_{ox} = \frac{d(\text{RH})}{dt} = \frac{-k_2 r_f^{1/2}}{k_t^{1/2}}\,[\text{RH}] = -\,k_3[\text{RH}] \tag{10}$$

The rate of reaction is a pseudo-first-order reaction, where k_3 is the overall reaction rate constant which is a function of r_f, the rate of peroxide formation. If the rate of peroxide formation is relatively constant (as expected in natural waters), then the free radical oxidation of the toxic organic can be computed as a pseudo-first-order reaction.

First-order oxidations of pesticides and organic chemicals have been reported in natural waters. However, these oxidations are often microbially mediated. Strictly chemical free radical oxidation of toxic organics in natural waters remains important for a few classes of compounds. Free radical oxidation is often a part of the photolytic cycle of reactions in natural waters and atmospheric waters.

Oxidations of organic chemicals by $O_{2(aq)}$ is generally slow, but it can be mediated by microorganisms. Cytochrome P450 monooxygenase is a well-studied enzyme with an iron porphyrin active site.[11] Methanotrophs and other organisms can use this pathway to oxidize organics in natural waters, a type of biological transformation.

7.2.3 Redox Reactions

Electron acceptors such as oxygen, nitrate, and sulfate can be reduced in natural waters while oxidizing trace organic contaminants. Oxidation reactions of toxic organic chemicals are especially important in sediments and groundwater, where conditions may be anoxic or anaerobic. The general scheme for utilization of electron acceptors in natural waters follows thermodynamics (Table 7.4).[10]

$$\text{organic} + \text{electron acceptor} \rightarrow \text{oxidized organic} + \text{reduced electron acceptor}$$

The sequence of electron acceptors is approximately:

$$O_2 \rightarrow NO_3^- \rightarrow \underbrace{\text{Mn(IV)}}_{(MnO_{2(s)})} \rightarrow \underbrace{\text{Fe(III)}}_{(Fe(OH)_{3(s)})} \rightarrow SO_4^{2-} \rightarrow CO_2$$

The organic chemical in Table 7.4 is represented as a simple carbohydrate (CH_2O such as glucose $C_6H_{12}O_6$) but other organics may be important reductants in natural waters and groundwaters.

Strict chemical reduction reactions that do not involve a biological catalyst (abiotic reactions) are common in groundwater but less important in natural waters and sediments, where a great complement of enzymes are available for redox transformations. In groundwater, H_2S is a common reductant. It can reduce nitrobenzene to aniline in

$$\text{organic} + \text{reductant} \rightarrow \text{reduced organic} + \text{oxidized reductant}$$

$$\phi\text{-}NO_2 + H_2S + H_2O \rightarrow \phi\text{-}NH_2 + H_2SO_3$$

Table 7.4 Redox Reactions in a Closed Oxidant System at 25 °C and pH 7.0 and Their Free Energies of Reaction.[a]

Reaction	Equation	$\Delta G^\circ(W)$,[b] kJ
Aerobic respiration	$CH_2O + O_2 = CO_2 + H_2O$	–500.59
Denitrification	$CH_2O + \frac{4}{5}NO_3^- + \frac{4}{5}H^+ = CO_2 + \frac{2}{5}N_2 + \frac{7}{5}H_2O$	–476.10
Mn(IV) reduction	$CH_2O + 2MnO_2 + 4H^+ = 2Mn^{2+} + 3H_2O + CO_2$	–339.83
Fe(III) reduction	$CH_2O + 8H^+ + 4Fe(OH)_3 = 4Fe^{2+} + 11H_2O + CO_2$	–107.19
Sulfate reduction	$CH_2O + \frac{1}{2}SO_4^{2-} + \frac{1}{2}H^+ = \frac{1}{2}HS^- + H_2O + CO_2$	–104.50
Methane fermentation	$CH_2O + \frac{1}{2}CO_2 = \frac{1}{2}CH_4 + CO_2$	–92.80
Nitrogen fixation	$CH_2O + H_2O + \frac{2}{3}N_2 + \frac{4}{3}H^+ = \frac{4}{3}NH_4^+ + CO_2$	–80.26

[a] Reactions are usually microbially mediated.
[b] $\Delta G^\circ(W) = \Delta G^\circ - RT \ln [H^+]^p$, where $[H^+] = 1.0 \times 10^{-7}$ mol L^{-1} and p is the stoichiometric coefficient for H^+. $\Delta G^\circ(W)$ is expressed in units of kilojoules per mole of CH_2O.
Source: Modified from Stumm and Morgan,[10] page 337.

homogeneous reactions.[11] Likewise, humic substances and their decay products (natural organic matter, NOM) are good reductants in homogeneous systems. Figure 7.4 is a structure–activity relationship demonstrating that, in homogeneous solution, the second-order kinetic rate constant k_{AB} is directly proportional to the one-electron reduction potential of the redox couple.[12]

$$ArNO_2 + H_2X \rightarrow ArNH_2 + \text{Products}$$

$$\frac{d[ArNO_2]}{dt} = -k_{AB}\,[ArNO_2]\,[H_2X] \tag{11}$$

where H_2X is the reductant. Schwarzenbach et al.[11] have shown that, in the case of juglone, it is not the diprotic dihydroquinone H_2JUG that is the reactant with nitroaromatics, but rather the anions $HJUG^-$ and JUG^{2-}. Reductants in natural waters include quinone, juglone (oak tree exudate), lawsone, and Fe-porphyrins. Nitroreduction is a two-electron, two-proton transfer reaction.

The reduction of nitroaromatic compounds in natural waters and soil water may be viewed as an electron transfer system that is mediated by NOM or its constituents. A bulk reductant is needed (in excess) to set the redox potential E_H of the system and to act as a redox buffer. H_2S often serves in this capacity in reduced sed-

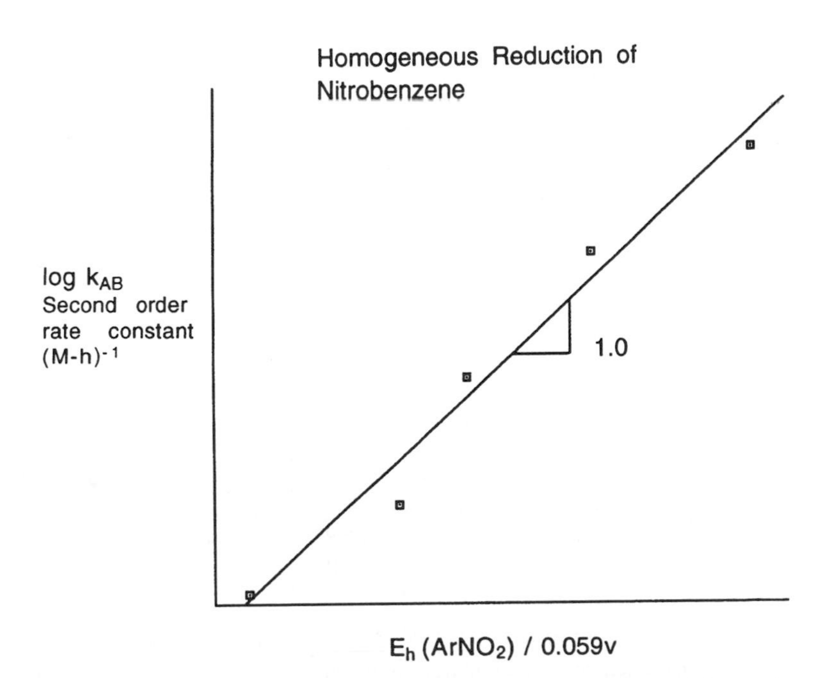

Figure 7.4 Linear free-energy relationship between second-order rate constant and the one electron potential for reduction of substituted nitrobenzenes with natural organic matter (Juglone). From Schwarzenbach, et al.[11]

iments and groundwater. Natural organic matter contains electron transfer mediators such as quinones, hydroquinones, and Fe-porphyrin-like substances.

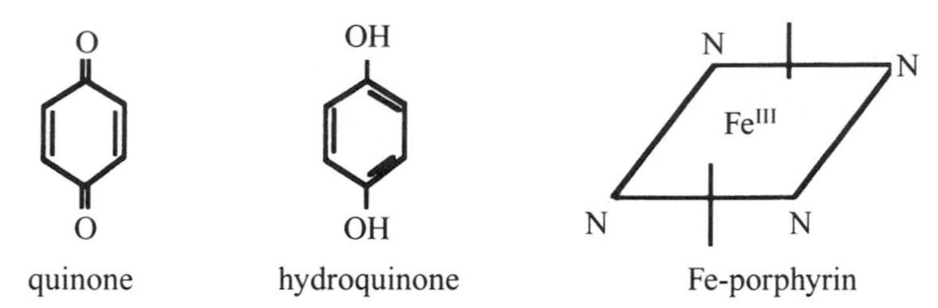

These mediators are reactants that are regenerated in the process by the bulk reductant, which is in excess.

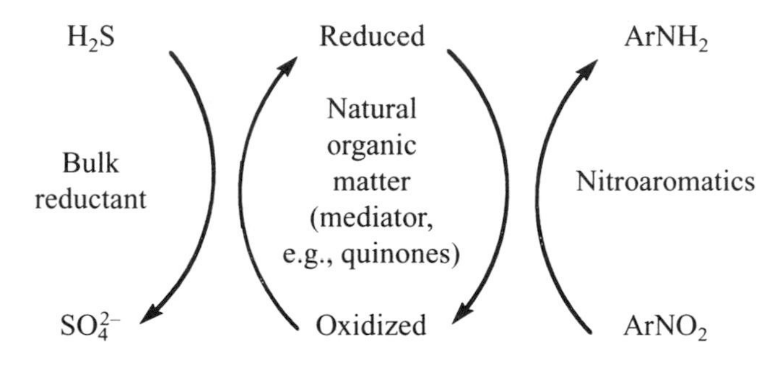

One can add half-reactions of xenobiotic organic oxidations with standard reductants in sediments and groundwater (H_2S, Fe^{2+}, and CH_4) to determine if the reaction is favored thermodynamically (Table 7.5). In the absence of bacteria, the reaction may be slow. However, it is difficult to differentiate *abiotic* from *biotic* reduction reactions because the reaction may occur in the absence of viable cells or on the surface of cells in the presence of extracellular enzymes.[13,14]

7.2.4 Photochemical Transformation Reactions

Direct photolysis, a light-initiated transformation reaction, is a function of the incident energy on the molecule and the quantum yield of the chemical.[15]

When light strikes the pollutant molecule, the energy content of the molecule is increased and the molecule reaches an excited electron state. This excited state is unstable and the molecule reaches a normal (lower) energy level by one of two paths: (1) it loses its "extra" energy through energy emission, that is, fluorescence or phosphorescence; or (2) it is converted to a different molecule through the new electron distribution that existed in the excited state. Usually the organic chemical is oxidized.

Photolysis may be direct or indirect. Indirect photolysis occurs when an intermediary molecule becomes energized, which then reacts with the chemical of interest. The basic equation for direct photolysis is of the form:

Table 7.5 Redox Half-Reactions Pertinent in Wastewater, Groundwater, and Sediment Reactions

Reaction Number	Half-Reaction		$\Delta G^{\circ}(W)$,[a] kcal per electron equivalent
	Reactions for Bacterial Cell Synthesis (R_c)		
	Ammonia as nitrogen source:		
1.	$\frac{1}{5}CO_2 + \frac{1}{20}HCO_3^- + \frac{1}{20}NH_4^+ + H^+ + e^-$	$= \frac{1}{20}C_5H_7O_2N + \frac{9}{20}H_2O$	
	Nitrate as nitrogen source:		
2.	$\frac{1}{28}NO_3^- + \frac{5}{28}CO_2 + \frac{29}{28}H^+ + e^-$	$= \frac{1}{28}C_5H_7O_2N + \frac{11}{28}H_2O$	
	Reactions for Electron Acceptors (R_a)		
	Oxygen:		
3.	$\frac{1}{4}O_2 + H^+ + e^-$	$= \frac{1}{2}H_2O$	−18.675
	Nitrate:		
4.	$\frac{1}{5}NO_3^- + \frac{6}{5}H^+ + e^-$	$= \frac{1}{10}N_2 + \frac{3}{5}H_2O$	−17.128
	Sulfate:		
5.	$\frac{1}{8}SO_4^{2-} + \frac{19}{16}H^+ + e^-$	$= \frac{1}{16}H_2S + \frac{1}{16}HS^- + \frac{1}{2}H_2O$	5.085
	Carbon dioxide (methane fermentation):		
6.	$\frac{1}{8}CO_2 + H^+ + e^-$	$= \frac{1}{8}CH_4 + \frac{1}{4}H_2O$	5.763
	Reactions for Electron Donors (R_d)		
	Organic Donors (Heterotrophic Reactions)		
	Domestic wastewater:		
7.	$\frac{9}{50}CO_2 + \frac{1}{50}NH_4^+ + \frac{1}{50}HCO_3^- + H^+ + e^-$	$= \frac{1}{50}C_{10}H_{19}O_3N + \frac{9}{25}H_2O$	7.6
	Protein (amino acids, proteins, nitrogenous organics):		
8.	$\frac{8}{33}CO_2 + \frac{2}{33}NH_4^+ + \frac{31}{33}H^+ + e^-$	$= \frac{1}{66}C_{16}H_{24}O_5N_4 + \frac{27}{66}H_2O$	7.7
	Carbohydrates (cellulose, starch, sugars):		
9.	$\frac{1}{4}CO_2 + H^+ + e^-$	$= \frac{1}{4}CH_2O + \frac{1}{4}H_2O$	10.0
	Grease (fats and oils):		
10.	$\frac{4}{23}CO_2 + H^+ + e^-$	$= \frac{1}{46}C_8H_{16}O + \frac{15}{46}H_2O$	6.6
	Acetate:		
11.	$\frac{1}{8}CO_2 + \frac{1}{8}HCO_3^- + H^+ + e^-$	$= \frac{1}{8}CH_3COO^- + \frac{3}{8}H_2O$	6.609
	Propionate:		
12.	$\frac{1}{7}CO_2 + \frac{1}{14}HCO_3^- + H^+ + e^-$	$= \frac{1}{14}CH_3CH_2COO^- + \frac{5}{14}H_2O$	6.664
	Benzoate:		
13.	$\frac{1}{5}CO_2 + \frac{1}{30}HCO_3^- + H^+ + e^-$	$= \frac{1}{30}C_6H_5COO^- + \frac{13}{20}H_2O$	6.892
	Ethanol:		
14.	$\frac{1}{6}CO_2 + H^+ + e^-$	$= \frac{1}{12}CH_3CH_2OH + \frac{1}{4}H_2O$	7.592
	Lactate:		
15.	$\frac{1}{6}CO_2 + \frac{1}{12}HCO_3^- + H^+ + e^-$	$= \frac{1}{12}CH_3CHOHCOO^- + \frac{1}{3}H_2O$	7.873
	Pyruvate:		
16.	$\frac{1}{5}CO_2 + \frac{1}{10}HCO_3^- + H^+ + e^-$	$= \frac{1}{10}CH_3COCOO^- + \frac{2}{5}H_2O$	8.545

(continued)

Table 7.5 *(continued)*

Reaction Number	Half-Reaction		$\Delta G°(W)$,[a] kcal per electron equivalent
	Methanol:		
17.	$\frac{1}{6}CO_2 + H^+ + e^-$	$= \frac{1}{6}CH_3OH + \frac{1}{6}H_2O$	8.965
	Inorganic Donors (Autotrophic Reactions)		
18.	$Fe^{3+} + e^-$	$= Fe^{2+}$	–17.780
19.	$\frac{1}{2}NO_3^- + H^+ + e^-$	$= \frac{1}{2}NO_2^- + \frac{1}{2}H_2O$	–9.43
20.	$\frac{1}{8}NO_3^- + \frac{5}{4}H^+ + e^-$	$= \frac{1}{8}NH_4^+ + \frac{3}{8}H_2O$	–8.245
21.	$\frac{1}{6}NO_2^- + \frac{4}{3}H^+ + e^-$	$= \frac{1}{6}NH_4^+ + \frac{1}{3}H_2O$	–7.852
22.	$\frac{1}{6}SO_4^{2-} + \frac{4}{3}H^+ + e^-$	$= \frac{1}{6}S + \frac{2}{3}H_2O$	4.657
23.	$\frac{1}{8}SO_4^{2-} + \frac{19}{16}H^+ + e^-$	$= \frac{1}{16}H_2S + \frac{1}{16}HS^- + \frac{1}{2}H_2O$	5.085
24.	$\frac{1}{4}SO_4^{2-} + \frac{5}{4}H^+ + e^-$	$= \frac{1}{8}S_2O_3^{2-} + \frac{5}{8}H_2O$	5.091
25.	$H^+ + e^-$	$= \frac{1}{2}H_2$	9.670
26.	$\frac{1}{2}SO_4^{2-} + H^+ + e^-$	$= \frac{1}{2}SO_3^{2-} + \frac{1}{2}H_2O$	10.595

[a]Reactants and products at unit activity except $[H^+] = 10^{-7}$.
Source: Sawyer and McCarty.[16]

$$\frac{dC}{dt} = -k_p C \tag{12}$$

where C is the concentration of organic chemical, and k_p is the rate constant for photolysis. Photolysis rate constants can be measured in the field with sunlight or under laboratory conditions. The first-order rate constant, k_p can be estimated directly:

$$k_p = \frac{2.303}{J} \phi \Sigma I_\lambda \varepsilon_\lambda \tag{13}$$

where k_p = photolysis rate constant, s^{-1}
$J = 6.02 \times 10^{20}$ = conversion constant
ϕ = quantum yield
I_λ = sunlight intensity at wavelength λ, photons $cm^{-2}\ s^{-1}$
ε_λ = molar absorbtivity or molar extinction coefficient at wavelength λ, $molarity^{-1}\ cm^{-1}$

The near-surface photolysis rate constants, quantum yields, and wavelengths at which they were measured are presented in Table 7.6. Photolysis will not be an important fate process unless sunlight is absorbed in the visible or near-ultraviolet wavelength ranges (above 290 nm) by either the organic chemical or its sensitizing agent. The quantum yield is defined by

$$\phi = \frac{\text{Number of moles of chemical reacted}}{\text{Number of einsteins absorbed}} \tag{14}$$

An einstein is the unit of light on a molar basis (a quantum or photon is the unit of light on a molecular basis). The quantum yield may be thought of as the efficiency of photoreaction. Incoming radiation is measured in units of energy per unit area per time (e.g., cal cm^{-2} s^{-1}). The incident light in units of einsteins cm^{-2} s^{-1} nm^{-1} can be converted to watts cm^{-2} nm^{-1} by multiplying by the wavelength (nm) and 3.03×10^{39}.

The intensity of light varies over the depth of the water column and may be related by

$$I_z = I_0 e^{-K_e z} \tag{15}$$

where I_z is the intensity at depth z, I_0 is the intensity at the surface, and K_e is an extinction coefficient for light disappearance. Light disappearance is caused by the scattering of light by reflection off particulate matter, and absorption by any mole-

Table 7.6 Photolysis Parameter Values for Selected Chemicals[6,7]

	Photolysis Reaction Parameters		
	Rate Constant (Near Surface) k_p, hr^{-1}	Quantum Yield	Wavelength, nm
Pesticides			
Carbaryl	0.002–0.004	0.01	Sunlight
DDT	$< 5 \times 10^{-7}$	0.16	254
Parathion	0.0024–0.003	< 0.001	> 280
Atrazine	0.0002	—	Sunlight
Trifluralin	0.03	0.002	313
2,4-D	0.00047	0.06	Sunlight
Monocyclic aromatics			
Pentachlorophenol	0.23–1.2	—	290–330
2,4-Dinitrotoluene	0.016	0.00075	313
Polycyclic aromatic hydrocarbons			
Anthracene	0.15	0.003	360
Benzo[*a*]pyrene	0.58	0.0009	313
Nitrosamines and miscellaneous			
Benzidine	11	—	280–380

[a]Near surface, summer, 40° N latitude.

cule. Absorbed energy can be converted to heat or can cause photolysis. Light disappearance is a function of wavelength and water quality (e.g., color, suspended solids, dissolved organic carbon).

Indirect or sensitized photolysis occurs when a nontarget molecule is transformed directly by light, which, in turn, transmits its energy to the pollutant molecule. Changes in the molecule then occur as a result of the increased energy content. The kinetic equation for indirect photolysis is

$$\frac{dC}{dt} = -k_2 CX = -k_p C \tag{16}$$

where k_2 is the indirect photolysis rate constant, X is the concentration of the nontarget intermediary, and k_p is the overall pseudo-first-order rate constant for sensitized photolysis. The important role of inducing agents (e.g., algae exudates and nitrate) has been demonstrated.[17,18]

Inorganics, especially iron, play an important role in the photochemical cycle in natural waters.[19] Hydrogen peroxide, a common transient oxidant, is a natural source of hydroxyl radicals in rivers, oceans, and atmospheric water droplets. Direct photolysis of H_2O_2 produces ·OH, but this pathway is relatively unimportant because H_2O_2 does not absorb visible light very strongly. The important source of ·OH involves hydrogen peroxide and iron (II) in a photo-Fenton reaction.[19]

$$\mathrm{Fe(III)\ L}_n + \mathrm{Light} \rightarrow \mathrm{Fe(II)}$$

$$\mathrm{Fe(II)} + \mathrm{H_2O_2} \rightarrow \cdot\mathrm{OH} + \mathrm{Fe(III)}$$

$$\cdot\mathrm{OH} + \mathrm{ArNO_2} \rightarrow \mathrm{Products}$$

As an example, nitobenzene forms products.

Hydroxyl radicals are a highly reactive and important transient oxidant of a wide range of organic xenobiotics in solution. They can be generated by direct photolysis of nitrate and nitrite in natural waters,[18] or they can be generated from H_2O_2 in the reaction shown above. Nitrobenzene, anisole, and several pesticides have been shown to be oxidized by hydroxyl radicals in natural waters. The indirect photooxidation of a wide variety of organic chemicals is expected, but natural dissolved organic matter also reacts with ·OH. This competitive parallel reaction likely scavenges the hydroxyl radicals in colored waters, preventing the oxidation of other organic compounds.

7.2.5 Chemical Hydrolysis

Chemical hydrolysis is that fate pathway by which an organic chemical reacts with water. Particularly, a nucleophile (hydroxide, water, or hydronium ions), N, displaces a leaving group, X, as shown.

$$\text{R-X} + \text{N} \rightarrow \text{RN} + \text{X}$$

Hydrolysis does not include acid–base, hydration, addition, or elimination reactions. The hydrolysis reaction consists of the cleaving of a molecular bond and the formation of a new bond with components of the water molecule (H^+, OH^-). It is often a strong function of pH (see Figure 7.5).

Three examples of a hydrolysis reaction are presented below.

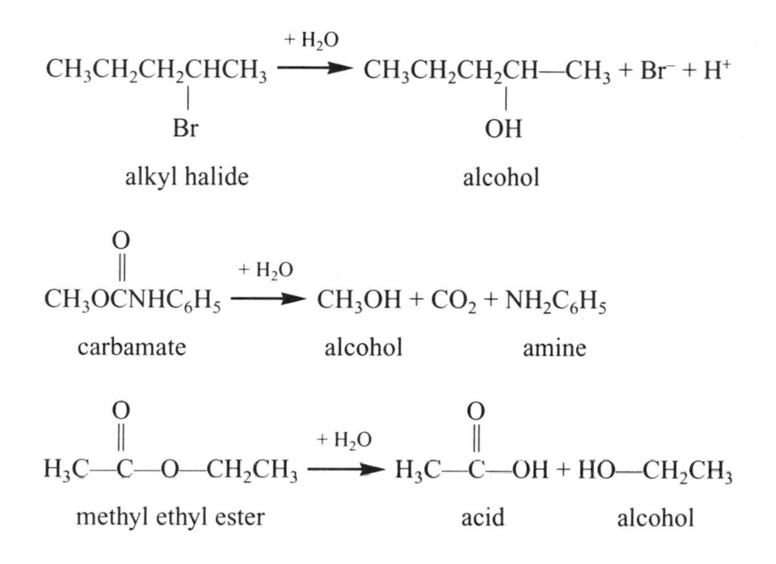

Types of compounds that are generally susceptible to hydrolysis are:

- Alkyl halides
- Amides
- Amines
- Carbamates
- Carboxylic acid esters
- Epoxides
- Nitriles
- Phosphonic acid esters
- Phosphoric acid esters
- Sulfonic acid esters
- Sulfuric acid esters

The kinetic expression for hydrolysis is

$$\frac{dC}{dt} = -k_a[H^+]C - k_n C - k_b[OH^-]C \tag{17}$$

where C is the concentration of toxicant; k_a, k_n, and k_b are the acid-, neutral-, and base-catalyzed hydrolysis reaction rate constants, respectively; and $[H^+]$ and $[OH^-]$ are the molar hydrogen and hydroxyl ion concentrations, respectively.

A summary of these data is presented in Table 7.7.

Hydrolysis experiments usually involve fixing the pH at some target value, eliminating other fate processes, and measuring toxicant disappearance over time. A sterile sample in a glass tube, filled to avoid a gas space, and kept in the dark eliminates the other fate pathways. In order to evaluate k_a and k_b, several non-neutral pH experiments must be conducted as depicted in Figure 7.5.

Often, the hydrolysis reaction rate expression in equation (17) is simplified to a pseudo-first-order reaction rate expression at a given pH and temperature (Table 7.7, 298 K and pH 7).

Table 7.7 Selected Chemical Hydrolysis Rate Constants, at 298 K and pH 7

Chemical	k_h, s^{-1}	$t_{1/2}$	Hydrolysis Rate Catalysis
FCH_3	7.4×10^{-10}	30 yr	Neutral
$ClCH_3$	2.4×10^{-8}	0.93 yr	Neutral
$BrCH_3$	4.1×10^{-7}	20 days	Neutral
ICH_3	7.3×10^{-8}	110 days	Neutral
$ClCH_2CH_3$	2.1×10^{-7}	38 days	Neutral
$CH_3CHClCH_3$	2.1×10^{-7}	38 days	Neutral
t-Chlorobutane	3.0×10^{-2}	23 sec	Neutral
CH_2CHCH_2Cl	1.2×10^{-7}	69 days	Neutral
p-$CH_3C_6H_4CH_2Cl$	4.5×10^{-4}	0.43 h	Neutral
$C_6H_5CHCl_2$	1.56×10^{-3}	0.1 h	Neutral
$CHCl_3$	6.9×10^{-10}	3500 yr	Base
CCl_4 [a]	4.8×10^{-7}	7000 yr	Base
H_3C—CH_3 (epoxide, O bridging)	18.3×10^{-7}	4.4 days	Neutral
$CH_3C(O)OCH_2CH_3$	1.1×10^{-8}	2.0 yr	Base
$CH_3SCH_2C(O)OCH_2CH_3$	9.2×10^{-8}	87 days	Base
$CH_3C(O)OC_6H5$	2.1×10^{-7}	38 days	Base, neutral
Acetamide	5.5×10^{-12}	3,950 yr	Base
Chloroacetamide	1.5×10^{-8}	1.46 yr	Base
Methoxychlor	3×10^{-8}	270 days	Base
Atrazine[b]	7.6×10^{-5}	2.5 h	Base
DDE	1×10^{-10}	1000 days	Base
Carbofuran	6×10^{-12}	> 30 yr	Base
Parathion	4.5×10^{-8}	180 days	Neutral
Bis(2-ethylhexyl)phthalate	1×10^{-11}	2000 yr	Base
PCB Aroclor 1248	~ 0	NA	NA

[a]Second order in CCl_4 concentration.
[b]Half-life is much longer in soil.
Source: Mabey and Mill[20] and Schnoor et al.[6]

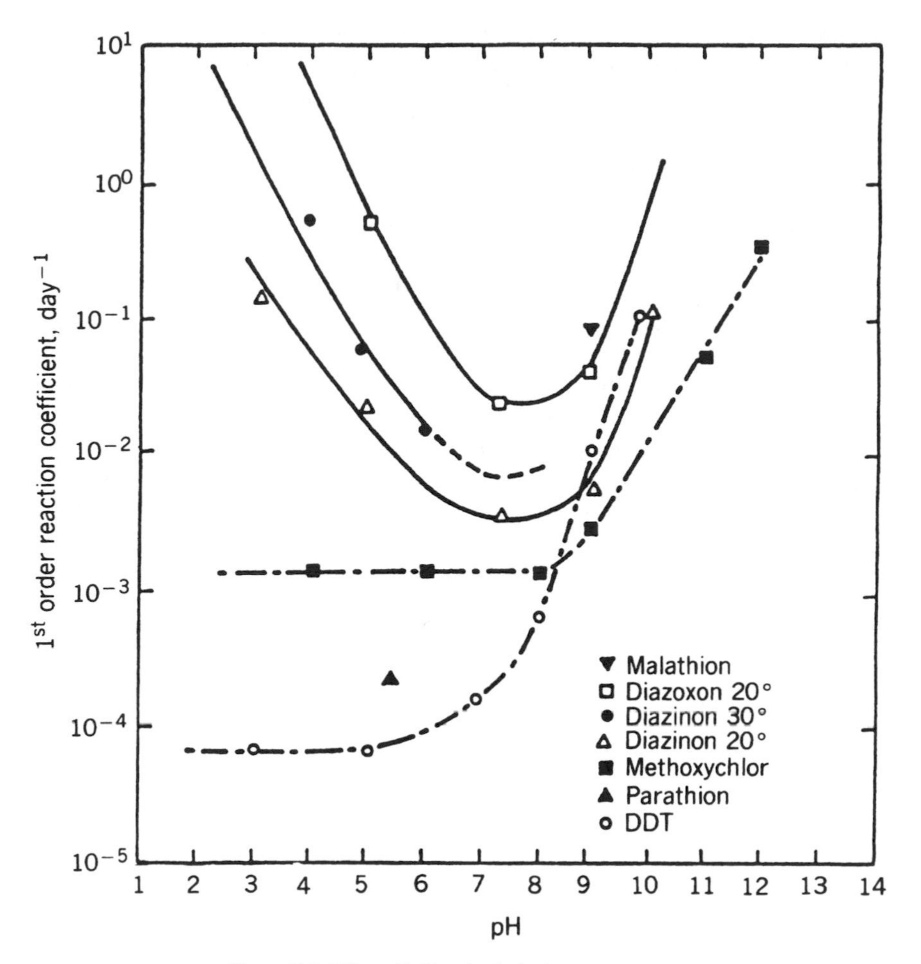

Figure 7.5 Effect of pH on hydrolysis rate constants.

$$\frac{dC}{dt} = -k_h C \tag{18}$$

where $k_h = k_b[OH^-] + k_a[H^+] + k_n$ and k_h is the pseudo-first-order hydrolysis rate constant, T^{-1}; k_b is the base-catalyzed rate constant, molarity^{-1} T^{-1}; k_a is the acid-catalyzed rate constant, molarity^{-1} T^{-1}; and k_n is the neutral rate constant, T^{-1}.

7.2.6 Volatilization/Gas Transfer

The transfer of pollutants from water to air or from air to water is an important fate process to consider when modeling organic chemicals. Volatilization is a transfer process; it does not result in the breakdown of a substance, only its movement from the liquid to gas phase, or vice versa. Gas transfer of pollutants is analogous to the reaeration of oxygen in surface waters and will be related to known oxygen transfer rates. The rate of volatilization is related to the size of the molecule (as measured by the molecular weight).

Gas transfer models are often based on two-film theory (Figure 7.6). Two-film theory was derived by Lewis and Whitman in 1923. Mass transfer is governed by molecular diffusion through a stagnant liquid and gas film. Mass moves from areas of high concentration to areas of low concentration. Transfer can be limited at the gas film or the liquid film. Oxygen, for example, is controlled by the liquid-film resistance. Nitrogen gas, although approximately four times more abundant in the atmosphere than oxygen, has a greater liquid-film resistance than oxygen.

Volatilization, as described by two-film theory, is a function of Henry's constant, the gas-film resistance, and the liquid-film resistance. The film resistance depends on diffusion and mixing. Henry's constant, H, is a ratio of a chemical's vapor pressure to its solubility. It is a thermodynamic ratio of the fugacity of the chemical (escaping tendency from air and water).

$$H = p_g / C_{s\ell} \tag{19}$$

where p_g is the partial pressure of the chemical of interest in the gas phase, and $C_{s\ell}$ is its saturation solubility. Henry's constant can be "dimensionless" [mg/L (in air)/mg/L (in water)] or it has units of atm m^3 mol^{-1}.

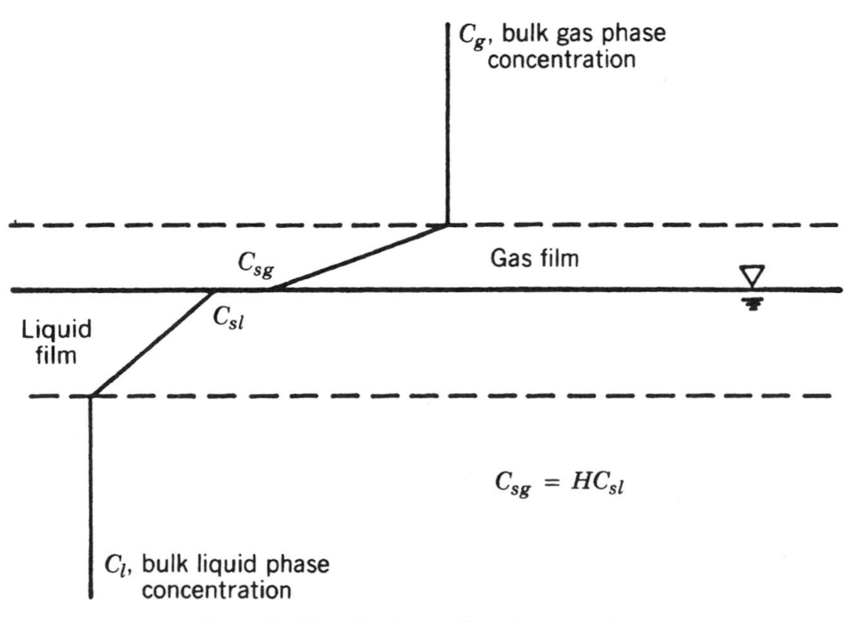

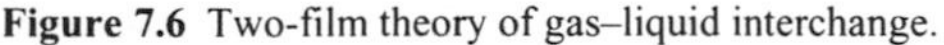
Figure 7.6 Two-film theory of gas–liquid interchange.

The value of H can be used to develop simplifying assumptions for modeling volatilization. If either the liquid-film or the gas-film controls—that is, one resistance is much greater than the other—the lesser resistance can be neglected. The threshold of Henry's constant for gas- or liquid-film control is approximately 0.1 for dimensionless H, or 2.2×10^{-3} atm m^3 mol^{-1}. Above this threshold value, the chemical is liquid-film controlled, and below it, it is gas-film controlled.

The flux of contaminants across the boundary can be modeled by Fick's first law of diffusion at equilibrium,

$$F = -D\frac{dC}{dx} \tag{20}$$

where D is the molecular diffusion coefficient and dC/dx is the concentration gradient in either the gas or liquid phase. If we consider the molecular diffusion to occur through a thin stagnant film, the mass flux is then

$$N = k\,\Delta C \tag{21}$$

where $k = D/\Delta z$ in which Δz is the film thickness and k is the mass transfer coefficient with units of LT^{-1}. At steady state, the flux through both films of Figure 7.6 must be equal:

$$N = k_g(C_g - C_{sg}) = k_\ell(C_{s\ell} - C_\ell) \tag{22}$$

If Henry's law applies exactly at the interface, we can express the concentrations in terms of bulk phase concentrations, which are measurable by substitution below:

$$N = k_g(C_g - H\,C_{s\ell}) = k_\ell(C_{s\ell} - C_\ell) \tag{23}$$

$$C_{s\ell} = \frac{N + k_\ell C_\ell}{k_\ell} \tag{24}$$

$$N = k_g\left[C_g - H\left(\frac{N + k_\ell C_\ell}{k_\ell}\right)\right] = k_\ell\left[\frac{N + k_\ell C_\ell}{k_\ell} - C_\ell\right] \tag{25}$$

By rearranging equation (25), we can solve for N in terms of bulk phase concentrations, mass transfer coefficients for each phase, and Henry's constant:

$$N = \frac{C_g/H - C_\ell}{\dfrac{1}{H\,k_g} + \dfrac{1}{k_\ell}} = K_L\left(\frac{C_g}{H} - C_\ell\right) \tag{26}$$

where K_L is the *overall* mass transfer coefficient derived for expression of the gas transfer in terms of a liquid phase concentration.

$$\frac{1}{K_L} = \frac{1}{k_\ell} + \frac{1}{H\,k_g} \tag{27}$$

We may think of the first term on the right-hand side of the equation as a liquid-film resistance and the second term as a gas phase resistance using an electrical resistance analogy. We can compare the two resistances to determine if the

$$R_t = r_\ell + r_g \tag{28}$$

gas phase resistance, r_g, or the liquid phase resistance, r_ℓ, predominates.

Equivalently, we could choose to write the overall mass transfer in terms of the bulk gas phase concentration.

$$N = K_G\,(C_g - HC_\ell) \tag{29}$$

where

$$\frac{1}{K_G} = \frac{1}{k_g} + \frac{H}{k_\ell} \tag{30}$$

If the gas is soluble, then H is small and the gas-film resistance controls mass transfer. Examples are SO_2, H_2O evaporation, NH_3, H_2S, most pesticides, and long chain organic molecules. If the gas is insoluble, then H is large and the liquid film controls (e.g., O_2, N_2, most volatile chlorinated aliphatics and solvents). If the liquid film controls gas transfer, stirring the liquid should increase gas exchange by increasing turbulence and k_ℓ. If the gas phase controls, stirring the liquid should not help greatly.

In terms of a differential equation, the overall gas transfer may be expressed in terms of a saturation concentration in equilibrium with the gas phase and an interfacial area for transfer.

$$\frac{dC_\ell}{dt} = K_L\,\frac{A}{V}\,(C_{\text{sat}} - C_\ell) \tag{31}$$

where $C_{\text{sat}} = p_g/H$, A is the interfacial surface area, and V is the volume of the liquid. In streams, A/V is the reciprocal depth of the water and the equation can be expressed as

$$\frac{dC_\ell}{dt} = \frac{K_L}{Z}\,(C_{\text{sat}} - C_\ell) = k_{li}(C_{\text{sat}} - C_\ell) \tag{32}$$

where Z is the mean depth and k_{li} is termed the volatilization rate constant (T^{-1}). Equations (31) and (32) apply for either gas absorption or gas stripping from the water body. It is a reversible process.

The mass transfer coefficients are dependent on the hydrodynamic characteristics of the air–water interface and flow regime. For flowing water, we may write

$$K_L = \left(\frac{D\bar{u}}{Z}\right)^{1/2} \tag{33}$$

where $\bar{u}$ is the mean stream velocity and Z is the mean depth.[21] Winds affect the gas transfer in natural waters greatly.[22] For smooth flow (no ripples or waves) and wind speed less than 5 ms^{-1}, $1/K_\delta$ predominates.

$$\frac{1}{K_L} = \frac{1}{K_\delta} + \frac{1}{K_\tau} \tag{34}$$

$$K_\delta = \frac{D}{\nu}\,\frac{\sqrt{C_D}\,W}{\Delta Z} \tag{35}$$

where C_D is the dimensionless drag coefficient, W is the wind speed, and ν is the kinematic viscosity.[22] The transfer term for aerodynamically rough flow with waves is

$$K_\tau = \left(\frac{Du_*}{\alpha d}\right)^{1/2} \tag{36}$$

where d is the diameter or amplitude of the waves, u_* is the surface shear velocity, and α is a constant dependent on the physics of the wave properties.

The diffusion coefficients in water and air have been related to molecular weight:

$$D_\ell = (2.2 \times 10^{-5}\ \mathrm{cm^2\ s^{-1}})\ (\mathrm{MW}^{-2/3}) \tag{37}$$

where D_ℓ is the diffusivity of the chemical in water and MW is the molecular weight, and

$$D_g = (1.9\ \mathrm{cm^2\ s^{-1}})\ (\mathrm{MW}^{-2/3}) \tag{38}$$

where D_g is the diffusivity of the chemical in air. The mass transfer rate constant, k_{li}, can then be related to the oxygen reaeration rate, k_a, by a ratio of the diffusivity of the chemical to that of oxygen in water:

$$k_{li}/k_a = (D_\ell/D_{O_2})^{1/2} \tag{39}$$

where D_{O_2} is 2.4×10^{-5} cm^2 s^{-1} at 20 °C. The reaeration rate, k_a, can be calculated from any of the formulas available, which were covered in Chapter 6, Table 6.1. In addition, the overall gas-film transfer rate may be calculated from

$$k_{gi} = \frac{0.001}{Z}\,(D_g/\nu_g)^{2/3}\,W \tag{40}$$

where ν_g is the kinematic viscosity of air (a function of temperature) as presented in Table 7.8, Z is the water depth, and W is the wind speed in m s^{-1}. k_{gi} has units of T^{-1}.

Solubility, vapor pressure, and Henry's constant data are presented in Table 7.9. (Henry's constant can be converted from atm m^3 mol^{-1} to a dimensionless number by multiplying by 44.64 = 1000 L mol/22.4 L m^3.) Dimensionless Henry's constants refer to a concentration ratio of mg/L air per mg/L in the water phase.

Yalkowsky[24] measured the solubility of 26 halogenated benzenes at 25 °C and developed the following relationship:

$$\log S_w = -0.01 \text{ MP} - 0.88 \log K_{ow} - 0.012 \tag{41}$$

where S_w is solubility (mol L^{-1}), MP is the melting point (°C), and K_{ow} is the estimated octanol/water partition coefficient.

Lyman et al.[26] compiled solubility data on 78 organic compounds and presented estimation methods based on K_{ow} for different classes of compounds. They also included a method based on the molecular structure. Mackay[27] measured Henry's constant for 22 organic chemicals as part of a study of volatilization characteristics. Transfer coefficients for the gas and liquid phases were correlated for environmental conditions as:

$$K_L = 34.1 \times 10^{-6} (6.1 + 0.6U_{10})^{0.5} U_{10} \text{ Sc}_L^{-0.5} \tag{42}$$

$$K_G = 46.2 \times 10^{-5} (6.1 + 0.63U_{10})^{0.5} U_{10} \text{ Sc}_G^{-0.67} \tag{43}$$

where U_{10} is the 10-m wind velocity (m s^{-1}), Sc_L and Sc_G are the dimensionless liquid and gas Schmidt numbers.

Volatile compounds such as those shown in Figure 7.3a are easily removed from water and wastewater by purging with air or by passing them through an air stripping tower. In natural waters, they are removed by stripping from the water to the atmosphere. The overall mass transfer coefficient K_L can be related to that of oxygen (Table 7.10) because so much information exists for oxygen transfer in nat-

Table 7.8 Kinematic Viscosity of Air

Temperature, °F	ν_g, cm^2 s^{-1}
0	0.117
20	0.126
40	0.136
60	0.147
80	0.156
100	0.167
120	0.176

Source: Rouse.[25]

Table 7.9 Summary Table of Volatilization Data at 20 °C

Chemical	Molecular Weight, g mol^{-1}	Density, g cm^{-3}	Solubility, mg L^{-1}	Vapor Pressure, atm	Henry's Law Constant, atm m^3 mol^{-1}	Henry's Law Constant, dimensionless	log K_{ow} 25 °C	Comments
Acetic acid	60.05	1.05	∞	—	—	—	—	
Aroclor 1254	325.06	1.50	1.2×10^{-2}	1×10^{-7}	2.7×10^{-3}	1.2×10^{-1}	6.5	Polychlorinated biphenyl mixture (PCB)
Aroclor 1260	371.22	1.57	2.7×10^{-3}	5.3×10^{-8}	7.1×10^{-3}	3.0×10^{-1}	6.7	Polychlorinated biphenyl mixture (PCB)
Atrazine	215.68		33	4×10^{-10}	3×10^{-9}	1×10^{-7}	2.68	Herbicide
Benzene	78.11	0.88	1780	1.25×10^{-1}	5.5×10^{-3}	2.4×10^{-1}	2.13	Gasoline constituent
Benz[*a*]anthracene	228.29		2.5×10^{-4}	6.3×10^{-9}	5.75×10^{-6}	2.4×10^{-4}	5.91	Polycyclic aromatic hydrocarbon (PAH)
Benzo[*a*]pyrene	252.32		4.9×10^{-5}	2.3×10^{-10}	1.20×10^{-6}	4.9×10^{-5}	6.50	PAH
Carbon tetrachloride	153.82	1.59	800	0.12	2.3×10^{-2}	9.7×10^{-1}	2.83	Chlorinated solvent
Chlorobenzene	112.56	1.11	472	1.6×10^{-2}	3.7×10^{-3}	1.65×10^{-1}	2.92	Industrial chemical
Chloroform	119.38	1.48	8,000	0.32	4.8×10^{-3}	2.0×10^{-1}	1.97	Chlorinated solvent
m-Cresol	108.14		2,780				1.96	Creosote
Cyclohexane	84.16	0.78	60	0.13	0.18	7.3	3.44	Industrial chemical
1,1-Dichloroethane	98.96	1.18	4,962	3.0×10^{-1}	6×10^{-3}	2.4×10^{-1}	1.79	Chlorinated solvent
1,2-Dichloroethane	98.96	1.24	8,426	9.1×10^{-2}	10^{-3}	4.1×10^{-2}	1.47	Chlorinated solvent
cis-1,2-Dichloroethene	96.94	1.28	3,500	0.26	3.4×10^{-3}	0.25	1.86	Chlorinated solvent
trans-1,2-Dichloroethene	96.94	1.26	6,300	0.45	6.7×10^{-3}	0.23	2.06	Chlorinated solvent
Ethane	30.07	—	2.4×10^{-3}	39.8	4.9×10^{-1}	20		Gaseous
Ethanol	46.07	0.79	∞	7.8×10^{-2}	6.3×10^{-6}		−0.31	Grain alcohol
Ethylbenzene	106.17	0.87	152	1.25×10^{-2}	8.7×10^{-3}	3.7×10^{-1}		Petroleum distillate
Lindane	290.9		7.3	1.2×10^{-8}	4.8×10^{-7}	2.2×10^{-5}		Pesticide

(continued)

Table 7.9 *(continued)*

Chemical	Molecular Weight, $g\ mol^{-1}$	Density, $g\ cm^{-3}$	Solubility, $mg\ L^{-1}$	Vapor Pressure, atm	Henry's Law Constant, $atm\ m^3\ mol^{-1}$	Henry's Law Constant, dimensionless	log K_{ow} 25 °C	Comments
Methane	16.04	—	6.7×10^{-3}	275	0.66	27	—	Natural gas
Methylene chloride	84.93	1.33	1.3×10^{4}	0.46	3×10^{-3}	1.3×10^{-1}	1.15	Also called dichloromethane
Naphthalene	128.17	1.03	33	3×10^{-4}	1.15×10^{-3}	4.9×10^{-2}	3.36	PAH
Nitrogen	28.01	—					—	Atmospheric gas
n-Octane	114.23	0.70	0.72	0.019	2.95	121	4.00	Alkane
Oxygen	32.00	—					—	Atmospheric gas
Pentachlorophenol	266.34	1.98	14	1.8×10^{-7}	3.4×10^{-6}	1.5×10^{-4}		Wood preservative
n-Pentane	72.15	0.63	40.6	0.69	1.23	50.3	3.62	Solvent
Perchloroethene	165.83	1.62	400	2×10^{-2}	8.3×10^{-3}	3.4×10^{-1}	2.88	Commonly used in dry cleaning; tetrachloroethene
Phenanthrene	178.23	0.98	6.2	8.9×10^{-7}	3.5×10^{-5}	1.5×10^{-3}	4.57	PAH
Styrene	104.15	0.91					2.95	Industrial chemical
Toluene	92.14	0.87	515	3.7×10^{-2}	6.6×10^{-3}	2.8×10^{-1}	2.69	Common solvent
1,1,1-Trichloroethane (TCA)	133.40	1.34	950	0.13	1.8×10^{-2}	7.7×10^{-1}	2.48	Common solvent
Trichloroethene (TCE)	131.39	1.46	1×10^{3}	8×10^{-2}	1×10^{-2}	4.2×10^{-1}	2.42	Common solvent
o-Xylene	106.17	0.88	175	8.7×10^{-3}	5.1×10^{-3}	2.2×10^{-1}	3.12	1,2-Dimethyl benzene
Vinyl chloride	62.50	0.91	90	3.4	2.4	99	0.60	Degradation product of TCE

Source: Modified from Hemond and Feichnor.[23]

Table 7.10 Estimated Henry's Constant and Mass Transfer Coefficients for Selected Organics at 20 °C

Chemical	MW, g mol^{-1}	VP, mm Hg	SOL, mg L^{-1}	H, $\frac{\text{mg/L air}}{\text{mg/L water}}$	$r_\ell \times 10^3$, h cm^{-1}	$r_g \times 10^3$, h cm^{-1}	K_L, cm h^{-1}	$\frac{K_L}{K_{O2}}$
Oxygen	32	—	9.2	25.6	50.0	—	20	1.0
trans-1,2-Dichloroethene	97	260	600	2.29	69.9	0.242	14.3	0.715
Carbon tetrachloride	154	96	800	1.01	80.0	0.625	12.4	0.619
Toluene	92	22	478	0.231	68.5	2.35	14.1	0.704
Benzene	78	72	1680	0.183	65.4	2.83	14.7	0.733
Chloroform	120	148	7800	0.124	74.1	4.74	12.7	0.633
Chlorobenzene	113	8.7	475	0.113	73.2	5.13	12.8	0.640
m-Dichlorobenzene	147	1.3	111	0.0935	78.7	6.67	11.7	0.584
p-Dichlorobenzene	147	0.6	69	0.0694	78.7	9.01	11.4	0.569
o-Dichlorobenzene	147	1	134	0.0605	78.7	10.3	11.2	0.559
Bromoform	253	5.4	2780	0.0268	92.6	27.5	8.3	0.415
Napthalene	128	0.09	30	0.021	75.8	28.5	9.6	0.479
Nitrobenzene	123	0.15	1900	0.0025	74.6	200	3.6	0.18
Ammonia	17			0.00073	41.3	448	1.9	0.0958
PCB (Aroclor 1254)	327	10^{-6}		0.000446	0.16	3.19	0.3	0.165
Dieldrin	385	2.7×10^{-6}	2.0	0.000035	0.196	48.4	0.021	0.00150
Alachlor	269.8	2.2×10^{-5}	242	1.3×10^{-6}	0.15	993	0.001	0.00005
Atrazine	215.7	1×10^{-6}	33	5.6×10^{-7}	0.13	2060	0.00048	0.00002
Water	18	17.5				3.3×10^{-4}	3000_a	

[a]For water, $K_G = 3000$ cm h^{-1}.

ural waters. In Table 7.10, most of the volatile organics from trans-1,2-dichloroethene through *o*-dichlorobenzene are liquid-film controlled ($r_\ell > r_g$) because they have a relatively high vapor pressure and low solubility that results in a relatively high Henry's constant. The overall mass transfer coefficient for oxygen, which is also liquid-film controlled, is assumed to be typical of ocean gas transfer, 20 cm h^{-1}. Relative to oxygen, the other chemicals have lower mass transfer rates. For chemicals that have relatively low Henry's constants (nitrobenzene down through atrazine in Table 7.10) the overall mass transfer coefficients are very small and the substances are considered semivolatile (nitrobenzene, PCB, and dieldrin) or nonvolatile (alachlor and atrazine).

7.2.7 Sorption Reactions

Soluble organics in natural waters can sorb onto particulate suspended material or bed sediments. The mechanism and the processes by which this occurs include: (1) physical adsorption due to van der Waals forces, (2) chemisorption due to a chemical bonding or surface coordination reaction, and (3) partitioning of the organic chemical into the organic carbon phase of the particulates. Physical adsorption is purely a surface electrostatic phenomenon. Surface coordination reactions result in a binding between the particulate binding site (e.g., an iron oxide edge of a mineral grain) and the dissolved organic molecule. Partitioning refers to the dissolution of hydrophobic organic chemicals into the organic phase of the particulate matter; it is an *ab*sorption phenomenon rather than a surface reaction, and it may occur slowly over time scales of minutes to days. The term *sorption* is purposely ambiguous to include all three phenomena in cases where the mechanism is not clear. Cationic and anionic organic compounds may also sorb by ion exchange reactions.

Adsorption isotherms refer to the equilibrium relationship of sorption between organics and particulates at constant temperature. Adsorption–desorption in natural waters is a reversible reaction in some cases. The chemical is dissolved in water in the presence of various concentrations of suspended solids. After an initial kinetic reaction, a dynamic equilibrium is established in which the rate of the forward reaction (sorption) is exactly equal to the rate of the reverse reaction (desorption). After chemical equilibrium is attained, samples are taken of the filtered water and/or solids and analyzed for chemical concentrations. The data for each batch reactor are plotted to make up the isotherm.

The sorption of toxicants to suspended particulates and bed sediments is a significant transfer mechanism. Partitioning of a chemical between particulate matter and the dissolved phase is not a transformation pathway; it only relates the concentration of dissolved and sorbed states of the chemical.

The octanol/water partition coefficient, K_{ow}, is related to the solubility of a chemical in water. The less soluble a chemical is in water, the more likely it is to sorb to the surfaces of sediments or suspended particles. Tables 7.9 and 7.11 provide log K_{ow} values for a number of organic chemicals of environmental interest. There is another list of log K_{ow} values for groundwater contaminants in Table 9.4.

The laboratory procedure for measuring K_{ow} is given by Lyman.[26]

Table 7.11 Octanol/Water Partition Coefficients of Selected Organics, 298 K

Chemical	log K_{ow}
Pesticides	
Carbofuran	1.60
DDT	6.4
Parathion	3.81
PCBs	
Aroclor 1248	5.75–6.11
Halogenated aliphatic hydrocarbons	
Chloroform	1.90–1.97
Halogenated ethers	
2-Chloroethyl vinyl ether	1.28
Monocyclic aromatics	
2,4-Dimethylphenol	2.42–2.50
Pentachlorophenol	5.01
Phthalate esters	
Bis(2-ethylhexyl)phthalate	8.73
Polycyclic aromatic hydrocarbons	
Anthracene	4.34–4.63
Benzo[*a*]pyrene	6.5
Nitrosamines & miscellaneous	
Benzidine	1.36–1.81
Dimethyl nitrosamine	0.06

1. Chemical is added to a mixture of pure octanol (a nonpolar solvent) and pure water (a polar solvent). The volume ratio of octanol and water is set at the estimated K_{ow}.
2. Mixture is agitated until equilibrium is reached.
3. Mixture is centrifuged to separate the two phases. The phases are analyzed for the chemical.
4. K_{ow} is the ratio of the chemical concentration in the octanol phase to chemical concentration in the water phase, and has no units. The logarithm of K_{ow} has been measured from -3 to +7.

If the octanol/water partition coefficient cannot be reliably measured or is not available in databases, it can be estimated from solubility and molecular weight information,[28]

$$\log S = -1.08 \log K_{ow} + 3.70 + \log \text{MW} \tag{44}$$

where MW is the molecular weight of the pollutant (g mol^{-1}) and S is in units of ppm for organics that are liquid in their pure state at 25 °C. For organics that are solid in their pure state at 25 °C,[28]

$$\log S = -1.08 \log K_{ow} + 3.70 + \log \text{MW} - \frac{\Delta S_f}{1360}(\text{MP} - 25) \quad (45)$$

where MP is the melting point of the pollutant (°C) and ΔS_f is the entropy of fusion of the pollutant (cal mol^{-1} deg^{-1}).

The octanol/water partition coefficient is dimensionless, but it derives from the partitioning that occurs in the extraction between the chemical in octanol and water.

$$K_{ow} = \frac{\text{mg chemical/L octanol}}{\text{mg chemical/L water}}$$

Octanol was chosen as a reference because it is a model solvent with some properties that make it similar to organic matter and lipids in nature.

For a wide variety of organic chemicals, the octanol water partition coefficient is a good estimator of the organic carbon normalized partition coefficient (K_{oc}).

$$K_{oc} = \frac{\text{mg chemical/kg organic carbon}}{\text{mg chemical/L } H_2O}$$

Karickhoff et al.[29] and Schwarzenbach and Westall[30] have published useful empirical equations for predicting K_{oc} as a function of K_{ow}.

$$\text{Karickhoff} \quad \log K_{oc} = 1.00 \log K_{ow} - 0.21 \quad (46)$$

$$\text{Schwarzenbach} \quad \log K_{oc} = 0.72 \log K_{ow} + 0.49 \quad (47)$$

Once an estimate of K_{oc} is obtained, the calculation of a sediment/water partition coefficient suitable for natural waters is straightforward because the

$$K_p = f_{oc} K_{oc} \quad (48)$$

where f_{oc} is the decimal fraction of organic carbon present in the particulate matter (mass/mass). Figure 7.7 is a schematic of how K_p, K_{oc}, and K_{ow} are interrelated. Figure 7.7a is the Langmuir adsorption isotherm for sorption of one chemical on particulate matter. In the low solute concentration range, the slope is constant, which is defined as K_p. But K_p is directly proportional to f_{oc} as shown in Figure 7.7b. There is a y-intercept in Figure 7.7b if other mechanisms in addition to absorption partitioning are important.

When many chemicals are considered, Figure 7.7c results, indicating the direct linear relationship on a log–log plot between K_{oc} and K_{ow}. In Chapter 9 on groundwater, values of log K_{oc} and log K_{ow} are given for a variety of organic chemicals (see Table 9.4).

The higher the value of K_{ow} for a chemical, the lower is its solubility in water and

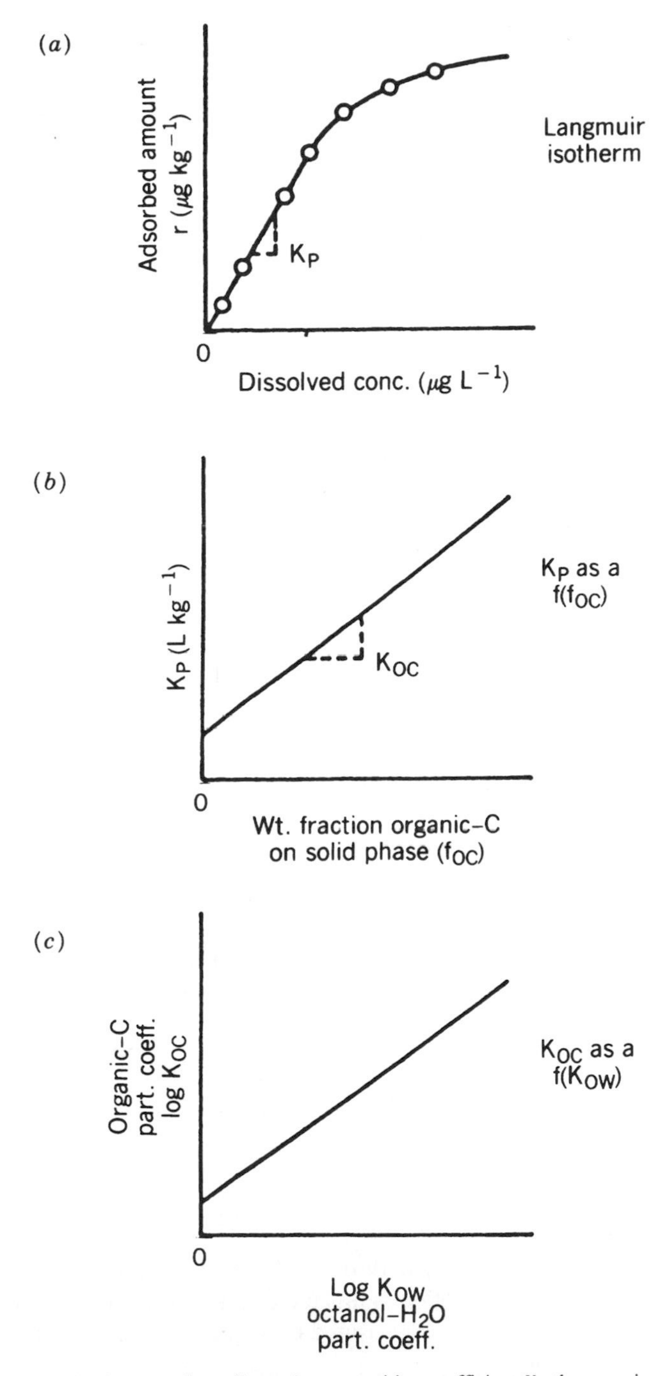

Figure 7.7 Relationships between the sediment/water partition coefficient K_p, the organic carbon partition coefficient K_{oc}, and the octanol/water partition coefficient K_{ow}. Plots (a) and (b) are for only one chemical and (c) is for many chemicals.

the more it "wants" to be in the organic phase. K_p is a measure of the actual partitioning in natural waters.

$$K_p = \frac{\text{mg chemical/kg solids}}{\text{mg chemical/L } H_2O}$$

The units on K_p are L kg^{-1}. (Some references will show it as dimensionless, which is correct if the water has a density of 1.0 kg L^{-1}, specific gravity = 1.00.)

The linear portion of the adsorption isotherm (Figure 7.7a) can be expressed by equation (49):

$$r = K_p\, C \tag{49}$$

where r is the mass of chemical sorbed per mass of particulate matter, mg kg^{-1}; and C is the mass of the chemical dissolved in water, mg L^{-1}.

The Langmuir isotherm in Figure 7.7a is derived from the kinetic equation for sorption–desorption:

$$\frac{dC}{dt} = -\, k_1 C\, [C_{pc} - C_p] + k_2 C_p \tag{50}$$

where C is the concentration of dissolved toxicant, C_p is the concentration of particulate toxicant, C_{pc} is the maximum adsorptive concentration of the solids, and k_1 and k_2 are the adsorption and desorption rate constants, respectively. A substitution can be made for C_p and C_{pc}. If m is the concentration of solids, r is the ratio of adsorbed toxicant to solids by mass, and r_c is the maximum adsorptive capacity of the solids, then

$$\frac{dC}{dt} = -\, k_1 Cm\, [r_c - r] + \mathrm{k}_2 rm \tag{51}$$

At steady-state, equation (51) reduces to a Langmuir isotherm in which the amount adsorbed is linear at low dissolved toxicant concentrations but gradually becomes saturated at the maximum value (r_c) at high dissolved concentrations.

$$r = \frac{C\, r_c}{k_2/k_1 + C} \tag{52}$$

Generally, the adsorption capacity of sediments is inversely related to particle size: clays > silts > sands. Sorption of organic chemicals is also a function of the organic content of the sediment, as measured by K_{oc}, and silts are most likely to have the highest organic content.

Sometimes a Freundlich isotherm is inferred from empirical data. The function is of the form

$$r = K\, C^{1/n} \tag{53}$$

where n is usually greater than 1. In dilute solutions, when n approaches 1, the Freundlich coefficient, K, is equal to the partition coefficient, K_p.

The partition coefficient is derived from simplification of the kinetic equations (50) and (51) if $r_c \gg r$ (the linear portion of the Langmuir isotherm). In this case, we may write

$$\frac{dC}{dt} = -k_f\, C + k_r\, C_p \tag{54a}$$

$$\frac{dC_p}{dt} = k_f\, C - k_r\, C_p \tag{54b}$$

where k_f is the adsorption rate constant and k_r is the desorption rate constant.

The total concentration of toxicant, C_T, is

$$C_T = C + C_p = f_d C_T + f_p C_T \tag{55}$$

where f_d and f_p are the dissolved and particulate fractions, respectively:

$$f_d = C_\infty / C_T = 1/(1 + K_p m) \tag{56}$$

$$f_p = C_{p\infty} / C_T = K_p m/(1 + K_p m) \tag{57}$$

and the ratio of the reaction rate constants is related by

$$\frac{k_r}{k_f} = \frac{C_\infty}{C_{p\infty}} = \frac{1}{K_p m} \tag{58}$$

where the ∞ subscripts indicate chemical equilibrium.

From kinetics experiments where dissolved and particulate concentrations are monitored over time, the ratio of steady-state concentrations can be read from the graph (Figure 7.8).

Sorption reactions usually reach chemical equilibrium quickly, and the kinetic relationships can often be assumed to be at steady-state. This is sometimes referred to as the "local equilibrium" assumption, when the kinetics of adsorption and desorption are rapid relative to other kinetic and transport processes in the system.

O'Connor and Connolly[31] first reported that, for organics and metals alike, the sediment/water partition coefficient K_p declines as sediment (solids) concentrations increase. It is a consistent phenomenon in natural waters that is particularly important for hydrophobic organic chemicals. For example, the K_p for a chemical in sediments is much lower than that observed in the water column. Most researchers attribute this fact to artifacts in the way that one attempts to measure K_p, including

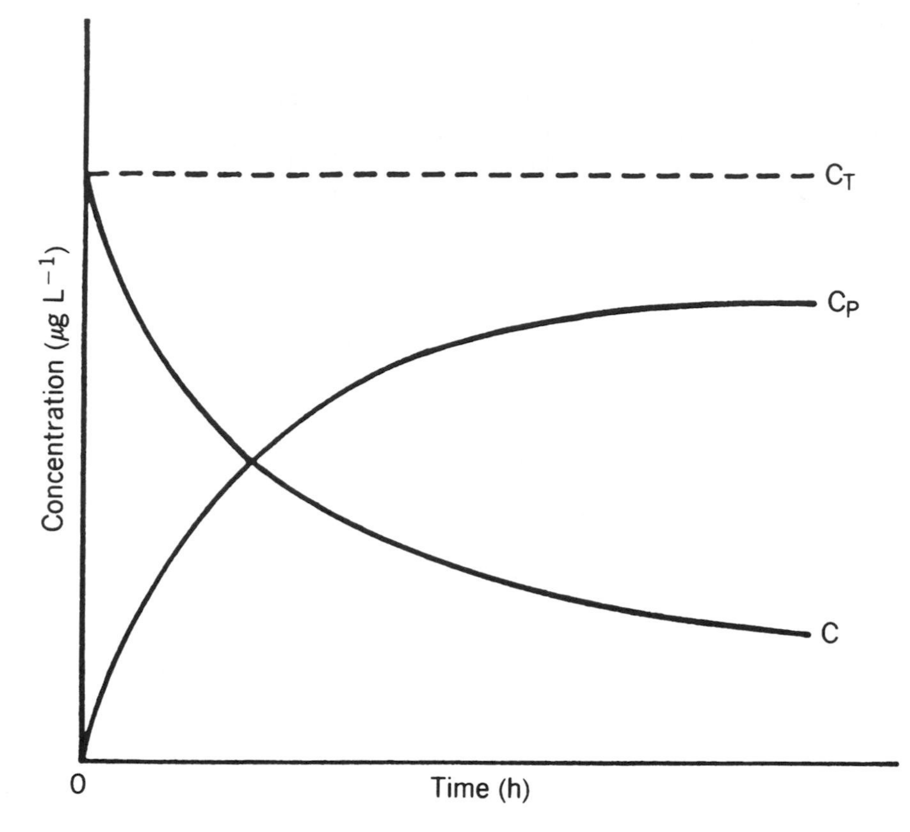

Figure 7.8 Kinetic sorption experiment in a batch reactor.

complexation of a chemical by colloids and dissolved organic carbon that pass a membrane filter.

7.2.8 Bioconcentration and Bioaccumulation

Bioconcentration of toxicants is defined as the direct uptake of aqueous toxicant through the gills and epithelial tissues of aquatic organisms. This fate process is of interest because it helps to predict human exposure to the toxicant in food items, particularly fish. Bioconcentration is part of the greater picture of bioaccumulation and biomagnification that includes food chain effects. Bioaccumulation refers to uptake of the toxicant by the fish from a number of different sources including bioconcentration from the water and biouptake from various food items (prey) or sediment ingestion. Biomagnification refers to the process whereby bioaccumulation increases with each step on the trophic ladder.

The terms bioconcentration, bioaccumulation, and biomagnification are sometimes mistakenly used interchangeably. It is useful to accept the following definitions for the sake of discussion.

Bioconcentration: the uptake of toxic organics through the gill membrane and epithelial tissue from the dissolved phase.

Bioaccumulation: the total biouptake of toxic organics by the organism from food items (benthos, fish prey, sediment ingestion, etc.) as well as via mass transport of dissolved organics through the gill and epithelium.

Biomagnification: that circumstance where bioaccumulation causes an increase in total body burden as one proceeds up the trophic ladder from primary producer to top carnivore.

Bioconcentration experiments measure the net bioconcentration effect after x days, having reached equilibrium conditions, by measuring the toxicant concentration in the test organism. The BCF (bioconcentration factor) is the ratio of the concentration in the organism to the concentration in the water.

The BCF derives from a kinetic expression relating the water toxicant concentration and organism mass:

$$\frac{dF}{dt} = ek_1C - k_2F \tag{59}$$

where e = efficiency of toxic absorption at the gill
k_1 = (L filtered/kg organism per day)
k_2 = depuration rate constant including excretion and clearance of metabolites, day^{-1}
C = dissolved toxicant, $\mu g\ L^{-1}$
B = organism biomass, $kg\ L^{-1}$
F = organism toxicant residue (whole body), $\mu g\ kg^{-1}$

The steady-state solution is

$$F = ek_1C/k_2 = (\text{BCF})\ (C) \tag{60}$$

where BCF has units of (μg/kg)/(μg/L). Bioconcentration is analogous to sorption of hydrophobic organics, which was discussed previously. Organic chemicals tend to partition into the fatty tissue of fish and other aquatic organisms, and BCF is analogous to the sediment/water partition coefficient, K_p. Bioconcentration also can be measured in algae and higher plants, where uptake occurs by adsorption to the cell surfaces or sorption into the tissues.

Many of the chemicals of interest are hydrophobic (lipophilic), which makes them more prone to bioconcentration. Several investigators observed strong correlations between the octanol/water partition coefficient and the BCF. Concentration of lipophilic toxicants in biological tissues is expected because the lipid concentration in cells is much higher than that in the water column. Because the chemical is more easily dissolved in a nonpolar solvent (e.g., oils and lipids), it will bioconcentrate into biological tissues. An empirical relationship for bioconcentration (BCF–K_{ow}) in bluegill sunfish in 28 days exposure for 84 organic priority pollutants was

$$\log \text{BCF} = 0.76 \log K_{ow} - 0.23 \tag{61}$$

and for rainbow trout with ten chlorobenzenes it was

$$\log \text{BCF} = 0.997 \log K_{ow} - 0.869 \tag{62}$$

for low-level exposures typical of natural waters.[32–35] Fathead minnow, bluegill, rainbow trout, brook trout, and mosquito fish are the species most frequently involved in bioconcentration tests.

Bioconcentration experiments, *per se,* do not measure the metabolism or detoxification of the chemical. Chemicals can be metabolized to more or less toxic products that may have different depuration characteristics. The bioconcentration experiment only measures the final body burden at equilibrium (although interim data that were used to determine when equilibrium was reached may be available). The fact that a chemical bioaccumulates at all is an indication that it resists biodegradation and is somewhat "biologically hard" or "nonlabile."

The kinetics of bioaccumulation are shown schematically in Figure 7.9. Fish can lose unmetabolized toxics via biliary excretion or "desorption" through the gill. On the other hand, toxic organics can undergo biotransformations and be eliminated as metabolic products. The rate constant, k_2, includes total depuration (both excretion of unmetabolized toxics, k_2', and elimination of metabolites, k_2''). Only a fraction of

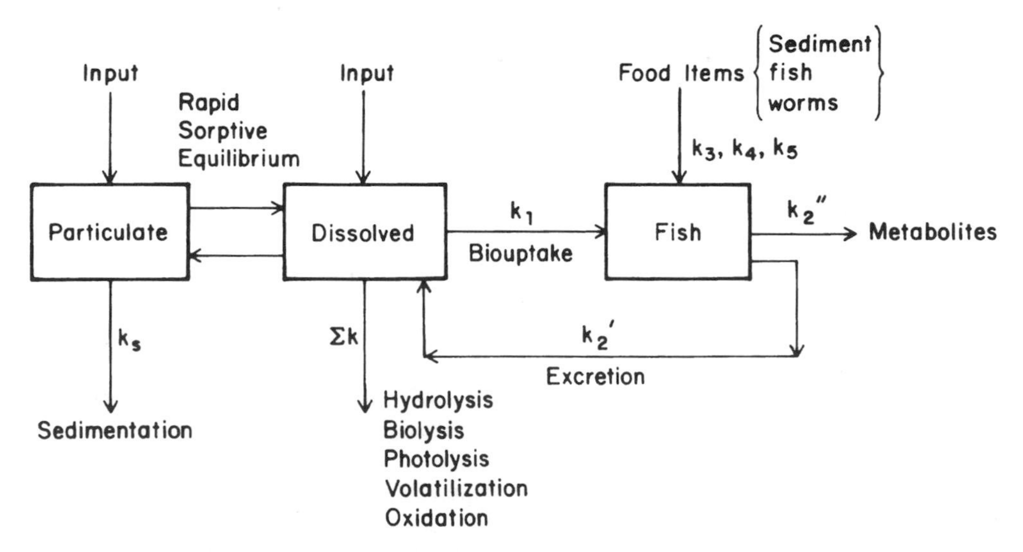

Figure 7.9 Bioaccumulation kinetics for hydrophobic organic chemicals in fish.

this elimination is returned to the water column as dissolved parent compound, designated as k_2' in Figure 7.9 (*Note:* $k_2 = k_2' + k_2''$.)

Hydrophobic organics tend to accumulate in fatty tissue of animals.[32–35] Lipid-normalized bioconcentration factors both in the laboratory and in the field have been correlated successfully with the hydrophobicity of toxic organics as measured by the octanol/water partition coefficient, K_{ow} (Table 7.12). But with large piscivorous fish and high octanol/water partitioning chemicals (log K_{ow} 5–7), Connolly and Thomann[36] have shown that biomagnification occurs in excess of the amount predicted by bioconcentration alone. Biomagnification occurs in lake trout for PCBs in the Great Lakes due to the contribution of alewife and small fish to the diet of these top carnivores.

Table 7.12 Bioconcentration Factors (BCF) for Selected Organic Chemicals in Fish (Units: μg/kg fish ÷ μg L water)

Chemical	BCF	Chemical	BCF
Chlorinated hydrocarbons		PCBs	
Aldrin	3,140	4-Chlorobiphenyl	590
Chlordane	8,260	4,4′-Dichlorobiphenyl	215
DDD	63,830	2,4,4′- and 2,2′,5-trichloro-biphenyl (Aroclor 1016, 1242)	49,000
DDE	27,400	2,2′,4,4′- and 2,2′,5,5′-Tetra-chlorobiphenyl (Aroclor 1248)	73,000
DDT	84,500	2,2′,4,5,5′-Pentachlorobiphenyl (Aroclor 1254)	46,000
Dieldrin	4,420	2,2′,4,4′,5,5′-Hexachlorobiphenyl	46,000
Endrin	1,360	PAHs	
Heptachlor	2,150	Anthracene	917
Lindane	560	Benzo[*a*]pyrene	~ 10,000
Methoxychlor	1,550	Other pesticides	
Toxaphene	4,250	Malathion	0
Kepone	8,400	Terbufos (Counter)	535
Mirex	220	Parathion	335
Pentachlorophenol	130	Carbaryl	< 1
Chlorobenzene	12	Carbofuran	0
p-Dichlorobenzene	215	Aldicarb	42
Hexachlorobenzene	8,600	2,4-D acid	0
Pentachlorobenzene	5,000	Dicamba	0
1,2,4,5-Tetrachlorobenzene	4,500	2,4,5-T	25
1,2,4-Trichlorobenzene	491	Trifluralin	926
Aniline	6	Atrazine	11
Carbon tetrachloride	18	Alachlor	0
Tetrachloroethylene	39	Propachlor	< 1

Source: Kenaga and Goring.[37]

7.2.9 Comparison of Pathways

Most of the transformations discussed in Section 7.2 are expressed as second-order reactions. It is difficult to compare the magnitudes of these reactions—the rate constants all have different units. Each of the transformations can be written as pseudo-first-order reactions assuming that the second concentration in the reaction rate expression can be assumed to be relatively constant. Then a fair comparison is possible to assess the relative magnitude of the various fate pathways and which, if any, may predominate.

The overall reaction rate may then be written as

$$\frac{dC}{dt} = -(k_b + k_o + k_r + k_p + k_h + k_v)C \tag{63}$$

where C = dissolved organic concentration, ML^{-3}
t = time, T
k_b = biotransformation rate constant, T^{-1}
k_o = oxidation rate constant, T^{-1}
k_r = reduction rate constant, T^{-1}
k_p = photolysis rate constant, T^{-1}
k_h = hydrolysis rate constant, T^{-1}
k_v = volatilization rate constant, T^{-1}

Equation (63) includes an assumption that the atmosphere has a negligible concentration (partial pressure) of the organic, so only volatilization occurs (stripping out of the water body). A direct comparison of the rate constants will reveal the maximum reaction that may predominate overall. All of the rate constants are pseudo-first order. Consideration of the reactions as pseudo-first order also allows estimation and comparison of their half-lives. For first-order reactions in a batch reactor without transport, the reaction rate can be written as

$$\frac{dC}{dt} = -\sum_{i=1}^{n} k_i C \tag{64}$$

Solving for the concentration as a function of time, we have

$$C = C_0 e^{-\Sigma k_i t} \tag{65}$$

or

$$\frac{C}{C_0} = 0.5 = e^{-\Sigma k_i t} \tag{66}$$

Taking the natural logarithm of both sides of equation (66) and solving for time (half-life) yields the well-known relationship below:

$$t_{1/2} = \frac{0.69}{\Sigma\, k_i} \tag{67}$$

where $t_{1/2}$ = overall half-life of the chemical due to all transformation reactions

$\sum_{i=1}^{n} k_i$ = the sum of all the pseudo-first-order reaction rate constants

Individual half-lives may be compared to determine which reaction predominates (gives the shortest half-life). Also, the overall half-life for the chemical can be estimated from equation (67), but this will be dependent on the transport regime into which the chemical is released (lake, river, estuary, sediment, ocean, etc.). It is also important to remember the assumptions that went into development of the pseudo-first-order reaction rate expressions.

7.3 ORGANIC CHEMICALS IN LAKES

7.3.1 Completely Mixed Systems

As an approximation, lakes can be represented as ideal completely mixed flow through reactors (CMF systems) or a network of CMF compartments. Then, it is possible to write a mass balance system of equations quite simply, as demonstrated in Chapter 2. We will follow the same basic approach as described in Chapter 2 and Chapter 5 on Eutrophication of Lakes.

Accumlation of mass within the control volume = Mass inputs – Mass outflow ± Rxns

Section 7.2 provided the reaction kinetics, and a schematic of the various reactions in the lake water column and sediment is shown in Figure 7.10. An assumption of local equilibrium may be used to relate the particulate adsorbed concentration to the dissolved concentration through the partition coefficient K_p.

$$C = r/K_p \tag{68a}$$

$$C_p = rM \tag{68b}$$

$$C_T = C + C_p = r\left(\frac{1}{K_p} + M\right) \tag{68c}$$

where K_p = sediment/water partition coefficient, L kg^{-1}

C = dissolved organic chemical concentration, μg L^{-1}

r = mass sorbed, μg kg^{-1}

M = suspended or bed solids concentration, kg L^{-1}

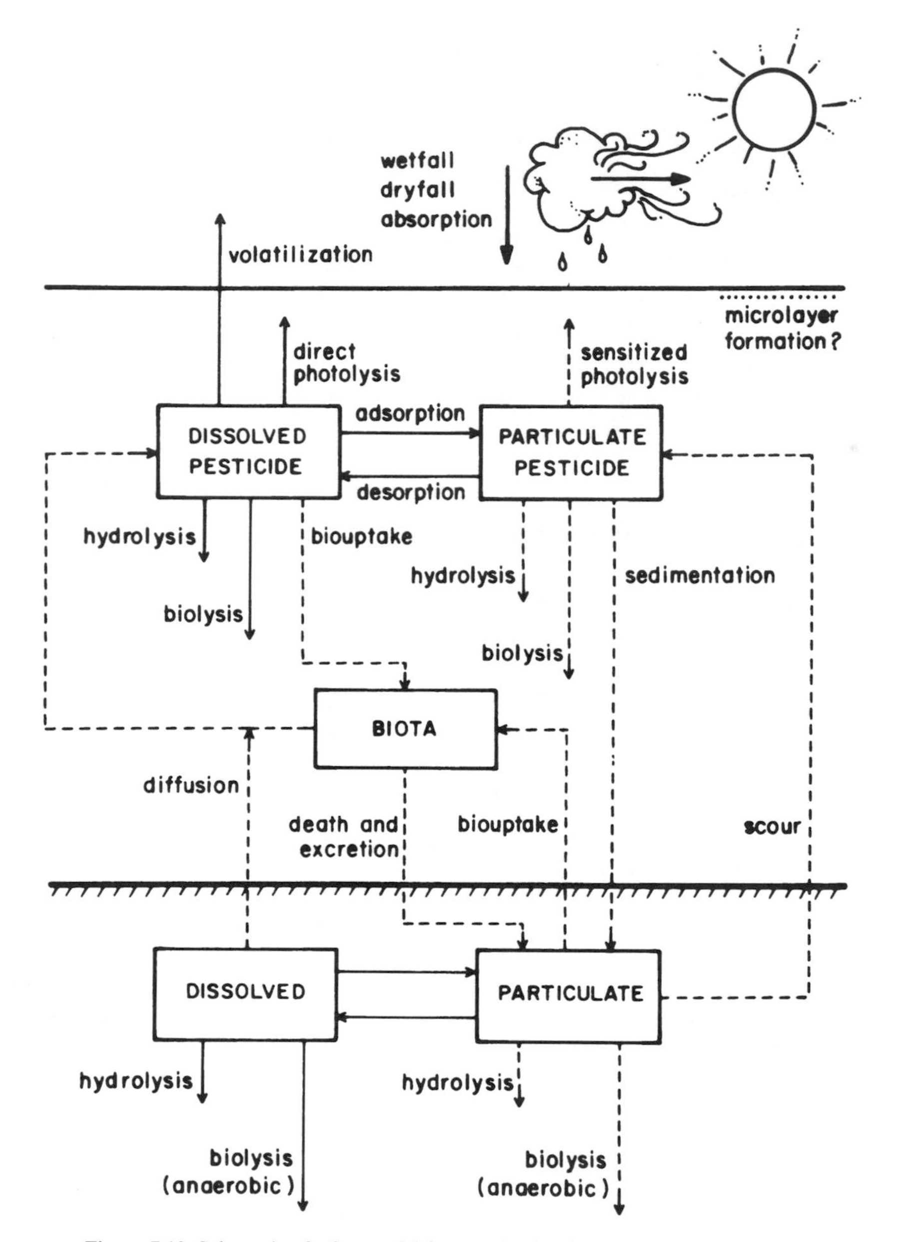

Figure 7.10 Schematic of a fate model for organic chemicals in water and sediment.

C_p = particulate adsorbed concentration, μg L^{-1}
C_T = total (dissolved plus particulate) concentration, μg L^{-1}

Sorptive equilibrium is usually a valid assumption in natural waters because the time scale for most sorption reactions (minutes to hours) is small compared to the time scale for reactions and transport (days to years). Nonetheless, there may be times when sorption kinetic reactions must be considered and the local equilibrium assumption is not valid.

Figure 7.10 also indicates a rapid local equilibrium assumption for bioconcentration. If uptake and depuration kinetics (hours to days) are fast relative to other reactions and time scales, this is a valid assumption. Use of the bioconcentration factor (BCF) helps to simplify the equations, and it is another partitioning coefficient that we may use similar to K_p.

$$F = \text{BCF}\,(C) \tag{69}$$

where BCF = bioconcentration factor, L kg^{-1}
C = dissolved chemical concentration, μg L^{-1}
F = residue concentration in whole fish, μg kg^{-1}

The total concentration of chemical C_T may be larger or smaller in the sediment than the overlying water depending on whether the water column or sediment was contaminated first. Partitioning of the chemical between the dissolved pore water C_2 and adsorbed sediment C_{p_2} may also be different due to the dependence of K_{p_2} on solids concentration.[31] Generally, $K_{p_2} < K_{p_1}$ because the sediment has a much higher solids concentration. Particulate adsorbed chemical settles to the sediment via the sedimentation rate constant $k_s = v_s/H$, where v_s is the mean sedimentation velocity and H is the mean depth of the water body.

A framework for a mass balance model for an organic chemical in a lake is given by Figure 7.10. Waste inputs, their fate and effects, can be assessed in this context. Anthropogenic inputs may also enter the water body from the atmosphere via wet precipitation and dry deposition. The concentration in rainfall is related to the gas phase concentration and Henry's constant, so the deposition mass is equal to the volume of rainfall times the aqueous phase concentration.

$$C_{\text{precip}} = C_g/H \tag{70}$$

where C_{precip} is the precipitation concentration, C_g is the gas phase concentration, and H is Henry's constant with the appropriate units. The flux of contaminants due to dry deposition is related to the depositional velocity and the gaseous concentration

$$J_d = v_d\,C_g \tag{71}$$

where v_d is the deposition velocity (LT^{-1}), C_g is the gas phase concentration (ML^{-3}), and J_d is the areal mass flux due to dry deposition ($ML^{-2}T^{-1}$). Equation (71) is empirical. Both gases and aerosol particles may contribute to dry deposition but the gas phase concentration should be proportional in either case, with v_d serving as the empirical proportionality constant.

The mass balance equation for a lake with toxic organic chemical inputs can be written assuming complete mixing, steady flow conditions, instantaneous local sorption equilibrium, and no atmospheric deposition.

$$\frac{d(VC_T)}{dt} = QC_{T_{\text{in}}} - QC_T - k_s C_p V - (k_b + k_o + k_r + k_p + k_h + k_v)\, CV \qquad (72)$$

Equation (72) has three unknown dependent variables—C_T, C_p, and C—but the assumption of local equilibrium allows us to write the equation entirely in terms of total (whole water, unfiltered) concentration.

$$\frac{d(VC_T)}{dt} = QC_{T_{\text{in}}} - QC_T - f_p k_s C_T V - \sum_{i=1}^{6} k_i f_d C_T V \qquad (73)$$

where C_T = total concentration = $C + C_p$, ML^{-3}
V = volume of the lake, L^3
t = time, T
Q = flowrate in and out, L^3T^{-1}
f_p = particulate fraction of total chemical concentration, dimensionless
= $C_p/C_T = K_pM/(1 + K_pM)$
f_d = dissolved fraction of total chemical concentration, dimensionless
= $C/C_T = 1/(1 + K_pM)$
C = dissolved chemical concentration, ML^{-3}
C_p = particulate chemical concentration, ML^{-3}
k_s = sedimentation rate constant, T^{-1}
k_i = sum of pseudo-first-order reaction rate constants [equation (63)], T^{-1}

Equation (73) is an ordinary differential equation with constant coefficients. It is solvable by first-order methods such as the integration factor method. Dividing through by the constant volume and rearranging, we have

$$\frac{dC_T}{dt} + \left(\frac{1}{\tau} + f_p k_s + \Sigma\, k_i f_d\right) C_T = \frac{C_{T_{\text{in}}}}{\tau} \qquad (74)$$

The final solution is

$$C_T = C_{T_0} e^{-\alpha t} + \frac{C_{T_{\text{in}}}}{\alpha\tau}\,(1 - e^{-\alpha t}) \qquad (75)$$

where C_{T_0} = initial total input concentration, ML^{-3}
α = integration factor = $1/\tau + f_p k_s + \Sigma k_i f_d$, T^{-1}
τ = mean hydraulic detention time = V/Q, T

We see that the solution to a continuous input of organic chemical to a lake is composed of two terms in equation (75). The first term is the die-away of initial conditions, and the second term is the asymptotic "hump" (the shape of a Langmuir isotherm), which builds to a steady-state concentration as $t \rightarrow \infty$.

$$C_{T_{ss}} = \frac{C_{T_{in}}}{\alpha\tau} \tag{76}$$

The steady-state concentration is directly proportional to the total concentration of organic inputs to the lake.

Because it takes an infinite amount of time for the lake to reach steady state in the strictest sense, we speak of time to *95%* of steady state, that is, the length of time required for the concentration in the lake to reach 95% of the value that it will ultimately achieve.

$$0.95 = \frac{C_T - C_{T_0}}{C_{T_{ss}} - C_{T_0}} \tag{77}$$

or

$$C_T = 0.05\ C_{T_0} + 0.95\ C_{T_{ss}} \tag{78}$$

By inspection, one can prove that equations (75) and (78) are equal when

$$t_{95\%\ ss} = \frac{3.0}{\alpha} \tag{79}$$

Equation (79) gives the time to 95% of steady state. For the simplest case of a nonadsorbing dissolved chemical undergoing first-order reaction decay, $\alpha = k + 1/\tau$. The greater is the flushing rate ($1/\tau$) and the reaction rate constant, the less is the time required to achieve steady state. Conservative substances ($k = 0$) take the longest time to reach steady state after a step function change in inputs.

Example 7.1 Slug Input of Parathion to a Pond: Comparison of Fate Pathways and Reactions

Parathion is an organic phosphorus insecticide that was added to an experimental fish pond in Israel. From hypothetical information given below, one can calculate the parathion concentration in the pond after a slug input. Estimate the pseudo-first-order rate constants for all the relevant removal processes including washout ($1/\tau$), photolysis, hydrolysis, volatilization, biotransformation, oxidation, and sedimentation (assuming a sedimentation velocity of particulate suspended solids of 0.1 m d^{-1}). Make a table of all the pseudo-first-order rate constants for each process in order to compare them. What is the half-life of parathion in the pond overall? Plot your model results (a continuous line) together with the observed parathion concentration measurements given below. The model that you will use is simply a completely mixed system (one-box) with outflow, inflow, and sedimentation. There is not a continuous input of parathion to the pond, however, only a slug input that gives an initial concentration of $C_0 = m/V = 0.091$ mg L^{-1}.

t, days	Observed Concentration, mg L^{-1}
0	0.091
1	0.041
2	0.028
5	0.013
7	0.010
9	0.004
14	0.002

Data	Definition
$C_0 = 0.091$ mg L^{-1}	Initial dosage of parathion
$Q = 20$ m^3 d^{-1}	Flowrate through the pond
$V = 400$ m^3	Pond volume
$Z = 1.0$ m	Mean depth of the pond
$k_p = 0.10$ d^{-1}	Sensitized photolysis
$k_B = 2.46 \times 10^3$ L mol^{-1} d^{-1}	Base-caalyzed hydrolysis rate constant
$k_N = 3.64 \times 10^{-3}$ d^{-1}	Uncatalyzed hydrolysis rate constant
$k_A = 1.28 \times 10^2$ mol^{-1} d^{-1}	Acid-catalyzed hydrolysis rate constant
pH = 9.0	pH
$k_v = 0.05$ d^{-1}	Overall volatilization rate constant
$k_b' = 2.35 \times 10^{-10}$ L $cell^{-1}$ d^{-1}	Second-order biolysis rate constant
$X = 10^9$ cell L^{-1}	Viable bacterial cells
$K_p = 50$ μg/kg per μg L^{-1}	Partition coefficient
$M = 20 \times 10^{-6}$ kg L^{-1}	Suspended solids concentration
$k_{ox} = 0.005$ d^{-1}	Chemical oxidation rate constant

Solution: The pond can be treated as a completely mixed system with a slug input of parathion at time $t = 0$.

$$V\frac{dC_T}{dt} = Q\cancelto{0}{C_{T_{in}}} - QC_T - k_s f_p C_T V - \Sigma\, k f_d C_T V$$

Accumulation = Inputs − Outflows − Sedimentation − Decay Rxns

$$\frac{dC_T}{dt} = -\alpha C$$

where

$$\alpha = \frac{1}{\tau} + \frac{k_s K_p M}{1 + K_p M} + \frac{\Sigma k}{1 + K_p M}$$

$$\boxed{C_T = C_{T_0} e^{-\alpha t}}$$

We use the data that were given to calculate the rate constants.

$$\tau = V/Q = 20 \text{ days}$$

$$f_d = \frac{1}{K_p M + 1} = 0.999$$

$$f_p = \frac{K_p M}{K_p M + 1} = 0.001 \quad \text{(only 0.1\% is in the adsorbed particulate phase)}$$

$$k_s f_p = (0.1 \text{ day}^{-1})(0.001) = 1 \times 10^{-4} \text{ day}^{-1}$$

$$k_b = k_b' X = (2.35 \times 10^{-10} \text{ L cell}^{-1} \text{ d}^{-1})(10^9 \text{ cell L}^{-1}) = 0.235 \text{ day}^{-1}$$

$$k_h = k_N + k_A[\text{H}^+] + k_B[\text{OH}^-] = 0.00364 + 128(10^{-9}) + 2460(10^{-5}) = 0.028 \text{ day}^{-1}$$

Comparison of Pathways

	$1/\tau$ Washout	k_p Photolysis	k_h Hydrolysis	k_v Volatility	k_b Biolysis	k_{ox} Oxidation	$k_s f_p$ Sediment	Σk Overall
Rate constant, day^{-1}	0.05	0.10	0.028	0.05	0.235	0.005	0.0001	0.47
Half-life, days	13.8	6.9	24.6	13.8	2.94	138	6900	1.5

$$C_T = (0.091 \text{ mg L}^{-1})\, e^{-0.47t}$$

Even without calibration, the model fits the experimental data quite well (Figure 7.11).

Example 7.2 PCB Concentrations in the Great Lakes

PCB concentrations in the Great Lakes have been slow to degrade and dissipate despite a ban on the manufacture of PCBs in 1972. Concentrations in the water are low (ng L^{-1}), but they bioaccumulate to large levels in fish and aquatic organisms.

Concentrations in Lake Michigan phytoplankton average 0.0025 mg kg^{-1}; in zooplankton, 0.123; rainbow smelt, 1.04; lake trout, 4.83; and herring gull eggs 124 mg kg^{-1} in 1988. There is a ban on commercial lake trout fishing in Lake Michigan. Figure 7.12 shows the estimated loadings in 1980. Atmospheric inputs account for a large percentage of the loadings (wet + dry deposition).

For the following data, make a dynamic computer model of the Great Lakes using ideal, completely mixed systems. Initial conditions are concentrations in 1980. For steady continuous loadings, run the model out to the year 2000. Assume sedimentation and gas transfer are the main reaction mechanisms. Loadings for the five lakes are the following: Superior—4300, Michigan—4800, Huron—3400, Erie—6000, and Ontario—2000 kg yr^{-1}. Dissolved PCB concentrations in 1980 were: Superior, 2.0; Michigan, 3.0; Huron, 1.5; Erie, 2.0; and Ontario, 2.5 ng L^{-1}. Sediment/water partition coefficient K_p is 100,000 L kg^{-1}.

a. Plot the dissolved and particulate PCB concentrations in each lake over time, 1980–2000.
b. If the loadings began to decrease by 3% per year in 1980, what are the predicted PCB concentrations in each lake? Plot them from 1980–2000. What is the time to steady state for a step function change in inputs in each lake?

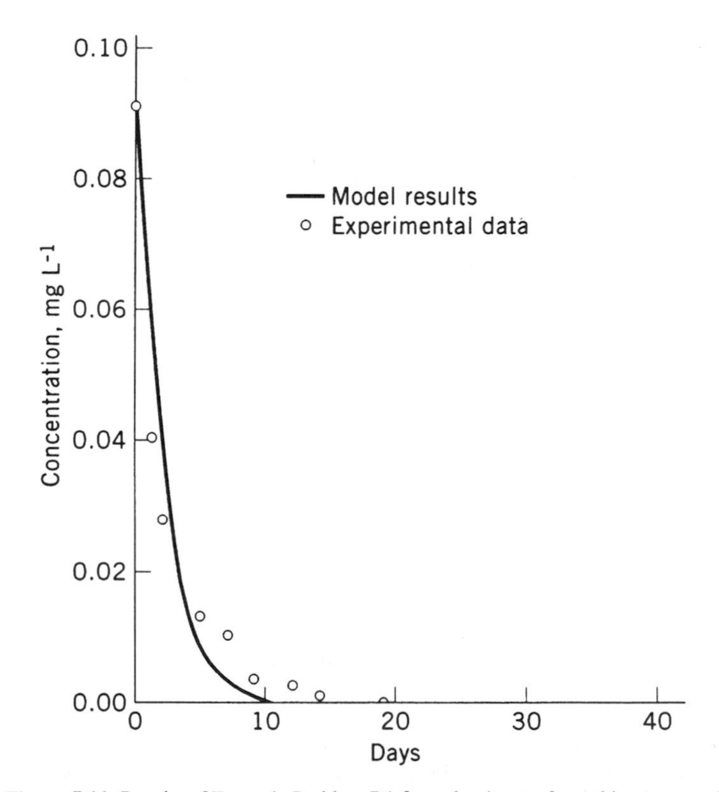

Figure 7.11 Results of Example Problem 7.1 for a slug input of parathion to a pond.

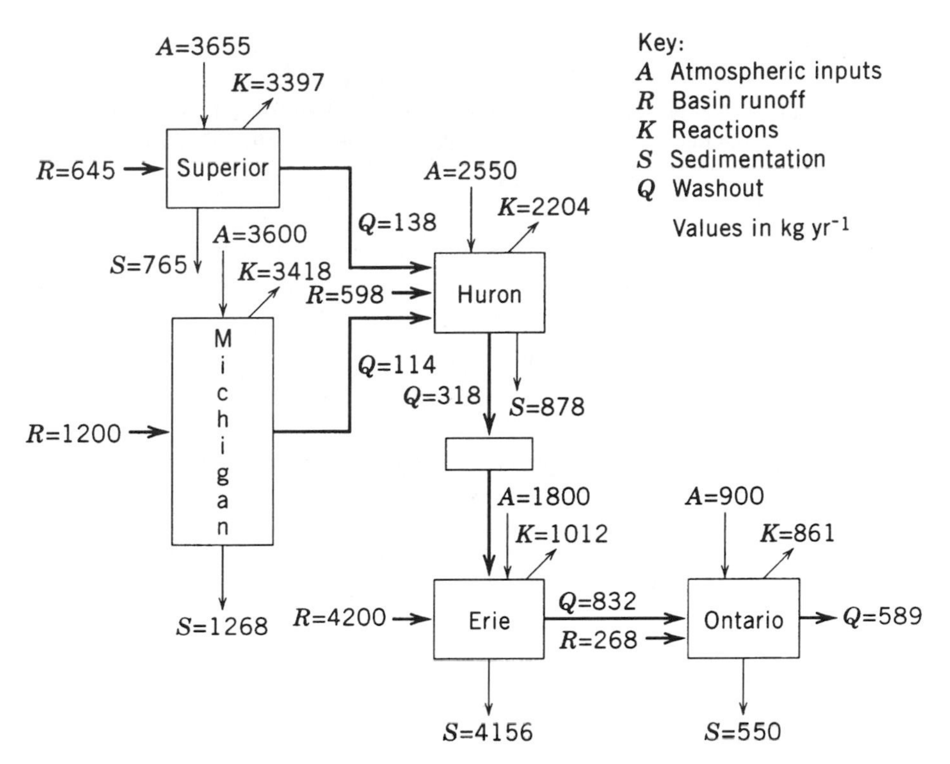

Figure 7.12 Estimated mass fluxes of PCBs through the U.S. Great Lakes in 1980.

c. Prepare a schematic mass balance for each lake in the entire Great Lake Basin showing the mass (in kg yr^{-1}) for each transport pathway (inputs, sedimentation, volatilization, and outflow).

Data

Item	Units	Superior	Michigan	Huron	Erie	Ontario
TSS	mg L^{-1}	3.00	4.00	4.00	20.00	6.00
Settling velocity	m yr^{-1}	19.60	24.20	26.00	53.60	27.80
Mean depth	m	145.20	85.00	59.00	18.60	86.00
Volume	km^3	11900.00	4910.00	3520.00	470.00	1630.00
Detention time	yr	177.00	136.00	21.90	2.60	7.70
Outflow	km^3 yr^{-1}	67.23	36.10	160.73	180.77	211.69
k_v rate constant	yr^{-1}	0.18	0.31	0.44	1.44	0.30
Atmospheric load	kg yr^{-1}	4300.00	4800.00	3400.00	6000.00	2000.00

Solution:

$$\text{Accumulation} = \text{Input} - \text{Output} \pm \text{Reactions}$$

Assumptions

1. All input is either atmospheric deposition or inflow from the upstream lakes. This means that all inflow from sources other than upstream lakes is assumed to have no PCBs.
2. Each lake is completely mixed.
3. All flows are steady. (This means lake volumes are constant.)
4. Dissolved and particulate fractions are always in equilibrium, and sediment scour and dissolution of PCBs are negligible.
5. PCB congeners can be treated as one chemical (not a good assumption in reality).

Discussion: The difference equation, which can be applied to each lake, is

$$\frac{\Delta C_T}{\Delta t} = \left(\frac{W_a}{V} + \frac{W_Q}{V} - \frac{C_T}{\tau} - (\Sigma k)\,(f_d)\,(C_T) - (k_s)\,(f_p)\,(C_T)\right)$$

or

$$\Delta C_T = \left\{\frac{W_a}{V} + \frac{W_Q}{V} - \left[\frac{1}{\tau} + (\Sigma k)\left(\frac{1}{1 + K_p M}\right) + \left(\frac{v_s}{H}\right)\left(\frac{K_p M}{1 + K_p M}\right)\right] C_T\right\} \Delta t$$

where C_T is the total concentration; W_a, V, τ, Σk, K_p, M, v_s, and H are all given; and W_Q is the loading from the upstream lakes.

Use a spreadsheet or a Runge–Kutta numerical solution with a time step on the order of 0.25 years (or less) to solve the system of five difference equations.

Results

Item	Units	Superior	Michigan	Huron	Erie	Ontario
f_d, dissolved fraction		0.77	0.71	0.71	0.33	0.63
C_0, initial PCB diss.	ng L^{-1}	2.00	3.00	1.50	2.00	2.50
C_{T_0}, initial total PCB	ng L^{-1}	2.60	4.20	2.10	6.00	4.00
Overall k	yr^{-1}	0.18	0.31	0.49	2.79	0.44
C_T at steady state	ng L^{-1}	2.08	3.15	2.14	4.84	4.02
Time to 95% of steady state	yr	16.2	9.45	5.60	0.94	5.26

Year 2000

Item	Units	Superior	Michigan	Huron	Erie	Ontario
Atm. + runoff	kg yr^{-1}	4300.00	4800.00	3400.00	6000.00	2000.00
Inflow	kg yr^{-1}	0.00	0.00	253.48	343.42	875.80
Sedimentation	kg yr^{-1}	−769.81	−1259.67	−946.94	−4374.62	−794.84

Item	Units	Superior	Michigan	Huron	Erie	Ontario
Volatilization	kg yr^{-1}	−3421.73	−3428.97	−2363.71	−1093.00	−1229.43
Outflow	kg yr^{-1}	−139.62	−113.87	−343.42	−875.80	−851.55
Net gain	kg yr^{-1}	−31.16	−2.51	−0.58	0.00	−0.01

During the 20-year simulation with steady loadings, the concentration of chemicals reached their 95%-of-steady-state values. All of the lake concentrations of PCBs declined slightly except for Lakes Huron and Ontario.

Assuming a 3% per year reduction in loading, the following C_T concentrations are simulated for PCBs in the Great Lakes. (The actual situation is for a 7–8% per year decline.) All concentrations are in units of ng L^{-1}.

Year	Superior	Michigan	Huron	Erie	Ontario
1980	2.6	4.2	2.1	6.0	4.0 ng L^{-1}
1985	2.15	3.15	1.9	4.2	3.65
1990	1.85	2.6	1.7	3.6	3.15
1995	1.6	2.2	1.45	3.1	2.7
2000	1.35	1.95	1.25	2.65	2.35

The reduction in loading rate is simulated as a time-variable input:

$$W_a = W_{a0} \exp(-0.03t)$$

where W_{a0} is the loading (in kg yr^{-1}) in 1980 and t is the time (in years). If the bioaccumulation factor in lake trout in Lake Michigan is log BCF = 6, it will still take 20 more years for the fishery to recover to less than 1.0 mg/kg.

7.3.2 Dieldrin Case Study in Coralville Reservoir, Iowa

The following case study is used to illustrate aspects of ecosystem recovery from a persistent hydrophobic organic pollutant. It also demonstrates the use of compartmentalization within a lake to simulate transport.

Figure 7.13 shows a schematic of water column, sediment, and fish concentrations following a period when large discharges of chemical were put into the system. Because the contaminant is hydrophobic and persistent, it remains in the system for a long time, accumulating in fish tissue and sediment. It disappears by washout (advection), burial into the deep sediment, and slow degradation reactions. Depending on the sediment dynamics of the system and the rate of chemical degradation, these can be slow processes taking years to decades. Some persistent insecticides (e.g. chlorinated hydrocarbons) used in the Midwest are given in Figure 7.14. These chemicals were banned in the 1970s and early 1980s because of their persistence and propensity to bioaccumulate in fish and wildlife. Also shown in Figure 7.14 are two replacement insecticides (ester compounds), which hydrolyze and break down in the environment. They are toxic but much less persistent.

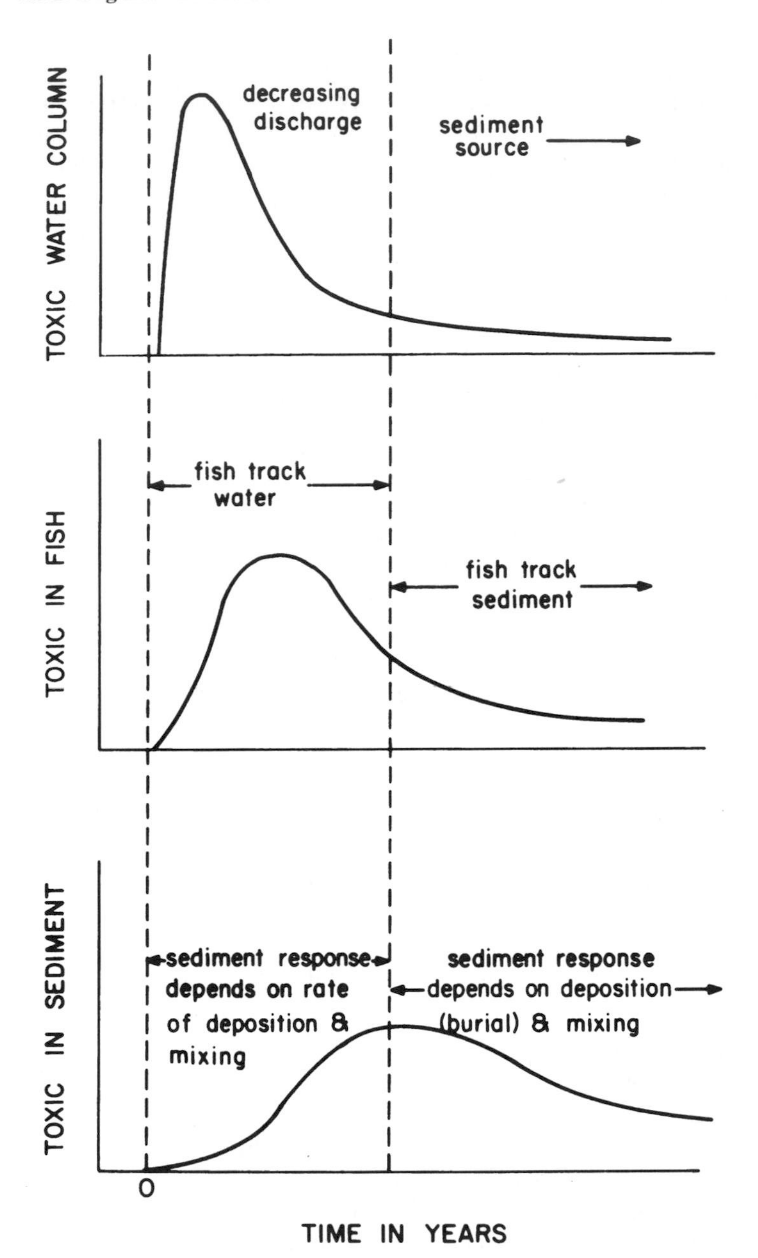

Figure 7.13 Schematic of lake recovery from a persistent hydrophobic pollutant.

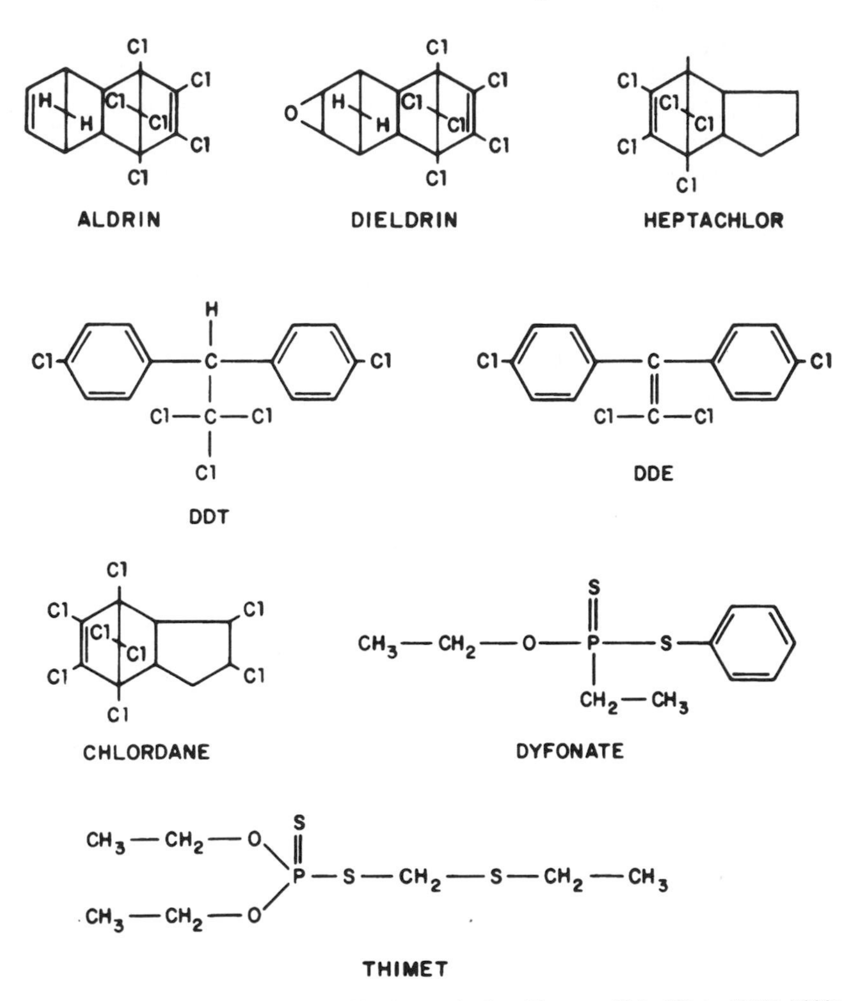

Figure 7.14 Selected insecticides used in the past in the midwestern United States (1945–1980).

Agricultural usage of pesticides in Iowa is widespread, particularly grass and broadleaf herbicides and row crop soil insecticides. One of the insecticides widely used for control of the corn rootworm and cutworm from 1960 to 1975 was the chlorinated hydrocarbon, aldrin. Aldrin is microbially metabolized to its persistent epoxide, dieldrin. Dieldrin is itself an insecticide of certain toxicity and is also a hydrophobic substance of limited solubility in water (0.25 ppm) and low vapor pressure (2.7×10^{-6} mm Hg at 25 °C). It is known to bioaccumulate to levels as high as 1.6 mg/kg wet weight in edible tissue of Iowa catfish.[38]

Coralville Reservoir is a mainstream impoundment of the Iowa River in eastern Iowa. It drains approximately 3084 square miles (7978 km^2) of prime Iowa farmland and receives extensive agricultural runoff with 90% of its drainage basin in intensive agriculture. It is a variable-level, flood control and recreational reservoir, which has undergone considerable sedimentation since it was created in 1958. At conservation pool (680 ft above mean sea level, msl), the reservoir has a capacity of 38,000 acre-ft (4.69×10^7 m^3), a surface area of 4900 acres (1.98×10^7m^2), a mean depth of approximately 8 ft (2.44 m), and a mean detention time of 14 days. In 1958, the capacity at conservation pool was 53,750 acre-ft (6.63×10^7 m^3).

Aldrin application in Iowa during the mid-1960s amounted to some 6.5×10^6 lb yr^{-1} (2.94×10^6 kg yr^{-1}) on 5.0 million acres (2×10^{10} m^2). However, the corn rootworm grew increasingly resistant to aldrin and, after 1967, usage decreased across the state by approximately one-half. Finally, the pesticide was banned in 1975, and little was applied after 1976. Although aldrin was no longer labeled, dieldrin residues in excess of the Food and Drug Administration "action" level (0.3 mg/kg wet edible tissue) were recorded for Coralville Reservoir fishes, and commercial fishing was banned in 1975.

The total pesticide concentration is the sum of the particulate plus the dissolved concentrations, with instantaneous sorptive equilibrium assumed.

$$\frac{dC_T}{dt} = \frac{W(t)}{V} - \frac{C_T}{\tau} - (\Sigma k) f_d\, C_T - k_s f_p C_T \tag{80}$$

where $f_d = C/C_T = 1/(1 + K_p M)$ = fraction of dissolved pesticide
$f_p = C_p/C_T = K_p M/(1 + K_p M)$ = fraction of particulate pesticide
$W(t)$ = time-variable loading of pesticide, M/T
C_T = total concentration in the water column, ML^{-3}
Σk = sum of the pseudo-first-order degradation rate constants
τ = mean hydraulic detention time
V = reservoir volume, L^3
k_s = sedimentation rate constant, T^{-1}.

The fish residue equation is

$$\frac{dF}{dt} = \frac{k_1 f_d C_T}{B} - k_d F \tag{81}$$

where k_1 = pesticide uptake rate by fish, T^{-1}
k_d = depuration rate constant, T^{-1}
F = whole-body fish residue level, M/M wet weight
B = fish biomass concentration, M/L^3 wet weight

The bioconcentration factor (BCF) between pesticide residue in whole fish and the dissolved concentration is the ratio of the biouptake rate constant to the depuration rate constant divided by the fish biomass, $k_1/k_d B$. If the pesticide is not metabolized in the fish, equations (80) and (81) are modified to reflect excretion of pesticide back into the dissolved phase.

Equations (80) and (81) may be solved analytically for constant coefficients and simple pesticide loading functions, $W(t)$, or they may be integrated numerically. In the case of a pesticide ban, the $W(t)$ might typically decline in an exponential manner due to degradation by soil organisms or a ban on application. For an exponentially declining loading function at rate ω, the analytical solutions to equations (80) and (81) are

$$C_T = C_{T_0} e^{-\delta t} + \frac{C_{T_{\text{in},0}}}{\varepsilon} (e^{-\omega t} - e^{-\delta t}) \tag{82}$$

$$F = F_0 e^{-k_d t} + \frac{k_1 f_d}{B} \left(\frac{C_{T_0}}{\gamma} e^{-\delta t} + \frac{C_{T_{\text{in},0}}}{\varepsilon\theta} e^{-\omega t} - \frac{C_{T_{\text{in},0}}}{\varepsilon\gamma} e^{-\delta t} \right)$$

$$- \frac{k_1 f_d}{B} e^{-k_d t} \left(\frac{C_{T_0}}{\gamma} + \frac{C_{T_{\text{in},0}}}{\varepsilon\theta} - \frac{C_{T_{\text{in},0}}}{\varepsilon\gamma} \right) \tag{83}$$

where C_{T_0} = initial total pesticide concentration in lake, ML^{-3}
$C_{T_{\text{in},0}}$ = initial total pesticide inflow concentration, ML^{-3}
ω = rate of exponentially declining inflow concentration, T^{-1}
$\alpha = (\Sigma k) f_d + k_1 f_d + k_s f_p$, T^{-1}
$\gamma = k_d - \alpha - (1/\tau)$, T^{-1}
$\delta = \alpha + (1/\tau)$, T^{-1}
$\varepsilon = \alpha\tau + 1 - \omega\tau$, dimensionless
$\theta = k_d - \omega$, T^{-1}.

For the fish residue level at steady state, the solution to equation (81) simply yields the fish partition coefficient times the equilibrium dissolved pesticide concentration.

Figure 7.15 is a schematic diagram of hypothetical pond or lake configurations that are possible for this problem. Each box is assumed to be completely mixed with bulk exchange between water compartments. There is dispersion in Coralville Reservoir that seems to be simulated best by the eight-compartment model based on dye studies. The choice is not arbitrary, but if dye studies are not available, one should choose a reasonable compartmental configuration based on physiographic features in the prototype, keeping in mind that rate constants used in such a hypothetical scheme are apparent values that depend on the model configuration.

Simulation of a two-compartment model (water and sediment) for dieldrin in Coralville Reservoir is shown in Figure 7.16. Model parameters based on calibration are given on the figure. The rate of declining inputs was 0.164 yr^{-1}; sedimentation rate constant was 0.18 day^{-1}; the rate of biodegradation of dieldrin in sediment was 0.005 day^{-1}; and the bioconcentration factor (BCF) was 70,000. Model results were within the range of field observations. Dieldrin residues in fish, sediment, and

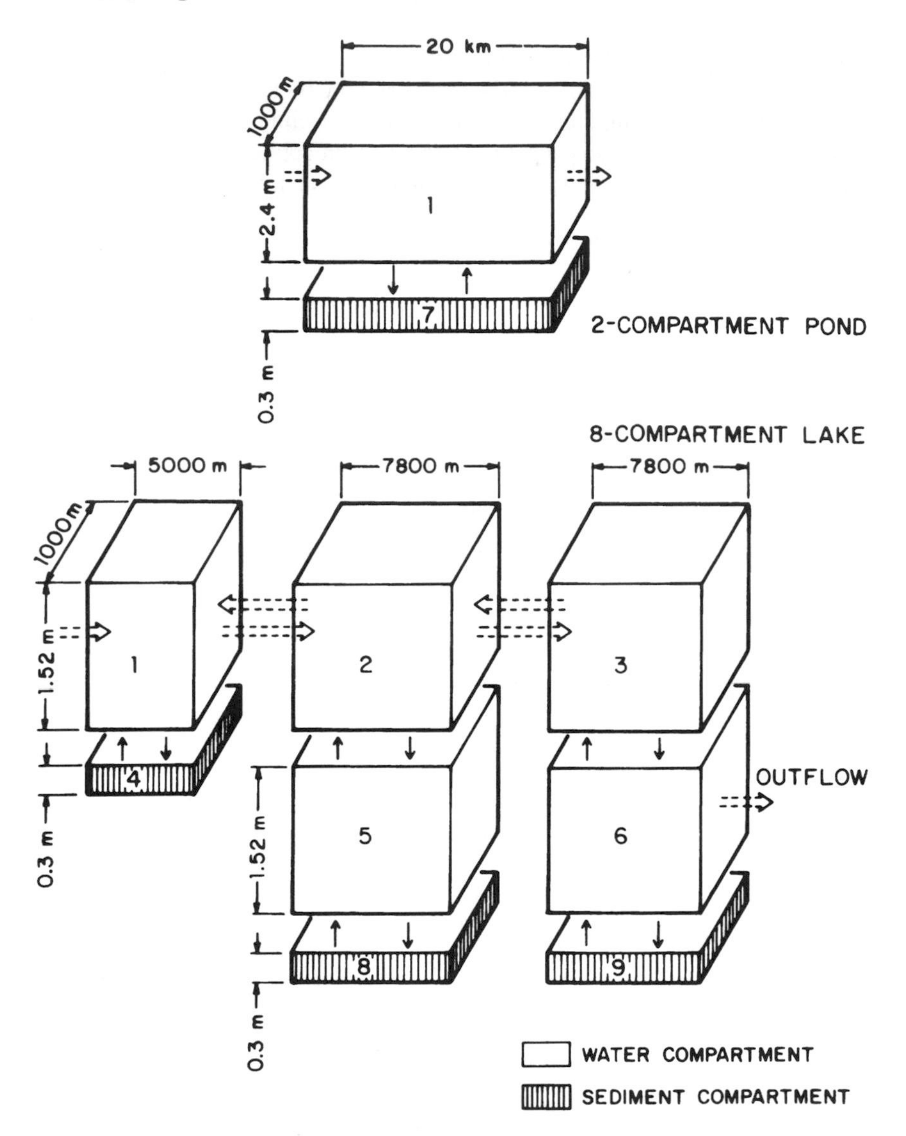

Figure 7.15 Compartmental configurations for a two-box pond model or an eight-box lake model.

water were all declining at ~15% per year. Approximately 50% of the pesticide load was exported from the reservoir, 40% underwent sedimentation, and 10% entered a huge biomass of bottom-feeding fish. The fishery was reopened in 1980 for commercial fishing of bigmouth buffalo fish. A post-audit study in 1989 showed that the model was quite robust in its predictions of fish residue levels 10 years later with no adjustments to model parameters (Figure 7.17).

Multicompartment solutions of equations (80) and (81) must include interflows

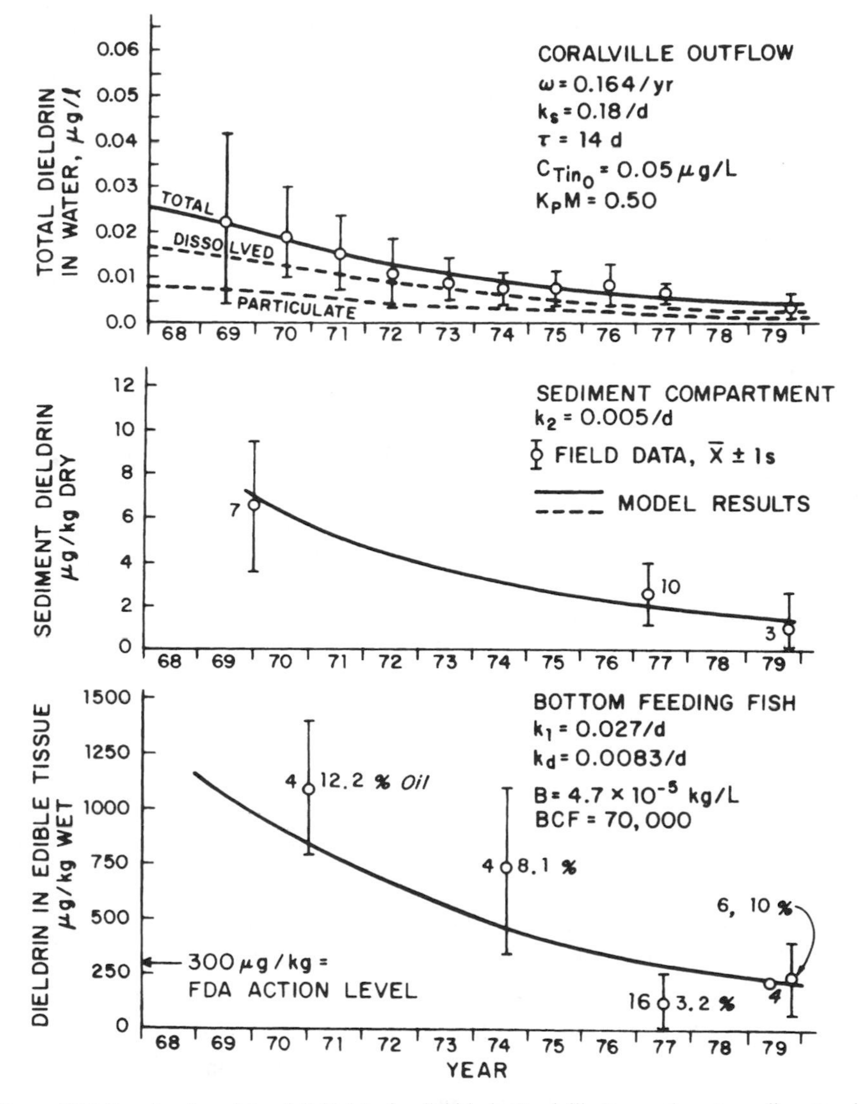

Figure 7.16 Results of model and field data for dieldrin in Coralville Reservoir water sediment and in fish. Reprinted with permission from the American Association for the Advancement of Science, Copyright (1981). Schnoor, J.L., *Science* 211, 840 (1981).

and bulk dispersion as well as an assumption regarding suspended solids and fish biomass distribution. For each constant-volume compartment,

$$V\frac{dC_T}{dt} = \Sigma Q_a C_a - \Sigma Q_b C_T - k_{da} f_d C_T V - (k_s + k_{pa}) f_p C_T V + k_{sa} f_p C_a V_a + \frac{EA}{\ell}(C_a - C) \tag{84}$$

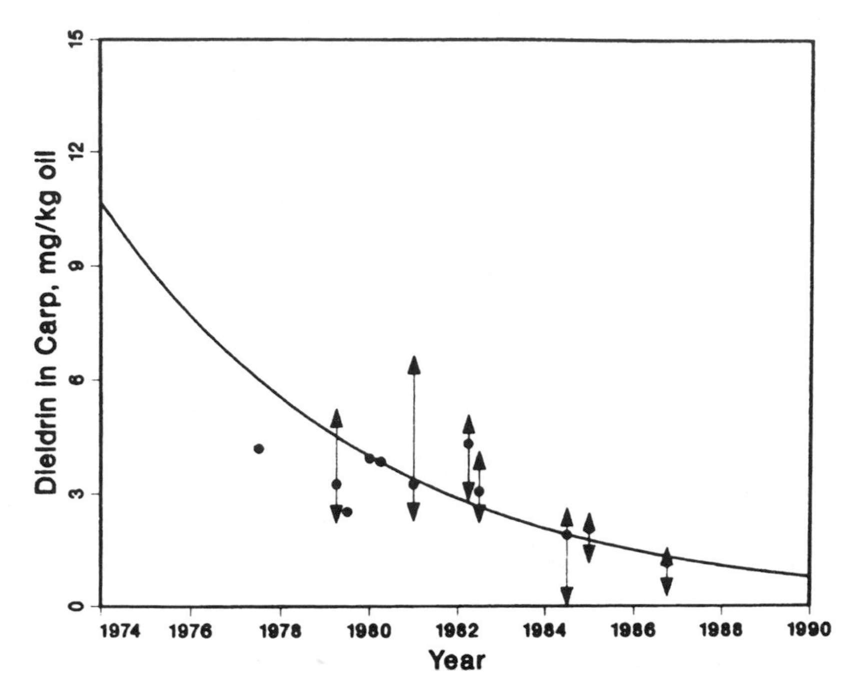

Figure 7.17 Post-audit study of dieldrin model for Coralville Reservoir showing utility of the model for forecasting fish residue levels. Reprinted with permission from the American Society of Civil Engineers. Copyright (1989). Mossman, D.K. and Schnoor, J.L. *J. Environ Eng., ASCE,* 115, 677 (1989).

where V = compartment volume, m^3

C_T = total pesticide concentration of the compartment, $\mu g\ L^{-1}$

t = time, days

Q_a = inflow of water from adjacent compartments, $m^3\ d^{-1}$

Q_b = outflow of water to adjacent compartments, $m^3\ d^{-1}$

C_a = total pesticide concentration in the adjacent compartment, $\mu g\ L^{-1}$

f_d = fraction of the total pesticide in the dissolved phase

f_p = fraction of the total pesticide in the particulate phase

k_{da} = reaction rate constant for the dissolved phase, day^{-1}

k_{pa} = reaction rate constant for the particulate phase, day^{-1}

k_s = settling rate constant of the compartment, day^{-1}

k_{sa} = settling rate constant from the above compartment, day^{-1}

E = bulk dispersion coefficient for adjacent compartments, $m^2\ d^{-1}$

A = surface area between two adjacent compartments, m^2

ℓ = mixing length between midpoints of adjacent compartments, m

V_a = volume of above compartment, m^3

The general mass balance equation for the jth compartment can be reduced to a general matrix equation:

$$\frac{d\{C_j\}}{dt} = \left[\frac{Q_{j,i}}{V_j}\right]\{C_i\} - \left[\frac{Q_{i,j}}{V_j}\right]\{C_j\} - [f_{d,j}\,k_{da} + f_{p,j}\,k_{s,j} + f_{p,j}\,k_{pa}]\,\{C_j\}$$

$$+ \left[f_{p,i}k_{s,i}\,\frac{V_i}{V_j}\right]\{C_i\} + \left[\frac{E_{i,j}\,A_j}{\ell_{i,j}V_j}\right]\{C_i - C_j\} \tag{85}$$

where i = subscript denoting adjacent compartments
j = subscript denoting the jth compartment
C_j = total pesticide concentration in the jth compartment, $\mu g\ L^{-1}$
C_i = total pesticide concentration in an adjacent compartment, $\mu g\ L^{-1}$
$Q_{i,j}$ = flow into compartment i from j, $m^3\ d^{-1}$
$Q_{j,i}$ = flow from compartment i to j, $m^3\ d^{-1}$
$f_{d,j}$ = dissolved fraction of a pesticide in compartment j
$f_{p,j}$ = particulate fraction of a pesticide in compartment j
k_{da} = sum of dissolved reaction rate constant, day^{-1}
k_{pa} = sum of particulate reaction rate constant, day^{-1}
$k_{s,i}$ = settling rate constant for compartment i, day^{-1}
$k_{s,j}$ = settling rate constant from compartment j, day^{-1}
$E_{i,j}$ = bulk dispersion coefficient between adjacent compartments, $m^2\ d^{-1}$
A_j = surface area of compartment j, m^2
$\ell_{i,j}$ = length between the midpoints of adjacent compartments, m
V_j = compartment volume, m^3
V_i = volume of adjacent compartment, m^3.

The equations comprise a set of linear, ordinary differential equations, which were numerically integrated via a fourth order Runge–Kutta approximation technique.

Results for the eight-compartment model are shown in Figures 7.18 and 7.19. The fit of model results to field data was similar to that which was obtained earlier. In this case, the more detailed eight-compartment model showed few advantages over the spatially aggregated two-compartment model. But if field measurements had been obtained throughout the reservoir, it is likely that the eight-compartment model would have performed better. More realistically, the vertical transport in a stratified lake or reservoir should be modeled using a finite difference scheme with depth (Appendix F).

7.4 ORGANIC CHEMICALS IN RIVERS AND ESTUARIES

Advection, dispersion, and reaction of chemicals may be simulated for large rivers in one, two, or three dimensions, depending on the application desired. A spill of chemical at the bank of a large river will be mixed laterally and vertically, and it will be transported downstream by current velocity (advection) and longitudinal dispersion. Regarding dispersion, Taylor's analogy for Fickian dispersion in the longitudi-

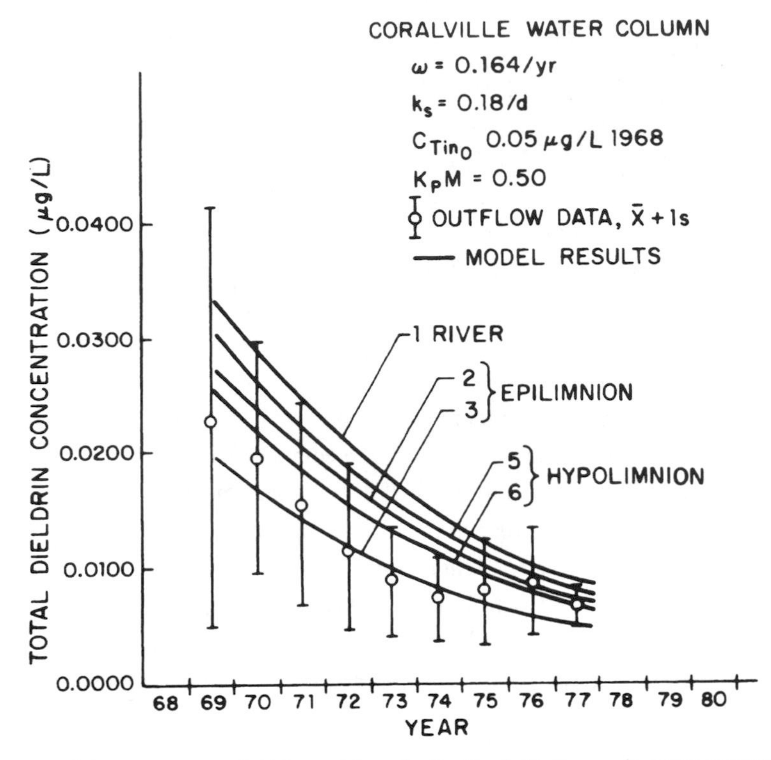

Figure 7.18 Eight-compartment dieldrin model results for Coralville Reservoir, water compartments. Reprinted with permission from the American Society of Civil Engineers. Copyright (1981).[38]

nal direction may result from vertical and lateral turbulence coupled with velocity gradients (due to shear forces at the air–water and sediment–water boundaries). After an initial mixing period, a three-dimensional advection-dispersion equation with Taylor's analogy may be applied for steady flow conditions and a uniform channel.

$$\frac{\partial C}{\partial t} = -u_i \frac{\partial C}{\partial x_i} + \frac{\partial}{\partial x_i} E_i \frac{\partial C}{\partial x_i} \pm \text{Reactions} \tag{86}$$

where C = chemical concentration, M L^{-3}

t = time, T

E = dispersion coefficients in the x-, y-, and z-directions, L^{2}T^{-1}

u_i = average velocities in the x-, y-, and z-directions, LT^{-1}

x = longitudinal distance, L

y = lateral (or transverse) distance, L

z = vertical distance, L

Because the conditions of steady flow and uniform cross section seldom occur, it is more useful to consider nonsteady-state flow conditions and a variable cross-sectional area of the river. Also, we will assume that the initial period is past, and that

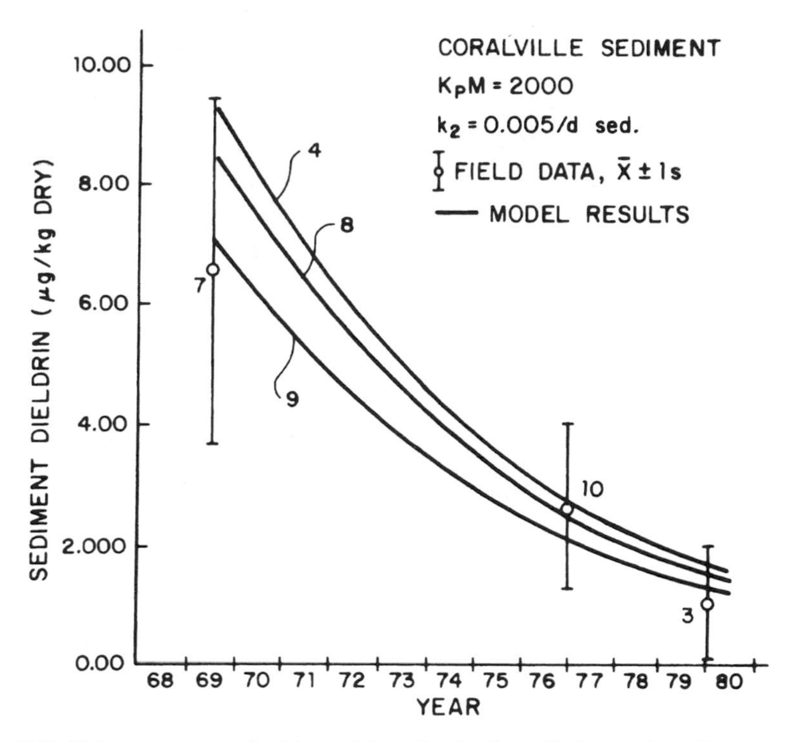

Figure 7.19 Eight-compartment dieldrin model results for Coralville Reservoir, sediment compartments. Reprinted with permission from the American Society of Civil Engineers. Copyright (1981).[38]

the chemical has been completely mixed in the lateral and vertical directions, so that terms for advection and dispersion in the y- and z-directions are zero. Under these assumptions, the mass balance equation in the longitudinal downstream dimension becomes

$$\frac{\partial C}{\partial t} = \frac{1}{A}\frac{\partial}{\partial x}\left(EA\frac{\partial C}{\partial x}\right) - \frac{1}{A}\frac{\partial (QC)}{\partial x} \pm \text{Reactions} \tag{87}$$

At this point one must consider the role of the sediment/water partitioning and sediment transport because chemicals that are adsorbed to suspended solids or bed sediment have a different fate and toxic effect than dissolved chemicals. Included for general applications should be kinetics of physical reactions (sedimentation, scour/resuspension, adsorption/desorption, and gas transfer), biological transformations (biological oxidation/reduction and co-metabolism), and chemical reactions (hydrolysis, oxidation, photolysis).

Figure 7.20 is a schematic of the reactions that occur in the water column and the bed sediment. It is assumed that chemical and biological transformation reactions occur predominantly in the soluble phase (C_s), although special transformation reactions may occur for chemical adsorped to the sediment under reducing conditions (λ'_b). Because environmental conditions differ in the sediment (e.g., photolysis and

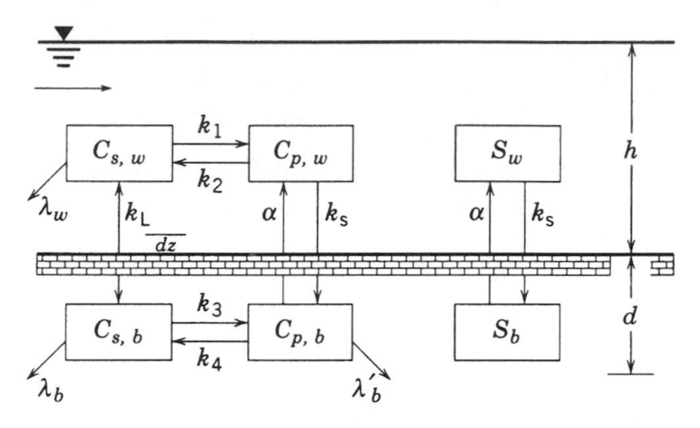

Figure 7.20 Schematic of reactions in the river water column and sediment including adsorption (k_1, k_3), desorption (k_2, k_4), sedimentation (k_s), scour and resuspension (α), and degradation (λ) for soluble chemical (C_s), particulate adsorbed chemical (C_p) and the concentration of solids in the water column (S_w) and in the bed (S_b). Mass transfer between the pore water of the bed and overlying water takes place via a mass transfer coefficient (k_L).

volatilization are not expected to occur from the sediment), the overall pseudo-first-order rate constant (λ_b) is different in the bed sediment from that in the water column (λ_w). Rate constants for adsorption and desorption also differ in the water column compared to the sediment because sorption processes and the sediment/water partition coefficient, in particular, have been reported to be a strong function of the solids concentration (S). This may be due to particle–particle interactions of adsorped chemical, but it is more likely due to binding of chemicals by macromolecules and colloids, which pass a 0.45-μm filter. Nevertheless, when calibrating the model based on such measurement techniques, it is important to include the sediment-dependent partition coefficient. There is some evidence that biological transformations, hydrolysis, and photolysis may occur in the adsorbed particulate form, especially when most of the chemical is in the adsorbed phase in bottom sediments, but in this development we assume that chemical and biological transformations occur only in the soluble aqueous phase ($\lambda'_b = 0$).

Figure 7.20 also includes sedimentation of suspended solids (S_w) in the river water and scour/resuspension of bed sediment (S_b) via the first-order rate constants k_s and α, respectively. Identical rate constants (k_s and α) can be applied to sedimentation and resuspension of adsorbed chemical (C_p). Mass transfer of contaminated river water to sediment pore water may occur in the initial stages of a chemical spill, or diffusion from contaminated sediment to overlying water may occur during the recovery phase. In either case, an overall mass transfer coefficient may be assumed (k_L) that gives the velocity at which dissolved chemical moves by molecular diffusion through a limiting film thickness Δz. Once the chemical has been transported through this limiting film, it is assumed to be mixed vertically with the remainder of the chemical in the water or sediment. This may be a good assumption during the recovery phase when the sediment is contaminated and chemical is diffusing to a vertically mixed water column, but it could cause difficulty with the proper choice of d, the depth of the active bed sediment layer, during the initial period of the spill. In reality, the sediment layer is not usually well-mixed, and the depth of the active layer may change during the period of interest.

To write the proper mass balance equations for the concentration of chemical in the dissolved and particulate-adsorbed phases, it is necessary to define the amount of contamination in the active sediment layer in terms of the particulate adsorbed concentration, $C_{p,b}$:

$$r_b = C_{p,b}/S_b \tag{88}$$

where r_b = amount of chemical adsorbed per mass of dry sediment, μg kg^{-1}
$C_{p,b}$ = particulate adsorbed chemical concentration, μg L^{-1}
S_b = bed solids concentration, kg L^{-1}.

The analytical determination of sediment contamination is measured on a sediment dry weight basis, so r_b is the state variable of interest in the bed sediment, not $C_{p,b}$. S_b and $C_{p,b}$ are based on environmental volumes, not water volumes. S_b is also known as the bulk density of the bed sediment if $C_{p,b}$ is expressed on the basis of environmental volume.

The final set of partial differential equations was developed from a mass balance around a control volume of $A\ \Delta x$. Six equations were solved simultaneously for soluble chemical, particulate adsorbed chemical, and suspended solids in the water column; and for soluble chemical in the sediment, adsorbed chemical per unit mass of sediment (dry weight), and bed solids concentration. The total concentration in the water column and in the bed sediment is the sum of the soluble and particulate adsorbed chemical,

$$C_{T,w} = C_{s,w} + C_{p,w} \quad \text{and} \quad C_{T,b} = C_{s,b} + C_{p,b} \tag{89}$$

where $C_{T,w}$ = total concentration, μg L^{-1}
$C_{s,w}$ = soluble chemical concentration in the water column, μg L^{-1}
$C_{p,w}$ = particulate adsorbed chemical concentration in the water column, μg L^{-1}
$C_{T,b}$ = total bed concentration, μg L^{-1}
$C_{s,b}$ = soluble chemical concentration in the bed sediment, μg L^{-1}
$C_{p,b}$ = adsorbed chemical concentration in the bed sediment, μg L^{-1}

All chemical concentrations in sediment and water refer to the mass per unit of total environmental volume (in liters), rather than on a basis of liquid water volume. The final set of six equations is:

$$\frac{\partial C_{s,w}}{\partial t} = -\frac{1}{A}\frac{\partial (QC_{s,w})}{\partial x} + \frac{1}{A}\frac{\partial}{\partial x}\left(EA\frac{\partial C_{s,w}}{\partial x}\right) - k_1\, C_{s,w}\, S_w - \lambda_w\, C_{s,w} + k_2 C_{p,w}$$

$$+ \frac{k_L}{h}(C_{s,b} - C_{s,w}) + F_s(x, t) \tag{90}$$

$$\frac{\partial C_{p,w}}{\partial t} = -\frac{1}{A}\frac{\partial (QC_{p,w})}{\partial x} + \frac{1}{A}\frac{\partial}{\partial x}\left(EA\frac{\partial C_{p,w}}{\partial x}\right) - k_s C_{p,w} + \alpha\frac{S_b r}{\gamma} + k_1 C_{s,w} S_w - k_2 C_{p,w} + F_p(x, t) \tag{91}$$

$$\frac{\partial S_w}{\partial t} = -k_s S_w + \frac{\alpha}{\gamma} S_b - \frac{1}{A}\frac{\partial (QS_w)}{\partial x} + \frac{1}{A}\frac{\partial}{\partial x}\left(EA\frac{\partial S_w}{\partial x}\right) + G_s(x, t) \tag{92}$$

$$\frac{\partial C_{s,b}}{\partial t} = -k_3 C_{s,b} S_b + k_4 r S_b - \lambda_b C_{s,b} - \frac{k_L}{d}(C_{s,b} - C_{s,w}) \tag{93}$$

$$\frac{\partial r}{\partial t} = k_3 C_{s,b} - k_r r + k_s\frac{C_{p,w}\gamma}{S_b} - \alpha r \tag{94}$$

$$\frac{\partial S_b}{\partial t} = +k_s S_w \gamma - \alpha S_b \tag{95}$$

where $C_{s,w}$ = soluble chemical concentration in the water column, μg L^{-1}
t = time, days
A = cross-sectional area of the river, m^2
Q = volumetric flowrate of the river, m^3 d^{-1}
x = longitudinal (downstream) distance, m
E = longitudinal dispersion coefficients, m^2 d^{-1}
k_1, k_3 = adsorption rate constants, L kg^{-1} d^{-1}
k_2, k_4 = desorption rate constants, day^{-1}
λ_w, λ_b = overall pseudo-first-order rate constant of photolysis, volatilization, biological transformation, chemical hydrolysis, and oxidation reactions in the water (subscript w) and bed (subscript b), day^{-1}
k_L = mass transfer coefficient between water column and pore water of bed sediment, m d^{-1}
h = depth of water column, m
$F_s(x, t)$ = distributed source of soluble chemical, μg L^{-1} d^{-1}
$C_{p,w}$ = particulate adsorbed chemical in the water column, μg L^{-1}
k_s = sedimentation rate constant (v_w/h, mean particle settling velocity divided by the mean depth), day^{-1}
α = scour/resuspension rate constant, day^{-1}
S_b = solids concentration in the bed, kg L^{-1}
S_w = solids concentration in the water column, kg L^{-1}
γ = ratio of water depth (or volume) to depth of the active bed sediment layer (h/d)
$F_p(x,t)$ = distributed source term for particulate adsorbed chemical to the water column, μg L^{-1} d^{-1}

$G(x, t)$ = distributed source term for suspended solids to the water column, kg L^{-1} d^{-1}

$C_{s,b}$ = soluble chemical concentration in the bed sediment, μg L^{-1}

r = amount of adsorbed chemical on sediment solids, μg kg^{-1} (dry weight)

In these equations, distributed (nonpoint) sources may be included as forcing functions on the right-hand side of the mass balance equations (F_s, F_p, G_s). Point sources may be handled as boundary conditions at the head end of each designated river segment (defined by a point source input, tributary input, or a change in river morphometry).

Equations (90)–(95) are applicable for hydrophobic chemicals, which may take a long time to adsorb or desorb (the kinetics of adsorption and desorption are considered explicitly). If the kinetics of transformation reactions or the time of transport (advection, dispersion, scour/resuspension, and sedimentation) are relatively slow compared to the kinetics of sorption, then an assumption of instantaneous equilibrium may be utilized. Under these conditions, the sediment/water partition coefficients are simply the ratio of the adsorption rate constant to the desorption rate constant:

$$K_{p,w} = \frac{k_1}{k_2} \quad \text{and} \quad K_{p,b} = \frac{k_3}{k_4} \tag{96}$$

where $K_{p,w}$ and $K_{p,b}$ are the partition coefficients (L kg^{-1}) in the water column and bed, respectively. We allow the possibility of a different sediment–water partition coefficient for the bed sediment than for the water column due to the dependence of K_p on solids concentration.

Quite often the solids concentrations in a river and the bed are rather constant during the period of interest. In this case, equation (95) may be assumed to be equal to zero (steady-state conditions, $dS_b/dt = 0$). Thus the right-hand side of equation (95) may be rearranged and solved for α, the scour coefficient:

$$k_s S_w \gamma = \alpha S_b \tag{97}$$

$$\alpha = k_s S_w \gamma / S_b \tag{98}$$

Given the assumption of an instantaneous local equilibrium for sorption and a steady-state solids concentration in the river water column, the set of six equations (90)–(95) can be reduced to a set of only two equations: one equation for the total concentration of chemical in the water column of the river and one equation for the amount of adsorbed chemical per unit mass of bed sediment.

$$\frac{\partial C_T}{\partial t} = -\frac{1}{A}\frac{\partial (QC_T)}{\partial x} + \frac{1}{A}\frac{\partial}{\partial x}\left(EA\frac{\partial C_T}{\partial x}\right) - \lambda_w\left(\frac{C_T}{1 + K_{p,w}S_w}\right) - k_s\left(\frac{K_{p,w}S_w}{1 + K_{p,w}S_w}\right)C_T$$

$$+ \frac{k_L}{h}\left(\frac{r}{K_{p,b}} - \frac{C_T}{1 + K_{p,w}S_w}\right) + \alpha\frac{S_b r}{\gamma} + F_T(x, t) \tag{99}$$

$$\frac{\partial r}{\partial t} = -\lambda_b \frac{r}{(1 + K_{p,b}S_b)} + \frac{k_s K_{p,w'} S_{w'}/(1 + K_{p,w'}S_{w'})C_T\gamma}{(S_b + 1/K_{p,b})} - \alpha S_b \frac{r}{(S_b + 1/K_{p,b})}$$

$$- \frac{k_L(r/K_{p,b} - C_T/(1 + K_{p,w'}S_{w'}))}{d(S_b + 1/K_{p,b})} \qquad (100)$$

where $C_T = C_{s,w'} + C_{p,w'}$ = total concentration in the water column, $\mu g\ L^{-1}$; and $F_T(x, t)$ is the distributed source for total chemical input, $\mu g\ L^{-1}\ d^{-1}$. In equation (99), C_T is abbreviated, but identical to $C_{T,w'}$ in equation (89). Equation (100) gives the change in sediment chemical concentration over time, so it is useful in predictions of recovery times for large rivers. The amount of chemical that is sorbed to bottom sediments can be related to the dissolved chemical concentration using a linear adsorption isotherm, under the assumption of instantaneous local equilibrium.

$$r = K_{p,b}C_{s,b} \qquad (101)$$

The concentration of the chemical in the dissolved phase in the water column and sediment pore water can be calculated below in terms of the total concentration in water $C_{T,w'}$ and in the bed $C_{T,b}$.

$$C_{s,w'} = \frac{C_{T,w'}}{(1 + K_{p,w'}S_{w'})} \qquad (102)$$

$$C_{s,b} = \frac{C_{T,b}}{(1 + K_{p,b}S_b)} \qquad (103)$$

Note that the concentration of pore water in the bed has been defined on a total environmental volume in the sediment ($\mu g\ L^{-1}$ total volume), not on a liquid water basis ($\mu g\ L^{-1}\ H_2O$). One must divide the concentration $C_{s,b}$ by the porosity of the sediment (H_2O volume/total volume) in order to obtain the pore water concentration that may be drained from a sediment core, for example.

Equations (99) and (100), coupled with the equilibrium relationships [equations (16), (17), (18)] provide a useful formulation for simulation of chemical spills, distributed source runoff, and point source problems under conditions of steady state for suspended solids and bed sediment with instantaneous sorption equilibrium.

The sets of equations (90)–(95) or (99), (100) can be solved simultaneously using an explicit numerical solution technique for the dispersion-reaction portion of the equation and an implicit technique for the advection term.

To solve numerically the set of equations (99) and (100), the model employs the scheme proposed in Marchuk.[39] The computational algorithm is based on a method of splitting the equations into different physical processes. For each incremental time interval between t_j and t_{j+1}, we consider the numerical scheme comprising three steps. At the first step the equation of chemical transport is solved:

$$\frac{\partial C}{\partial t} + \frac{1}{A}\frac{\partial (QC)}{\partial x} = 0 \tag{104}$$

At the second step we solve the diffusion equation:

$$\frac{\partial C}{\partial t} = \frac{1}{A}\frac{\partial}{\partial x}\left(EA\frac{\partial C}{\partial x}\right) \tag{105}$$

The third step solves the reaction rate equations for local transformations of chemicals, their interaction with the bottom sediments, and source influence. This representation of the chemical transport model simplifies its computation and allows for optimal solution algorithms at each step. The equations are treated as separate solutions at the first two steps and combined with each other at the third.

The third-step equations can be considered at each point of the integration domain as a set of ordinary difference equations with the coefficients dependent on the spatial coordinates. The solution of the previous step, at an instant of time $t = t_j$, is taken to be the initial condition for the following step at the time $t = t_{j+1}$. To solve equation (104) we use the hybrid numerical scheme quoted in Marchuk.[39] The diffusion equation of the second step is solved with the use of implicit approximations. The set of equations of the third step is approximated in time with an accuracy not less than second order. The third step is implemented with the aid of standard algorithms for ordinary differential equations.

Example 7.3 Pesticide Degradation in an Irrigation Canal

Acrolein is a toxic herbicide that is used for submersed weed control in irrigation canals. The data given below are from Bartley and Gangstad.[40] Develop a steady-state model to calculate the acrolein concentration in the downstream receiving water below the treated area of the Wahluke Branch Canal of the Columbia River Basin in Washington. Dosages required are typically 100 ppb acrolein.

$k_v' = 0.305$ m d^{-1}	Volatilization mass transfer coefficient
$k_b' = 8.9 \times 10^{-9}$ L cells^{-1} d^{-1}	Second-order biolysis rate constant
$X = 10^8$ cells L^{-1}	Bacterial cells
$U_x = 0.305$ m s^{-1}	Mean velocity
$H = 0.91$ m	Mean depth

Field Data

Concentration, ppb	105	95	106	92	54	50	33	33	31	30
Distance, km	0.805	1.61	4.43	8.05	12.1	16.1	20.1	32.2	48.3	64.4

Calculate the overall pseudo-first-order reaction rate constant. Since acrolein is nearly totally soluble, the problem then becomes analogous to BOD degradation in a stream. The primary loss mechanism is apparently an initial hydration to β-hydroxypropionaldehyde and subsequent biotransformation. Assume plug-flow conditions and steady state.

Solution: For a plug-flow stream at steady state,

$$C = C_0 e^{-\Sigma kx/U}$$

A linear regression equation was used for model calibration to obtain the parameter Σk (Figure 7.21). The pseudo-first-order rate constant obtained by this method was 0.57 day^{-1}. Then the pseudo-first-order rate constant was calculated from the measured rate constants given for volatilization and biodegradation. The result using this method was 1.23 day^{-1}, about two times larger. The agreement between the two estimates is probably acceptable given large uncertainties in measuring rate constants. If the last two field data points at km 48.3 and 64.4 are ignored, then the best fit regression line yields a pseudo-first-order rate constant of 1.2 day^{-1}, in close agreement to the measured rate constants.

Model Calibration

$$k = \frac{0.0215}{\text{km}} \quad \text{best fit from least squares plot}$$

$$k \cdot \bar{u} = \Sigma k = \left(\frac{0.0215}{\text{km}}\right)\left(\frac{26.35 \text{ km}}{\text{day}}\right) = \boxed{0.57 \text{ day}^{-1}}$$

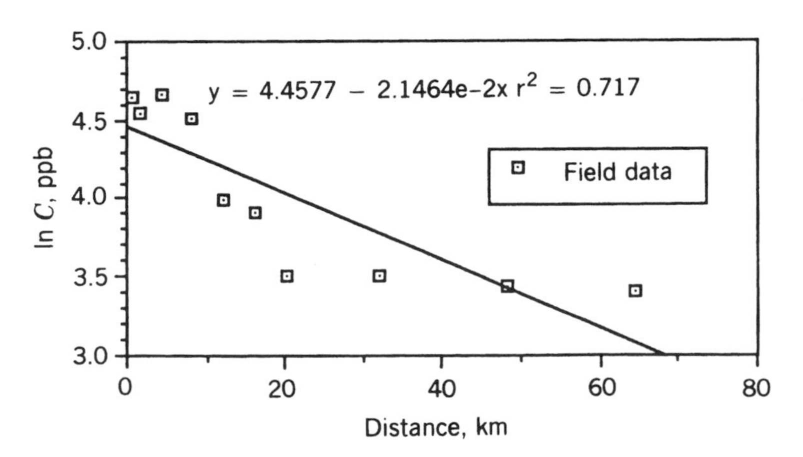

Figure 7.21 Acrolein in an irrigation canal (Wahluke Canal). Best fit of model to field data for Example 7.3.

Measured Rate Constants Given

$$\frac{k_v'}{H} + k_b'X \text{ should equal } \Sigma k$$

$$\frac{k_v'}{H} = \frac{0.305 \dfrac{\text{m}}{\text{day}}}{0.91 \text{ m}} = 0.34 \text{ day}^{-1}$$

$$k_b'X = \left(8.9 \times 10^{-9} \frac{\text{L}}{\text{cells day}}\right)\left(10^8 \frac{\text{cells}}{\text{L}}\right) = 0.89 \text{ day}^{-1}$$

$$\Sigma k = \frac{k_v'}{H} + k_b'X = 0.34 + 0.89 = \boxed{1.23 \text{ day}^{-1}}$$

Example 7.4 Rhine River Chemical Spill Model[41]

A pulse input of pollutants, which were primarily organophosphate pesticides and organic mercurial compounds, to the Rhine River at Basel, Switzerland, was caused by a fire at a chemical warehouse on November 1, 1986. An estimated 7 metric tons of contaminants were washed into the Rhine by fire-fighting runoff. A fish kill extended over 250 km following this spill. Subsequent monitoring of the pollutant plume by Swiss, German, French, and Dutch environmental agencies provided an excellent database for analyzing pollutant fate and transport.

Use the data given below and model equations (99) and (100) to estimate the fate and transport of the sum of the phosphoester pesticides in the Rhine River. Field data for model calibration are given in Figure 7.22.

Rate Constants and Coefficients for the Sum of Organo-P Ester Pesticides in the Rhine River Chemical Spill at Basel, Switzerland

$\lambda_w = 0.20 \text{ day}^{-1}$	$\alpha = 5 \times 10^{-5} \text{ day}^{-1}$
$\lambda_b = 0.20 \text{ day}^{-1}$	$S_w = 10 \text{ mg L}^{-1}$
$k_s = 0.5 \text{ m d}^{-1}$	$S_b = 0.1 \text{ kg L}^{-1}$
$d = 0.02 \text{ m}$	$A = 1500 \text{ m}^2$
$k_L = 10^{-6} \text{ cm s}^{-1}$	$h = 5 \text{ m}$
$E_x = 4 \times 10^{-6} \text{ cm}^2 \text{ s}^{-1}$	$Q = 1350 \text{ m}^3 \text{ s}^{-1}$
$K_{p,w} = 200 \text{ L kg}^{-1}$	$F = 7000$ kg over 12 h
$K_{p,b} = 100 \text{ L kg}^{-1}$	

Solution: Equations (99) and (100) were solved with a split operator method under the steady flow assumption, time-variable concentrations.

The measured mass of material passing each of the four locations decreased with downstream distance (Figure 7.22). The sum of phosphoester pesticides was approximately 4700 kg at Maximiliansau (362 km), 3700 kg at Mainz (496 km), 3200

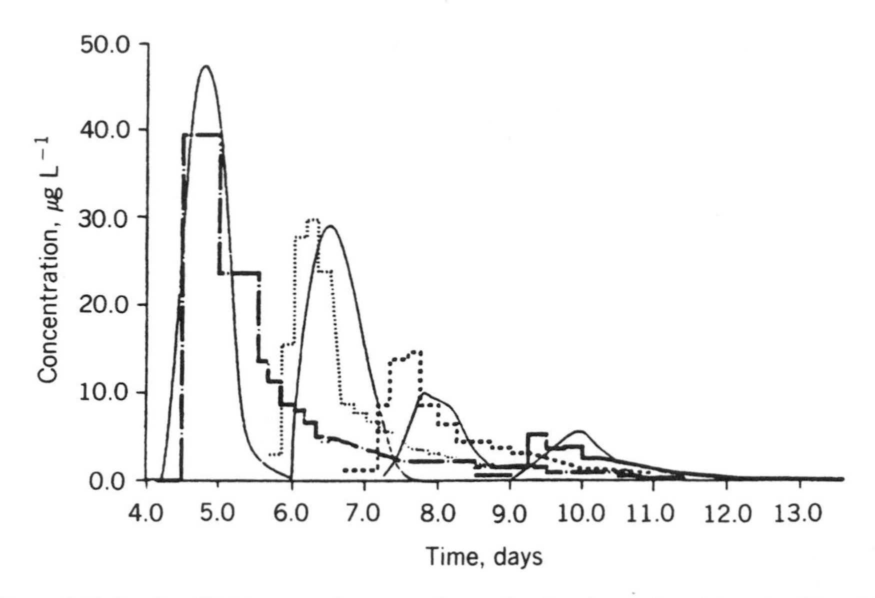

Figure 7.22 Results of field measured concentrations at four locations and model results (thin solid lines) versus time in days of November 1986.

kg at Bad Honnef (640 km), and 1400 kg at Lobith (865 km). An overall pseudo first-order transformation rate constant of 0.20 day^{-1} was used in order to reproduce the estimated mass fluxes. Effects of the accident would have occurred over a much longer duration if the pesticide chemicals had been hydrophobic, persistent, and trapped in the sediments, for example, DDT. Figure 7.22 shows the results of model calibration with little "tuning" of the parameters.

7.5 REFERENCES

1. Organization for Economic Cooperation and Development, *Guidelines for Testing of Chemicals,* OECD, Paris (1981).
2. Morrison, R.T., and Boyd, R.N., *Organic Chemistry,* 2nd ed., Allyn and Bacon Publishers, Boston (1966).
3. Horvath, R.S., *Bacteriol. Rev.* 36:2, 145 (1972).
4. Higgins, I.J., and Burns, R.G., *The Chemistry and Microbiology of Pollution,* Academic Press, New York (1975).
5. Schnoor, J.L., in *Fate of Pesticides and Chemicals in the Environment,* J.L. Schnoor, Ed., Wiley, New York, pp. 1–24 (1992).
6. Schnoor, J.L., Sato, C., McKechnie, D., and Sahoo, D., Processes, Coefficients, and Models for Simulating Toxic Organics and Heavy Metals in Surface Waters. EPA/600/3-87/015, Athens, GA (1987).

7. Mabey, W.R., et al., Aquatic Fate Process Data for Organic Priority Pollutants. EPA/440/4-81-014, Washington, DC (1982).
8. Smith, J.H., Mabey, W.R., Bohonos, N., Holt, B.R., Lee, S.S., Chou, T.W., Bomberger, D.C., and Mill, T., Environmental Pathways of Selected Chemicals in Freshwater Systems, Part II: Laboratory Studies. EPA/600/7-78-074, Washington, DC (1978).
9. Bader, H., Sturzenegger, V., and Hoigné J., *Water Res.*, 22:9, 1109 (1988).
10. Stumm, W., and Morgan, J.J., *Aquatic Chemistry,* Wiley-Interscience, New York, pp. 337–338 (1970).
11. Schwarzenbach, R., Gschwend, P., and Imboden, D., *Environmental Organic Chemistry,* Wiley, New York (1993).
12. Dunnivant, F.M., Schwarzenbach, R.P., and Macalady, D.L., *Environ. Sci. Technol.,* 26, 2133 (1992).
13. Zepp, R.G., in *Fate of Pesticides and Chemicals in the Environment,* J.L. Schnoor, Ed., Wiley-Interscience, New York, pp. 127–128 (1992).
14. Alvarez, P.J.J., Anid, P.J., and Vogel, T.M., *Biodegradation,* 2, 1191 (1991).
15. Wolfe, N.L., Chemical and Photochemical Transformation of Selected Pesticides in Aquatic Systems. EPA/600/3-76-067, Athens, GA (1976).
16. Sawyer, C.N., and McCarty, P.L., *Chemistry for Environmental Engineering,* 3rd ed., McGraw-Hill, New York, p. 242 (1978).
17. Zepp, R.G., Schlotzhauer, P.F., and Sink, R.M., *Environ. Sci. Technol.,* 19, 48 (1985).
18. Zepp, R.G., Hoigné, J., and Bader, H., *Environ. Sci. Technol.,* 21, 443 (1987).
19. Zepp, R.G., Faust, B.C., and Hoigné, J., *Environ. Sci. Technol.,* 26, 313 (1992).
20. Mabey, W., and Mill, T., *J. Phys. Chem. Ref. Data,* 7, 383 (1978).
21. O'Connor, D.J., *Trans Am. Soc. Civ. Eng.,* 123, 655 (1958).
22. O'Connor, D.J., *J. Environ. Eng., ASCE,* 109:3, 731 (1983).
23. Hemond, H.F., and Feichner, E.J., *Chemical Fate and Transport in the Environment,* Academic Press, San Diego, CA, pp. 36–37 (1994).
24. Yalkowsky, S.H., Valvani, S.C., and Mackay, D., *Residue Reviews,* 85, 43 (1983).
25. Rouse, H., *Elementary Mechanics of Fluids,* Dover, New York (1946).
26. Lyman, W.J., et al., *Handbook of Chemical Property Estimation Methods,* McGraw Hill, New York (1982).
27. Mackay, D., Volatilization of Organic Pollutants from Water. EPA/600/3-82/019, Athens, GA (1982).
28. Yalkowsky, S.H., and Valvani, S.C., *J. Chem. Eng. Data,* 24, 127 (1979).
29. Karickhoff, S.W., Brown, D.S., and Scott, T.A., *Water Res.,* 13, 241 (1979).
30. Schwarzenbach, R.P., and Westall, J., *Environ. Sci. Technol.,* 15, 1360 (1981).
31. O'Connor, D.J., and Connolly, J.P., *Water Res.,* 14, 1517 (1980).
32. Neely, W.G., Branson, D.R., and Blau, G.E., *Environ. Sci. Technol.,* 8, 1113 (1974).
33. Metcalf, R.L., Sanborn, J.R., and Lu, P.Y., *Arch. Environ. Contam. Toxicol.,* 3, 151 (1975).
34. Chiou, C.T., Freed, V.H., Schmedding, D.W., and Kohnert, R.L., *Environ. Sci. Technol.,* 11, 457 (1977).
35. Veith, G.D., DeFoe, D.L., and Bergstedt, B.V., *J. Fish. Res. Board Can.,* 36, 1040 (1979).

36. Connolly, J.P., and Thomann, R.V., in *Fate of Pesticides in Chemicals in the Environment,* J.L. Schnoor, Ed., Wiley-Interscience, New York (1992).
37. Kenaga, E.E., and Goring, C.A.I., in *Aquatic Toxicology,* ASTM STP 707, J.G. Eaton, P.R. Parrish, and A.C. Hendricks, Eds., American Society for Testing and Materials, Philadelphia (1980).
38. Schnoor, J.L., and McAvoy, D.C., *J. Environ. Eng., ASCE,* 107, 1229 (1981).
39. Marchuk, G.I., *Mathematical Modeling for Problems of the Environment,* Science Publishers (in Russian), Moscow, (1982).
40. Bartley, J., and Gangstad, F., *J. Irrig. Drain., ASCE,* 100, 231 (1974).
41. Schnoor, J.L., Mossman, D.J., Borzilov, V.A., Novitsky, M.A., Voszhennikov, O.I., and Gerasimenko, A.K., in *Fate of Pesticides and Chemicals in the Environment,* J.L. Schnoor, Ed., Wiley-Interscience, New York (1992).

7.6 PROBLEMS

1. For the following reactions and rate constants, estimate the overall half-life of the chemical in a batch reactor. Which reaction pathway predominates, if any?

$k_b = 7.0 \times 10^{-9}$ mL cells^{-1} d^{-1}	Biodegradation
$X = 1 \times 10^7$ cells mL^{-1}	Cell concentration
$k_{ox} = 1 \times 10^9$ M^{-1} d^{-1}	Oxidation
$^1O_2 = 1 \times 10^{-12}$ M	Singlet oxygen
$k_p = 0.01$ day^{-1}	Photolysis
$\phi = 0.1$	Quantum yield

2. If the suspended solids concentration in Coralville Reservoir is 200 mg L^{-1} and the partition coefficient for sorption is 100 L kg^{-1} for atrazine, calculate the fraction of atrazine that is in the dissolved phase, f_d. What is the fraction in the dissolved pore water if the sediment solids concentration is 0.4 kg L^{-1} of H_2O?

3. If the BCF (bioconcentration factor) for PCB (Aroclor 1260) is 100,000 L kg^{-1}, and the dissolved phase concentration of organic hydrophobic chemical is 2 ng L^{-1}, what is the concentration expected in fish (in mg kg^{-1})? (*Note:* This is such a low concentration in the water. Two parts per trillion is equivalent to a 1-foot journey on the way to the sun, 93 million miles away!)

4. For a toxic organic chemical in a small completely mixed lake, write a mass balance differential equation for the total chemical concentration in the lake water C_T and the total sediment concentration B_T. Assume instantaneous local equilibrium for sorption, first-order decay reactions in the dissolved phase and pore water, and sedimentation of particulate-adsorbed organic chemical. Use the following variables:

   k_s = sedimentation rate constant, T^{-1}

k = sum of pseudo-first-order decay rate constants, T^{-1}
τ = detention time, T
f_d, f_p = dissolved and particulate fractions in water column
f_{db}, f_{pb} = dissolved and particulate fractions in the bed
k_{bed} = sum of pseudo-first-order rate constants in the bed, T^{-1}

5. The kinetics of adsorption and desorption are shown below:

A	+	B	$\underset{k_2}{\overset{k_1}{\rightleftharpoons}}$	D
dissolved chemical concentration $\mu g\ L^{-1}$		solids concentration kg L^{-1}		adsorbed chemical concentration $\mu g\ kg^{-1}$

a. Sketch a graph of the concentration of A and D with time assuming that B is constant.

b. Write the rate expression for the concentration of A with respect to time (dA/dt) and for the concentration of D with respect to time (dD/dt).

c. Give the ratio of D/A at sorptive equilibria. (This ratio is the sediment-to-water partition coefficient of the chemical).

6. Use equations (99) and (100) to simulate the problem of kepone contamination of the James River, Virginia, in 1980. Some parameters are:

$K_p = 6000\ L\ kg^{-1}$
$k_s = 0.34\ day^{-1}$
$\alpha = 2.5 \times 10^{-5}\ day^{-1}$
$S_w = 40\ mg\ L^{-1}$
$S_b = 0.35\ kg\ L^{-1}$
$u = 0.056\ m\ s^{-1}$ (net nontidal)
$r_b = 1.0\ \mu g\ g^{-1}$ initial bed sediment concentration
$\lambda = 0$ no degradation
$E_x = 4 \times 10^{-4}\ cm^2\ s^{-1}$

7. Same as Example 7.1 for parathion except the discharge is continuous to a plug-flow stream with a pH of 7.0, a suspended solids concentration of 1000 mg L^{-1} (turbid), and a bacteria concentration of 5×10^8 cells L^{-1}. The net sedimentation rate constant is $k_s = 0.2\ day^{-1}$. The initial concentration in the stream is 100 $\mu g\ L^{-1}$ ($C_0 = W/Q = 100\ \mu g\ L^{-1}$). You may assume that the volatilization rate constant is 0.05 day^{-1} and that there is no background atmospheric concentration of parathion. The velocity of the stream is 0.5 m s^{-1} under steady flow, steady state conditions. Compare the fate pathways (as was done in Example 7.1) and plot the concentration of parathion with distance for 100 km.

8. Same as Example 7.2 for PCBs in the Great Lakes. All data are the same as given in the example problem except that the rate of decrease of the atmospheric load is more rapid than expected.

$$W_a = W_{a_0} \exp(-0.07\,t) \quad \text{where } t \text{ is in years}$$

ANSWERS TO SELECTED PROBLEMS—CHAPTER 7

1. Biodegradation predominates.

2. $f_d = 0.98$ in water column
$f_d = 0.024$ in sediment

3. $F = 0.2$ mg/kg

7. $f_d = 0.952$; $\Sigma k = 0.2764$ day^{-1}

$C_T = 100$ µg L^{-1} at $x = 0$ km
$= 93.9$ at $x = 10$
$= 72.9$ at $x = 50$
$= 53.2$ at $x = 100$

8

MODELING TRACE METALS

Art is the lie that helps us to see the truth.
—Pablo Picasso

8.1 INTRODUCTION

Modeling is a little like art in the words of Pablo Picasso. It is never completely realistic; it is never the truth. But it contains enough of the truth, hopefully, and enough realism to gain understanding about environmental systems.

Improvements in analytical quantitation have affected markedly the basis for governmental regulation of trace metals and increased the importance of mathematical modeling. Figure 8.1 was compiled from manufacturers' literature on detection limits for lead in water samples. As we changed from using wet chemical techniques for lead analysis (1950–1960), to flame atomic absorption (1960–1975), to graphite furnace atomic absorption (1975–1990), to inductively coupled plasma mass spectrometry with preconcentration (1985–present), the limit of detection has improved from approximately 10^{-5} (10 mg L^{-1}) to 10^{-12} (1.0 ng L^{-1}).[1] That is seven orders of magnitude in just 35 years! As our limits of detection and clean analytical techniques have improved, what is defined as an unpolluted "background concentration" has decreased dramatically.[1] In some cases, water quality standards have become more stringent when there are demonstrated effects at low concentrations.

Some perspective on Figure 8.1 is provided by Table 1.3. The freshwater Pb chronic water quality criterion is 3.2 $\mu g\ L^{-1}$ and the maximum contaminant level (MCL) drinking water standard is 15 $\mu g\ L^{-1}$. Only since 1980 have we been able to detect such low levels in water samples. Drinking water standards are given in Table 8.1 for metals and inorganic chemicals. (Refer to Table 1.3 for scientific determined water quality criteria. Enforceable water quality standards for surface waters are governed state by state.)

"Heavy metals" usually refer to those metals between atomic number 21 (scandium) and atomic number 84 (polonium), which occur either naturally or from anthropogenic sources in natural waters. These metals are sometimes toxic to aquatic organisms depending on their concentration and chemical speciation. Aluminum (atomic number 13) and the metalloids arsenic and selenium (atomic numbers 33 and 34) often are included in this broad class of pollutants. Other metals that are

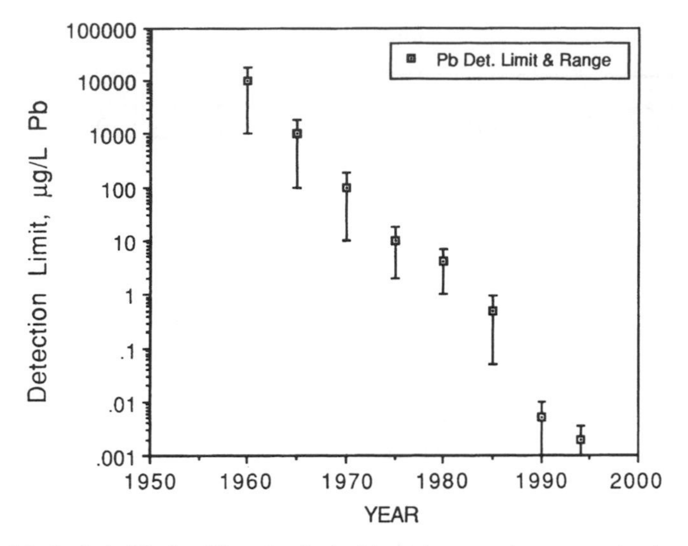

Figure 8.1 Analytical limits of detection for lead in environmental water samples since 1960.

lighter than scandium (i.e., Li, Be, Na, Mg, K, and Ca) are also important in the chemistry of natural waters.

Heavy metals differ from xenobiotic organic pollutants in that they frequently have natural background sources from dissolution of rocks and minerals. The mass of total metal is conservative in the environment; although the pollutant may change its chemical speciation, the *total* remains constant. It cannot be "mineralized" to innocuous end-products as is often the case with toxic organic chemicals. Heavy metals are a serious pollution problem when their concentration exceeds water quality

Table 8.1 Maximum Contaminant Level (MCL) Drinking Water Standards for Inorganic Chemicals and Metals (mg L^{-1} except as noted)

Contaminant	MCL, mg L^{-1}
Arsenic	0.05
Asbestos	7.0 MFL[a]
Barium	2.0
Cadmium	0.005
Chromium	0.100
Copper	1.3
Fluoride	4.0
Lead	0.015
Mercury	0.002
Nitrate-N	10.
Nitrite-N	1.0
Selenium	0.05

[a]Units for asbestos MCL are millions fibers per liter.

standards. Thus waste load allocations are needed to determine the permissible discharges of heavy metals by industries and municipalities. In addition, volatile metals (Cd, Zn, Hg, Pb) are emitted from stack gases, and fly ashes contain significant concentrations of As, Se, and Cr, all of which can be deposited to water and soil by dry deposition.

8.1.1 Hydrolysis of Metals

All metal cations in water are hydrated, that is, they form aquo complexes with H_2O. The number of water molecules that are coordinated around a water is usually four or six. We refer to the metal ion with coordinated water molecules as the "free aquo metal ion." Usually, one neglects to write the H_2O hydration molecules ($M^{n+} \cdot x\, H_2O$) in chemical reaction equations, but they are always present on the free metal ion. The analogous situation is when one writes H^+ rather than H_3O^+ for a proton in water; the coordination number is one for a proton.

The acidity of H_2O molecules in the hydration shell of a metal ion is much larger than the bulk water due to the repulsion of protons in H_2O molecules by the + charge of the metal ion. Thus the acidity of the aquo metal ion is expected to increase with a decrease in the ionic radius and an increase in the charge of the metal ion. Trivalent aluminum ions (monomers) coordinate six waters. Al^{3+} is quite acidic,

$$Al^{3+} + 6\, H_2O \rightleftharpoons Al^{3+} \cdot 6\, H_2O \qquad (1)$$

because the charge to ionic radius ratio of Al^{3+} is large. $Al^{3+} \cdot 6\, H_2O$ can also be written as $Al(H_2O)_6^{3+}$.

The maximum coordination number of metal cations for ligands in solution is usually two times its charge. Metal cations coordinate 2, 4, 6, or sometimes 8 ligands. Two ligands are arranged linearly; four ligands are in a square (planar) or tetrahedron configuration; and six ligands form an octahedron around the central metal ion. Copper(II) is an example of a metal cation with a maximum coordination number of 4.

$$Cu^{2+} + 4\, H_2O \rightleftharpoons Cu(H_2O)_4^{2+} \qquad (2)$$

$$Cu(H_2O)_4^{2+} + 4\, NH_3 \rightleftharpoons Cu(NH_3)_4^{2+} + 4\, H_2O \qquad (3)$$

When free cupric ions are complexed with four ammonia molecules, the Cu2+ ion loses the water molecules in its hydration shell and coordinates with the nitrogen atoms in each ammonia molecule. Metal ions can hydrolyze in solution, demonstrating their acidic property. This discussion follows that of Stumm and Morgan.[2]

$$Cu(H_2O)_4^{2+} \rightleftharpoons Cu(H_2O)_3(OH)^{+} + H^{+} \qquad (4)$$

Equation (4) can be written more succinctly below. Hydrolysis of metal ions is the donation of a proton or the acceptance of OH^- ion by the metal ion.

$$Cu^{2+} + H_2O \rightleftharpoons CuOH^+ + H^+ \qquad \log {}^*K_1 = -8.0 \tag{5}$$

$^*K_1$ is the equilibrium constant, the first acidity constant, for the free aquo cupric ion reacting with a protonated ligand HL (in this case it is reacting with H_2O). If hydrolysis leads to supersaturated conditions of metal ions with respect to their oxide or hydroxide precipitates, polynuclear species can be formed as intermediates, such as the dimer for Cu(II) ions given by equation (6):

$$2\,CuOH^+ \rightleftharpoons Cu_2(OH)_2^{2+} \qquad \log {}^*K_{22} = +1.5 \tag{6}$$

Formation of precipitates can be considered as the final stage of polynuclear complex formation:

$$Cu_2(OH)_2^{2+} \rightleftharpoons Cu^{2+} + Cu(OH)_{2(s)} \qquad \Delta G^\circ = -2.6 \text{ kcal} \tag{7}$$

In Section 8.3, we will discuss the equilibrium modeling of metals, complexation, and precipitation.

8.1.2 Chemical Speciation

Speciation of metals determines their toxicity. Often, free aquo metal ions are toxic to aquatic biota and the complexed metal ion is not. Chemical speciation also affects the relative degree of adsorption or binding to particles in natural waters, which affects its fate (sedimentation, precipitation, volatilization) and toxicity. Figure 8.2 is a periodic chart that has been abbreviated to emphasize the chemistry of metals into two categories of behavior: A-cations (hard) or B-cations (soft).[3]

The classification of A- and B-type metal cations is determined by the number of electrons in the outer *d* orbital. Type-A metal cations have an inert gas type of arrangement with no electrons (or few electrons) in the *d* orbital, corresponding to "hard sphere" cations. Their electron sheath is not readily deformable by the influence of electronic fields from approaching ligands. Type-A metals form complexes with F^- and oxygen donor atoms in ligands such as OH^-, CO_3^{2-}, and HCO_3^{2-}. They coordinate more strongly with H_2O molecules than NH_3 or CN^-.

Type-B metals (e.g., Ag^+, Au^+, Hg^{2+}, Cu^+, Cd^{2+}) have many electrons in the outer *d* orbital. They are not spherical, and their electron cloud is easily deformed by ligand fields. They form strong complexes and precipitates with HS^- and S^{2-}; they prefer I^- or Cl^- to F^- ligands; and they bind ammonia more strongly than water molecules.[3]

Transition metal cations are in-between Type A and Type B on the periodic chart (Figure 8.2). They have some of the properties of both hard and soft metal cations, and they employ 1–9 *d* electrons in their outer orbital. A qualitative generalization of the stability of complexes formed by the transition metal ions is the Irving–Williams series:

$$Mn^{2+} < Fe^{2+} < Co^{2+} < Ni^{2+} < Cu^{2+} > Zn^{2+} \tag{8}$$

increasingly stable complexes →

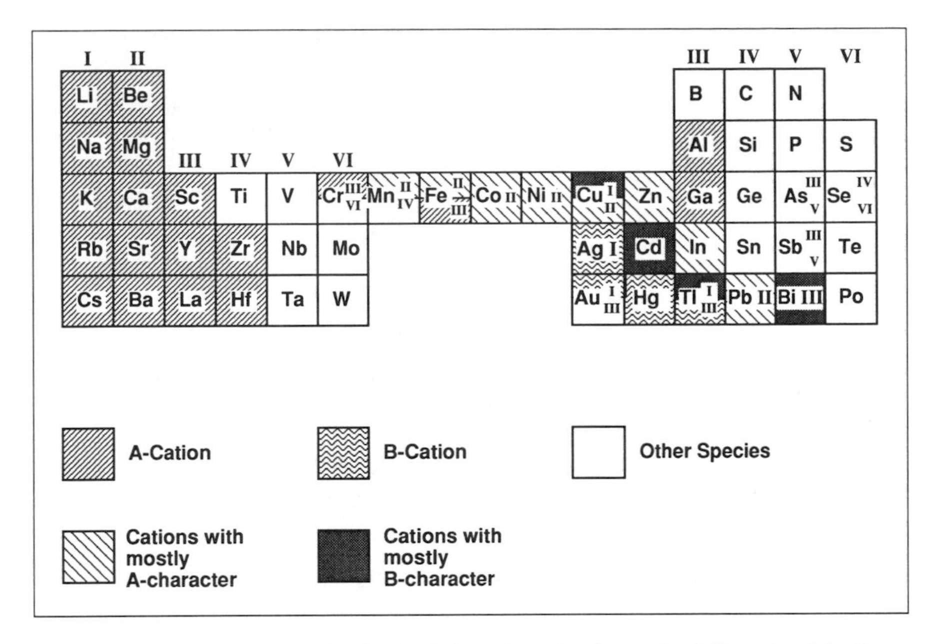

Figure 8.2 Partial periodic table showing hard (A-type) metal cations and soft (B-type) metal cations.

These transition metal cations show mostly Type-A (hard sphere) properties. Zinc(II) is variable in the strength of its complexes; it forms weaker complexes than copper(II), but stronger complexes than several of the others in the transition series.

Table 8.2 gives the typical species of metal ions in fresh water and seawater. This is a generalization because the species may change with the chemistry of an individual natural water. For example, colored waters containing significant humic and fulvic acids that can complex Cu(II) and Al(III) very strongly and dramatically alter both the toxicity and speciation of the metal. The presence of colloids and suspended solids can decrease the concentration of free aquo metal ions and decrease toxicity in most cases. Table 8.2 is applicable to aerobic waters—anaerobic systems would have large HS^- concentrations that would likely precipitate Hg(II), Cd(II), and several other Type-B metal cations.

8.1.3 Dissolved Versus Particulate Metals

Heavy metals that have a tendency to form strong complexes in solution also have a tendency to form surface complexes with those same ligands on particles. Difficulty arises because chemical equilibrium models must distinguish between what is actually dissolved in water versus that which is adsorbed. Analytical chemists have devised an operational definition of "dissolved constituents" as all those which pass a 0.45 μm membrane filter. This operational definition is problematic because it is known that many colloids and macromolecules are able to pass the filter, resulting in a high bias of truly dissolved metals concentrations. But the definition has been

Table 8.2 Major Species of Metal Cations in Aerobic Fresh Water and Seawater

	Element	Fresh Water[a] Major Species	Seawater[b] Additional Species
Predominantly free aquo ions	Li	Li^+	
	Na	Na^+	
	Mg	Mg^{2+}	$MgCO_3^0$
	K	K^+	
	Ca	Ca^{2+}	$CaSO_4^0$
	Sr	Sr^{2+}	
	Cs	Cs^+	
	Ba	Ba^{2+}	
Complexation with OH^-, CO_3^{2-}, HCO_3^-, Cl^-	Be(II)	$BeOH^+$, $Be(OH)_2^0$	
	Al(III)	$Al(OH)_{3(s)}$, $Al(OH)_2^+$, $Al(OH)_4^-$	
	Ti(IV)	$TiO_{2(s)}$, $Ti(OH)_4^0$	
	Mn(IV)	$MnO_{2(s)}$	
	Fe(III)	$Fe(OH)_{3(s)}$, $Fe(OH)_2^+$, $Fe(OH)_4^-$	
	Co(II)	Co^{2+}, $CoCO_3^0$	
	Ni(II)	Ni^{2+}, $NiCO_3^0$	$NiCl^+$
	Cu(II)	$CuCO_3^0$, $Cu(OH)_2^0$	$CuCl^+$
	Zn(II)	Zn^{2+}, $ZnCO_3^0$	$ZnCl^+$
	Ag(I)	Ag^+, $AgCl^0$	$AgCl_2^-$
	Cd(II)	Cd^{2+}, $CdCO_3^0$	$CdCl^+$, $CdCl_2^0$
	Hg(II)	$Hg(OH)_2^0$	$HgCl_4^{2-}$
	Pb(II)	$PbCO_3^0$	$PbCl^+$
Hydrolyzed, anionic	B(III)	H_3BO_3, $B(OH)_4^-$	
	V(V)	HVO_4^{2-}, $H_2VO_4^-$	
	Cr(VI)	CrO_4^{2-}	
	As(V)	$HAsO_4^{2-}$	
	Se(VI)	SeO_4^{2-}	
	Mo(VI)	MoO_4^{2-}	
	Si(IV)	$Si(OH)_4$	

[a]Fresh water: pH 8, alk = 2×10^{-3} M, $Cl^- = 2.5 \times 10^{-4}$ M, O_2 = satd., $I = 5 \times 10^{-3}$ M.
[b]Seawater: pH 8.2, alk = 2.5×10^{-3} M, Cl^- = 0.4 M, O_2 = satd., I = 0.7 M.
Source: Stumm and Morgan.[3]

accepted by governmental regulators, and it is used to generate comparable field data for environmental models.

Windom et al.[4] and Cohen[5] have demonstrated that filtered samples can become easily contaminated in the filtration process. As per Horowitz,[6] there are two philosophies for processing a filtered sample for quantitation of dissolved trace elements: (1) treat the colloidally associated trace elements as contaminants and attempt to exclude them from the sample as much as possible, or (2) process the sample in such a way that most colloids will pass the filter (large particles will not) and try to standardize the technique so that results are operationally comparable. The

United States Geological Survey has chosen the second philosophy of maintaining 0.45-μm membrane filtration as the operational definition for dissolved constituents, but it recommends that a capsule filter be used that has 600 cm^2 of surface area (as opposed to the smaller 17.5-cm^2 Nucleopore plate filter). The larger surface area is to assure that most colloids will pass the filter and be included consistently as "dissolved" constituents.

There is a school of thought that the first philosophy is the best approach from the standpoint of chemical equilibrium modeling. Shiller and Taylor[7] and Buffle[8] have proposed ultrafiltration, sequential flow fractionation, or exhaustive filtration as methods to approach the goal of excluding colloidally associated metals from the water sample entirely.

Flegal and Coale[9,10] were among the first to report that previous trace metal data sets suffered from serious contamination problems. Virtually all marine trace metal data from before 1985 are now considered invalid.[1] The modeler needs to understand how water samples were obtained because it can affect results when simulating trace metals concentrations in the nanogram per liter range, especially.

According to Benoit,[1] ultraclean procedures have three guiding principles: (1) samples contact only surfaces consisting of materials that are low in metals (Teflon or low-density polyethylene) and that have been extensively acid-cleaned in a filtered air environment; (2) samples are collected and transported taking extraordinary care to avoid contamination from field personnel or their gear; and (3) all other sample handling steps take place in a filtered air environment and using ultrapure reagents.

The "sediment-dependent" partition coefficient for metals (K_d) is an artifact, to large extent, of the filtration procedure in which colloid-associated trace metals pass the filter and are considered "dissolved constituents." The greater the suspended solids concentration in the original whole water sample, the greater will be the concentration of colloidal metal that passes the filter, and the less will be the corresponding determination of K_d, that is, the distribution coefficient between particulate and operationally defined dissolved phases.

8.2 MASS BALANCE AND WASTE LOAD ALLOCATION FOR RIVERS

We will begin with a model for trace metals in rivers. Rivers are open channels that have flowrates that vary with time and longitudinal distance, and cross-sectional areas that vary with longitudinal distance.

8.2.1 Open-Channel Hydraulics

A one-dimensional model is the St. Venant equations[11,12] for nonsteady, open-channel flow. We must know where the water is moving before we can model the trace metal constituents.

$$\frac{\partial z}{\partial t} = -\frac{1}{b}\frac{\partial Q}{\partial x} + \frac{1}{b} q_i \tag{9}$$

$$\frac{\partial Q}{\partial t} = \left(\frac{Q^2 b}{A^2} - gA\right)\frac{\partial z}{\partial x} - \frac{2Q}{A}\frac{\partial Q}{\partial x} + \frac{Q^2}{A^2}\frac{\partial A}{\partial x} - gAS_f \tag{10}$$

where Q = discharge, L^3T^{-1}
t = time, T
z = absolute elevation of water level above sea level, L.
A = total area of cross section, L^2
b = width at water level, L
g = gravitational acceleration, LT^{-2}
x = longitudinal distance, L
q_i = lateral inflow per unit length of river, L^2T^{-1}
S_f = the friction slope

$$S_f = \frac{f}{8g}\frac{P}{A}\frac{Q^2}{A^2} \tag{11}$$

where P = wetted perimeter, L
f = Darcy–Weisbach friction factor

Only a few numerical codes that are available solve the fully dynamic open-channel flow equations.[13,14] It requires detailed cross-sectional area/stage relationships, stage/discharge relationships, and stage and discharge measurements with time. During storm events, measurements would need to be hourly at a number of stations, so a large data set is needed. Trace metals are often associated with bed sediments, and sediment transport is very nonlinear. Bed sediment is scoured during flood events when critical shear stresses at the bed–water interfaces are exceeded. Thus, for some problems, it is necessary to solve the fully dynamic St. Venant equations.

In most cases, equations (9) and (10) can be simplified: (1) lateral inflow q_i can be neglected and accounted for at the beginning of each stream segment; (2) the river can be segmented into reaches where $Q(x)$ is approximately constant within each segment; and/or (3) the river can be segmented into reaches where $A(x)$ is approximately constant within each segment. If Q and A can be approximated as constants in (x, t), then the velocity is constant and the problem reduces to one of steady flow. Codes that are widely used for river water quality modeling generally assume steady-flow conditions and constant values for Q and A within each stream segment. QUAL-2E by Brown and Barnwell[15] solves the time-variable water quality equations for dissolved constituents, but it assumes steady flow. NONEQUI by Fontaine[16] was specifically developed for trace metals modeling and speciation, but steady flow was also assumed.

8.2.2 Mass Balance Equation

A schematic of trace metal transport in a river is given in Figure 8.3. Metals can be in the particulate-adsorbed or dissolved phases in the sediment or overlying water column. Interchange between sorbed metal ions and aqueous metal ions occurs via adsorption/desorption mechanisms. Within the dissolved phase, the metal can be complexed with a number of different ligands in natural waters. Likewise, metal ions form surface complexes at a number of different sites on particulate matter. Bed sediment can be scoured and can enter the water column, and suspended solids can undergo sedimentation and be deposited on the bed. Metal ions in pore water of the sediment can diffuse to the overlying water column and vice versa, depending on the concentration driving force.

Under steady flow conditions ($dQ/dt = 0$), we can consider the time-variable concentration of metal ions in sediment and overlying water. Here, we will assume that adsorption and desorption processes are fast relative to transport processes and that the suspended solids concentration is constant within the river segment. These assumptions are not required and they are not always applicable, but it allows us to reduce the number of mass balance equations from six (dissolved metal concentration, adsorbed metal concentration and suspended solids concentration in the overlying water column and the sediment) down to only two equations (total metal concentration in the unfiltered water and adsorbed metal concentration in the sediment). Sorption kinetics are typically complete within 1 hour, which is fast compared to transport processes, except perhaps during high-flow events. Suspended solids concentrations can be assumed as constant within a segment under steady flow conditions.

The modeling approach is to couple a chemical equilibrium model with the mass balance equations for total metal concentration in the water column as depicted in Figure 4.1. In this way, we do not have to write a mass balance equation for every chemical species (which is impractical). The approach assumes that chemical equilibrium among metal species in solution and between phases is valid. Acid–base reactions and complexation reactions are fast (on the order of minutes to microseconds). Precipitation and dissolution reactions can be slow; if this is the case, the chemical equilibrium submodel will flag the relevant solid phase reaction (provided that the user has anticipated the possibility). Redox reactions can also be slow. The modeler must have some knowledge of aquatic chemistry in order to make an intelligent choice of chemical species in the equilibrium model. If the assumption of local chemical equilibrium is not valid for a particular problem, then the modeler must go back to the set of six differential equations including the kinetics of adsorption and desorption and time-variable solids concentrations. Because the equations will be solved with a computer using a numerical technique, it is not that much more difficult to include the kinetics of adsorption and desorption if it is necessary.

Assuming local equilibrium for sorption, we may write the mass balance equations similarly to those given in Chapter 7 for toxic organic chemicals in a river. The only difference is that metals do not degrade—they may undergo chemical reactions or biologically mediated redox transformations, but the total amount of metal re-

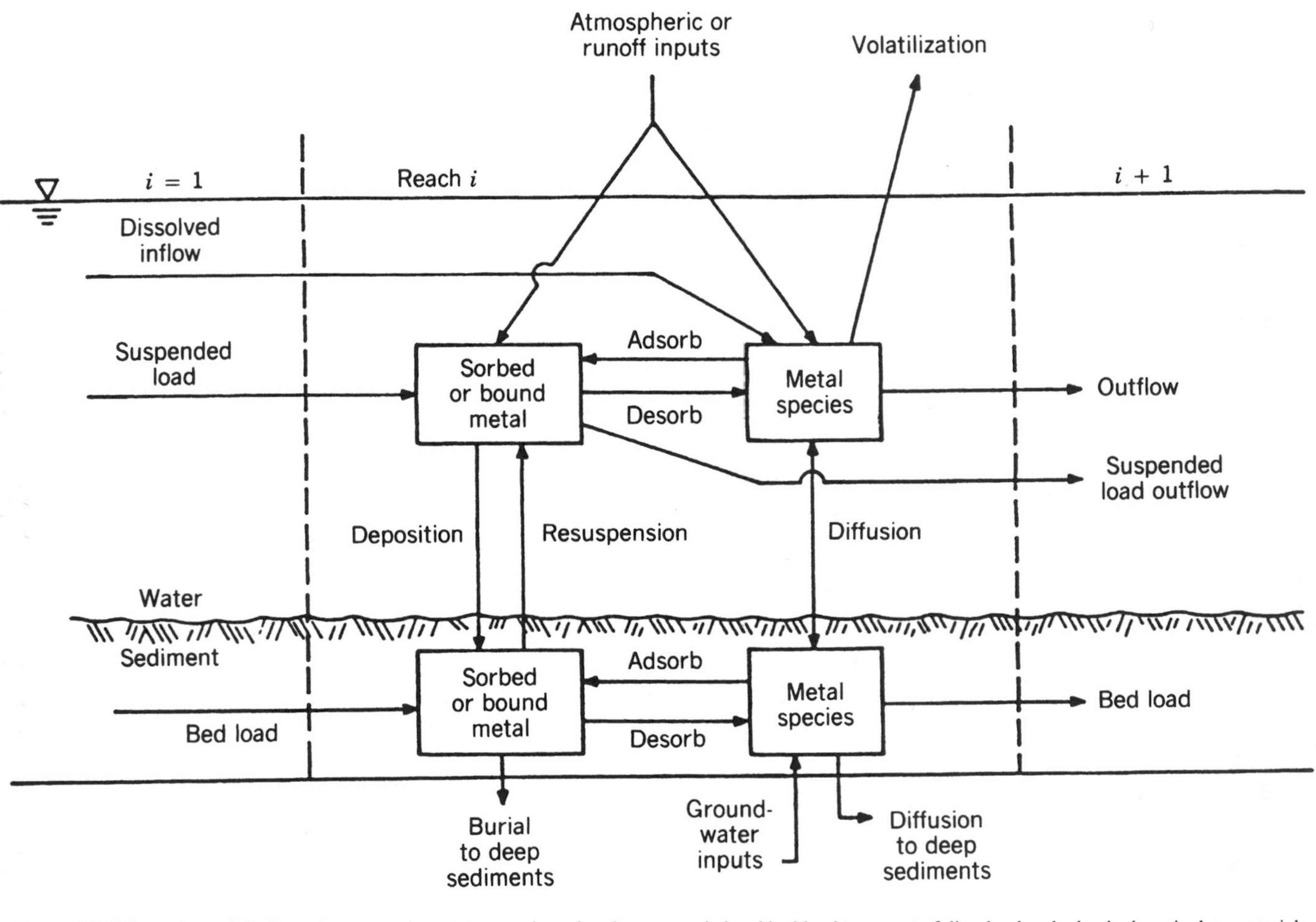

Figure 8.3 Schematic model of metal transport in a stream or river showing suspended and bed load transport of dissolved and adsorbed particulate material.

mains unchanged. If chemical equilibrium does not apply, each chemical species of concern must be simulated with its own mass balance equation. An example would be the methylation and dimethylation of Hg and Hg^0 volatilization, which are relatively slow processes, and chemical equilibrium would not apply.

$$\frac{\partial C_T}{\partial t} = -\frac{1}{A}\frac{\partial(QC_T)}{\partial x} + \frac{1}{A}\frac{\partial}{\partial x}\left(EA\frac{\partial C_T}{\partial x}\right) - k_s f_{p,w} C_T + \frac{k_L}{h}\left(\frac{r}{K_{p,b}} - f_{d,w} C_T\right) + \frac{\alpha S_b r}{\gamma} \tag{12}$$

$$\frac{dr}{dt} = +\underbrace{\frac{k_s f_{p,w} C_T \gamma}{\left(S_b + \frac{1}{K_{p,b}}\right)}}_{\text{sedimentation}} - \underbrace{\frac{k_L}{d}\left(\frac{\frac{r}{K_{p,b}} - f_{d,w} C_T}{S_b + \frac{1}{K_{p,b}}}\right)}_{\text{pore-water diffusion}} - \underbrace{\alpha S_b \frac{r}{\left(S_b + \frac{1}{K_{p,b}}\right)}}_{\text{scour from bed}} \tag{13}$$

where C_T = total whole water (unfiltered) concentration in the water column, M L^{-3}
t = time, T
A = cross sectional area, L^2
Q = flowrate, L^3T^{-1}
x = longitudinal distance, L
E = longitudinal dispersion coefficient, L^2T^{-1}
k_s = sedimentation rate constant, T^{-1}
$f_{p,w}$ = particulate adsorbed fraction in the water column, dimensionless
$f_{d,w}$ = dissolved chemical fraction in the water column, dimensionless
$K_{p,b}$ = sediment–water distribution coefficient in the bed, L^3M^{-1}
r = adsorbed chemical in bed sediment, MM^{-1}
k_L = mass transfer coefficient between bed sediment and water column, LT^{-1}
h = depth of the water column (stage), L
α = scour coefficient of bed sediment into overlying water, T^{-1}
S_b = bed sediment solids concentration, ML^{-3} bed volume
γ = ratio of water depth to active bed sediment layer, dimensionless
d = depth of active bed sediment, L

The model is similar to that depicted in Figure 8.3 except that we assume instantaneous sorption equilibrium in the water column and sediment between the adsorbed and dissolved chemical phases. The first two terms of equation (12) are for advective and dispersive transport for the case of variable Q, E, and A with distance. Sedimentation of the particulate fraction occurs via a first-order sedimentation rate constant, k_s. Pore-water diffusion is based on a driving force between the concentration of dissolved metal in the pore water minus the dissolved concentration in the overlying water ($C_{d,b} - C_{d,w}$). [All other parameters in the pore water diffusion term of equation (13) simply allow conversions between sediment adsorbed concentration, r, and total unfiltered water concentration, C_T.] The last term in equations (12) and (13) accounts for scour of particulate adsorbed material from the bed sediment to the overlying water using a first-order coefficient for scour, α, that would depend on flow and sediment properties. Equations (12) and (13) are a set of partial differential equations that can be solved by the method of characteristics[17,18] or operator splitting methods.[19] Generally, the reactions terms (sedimentation, diffusion, and scour) are solved by a Runge–Kutta or implicit–explicit scheme, the advection term by the method of characteristics, and the dispersion term by Crank–Nicholson operator.

This set of two equations is very powerful because it can be used to model situations when the bed is contaminated and diffusing into the overlying water or when there are point source discharges that adsorb and settle out of the water column to the bed. It is a dynamic (time-variable) system of equations that allows investigation of the time required for recovery and natural restoration of a metals-contaminated river.

Local equilibrium was assumed in equations (12) and (13) for adsorption and desorption. If a simple equilibrium distribution coefficient is applicable, one can substitute for the fraction of chemical that is dissolved and particulate according to the following relationships:

$$f_{d,w} = \frac{1}{1 + K_{p,w} S_w}$$

$$f_{p,w} = \frac{K_{p,w} S_w}{1 + K_{p,w} S_w}$$

where $f_{d,w}$ = fraction dissolved in the water column
$f_{p,w}$ = fraction in particulate adsorbed form in the water column
$K_{p,w}$ = distribution coefficient in the water, L kg^{-1}
S_w = suspended solids concentration, kg L^{-1}.

Alternatively, one can use a chemical equilibrium submodel to compute $K_{p,b}$, $f_{d,w}$, and $f_{p,w}$ in the water and sediment. In this case, several options are available including Langmuir sorption, diffuse double-layer model, constant capacitance model, or triple-layer model. The fractions dissolved and particulate would be updated in the mass balance equation at every time step when the chemical equilibrium submodel was called (see Figure 4.1).

8.2.3 Waste Load Allocations

Point source discharges that continue for a long period will eventually come to steady state with respect to river water quality concentrations. The steady state will be perturbed by rainfall runoff or discharge events, but the river may tend toward an "average condition." If metals concentrations from point source and/or nonpoint source discharges violate a stream water quality standard, a waste load allocation model must be used to determine an allowable discharge level that will comply with standards. Under steady-state conditions ($\partial C_T/\partial t = 0$ and $\partial r/\partial t = 0$), equations (12) and (13) reduce to the following:

$$u\frac{dC_T}{dx} = E_x\frac{d^2C_T}{dx^2} - k_s f_{p,w} C_T + \frac{k_L}{h}\left(\frac{r}{K_{p,b}} - f_{d,w}C_T\right) + \frac{\alpha S_b r}{\gamma} \tag{14}$$

Equation (14) can be solved directly because r, the adsorbed sediment concentration, is a constant under steady-state conditions. It is an input parameter that specifies the contamination level in the bed sediment. The model must be segmented to reflect variations in r with longitudinal distance.

For swift-flowing streams and rivers, the longitudinal dispersion term ($E_x\, d^2C_T/dx^2$) in equation (14) can be neglected. For large rivers, after an initial mixing period, dispersion can be estimated by Fischer's equation[12]:

$$E_x = \frac{\beta u_x^2 B_x^2}{h u_*} \tag{15}$$

where E_x is the longitudinal dispersion coefficient (L^2T^{-1}), β is a constant typical of the river in question, u_x is the mean flow velocity (LT^{-1}), B_x is the mean width (L), h is the mean depth (L), and u_* is the mean shear velocity (LT^{-1}) at the sediment–water interface, often taken to be $0.1\ u_x$ or

$$u_* = \sqrt{ghS} \tag{16}$$

where g is the gravitational constant (LT^{-2}), h is the mean depth (L), and S is the energy slope of the channel (LL^{-1}).

A mixing zone in the river is allowed in the vicinity of the wastewater discharge where locally high concentrations may exist before there is adequate time for mixing across the channel and with depth. The mixing zone can be estimated from the empirical equation (17):

$$X_\ell = \frac{0.4\ B^2}{E_y} \tag{17}$$

where X_ℓ = mixing length for a side bank discharge, ft

B = river width, ft

E_y = lateral dispersion coefficient, $ft^2\ s^{-1} = \epsilon h u_*$

ϵ = proportionality coefficient = 0.3–1.0

u_* = shear velocity as defined in equation (16), ft s^{-1}

h = mean depth, ft

[Equation (17) is empirical and the units do not cancel.] Allowable mixing zones vary by state. Generally, at least 500–1000 feet are allowed as a mixing zone from the end of the property line. Mixing across large rivers takes a long distance, on the order of tens of miles in the Mississippi and Missouri Rivers! Such large distances are not considered as allowable mixing zones, however.

The most basic waste load allocation is a simple dilution calculation assuming complete mixing of the waste discharge with the upstream metals concentration (a flow-weighted average concentration). One assumes that the total metal concentration (dissolved plus particulate adsorbed concentrations) are bioavailable to biota as a worst case condition. In most cases this is a very conservative assumption, but particulate metal may become bioavailable to aquatic organisms at a later time in the water body, or it may be available to benthic organisms after it is deposited in the sediment, so there is a rationale for the assumption. We use water quality criteria to protect aquatic organisms that come into contact with sediments because there is not a separate regulatory requirement for sediment quality criteria at the present time.

At the next level of complexity, the waste load allocation may become a probalistic calculation using a Monte Carlo analysis. This approach has several advantages because low-flow conditions are usually assumed as the worst case condition for acute exposures, and low-flow conditions are probalistic, such as the low flow condition in a seven day running average that occurs only once in every ten years (7 day, 1-in-10 year low flow). Also, it has the advantage of being extended to any recurrence interval that is desired for acute or chronic exposures. Because spreadsheets and most workstation compilers have random number generators, it is easy to perform a Monte Carlo analysis using the following equation.

$$C_T = \frac{Q_u C_u + Q_w C_w}{Q_u + Q_w}$$

where C_T = total metals concentration, ML^{-3}
Q_u = upstream flowrate, L^3T^{-1}
C_u = upstream metals concentration, ML^{-3}
Q_w = wastewater discharge flowrate, L^3T^{-1}
C_w = wastewater metals concentration, ML^{-3}

The steps in the procedure include estimating a frequency histogram (probability density function) for each of the four model parameters Q_u, C_u, Q_w, and C_w. Usually, normal or lognormal distributions are assumed or estimated from field data. A mean and a standard deviation are specified. Then the Monte Carlo program estimates 250-1000 realizations of the model equation; this procedure provides an estimate of the frequency distribution of the total metals concentration in the stream, C_T, assuming complete mixing. Excursions of acute water quality criteria are allowed by state agencies only one in every three years (a frequency of 1 day in 3 years or 0.000913).

A more detailed waste load allocation was prepared for the Deep River in Forsyth County, North Carolina.[20] A steady-state mass balance equation was used that neglected longitudinal dispersion and scour.

$$u\frac{dC_T}{dx} = -k_s f_{p,w} C_T + \frac{k_L}{h}\left(\frac{r}{K_{p,b}} - f_{d,w} C_T\right) \tag{18}$$

[The terms are defined the same as in equation (14).] Low-flow conditions were simulated (worst case scenario) during August–September 1983, when a stream survey of copper and other toxic metals as conducted.[21] Equation (18) was coupled with a MINTEQ[22] chemical equilibrium model to determine chemical speciation of dissolved metal cations.

The upper Deep River from High Point Lake to the town of Randleman was the segment of stream used in the waste load allocation. The small river received 41 permitted wastewater discharges at the time of the study, the majority of which were small municipal discharges. An average width of 12.2 m, depth of 0.2 m, and velocity of 0.021 m s^{-1} were characteristic of low-flow conditions. Under these conditions, almost half of the flow in the river was from wastewater discharges. Thus, based on a water quality standard of 20 μg L^{-1} for dissolved copper, wastewater discharges had to be reduced considerably to meet the standard using a waste load allocation.

Figure 8.4 is the result of the waste load allocation. Steps in the procedure were the following:

1. Select critical conditions when concentrations and/or toxicity are expected to be greatest. Usually 7-day, 1-in-10-year low-flow periods are assumed.
2. Develop an effluent discharge database. If measured flow and concentration data are not available for each discharge, the maximum permitted values are usually assumed, or some fraction thereof.
3. Obtain stream survey concentration data during critical conditions. Suspended solids, bed sediment, and filtered and unfiltered water samples should be analyzed for each trace metal of concern. Dye studies for time of travel, gage station flow data (stage-discharge), and cross-sectional areas of the river reach should be obtained. Measurement of nonpoint source loads should also be accomplished at this time.
4. Perform model simulations. First, calibrate the model by using the effluent discharge database and the stream survey results. Adjustable parameters include k_s and k_L. All other parameters and coefficients should have been measured. k_s and k_L should be varied within a reasonable range of reported literature values.
5. Determine how much the measured trace metal concentrations exceed the water quality standard. (If they do not exceed the standard, there is no need for a waste load allocation.) Reduce the effluent discharges proportionally until the entire concentration profile is within acceptable limits.

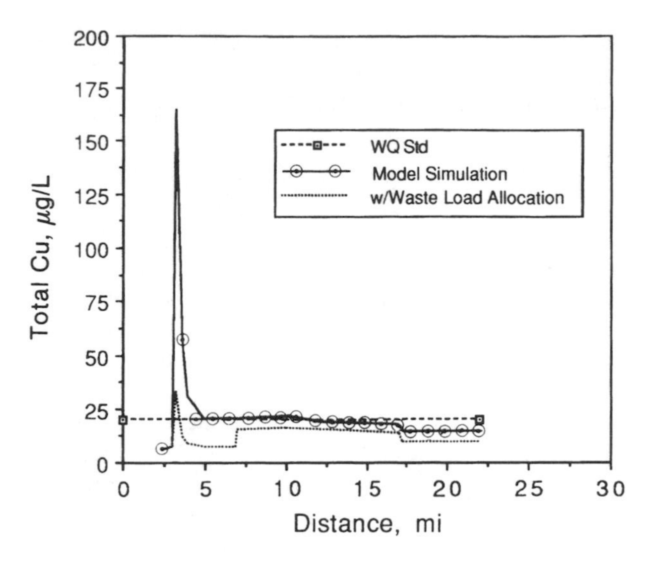

Figure 8.4 Waste load allocation for copper in the Deep River, North Carolina. Total dissolved copper concentrations in field samples and model simulation are shown. The water quality standard is 20 μg L ¹.

In Figure 8.4, model calibration (the solid line) fit the field survey data points quite well. A major discharge occurred from the Jamestown wastewater treatment plant at mile point 3. It caused a large peak in the simulated copper concentration of 165 μg L^{-1}. Unfortunately, there were no field measurements at the point of the predicted maximum in copper concentrations. The next data point was at approximately mile point 4 (55 μg L^{-1}), and the simulated concentration matched the measured value quite closely. The simulated total copper concentration declined dramatically between mile point 3 and 5 due to sedimentation of a significant suspended particulate load that entered at Jamestown. After mile point 7, field survey data were at or slightly below the water quality standard of 20 μg L^{-1}, as were the simulated values. The k_s value, sedimentation rate constant was 1.2 day^{-1}, which corresponds to a settling velocity ($v_s = k_s h$) of 0.24 m d^{-1}, a typical rate for algae and fine particles to settle from natural waters. There were no further significant discharges after Jamestown except a small one at mile point 7. The remainder of the water quality samples and the model results were slightly less than the water quality standard.

By reducing the waste discharge concentration at Jamestown from 250 μg L^{-1} to 48 μg L^{-1} copper, the waste load allocation succeeded in lowering concentrations to within water quality standards except for ½ mile immediately downstream. It was not necessary to reduce effluent concentrations elsewhere (Figure 8.4). The treatment plant at Jamestown (mile point 3) might be given the alternatives of (1) eliminating industrial wastes from their plant, (2) requiring pretreatment of the industrial waste, or (3) precipitating copper and other metals at high pH from the effluent.

One might ask the question, How toxic is the effluent? There are at least two approaches to the question: a whole effluent approach where toxicity tests are performed on each effluent until test organisms can survive and dilution requirements are determined[23]; or chemical speciation and equilibrium modeling of the metal ions to determine if they are in their toxic (bioavailable) form. The second approach requires a chemical equilibrium model coupled with the mass balance equation. For the case of the Deep River, MINTEQ[22] indicated that most of the dissolved metal near the Jamestown discharge (mile point 3) was in the form of copper–fulvate complexes due to dissolved organic matter in the water, $10^{-4.5}$ M fulvic acid (FA). Results indicated that the following Cu(II) species were prevalent in the Deep River, North Carolina: Cu–FA = 36%, $Cu(OH)_{2(aq)}$ = 42%, $CuCO_{3(aq)}$ = 19%, and Cu^{2+} = 1.4%. Cu–FA complexes are not expected to be toxic, and the low percentage of free copper (presumably the toxic species) means that the concentration of Cu^{2+} was only 0.8 $\mu g\ L^{-1}$, well below most toxicity thresholds.

8.3 COMPLEX FORMATION AND SOLUBILITY

8.3.1 Formation Constants

Complexation of free aquo metal ions in solution can affect toxicity markedly. Stability constants are equilibrium values for the formation of complexes in water. They can be expressed as either stepwise constants or overall stability constants. Stepwise constants are given by equations (19) to (21) as general reactions, neglecting charge.

$$\mathrm{M} + \mathrm{L} \rightleftharpoons \mathrm{ML} \qquad K_1 = \frac{\{\mathrm{ML}\}}{\{\mathrm{M}\}\ \{\mathrm{L}\}} \tag{19}$$

$$\mathrm{ML} + \mathrm{L} \rightleftharpoons \mathrm{ML}_2 \qquad K_2 = \frac{\{\mathrm{ML}_2\}}{\{\mathrm{ML}\}\ \{\mathrm{L}\}} \tag{20}$$

$$\vdots$$

$$\mathrm{ML}_{i-1} + \mathrm{L} \rightleftharpoons \mathrm{ML}_i \qquad K_i = \frac{\{\mathrm{ML}_i\}}{\{\mathrm{ML}_{i-1}\}\ \{\mathrm{L}\}} \tag{21}$$

Overall stability constants are reactions written in terms of the free aquo metal ion as reactant:

$$\mathrm{M} + 2\mathrm{L} \rightleftharpoons \mathrm{ML}_2 \qquad \beta_2 = \frac{\{\mathrm{ML}_2\}}{\{\mathrm{M}\}\ \{\mathrm{L}\}^2} \tag{22}$$

$$\mathrm{M} + i\mathrm{L} = \mathrm{ML}_i \qquad \beta_i = \frac{\{\mathrm{ML}_i\}}{\{\mathrm{M}\}\ \{\mathrm{L}\}^i} \tag{23}$$

The product of the stepwise formation constants gives the overall stability constant:

$$\beta_i = K_1 \times K_2 \times K_3 \times \cdots K_i \tag{24}$$

For polynuclear complexes, the overall stability constant is used, and it reflects the composition of the complex in its subscript.[2] An example would be $Fe_2(OH)_2^{4+}$.

$$m\text{M} + n\text{L} \rightleftharpoons \text{M}_m\text{L}_n \qquad \beta_{nm} = \frac{\{\text{M}_m\text{L}_n\}}{\{\text{M}\}^m \{\text{L}\}^n} \tag{25}$$

$$2\ \text{Fe}^{3+} + 2\ \text{OH}^- \rightleftharpoons \text{Fe}_2(\text{OH})_2^{4+} \qquad \beta_{22} = \frac{\{\text{Fe}_2(\text{OH})_2^{4+}\}}{\{\text{Fe}^{3+}\}^2 \{\text{OH}^-\}^2} \tag{26}$$

Table 8.3 is adapted from Morel and Hering,[24] and it provides some common overall stability constants and solubility constants for 19 metals and 10 ligands. Note that the values are given as log β values for the *formation* constants. (It is the inverse of the solubility product. For example, the log β_1 value for $CuS_{(s)}$ is given as 36.1.)

$$\text{Cu}^{2+} + \text{S}^{2-} \rightleftharpoons \text{CuS}_{(s)} \qquad \beta_1 = \frac{\{\text{CuS}\}}{\{\text{Cu}^{2+}\}\{\text{S}^{2-}\}} = 10^{36.1} \tag{27}$$

Note also that the stability and solubility constants are given at 25 °C and zero ionic strength (infinite dilution). At low ionic strengths, the constants may be used with concentrations rather than the activities indicated by the brackets in equations (19)–(27). However, at high ionic strengths, activity corrections should be made as discussed in Chapter 4, or a conditional stability constant should be determined for those conditions.

Complexation reactions are normally fast relative to transport reactions, but precipitation and dissolution reactions can be quite slow (mass transfer or kinetic limitations). Use of chemical equilibrium models is not always warranted for precipitation and dissolution in natural waters.

Under anaerobic conditions, sulfide is generated; it is a microbially mediated process of sulfate reduction in sediments and groundwater. Sulfide precipitates control the solubility of many trace metals under anaerobic conditions. In contrast, under aerobic conditions, trace metals solubility is controlled by amorphous hydroxide precipitation at neutral pH and above. Because the hydroxides are amphoteric, they show minimum solubility at the pH where the neutral species predominates. Figure 8.5 shows the solubility of a few metal sulfides and hydroxides. Solubility calculations are highly dependent on the K_{s0} values employed, which are somewhat variable depending on the mineral formed and its crystallinity. $Pb(OH)_3^-$, $Zn(OH)_3^-$, $Ni(OH)_3^-$, and $Cu(OH)_4^{2-}$ increase the solubility of the metals at high pH.

At pH 7 and anaerobic conditions, the solubility of Cu^{2+}, Pb^{2+}, Cd^{2+}, Zn^{2+}, and Ni^{2+} is 1 μg L^{-1} or less; it is controlled by anaerobic generation of HS^- and S^{2-} in

Table 8.3 Logarithm of Overall Stability Constants for Formation of Complexes and Solids from Metals and Ligands at 25 °C and $I = 0.0$ M

log β Values	OH^-	CO_3^{2-}	SO_4^{2-}	Cl^-	Br^-	F^-	NH_3	CN^-	S^{2-}	PO_4^{3-}
H^+		HL 10.33	HL 1.99			HL 3.2	HL 9.24	HL 9.2	HL 13.9	HL 12.35
	$HL_{(w)}$ 14.00	H_2L 16.68					$L_{(g)}$ −1.8		H_2L 20.9	H_2L 19.55
		$H_2L_{(g)}$ 18.14							$H_2L_{(g)}$ 21.9	H_3L 21.70
Na^+		NaL 1.27	NaL 1.06							NaHL 13.5
		NaHL 10.08								
K^+			KL 0.96							KHL 13.4
Ca^{2+}	CaL 1.15	CaL 3.2	CaL 2.31			CaL 1.1				CaL 6.5
		CaHL 11.59	$CaL_{(s)}$ 4.62			$CaL_{2(s)}$ 10.4				CaHL 15.1
	$CaL_{2(s)}$ 5.19	$CaL_{(s)}$ 8.22								CaH_2L 21.0
		$CaL_{(s)}$ 8.35								$CaHL_{(s)}$ 19.0
Mg^{2+}	MgL 2.56	MgL 3.4	MgL 2.36			MgL 1.8				MgL 4.8
	Mg_4L_4 16.28	MgHL 11.49				$MgL_{2(s)}$ 8.2				MgHL 15.3
	$MgL_{2(s)}$ 11.16	$MgL_{(s)}$ 4.54								MgH_2L 20.0
		$MgL_{(s)}$ 7.45								$Mg_3L_{2(s)}$ 25.2
										$MgHL_{(s)}$ 18.2
Sr^{2+}		$SrL_{(s)}$ 9.0	SrL 2.6			$SrL_{2(s)}$ 8.5				SrL 5.5
			$SrL_{(s)}$ 6.5							SrHL 14.5
										SrH_2L 20.3
										$SrHL_{(s)}$ 19.3
Ba^{2+}		BaL 2.8	BaL 2.7			$BaL_{2(s)}$ 5.8				$BaHL_{(s)}$ 19.8
		$BaL_{(s)}$ 8.3	$BaL_{(s)}$ 10.0							

(continued on next page)

Table 8.3 *(Continued)*

log β Values	OH^-	CO_3^{2-}	SO_4^{2-}	Cl^-	Br^-	F^-	NH_3	CN^-	S^{2-}	PO_4^{3-}
Cr^{3+}	CrL 10.0		CrL 3.0	CrL 0.23		CrL 5.2				
	CrL_2 18.3					CrL_2 9.2				
	CrL_3 24.0					CrL_3 12.0				
	CrL_4 28.6									
	Cr_3L_4 47.8									
	$CrL_{3(s)}$ 30.0									
Al^{3+}	AlL 9.0									
	AlL_2 18.7					AlL 7.0				
	AlL_3 27.0					AlL_2 12.6				
	AlL_4 33.0					AlL_3 16.7				
	Al_3L_4 42.1					AlL_4 19.1				
	$AlL_{3(s)}$ 33.5									
Fe^{3+}	FeL 11.8		FeL 4.0	FeL 1.5	FeL 0.6	FeL 6.0		FeL_6 43.6		$FeHL$ 22.5
	FeL_2 22.3		FeL_2 5.4	FeL_2 2.1		FeL_2 10.6				FeH_2L 23.9
	FeL_4 34.4					FeL_3 13.7				$FeL_{(s)}$ 26.4
	Fe_2L_2 25.0									
	$FeL_{3(s)}$ 42.7									
	$FeL_{3(s)}$ 38.8									
Mn^{2+}	MnL 3.4	$MnHL$ 12.1	MnL 2.3	MnL 0.6		MnL 1.3	MnL 1.0		$MnL_{(s)}$ 10.5	
	MnL_2 5.8	$MnL_{(s)}$ 9.3					MnL_2 1.5		13.5	
	MnL_3 7.2									
	MnL_4 7.7									
	$MnL_{2(s)}$ 12.8									

log β Values	OH⁻		CO_3^{2-}		SO_4^{2-}		Cl^-		Br^-	F^-		NH_3		CN^-		S^{2-}		PO_4^{3-}	
Fe^{2+}	FeL	4.5	$FeL_{(s)}$	10.7	FeL	2.2				FeL	1.4			FeL_6	35.4	$FeL_{(s)}$	18.1	FeHL	16.0
	FeL_2	7.4																FeH_2L	22.3
	FeL_3	11.0																$Fe_3L_{2(s)}$	36.0
	$FeL_{2(s)}$	15.1																	
Co^{2+}	CoL	4.3	$CoL_{(s)}$	10.0	CoL	2.4	CoL	0.5		CoL	1.0	CoL	2.0			$CoL_{(s)}$	21.3	CoHL	15.5
	CoL_2	9.2										CoL_2	3.5			$CoL_{(s)}$	25.6		
	CoL_3	10.5										CoL_3	4.4						
	$CoL_{2(s)}$	15.7										CoL_4	5.0						
Ni^{2+}	NiL	4.1	$NiL_{(s)}$	6.9	NiL	2.3	NiL	0.6		NiL	1.1	NiL	2.7	NiL	7.3	$NiL_{(s)}$	19.4	NiHL	15.4
	NiL_2	9.0										NiL_2	4.9	NiL_4	30.2	$NiL_{(s)}$	24.9		
	NiL_3	12.0										NiL_3	6.6	NiH_2L_4	40.8	$NiL_{(s)}$	26.6		
	$NiL_{2(s)}$	17.2										NiL_4	7.7	$NiHL_4$	36.1				
												NiL_5	8.3						
Cu^{2+}	CuL	6.3	CuL	6.7	CuL	2.4	CuL	0.5		CuL	1.5	CuL	4.0	CuL		$CuL_{(s)}$	36.1	CuHL	16.5
	CuL_2	11.8	CuL_2	10.2	$Cu_4(OH)_6L_{(s)}$	68.6						CuL_2	7.5	CuL_2	16.3			CuH_2L	21.3
	CuL_4	16.4	$CuL_{(s)}$	9.6								CuL_3	10.3	CuL_3	21.6				
	Cu_2L_2	17.7	$Cu_2(OH)_2L_{(s)}$	33.8								CuL_4	11.8	CuL_4	23.1				
	$CuL_{2(s)}$	19.3	$Cu_3(OH)_2L_{2(s)}$	46.0															
	$CuL_{2(s)}$	20.4																	
Zn^{2+}	ZnL	5.0	$ZnL_{(s)}$	10.0	ZnL	2.1	ZnL	0.4		ZnL	1.2	ZnL	2.2	ZnL	5.7	ZnL	16.6	ZnHL	15.7
	ZnL_2	11.1			ZnL_2	3.1	ZnL_2	0.2				ZnL_2	4.5	ZnL_2	11.1	$ZnL_{(s)}$	24.7	ZnH_2L	21.2
	ZnL_3	13.6					ZnL_3	0.5				ZnL_3	6.9	ZnL_3	16.1			Zn_3L_2	35.3
	ZnL_4	14.8					$Zn_2(OH)_3L_{(s)}$	26.8				ZnL_4	8.9	ZnL_4	19.6				
	$ZnL_{2(s)}$	15.5												$ZnL_{2(s)}$	15.9				
	$ZnL_{2(s)}$	16.8																	

(continued on next page)

Table 8.3 *(Continued)*

log β Values	OH^-	CO_3^{2-}	SO_4^{2-}	Cl^-	Br^-	F^-	NH_3	CN^-	S^{2-}	PO_4^{3-}
Pb^{2+}	PbL 6.3	PbL 6.3	PbL 2.8	PbL 1.6	PbL 1.8	PbL 2.0			$PbL_{(s)}$ 27.5	$PbHL$ 15.5
	PbL_2 10.9	$PbL_{(s)}$ 13.1	$PbL_{(s)}$ 7.8	PbL_2 1.8	PbL_2 2.6	PbL_2 3.4				PbH_2L 21.1
	PbL_3 13.9			PbL_3 1.7	PbL_3 3.0	$PbL_{2(s)}$ 7.4				$Pb_3L_{2(s)}$ 43.5
	$PbL_{2(s)}$ 15.3			PbL_4 1.4	$PbL_{2(s)}$ 5.7					$PbHL_{(s)}$ 23.8
				$PbL_{2(s)}$ 4.8						
Hg^{2+}	HgL 10.6	$HgL_{(s)}$ 16.1	HgL 2.5	HgL 7.2	HgL 9.6	HgL 1.6	HgL 8.8	HgL 17.0	HgL 7.9	
	HgL_2 21.8		HgL_2 3.6	HgL_2 14.0	HgL_2 18.0		HgL_2 17.4	HgL_2 32.8	HgL_2 14.3	
	HgL_3 20.9			HgL_3 15.1	HgL_3 20.3		HgL_3 18.4	HgL_3 36.3	$HgOHL$ 18.5	
	$HgL_{2(s)}$ 25.4			HgL_4 15.4	HgL_4 21.6		HgL_4 19.1	HgL_4 39.0	$HgL_{(s)}$ 52.7	
				$HgOHL$ 18.1	$HgL_{2(s)}$ 19.8			$HgOHL$ 29.6	$HgL_{(s)}$ 53.3	
Cd^{2+}	CdL 3.9	$CdL_{(s)}$ 13.7	CdL 2.3	CdL 2.0	CdL 2.1	CdL 1.0	CdL 2.6	CdL 6.0	CdL 19.5	
	CdL_2 7.6		CdL_2 3.2	CdL_2 2.6	CdL_2 3.0	CdL_2 1.4	CdL_2 4.6	CdL_2 11.1	$CdHL$ 22.1	
	$CdL_{2(s)}$ 14.3		CdL_3 2.7	CdL_3 2.4			CdL_3 5.9	CdL_3 15.7	CdH_2L_2 43.2	
				CdL_4 1.7			CdL_4 6.7	CdL_4 17.9	CdH_3L_3 59.0	
									CdH_4L_4 75.1	
									$CdL_{(s)}$ 27.0	
Ag^+	AgL 2.0	$Ag_2L_{(s)}$ 11.1	AgL 1.3	AgL 3.3	AgL 4.7	AgL 0.4	AgL 3.3		AgL 19.2	$Ag_3L_{(s)}$ 17.6
	AgL_2 4.0		$Ag_2L_{(s)}$ 4.8	AgL_2 5.3	AgL_2 6.9		AgL_2 7.2	AgL_2 20.5	$AgHL$ 27.7	
	$AgL_{(s)}$ 7.7			AgL_3 6.4	AgL_3 8.7			AgL_3 21.4	$AgHL_2$ 35.8	
				$AgL_{(s)}$ 9.7	AgL_4 9.0			$AgOHL$ 13.2	AgH_2L_2 45.7	
					$AgL_{(s)}$ 12.3			$AgL_{(s)}$ 15.7	$Ag_2L_{(s)}$ 50.1	

Source: Modified from Morel and Hering.[24]

sediments (Figure 8.5). Under aerobic conditions at pH 7, these trace metals have solubilities in excess of 100 μg L^{-1} based on their hydroxides. In this case, the role of particles (adsorption) and complexation with other ligands (such as natural organic matter) are important in controlling the dissolved phase concentration and toxicity. Cu(II) and Ag(I) exceed water quality standards most often.

Example 8.1 Speciation of Copper(II) with Inorganic Ligands

Copper(II) is toxic in natural waters. To most aquatic biota, it is the free aquo copper ion Cu^{2+} that is the most toxic. Use a chemical equilibrium model to determine the speciation of dissolved Cu(II) in a fresh water with the following conditions: TOT Cu = 1×10^{-7} M, TOT Cl^- = 7×10^{-4} M, TOT SO_4^{2-} = 2×10^{-4} M, and TOT CO_3^{2-} = 3×10^{-3} M. Use the complexation constants given below, which have been corrected for these conditions. Vary the pH from 2 to 8. Assume a closed system (no CO_2 gas exchange).

Solution: First we must decide on the species and the components. By examining Table 8.3, we determine that the carbonate, sulfate, and hydroxy complexes may be important.

Species: Cu^{2+}, Cl^-, $CuCl^+$, CO_3^{2-}, HCO_3^-, H_2CO_3, $CuCO_3^0$, $Cu(CO_3)_2^{2-}$, SO_4^{2-}, HSO_4^-, $CuSO_4^0$, $CuOH^+$, $Cu(OH)_2^0$, OH^- and H^+ (15 species)

Components: Cu^{2+}, Cl^-, CO_3^{2-}, SO_4^{2-}, H^+ (5 components)

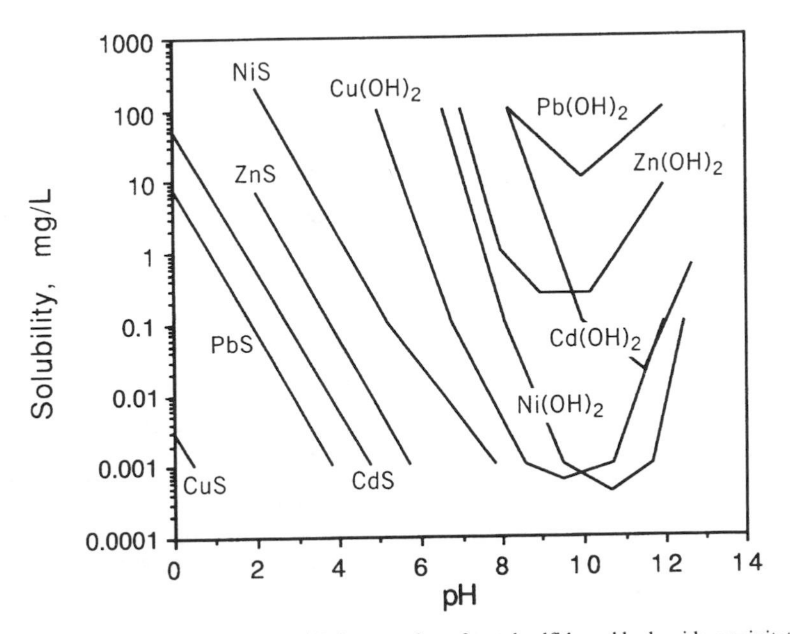

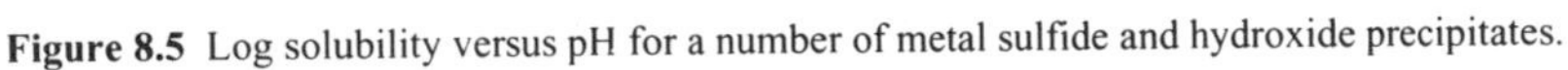
Figure 8.5 Log solubility versus pH for a number of metal sulfide and hydroxide precipitates.

The components are determined in large part as the metal ion plus each ligand for which a mass balance (total concentration) has been specified. Chapter 4 gives details on how to set up the matrix needed to solve this problem (see Examples 4.7 and 4.8).

The matrix is given by the rows of species and the columns of components related by their mass law equations. Note that the log *K* values for $CuOH^+$ and $Cu(OH)_2$ were rearranged and written in terms of H^+ ions rather than OH^- ions (if constants from Table 8.3 are to be used).

	Components					
Species	Cu^{2+}	Cl^-	CO_3^{2-}	SO_4^{2-}	H^+	log *K*
Cu^{2+}	1	0	0	0	0	0
Cl^-	0	1	0	0	0	0
$CuCl^+$	1	1	0	0	0	0.221
CO_3^{2-}	0	0	1	0	0	0
HCO_3^-	0	0	1	0	1	10.15
H_2CO_3	0	0	1	0	2	16.59
$CuCO_3$	1	0	1	0	0	6.39
$Cu(CO_3)_2^{2-}$	1	0	2	0	0	9.56
SO_4^{2-}	0	0	0	1	0	0
HSO_4^-	0	0	0	1	1	1.81
$CuSO_4^0$	1	0	0	1	0	2.00
$CuOH^+$	1	0	0	0	−1	−7.79
$Cu(OH)_2$	1	0	0	0	−2	−15.29
OH^-	0	0	0	0	−1	−14
H^+	0	0	0	0	1	0
TOT	1×10^{-7}	7×10^{-4}	3×10^{-3}	2×10^{-4}		

Results of MacμQL are given below. At pH 2, almost all of the Cu(II) is as free aquo metal ion, Cu^{2+}. At pH 6.6 and above, $CuCO_{3(aq)}$ becomes the dominant species. Carbonate complexes are very important for copper complexation in natural waters with significant alkalinity. In seawater, $CuSO_4^0$ and $CuCl^+$ also become important.

Solution Speciation

pH	Cu^{2+} Log *C*	$CuCl^+$ Log *C*	$CuCO_3^0$ Log *C*	$Cu(CO_3)_2^{2-}$ Log *C*	$CuSO_4^0$ Log *C*	$CuOH^+$ Log *C*	$Cu(OH)_2^0$ Log *C*
2	−7.00575	−9.93965	−15.7286	−27.6715	−8.92106	−12.7957	−18.2957
3	−7.00859	−9.94249	−13.7316	−23.6757	−8.37473	−11.7986	−16.2986
4	−7.00912	−9.94302	−11.7336	−19.678	−8.71089	−10.7991	−14.2991
5	−7.01056	−9.94446	−9.74892	−15.7073	−8.70981	−9.80056	−12.3006

(continued)

Solution Speciation

pH	Cu^{2+} Log C	$CuCl^{+}$ Log C	$CuCO_3^0$ Log C	$Cu(CO_3)_2^{2-}$ Log C	$CuSO_4^0$ Log C	$CuOH^{+}$ Log C	$Cu(OH)_2^0$ Log C
6	–7.07069	–10.0046	–7.9281	–12.0055	–8.76969	–8.86069	–10.3607
7	–7.72672	–10.6606	–7.1155	–9.72429	–9.42569	–8.51672	–9.01672
8	–8.7759	–11.7098	–7.07357	–8.59124	–10.4749	–8.5659	–8.0659

8.3.2 Complexation by Humic Substances

Humic substances comprise most naturally occurring organic matter (NOM) in soils, surface waters, and groundwater. They are the decomposition products of plant tissue by microbes. Humic substances are separated from a water sample by adjusting the pH to 2.0 and capturing the humic materials on an XAD-8 resin; subsequent elution in dilute NaOH is used to recover the humic substances from the resin. In nature, soils are leached by precipitation, and runoff carries a certain amount of humic and fulvic compounds into the water.

$$\text{Plant residues} \rightarrow \text{Humic compounds} \rightarrow \text{Fulvic compounds} \rightarrow CO_2 \qquad (28)$$

The term fulvic acids (FA) refers to a mixture of natural organics with MW $\simeq$ 200–5000, the fraction that is soluble in acid at pH 2 and also soluble in alcohol. The median molecular weight of FA is ~500 Daltons. It consists primarily of polyphenols and benzene carboxylic acids with 2–20 meq per gram of carbon (meq g^{-1} C) base-titrateable groups. The fulvic fraction (FF) includes carbohydrates, fulvic acids, and polysaccharides.

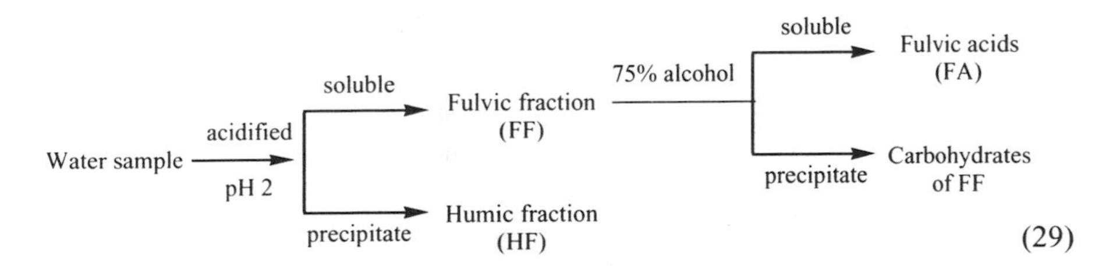

(29)

Buffle[25] has shown that fulvic acids, in particular, are good complexing agents for trace metals in natural waters. The same total concentration of trace metals, one with fulvic acid present (a colored water) and one without, often give remarkably different measures of toxicity. Fulvic acid-metal complexes are presumably too large to pass through organism membranes without dissociating. After dissociation, the free aquo metal ion or a metal–porter complex ("porter" refers to an active transporting agent) is transported into the cell. Thermodynamically, the transport process is less favorable if the metal ion is initially complexed by fulvic acid functional groups (phenolic, carboxylic, and quinones). Thus chemical equilibrium modeling should take fulvic acid complexation into account, especially for metals with large log K values for metal–fulvate (M–FA) complexes such as mercury, copper, lead, and cadmium. Table 8.4 is a summary of metal–fulvic acid (M–FA) complexation constants.

One approach is to use conditional stability constants (such as Table 8.4) for the natural water that is being modeled

$$M + FA \rightleftharpoons M\text{–}FA \qquad {}^{c}K = \frac{[M\text{–}FA]}{[M]\,[FA]} \tag{30}$$

where ${}^{c}K$ is the conditional stability constant at a specified pH, ionic strength, and chemical composition. One does not always know the stoichiometry of the reaction in natural waters, whether 1:1 complexes or 1:2 complexes (M/FA) are formed, for example. Also, the complexing agent is not homogeneous—we measure the organic complexant in units of mol L^{-1} of sites or mmol g^{-1} C of dissolved organic carbon (DOC). DOC, the complexant, is a macromolecule of organic carbon in which there are many different sites with varying binding energies. So equation (30) is simply a lumped-parameter conditional stability constant, but it does allow intercomparisons of the relative importance of complexation among different metal ions and DOC at specified conditions.

The conditional complexation constant (stability constant) should be measured for each site. This is most easily accomplished by ion selective electrodes (ISEs), anodic stripping voltammetry (ASV), or adsorption of the complex onto XAD resin. Total dissolved concentrations of metals (including M–FA) are measured using atomic absorption spectroscopy or inductively coupled plasma emission spectroscopy, and the aquo free metal ion is measured by ISEs or ASV. Metals can be preconcentrated using XAD resin or similar commercial preparations. In the laboratory, titration curves can be measured of total metal titrated (on the x-axis) versus aquo free metal ion (on the y-axis) to estimate the complexation constant. Often, one complexation constant is not sufficient to model the entire pH range.

Table 8.4 Freshwater Conditional Stability Constants for Formation of M–FA Complexes[25] (Roughly in Order of Binding Strength)

Metal Ion	DOC, mg L^{-1}	I, M	pH Range	log ${}^{c}K$ Range
Hg^{2+}	5–20	0.01–0.10	5–8	5–20
Cu^{2+}	5–20	0.01–0.04	5–8	4–10[b]
Pb^{2+}	5–100[a]	0.02–0.10	5–7	4.7–6.1
Cd^{2+}	5–300[a]	0.01–0.10	4–6	3–6
Ni^{2+}	70–200[a]	0.02	8	5.1
Co^{2+}	70–200[a]	0.01–0.10	5–8	4–5
Zn^{2+}	70–200[a]	0.01–0.10	5–8	2–5

[a]Fractionated from whole water sample to increase the concentration of DOC and the accuracy of measurement.

[b]Xue and Sigg[26] have shown that algae exudates exhibit log ${}^{c}K = 15$ at pH 8

Conditional stability constants for metal–fulvic acid complexes vary in natural waters because[3]:

- There are a range of affinities for metal ions and protons in natural organic matter resulting in a range of stability constants and acidity constants.
- Conformational changes and changes in binding strength of M–FA complexes in natural organic matter result from electrostatic charges (ionic strength), differential and competitive cation binding, and, most of all, pH variations in water.

The stability constants given in Table 8.5 vary markedly with pH. In general, the lower the pH of the solution, the lower is the stability constant due to competition of the metal ion with protons for organic ligands. The equilibrium constant ${}^{c}K$ is defined by equation (30).

Best estimates of conditional stability constants at a specific pH in fresh water are on the order of: Hg–FA log ${}^{c}K = 5.1$ at pH 5, 15 at pH 8; Cu–FA log ${}^{c}K = 6.0$ at pH 6; Pb–FA log ${}^{c}K = 6.1$ at pH 6; Cd–FA log ${}^{c}K = 4.7$ at pH 6.5; Zn–FA log ${}^{c}K = 2.5$ at pH 5. More accurately, the spread in binding strength of functional groups of a fulvic acid can be represented by a continuous distribution of equilibrium constants.[25,28]

The total number of metal-titrateable groups is in the range of 0.1–5 meq per gram of carbon, DOC complexation capacity. Complexation of metals with fulvic acid is most significant for those ions that are appreciably complexed with CO_3^{2-} and OH^- (functional groups like those contained in fulvic acid that contains carboxylic acids and catechol with adjacent —OH sites). These include mercury, copper, and lead.

Cabaniss and Shuman[27] have demonstrated a second approach to modeling the complexation of dissolved fulvic acid and copper ions in natural water. Rather than use a conditional complexation constant or a distribution of discrete complexation constants, they defined a five-site model with five different ligands (sites), five separate stability constants, and five different acidity constants (Table 8.5). Many investigators have used a two site model representing fulvic acid functional groups

Table 8.5 Fitting Parameters for Five-Site Model

Site	Complexation Capacity, mol mg^{-1} C	log K	Reaction	Proton Dependence
1. L_a	5×10^{-6}	3.9	$Cu^{2+} + L_a \rightleftharpoons CuL_a$	0
2. L_b	1.9×10^{-7}	1.5	$Cu^{2+} + HL_b \rightleftharpoons CuL_b + H^+$	1
3. L_c	1.1×10^{-6}	−0.36	$Cu^{2+} + HL_c \rightleftharpoons CuL_c + H^+$	1
4. L_d	1.4×10^{-7}	−7.48	$Cu^{2+} + H_2L_d \rightleftharpoons CuL_d + 2\,H^+$	2
5. L_e	9.6×10^{-6}	−10.05	$Cu^{2+} + H_2L_e \rightleftharpoons CuL_e + 2\,H^+$	2

Source: Cabaniss and Shuman.[27]

with acidity constants $pK_{a1} \simeq 3.5$ and $pK_{a2} \simeq 5.0$, and metal complexation constants similar to those in Table 8.5.

Using only one conditional stability constant for Cu–FA, Figure 8.6 demonstrates the importance of including organics complexation in chemical equilibrium modeling for metals. Filson Creek is located in northernmost Minnesota on the edge of the Boundary Waters Canoe Area. It was historically a copper–nickel–zinc mining area, and the waters are high in metals concentrations. It is a highly colored stream due to drainage through wetlands with a dissolved organic carbon concentration of 10 mg L^{-1} DOC (approximately 4.6×10^{-5} M FA sites for metal binding), a $\log {}^cK = 6.2$ for Cu–FA, and a pH of 6.1. The MINTEQ[22] chemical equilibrium program was used to solve the equilibrium problem, and the distribution diagrams for chemical speciation are given in Figure 8.6. Ion chemistry was measured as the following dissolved concentrations in mg L^{-1} at pH 6.1: $Ca^{2+} = 2.1$, $Na^+ = 0.7$, $Mg^{2+} = 1.7$, $K^+ = 0.20$, $NH_4^+ = 0.025$, $SO_4^{2-} = 4.5$, $NO_3^- = 0.079$, $PO_4^{3-} = 0.95$, alkalinity = 2.59 as $CaCO_3$, Zn(II) = 0.006, Pb(II) = 0.1, Ni(II) = 0.4, Al(III) = 0.2, Mn(T) = 0.01, Fe(T) = 0.49, Cd(II) = 0.005, and $H_3AsO_4 = 0.076$ mg L^{-1}. The total dissolved copper concentration was 0.006 mg L^{-1} (6 μg L^{-1}). MINTEQ was used to find the percentage of each species of Cu(II) at pH increments of 0.5 units. An acidity constant for HFA of $\log K = -4.5$ was utilized. (A more accurate representation of the system would be to vary the $\log {}^cK$ for Cu–FA with pH; lower pH waters would have lower $\log {}^cK$ stability constants for Cu–FA.)

Results in Figure 8.6a show that at the pH of the sample (pH 6.1) and below, the copper is almost totally as free aquo metal ion (hypothetically, in the absence of FA complexing agent). But Figure 8.6b shows that, in reality, almost all of the dissolved Cu(II) is present as the Cu–FA complex. Only 2% of the copper is present as the free aquo metal ion (~ 0.12 μg L^{-1} or 1.9×10^{-9} M). This is 100 times below the chronic threshold water quality criterion (based on toxicity) of 12 μg L^{-1}.

Results for cadmium are more problematic, however. In Figure 8.7a, assuming an absence of FA complexing agent, the free aquo cadmium concentration would be nearly 5 μg L^{-1} at pH 6.1. The chronic threshold water quality criterion for fresh water is 1.1 μg L^{-1} (see Table 1.1). But in this case, the stability constant between Cd(II) and fulvic acid is not as great as for Cu(II). Thus the free aquo Cd^{2+} concentration is modeled to be 4.3 μg L^{-1} (3.9×10^{-8} M) with fulvic acid present, about 87% of the total dissolved cadmium concentration at pH 6.1 (Figure 8.7b). This is almost four times greater than the chronic toxicity threshold as specified in the water quality criterion for cadmium. There exists a population of small fish (yellow perch and blue gill) in Filson Creek, but it is not known whether the population is stressed by metals toxicity.

Example 8.2 Lead Chemical Speciation

A chemical equilibrium program with a built-in database can be used to model the chemical speciation for Pb in natural waters. Lake Hilderbrand, Wisconsin, is a small seepage lake in northern Wisconsin with some drainage from a bog. It receives acid deposition from air emissions in the Midwest, and its chemistry reflects

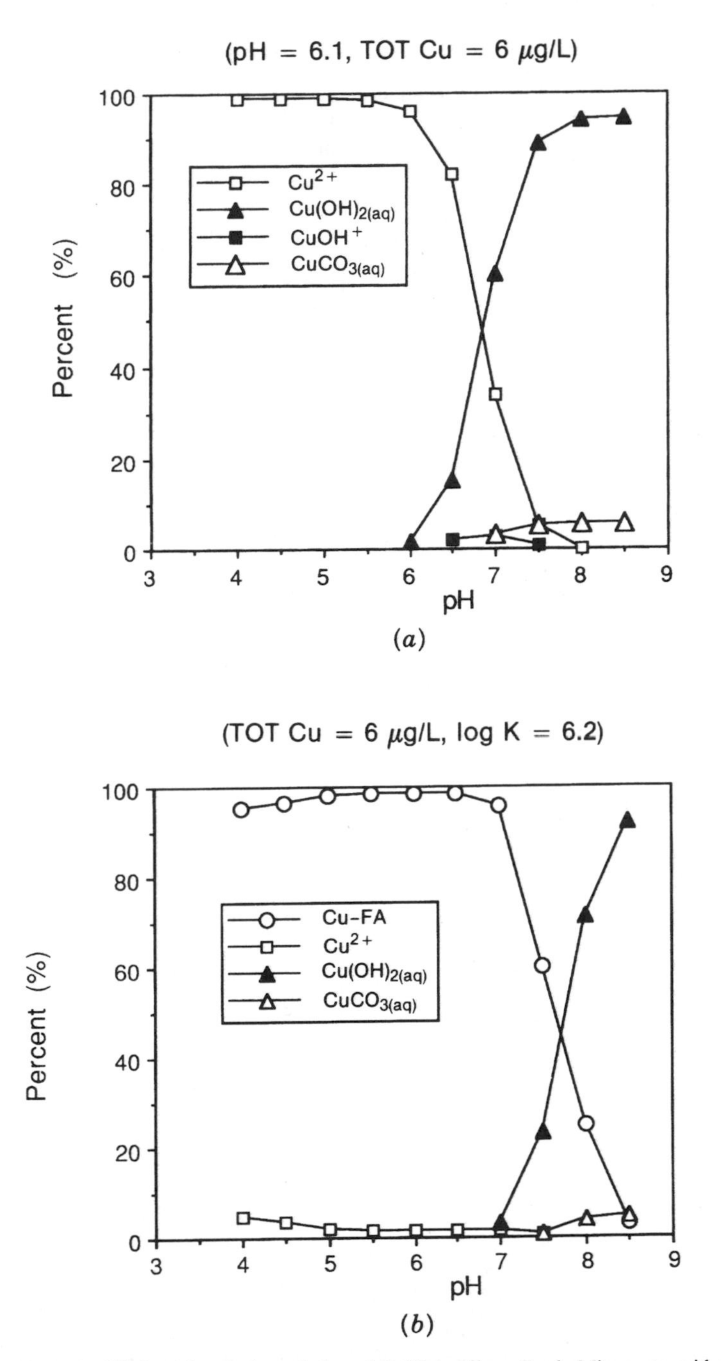

Figure 8.6 (a) Equilibrium chemical speciation of Cu(II) in Filson Creek, Minnesota, without consideration of complexation by dissolved organic carbon (fulvic acid). (b) Equilibrium chemical speciation of Cu(II) in Filson Creek, Minnesota, with copper–fulvic acid formation constant of log K = 6.2. [$Cu^{2+} + FA^{2-} \rightleftharpoons Cu\text{–}FA_{(aq)}$.]

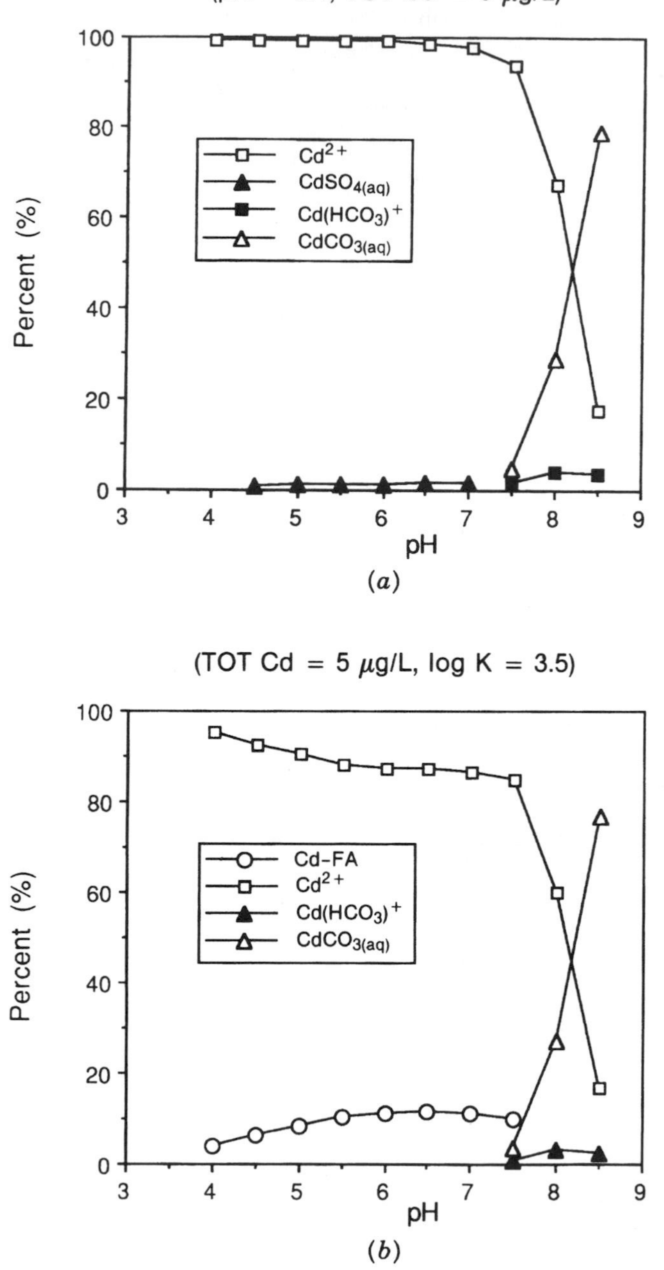

Figure 8.7. (a) Equilibrium chemical speciation of Cd(II) in Filson Creek, Minnesota, without consideration of complexation by dissolved organic carbon (fulvic acid). (b) Equilibrium chemical speciation of Cd(II) in Filson Creek, Minnesota, with cadmium–fulvic acid formation constant of log $K = 3.5$. [$Cd^{2+} + FA^{2-} \rightleftharpoons Cd\text{-}FA_{(aq)}$.]

acid precipitation, weathering/ion exchange with sediments, heavy metals deposition, and fulvic acids. It was probably never an alkaline lake because much of its water comes directly onto the surface of the lake via precipitation and from drainage of bogs; the current pH is 5.38. It has 12.0 units of color (Pt–Co units), and we can assume a FA concentration of 2.9×10^{-5} M. The log ^{c}K for Pb–FA in the system is in the range of 4–5. Use a chemical equilibrium model such as MINTEQA2 to solve for all the Pb species in solution from pH 4 to 8 with increments of 0.5 units. The temperature of the sample was 13.7 °C. Its ion chemistry (in mg L^{-1}) was the following: Ca^{2+} = 1.37, Na^{+} = 0.21, Mg^{2+} = 0.31, K^{+} = 0.64, ANC (titration alkalinity as $CaCO_3$) = 0.03, SO_4^{2-} = 5.25, Cl^{-} = 0.38, Cd(II) = 0.000037, and Pb(II) = 0.00035 mg L^{-1}. Is the Pb(II) expected to be present as the free aquo metal ion?

Solution: All the ions must be converted to mol L^{-1} concentration units. The ion balance should be checked; the sum of the cations minus the anions should be within 5% of the sum of the anions. At pH 5.38 some of the FA is anionic (fulvate), so it might need to be included as an anion in the charge balance. In this case, there is a good charge balance so the fulvic acid either (1) has an acidity constant near the pH of the solution and it does not contribute much to the charge, and/or (2) a portion of the Ca^{2+} is actually complexed by FA^{-}, which fortuitously causes the calculated $\Sigma(+)$ charges to be nearly equal to the $\Sigma(-)$ charges. The ion balance checks to within 2.4%.

H^{+}	4.2 μeq L^{-1}		
Ca^{2+}	68.4	SO_4^{2-}	109.4 μeq L^{-1}
Na^{+}	9.1	Cl^{-}	10.7
Mg^{2+}	25.5	Alk	0.6
K^{+}	16.4	FA^{-}	—
$\Sigma(+)$	123.6	$\Sigma(-)$	120.7

Next, we must decide what species and components are necessary to solve the problem as described in Chapter 4. At high pH we expect $PbCO_{3(aq)}$ and $PbOH^{+}$; at low pH we expect Pb^{2+} and possibly Pb–FA.

Species: Pb^{2+}, $PbOH^{+}$, $PbCO_3^0$, Pb–FA, CO_3^{2-}, HCO_3^{-}, H_2CO_3, HFA, FA^{-}, OH^{-}, H^{+}

Components: Pb^{2+}, HFA, CO_3^{2-}, H^{+}

The acidity constant for HFA is log K = –4.8.

We neglect the complex of $PbSO_4^0$ because the stability constant is rather small (Table 8.3). As we increase pH from 4 to 8, we can assume an open or a closed system with respect to $CO_{2(g)}$. In this case, a closed system was assumed.

Results for three different cases are given in Figure 8.8. The model is quite sensitive to the log ^{c}K that is assumed for the Pb–FA complex. Given a log ^{c}K = 5.0, most of the total lead that is present in solution is complexed with fulvic acid at pH 5.38.

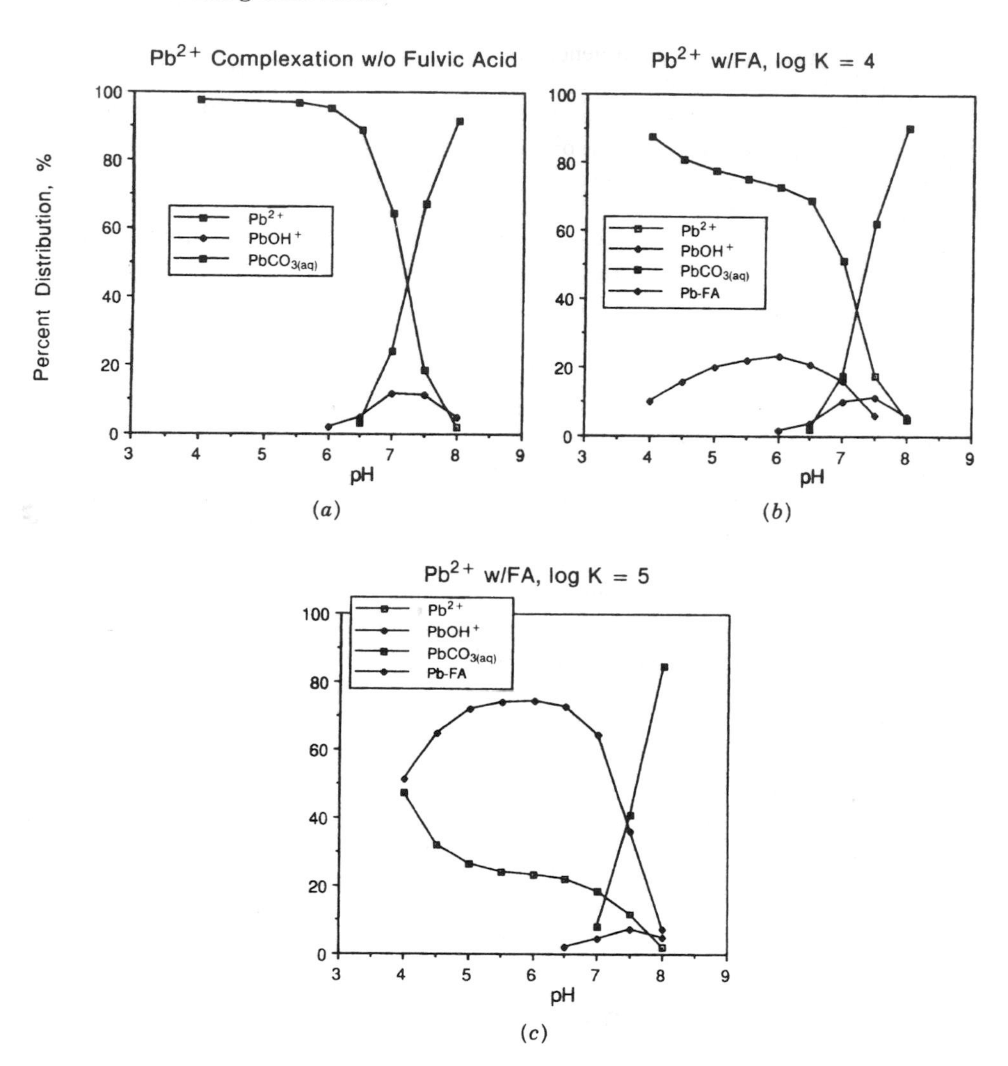

Figure 8.8 Lead speciation in Lake Hilderbrand, Wisconsin. (a) Without consideration of Pb–FA complexation. (b) With Pb–FA complexation, log $K = 4$. (c) With Pb–FA complexation, log $K = 5$

Schnitzer and Skinner[29] suggest an even stronger Pb–FA complex at pH 5 with log $^{c}K = 6.13$ in soil water. Most of the Pb(II) is apparently complexed with fulvate at pH 5.38 in Lake Hilderbrand. It is not considered toxic; the chronic water quality criterion for fresh water is 3.2 $\mu g\ L^{-1}$.

Example 8.2 brings up an important model limitation of fulvic acid binding for metal ions in natural waters. It is not clear how much competition exists among divalent cations. For example, the reaction between Ca–FA and Pb^{2+} in solution is certainly less favorable thermodynamically than FA with Pb^{2+}. Yet, these competition reactions (ligand exchange) are generally not considered in chemical equilibrium modeling.

$$\text{Ca–FA} + \text{Pb}^{2+} \rightleftharpoons \text{Pb–FA} + \text{Ca}^{2+} \tag{31}$$

The log ${}^{c}K$ for Al–FA is approximately 5.0 (quite a strong complex), and this competition was not considered in Example 8.2. Also, the affinity of Fe(III) for FA is almost equal to that of Pb–FA.[29] Thus, if a natural water has many divalent and trivalent metal ions, they all may need to be considered.

Fulvic acid binding of trace metals is complicated. Not only is FA a distribution of molecules from molecular weight 200 to 10,000, it has a variety of binding sites that are affected by pH, ionic strength, and the chemistry of ions in solution. Figure 8.9 shows the nature of those binding sites where adjacent carboxylic acid and phenolic groups are hydrogen bonded to form a polymeric structure with considerable stability. It is these same sites that bind trace metal ions. If fulvic acid is approximately 50% carbon and has an average of one site per 700 molecular weight, then we can estimate (crudely) a metals binding-site density of 1.43 meq g^{-1} C, as dissolved organic carbon. Binding site measurements and complexation titrations are needed to advance the field of chemical equilibrium modeling. In most cases, two or three types of sites with different stability constants and measurement of the major metal ions will be sufficient to estimate chemical speciation over a wide range of conditions. One operational stability constant can be utilized for specific conditions.

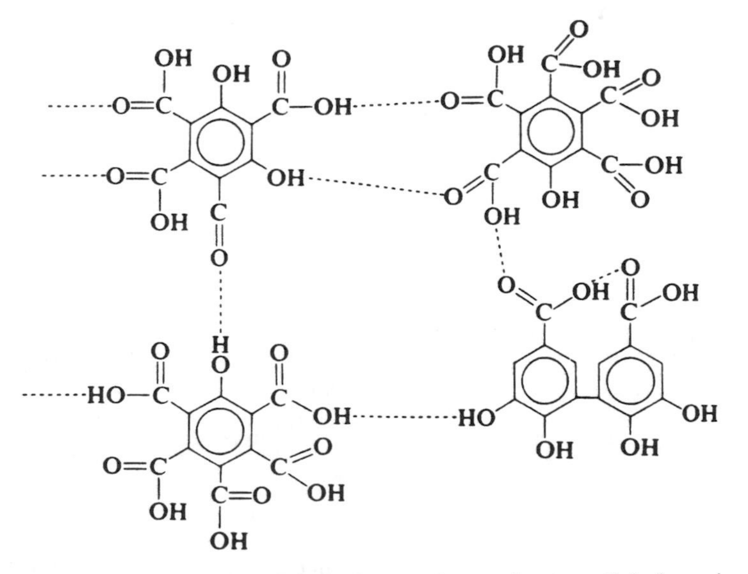

Figure 8.9 Typical structure of dissolved organic matter in natural waters with hydrogen bonding and bridging among functional groups that also make binding sites for metals.

8.4 SURFACE COMPLEXATION/ADSORPTION

8.4.1 Particle–Metal Interactions

In addition to aqueous phase complexation by inorganic and organic ligands, metal ions can form surface complexes with particles in natural waters. This phenomenon is variously named in scientific literature as surface complexation, ion exchange, adsorption, metal–particle binding, and metals scavenging by particles. Complexation at the surface of particles is analogous to aqueous phase complexation, and it is the subject of this section.

Particles in natural water are many and varied, including hydrous oxides, clay particles, organic detritus, and microorganisms. They range a wide gamut of sizes (Figure 8.10). Colloids less than 1 μm are generally stable in suspension and will not settle out in natural waters except following aggregation. These are often clay particles that have become very small by chemical and mechanical weathering processes. Suspended particles larger than 4 μm settle out easily, including some algae, calcite particles, hydrous oxides, and fine quartz.

Particles provide surfaces for complexation of metal ions, acid–base reactions, and anion sorption.

Metal ions:
$$SOH + M^{2+} \rightleftharpoons SOM^{+} + H^{+} \tag{32}$$
$$(SOH)_2 + M^{2+} \rightleftharpoons (SO)_2M + 2H^{+} \tag{33}$$

Acid–base:
$$SOH \rightleftharpoons SO^{-} + H^{+} \tag{34}$$
$$SOH + H^{+} \rightleftharpoons SOH_2^{+} \tag{35}$$

Organic acids:
$$SOH + H_2A \rightleftharpoons SOAH^{2-} + 2\,H^{+} \tag{36}$$

Anions:
$$SOH + A^{2-} \rightleftharpoons SA^{-} + OH^{-} \tag{37}$$
$$(SOH)_2 + A^{2-} \rightleftharpoons S_2A + 2\,OH^{-} \tag{38}$$

where SOH represents hydrous oxide surface sites, M^{2+} is the metal ion, and A^{2-} is the anion.

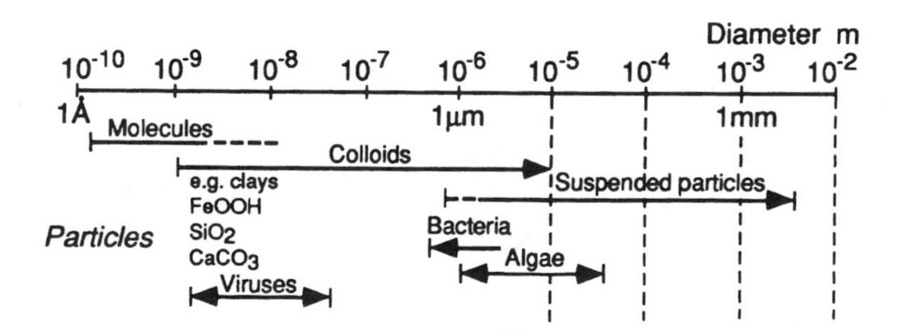

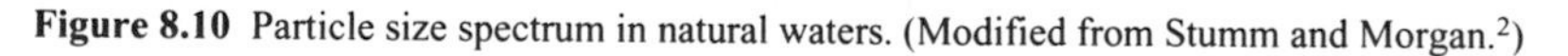

Figure 8.10 Particle size spectrum in natural waters. (Modified from Stumm and Morgan.[2])

Figure 8.11 shows the "sorption edges" for three hypothetical metal cations forming surface complexes on hydrous oxides (e.g., amorphous iron oxides $FeOOH_{(s)}$ or aluminum hydroxide $Al(OH)_{3(s)}$). Equations (32) and (33) indicate that the higher is the pH of the solution, the greater the reactions will proceed to the right.

The adsorption edge culminates in nearly 100% of the metal ion bound to the surface of the particles at high pH. At exactly what pH that occurs depends on the acidity and basicity constants for the surface sites, equations (34) and (35), and the strength of the surface complexation reaction, equations (32) and (33).

Similarly, anions can form surface complexes on hydrous oxides (Figure 8.11). Most particles in natural waters are negatively charged at neutral pH. At low pH, the surfaces become positively charged and anion adsorption is facilitated, equations (37) and (38). Weak acids (such as organic acids, fulvic acids) typically have a maximum sorption near the pH of their first acidity constant ($pH = pK_a$), a consequence of the equilibrium equations (34)–(36).

In Figure 8.12, sorption edges for surface complexation of various metals on hydrous ferric oxide (HFO) are depicted.[30,31] Metal ions on the left of the diagram form the strongest surface complexes (most stable) and those on the right at high pH form the weakest. Of course, the order may change in natural waters depending on the aqueous phase complexants (ligands) and the nature of the particles. Because the surface coordinating ligands are —OH groups, the sequence of metal binding strength on hydrous oxides (Figure 8.12) is related to the complex formation constants with OH^- in Table 8.3: $Pb^{2+} > Cd^{2+} > Zn^{2+} > Ni^{2+}$. Particles including hydrous

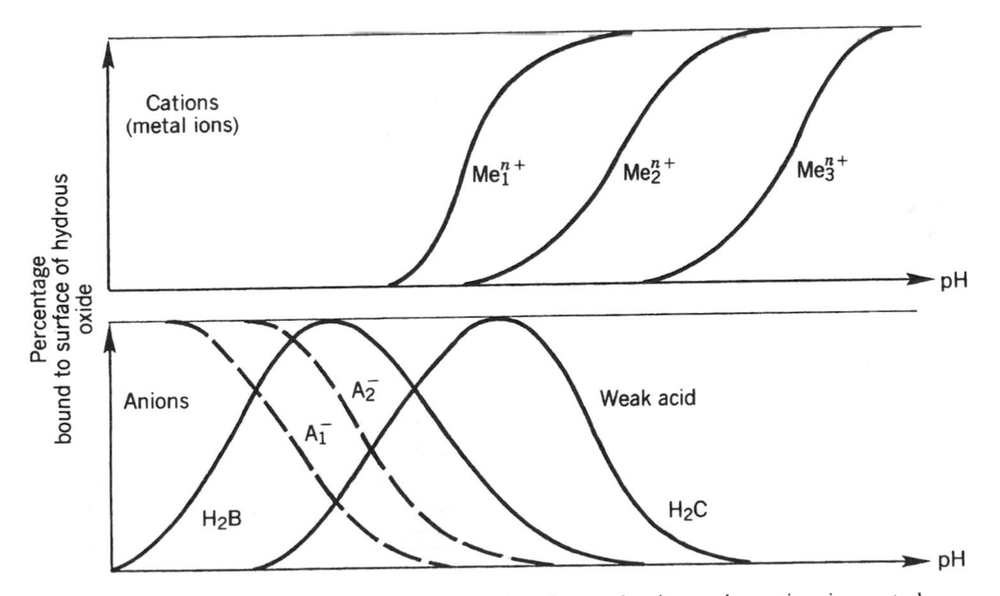

Figure 8.11 Surface complexation and sorption edges for metal cations and organic anions onto hydrous oxide surfaces.

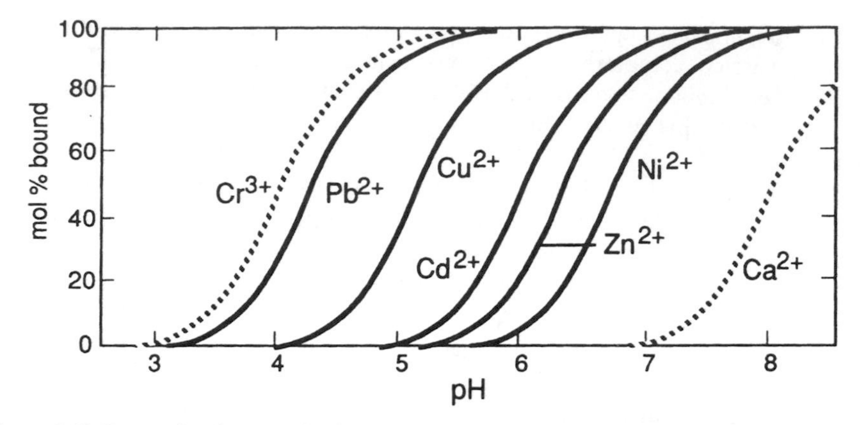

Figure 8.12 Extent of surface complex formation as a function of pH (measured as mol % of the metal ions in the system adsorbed or surface bound). [TOT Fe] = 10^{-3} M (2×10^{-4} mol reactive sites L^{-1}); metal concentrations in solution = 5×10^{-7} M; $I = 0.1$ M $NaNO_3$. (Modified from Stumm[30]; the curves are based on data compiled by Dzombak and Morel.[31])

ferric oxides ($FeOOH_{(s)}$) exert a large influence in controlling trace metal concentrations in natural waters. $FeOOH_{(s)}$ is ubiquitous in the environment, and it can form precipitates on the surface of other particles creating many surface sites for metals scavenging, disproportionate to the fraction of elemental iron that is present in the suspended sediment (Figure 8.13).

8.4.2 Natural Waters

Particles from lakes and rivers exert control on the total concentration of trace metals in the water column. The adsorption–sedimentation process scavenges metals from water and carries them to the sediments. Müller and Sigg[32] showed that particles in the Glatt River, Switzerland, adsorbed more than 30% of Pb(II) in whole water samples (Figure 8.14). Also, more than half of the Pb(II) was present as $PbCO_{3(aq)}$, a common occurrence in natural waters with high alkalinity.

It is interesting that small quantities of chelates in natural waters and groundwater can exert a significant influence on trace metal speciation. At concentrations typically measured in rivers receiving municipal wastewater discharges (on the order of 10^{-8} to 10^{-7} M for EDTA and NTA), chelation of Pb(II) accounts for more than half of dissolved lead (Figure 8.14). Ethylene diamine tetraacetic acid (EDTA) and nitrilotriacetic acid (NTA) are chelates that are used in foods, pharmaceuticals, cosmetics, water softening, and detergents as stabilizers. They form stable chelate complexes (with multidentate ligands) on trace metal ions, rendering them mobile and nonreactive. In natural waters, chelates such as EDTA are biodegraded or photolyzed slowly, and steady-state concentrations can be sufficient to affect trace metal speciation.

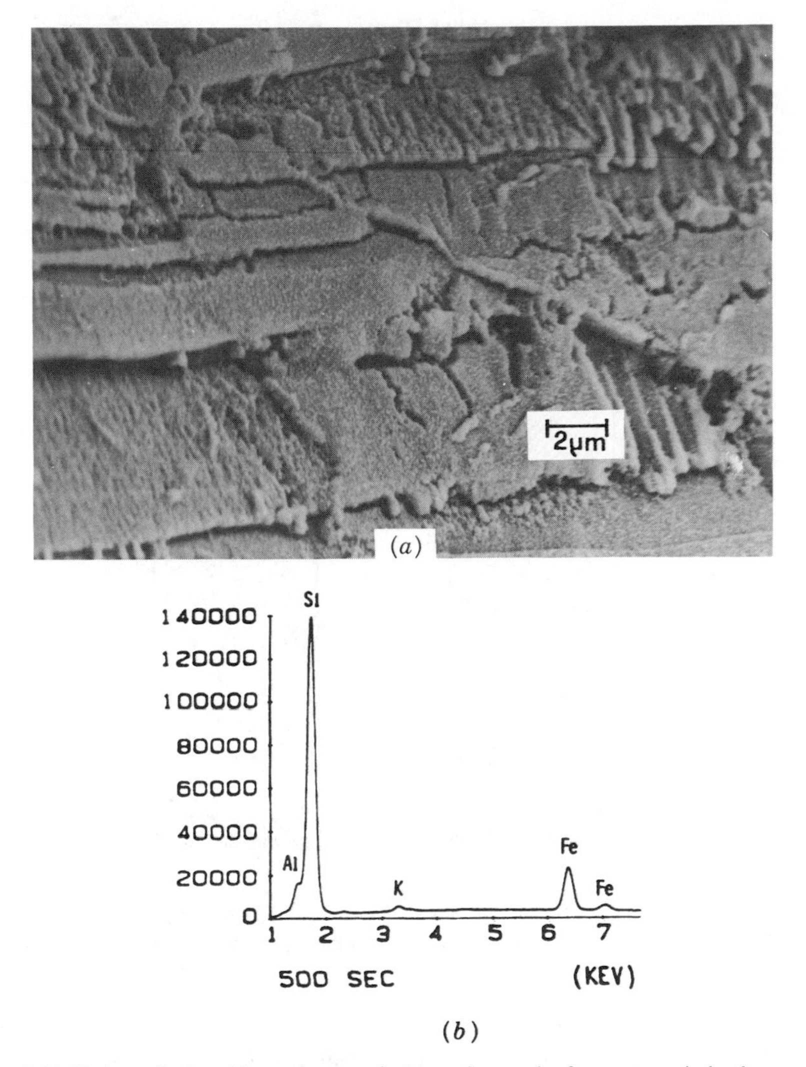

Figure 8.13 Hydrous ferric oxide coating seen in (a) a micrograph of a quartz grain by the scanning electron microscope and (b) EDS spectra of the surface chemistry showing Fe peaks and underlying quartz grain. (Micrograph courtesy of Prof. Rudolph Giovanoli, University of Berne.)

Example 8.3 Pb(II) Complexation by NTA and EDTA

Use a chemical equilibrium model to solve for all species in solution for the natural water depicted in Figure 8.14b. EDTA concentration is 7×10^{-8} M; NTA is 4×10^{-8} M; and Pb(II) is 1.8×10^{-8} M at pH 8. To simplify, we will ignore the carbonate complexes. (See Müller and Sigg.[32])

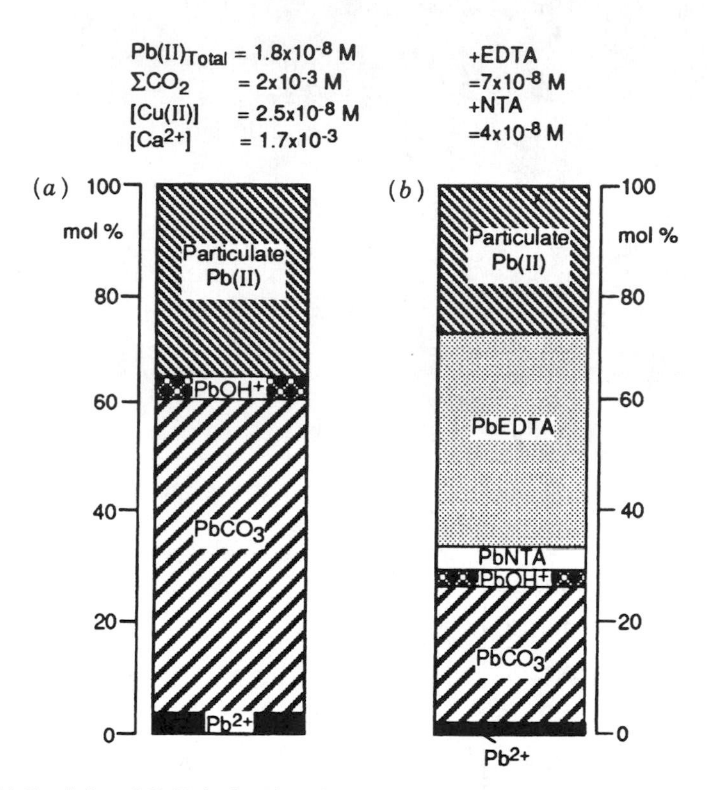

Figure 8.14 Speciation of Pb(II) in the Glatt River, Switzerland, including (a) adsorption onto particulates and (b) complexation by EDTA and NTA chelates, which are measured in the environment. (From Müller and Sigg.[32])

Solution: Make a matrix of all the pertinent equilibrium reactions and solve using a chemical equilibrium model. Morel and Hering[24] provide a compilation of stability constants for EDTA and NTA with trace metals including lead. EDTA is tetraprotic, H_4EDTA. The first two stability constants are sufficient because concentrations become very low after that at pH 8.

$$Pb^{2+} + H_4EDTA \rightleftharpoons PbEDTA^{2-} + 4\,H^+ \qquad \log K = -3.96$$

$$Pb^{2+} + H_4EDTA \rightleftharpoons PbHEDTA^{-} + 3\,H^+ \qquad \log K = -0.76$$

NTA is triprotic. The first stability constant is sufficient.

$$Pb^{2+} + NTA^{3-} \rightleftharpoons PbNTA^{-} \qquad \log K = 12.6$$

Each species must be written in terms of the components. We must choose one species of lead, one species of EDTA, one species of NTA, and H^+ as components.

Components: Pb^{2+}, H_4EDTA, NTA^{3-}, H^+

Of course, we could have chosen other species of EDTA and NTA as components as long as the equilibrium expressions are written for each species in terms of its components. Once the following matrix is assembled, the problem is solved on a computer with a Newton–Raphson numerical technique.

	Pb^{2+} Total	H_4EDTA Total	NTA^{3-} Total	H^+ Total	{log K}
Pb^{2+}	1	0	0	0	0
$PbOH^+$	1	0	0	–1	–7.7
$Pb(OH)_2$	1	0	0	–2	–17.1
$Pb(OH)_3^-$	1	0	0	–3	–28.1
$PbNTA^-$	1	0	1	0	12.6
$PbEDTA^{2-}$	1	1	0	–4	–3.96
$PbHEDTA^-$	1	1	0	–3	–0.76
$EDTA^{4-}$	0	1	0	–4	–23.76
$HEDTA^{3-}$	0	1	0	–3	–12.64
H_2EDTA^{2-}	0	1	0	–2	–3.66
H_3EDTA^-	0	1	0	–1	–2.72
H_4EDTA	0	1	0	0	0
NTA^{3-}	0	0	1	0	0
$HNTA^{2-}$	0	0	1	1	9.71
H_2NTA^-	0	0	1	2	12.19
H_3NTA	0	0	1	3	13.99
H_4NTA^+	0	0	1	4	14.79
OH^-	0	0	0	–1	–14
H^+	0	0	0	1	0
TOT (C_T)	1.8×10^{-8}	7×10^{-8}	4×10^{-8}	1×10^{-8}	

Example 8.4 Pb Binding on Hematite (Fe_2O_3) Particles

(This example is from Sigg and Stumm.[33]) Iron oxides such as hematite are important particles in rivers and sediments. Determine the speciation of Pb(II) adsorption onto hematite particles from pH 2 to 9 for the following conditions. The surface of the oxide becomes hydrous ferric oxide in water with —OH functional groups that coordinate Pb(II).

Specific surface area = 40,000 $m^2\ kg^{-1}$
Site density = 0.32 mol kg^{-1}
Hematite particle concentration = 8.6×10^{-6} kg L^{-1}

Ionic strength = 0.005 M
Inner capacitance = 2.9 F m^{-2} (double-layer model)

Solution: The relevant equations are the intrinsic acidity constants for the surface of the hydrous oxide and the surface complexation constant with Pb^{2+}.

$$\exists FeOH_2^+ \rightleftharpoons \exists FeOH + H^+ \qquad \log K_{a1}^{intr} = -7.25$$

$$\exists FeOH \rightleftharpoons FeO^- + H^+ \qquad \log K_{a2}^{intr} = -9.75$$

$$\exists FeOH + Pb^{2+} \rightleftharpoons \exists FeOPb^+ + H^+ \qquad \log K^{intr} = +4.0$$

The problem is similar to that of Example 4.9. Recall that the electrostatic effect of bringing a Pb^{2+} atom to the surface of the oxide must be considered (the coulombic correction factor). This is accounted for in the program by correcting the intrinsic surface equilibrium constants.

Species: $\exists FeOH$, $\exists FeOH_2^+$, $\exists FeO^-$, $\exists FeOPb^+$, Pb^{2+}, $PbOH^+$, H^+ (7 species)
Components: $\exists FeOH$, Pb^{2+}, surface charge, H^+ (4 components)
Surface charge = $\exists FeOH_2^+ + \exists FeOPb^+ - \exists FeO^- = \sigma_T$.

$$K^{app} = K^{intr} e^{-F\psi/RT}$$

The column in the matrix below, "surface charge" refers to the coulombic correction factor $\exp(-F\psi/RT)$, which multiplies the intrinsic equilibrium constant to obtain the apparent equilibrium constant in the Newton–Raphson solution. The total concentration of binding sites for the mass balance is the product of the site density and the hematite particle concentration, 2.75×10^{-6} M.

	$\exists FeOH$	Pb^{2+}	Surface Charge	H^+	{log K}
$\exists FeOH$	1	0	0	0	0
$\exists FeOH_2^+$	1	0	1	1	+7.25
$\exists FeO^-$	1	0	−1	−1	−9.75
$\exists FeOPb^+$	1	1	1	−1	4.0
Pb^{2+}	0	1	0	0	0
$PbOH^+$	0	1	0	−1	−7.7
H^+	0	0	0	1	0
TOT	2.75×10^{-6}	1×10^{-7}		1×10^{-7} (initial)	

The matrix can be used to solve the problem using a chemical equilibrium program. Results are shown in Figure 8.15. Interactive models such as MacμQL will prompt the user to provide information about the adsorption model (double-layer model in this case) and the inner capacitance. Note how sharp the "sorption edge" is in Figure 8.15 between pH 4 and 6. Above pH 6, almost all of the Pb(II) is adsorbed under these conditions. Lines in the pC versus pH plot of Figure 8.15 are "wavy" due to the coulombic correction factor.

Of course, the nature of the particles has a great effect on surface complexation constants for trace metals. Following are a few types of particles suspended in natural waters and their surface groups.

- Clays (—OH groups and H^+ exchange).
- $CaCO_3$ (carbonato complexes and coprecipitation).
- Algae + bacteria (binding and active transport for Cu, Zn, Cd, AsO_4^{3-}, ion exchange).
- FeOOH (oxygen donor atoms, high specific surface area).
- MnO_2 (oxygen donor atoms, high specific surface area).
- Organic detritus (phenolic and carboxylic groups).
- Quartz (oxide surface with coatings).

Quartz particles do not bind trace metals to the extent of FeOOH and MnO_2 particles, but quartz grains can be coated by iron oxides (Figure 8.13) or organic coatings. Organic coatings and algal exudates are strong complexants for Cu(II). $MnO_{2(s)}$ is a great surface for scavenging most trace metals, especially Cd(II) and Zn(II). Solid/water distribution coefficients (K_d values) vary depending on the particle type. In natural waters, the determination of *one* K_d value assumes that the composition of particles will remain constant, but particles are highly variable with season and due to storm events.

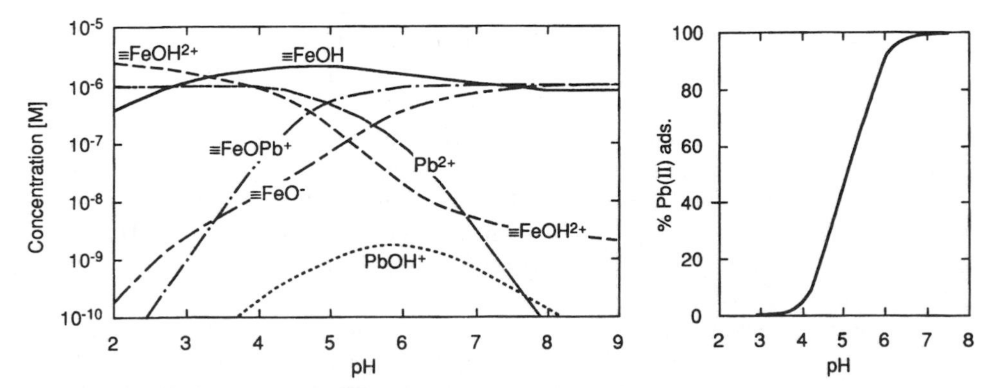

Figure 8.15 Chemical equilibrium modeling of Pb(II) on geothite with sorption edge from pH 4 to 6 (From Sigg and Stumm[33]).

8.4.3 Mercury as an Adsorbate

One of the most difficult metals to model is Hg(II) because it undergoes many complexation reactions, both in solution and at the particle surface, and it is active in redox reactions as well. Hg(II) forms strong complexes with Cl^- in seawater, OH^-, and organic ligands in fresh water. On surfaces, it seeks to coordinate with oxygen donor atoms in inner-sphere complexes. Tiffreau et al.[34] have used the diffuse double-layer model with two binding sites to describe the adsorption of Hg(II) onto quartz (α-SiO_2) and hydrous ferric oxide in the laboratory over a wide range of conditions.

It was necessary to invoke ternary complexes, that is, the complexation of three species together, in this case SOH, Hg^{2+}, and Cl^-. For hydrous ferric oxides (HFO), all parameters were taken from Dzombak and Morel[31] except the weak binding site density. The HFO was prepared in the laboratory, 89 g mol^{-1} Fe (298.15 K, $I = 0$ M).

Surface area = 600 m^2 g^{-1}
Site density$_1$ = 0.205 mol mol^{-1} Fe (weak binding sites)
Site density$_2$ = 0.029 mol mol^{-1} Fe (strong binding sites)
Acidity constant$_1$ $\exists FeOOH_2^+$ $pK_{a1}^{intr} = 7.29$
Acidity constant$_2$ $\exists FeOOH$ $pK_{a2}^{intr} = 8.93$

Quartz grains have a much lower surface area and are generally less important for sorption in natural waters.

Surface area = 4.15 m^2 g^{-1}
Site density = 4.5 sites nm^{-2}
Acidity constant$_1$ $\exists SiOH_2^+$ $pK_{a1}^{intr} = -0.95$
Acidity constant$_2$ $\exists SiOH$ $pK_{a2}^{intr} = 6.95$
Solids concentration = 40 g L^{-1} SiO_2

Results of the simulation for adsorption ($(Hg(II))_T = 0.184$ μM) onto quartz at $I = 0.1$ M $NaClO_4$ is shown in Figure 8.16. The laboratory data are from McNaughton[35] and statistically fitted with the nonlinear program FITEQL.[36] Equilibrium constants for the surface complexation reactions are given in Table 8.6.

Figure 8.16 shows that metal sorption can be "amphoteric" just like metal oxides and hydroxides. The pH of zero charge for the quartz is pH ~3. As the surface becomes more negatively charged above pH 3, the sorption edge for Hg(II) on α-SiO_2 is observed. However, at pH > 7, solute complexes $Hg(OH)_2$, $Hg(OH)_3^-$, and $Hg(OH)_4^{2-}$ become important, and the percent adsorption decreases once again (Figure 8.16).

Even in pure laboratory systems, modeling the sorption of trace metals on particles becomes complicated. That is why many investigators use a conditional equilibrium constant approach, that of the solids/water distribution coefficient, K_d. The approach is fine as long as the modeler recognizes its limitations and the specific conditions where it applies.

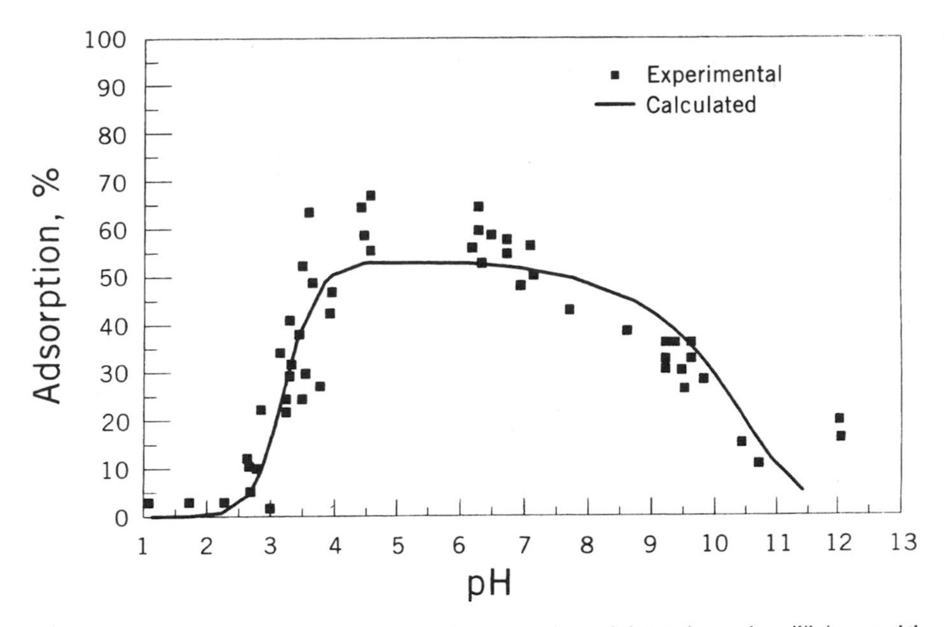

Figure 8.16 Adsorption of Hg(II) onto silica (SiO_2). Experimental data points and equilibrium model line are from Tiffreau et al.[34]

Table 8.6 Summary of Surface Reaction Constants for Hg(II) Adsorption onto Quartz Grains and Hydrous Ferric Oxide (HFO) (298.15 K, $I = 0$ M)

Constants	Reactions	log K^{int} (α-SiO_2)	log K^{int} (HFO)
$K^{int}_{\exists SOHg^+}$	$\exists$ S-OH^0 + Hg^{2+} $\Leftrightarrow$ S-OHg^+ + H^+	—	6.9 ± 0.2
$K^{int}_{\exists SOHgOH}$	$\exists$ S-OH^0 + Hg^{2+} + H_2O $\Leftrightarrow$ $\exists$ S-OHgOH + $2H^+$	–3.2 ± 0.1	–0.9 ± 0.2
$K^{int}_{\exists SOHgCl}$	$\exists$ S-OH^0 + Hg^{2+} + Cl^- $\Leftrightarrow$ $\exists$ S-OHgCl + H^+	7.0 ± 0.1	9.8 ± 0.4

8.5 STEADY-STATE MODEL FOR METALS IN LAKES

8.5.1 Introduction

Surface coordination of metals with particles in lakes may be described in the simplest case as

$$SOH + M^{2+} = SOM^+ + H^+ \tag{39}$$

$$K_{eq} = \frac{[SOM^+]\,[H^+]}{[SOH]\,[M^{2+}]} \tag{40}$$

where SOH are the binding sites on the solid hydrous oxide particle, M^{2+} is the divalent metal ion, SOM^+ is the adsorbed metal species, and K_{eq} is the equilibrium constant for the surface coordination reaction.

Traditionally, researchers have described metals adsorption in natural waters with an empirical distribution coefficient defined as the ratio of the adsorbed metal concentration to the dissolved metal fraction. (Dissolved fraction is operationally defined in most field studies as filtration through a 0.45-μm membrane filter; it may include some colloidal material.) For the case described by equation (39), the distribution coefficient would be defined as the following:

$$K_d = \frac{\text{adsorbed metal}}{\text{dissolved metal}} = \frac{[SOM^+]}{[SOH]\,[M^{2+}]} \tag{41}$$

where $[SOM^+]/[SOH]$ is expressed in μg kg^{-1} solids and M^{2+} is in μg L^{-1}. By combining equations (40) and (41), we see that K_d is inversely proportional to the H^+ ion concentration.

$$K_d = K'_{eq}/[H^+] \tag{42}$$

K_d is expressed in units of μg kg^{-1} adsorbed metal ion per μg L^{-1} dissolved metal ion (L kg^{-1}). The inverse linear dependence of K_d on H^+ concentration is a consequence of the way we formulated equation (39). If a binuclear complex was hypothesized, we would have produced a different result. Nevertheless, formulation of the problem in this manner produces a useful model, and K_d values in nature are often inversely proportional to $[H^+]$.

8.5.2 Steady-State Model

We can formulate a simple steady-state mass balance model to describe the input of metals to a lake.

$$V\frac{dC_T}{dt} = IAC_{T\,\text{precip}} + v_d AC_{T\,\text{air}} - k_s V f_p C_T - QC_T \tag{43}$$

where V = lake volume, L^3

C_T = total (unfiltered) metal concentration suspended in the lake, ML^{-3}

I = precipitation rate, LT^{-1}

$C_{T\,\text{precip}}$ = total metal concentration in precipitation, ML^{-3}

A = surface area of the lake, L^2

$C_{T\,\text{air}}$ = metal concentration in air, ML^{-3}

v_d = dry deposition velocity, LT^{-1}

k_s = sedimentation rate constant, T^{-1}

Q = outflow discharge rate, L^3T^{-1}

f_p = particulate fraction of total metal concentration in lake, dimensionless
$= K_d M/(1 + K_d M)$
M = suspended solids in the lake, ML^{-3}
K_d = solids/water distribution coefficient, L^3M^{-1}

Under steady-state conditions, equation (43) can be solved for the average total metals concentration (unfiltered) in the lake.

$$C_T = (IC_{T\ \text{precip}} + v_d C_{T\ \text{air}})/H(k_s f_p + 1/\tau) \tag{44}$$

where τ = mean hydraulic detention time of the lake = V/Q, T
H = mean depth of the lake, L

or

$$C_T = C_{T\ \text{in}}/(1 + k_s f_p \tau) \tag{45}$$

where $C_{T\ \text{in}}$ = total input concentration of wet plus dry deposition, ML^{-3}
$= (I\tau C_{T\ \text{precip}} + v_d \tau C_{T\ \text{air}})/H$

From equation (45), one can estimate the fraction of metal removed by the lake due to sedimentation.

$$R = 1 - (C_T/C_{T\ \text{in}}) = k_s f_p \tau/(1 + k_s f_p \tau) \tag{46}$$

Equation (46) can be used to estimate the "trap efficiency" of the lake for metals. It is entirely dependent on only two dimensionless parameters: $k_s\tau$, the sedimentation rate constant times the mean hydraulic detention time of the lake; and $K_d M$, the distribution coefficient times the suspended solids concentration. The dimensionless number $K_d M$ is involved in equation (46) through the particulate adsorbed fraction of metal, $f_p = K_d M/(1 + K_d M)$. Figure 8.17 shows that the fraction of metal that is removed from the lake by sedimentation increases with $K_d M$ and $k_s\tau$. Lakes with very short hydraulic detention times, τ, are not able to remove all of the adsorbed metal by sedimentation even if the $K_d M$ parameter is very large because the residence time is not long enough.

Model equation (46) provides a steady-state estimate of the fraction of metals that are adsorbed and sedimented to the bottom of the lake. To use the model, we must know the following parameters: k_s, τ, K_d, and the suspended solids M (Table 8.7). The sedimentation constant can be estimated from the mean settling velocity of particles divided by the mean depth ($k_s = v_s/H$).

Values for K_d vary considerably from lake to lake because the nature of particles differs and aqueous phase complexing agents may vary. Most investigators report a solid concentration-dependent K_d value as per O'Connor and Connolly.[37] K_d values decrease with increasing concentration of total suspended solids according to empirical relationships such as K_d = 250,000/mg L^{-1} TSS) for copper, zinc, cadmium,

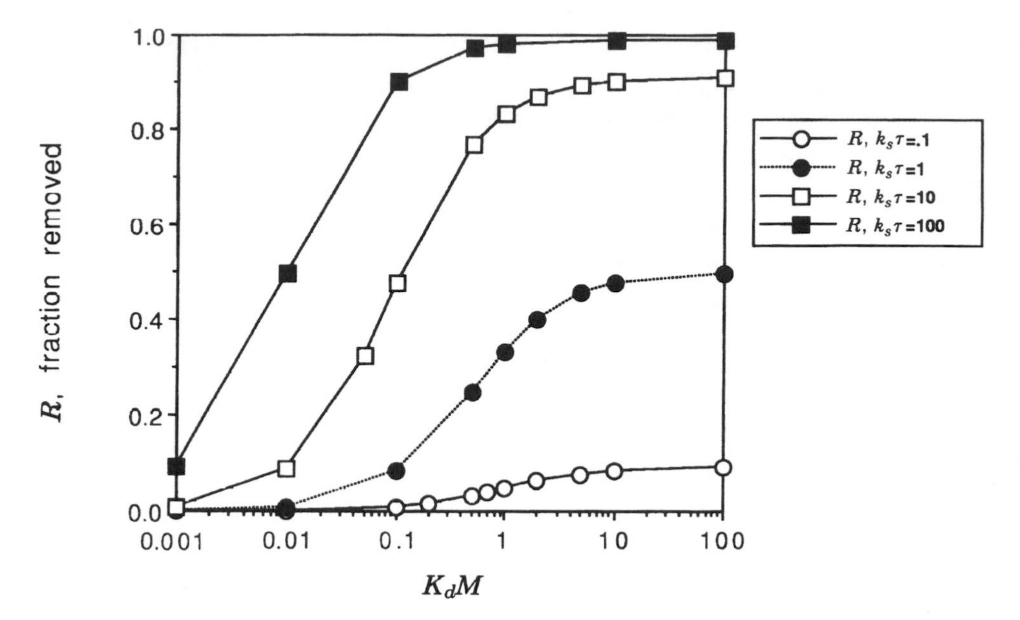

Figure 8.17 Effect of two dimensionless numbers on the fraction of metal inputs that are removed by sedimentation in lakes. K_dM is the distribution coefficient times the suspended solids concentration. $k_s\tau$ is the sedimentation rate constant times the mean hydraulic detention time. Theoretical curves are based on a simple model with completely mixed reactor assumption under steady-state conditions.

chromium, lead, and nickel.[38] This is likely due to colloids passing the filter in samples with large suspended solids concentrations. Or organic complexing ligands could be present at higher concentrations in samples with large suspended solids concentrations. In samples with less than 10 mg L^{-1} total suspended solids, K_d values for Pb^{2+} may be as large as 10^6 L kg^{-1} and Cu^{2+}, Cd^{2+}, Zn^{2+}, and Ni^2 will be on the order of 10^4–10^5 L kg^{-1} at pH 7.

Table 8.7 Parameters in Steady-State Model [equation (46)]

Parameter	Units	Estimation Method
τ	days	V/Q (annual average)
v_s	m d^{-1}	Settling column test
H	m	Bathymetric map
K_d	L kg^{-1}	Adsorption isotherm or filtered/unfiltered samples[a]
M	kg L^{-1}	Gravimetric determination

[a]Operational definition can give variable results. See Section 8.1.3

Example 8.5 Steady-State Model for Pb in Lake Cristallina

Lake Cristallina is a small, oligotrophic lake at 2400 m above sea level in the southern Alps of Switzerland. It has a mean hydraulic detention time of only 10 days, a particle settling velocity of 0.1 m d^{-1}, a mean depth of 1 m, and a K_d value of 10^5 L kg^{-1} for Pb at pH 6. The suspended solids concentration is 1 mg L^{-1}. Estimate the fraction of Pb(II) that is adsorbed to suspended particles and the fraction that is sedimented. Hypothetically, what would be the variation in the results if the pH of the lake varied between pH 5 and 8, but other parameters remained constant?

Solution: We can calculate two dimensionless parameters $k_s\tau$ and $K_d M$ and estimate the fraction in the particulate adsorbed phase, f_p, and the fraction removed to the sediment [equation (46)]. At pH 6,

$$K_d = 10^5 \text{ L kg}^{-1}$$

$$K_d M = (10^5 \text{ L kg}^{-1})\,(1 \times 10^{-6} \text{ kg L}^{-1}) = 0.1$$

$$k_s = v_s/H = 0.1/1 = 0.1 \text{ day}^{-1}$$

$$k_s\tau = (0.1)(10) = 1.0$$

$$f_p = K_d M/(1 + K_d M) = 0.0909 \quad \text{at pH 6}$$

$$R = k_s \tau f_p/(1 + k_s \tau f_p) = 0.0833 \text{ at} \quad \text{pH 6}$$

Assuming a linear dependence of K_d on $1/[H^+]$ as indicated by equation (42), one can complete the calculations for pH 5, 7, and 8. Results are plotted in Figure 8.18. In reality, the relationship between K_d and $[H^+]$ is complicated by the stoichiometry of surface complexation and aqueous phase complexing ligands. The best way to establish the effect of pH on K_d is experimentally at each location.

Figure 8.18 shows a steep sorption edge for Pb in Lake Cristallina as pH increases. Because Lake Crystallina has such a short hydraulic detention time, little of the Pb(II) is removed at pH 5–6. Thus acidic lakes have significantly higher concentrations of toxic metals because less is removed by particle adsorption and sedimentation. Also, when the pH is low, there is a greater fraction of the total Pb(II) that is present as the free aquo metal ion (more toxicity). For these reasons, small alpine lakes like Lake Cristallina have higher concentrations of trace metals and more toxicity to fish than large lakes receiving wastewater discharges of metals. Table 8.8 is a compilation of some European lakes and trace metal concentrations relative to Lake Cristallina.[39]

The pH dependence of trace metal binding on particles affects a number of acid lakes receiving atmospheric deposition. Lower pH waters have significantly higher concentrations of zinc, for example (Figure 8.19).[40] This is attributed to lower K_d

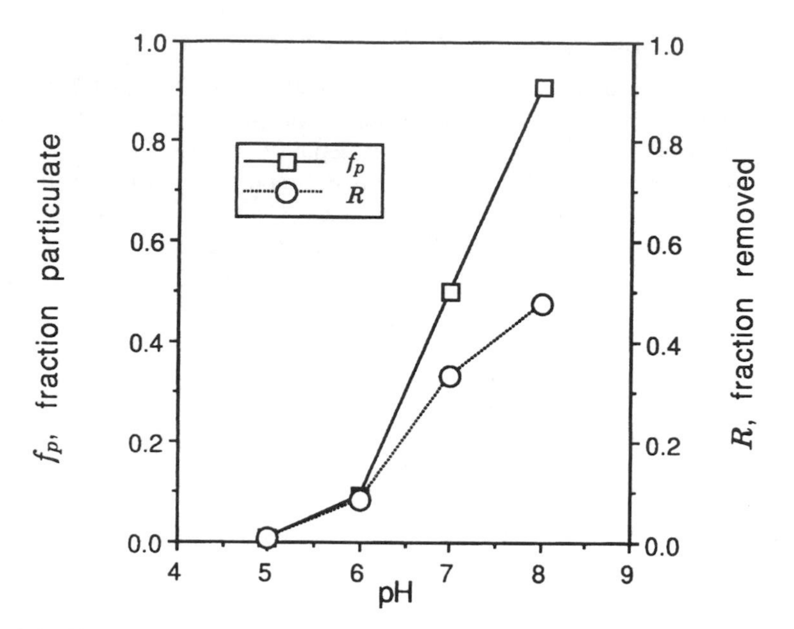

Figure 8.18 Simple steady-state model for metals in lakes applied to Pb(II) binding by particles in Lake Cristallina. K_d value was 10^5 at pH 6. Log K_d was directly proportional to pH (log $K_d = 4$ at pH 5, log $K_d = 5$ at pH 6, etc.).

values in acidic systems and less adsorption of the total dissolved Zn concentration by particles. Often lakes with pH < 5.5 have poor fisheries or are barren of fish. The toxic effect is attributed mostly to aluminum ions as we shall see in Chapter 10.

What controls trace metal concentrations in lakes? In Section 8.3, we saw that soluble ligands and fulvic acids are important complexing agents that can stabilize metals in solution (with less toxicity perhaps). Solubility exerts some control on metals, particularly under anaerobic conditions of sediments (sulfide solubility

Table 8.8 Concentrations of Trace Metals in Alpine Lakes and in Lower Elevation Lakes Receiving Runoff and Wastewater (Total Concentrations of Cu, Zn, Cd, Pb, and Dissolved Concentrations of Al)

		Total Concentration, nmol L^{-1}				
Lake	pH	Cu, nM	Zn, nM	Cd, nM	Pb, nM	Al (dissolved), μM
Cristallina[a]	5.9	5	42	0.4	3.4	0.65
Zotta[a]	5.7	5	50	0.5	3.0	1.04
Val Sabbia[a]	6.9	4	11	0.2	1.5	0.24
Constance[b]	7.8–8.5	5–15	10–30	0.05–0.1	0.2–0.5	—
Zurich[b]	7.5–8.5	6–12	5–45	0.04–0.1	0.05–1	—

[a]Small alpine lakes in the southern part of the Swiss Alps receiving atmospheric deposition.
[b]Large lakes receiving wastewater discharges.

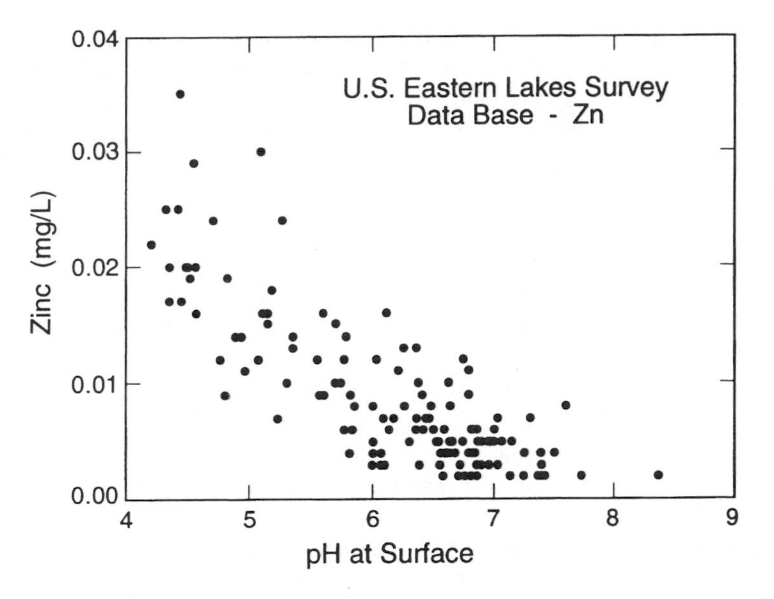

Figure 8.19 Total dissolved zinc concentration in lakes of the Eastern United States as a function of the pH value of the surface sample.

shown in Figure 8.5). And particles can scavenge trace metals, decreasing the total dissolved concentration remaining in the water column. We have seen that for Pb(II) both surface complexation on particles and aqueous phase complexation are important. For Cu(II), organic complexing agents are important, especially algal exudates and organic macromolecules.[26] So, the modeler must be aware of factors affecting chemical speciation in natural waters, particularly when toxicity is the biological effect under investigation. The controls on trace metals are threefold:

- Particle scavenging (adsorption).
- Aqueous phase complexation.
- Solubility.

They all can be important under different circumstances and the relative importance of each can be modeled. As evidenced in Figure 8.5, solubility of trace metal hydroxides and oxides is relatively large at neutral pH in natural aerobic waters (pH 6–8), so adsorption and complexation are usually the most important. The first process, particle scavenging, is affected by the aggregation of particles in natural waters.

8.5.3 Aggregation, Coagulation, and Flocculation

Coagulation in natural waters refers to the aggregation of particles due to electrolytes. The classic example is coagulation of suspended solids as salinity increases toward the mouth of an estuary. Polymers are also known to "bridge" particles causing flocculation. Flocculation is especially important in water treatment plants when iron salts or aluminum salts [$FeSO_4$, $FeCl_3$, and $Al_2(SO_4)_3$] are used to destabilize colloids and to form polymers and precipitates [$Fe(OH)_3$ and $Al(OH)_3$] that promote flocculation.[2] Organic macromolecules may be effective in aggregating particles in natural waters, especially near the sediment–water interface where particle concentrations are high.

Particles less than 1.0 μm are usually considered to be in colloidal form, and they may be suspended for a long time in natural waters because their gravitational settling velocities are less than 1 m d^{-1}. Colloidal suspensions are "stable," and they do not readily settle or aggregate in solution. Fine particles are usually negatively charged at neutral pH [with the exception of calcite, aluminum oxide, and other nonhydrated (crystalline) metal oxides]. Algae, bacteria, and biocolloids are negatively charged and emit organic macromolecules (e.g., polysaccharides) that may serve to stabilize the colloidal suspension and to disperse particles in the solution (see Figure 4.4).

Perikinetic coagulation refers to the collision of particles due to their thermal energy caused by Brownian motion. Von Smoluchowski showed in 1917 that perikinetic coagulation of a monodisperse colloidal suspension followed second-order kinetics.

$$dN/dt = -\, k_p\, N^2 \tag{47}$$

where k_p is the rate constant for perikinetic coagulation and N is the number of particles per unit volume. The rate constant k_p is on the order of 2×10^{-12} cm^3 s^{-1} particle^{-1} for water at 20 °C and for a sticking coefficient of 1.0 (the fractional efficiency of a particle collision successfully leading to agglomeration).[2] It results in relatively long periods required for particle coagulation; 6 days would be required for 10^6 particles cm^{-3} to be reduced by half by perikinetic coagulation in a batch system.

In streams and estuaries, the fluid shear rate is high due to velocity gradients, and orthokinetic coagulation exceeds perikinetic processes. The mixing intensity $G(s^{-1})$ is used as a measure of the shear velocity gradient, du/dz. Under these conditions, the rate of decrease in particle concentrations in natural waters follows first-order kinetics.

$$\frac{dN}{dt} = -k_0\, N \tag{48}$$

and

$$k_0 = 4\alpha_0\, \phi\, G/\pi$$

where k_0 is the rate constant due to orthokinetic coagulation (s^{-1}), α_0 is the sticking coefficient, ϕ is the volume of colloids per unit volume (volume fraction), and G is the mixing intensity (s^{-1}). Stumm and Morgan[2] show that a slow stirring beaker at one revolution per second would create a mixing intensity $G = 5\ s^{-1}$. For colloids of 1-μm diameter, the volume fraction would be $\phi = 5 \times 10^{-7}\ cm^3\ cm^{-3}$, resulting in a first-order rate constant $k_0 = 3.2 \times 10^{-6}\ s^{-1}$. A particle number half-life of 2.5 days would result from orthokinetic coagulation under these conditions. Most of the chemical and electrostatic factors (surface coordination, zero point of charge, electrolytes in solution) for coagulation are included in the sticking coefficient, α_0 in equation (48), and the physical factors are contained in the parameters ϕG. Stable colloidal suspensions ($\alpha \ll 1.0$) can result in greatly reduced coagulation rates. Differential settling and agglomeration may increase coagulation rates when particles size distributions are widely nonuniform.

The particle size distribution shifts during the course of coagulation and agglomeration. The number of particles in each size range follows a power law relationship:

$$N = A d_p^{-\beta} \tag{49}$$

in which N is the number of particles per unit volume, A is a proportionality constant related to the total concentration of particles in the water, d_p is the diameter of the particles, and β is the empirical exponent of the relationship. β is usually between 3.5 and 4.5 in natural waters, and in oceans β has been found to be ~4. If β = 4, the volume of solids is equally distributed in each logarithmic size interval, so the volume fraction ϕ does not change during coagulation (although the surface area and the number of particles decreases as coagulation proceeds).

8.6 REDOX REACTIONS AND TRACE METALS

There are a variety of trace metals and metalloids that undergo important redox reactions in soils, sediments, and water. These include As, Se, Cr, Hg, and, of course, Fe and Mn. Iron and manganese are important scavengers of trace metals, and their redox reactions are important because they affect solubility of Fe and Mn, which, in turn, exerts controls on toxic metals. Arsenic, cadmium, selenium, chromium, and mercury are toxic to aquatic biota and to humans. In trace quantities, selenium and chromium are considered to be nutrients, but at high concentrations they are toxicants.

8.6.1 Mercury and Redox Reactions

Mercury has valence states of +2, +1, and 0 in natural waters. In anaerobic sediments, the solubility product of HgS is so low that $Hg(II)_{(aq)}$ is below detection limits, but in aerobic waters a variety of redox states and chemical transformations are of environmental significance. Mercury is toxic to aquatic biota as well as humans. In humans, it binds sulfur groups in recycling amino acids and enzymes, rendering them inactive. Bioconcentration in fish, shellfish, and humans is a major problem, as evidenced by the Minimata Bay disaster in Japan where 798 people died or suffered permanent neurological damage.[41] Mercury bioconcentration in fish in oligotrophic, poorly buffered lakes in the United States, Canada, and Scandinavia has caused hundreds of lakes to be closed to commercial and sport fishing.

Mercury is released into air by outgassing of soil, by transpiration and decay of vegetation, and by anthropogenic emissions from coal combustion. Most mercury is adsorbed onto atmospheric particulate matter or in the elemental gaseous state. This is removed from air by dry fallout, gas exchange, and rainout. Humic substances form complexes, which are adsorbed onto particles, and only a small soluble fraction of mercury is taken up by biota. In the marine environment, submicron clay particles are distributed throughout the oceans because of slow settling. Pelagic organisms agglomerate and excrete the mercury-bearing clay particles, thus promoting sedimentation as one sink of mercury from the midoceanic food chain. Another source of mercury to biota is uptake of dissolved mercury by phytoplankton and algae.

Figure 8.20 shows the inorganic equilibrium E_h–pH stability diagram for Hg in fresh water and groundwater. Mercury exists primarily as the hydroxy complex $Hg(OH)_2^0$ in aqueous form in surface waters.[42] At chemical equilibrium in soil water and groundwater, inorganic mercury is converted to $Hg^0_{(aq)}$, although the process may occur slowly. Organic complexation, methylation, and/or anaerobic conditions in the presence of HS^- can curtail the formation of $Hg^0_{(aq)}$. A purely inorganic chemical equilibrium model should consider the many hydroxide and chloride complexes of mercury given by Table 8.9 The chloro complexes are important in seawater.

By using the REDEQL 2 program, Vuceta and Morgan[43] studied the oxic freshwater system (pE = 12, $P_{CO_2} = 10^{-3.5}$ atm), containing major cations (Ca, Mg, K, Na), nine trace metals (Pb, Cu, Ni, Zn, Cd, Co, Hg, Mn, Fe), eight inorganic ligands (CO_3, SO_4, Cl, F, Br, NH_3, PO_4, OH), and a surface with the adsorption characteristics of $SiO_{2(s)}$. They found that Hg(II) was present mainly in chlorocomplexes (pH < 7.1) or in hydroxy complexes (pH > 7.1). They also modeled the conditions at pH 7 with the presence of organic ligands (EDTA, citrate, histidine, aspartic acid, and cysteine).

It is apparent that most mercury compounds released into the environment can be directly or indirectly transformed into monomethylmercury cation or dimethylmercury via bacterial methylation.[44] At low mercury contamination levels, dimethylmercury is the ultimate product of the methyl transfer reactions, whereas if higher concentrations of mercury are introduced, monomethylmercury is predominant. Moreover, it appears that neutral and alkaline environments favor the formation of dimethylmercury, which readily decomposes into monomethylmercury in mildly acid environments. This explains why some acidic lakes in remote areas with no direct sources of mercury pollution still contain fish with relatively high mercury concentrations. If the lakes are acid, mercury is methylated to CH_3Hg^+ and bioconcentrated by fish. The greater is the organic matter content of sediments, the greater is the tendency for methylation (formation of CH_3Hg^+) and bioconcentration by fish.[42]

Kinetics control the bacterial methylation, demethylation, and bioconcentration of mercury in the aquatic environment (Figure 8.21). Thus an equilibrium approach

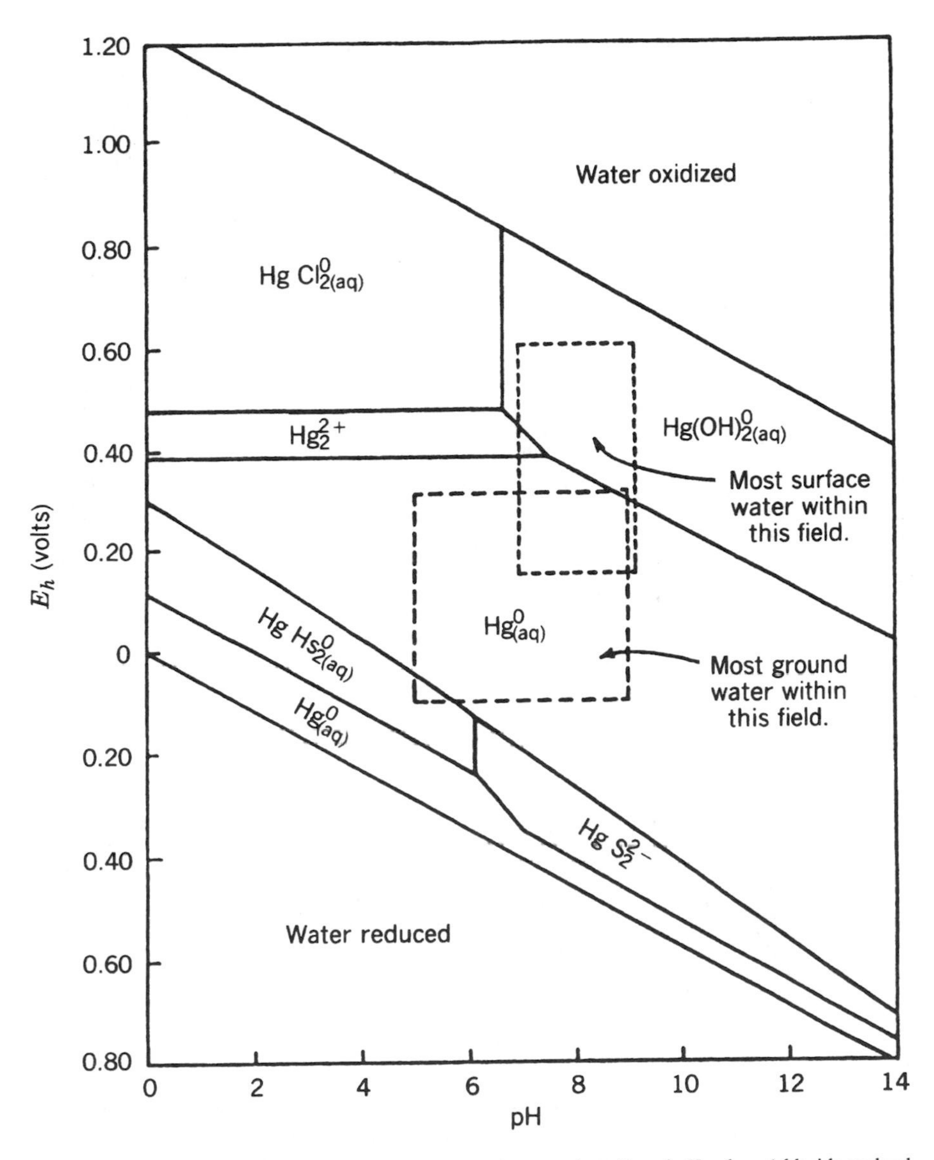

Figure 8.20 Stability fields for aqueous mercury species at various E_h and pH values (chloride and sulfur concentrations of 1 mM each were used in the calculation; common E_h–pH ranges for groundwater are also shown). (Modified from Gavis and Ferguson.[42])

such as that given by Table 8.9 is not sufficient. We must couple chemical equilibrium expressions with mass balance kinetic equations and solve simultaneously in order to model such problems. The basic approach was discussed in Chapter 4 and shown in Figure 4.1.

As a first approach to the problem, let us assume that most of the processes depicted in Figure 8.21 can be modeled using first-order reaction kinetics. Rate con-

Table 8.9 Stability Constants Used in the Inorganic Complexation Model for Hg(II), 298.15 K, $I = 0$ M

Species	Equilibrium Expression	log K
OH^-	$K_1 = [H^+][OH^-]$	−14.00
$HgOH^+$	$K_2 = [HgOH^+][H^+]/[Hg^{2+}]$	−3.40
$Hg(OH)_2$	$K_3 = [Hg(OH)_2][H^+]^2/[Hg^{2+}]$	−6.20
$Hg(OH)_3^-$	$K_4 = [Hg(OH)_3^-][H^+]^3/[Hg^{2+}]$	−21.10
$Hg(OH)_4^{2-}$	$K_5 = [Hg(OH)_4^{2-}][H^+]^4/[Hg^{2+}]$	−36.80
Hg_2OH^{3+}	$K_6 = [Hg_2OH^{3+}][H^+]/[Hg^{2+}]^2$	−3.30
$Hg_3(OH)_3^{3+}$	$K_7 = [Hg_3(OH)_3^{3+}][H^+]^3/[Hg^{2+}]^3$	−6.40
$HgCl^+$	$K_8 = [HgCl^+]/([Hg^{2+}][Cl^-])$	7.30
$HgCl_2$	$K_9 = [HgCl_2]/([Hg^{2+}][Cl^-]^2)$	14.00
$HgCl_3^-$	$K_{10} = [HgCl_3^-]/([Hg^{2+}][Cl^-]^3)$	15.00
$HgCl_4^{2-}$	$K_{11} = [HgCl_4^{2-}]/([Hg^{2+}][Cl^-]^4)$	15.60
$HgOHCl$	$K_{12} = [HgOHCl][H^+]/([Hg^{2+}][Cl^-])$	4.10
$HgOHCl_2^-$	$K_{13} = [HgOHCl_2^-][H^+]/([Hg^{2+}][Cl^-]^2)$	3.45
$HgOHCl_3^{2-}$	$K_{14} = [HgOHCl_3^{2-}][H^+]/([Hg^{2+}][Cl^-]^3)$	3.10
$Hg(OH)_2Cl^-$	$K_{15} = [Hg(OH)_2Cl^-][H^+]^2/(Hg^{2+}][Cl^-])$	−8.60
$Hg(OH)_3Cl^{2-}$	$K_{16} = [Hg(OH)_3Cl^{2-}][H^+]^3/([Hg^{2+}][Cl^-])$	−23.10
$Hg(OH)_2Cl_2^{2-}$	$K_{17} = [Hg(OH)_2Cl_2^{2-}][H^+]^2/([Hg^{2+}][Cl^-]^2)$	−10.10
$HgNO_3^+$	$K_{18} = [HgNO_3^+]/([Hg^{2+}][NO_3^-])$	0.39
$Hg(NO_3)_2$	$K_{19} = [Hg(NO_3)_2]/([Hg^{2+}][NO_3^-]^2)$	0.42
HCO_3^-	$K_{20} = [HCO_3^-][H^+]/[H_2CO_3^*]$	−6.35
CO_3^{2-}	$K_{21} = [CO_3^{2-}][H^+]^2/[H_2CO_3^*]$	−16.68
$HgCO_3$	$K_{22} = [HgCO_3][H^+]^2/([Hg^{2+}][H_2CO_3^*])$	−4.50
$Hg(CO_3)_2^{2-}$	$K_{23} = [Hg(CO_3)_2^{2-}][H^+]^4/([Hg^{2+}][H_2CO_3^*]^2)$	−17.70
$HgHCO_3^+$	$K_{24} = [HgHCO_3^+][H^+]/([Hg^{2+}][H_2CO_3^*])$	−0.35
$HgOHCO_3^-$	$K_{25} = [HgOHCO_3^-][H^+]^3/([Hg^{2+}][H_2CO_3^*])$	−11.40
$Hg(OH)_{2(s)}$	$K_{spHg} = [Hg^{2+}]/([H^+]^2\ \{HgO_{(s)}\})$	2.51
$2HgO \cdot HgCO_3(s)$	$K_{spCO_3} = [Hg^{2+}]^3\ P_{CO_2}/(\{2HgO \cdot HgCO_{3(s)}\}\ [H^+]^6)$	7.00
$H_2CO_3^*$	$K_H = [H_2CO_3^*]/P_{CO_2}$	−1.47

Source: Modified from Tiffreau et al.[34]

stants are conditional and site specific because they depend on environmental variables such as organic matter content of sediments, pH, and fulvic acid concentration. We can neglect the disproportionation reaction

$$Hg^0 + Hg^{2+} \rightleftharpoons 2\ Hg^+ \tag{50}$$

Also, the formation of dimethyl mercuric sulfide ($CH_3S–HgCH_3$) is known to be important in marine environments, but it is neglected here.

A chemical equilibrium submodel would be used to estimate the fraction of total Hg(II) that was adsorbed onto hydrous oxides and particulate organic matter, $[Hg^{2+}]_{ads}$. It would also calculate the amount of Hg(II) that precipitates as $HgS_{(s)}$. The mass balance model is used to calculate the kinetics of redox reactions among

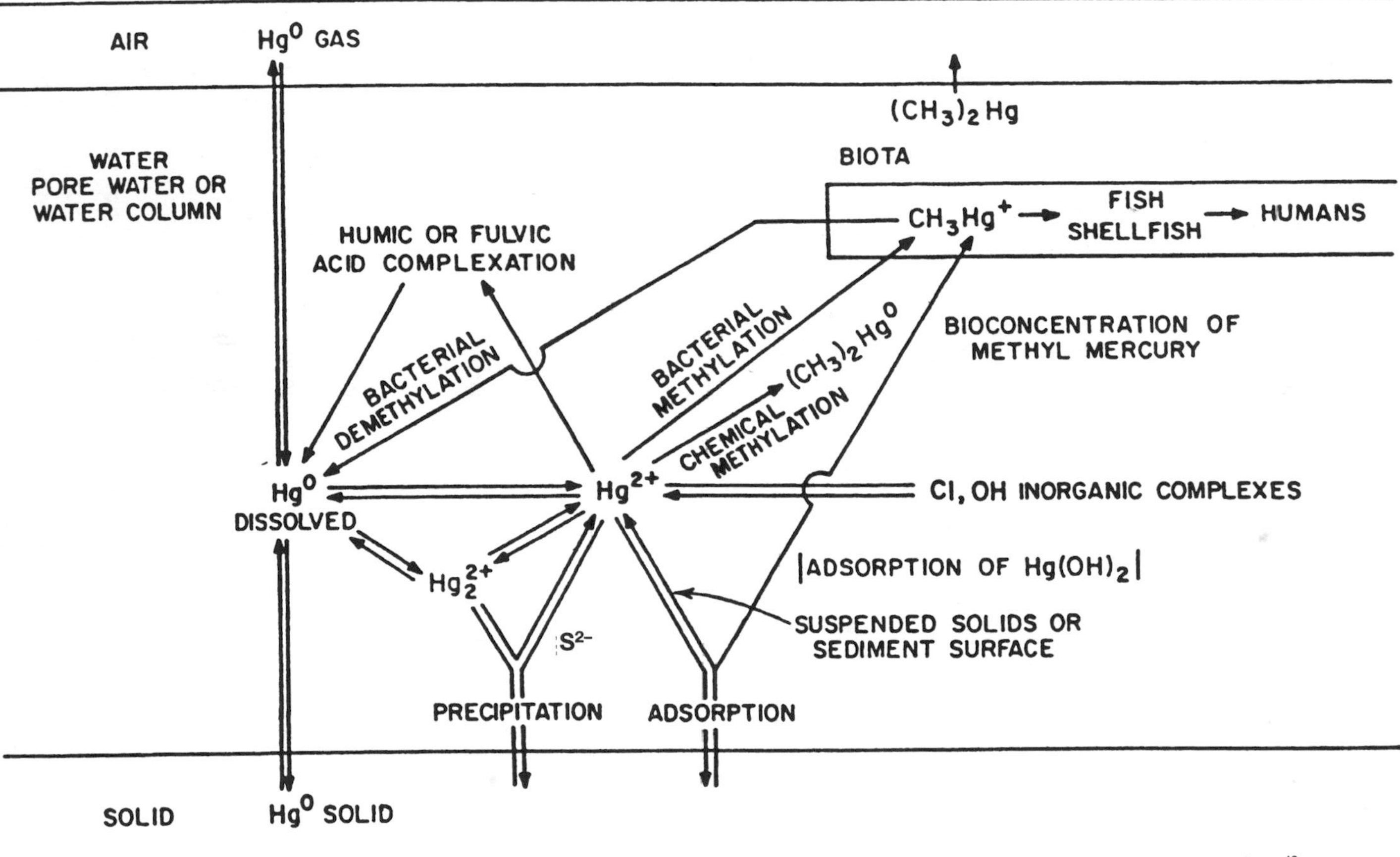

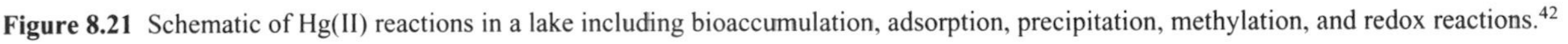

Figure 8.21 Schematic of Hg(II) reactions in a lake including bioaccumulation, adsorption, precipitation, methylation, and redox reactions.[42]

the species Hg^{2+}, Hg^0, CH_3Hg^+, and $(CH_3)_2Hg^0$, as well as the rate of nonredox processes such as uptake by fish, volatilization to the atmosphere, and sedimentation of particulate-adsorbed mercury and precipitated solids (if any). Simplified mass balance equations for a completely mixed system (such as a lake) are given by equations (51)–(54).

$$\frac{d[\mathrm{Hg(II)}]_T}{dt} = \underbrace{F_{d2}/H}_{\text{deposition inputs}} \; \underbrace{+\,k_{\mathrm{ox}}[\mathrm{Hg}^0]}_{\text{oxidation of Hg}^0} \; \underbrace{-\,[\mathrm{Hg(II)}]_T/\tau}_{\text{outflows}} \; \underbrace{-\,k_m[\mathrm{Hg(II)}][\mathrm{Bacteria}]}_{\text{methylation}} \; \underbrace{-\,k_{\mathrm{FA}}[\mathrm{Hg(II)}][\mathrm{FA}]}_{\text{reduction by organics}} \; \underbrace{-\,k_{\mathrm{di}}[\mathrm{Hg(II)}]}_{\text{chemical dimethylation}} \; \underbrace{-\,k_{s1}[\mathrm{Hg(II)}]_{\mathrm{ads}}}_{\text{adsorption \& sediment.}} \; \underbrace{-\,k_{s2}[\mathrm{HgS}]}_{\text{precip. \& sediment.}} \quad (51)$$

$$\frac{d[\mathrm{CH_3Hg^+}]_T}{dt} = \underbrace{+\,k_m[\mathrm{Hg(II)}][\mathrm{Bacteria}]}_{\text{methylation}} \; \underbrace{-\,k_{\mathrm{bio}}[\mathrm{CH_3Hg^+}][\mathrm{Fish}]}_{\text{uptake by fish}} \; \underbrace{-\,k_{\mathrm{de}}[\mathrm{CH_3Hg^+}]}_{\text{demethylation}} \; \underbrace{-\,[\mathrm{CH_3Hg^+}]/\tau}_{\text{outflows}} \quad (52)$$

$$\frac{d[(\mathrm{CH_3})_2\mathrm{Hg}^0]}{dt} = \underbrace{+\,k_{\mathrm{di}}[\mathrm{Hg(II)}]}_{\text{chemical dimethylation}} \; \underbrace{-\,k_{a2}[(\mathrm{CH_3})_2\,\mathrm{Hg}^0]}_{\text{volatilization}} \; \underbrace{-\,[(\mathrm{CH_3})_2\,\mathrm{Hg}^0]/\tau}_{\text{outflows}} \quad (53)$$

$$\frac{d[\mathrm{Hg}^0]}{dt} = \underbrace{F_{d0}/\mathrm{H}}_{\text{deposition inputs}} \; \underbrace{+\,k_{\mathrm{FA}}[\mathrm{Hg(II)}][\mathrm{FA}]}_{\text{reduction of Hg(II)}} \; \underbrace{-\,k_{a1}[\mathrm{Hg}^0]}_{\text{volatilization}} \; \underbrace{-\,k_{\mathrm{ox}}[\mathrm{Hg}^0]}_{\text{oxidation}} \; \underbrace{-\,k_{s3}[\mathrm{Hg}^0]}_{\text{formation of metallic Hg}^0} \quad (54)$$

$$\frac{d[\mathrm{CH_3Hg\text{-}Fish}]}{dt} = \underbrace{+\,k_{\mathrm{bio}}[\mathrm{CH_3Hg^+}]}_{\text{uptake by fish}} \; \underbrace{-\,k_d[\mathrm{CH_3Hg\text{-}Fish}]}_{\text{depuration}} \quad (55)$$

where $[\mathrm{Hg(II)}]_T$ = total Hg(II) concentration (dissolved and particulate), ML^{-3}

t = time, T

F_{d2} = total inputs including wet and dry deposition of Hg(II), $ML^{-2}T^{-1}$

F_{d0} = total inputs of Hg^0, $ML^{-2}T^{-1}$

H = mean depth of the lake, L
k_{ox} = oxidation rate constant for $Hg^0 \rightarrow Hg(II)$, T^{-1}
$[Hg^0]$ = dissolved elemental mercury concentration, ML^{-3}
τ = mean hydraulic detention time, T
k_m = methylation rate constant, L^3 $cells^{-1}$ T^{-1}
$[Hg(II)]$ = Hg(II) aqueous concentration, ML^{-3}
[Bacteria] = methylating bacteria concentration, cells L^{-3}
k_{FA} = rate constant for reduction of Hg(II), L^{-3} mol^{-1} T^{-1}
[FA] = fulvic acid (DOC) concentration, mol L^{-3}
k_{di} = rate constant for chemical dimethylation, T^{-1}
$[Hg(II)]_{ads}$ = adsorbed Hg(II) concentration on particles, ML^{-3}
k_{s1} = sedimentation rate constant of adsorbed Hg(II), T^{-1}
k_{s2} = sedimentation rate constant of precipitated HgS, T^{-1}
k_{s3} = sedimentation rate constant for metallic mercury, T^{-1}
$[CH_3Hg^+]$ = dissolved methylmercury concentration, ML^{-3}
k_{bio} = biouptake by fish, (L^3 filtered) M_{fish}^{-1} T^{-1}
[Fish] = concentration of fish in the lake, (M_{fish}) L^{-3}
k_{de} = demethylation rate constant, T^{-1}
$[(CH_3)_2 Hg^0]$ = dimethylmercury dissolved concentration, ML^{-3}
k_{a1} = volatilization of Hg^0 rate constant, T^{-1}
k_{a2} = volatilization of dimethylmercury rate constant, T^{-1}
$[CH_3Hg\text{-}Fish]$ = mass of mercury per mass of fish, MM^{-1}
k_d = depuration rate constant of mercury from fish, T^{-1}

Equations (51)–(55) can be used to model the contamination and recovery of mercury in a lake and fish. Remedial action alternatives can be evaluated such as aeration to volatilize mercury from the lake, natural recovery, or the effectiveness of mitigating inputs to the lake. As in all models, calibration and verification data are needed for state variables being modeled. As an additional check on the model rate constants, microcosms should be set up in the laboratory with actual lake water and sediment to evaluate the methylation rate constant. Radioisotopes, ^{203}Hg, can be used in microcosm studies or "cold" Hg inputs to enclosure experiments in the lake to obtain model parameters.

Example 8.6 Equilibrium Model for Hg(II) with K_d

What is the distribution of Hg(II) species in the Mississippi River at Memphis, Tennessee, if the distribution coefficient is K_d = 19,000 L kg^{-1}, the total suspended solids concentration is 200 mg L^{-1}, the chloride concentration is 3×10^{-3} M, the $Hg(II)_T$ concentration is 1×10^{-9} M and the pH is 8.3?

Solution: The $Hg(OH)_2$ complex is expected to be predominant for Hg(II) in fresh waters. Chloro complexes are crucial for seawater with 0.7 M Cl^-. The equilibrium constants are available in tables of stability constants. The distribution coefficient K_d can be handled just like any other equilibrium constant, assuming that the dissolved species is $Hg(OH)_2$. Units for K_d need to be converted to mol L^{-1} from L kg^{-1}.

$$K_d = \frac{[\text{Hg(II)}]_{\text{ads}}}{[\text{Hg(OH)}_2]} = 19{,}000\,\frac{\mu\text{g kg}^{-1}}{\mu\text{g L}^{-1}} = 19{,}000\,\frac{\text{mol kg}^{-1}}{\text{mol L}^{-1}}$$

$$K_d \times M = 19{,}000\,\frac{\text{mol kg}^{-1}}{\text{mol L}^{-1}} \times 200 \times 10^{-6}\ \text{kg L}^{-1} = 3.8\,\frac{\text{mol L}^{-1}}{\text{mol L}^{-1}}$$

where M is the suspended solids concentration in kg L^{-1}.

We must write the equilibrium equation for $Hg(II)_{ads}$ in terms of the components $Hg(OH)_2$ and Hg^{2+}.

$$\text{Hg(OH)}_2 = 10^{-6.2}\,[\text{Hg}^{2+}]/[\text{H}^+]^2$$

$$K_d \times M = 3.8 = \frac{[\text{Hg(II)}]_{\text{ads}}}{10^{-6.2}[\text{Hg}^{2+}]/[\text{H}^+]^2}$$

$$\therefore \quad [\text{Hg(II)}]_{\text{ads}} = 10^{-5.62}\,[\text{Hg}^{2+}]/[\text{H}^+]^2$$

Now the problem can be formulated like any other equilibrium problem using the matrix method and a Newton–Raphson numerical solution technique. At high pH and mercury concentration, the dimer Hg_2OH^{3+} is important and will be included here for completeness.

Species: Hg^{2+}, $HgOH^+$, $Hg(OH)_{2(aq)}$, $Hg(OH)_3^-$, Hg_2OH^{3+}, $HgCl^+$, $HgCl_{2(aq)}$, $HgCl_3^-$, $HgCl_4^{2-}$, $Hg(II)_{ads}$, Cl^-, OH^-, H^+ (13 species)

Only three components are required for the solution. We may choose:

Components: Hg^{2+}, Cl^-, and H^+ (3 components)

	Hg^{2+}	Cl^-	H^+	log K
Hg^{2+}	1	0	0	0
$HgOH^+$	1	0	–1	–3.4
$Hg(OH)_2^0$	1	0	–2	–6.2
$Hg(OH)_3^-$	1	0	–3	–21.1
Hg_2OH^{3+}	2	0	–1	–3.3

(continued)

	Hg^{2+}	Cl^-	H^+	log K
$HgCl^+$	1	1	0	6.72
$HgCl_2^0$	1	2	0	13.23
$HgCl_3^-$	1	3	0	14.2
$HgCl_4^{2-}$	1	4	0	15.3
$Hg(II)_{ads}$	1	0	–2	–5.62
Cl^-	0	1	0	0
OH^-	0	0	–1	–14
H^+	0	0	1	0
TOT, C_T	1×10^{-9} M	3×10^{-3} M	5×10^{-9} M	

Results of the calculation show that 79% of the $Hg(II)_T$ is adsorbed to suspended solids under these conditions. The remainder is primarily as $Hg(OH)_{2(aq)}$. Thus the assumption that the dissolved $Hg(OH)_2$ species represents $Hg(II)_{(aq)}$ in the distribution coefficient was valid.

Answer:

$$Hg^{2+} = 8.24 \times 10^{-21}\ M$$

$$Hg(OH)_2 = 2.08 \times 10^{-10}\ M$$

$$HgCl_2 = 1.26 \times 10^{-12}\ M$$

$$Hg(II)_{ads} = 7.91 \times 10^{-10}\ M$$

Example 8.7 Steady-State Model for Hg in the Mississippi River

Develop a simple steady-state model for Hg in the Mississippi River based on the equilibrium model of Example 8.6.

Approach: The kinetic equations (51)–(55) can be used as the basis for reactions in the model. We will assume plug-flow conditions (the Mississippi is quite fast from Memphis to Vicksburg). The approach is to solve the mass balance equation coupled with the equilibrium model of Example 8.6. Methylation, reduction, and sedimentation are the primary sinks of Hg(II) in the river; and wastewater discharges and scour of contaminated sediment are the primary sources of Hg(II) to the water column. Wastewater discharges should be handled as boundary conditions at the beginning of each river segment where they occur. Flux of mercury by scour from the sediments is accounted for as a first-order process, where the scour coefficient, α, depends on the shear velocity u_* at the sediment–water interface. An example of a mass balance equation for $Hg(II)_T$ would be:

$$u\frac{d[\mathrm{Hg(II)}]_T}{dx} = -\underbrace{k_m[\mathrm{Hg(II)}]\,[\mathrm{Bacteria}]}_{\text{methylation}} - \underbrace{k_s\,[\mathrm{Hg(II)_{ads}}]}_{\text{sedimentation}} - \underbrace{k_{\mathrm{FA}}[\mathrm{Hg(II)}][\mathrm{FA}]}_{\text{reduction to Hg}^0} + \underbrace{\alpha[\mathrm{Hg(II)}]_{\mathrm{sed}}\,S_{\mathrm{sed}}\,d/h}_{\text{scour of sediment}}$$

where $[\mathrm{Hg(II)}]_T$ = total (whole water) Hg(II) concentration, ML^{-3}
x = longitudinal distance, L
u = mean velocity ~ 0.7 m s^{-1}, LT^{-1}
k_m = methylation rate constant, L^3 $cells^{-1}$ T^{-1}
k_s = sedimentation rate constant, T^{-1}
$[\mathrm{Hg(II)_{ads}}]$ = adsorbed particulate concentration, ML^{-3}
k_{FA} = reduction by fulvic acid, $L^3M^{-1}T^{-1}$
α = scour coefficient, T^{-1}
$[\mathrm{Hg(II)}]_{\mathrm{sed}}$ = sediment Hg concentration (an input parameter), MM^{-1}
S_{sed} = sediment solids concentration, ML^{-3}
d = depth of active bed in exchange with the water column, L
h = water column depth (mean depth of river), L

The equation(s) can be solved by a Runge–Kutta or predictor–corrector numerical technique. The sediment concentration would be constant for each segment of the model. (Each segment of the river model has constant coefficients.) If parameters vary, then more segments would need to be created. Sediment concentration is expressed in units of μg kg^{-1} or μmol kg^{-1} of solids, consistent with the other coefficients.

To save computation time, chemical speciation using the equilibrium model could be updated every 10 time steps or so. Total suspended solids concentrations would affect the log K value for $[\mathrm{Hg(II)_{ads}}]$ in the chemical equilibrium model as demonstrated in Example 8.6. These concentrations would be constant within each model segment of the river. The advantage of this approach is that both chemical speciation and fate and transport are considered in the model.

8.6.2 Arsenic and Redox Reactions

The metalloid arsenic provides a challenge for modelers due to its redox reactions and relatively complex adsorption chemistry. It can be modeled in a manner similar to mercury even though it exists primarily as an anion in natural waters, soils, and sediments. Arsenate, $HAsO_4^{2-}$, is the primary anion in aerobic surface waters, and arsenite (H_3AsO_3 or $H_2AsO_3^-$) is the primary species in groundwater (Figure 8.22). Arsenate has a valence state As(V), while arsenite is As(III). Arsenite (the reduced form) is considerably more mobile and toxic than arsenate. Thus it is important to be able to differentiate arsenic species, but chemical equilibrium is slow to be

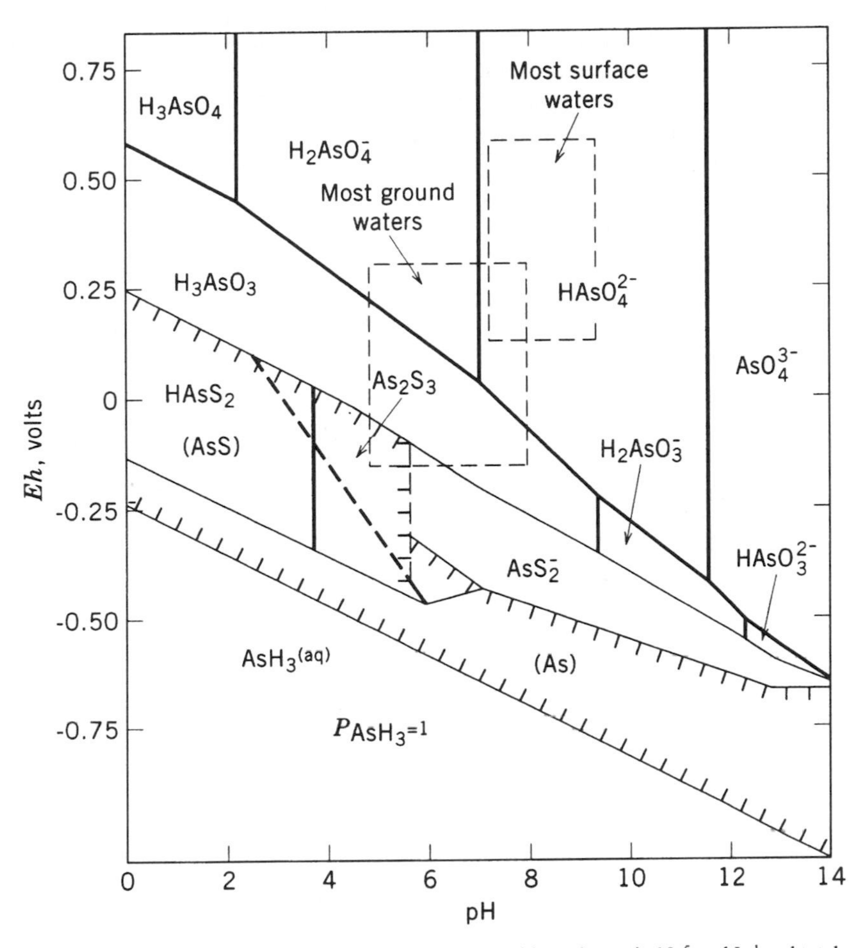

Figure 8.22 The E_h–pH diagram for As at 25 °C and 1 atm with total arsenic 10^{-5} mol L^{-1} and total sulfur 10^{-3} mol L^{-1}. Solid species are enclosed in parentheses in cross-hatched area, which indicates solubility less than $10^{-5.3}$ mol L^{-1} (Modified from Ferguson and Gavis.[45])

achieved. A chemical equilibrium model coupled with the kinetic mass balance equation is necessary.

Arsenate adsorbs to iron oxides and hydroxides especially under low pH conditions when hydrous oxides have a positive surface charge. Coprecipitation of $FeAsO_4$ with $Fe(OH)_3$ has also been reported at high arsenate concentrations, although $FeAsO_4$ is quite soluble under most conditions ($pK_{s0} = 20.2$ at 25 °C).[45] Oxidation of arsenite to arsenate by dissolved oxygen occurs slowly at neutral pH, but it can be acid or base catalyzed. If arsenite adsorbs to MnO_2 mineral grains, it is rapidly oxidized and Mn(IV) is reduced to Mn(II). Heterogeneous oxidation of arsenite on hydrous ferric oxides is also possible. Photoreduction of arsenate to arsenite in the euphotic zone and in the presence of algae is possible. Arsenate behaves similarly to phosphate in some biological reactions, like a nutrient.

Under extremely reducing conditions it is possible to form arsine, AsH_3, with arsenic in the –3 valence state. Arsine is very toxic to aquatic biota, but it is rarely measured in natural waters. More common under reducing conditions is the formation of polysulfide precipitates such as As_2S_3 or AsS_2^- shown in Figures 8.22 and 8.23. Arsenic is a frequent pollutant in acid mine drainage stemming from the oxidation of minerals such as arsenopyrite FeAsS, where bacteria mediate arsenic and sulfur oxidation.

Braman and Foreback[46] were the first to report widespread methylation of arsenic in natural waters and sediments. The process is bacterially mediated. Methylarsenic acid, dimethyl arsenic acid, and trimethyl arsenic acid have all been reported. These organic arsenicals are less toxic than arsenite to aquatic biota, but they can bioaccumulate. Figure 8.23 is suggestive of some of the kinetic processes (not at chemical equilibrium) that would need to be included in an arsenic lake model. For example, a mass balance equation for arsenate would include reduction to arsenite, adsorption onto hydrous oxides and algae, plus sedimentation, methylation by bacteria, and oxidation of arsenite to arsenate.

$$\frac{d[\mathrm{HAsO_4^{2-}}]}{dt} = \underbrace{-k_r[\mathrm{HAsO_4^{2-}}]}_{\text{reduction}} \underbrace{- k_s[\mathrm{HAsO_4^{2-}}]_{\mathrm{ads}}}_{\text{sedimentation}}$$

$$\underbrace{- k_m[\mathrm{HAsO_4^{2-}}]\,[\mathrm{Bacteria}]}_{\text{methylation}} + \underbrace{k_0[\mathrm{H_3AsO_3}]}_{\text{oxidation of arsenite}} \tag{56}$$

± Other sources and sinks (weathering, atmospheric deposition, waste discharges)

Equation (56) represents the kinetic terms of the model; transport would need to be added. As in the mercury model, the mass balance equation would be solved numerically; and acid–base reactions, adsorption, and complexation reactions could be

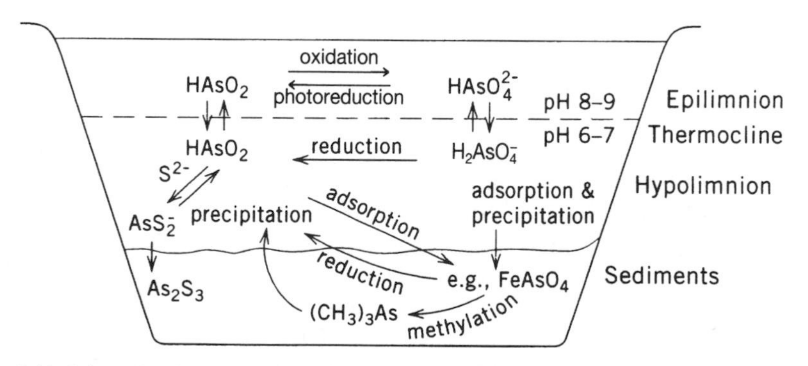

Figure 8.23 Schematic of arsenic speciation and redox reactions in a thermally stratified lake ecosystem.[45]

solved by a chemical equilibrium submodel. Concentrations of species would be updated periodically.

8.7 METALS MIGRATION IN SOILS

Trace metals with large K_d values accumulate in soils and, even after anthropogenic inputs to soils have been curtailed, recovery times can be lengthy. Accumulation may result from (1) atmospheric deposition of metals such as Cd, Hg, Pb, and Zn (Chapter 10); (2) industrial discharges from landfilling, spray irrigation of wastes, mine wastes, compost, and sewage sludge applications; and (3) unintentional addition of metals to land by fertilization, such as Cd in phosphate fertilizer, or selenium and arsenic in irrigation return water. In addition, toxic metals such as Al, Zn, and Be that occur naturally in soil minerals can be liberated due to acidification of soils by acid deposition.

A detailed 1-D transport equation for soil moisture is needed in the root zone and unsaturated zone to adequately trace metals migration. Transport and reaction equations are covered in Chapter 9 on Groundwater Contaminants, but in the simplest case, we may neglect overland flow and dispersion to formulate a transport equation for metals migration in the unsaturated zone.

$$\frac{\partial C}{\partial t} = \underbrace{-\frac{v_z}{n}\frac{\partial C}{\partial z}}_{\text{percolation}} \underbrace{-\frac{\rho_b}{n}\frac{\partial r}{\partial t}}_{\text{sorption}} \tag{57}$$

where C = total dissolved metal concentration, ML^{-3}

t = time, T

v_z = average rainfall percolation rate, LT^{-1}

z = depth below soil surface, L

ρ_b = dry soil bulk density, ML^{-3}

n = effective soil porosity, voids volume/total volume, dimensionless

r = adsorbed total metal concentration, MM^{-1}

No reaction terms are necessary if *total* dissolved metal is being modeled. When redox reactions are important and the various valence states of the metals have different K_d values, then a coupled set of equations is necessary, one equation for each valence state of relevance. The rainfall percolation rate is defined as the precipitation rate minus evapotranspiration.

Assuming a linear adsorption isotherm for low concentrations of trace metals ($r = K_dC$), it is possible to substitute for the adsorbed concentration, r, in terms of dissolved total metal, C:

$$\frac{\partial C}{\partial t} = -\frac{v_z}{n}\frac{\partial C}{\partial z} - \frac{\rho_b K_d}{n}\frac{\partial C}{\partial t} \tag{58}$$

$$\frac{\partial C}{\partial t}\left(1 + \frac{\rho_b K_d}{n}\right) = -\frac{v_z}{n}\frac{\partial C}{\partial z} \tag{59}$$

$$\frac{\partial C}{\partial t} = -\frac{v_z}{nR_f}\frac{\partial C}{\partial z} \tag{60}$$

where R_f = retardation factor due to adsorption of metal, dimensionless

$$= \left(1 + \frac{\rho_b K_d}{n}\right)$$

K_d = soil/water distribution coefficient, L^3M^{-1} (liters kg^{-1})

Binding of trace metals by adsorption to soil particles retards the migration of metals—in essence, slowing down the movement of dissolved metal by the magnitude of the retardation factor R.

Equation (60) admits various numerical solutions depending on the boundary equations employed and the temporal variations in v_z, the rainfall percolation rate. Metals will migrate vertically downward in soils as the flushing rate increases (v_z increases) and if the retardation factor decreases (K_d decreases). Acidification of soils over decades (see Chapter 10) tends to decrease K_d, making metals migrate. On the other hand, rich soils with plenty of organic matter tend to bind metals with large K_d values and retardation factors. The migration of metals can be viewed as a beneficial process from the standpoint of "cleansing" the soils of toxic metals but, at the same time, it causes metals to contaminate groundwater and streams that interchange with surficial groundwater. Rich soils with deep organic matter can bind metals and retard their movement for decades, as discussed in Chapter 10 concerning the buildup of Pb in soils from combustion of leaded gasoline.

Speciation models of trace metals in soils should include adsorption of metals to binding sites. Both organic and inorganic binding sites may be necessary for a general model that could be used at many different sites. In the case of cadmium, the following chemical reactions are important: adsorption to hydrous oxides, hydrolysis, chloride complexation in saline and dry soils, sulfate complexation in soils receiving acid rain H_2SO_4, and organic complexation by fulvic acids. A chemical equilibrium model is given by the matrix of reactions below. There are 15 species and 7 components (including the coulombic correction factor, the dummy component recorded as surface charge). Local environmental conditions are needed for the site density of SOH (mol L^{-1} of soil water) and the total concentration of cadmium and ligands (v, w, x, y, and z).

Results of chemical equilibrium models for Cd in soils indicate that organic acids (H_2FA) can be effective in mobilizing Cd in soils. The K_d for cadmium adsorption by iron oxides and aluminum oxides is defined by the equation for species 6 in Table 8.10 when free aquo metal ion is the predominant dissolved species.

$$[SOCd^+] = 10^{0.3}[SOH]\,[Cd^{2+}]/[H^+] \tag{61}$$

Table 8.10 Matrix of Reactions for Chemical Equilibrium Model of Cadmium in Soils

	Components							
Species	Cd^{2+}	SOH	FA^{2-}	Cl^-	SO_4^{2-}	Surface Charge	H^+	log *K*
1. Cd^{2+}	1	0	0	0	0	0	0	0
2. $CdOH^+$	1	0	0	0	0	0	–1	3.9
3. SOH	0	1	0	0	0	0	0	0
4. SOH_2^+	0	1	0	0	0	+1	1	4.0
5. SO^-	0	1	0	0	0	–1	–1	–9.0
6. $SOCd^+$	1	1	0	0	0	+1	–1	0.3
7. CdFA	1	0	1	0	0	0	0	3.5
8. $CdCl^+$	1	0	0	1	0	0	0	2.0
9. $CdSO_4^0$	1	0	0	0	1	0	0	2.3
10. H_2FA	0	0	1	0	0	0	2	11.8
11. FA^{2-}	0	0	1	0	0	0	0	0
12. HFA^-	0	0	1	0	0	0	1	7.0
13. Cl^-	0	0	0	1	0	0	0	0
14. SO_4^{2-}	0	0	0	0	1	0	0	0
15. H^+	0	0	0	0	0	0	1	0
C_T	*v*	1.0	*w*	*x*	*y*		*z*	

Rearranging

$$K_d = \frac{[SOCd^+]}{[Cd^{2+}]\,[SOH]} = \frac{10^{0.3}}{[H^+]} \tag{62}$$

K_d decreases with increasing H^+ ion concentration. The more acidic the soil solution, the faster Cd will desorb from the soil matrix and be transported to groundwater.

Nikolaidis et al.[47] have demonstrated that Cd sorption in soils is often irreversible, and desorption has mass transfer limitations that cause immobile water in fine pore spaces to exchange slowly with water in primary pores. Results of column experiments with Cd-contaminated soil from an electroplating operation are shown in Figure 8.24. Higher flushing rates cause Cd concentrations in leachate water to decrease faster and to lower threshold concentrations. A chemical equilibrium model similar to Table 8.10 was used for speciation, and dissolved cadmium in the leachate water was primarily as the free aquo metal ion at pH 4 and $I = 0.1$ M. After an equivalent of 50 years of precipitation was flushed through the columns, 50–90% of the metal remained in the soil matrix.

Example 8.8 Equilibrium Model for Metal Adsorption in Soil

Develop a simple equilibrium model for a trace metal in soil water given the following information and predominant reactions:

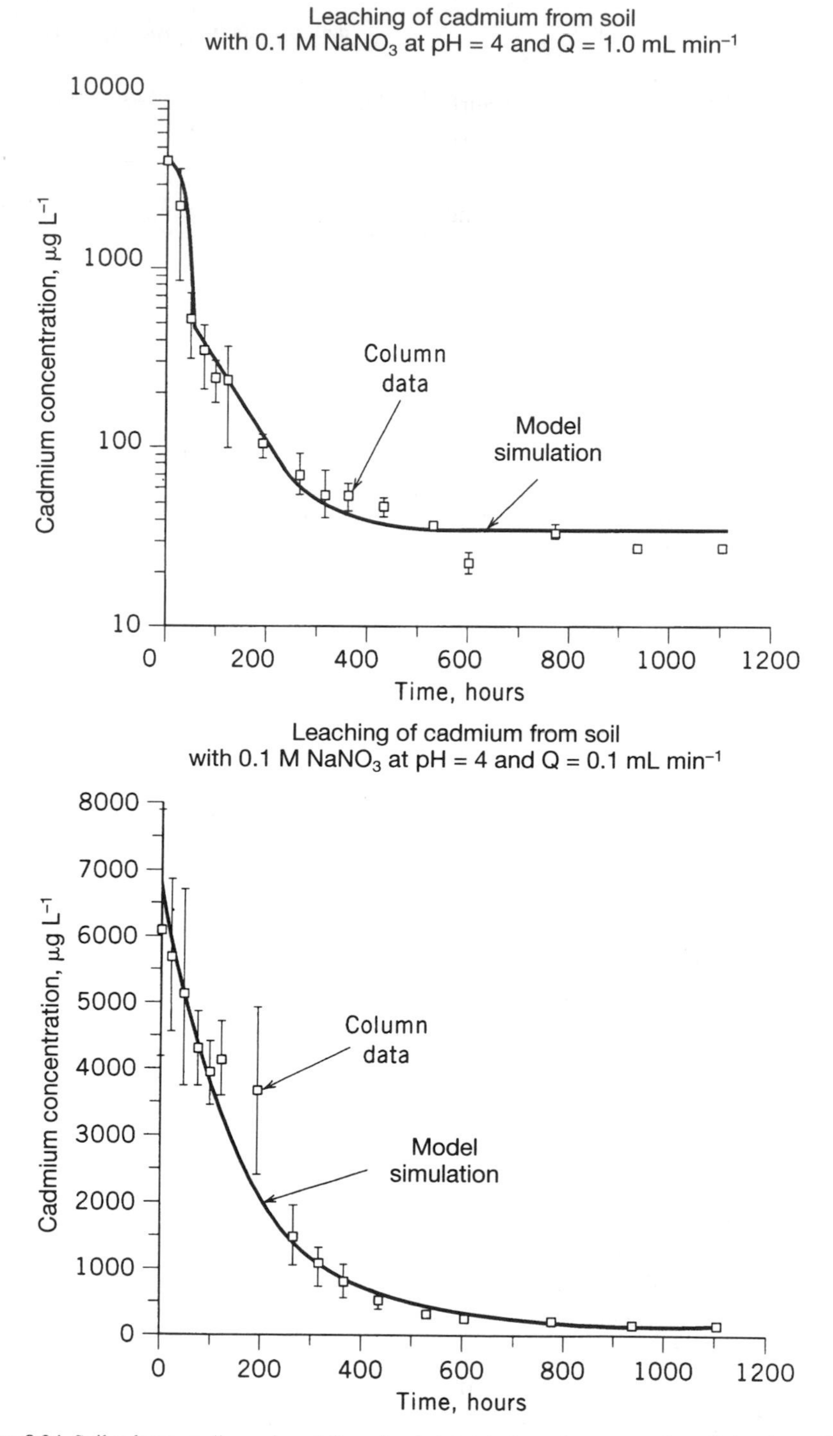

Figure 8.24 Soil column studies and modeling of cadmium at a hazardous waste site, pH 4 and two different flushing rates. (From Nikolaidis et al.[47])

SOH_2^+	$\log K_{a1} = -4.0$	Surface acidity constant 1
SOH	$\log K_{a2} = -9.0$	Surface acidity constant 2
Me^{2+}	$\log K = -8.0$	Hydrolysis 1
$SOMe^+$	$\log K = -1.0$	Binding constant (stability of formation constant)

The surface area of adsorbing hydrous oxides is 10 $m^2\ g^{-1}$, the number of binding sites is 10^{-8} mol m^{-2}, and the concentration of adsorbing solids is 1000 g L^{-1}. Total dissolved metal concentration is 10^{-7} M, and the pH of the soil is 5.0.

a. Calculate the concentration of all species from pH 3 to 10.
b. Plot the adsorption edge of metal onto soil particles. At what pH is essentially all of the free aquo metal ion bound?
c. Plot the surface charge as a function of pH. At pH 5, are the soil particles positively or negatively charged?
d. Estimate the K_d value at pH 5 and pH 7.

Solution: Use the diffuse double layer model with an inner capacitance of 2 F m^{-2}. Assume an ionic strength of 0.01 M. There are seven species and four components including the coulombic correction factor (surface charge). The concentration of surface sites is 10^{-4} M even though we must enter its activity as 1.0 in the chemical equilibrium program (activity of a solid phase).

$$[SOH] = (10\ m^2/g)\ (10^{-8}\ mol/m^2)\ (10^3\ g/L) = 10^{-4}\ M$$

The matrix of chemical reactions must be rearranged to express all of the species in terms of the components. Components should include Me^{2+}, the concentration of surface sites SOH, and H^+.

	Components				
Species	Me^{2+}	SOH	Surface Charge	H^+	log K
Me^{2+}	1	0	0	0	0
$MeOH^+$	1	0	0	−1	− 8.0
SOH	0	1	0	0	0
SOH_2^+	0	1	+1	1	4.0
SO^-	0	1	−1	−1	− 9.0
$SOMe^+$	1	1	+1	−1	− 1.0
H^+	0	0	0	1	0
C_T	1×10^{-7}	1.0		1×10^{-5}	

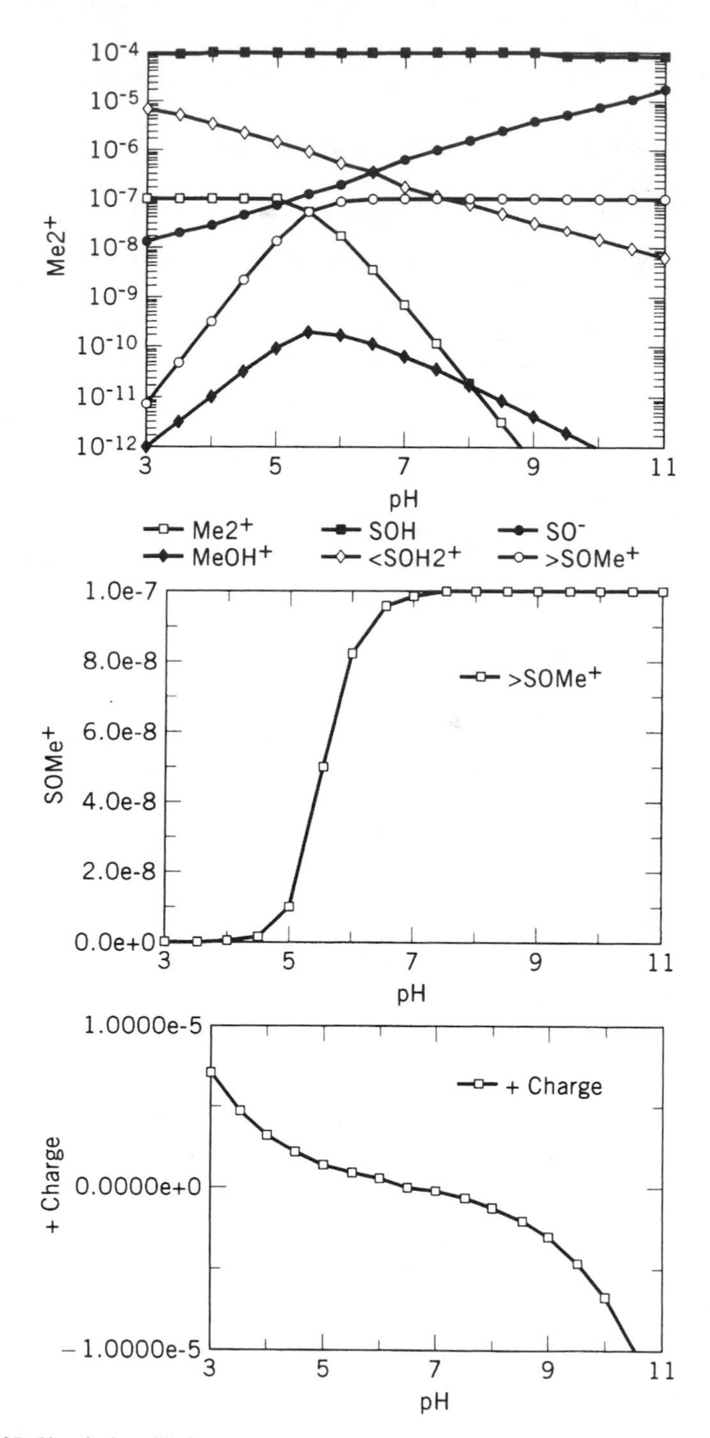

Figure 8.25 Chemical equilibrium model for metals speciation in a contaminated soil. *Top*: Equilibrium concentrations (M) as a function of pH. *Middle*: Adsorption edge from pH 5 to 7. *Bottom*: Surface charge on the soil particles as a result of the surface complexation of hydrous oxides for metal cations.

Figure 8.25 shows graphical output from the model.

a. The concentration of free aquo metal ion (Me^{2+}) predominates at pH $\leq$ 5. Adsorbed metal concentrations are predominant above pH 6. At pH 5.5, the adsorbed and dissolved concentrations are equal.
b. The adsorption edge rises sharply from pH 5 to pH 7. Essentially all of the metal ion is bound at pH 7, so liming the soil would be an effective way to immobilize the metal ions.
c. The surface charge is zero at pH 6.5. Soil particles are positively charged due to adsorption of metal ions at pH 5.
d. The K_d values are strong functions of pH.

$$\frac{[SOMe^+]}{[Me^{2+}]} \simeq 0.1 \quad \text{at pH 5}$$

$$\simeq 100 \quad \text{at pH 7}$$

8.8 CLOSURE

In this chapter, we have learned that chemical speciation drives the fate, transport, and toxicity of metals in natural systems. We must combine mass balance equations with chemical equilibrium models to address these problems. It is difficult to obtain all the desired model parameters and equilibrium constants, but a multiple approach of model calibration, collection of field data, microcosm experiments, and laboratory measurements will prove to be successful. Considerable improvements in the state-of-the-art for waste load allocations of metals are needed to put regulatory actions on a more scientifically defensible basis.

8.9 REFERENCES

1. Benoit, G., *Environ. Sci. Technol.*, 28, 1987 (1994).
2. Stumm, W., and Morgan, J.J., *Aquatic Chemistry*, 2nd ed., Wiley, New York (1981).
3. Stumm, W., and Morgan, J.J., *Aquatic Chemistry*, 3rd ed., Wiley, New York (1995).
4. Windom, H.L., Byrd, J.T., Smith, R.J. Jr., and Huan, F., *Environ. Sci. Technol.*, 25, 1137 (1991).
5. Cohen, P., *Environ. Sci. Technol.*, 25, 1940 (1991).
6. Horowitz, A.J., Lum, K.R., Garbarino, J.R., Hall, G.E.M., Lemieux, C., Demas, C.R., *Environ. Sci. Technol.*, 30, 954 (1996).
7. Shiller, A.M., and Taylor, H., *Environ. Sci. Technol.*, 29, 1313 (1995).
8. Buffle, J., in *Determination of Trace Metals in Natural Waters*, T.W. West and W. Nürnberg, Eds., Blackwell, Oxford (1988).
9. Flegal, A.R., and Coale, K.H., *Water Resour. Bull.*, 25, 1275 (1989).

10. Coale, K.H., and Flegal, A.R., *Sci. Total Environ.*, 87/88, 297 (1989).
11. Yen, B.C., in *Modeling of Rivers*, H.W. Shen, Ed., Wiley, New York (1979).
12. Fischer, H.B., List, E.J., Koh, R.C.Y., Imberger, J., and Brooks, N.H., *Mixing in Inland and Coastal Waters*, Academic Press, New York (1979).
13. Raytheon Company, Oceanographic and Environmental Services *RECEIV-II Water Quantity and Quality Model*, Office of Water, U.S. Environmental Protection Agency, Washington, DC (1974).
14. Johanson, R.C., Imhoff, J.C., and Davis, H.H., *User's Manual for Hydrological Simulation Program Fortran (HSPF)*, EPA-600/9-80-015, U.S. Environmental Protection Agency, Athens, GA (1980).
15. Brown, L.C., and Barnwell, T.O. Jr., *Computer Program Documentation for the Enhanced Stream Water Quality Model QUAL-2E*, EPA-600/3-85/065, U.S. Environmental Protection Agency, Athens, GA (1985).
16. Fontaine, T.D., *Ecol. Modeling*, 21, 287 (1984).
17. Holly, F.M. Jr., and Preissmann, A., *J. Hydrol. Div., ASCE*, 103, 1259 (1977).
18. Mossman, D.J., Schnoor, J.L., and Stumm, W., *J. Water Pollut. Control Fed.*, 60(10), 1806 (1988).
19. Schnoor, J.L., Ed., *Fate of Pesticides and Chemical in the Environment*, Chapter 11, Wiley, New York (1992).
20. Schnoor, J.L., Sato, C., McKechnie, D., and Sahoo, D., Processes, Coefficients, and Models for Simulating Toxic Organics and Heavy Metals in Surface Waters, EPA/600/3-87/015, U.S. EPA Environmental Research Laboratory, Athens, GA (1987).
21. North Carolina Department of Natural Resources and Community Development, Water Quality Evaluation Upper Deep River, Cape Fear River Basin, Division of Environmental Management Water Quality Section, Raleigh, NC (1985).
22. Allison, J.D., Brown, D.S., and Novo-Gradac, K.J., MINTEQA2/PRODEFA2 A Geochemical Assessment Model for Environmental Systems, EPA/600/3-91/021, U.S. Environmental Protection Agency, Washington, DC (1991).
23. Mills, W.B., Bowie, G.L., Grieb, T.M., and Johnson, K.M., Stream Sampling Handbook for Waste Load Allocation Applications, EPA/625/6-86/013 U.S. Environmental Protection Agency, Washington, DC (1986).
24. Morel, F.M.M., and Hering, J.G., *Principles and Applications of Aquatic Chemistry*, Wiley, New York (1993).
25. Buffle, J., *Complexation Reactions in Aquatic Systems: An Analytical Approach*, Ellis Harwood Limited, Chichester, England (1988).
26. Xue, H.B., and Sigg, L., *Limnol. Oceanogr.*, 38, 1200 (1993).
27. Cabaniss, S.E., and Shuman, M.S., *Geochim.Cosmochim. Acta*, 52, 185 (1988).
28. Perdue, E.M., and Lytle, C.R., in *Aquatic and Terrestrial Humic Materials*, R.F. Christman and E.T. Gjessing, Eds., Ann Arbor Science, Ann Arbor, MI (1983).
29. Schnitzer, M., and Skinner, S.I.M., *Soil Sci.*,103, 247 (1967).
30. Stumm, W., *Chemistry of the Solid–Water Interface*, Wiley, New York (1992).
31. Dzombak, D.A., and Morel, F.M.M., *Surface Complexation Modeling*, Wiley, New York (1990).
32. Müller, B., and Sigg, L., *J. Aquatic Sci.*, 52, 75 (1990).
33. Sigg, L., and Stumm, W., *Aquatische Chemische*, VDF Publishers, Zürich (1989).

34. Tiffreau, C., Lützenkirchen, J., and Behra, P., *J. Colloid Interface Sci.*, 172, 82–93 (1995).
35. MacNaughton, M.G., Adsorption of Hg(II) at the Solid–Water Interface, Ph.D. Thesis, Stanford University, Stanford, CA (1973).
36. Westall, J.C., FITEQL (version 2.0) A Program for Determination of Chemical Equilibrium Constants from Experimental Data, Oregon State University, Corvallis, OR (1982).
37. O'Connor, D.J., and Connolly, J.P., *Water Res.*, 14, 1517 (1980).
38. Thomann, R.V., and Mueller, J.A., *Principles of Surface Water Quality Modeling and Control*, Harper & Row, New York (1987).
39. Schnoor, J.L., Giovanoli, R., Sigg, L., Stumm, W., Sulzberger, B., and Zobrist, J., *EAWAG News* 26/27, 5 (1989).
40. Linthurst, R.A., Landers, D.A., Eilers, J.M., Brakke, D.F., Overton, W.S., Meier, E.P., and Crowe, R.E., Characteristics of Lakes in the Eastern United States, EPA 600/4-96/007a, U.S. Environmental Protection Agency, Washington, DC (1988).
41. Francis, B.M., *Toxic Substances in the Environment*, John Wiley & Sons, New York, p. 28 (1994).
42. Gavis, J., and Ferguson, J.F., *Water Res.*, 6, 989 (1972).
43. Vuceta, J., and Morgan, J.J., *Environ. Sci. Technol.*, 12, 1302 (1978).
44. Fagerström, T., and Jernelöv, A., *Water Res.*, 6, 1193 (1972).
45. Ferguson, J.F., and Gavis, J., *Water Res.*, 6, 1259 (1972).
46. Braman, R.S., and Foreback, C.C., *Science*, 182, 1247 (1973).
47. Nikolaidis, N. P., Cheeda, P., Lackovic, J., and Carley, R. A. Framework for the Development of Mobility Based Cleanup Standards for Heavy Metal Contaminated Soils. Technical Report ERI-96.05, Environmental Research Institute, The University of Connecticut, Storrs, CT (1996).

8.10 PROBLEMS

1. Use a chemical equilibrium model to compute and plot pC versus pH for the solubility of $Al(OH)_{3(s)}$, amorphous aluminum hydroxide using log K_{s0} = 8.0 at 25 °C and a table for stability constants of aluminum hydroxy complexes.

$$Al(OH)_{3(s)} + 3H^+ \rightleftharpoons Al^{3+} + 3H_2O$$

$$\frac{[Al^{3+}]}{[H^+]^3} = 10^{8.0}$$

Give two reasons why the solubility of amorphous aluminum hydroxide is important in acidic natural waters. Rework the problem for 5 °C. Is it more or less soluble at 5 °C than at 25 °C?

2. Use a chemical equilibrium model to compute and plot pC versus pH for the solubility of $Fe(OH)_{3(s)}$, amorphous iron hydroxide. Consider the following equilibrium reactions at 25 °C. Why is iron hydroxide important in natural waters?

$$Fe^{3+} + H_2O \rightleftharpoons FeOH^{2+} + H^+ \qquad \log K_1 = -2.16$$

$$Fe^{3+} + 2H_2O \rightleftharpoons Fe(OH)_2^+ + 2H^+ \qquad \log K = -6.74$$

$$Fe(OH)_{3(s)} \rightleftharpoons Fe^{3+} + 3OH^- \qquad \log K_{s0} = -38$$

$$Fe^{3+} + 4H_2O \rightleftharpoons Fe(OH)_4^- + 4H^+ \qquad \log K = -23$$

$$2Fe^{3+} + 2H_2O \rightleftharpoons Fe_2(OH)_2^{4+}\ 2H^+ \qquad \log K = -2.85$$

3. Lake Clara is a clear water oligotrophic seepage lake (no inlets or outlets) in northern Wisconsin with a pH of 6.1. It receives H_2SO_4 and heavy metals (Pb, Cd, and Hg) from atmospheric deposition. The dissolved organic carbon concentration is 6 mg L^{-1}. Assume a fulvic acid concentration of 1.43 meq g^{-1} C. Given the following lake chemistry, use a fixed-pH chemical equilibrium model to estimate the Cd species in solution. The binding constant for Cd–FA is $\log {}^cK = 3.5$ from metals titration analyses. Is cadmium considered toxic at this concentration? Dissolved concentrations in mg L^{-1}: Ca = 1.98, Na = 2.6, Mg = 0.54, K = 0.88, alkalinity = 1.89, sulfate = 4.33, chloride = 4.5, Pb(II) = 0.00012, and Cd(II) = 0.000014 mg L^{-1}. Temperature of the sample was 14.3 °C.

4. Repeat Problem 3 for Pb speciation at pH 6.1 and 14.3 °C. Assume $\log {}^cK = 5$ for Pb–FA complexation.

5. Lake Ontario has a mean hydraulic detention time of 8 years and a mean depth of 80 m. Most of the particles in the pelagic zone (open water) are from phytoplankton, 6 mg L^{-1} total suspended solids. Assume a solid/water distribution coefficient of $\log K_d = 7$ at pH 8 for Pb adsorption onto particles. Estimate the fraction of lead that is in the particulate adsorbed phase, f_p, and the fraction removed, R, if phytoplankton sinking velocities are 0.5 m d^{-1}.

6. Trace metal binding to particles is complicated by organic coatings or inorganic particle nonhomogeneities. For example, humic and fulvic acids adsorb to hydrous aluminum oxides, such as amorphous gibbsite, $Al(OH)_{3(s)}$. Given the following equilibrium constants, use the diffuse double-layer model to simulate the adsorption of fulvic acid on amorphous $Al(OH)_3$.

Acidity constant$_1$ for $\exists\ AlOH_2^+$	$\log K_{a1}^{intr} = 7.20$
Acidity constant$_2$ for $\exists\ AlOH$	$\log K_{a2}^{intr} = -9.50$
Acidity constant$_1$ for H_2FA	$\log K_{FA1} = -5.00$
Acidity constant$_2$ for HFA^-	$\log K_{FA2} = -14.00$
Surface complexation $\exists\ AlFA^-$	$\log K^{intr} = -2.00$

Assume a total concentration of fulvic acid of 1×10^{-5} M, a pH of 6, a surface area of 40 $m^2\ g^{-1}$, a surface site density of 3.2×10^{-4} mol g^{-1}, an ionic strength of

10^{-3} M, and 8.6 mg L^{-1} of $Al(OH)_{3(s)}$ particles. What is the concentration of $\exists$ $AlFA^-$ at pH 6? What fraction of the total sites ($\exists$ AlOH) are occupied by FA?

7. Make a plot of the time required to reduce a colloidal suspension of particles by half its number under quiescent conditions (perikinetic coagulation at 20 °C with a sticking coefficient of 1.0) versus the number of particles per cm^3 initially. See equation (47).

8. Waste Load Allocation Problem. The Iowa River at Iowa City has an average flow of approximately 2000 cfs. The 7 day, 1-in-10-year flow (7Q10) can be taken to be 50 cfs which is the lowest flow that is recommended by the U.S. Army Corps of Engineers, who run the dam at Coralville upstream (low flow augmentation). A new industry wants to locate a specialty metals finishing plant at the location, and they have applied for a new waste discharge permit for copper. The design flow for the plant wastewater discharge is 1×10^6 gallons per day with a technology-based maximum copper discharge concentration of 1.0 mg/L that is expected. The water quality standard for the State of Iowa for copper is 12 μg/L that can be exceeded no more frequently than one day in every three years.

a. Assuming an upstream concentration of copper of 5 μg/L, calculate the allowable discharge permit concentration for the new industry using the 7Q10 low flow value of 50 cfs. Use a simple flow-weighted average mass balance model to set the allowable discharge concentration. (The permit is based on total recoverable metal, dissolved + particulate forms of Cu.)

b. Now assume probability density functions for each of the key variables in the mass balance equation. Calculate C_T, the total copper concentration in the Iowa River after mixing, and the frequency with which it exceeds the standard of 12 μg/L if the following parameters apply. The data are mean and standard deviations.

$$Q_u = 2000 \pm 1000 \text{ cfs (river discharge)}$$

$$C_u = 5 \pm 2\ \mu\text{g/L (upstream copper concentration)}$$

$$Q_w = 1{,}000{,}000 \pm 200{,}000 \text{ gal/day (wastewater discharge)}$$

$$C_w = 0.5 \pm 0.2 \text{ mg/L (wastewater copper concentration)}$$

c. More realistically, the variables are lognormally distributed. Redo part b with the following lognormal distributions (geometric mean ± geometric deviations).

$\log Q_u = 3.26 \pm 0.5$
$\log C_u = 0.6 \pm 0.2$
$\log Q_w = 5.95 \pm 0.3$
$\log C_w = -0.4 \pm 0.2$

ANSWERS TO SELECTED PROBLEMS—CHAPTER 8

1. Species: Al^{3+}, $AlOH^{2+}$, $Al(OH)_2^+$, $Al(OH)_4^-$, H^+

Components: $Al(OH)_{3(s)}$, H^+

	$Al(OH)_{3(s)}$	H^+	log K
Al^{3+}	1	3	8.00
$AlOH^{2+}$	1	2	3.00
$Al(OH)_2^+$	1	1	–2.00
$Al(OH)_4^-$	–1	–1	–14.20
H^+	0	1	0.00
TOT (C_T)	1.0	1.0×10^{-6} initial fixed value	

Two reasons: (1) surface adsorption is one of the controls on trace metal concentrations in natural waters, and (2) inorganic monomeric aluminum is toxic to fish. Aluminum is more soluble at cold temperatures; dissolution of $Al(OH)_{3(s)}$ is exothermic. For solubility, sum the concentrations of soluble species.

3. $[Cd^{2+}] = 1.25 \times 10^{-5}$ mg L^{-1} = 1.1×10^{-10} M

[Cd–FA] = 1.2×10^{-6} mg L^{-1} = 1.05×10^{-11} M

It is mostly in the free aquo metal ion form, but cadmium is not considered toxic at such low concentrations.

6. The matrix for computer modeling is given below. The key reaction is

$$\exists\, AlOH + H_2FA \rightleftharpoons \exists\, AlFA^- + H^+ + H_2O \qquad \log K^{intr} = -2.0$$

At pH 6, $\exists$ $AlFA^-$ concentration is 5×10^{-7} M, which is about 25% of the total site concentration $\exists$ AlOH of 2.2×10^{-6} M.

	H_2FA	$\exists$ AlOH	Surface Charge	H^+	log K
$\exists$ AlOH	0	1	0	0	0.00
$\exists$ $AlOH_2^+$	0	1	1	1	7.20
$\exists$ AlO^-	0	1	–1	–1	–9.50
H_2FA	1	0	0	0	0.00
HFA^-	1	0	0	–1	–5.00
FA^{2-}	1	0	0	–2	–14.00
$\exists$ $AlFA^-$	1	1	–1	–1	–22.00
H^+	0	0	0	1	0.00
TOT	10^{-5} M	1.0 unit activity		10^{-6} M	

9

GROUNDWATER CONTAMINANTS

Make things as simple as possible, but not any simpler.
—Albert Einstein

9.1 INTRODUCTION

Groundwater contamination and soil pollution have become recognized as important environmental problems over the last 20 years. Recognition of the problem was followed by legislation to remediate the sites and to prevent further problems. Monitoring and remediation of these sites have proved to be costly and time consuming, and sometimes contaminant modeling of the subsurface is necessary to aid decision-making.

A wide variety of pollutants contaminate soils and groundwaters including toxic metals, organics, and radionuclides. Usually, the desired remediation criterion is to clean up groundwater contaminants to drinking water standards (see Chapter 1), a very stringent goal. Spills and leaks are often the source of contaminants that enter the soil (unsaturated zone) and eventually are transported to the groundwater table (saturated zone).

Modeling groundwater contaminants is difficult because of the inaccessibility of the plume below the ground surface and the heterogeneity of porous media. Transport of contaminants is complicated by the tortuous path through heterogeneous porous media and by the possibility of several contaminant phases (dissolved, adsorbed, gaseous, and pure organic phase). It is difficult to probe the site and expensive to obtain samples for analyses—monitoring wells and piezometers must be installed.

The first rule of groundwater modeling is to "know the geology"! You must determine the vertical stratigraphy, that is, the vertical layers of soil, sand, and rock units that comprise the site. From a good stratigraphic record (core borings or driller's logbook records) and a knowledge of the hydraulic gradient at the site, one can hypothesize the probable movement of contaminants and develop a plan to test that hypothesis. Engineers, environmental scientists, and geologists must work together to understand hazardous waste sites and remediation strategies. It is an interdisciplinary field. Table 9.1 gives a glossary of hydrogeology terms that are useful.

Table 9.1 Glossary of Hydrogeology Terms for Groundwater Contaminant Modeling

Aquiclude—a saturated geologic unit that is not capable of transmitting water.

Aquifer—a saturated permeable geologic unit that can transmit significant quantities of water.

Aquitard—a geologic unit of low permeability that precludes much water flow.

Anisotropic—an aquifer with different hydraulic conductivities and dispersivities in each direction due to preferential flow directions through the porous media.

Bailer test—an in situ, rapid drawdown of the water in a single well to measure the rate that it refills, to estimate the saturated hydraulic conductivity in the immediate vicinity of the well.

Capillary fringe—the rise of water just above the water table to fill pores due to surface tension and capillary action in the porous media; tension saturated zone with pressure head less than atmospheric.

Confined aquifer—an aquifer that is confined between two aquitards.

Darcy's law—the empirical relationship derived by Darcy in 1856 that superficial velocity (specific discharge) is directly proportional to the hydraulic gradient in the direction of flow.

Head—the unit of groundwater energy per unit volume of water that includes pressure, elevation (potential energy), and kinetic energy (often negligible), given in meters of water.

Hydraulic conductivity—the rate of flow per unit time per unit cross-sectional area; it is a property of the fluid (water) and the porous media (in units such as cm s^{-1} or gal d^{-1} ft^{-2}).

Homogeneous—the same porous media in all directions in the subsurface environment.

Monitoring well—a constructed well that is screened over an interval and used to collect groundwater samples from a given aquifer.

Permeability—the relative degree of flow through porous media; it depends on grain size and media properties only; units are cm^2.

Permeameter—a device to estimate saturated hydraulic conductivity of aquifer material in the laboratory.

Piezometer—a small diameter well that is open at the top and at the very bottom; it is used to measure head at the point where it is inserted at the bottom.

Porosity—ratio of pore volume to total volume, dimensionless.

Porous media—the soil, sand, silt, clay, or rock through which groundwater flows.

Pump test—the in situ drawdown of a well in a field of monitoring wells to measure the change in head of the monitoring wells and to estimate the hydraulic conductivity and specific storage of the aquifer in the area of the well field.

Saturated zone—geologic unit that is saturated with water in the pore spaces (includes the capillary fringe).

Slug test—the in situ, rapid input of water into a single well to measure the rate at which water escapes and to estimate the saturated hydraulic conductivity.

Specific storage—volume of water that a unit volume of aquifer releases from storage under a unit decline in hydraulic head; units are L^{-1}.

Specific yield—storage in unconfined aquifers, that is, the volume of water that will drain freely per unit decline in water table per unit surface area of an aquifer, dimensionless.

Storativity—same as storage coefficient, the volume of water that an aquifer releases from storage per unit surface area of aquifer per unit decline in hydraulic head, dimensionless.

Stratigraphy—the record of each geologic unit with depth, the depth of each unit, and the order in which they occur.

(continued)

Table 9.1 *(continued)*

Suction head—negative pressure head due to capillary forces under unsaturated conditions, L.
Surficial aquifer—same as water table aquifer or phreatic aquifer.
Transmissivity—the saturated hydraulic conductivity times the aquifer thickness, gal d^{-1} ft^{-1}.
Unconfined aquifer—same as water table aquifer, no confining units above this aquifer.
Unsaturated zone—the subsurface down to the top of the capillary fringe.
Vadose zone—the unsaturated zone below the root zone and above the saturated zone.
Water table aquifer—an aquifer in which the water table forms the upper boundary; same as surficial aquifer.
Water table—the surface on which the fluid pressure in the porous medium is exactly atmospheric; the top of the saturated zone below the capillary fringe.

Groundwater moves very slowly, on the order of 1 cm per day, so it takes a long time for contaminants to reach a drinking water aquifer. The residence time of the water in the surficial aquifer is likely to be on the order of decades, and deep aquifer waters are thousands of years old. So the good news is that it takes a long time to pollute an aquifer, but the bad news is that once the aquifer is contaminated, it will probably take a very long time for natural restoration.

Sources of groundwater and soil contamination are many and varied:

- Agricultural infiltration (N-fertilizers, pesticides).
- Leachate from mounds of chemicals stored aboveground.
- Infiltration from pits, ponds, and lagoons.
- Landfill leachate and infiltration (municipal, industrial, fly ash).
- Leaks from underground storage tanks, septic tanks, or sewer lines.
- Spills (jet fuel, industrial chemicals, waste oil, etc.).
- Hazardous industrial wastes in buried drums or landfills (petrochemicals, chlorinated solvents, munitions, manufactured gas plant PAHs, creosote plants, electroplating, etc.).
- Nuclear wastes (low-level, intermediate, and mixed wastes).

By the time that the environmental scientist or engineer is called upon, the source of the waste has usually abated. The company has moved, closed, or gone bankrupt, and the record of what was done at the site is sketchy. It is the job of the mathematical modeler to use existing information to reconstruct the pollution history and to devise remediation strategies. The first step is to construct a good flow equation for the movement of water.

9.2 DARCY'S LAW

In 1856, a French hydraulic engineer, Henry Darcy, developed an empirical relationship for water flow through porous media. He found that the specific discharge was directly proportional to the energy driving force (the hydraulic gradient) according to the following relationship:

$$\nu_x \propto \frac{\Delta h}{\Delta x} \tag{1}$$

where ν_x = specific discharge in the x-direction, LT^{-1}
Δh = the change in head from point 1 to point 2, L
Δx = the distance between point 1 and point 2, L
$\Delta h/\Delta x$ = the hydraulic gradient in the x-direction, dimensionless

Figure 9.1 is a schematic illustrating Darcy's law.[1] It represents a tube filled with sand or suitable porous media with cross-sectional area A. The specific discharge ν_x is defined as the flow volume per time per unit area (cm s^{-1} or gal d^{-1} ft^{-2}). Darcy found that if he doubled the hydraulic gradient, $\Delta h/\Delta x$, by raising the water reservoir, then the specific discharge would also double. But equation (1) needed a proportionality constant to be dimensionally consistent.

$$\nu_x = - K_x \frac{dh}{dx} \tag{2}$$

where K_x = saturated hydraulic conductivity in the x-direction, LT^{-1}

The negative sign is necessary in Darcy's equation (2) because water flows from areas of high head to low head (a negative hydraulic gradient). To achieve a positive

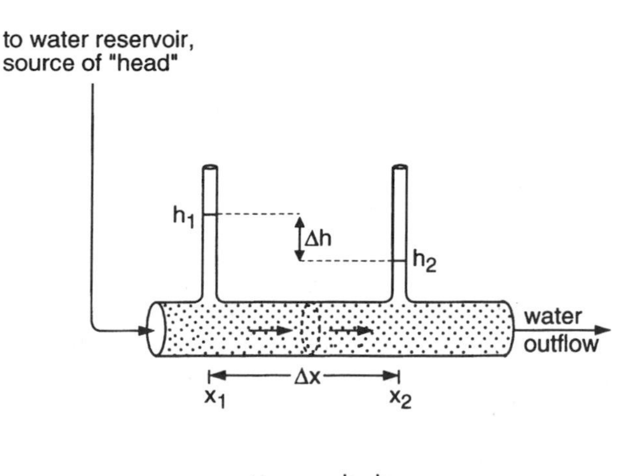

Figure 9.1 Schematic illustrating Darcy's law for flow through porous media, which is proportional to the hydraulic gradient, $\Delta h/\Delta x$. [$\Delta h/\Delta x = (h_2 - h_1)/(x_2 - x_1)$.]

value for v_x, a negative hydraulic gradient is required. The saturated hydraulic conductivity K_x is a measure of both fluid properties and aquifer media properties. In general, it is greatest for coarse sands and gravel (Table 9.2).

Specific discharge, v_x, is sometimes referred to as the superficial velocity or the Darcy velocity. It has units of velocity (length per time), but it is not the actual velocity that water would move through the aquifer if we injected a conservative tracer. The actual velocity is the specific discharge divided by the fractional porosity under saturated conditions. It is greater than the specific discharge because water is "throttled" through the narrow pore spaces, creating faster movement.

$$u_x = \frac{v_x}{n} \quad \text{or} \quad \frac{v_x}{n_e} \tag{3}$$

where u_x = actual fluid velocity, LT^{-1}

n = porosity or void volume/total volume

n_e = effective porosity

In consolidated aquifers, the effective porosity can be smaller than the total porosity (void volume/total volume). Effective porosity reflects the interconnected pore volume through which water actually moves, and it is the proper parameter to use in equation (3).

Example 9.1 Velocity and Time of Travel for Groundwater Flow

Piezometers are installed 100 m apart along a transect of a surficial aquifer. The following hydraulic heads are recorded relative to mean sea level.

Table 9.2 Typical Hydrogeologic Parameters for Various Aquifer Units and Materials

Material	Porosity n, %	Specific Yield, %	Hydraulic Conductivity, K, cm s^{-1}	Permeability, k, cm^2
Unconsolidated deposits				
Gravel	25–35	25–35	1–100	10^{-5}–10^{-3}
Sand	30–45	25–40	10^{-4}–10^{-1}	10^{-9}–10^{-6}
Silt	35–45	20–35	10^{-6}–10^{-4}	10^{-11}–10^{-9}
Clay	40–55	2–10	10^{-9}–10^{-6}	10^{-14}–10^{-11}
Rocks				
Karst limestone	15–40	10–35	10^{-4}–10^{-1}	10^{-8}–10^{-6}
Limestone, non-Karst	5–15	2–10	10^{-6}–10^{-4}	10^{-11}–10^{-9}
Sandstone	10–25	5–15	10^{-7}–10^{-4}	10^{-12}–10^{-9}
Shale	0–10	0–5	10^{-11}–10^{-7}	10^{-16}–10^{-13}
Crystalline rock (fractured)	1–10	1–10	10^{-6}–10^{-4}	10^{-11}–10^{-9}
Crystalline rock (unfractured)	0–2	0–1	10^{-11}–10^{-9}	10^{-16}–10^{-14}

Piezometer	Head, m	Distance, m
A1	150.0	0
A2	149.5	100
A3	149.0	200

The surficial aquifer is composed of fine sand with a porosity of 0.33 and a saturated hydraulic conductivity of 0.001 cm s^{-1}. Calculate the specific discharge, the actual velocity, and the time of travel for groundwater to move from point A1 to A3.

Solution: The hydraulic gradient is constant along the three-piezometer transect.

$$\frac{\Delta h}{\Delta x} = \frac{h_3 - h_1}{x_3 - x_1} = \frac{149.0 - 150.0}{200 - 0} = -0.005 \text{ m m}^{-1}$$

Using Darcy's law, we can find the specific discharge. Groundwater is moving from piezometer A1 (high head) to A3 (low head).

$$v_x = -K_x \frac{dh}{dx} = -\left(\frac{0.001 \text{ cm}}{\text{s}}\right)\left(\frac{-0.005 \text{ m}}{\text{m}}\right)$$

$$v_x = 5.0 \times 10^{-6} \text{ cm s}^{-1}$$

Then the actual velocity is only about 1 cm per day (typical).

$$u_x = v_x/n = 1.51 \times 10^{-5} \text{ cm s}^{-1} = 1.31 \text{ cm d}^{-1}$$

The travel time for 200 m is 41.8 years!

$$t = x/u_x = 200 \text{ m}/0.0131 \text{ m d}^{-1} = 15{,}300 \text{ days} = 41.8 \text{ years}$$

Sometimes it is desirable to estimate the flowrate of groundwater moving in the aquifer, and it is necessary to define a cross-sectional area through which the aquifer is flowing (Figure 9.2).

$$Q = -K_x A \frac{dh}{dx} \tag{4}$$

where Q = volumetric flowrate, $L^3 T^{-1}$
K_x = saturated hydraulic conductivity, LT^{-1}
A = cross-sectional area, L^2
dh/dx = hydraulic gradient, LL^{-1}

The specific discharge and the flowrate are dependent on both the properties of the fluid (water) and properties of the media (aquifer materials) contained in the hydraulic conductivity parameter.

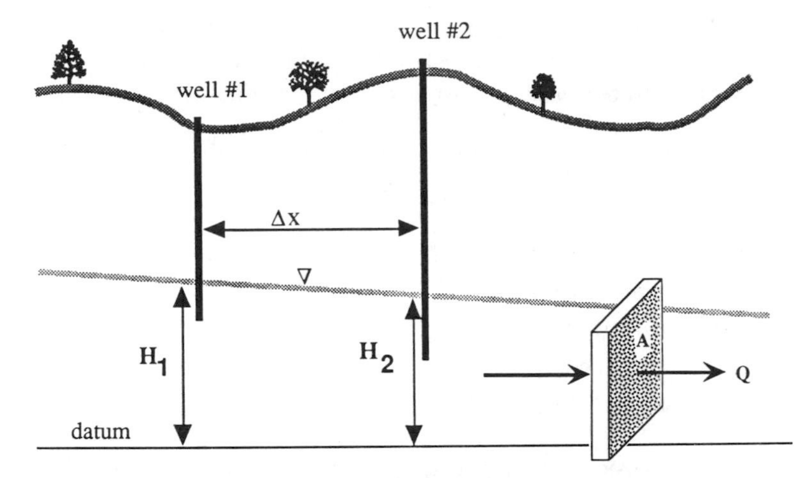

Figure 9.2 Flowrate Q in an aquifer is defined through a cross-sectional area A and by the hydraulic gradient between two observation wells. $Q = -K_x A\ (\Delta h/\Delta x)$.

$$K = \frac{Cd^2 \rho g}{\mu} = \frac{k \rho g}{\mu} \tag{5}$$

where K = saturated hydraulic conductivity, LT^{-1}
C = proportionality constant, dimensionless
d = particle diameter, L
ρ = fluid density, ML^{-3}
g = gravitational constant, LT^{-2}
μ = viscosity, $ML^{-1}T^{-1}$ ($N\ s\ m^{-2} \equiv kg\ m^{-1}\ s^{-1}$)
k = intrinsic permeability, L^2 ($k = Cd^2$)

The properties of the porous media are contained in the parameters Cd^2, which is called the specific or intrinsic permeability (k). The properties of the fluid are embedded in the parameters ρ and μ. Water density depends on temperature and salinity, while the viscosity is dependent on temperature.

Hazen related the saturated hydraulic conductivity empirically to the grain size diameter for uniformly graded sands

$$K = ad_{10}^2 \tag{6}$$

where a = Hazen's constant
d_{10} = particle diameter from the standard sieve analysis where 10% (by mass) of the particles are smaller than this diameter

Hazen's constant, a, is 1.0 if d_{10} is expressed in units of mm and K is expressed in units of cm s^{-1} for water.

The Kozeny–Carmen equation is another relationship that can be used to estimate saturated hydraulic conductivity.[2] It is similar in form to equation (5), except that it replaces the proportionality constant C with a relationship based on porosity:

$$K = \frac{n^3}{(1-n)^2}\left(\frac{d_m^{\ 2}}{180}\right)\left(\frac{\rho g}{\mu}\right) \tag{7}$$

where n = porosity, dimensionless
d_m = median particle diameter, L
ρ = fluid density, ML^{-3}
g = gravitational constant, LT^{-2}
μ = viscosity, $ML^{-1}T^{-1}$

Equation (7) is dimensionally consistent, so any set of consistent units will result in a hydraulic conductivity with dimensions LT^{-1}.

The best way to estimate K is from pump tests with multiple observation wells in the field. Slug or bailer tests may also be used for estimates in the vicinity of a single well. Often, lab permeater tests are run on porous material from drill borings, but disturbed point samples should not be expected to yield accurate measurements of K.

Darcy's law applies in all three directions [equations (8)–(10) for horizontal and vertical flow in aquifers], but it is an empirical relationship with some limitations. At very low velocities in fine-grained materials, there may be a hydraulic gradient threshold below which flow does not occur. Also, at very high velocities (turbulent flow), the kinetic energy term causes the relationship between specific discharge and hydraulic gradient to be nonlinear. Nonlaminar flow does not often occur in groundwater, but flow in fractured rock can be rapid. It depends on the connectivity of the fractures, and the aquifer may not behave as a Darcy continuum.

$$\text{Longitudinal} \qquad v_x = -K_x \frac{dh}{dx} \tag{8}$$

$$\text{Transverse} \qquad v_y = -K_y \frac{dh}{dy} \tag{9}$$

$$\text{Vertical} \qquad v_z = -K_z \frac{dh}{dz} \tag{10}$$

When K_x differs from K_y and K_z at a point in an aquifer, the flow is anisotropic. When the hydraulic conductivity varies from point to point within an aquifer, it constitutes nonhomogeneous conditions. Figure 9.3 illustrates the point. Often, we can assume homogeneous, anisotropic conditions for horizontally bedded sedimentary units and sand/gravel deposits. Usually $K_x = K_y > K_z$ because flow in the horizontal plane is preferred over vertical flow through bedded deposits.

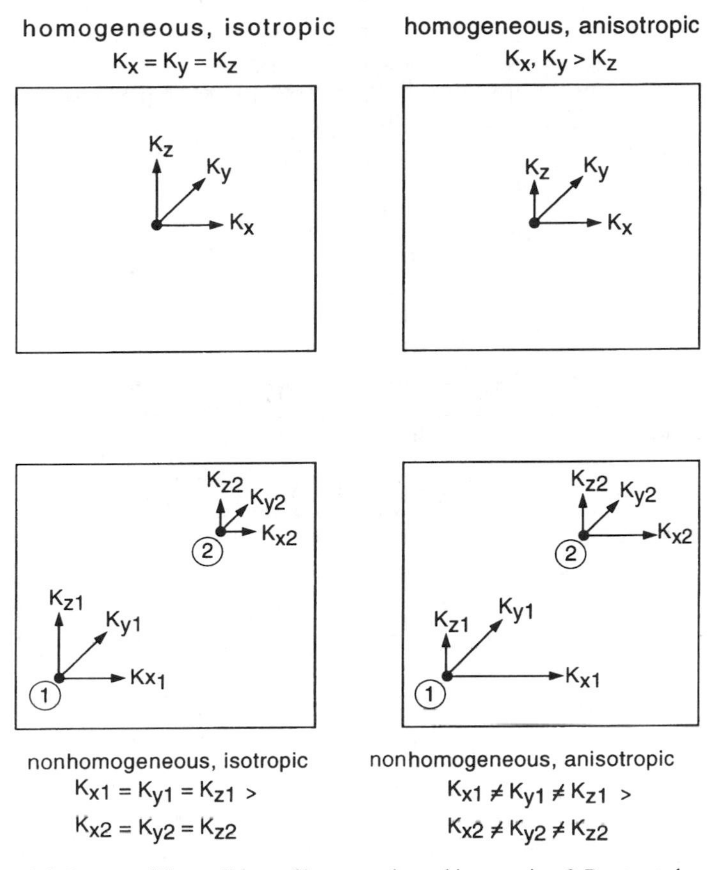

Figure 9.3 Four possible conditions of homogeneity and isotropy in a 3-D saturated aquifer.

Hydraulic gradients in the *x*- and *y*-directions are obtained from piezometers that are completed into the same aquifer unit in the horizontal plane (see Example 9.1). But vertical hydraulic gradients must be obtained from nests of piezometers completed at different depths to estimate the potentiometric difference (Example 9.2). Figure 9.4 illustrates the use of piezometers to measure hydraulic gradients. In Figure 9.4, there are three nests of piezometers; each nest is located at one location on the land surface (there is essentially no horizontal distance between the piezometers located at position A, for example). Piezometers A2, B2, and C2 are completed within the surficial sand aquifer as are A3, B3, and C3. They also measure the head at the water table elevation. Piezometers A1, B1, and B2 measure the head in the confined aquifer 1, which defines the potentiometric surface there. Each aquifer unit has its own potentiometric surface.

Piezometers measure the head (the hydraulic potential) at the point at the very bottom of the tube where they are open (or screened). There is no vertical flow between the points at the bottom of piezometers A2 and A3, for example, because

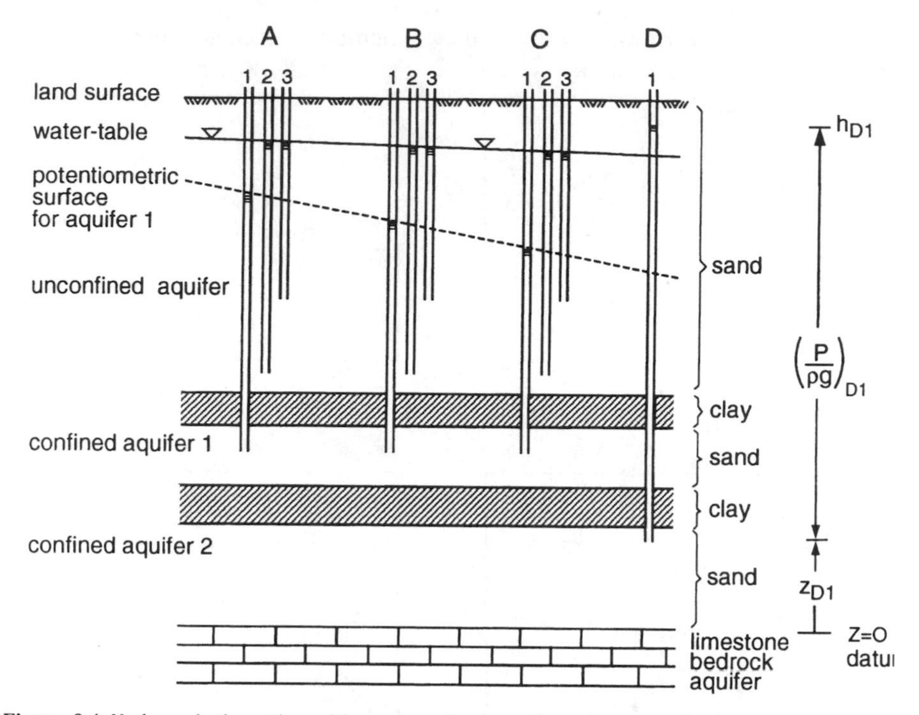

Figure 9.4 Hydrogeologic setting with an unconfined aquifer and two confined aquifers. Nests of piezometers located at A, B, and C along a transect define the vertical and longitudinal hydraulic gradient.

there is no head difference between them. There is a head differential between piezometers A2 and B2 (and A3 and B3), indicating longitudinal flow between points A and B. The water table defines the hydraulic gradient for the surficial sand aquifer. Also, there is a head difference between piezometers A1 and A2, indicating vertical flow of water in the downward direction from the sand aquifer through the clay confining layer to the confined aquifer 1. Because the clay unit is an aquitard, not much water is flowing and K_z is very small, but nevertheless there is a tendency for water to flow downward from the sand aquifer to the confined aquifer 1. In effect, the surficial sand aquifer is recharging the confined aquifer 1. Piezometer D1 is completed into confined aquifer 2, and water rises in the tube to an elevation greater than the water table. This makes the confined aquifer 2 an "artesian well" because it has a greater total head than the surficial sand aquifer. If the pressure head had been above the ground elevation, piezometer D would be a "flowing artesian" well. There is vertical flow upward from confined aquifer 2 to confined aquifer 1.

Hydraulic head is a measure of the energy per unit volume of water:

$$h = z + \frac{p}{\rho g} + \frac{v^2}{2g} \tag{11}$$

where h = hydraulic head, L
z = elevation of the water above a reference point, L
p = the gage pressure (above atmospheric), $ML^{-1}T^{-2}$
ρ = water density, ML^{-3}
g = acceleration due to gravity, LT^{-2}
v = velocity of the water, LT^{-1}

The three terms on the right-hand side of equation (11) correspond to elevation head, pressure head, and kinetic energy. The kinetic energy term can be neglected in most cases because flow in groundwater aquifers is laminar and velocities are very small. (An exception is the flow in the immediate vicinity of production wells where the pumping rate is high.) In Figure 9.4, the elevation head and the pressure head are shown for piezometer D1. Elevation head is relative to a datum reference point that is arbitrarily chosen; often mean sea level is used. The elevation head z is measured at the point where the measurement is taken, that is, the bottom opening of the piezometer.

Example 9.2 Vertical Flow Calculation from Nested Piezometers

Estimate the vertical velocity between the surficial sand aquifer (n = 0.30) and the confined aquifer 1 (n = 0.45) at point B in Figure 9.4. The hydraulic conductivity K_z in the sand is 10^{-4} cm s^{-1}, and the K_z of the confining clay layer is 10^{-7} cm s^{-1}. The piezometer elevation is at the bottom of the piezometer (where the measurement is made).

Piezometer	Piezometer Elevation, z	Measured Head, h
B1	220 m	460 m
B2	245 m	470 m

Solution: Calculate the vertical hydraulic gradient

$$\frac{\Delta h}{\Delta z} = \frac{h_2 - h_1}{z_2 - z_1} = \frac{470 - 460}{245 - 220} = 0.40 \text{ m m}^{-1}$$

As a first approximation, the proper hydraulic conductivity to use is the K_z for the clay unit because it has the highest resistance to flow. Calculate the specific discharge and then the velocity.

$$v_z = -K_z \frac{dh}{dz}$$

$$v_z = -(10^{-7} \text{ cm s}^{-1})(0.40 \text{ m m}^{-1}) = -4.0 \times 10^{-8} \text{ cm s}^{-1}$$

$$u_z = -\frac{4 \times 10^{-8}}{0.45} = -8.9 \times 10^{-8} \text{ cm s}^{-1} = -2.8 \text{ cm yr}^{-1}$$

Water is moving downward from the sand unconfined aquifer to the confined aquifer 1 at a very slow rate. If contamination occurred in the sand aquifer, it would take decades or longer to reach the confined aquifer 1.

9.3 FLOW EQUATIONS

We cannot determine contaminant transport in an aquifer if we do not know where the water is moving. Groundwater must satisfy the equation of continuity; water is conserved in flow through porous media. The equation of continuity is given below for nonsteady-state conditions in a confined or an unconfined aquifer.

$$\frac{\partial (\rho v_x)}{\partial x} + \frac{\partial (\rho v_y)}{\partial y} + \frac{\partial (\rho v_z)}{\partial z} = \frac{\partial (\rho n)}{\partial t} \tag{12}$$

where ρ = water density, ML^{-3}
$v_{x,y,z}$ = specific discharge in longitudinal, lateral, and vertical directions, LT^{-1}
n = porosity of the porous medium
t = time, T

The units of each term of equation (12) are $ML^{-3}T^{-1}$ or mass change of water per unit volume (in an elementary control volume) due to any changes of flow in three dimensions. The term on the right-hand side of equation (12) is for the change in mass of water due to expansion or compression (a change in density) or due to compaction of the porous medium (a change in porosity).

$$\frac{\partial (\rho n)}{\partial t} = n\frac{\partial \rho}{\partial t} + \rho\frac{\partial n}{\partial t} = \rho\, S_s \frac{\partial h}{\partial t} \tag{13}$$

where S_s = specific storage = $\rho g(\alpha + n\beta)$, L^{-1}
α = aquifer compressibility, LT^2M^{-1}
β = fluid compressibility, LT^2M^{-1}

Units on compressibility are generally $m^2\ N^{-1}$ or $m\ s^2\ kg^{-1}$. Substituting equation (13) into equation (12) results in

$$\frac{\partial (\rho v_x)}{\partial x} + \frac{\partial (\rho v_y)}{\partial y} + \frac{\partial (\rho v_z)}{\partial z} = \rho\, S_s \frac{\partial h}{\partial t} \tag{14}$$

Expanding terms on the left-hand side of equation (14) results in terms with $\partial\rho/\partial x$,

$\partial\rho/\partial y$, and $\partial\rho/\partial z$, which can be neglected in comparison to those terms shown below.

$$\rho\frac{\partial v_x}{\partial x}+\rho\frac{\partial v_y}{\partial y}+\rho\frac{\partial v_z}{\partial z}=\rho S_s\frac{\partial h}{\partial t} \tag{15}$$

We may divide by the fluid density and substitute Darcy's law into the left-hand side of equation (15).

$$\frac{\partial}{\partial x}\left(K_x\frac{\partial h}{\partial x}\right)+\frac{\partial}{\partial y}\left(K_y\frac{\partial h}{\partial y}\right)+\frac{\partial}{\partial z}\left(K_z\frac{\partial h}{\partial z}\right)=S_s\frac{\partial h}{\partial t} \tag{16}$$

Equation (16) is a second-order partial differential equation in three dimensions that requires a numerical solution. Boundary conditions and an initial condition are needed, as well as aquifer parameters K and S_s. It is the basic equation for unsteady flow in aquifers.

If the aquifer material is homogeneous and isotropic, we may divide equation (16) by the saturated hydraulic conductivity and obtain an expression with second-order differentials on the left-hand side:

$$\frac{\partial^2 h}{\partial x^2}+\frac{\partial^2 h}{\partial y^2}+\frac{\partial^2 h}{\partial z^2}=\frac{S_s}{K}\frac{\partial h}{\partial t} \tag{17}$$

where K is now the hydraulic conductivity in all three directions. For flow to a well and the flow net that is created, equation (17) is usually expressed in terms of transmissivity and storage. Storage in a horizontal confined aquifer of thickness (depth) b is defined as $S = S_s\, b$. Transmissivity of a confined aquifer is defined as $T = Kb$. Therefore, we can modify the right-hand side of equation (17) for confined aquifers of thickness b and for horizontal flow to a well.

$$\frac{\partial^2 h}{\partial x^2}+\frac{\partial^2 h}{\partial y^2}=\frac{S}{T}\frac{\partial h}{\partial t} \tag{18}$$

where S = storage coefficient, dimensionless
T = transmissivity, L^2T^{-1}

Note that each term in the equation has units of L^{-1}. If steady-state conditions exist, we have the well-known Laplace equation in two dimensions.

$$\frac{\partial^2 h}{\partial x^2}+\frac{\partial^2 h}{\partial y^2}=0 \tag{19}$$

The Laplace equation governs flow in a unit volume of saturated, homogeneous, and isotropic porous medium under steady-state conditions. Boundary conditions are required. The solution to equation (19) is a value of head at each point in the flow system.

Equation (19) can be solved by numerical methods such as the four-star or five-point operator routine to obtain the potentiometric surface of a two-dimensional aquifer. Development of these equations has followed that of the classical treatment by Freeze and Cherry,[1] and it is essentially that of Bear[2] and Jacob.[3] Other groundwater reference texts that are recommended include Strack,[4] Domenico and Schwartz,[5] and Hemond and Fechner.[6]

Example 9.3 Numerical Solution to Laplace Equation

Calculate the heads for a simple two-dimensional aquifer (depth and longitude) assuming no flow boundaries at the bottom. Use the five-point central differencing method outlined below. A FORTRAN program is listed to help with the programming.

Aquifer Data

K = isotropic hydraulic conductivity = 100 gal d^{-1} ft^{-2}
H_0 = initial head = 500.00 ft
Δx = numerical spatial increment = 7.00 ft
dh/dx = hydraulic gradient (slope) = 0.05 ft ft^{-1}

Numerical Method: We use the five-point method, central differencing technique. In this case, we will use 7 nodes in the vertical j-direction and 13 nodes in the longitudinal i-direction.

$$h_{i,j} = \frac{h_{i,j+1} + h_{i,j-1} + h_{i-1,j} + h_{i+1,j}}{4}$$

The above equation pertains to an isotropic aquifer, or the following equation is needed for an anisotropic aquifer where $K_x \neq K_y$. (Here the vertical hydraulic conductivity is defined as K_y.)

$$h_{i,j} = \frac{(\mathrm{KX}_{i,j})(h_{i+1,j}) + (\mathrm{KX}_{i-1,j})(h_{i-1,j}) + (\mathrm{KY}_{i,j})(h_{i,j+1}) + (\mathrm{KY}_{i,j-1})(h_{i,j-1})}{\mathrm{KX}_{i-1,j} + \mathrm{KX}_{i,j} + \mathrm{KY}_{i,j-1} + \mathrm{KY}_{i,j}}$$

Solution

```
PROGRAM FLOWNET
C      EXAMPLE TO CALCULATE FLOW 'Q' WITH VARIABLE CONDUCTIVITY
       REAL*4 H(13,7),KX(13,7), KY(13,7),Q(13), HO,DX,SLOPE,KK,KO
    1  WRITE (6,5)
    5  FORMAT ('ENTER INITIAL HEAD, VALUE FOR DX, AND SLOPE:')
       READ (5, 10) HO, DX, SLOPE
   10  FORMAT (F10.2/F10.2/F10.2)
```

```
C     DEFINE THE KX AND KY ARRAYS
      KO=100
      DO 15 J=1,7
      DO 15 I=1,13
      KX(I,J)=KO
      KY(I,J)=KO
 15   CONTINUE
C     INITIALIZE ALL HEADS TO HO
      DO 20 J=1,7
      DO 20 I=1,13
      H(I,J)=HO
 20   CONTINUE
C     WATER TABLE BOUNDARY: H(X,Y)=HO-(HO-H)*(X/XL)
      DO 30 I=2,12
      H(I,1)=SLOPE*DX*(I-2)+HO
 30   CONTINUE
C     KEEP TRACK OF THE NUMBER OF ITERATIONS AND OF LARGEST ERROR
C     NO FLOW BOUNDARIES NEED TO BE RESET WITHIN EACH INTERATION LOOP
      NUMIT=0
 40   AMAX=0.0
      NUMIT=NUMIT+1
C     LEFT AND RIGHT NO-FLOW BOUNDARIES
      DO 50 J=1,7
      H(1,J)=H(3,J)
      H(13,J)=H(11,J)
 50   CONTINUE
C     BOTTOM NO FLOW BOUNDARY
      DO 60 I=2,12
      H(I,7)=H(I,5)
 60   CONTINUE
C     SWEEP INTERIOR POINTS WITH 5-POINT OPERATOR
      DO 70 J=2,6
      DO 70 I=2,12
      OLDVAL=H(I,J)
      KK=KX(I-1,J)+KX(I,J)+KY(I,J-1)+KY(I,J)
      H(I,J)=KX(I,J)*H(I+1,J)+KX(I-1,J)*H(I-1,J)+KY(I,J)*H(I,J+1)
     #  +KY(I,J-1)*H(I,J-1)
      H(I,J)=H(I,J)/KK
      ERR=ABS(H(I,J)-OLDVAL)
      IF (ERR.GT.AMAX) AMAX=ERR
 70   CONTINUE
C     DO ANOTHER ITERATION IF LARGEST ERROR AFFECTS 1ST DECIMAL POINT
      IF (AMAX.GT.0.001) GO TO 40
C     WE ARE DONE.
      DO 75 I=2,12
      Q(I)=(H(I,2)-H(I,1))*KY(I,1)
 75   CONTINUE
      WRITE (6,80) NUMIT, ((H(I,J),I=2,12),J=1,6)
 80   FORMAT (//1X, 'NUMBER OF ITERATIONS IS: ',I5,///6(11F8.1///))
```

```
      WRITE (6,90) (Q(I), I=2,12)
  90  FORMAT ('FLOW RATE ACROSS THE WATER @TN:Table IS: ',/11F8.2)
      WRITE (6,100)
 100  FORMAT (//'DO YOU WISH ENTER MORE DATA? IF SO, ENTER ''YES''.')
      READ (5,105) ANS
 105  FORMAT (A2)
      IF (ANS.EQ.'YE') GO TO 1
      CALL EXIT
      END
```

```
OK, R *AQU
ENTER INITIAL HEAD, VALUE FOR DX, AND SLOPE:
500.00
7.00
0.05

NUMBER OF ITERATIONS IS: 106
   500.0 500.3 500.7 501.0 501.4 501.7 502.1 502.4 502.8 503.1 503.5
   500.6 500.7 500.9 501.2 501.5 501.7 502.0 502.3 502.6 502.8 502.9
   500.9 500.9 501.1 501.3 501.5 501.7 502.0 502.2 502.4 502.6 502.6
   501.0 501.1 501.2 501.3 501.5 501.7 501.9 502.1 502.3 502.4 502.4
   501.1 501.1 501.2 501.4 501.5 501.7 501.9 502.1 502.2 502.3 502.3
   501.1 501.2 501.3 501.4 501.6 501.7 501.9 502.1 502.2 502.3 502.3
FLOW RATE ACROSS THE WATER TABLE IS:
   55.17 32.91 20.20 11.63 5.04 -0.76 -6.56 -13.17 -21.73 -34.56 -56.73
```

Draw in the equipotential lines (constant head lines) and the orthogonal flow lines above. The flow lines show that water moves down and through the aquifer from right to left, and it exits from the upper left hand corner of the aquifer. The bottom is an impermeable, no-flow boundary.

9.4 CONTAMINANT SOLUTE TRANSPORT EQUATION

Solute transport and reactions can be modeled with a similar approach to that of organic contaminants in the estuary (Chapter 7) using the advection–dispersion equation coupled with the flow equation from the previous Section 9.3. In tensor notation, the three-dimensional transport of solute is written

$$\frac{\partial C}{\partial t} = -\frac{\partial}{\partial x_i}\left(u_i C\right) + \frac{\partial}{\partial x_i}\left(D_{ij}\frac{\partial C}{\partial x_j}\right) \pm \sum_{m=1}^{n} r_m \tag{20}$$

where C = solute concentration, ML^{-3}

t = time, T

u_i = velocity in three dimensions, LT^{-1}

x_i = longitudinal, lateral, and vertical distance, L
D_{ij} = dispersion coefficient tensor, L^2T^{-1}
r_m = physical, chemical, and biological reaction rates, $ML^{-3}T^{-1}$

If the velocity and dispersion coefficients are constant in space and time, and if the principal components of dispersion are D_{xx}, D_{yy}, and D_{zz} for longitudinal, lateral, and vertical flow, equation (20) simplifies to a partial differential equation in three dimensions with constant coefficients.

$$\frac{\partial C}{\partial t} = -\left[u_x \frac{\partial C}{\partial x} + u_y \frac{\partial C}{\partial y} + u_z \frac{\partial C}{\partial z}\right] + \left[D_x \frac{\partial^2 C}{\partial x^2} + D_y \frac{\partial^2 C}{\partial y^2} + D_z \frac{\partial^2 C}{\partial z^2}\right] \pm \Sigma\, r_m \tag{21}$$

Using gradient operator notation, the equation is

$$\frac{\partial C}{\partial t} = -\, u_i\, \nabla C + D_i\, \nabla^2 C \pm \Sigma\, r_m \tag{22}$$

9.4.1 Velocity Parameters

Usually, it is necessary to solve equation (22) in only one or two dimensions. Velocity values can be obtained from the solution of the flow equation (16), so it is desirable to solve the flow balance before contaminant transport modeling begins. At the very least, it is necessary to estimate the horizontal and/or vertical velocities for the aquifer from knowledge of K, n, and $dh/d\ell$. Estimates from Darcy's law in three dimensions for u_x, u_y, and u_z can be substituted into equation (22) to solve the contaminant transport equation.

9.4.2 Dispersion Coefficient

Dispersion coefficients are difficult to determine for use in equation (22). Normally, one would obtain a range of dispersion coefficients from the literature and fix D_x, D_y, and D_z by model calibration. The best method to obtain them is from injection of a conservative tracer and monitoring its arrival at observation wells. Potassium bromide, sodium chloride, tritium, fluoroscein, and Rhodamine WT® dyes are occasionally used. Such tracers are useful to determine the velocity of water as well as the dispersion coefficient. Because the subsurface is so heterogeneous, it is difficult to get good results from tracer tests on the first experiment. Tracer tests are costly and only large field or research projects generally make use of them.

Dispersion occurs in groundwater, not due to turbulent flow (groundwater flow is laminar), but rather due to mechanical dispersion and the tortuous path that groundwater must follow through porous media. Nonhomogeneity causes microscopic and macroscopic variations in hydraulic conductivity that ultimately lead to dispersion. Figure 9.5 shows how groundwater flow paths become more numerous with greater distance from the source (length scale). Flow trajectories that go through narrow pore spaces will speed up relative to other parcels of water. Thus a tracer will spread in all three dimensions, and not at equal rates unless the geologic formation is isotropic and homogeneous.

Dispersion occurs by a second mechanism, "stored" water. Table 9.2 gives typical specific yield and porosity values for a variety of aquifer materials. Clay has a very large porosity but a small specific yield, indicating that water will not drain freely from clay in unconfined aquifers. This is because water becomes stored in interlayers of the clay particles and bound there. Molecular diffusion and other slow processes will eventually displace the stored water, but this mechanism can be a source of dispersion at the macroscopic scale for clays and other media also. Later in this chapter, we will see how related processes can cause hysteresis in contaminant sorption–desorption processes that sometimes make remediation efforts difficult.

Dispersion coefficients in contaminant transport models are empirical, and they are a strong function of scale. When setting initial estimates of dispersion coefficients, it is best to begin from experience and/or literature values. Table 9.3 may provide some guidance. Following the treatment of Scheidegger,[7] a scaling factor is used that correlates with a length scale in laboratory soil columns and field tracer tests. The scaling factor is called dispersivity, α.

The dispersion coefficient is directly related to the velocity in the porous media. Larger velocities will cause more spread of a tracer due to realization of more flow paths. Dispersivity, α, has units of length analogous to a mixing length.

$$D_x = \alpha_x u_x + D_* \tag{23a}$$

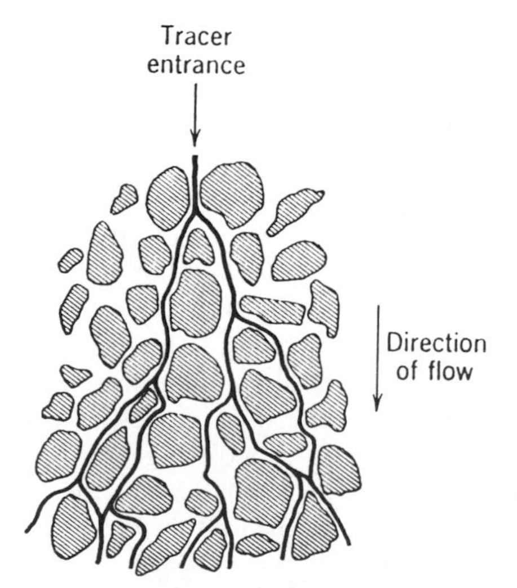

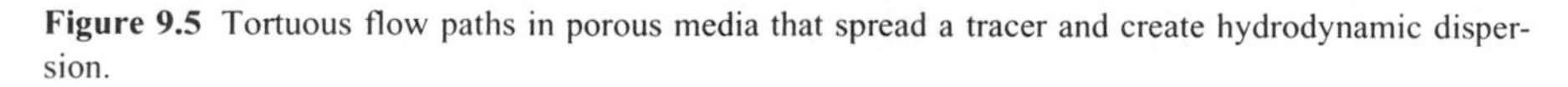

Figure 9.5 Tortuous flow paths in porous media that spread a tracer and create hydrodynamic dispersion.

Table 9.3 Empirical Value of Longitudinal Dispersivity, α, as a Function of Experimental Scale in Unconsolidated Porous Media

	Scale, m	Longitudinal Dispersitivity, α, m	
		Average	Range
Laboratory	< 1	0.001–0.01	0.0001–0.01
Field, small scale	1–10	0.1–1.0	0.001–1.0
Field, large scale	10–100	25	1–100

$$D_y = \alpha_y u_x + D_* \tag{23b}$$

where D_x = longitudinal dispersion coefficient, L^2T^{-1}
D_y = lateral dispersion coefficient, L^2T^{-1}
α_x = longitudinal dispersivity, L
α_y = lateral dispersivity, L
u_x = longitudinal velocity, LT^{-1}
D_* = molecular bulk diffusion coefficient, L^2T^{-1}

The dispersion coefficients in equations (23a) and (23b) are the summation of two terms: hydrodynamic (mechanical) dispersion and bulk molecular diffusion. The molecular diffusion coefficient, D_*, is on the order of 10^{-5} cm^2 s^{-1}, and it is generally not important in field situations. Dispersivity is greatest in the direction of flow, and vertical dispersivity is usually small, especially if fluvial, glacial, or sedimentary deposits have caused horizontal bedding planes.

$$\alpha_x \geq \alpha_y \gg \alpha_z \tag{24}$$

Estimates of the dispersion coefficient are required to solve the contaminant transport equation (22). Reactions, such as sorption of chemicals to the aquifer medium, may also play an important role.

9.5 SORPTION, RETARDATION, AND REACTIONS

Suppose a contaminant is released into the groundwater as in Figure 9.6. If the contaminant is well-mixed with depth, and if the hydraulic gradient is from left to right, the step function input of pollutant will form a one-dimensional plume. The plume will gradually spread at the edges due to hydrodynamic (mechanical) dispersion. It will form a broader contaminant "edge" as it travels through the aquifer from time t_1 to time t_2 (Figure 9.6a). If monitoring wells are installed at points x_1 and x_2 in the aquifer, one can measure the concentration of contaminant as it moves through each point. Figure 9.6b shows an S-shaped curve in time as the pollutant moves through the monitoring well at x_1. At x_2, the "breakthrough curve" is even more S-shaped

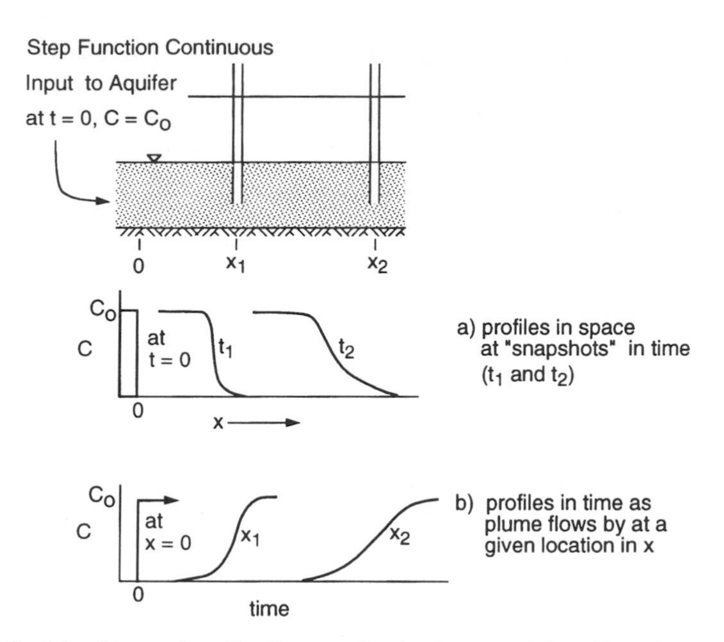

Figure 9.6 Spatial and temporal profiles for a step function input to a 1-D aquifer and a *conservative* substance. Hydrodynamic dispersion is responsible for spreading out the signal with space and time.

because the contaminant has had more time to disperse. The more dispersion in the porous media, the greater will be the spread of the breakthrough curve.

For this case of a conservative contaminant transported through a homogeneous, one-dimensional aquifer, we can simplify equation (21). Velocity vectors u_y and u_z and dispersion coefficients D_y and D_z can be neglected. Transport and reaction of the contaminant in the porous medium can be represented as a second-order partial differential equation with constant coefficients in one dimension.

$$\frac{\partial C}{\partial t} = -u_x \frac{\partial C}{\partial x} + D_x \frac{\partial^2 C}{\partial x^2} \pm \sum_{m=1}^{n} r_m \tag{25}$$

One of the most important reactions that contaminants can undergo in the subsurface is sorption. Here, we will refer to the general term "sorption" indicating a number of possible mechanisms for how pollutants can be removed from the aqueous phase and immobilized onto particles in the porous medium. Mechanisms for sorption of solutes to solid particles include:

- Hydrophobic partitioning of organic chemicals (absorption) in the organic coatings or organic matter contained in the subsurface.
- Adsorption of organics and metals to the surface of particles by electrostatic and/or suface coordination chemistry.

- Ion exchange of metal ions and ligands at exchange sites and in the interlayers of clays.

The reaction term in equation (25) refers to a number of possible reactions. For organic contaminants, most of the reactions discussed in Chapter 7 pertain, including sorption, biodegradation, hydrolysis, redox reactions, and volatilization (in the unsaturated zone). For metals, principal reactions are sorption, ion exchange, and redox reactions for elements with multiple valence states. Metal contaminants such as Cd^{2+}, Pb^{2+}, Zn^{2+}, Ni^{2+}, and Cu^{2+} also may be released from sorption sites on iron and manganese oxides when reducing conditions result in dissolution and reduction of the oxides.

Sorption of organic chemicals due to hydrophobic partitioning has already been covered in Section 7.2.7. For low concentrations of contaminants typical in groundwater, one may assume a linear sorption isotherm. We will generalize the discussion for all sorption reactions (organics and metals) to a linear equilibrium constant, K_d, that is derived from field or laboratory measurements.

$$S = K_d C \tag{26}$$

where S = amount sorbed onto porous medium, MM^{-1} (mg kg^{-1})
K_d = distribution coefficient, L^3M^{-1} (L kg^{-1})
C = solute concentration, ML^{-3} (mg L)

Equation (25) can be expanded to include sorption explicitly.

$$\frac{\partial C}{\partial t} = -u_x \frac{\partial C}{\partial x} + D_x \frac{\partial^2 C}{\partial x^2} - \frac{\partial S}{\partial t}\frac{\rho_s(1-n)}{n} \pm \Sigma r_i$$

Change in concentration = Advection + Dispersion − Sorption ± Rxns (27)

where ρ_s = the solid density of the particles, ML^{-3}
n = effective porosity, dimensionless
r_i = chemical and biological reactions, $ML^{-3}T^{-1}$

Often it is more convenient to use bulk density of the porous medium rather than solid particle density.

$$\frac{\partial C}{\partial t} = -u_x \frac{\partial C}{\partial x} + D_x \frac{\partial^2 C}{\partial x^2} - \frac{\partial S}{\partial t}\frac{\rho_b}{n} \pm \Sigma r_i \tag{28}$$

where ρ_b = bulk density of the porous medium, ML^{-3}

Typical conversion units are given below that cause the sorption term to reflect the loss in concentration from the aqueous phase.

$$\frac{\partial S}{\partial t}\frac{\rho_b}{n} = \frac{\text{mg chemical}}{\text{kg solids-day}} \frac{\text{kg solids/liter total}}{\text{liters H}_2\text{O/liter total}} \tag{29}$$

Taking the derivative of both sides of equation (26) and substituting into equation (28) for $\partial S/\partial t$ yields:

$$\frac{\partial C}{\partial t} = -u_x\frac{\partial C}{\partial x} + D_x\frac{\partial^2 C}{\partial x^2} - K_d\frac{\rho_b}{n}\frac{\partial C}{\partial t} \pm \Sigma\, r_i \tag{30}$$

Rearranging terms, we find

$$\frac{\partial C}{\partial t}\left(1 + K_d\frac{\rho_b}{n}\right) = -u_x\frac{\partial C}{\partial x} + D_x\frac{\partial^2 C}{\partial x^2} \pm \Sigma\, r_i \tag{31}$$

We will define the dimensionless retardation factor as

$$R = 1 + \frac{K_d\,\rho_b}{n} = 1 + \frac{K_d\,\rho_s\,(1-n)}{n} \tag{32}$$

Let us assume a first-order degradation reaction (such as biodegradation) for the reaction term in equation (31), and dividing through by the retardation factor, one obtains equation (33):

$$\frac{\partial C}{\partial t} = -\frac{u_x}{R}\frac{\partial C}{\partial x} + \frac{D_x}{R}\frac{\partial^2 C}{\partial x^2} - \frac{k}{R}C \tag{33}$$

where R = retardation factor, dimensionless
k = first-order degradation rate constant, T^{-1}

The retardation factor is a dimensionless number that is equal to 1.0 in the absence of sorption, $K_d = 0$, or it is a number that is greater than 1.0, which serves to slow down or "retard" the actual contaminant velocity. Note that the diffusion coefficient is also affected such that the "apparent diffusion" coefficient is divided by a factor of R, as is the first-order degradation rate constant. Here we assume that the degradation reaction occurs only in the aqueous phase. If the adsorbed chemical is also available for degradation reactions, then equation (33) would need to be modified. The last term would be simply $-kC$. Equation (33) is the general mass balance equation for one-dimensional contaminant migration with advection dispersion, sorption, and a first-order degradation reaction.

The retardation factor has the effect of slowing down the entire process of pollutant migration.

$$R = \frac{\text{mean velocity of water}}{\text{mean velocity of contaminant}} = \frac{u_x}{u_R} \geq 1.0 \tag{34}$$

In Figure 9.7, the first curve (1) is the breakthrough curve at a monitoring well for a continuous input of a conservative substance that does not sorb, such as KBr. It is slightly S-shaped due to dispersion, but the mean time of travel is equal to the mean residence time of water molecules in the porous medium ($t/\tau = 1$). The second curve is more S-shaped because greater time has elapsed for dispersion, and the retardation factor is approximately 3.7. Curve 2 has $k = 0$, no degradation reactions.

In Figure 9.8, the third curve illustrates breakthrough for a contaminant that disperses, sorbs, and undergoes a first-order decay reaction as in equation (33). The steady-state concentration is less than the initial contaminant concentration C_0 because of degradation. Biodegradation often occurs in groundwater after a lag time, or the rate of biotransformation accelerates with time due to adaptation of the microorganisms (enzyme induction) and/or an increase in their numbers (increased enzyme activity). Curve 4 demonstrates that the contaminant may continue to be biodegraded at an increasing rate, $k = f(t)$, if adaptation occurs. Some reports in the literature indicate a threshold concentration of organic contaminants in groundwater below which further biodegradation is difficult.[8]

9.5.1 One-Dimensional (1-D) Contaminant Equations

Equation (33) yields an analytical solution for simple boundary conditions (BC) and an initial condition (IC). For conditions of a 1-D aquifer with a step function input of contaminant at $t = 0$, the following solution applies:[9]

$$\text{BC1:} \quad C(0, t) = C_0 \quad \text{for } t > 0$$

$$\text{BC2:} \quad \partial C/\partial x = 0 \quad \text{for } x = \infty$$

$$\text{IC:} \quad C(x, 0) = 0 \quad \text{for } x \geq 0$$

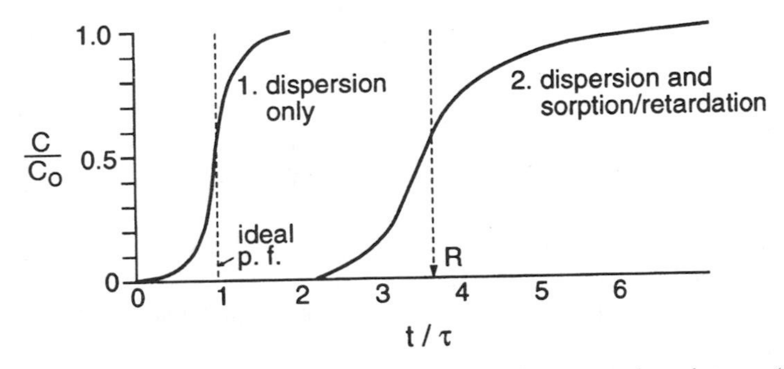

Figure 9.7 Temporal breakthrough curves in porous medium for conservative substances (one is nonsorbing and the other is sorbing). C/C_0 is the concentration relative to the continuous input concentration at the source of the contamination. t/τ is the time relative to the mean residence time. The retardation factor for the sorbing substance is approximately 3.7.

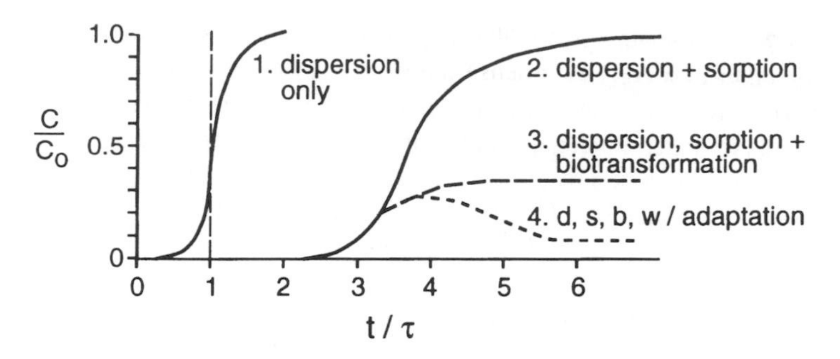

Figure 9.8 Normalized concentration versus time breakthrough curves showing the effect of biotransformation and adaptation on contaminant fate and transport in porous medium. (Modified from McCarty.[8])

$$C = \frac{C_0}{2} \exp \frac{(u_x - v)x}{2D_x} \operatorname{erfc}\left[\frac{Rx - vt}{2\sqrt{D_x Rt}}\right] + \frac{C_0}{2} \exp \frac{(u_x + v)x}{2D_x} \operatorname{erfc}\left[\frac{Rx + vt}{2\sqrt{D_x Rt}}\right] \tag{35}$$

$$v = u_x \left(1 + 4kD_x/u_x^2\right)^{1/2} \tag{36}$$

where

$$\operatorname{erfc}(y) = 1 - \frac{2}{\sqrt{\pi}} \sum_{n=0}^{\infty} \frac{(-1)^n y^{2n+1}}{n!(2n+1)} \tag{37}$$

The complementary error function is an infinite series solution that arises upon integration of equation (33). The easiest way to compute the complementary error function is to use a table of values found in Appendix C or to call the function erfc (y) from a computer code. Most spreadsheets or compilers have routines that compute the complementary error function for you.

Equation (35) can be used to analyze the case of a continuous input of contaminant to a 1-D aquifer, similar to the schematic in Figure 9.6, but including sorption and first-order decay. Curves 1, 2, and 3 in Figure 9.8 can be generated using equation (35).

Occasionally, the source of contamination is not continuous but is rather an impulse input to an aquifer, such as a jet fuel spill. For a 1-D aquifer, the solution to equation (33) is given below with the initial condition that a slug of mass M is injected at $x = 0$ and $t = 0$.

$$C = \frac{MR}{2A\sqrt{\pi D_x Rt}} \exp\left[\frac{-\left(x - \frac{u_x}{R}t\right)^2}{\frac{4D_x t}{R}}\right] \exp\left(\frac{-k}{R}t\right) \tag{38}$$

where M = mass impulse input to an aquifer at $t = 0$ (planar source), M
R = retardation factor, dimensionless
A = cross-sectional flow area of mass input, L^2
D_x = longitudinal dispersion coefficient, L^2T^{-1}
t = time, T
u_x = longitudinal actual velocity of water, LT^{-1}
k = first-order degradation rate constant, T^{-1}

If the adsorbed contaminant phase undergoes degradation, then the last multiplier in equation (38) should be exp $(-kt)$ rather than exp $(-kt/R)$.

9.5.2 Two-Dimensional (2-D) and Three-Dimensional Contaminant Equations

Two-dimensional and three-dimensional contaminant transport equations have been developed for continuous discharges with similar boundary conditions as for equation (35).[9] The following solutions are adapted assuming linear sorption kinetics and first-order degradation. Figure 9.9 gives a graphical depiction of the plume. As in the case of the 1-D equation (38), the source of contamination is a plane, orthogonal to the flow direction (i.e., in the *yz* plane). The partial differential contaminant transport equation is:

$$R\frac{\partial C}{\partial t} = D_x\frac{\partial^2 C}{\partial x^2} + D_y\frac{\partial^2 C}{\partial y^2} + D_z\frac{\partial^2 C}{\partial z^2} - u_x\frac{\partial C}{\partial x} - kRC \tag{39}$$

Here we assume that both the dissolved and particulate adsorbed chemical fractions are available for biotransformation. The solution to equation (39) is provided by Domenico and Schwartz.[5] Equation (40) describes the development of cigar-shaped plumes as depicted in Figure 9.9.

$$\begin{aligned} C(x, y, z, t) = {} & \left(\frac{C_0}{8}\right) \exp\left\{\frac{xu_x}{2D_x}\left[1 - \left(\frac{1 + 4kD_x/u_x}{u_x/R}\right)^{1/2}\right]\right\} \\ & \times \operatorname{erfc}\left[\frac{x - \frac{u_x}{R}t\left(1 + \frac{4kD_xR}{u_x^2}\right)^{1/2}}{2(D_xt/R)^{1/2}}\right] \\ & \times \left\{\operatorname{erf}\left[\frac{y + \frac{Y}{2}}{2\left(\frac{D_yx}{u_x}\right)^{1/2}}\right] - \operatorname{erf}\left[\frac{y - \frac{Y}{2}}{2\left(\frac{D_yx}{u_x}\right)^{1/2}}\right]\right\} \\ & \times \left\{\operatorname{erf}\left[\frac{z + Z}{2\left(\frac{D_zx}{u_x}\right)^{1/2}}\right] - \operatorname{erf}\left[\frac{z - Z}{2\left(\frac{D_zx}{u_x}\right)^{1/2}}\right]\right\} \end{aligned} \tag{40}$$

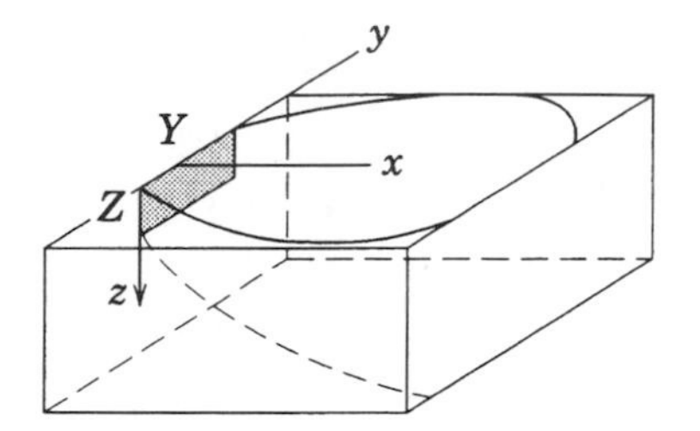

Figure 9.9 Two-dimensional contaminant transport with impulse input and step function input. Isoconcentration contour lines are shown. Sorption and degradation reactions would shrink the plumes and retard them back toward the origin.

where Y = width of waste source in saturated zone, L

Z = depth of waste source in saturated zone, L

D_x, D_y, D_z = dispersion coefficients in x, y, and z directions, L^2T^{-1}

Most realistic situations in the environment do not have simple boundary conditions and intitial conditions. For these cases, it is necessary to do a finite difference or a finite element numerical solution to solve the problem.

We have discussed how to obtain estimates of parameters u_x and D_x. D_y must be obtained by model calibration or by initially setting it at 0.1–1.0 times D_x. Dispersivity estimates may be utilized [equation (23)]. The retardation factor R should be estimated from equation (32). Tables 9.4 and 9.5 provide some literature values for hydrophobic sorption of organics and binding of metals, respectively. There are at least seven different equations that can be used to estimate K_{oc} ($K_d = K_{oc} f_{oc}$) for hydrophobic organics in Table 9.4, so you should be careful to choose one that is closest to the chemical of interest.

Values of K_d for metals are even more empirical because the mechanisms (ion exchange, surface coordination to oxides, and organic binding) are variable in different porous media. Mineralogy (feldspars, quartz, limestone, etc.) affects the K_d value dramatically as well as the presence of chelating agents such as EDTA at nuclear waste repositories (Table 9.5). Distribution coefficients for metals are affected by mineralogy, organic binding (coatings), redox state, and chemical speciation. Equations to predict K_d values for metals in aquifers do not exist because of the variety of environmental conditions that affect sorption.

Example 9.4 Estimation of Retardation Factor for Organic Chemicals from K_{ow} and f_{oc}

Lyman et al.[10] provide several relationships to predict the organic carbon normalized partition coefficient K_{oc} from octanol/water partition coefficients for hydrophobic chemicals in groundwater. This is the standard method to estimate distribution coefficients (partition coefficients) *a priori* for hydrophobic organics. Of course, the best method is to perform an adsorption isotherm in the laboratory with actual aquifer medium.

Table 9.4 (A) Compilation of K_{ow} and Solubility Values from Schnoor et al.[12]; (B) Equations to Estimate K_{oc} from Lyman.[10]

A. Compound	log K_{ow}	log K_{oc}	Water Solubility at 20 °C, mg L^{-1}	Equation Number
Halogenated aliphatics				
Bromoform	2.30	2.16	—	7
Carbon tetrachloride	2.64	2.40	800	7
Chloroethane	1.49	1.57	5,740	7
Chloroform	1.97	1.92	8,200	7
Chloromethane	0.95	1.18	6,450	7
Dichlorodifluoromethane	2.16	2.05	280 (25 °C)	7
Dichloromethane	1.26	1.41	20,000	7
Hexachloroethane	3.34	2.81	50	7
Tetrachloroethylene	2.88	2.57	200	7
Trichloroethylene	2.29	2.15	1,100	7
Vinyl chloride	0.60	0.93	90	7
1,1-Dichloroethane	1.80	1.80	400	7
1,1-Dichloroethylene	1.48	1.57	400	7
1,2-Dichloroethane	1.48	1.57	8,000	7
1,2-*trans*-Dichloroethylene	2.09	2.00	600	7
1,1,1-Trichloroethane	2.51	2.30	4,400	7
1,1,2-Trichloroethane	2.07	1.99	4,500	7
Aromatics				
Benzene	2.13	1.92	1,780	3
Benzo[*a*]pyrene	6.06	5.85	0.0038	3
Chlorobenzene	2.84	2.63	500	3
Ethylbenzene	3.34	3.13	152	3
Hexachlorobenzene	6.41	6.20	0.006	3
Naphthalene	3.29	3.08	31	2
Nitrobenzene	1.87	1.66	1,900	2
Pentachlorophenol	5.04	4.83	14	3
Phenol	1.48	1.27	93,000	3
Toluene	2.69	2.48	535	3
1,2-Dichlorobenzene	3.56	3.35	100	3
1,3-Dichlorobenzene	3.56	3.35	123	3
1,4-Dichlorobenzene	3.56	3.35	79	3
1,2,4-Trichlorobenzene	4.28	4.07	30	3
2-Chlorophenol	2.17	1.96	28,500	3
2-Nitrophenol	1.75	1.54	2,100	3
2,4,5,2′,4′,5′-PCB	6.72	6.51	—	3
Pesticides				
Acrolein	0.01	0.86	210,000	6
Alachlor	2.92	2.96	242	1
Atrazine	2.69	2.55	33	4
Dieldrin	3.54	3.33	0.2	3
DDT	6.91	6.70	0.0055	3
Lindane	3.72	3.51	7.52	3
2,4-D	1.78	1.65	900	5

(continued)

Table 9.4 ***(continued)***

B. Equation	Number	r^2	Chemical Classes Represented
1. $\log K_{oc} = 0.544 \log K_{ow} + 1.377$	45	0.74	Wide variety, mostly pesticides
2. $\log K_{oc} = 0.937 \log K_{ow} - 0.006$	19	0.95	Aromatics, polynuclear aromatics, triazines, and dinitroaniline herbicides
3. $\log K_{oc} = 1.00 \log K_{ow} - 0.21$	10	1.00	Mostly aromatic or polynuclear aromatics; two chlorinated
4. $\log K_{oc} = 0.94 \log K_{ow} + 0.02$	9	NA	*s*-Triazines and dinitroaniline herbicides
5. $\log K_{oc} = 1.029 \log K_{ow} - 0.18$	13	0.91	Variety of insecticides, herbicides, and fungicides
6. $\log K_{oc} = 0.524 \log K_{ow} + 0.855$	30	0.84	Substituted phenylureas and alkyl-*N*-phenylcarbamates
7. $\log K_{oc} = 0.72 \log K_{ow} + 0.5$	14	0.95	Halogenated hydrocarbons, both aliphatic and aromatic[13]

The aquifer is contaminated with toluene from a petrochemical spill. Given the following information, estimate the K_d and the retardation factor R. We will use the equation of Schwarzenbach and Westall[11] to estimate K_{oc}.

$$\log K_{oc} = 0.72 (\log K_{ow}) + 0.5$$

$$\rho_b = 1.5 \text{ g cm}^{-3}$$

Table 9.5 Order-of-Magnitude Compilation of Distribution Coefficients (K_d Values) in Sandy Aquifers for Toxic Metals.[a]

	K_d, L kg^{-1}
As	1–10
Cd	5–50
Se	1–10
Cr	10–100
Cu	5–50
Ni	5–50
Pb	2–25
Zn	5–50
^{239}Pu	50–500
^{60}Co	10–100
^{90}Sr	5–50

[a]Exact value depends on mineralogy, redox, organics (f_{oc}), pH, and chemical speciation.

$$n = 0.4$$

$$f_{oc} = 0.001$$

$$\log K_{ow} = 2.69 \quad \text{for toluene}$$

Solution

$$\log K_{oc} = 0.72\,(2.69) + 0.5$$

$$K_{oc} = 273.4$$

$$K_{oc}\, f_{oc} = K_d = 0.273 \text{ L kg}^{-1}$$

$$R = 1 + K_d \frac{\rho_b}{n} = 2.03$$

Toluene is retarded twofold in the aquifer, and the mean velocity of toluene is one-half that of the average H_2O molecule because of hydrophobic sorption of the chemical into the organic phase of the porous medium.

Following are some additional examples for two different porous media. Some estimates of retardation factors are given for measured K_{oc} values in two different porous media. [*Note:* $\rho_b = (1 - n)\,\rho_s$, where ρ_s is the average density of the aquifer medium.]

			Sandy Aquifer $\rho_b = 1.8$, $n = 0.32$, $f_{oc} = 0.001$		Sandy Loam $\rho_b = 1.25$, $n = 0.40$, $f_{oc} = 0.01$	
	$\log K_{ow}$	Measured $\log K_{oc}$	K_d	R	K_d	R
Toluene[12]	2.69	2.44	0.273	2.54	2.73	9.53
Benzene[12]	2.13	1.92	0.083	1.47	0.83	3.59
Trichloroethylene[12]	2.29	2.15	0.141	1.79	1.41	5.41
Atrazine[12]	2.69	2.55	0.355	3.00	3.55	12.10
2,4,5,2′,4′,5′-PCB[13]	6.72	6.51	3240	18,200	32,400	101,000

PCB is retarded ~100,000 times in sandy loam soil! Except for nonheterogeneities such as macropore flow through root channels, which is a possibility, PCB should not migrate to the groundwater.

Example 9.5 Contaminant Transport and Reaction, One Dimension

A one-dimensional surficial aquifer with properties given in the previous example has received a continuous input of toluene from a leaking underground storage tank.

The mean longitudinal velocity of the aquifer is 2 cm d^{-1} and the dispersivity is estimated to be ~1.0 m. How long will it take the toluene to reach the neighbors across the street, down gradient 25 m away? The concentration at the source is 1.0 mg L^{-1}. Toluene degrades aerobically by indigenous microorganisms with $k = 0.03$ day^{-1}.

Solution: Use equation (35) and solve for the concentration as a function of distance for several choices of time.

$$R = 2.0$$

$$u_x = 0.02 \text{ m d}^{-1}$$

$$D_x = \alpha u_x + D_* = (1.0)(0.02) + 10^{-4} \text{ m}^2 \text{ d}^{-1}$$

$$= 0.02 \text{ m}^2 \text{ d}^{-1}$$

Toluene is not very toxic to humans, but it has a taste and odor threshold of 20 μg L^{-1}, and it is indicative of other potentially more toxic contaminants in the petrochemical mixture.

9.6 BIOTRANSFORMATIONS

Assuming a first-order decay constant to account for all reactions other than sorption in the subsurface environment is, of course, simplistic. There are many microorganisms (bacteria, fungi) in the subsurface environment that are capable of degrading a large variety of organic compounds. These microorganisms exist even in surprisingly deep aquifers (> 50 m) although fewer in number. Ghiorse and Balkwill[14] have discussed the types of microorganisms in the subsurface and methods of identifying them. The most common method to measure microbial biomass in the subsurface is by epifluorescence microscopy using acridine orange as a fluorescent stain for double-stranded DNA. The bacteria take on the stain and, in most cases, are readily recognizable. However, one cannot be sure whether nonviable cells are also staining, and it is difficult to search enough fields of the microscope slide to get an accurate estimate of cell count for cases where the microbial population is small (deep groundwater). Cells that are counted by this method are termed AODCs (acridine orange direct counts). If estimates of biomass are needed, the AODC cells must be multiplied by an average cell volume and density (1.01–1.03 g cm^{-3}) to obtain the biomass (in mg L^{-1}). Imaging techniques and computer software can speed the time-consuming counting procedure. Table 9.6 gives some methods for measuring biomass and their relative advantages and disadvantages.[14] If nutrient agars are to be used, they must be diluted 10–20×. Otherwise, special oligotrophic media are needed to obtain accurate estimates of biomass for these nutrient-poor subsurface conditions. Less than 10% of the viable cells are thought to be counted by these

Table 9.6 Methods for Determining Bacterial Biomass in the Subsurface

Methods	Advantages	Disadvantages
Scanning electron microscopy	direct counts	cells difficult to find, sensitivity at low densitites
Epifluorescence Microscopy	direct counts	viability ? sensitivity
Pour-plates, nutrient media	viable cells counted	more than one cell per CFU; difficult to find proper nutrient-poor media
Chloroform-extractable lipid phosphate	easy, greater sensitivity due to increased sample size	viability, activity?
Muramic acid extraction (cell wall extract)	greater sensitivity due to increased sample size	viability, activity?
ATP	ease of determination	reliability, ATP/cell?
16s r-RNA & gene probes	identifies specific bacteria	are they active?
Immunochemical probes	identifies specific bacteria	reliability, cross-reactivity

Source: Compiled from Ghiorse and Balkwill.[14]

methods because we do not have the optimum medium to grow the various microorganisms.

Figure 9.10 is illustrative of some microorganisms in the subsurface environment. In deep groundwater, the density of microorganisms is so low that it is difficult to find the organisms amidst the particles and the debris in porous media when their densities are low. Critical point drying may help. Figures 9.10a and 9.10b were from a recharge area of an infiltration zone, where bacterial densities were $\sim 10^6$ g^{-1}.[15] Densities between 10^4 and 10^7 g^{-1} are typical in the subsurface with the higher AODCs in shallow surficial aquifers. Biofilms can form in areas where high organic concentrations are prevalent such as near hazardous waste dump sites (Figure 9.10b).

Molecular biology has added a whole new realm of gene probe and immunochemical techniques that can aid the modeler. It is possible to identify known "active degraders" by using gene probes. Figure 9.10d shows how fluorescent agents can be bound to target bacteria to produce an easily identifiable image. A polyclonal antibody was developed by injecting a rabbit with *T. ferrodoxans* and the blood of the rabbit produced antibodies to the bacteria.[16] Then the antibodies were complexed with anti-goat antibody, which contains the fluorescing agent. A relatively easy to use field kit to identify *T. ferrodoxans* can be developed using ELISA plates.

Environmental factors such as dissolved oxygen concentration, moisture content of unsaturated soils, organic carbon content, and electron acceptor (redox conditions) strongly influence the rate of microbial transformation of organic chemicals. Borden and Bedient[17] were the first to quantify the Monod-type limitation on rate

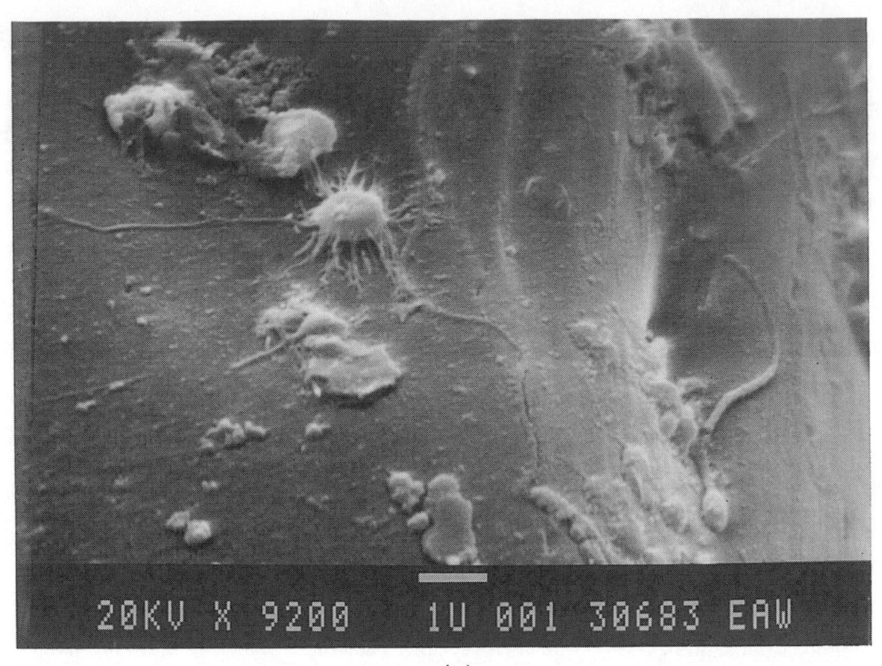

(a)

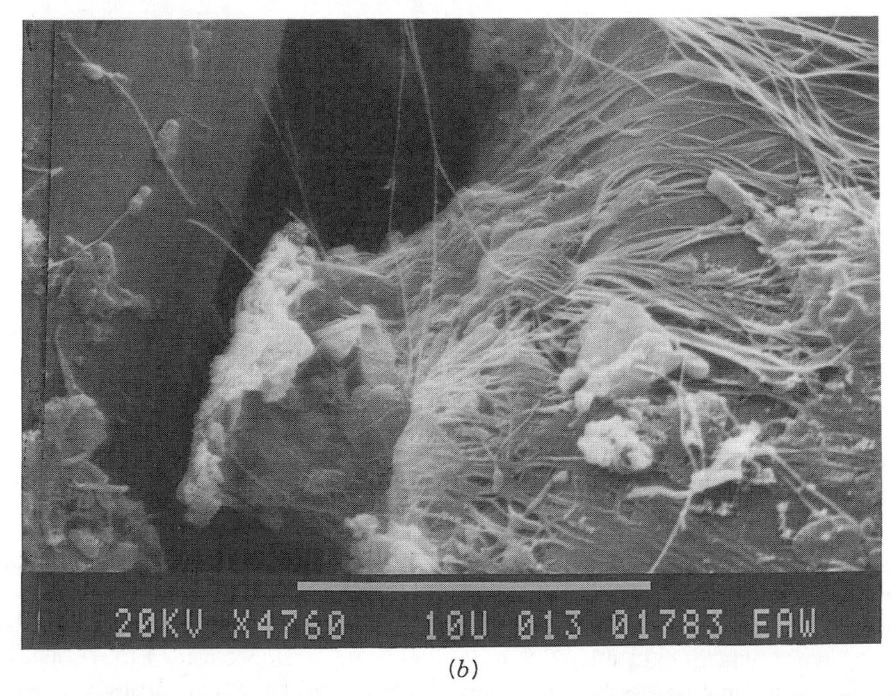

(b)

Figure 9.10 Bacteria in the subsurface environment. (a) Bacterial micrograph in porous media collected from the Glatt River infiltration area. (Ref.: Kuhn et al.[15]). (b) Glatt River infiltration aquifer media fed in column studies with simulated groundwater and 2 mg L^{-1} of acetic acid. Note the production of polysaccharide material.

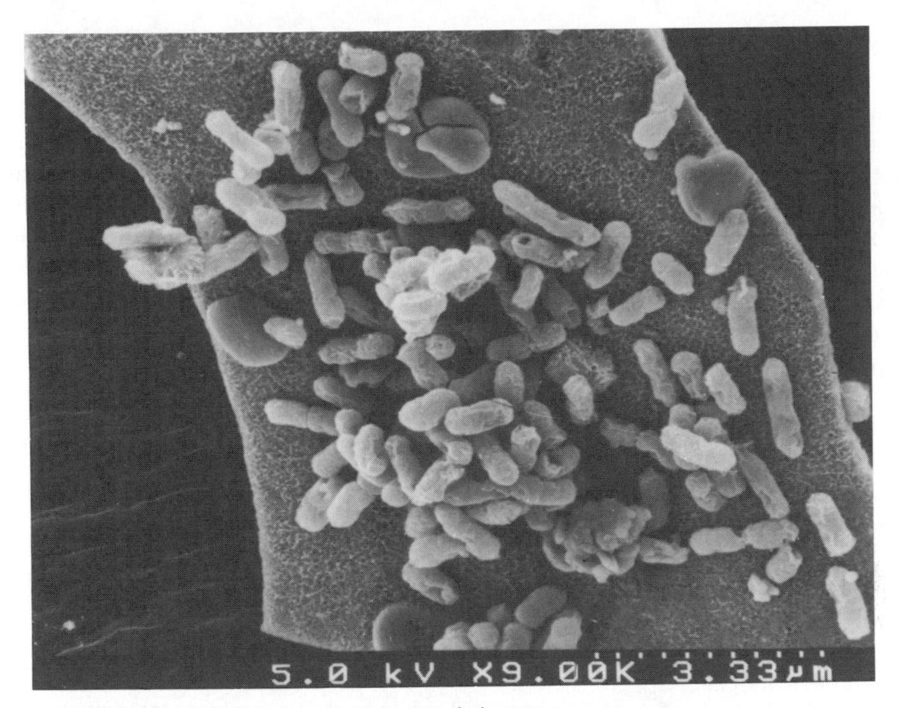

(c)

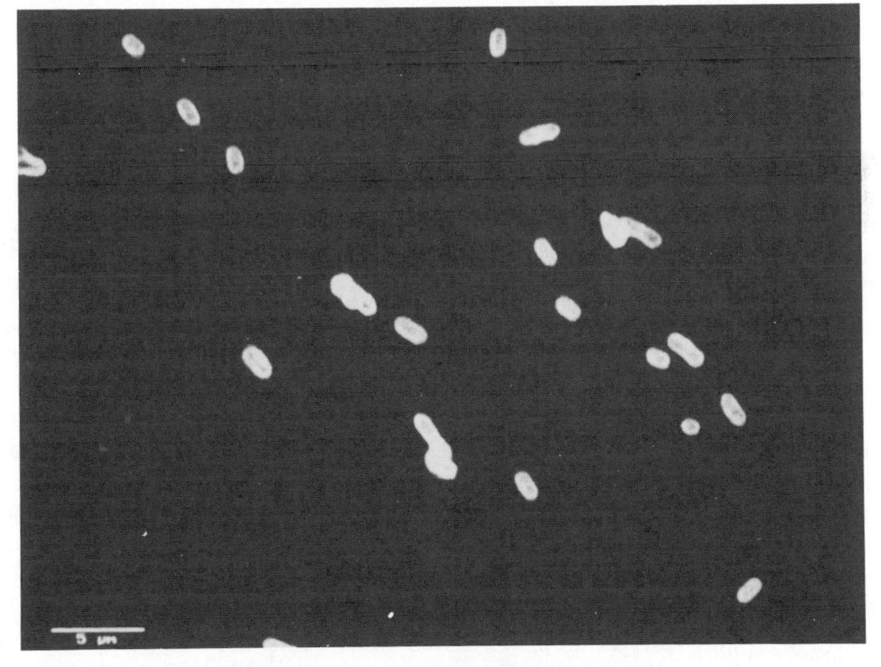

(d)

(c) Micrographs of *Thiobacillus ferrodoxans* in sediments of a stream in the Rocky Mountains, Colorado, from dissertation of Barry.[16] (d) Same bacteria as in (c) but stained with fluorescent polyclonal antibody from a rabbit in this immunochemical assay for *T. ferrodoxans* by Barry.[16]

that low dissolved oxygen concentrations cause for aerobic biological transformations. Electron acceptor conditions are critical, especially when redox conditions change from aerobic to anaerobic in the subsurface.[18,19] Nair and Schnoor[20] showed that atrazine mineralization rates in the unsaturated zone were directly proportional to soil moisture content and to f_{oc}. Microorganisms need soil moisture to thrive and, presumably, to solubilize substrates for transport. Too much soil moisture results in water-logged soils and low D.O. An empirical equation was given that modelers could use to adjust biotransformation rates for various environmental conditions in the case of atrazine mineralization to carbon dioxide.[21]

$$k = k_0\, \theta^{(T\text{-}20)} \frac{[O_2]}{K_{O2} + [O_2]} f_{om}\, \phi \tag{41}$$

where k = first-order mineralization rate constant, day^{-1}

k_0 = reference rate constant = 0.047 day^{-1} for atrazine mineralization

θ = temperature response factor = 1.045

$[O_2]$ = partial pressure of oxygen, atm

K_{O_2} = half-saturation constant for oxygen as an electron acceptor = 0.1 atm

f_{om} = fraction of organic matter in the porous media (mass/mass), dimensionless

ϕ = soil water content measured as mass fraction of field capacity, dimensionless

T = temperature in °C

When soils went anoxic and nitrate was respired as the electron acceptor (denitrifying conditions), atrazine mineralization rates decreased by more than tenfold. Equation (41) demonstrates that environmental factors can be critical in biodegradation reactions, and reported rate constants must be considered as site specific.

Table 9.7 is a qualitative summary of selected groundwater organic contaminants and their potential for microbial transformation. In general, aromatic compounds are readily degraded aerobically, but an acclimation period of days to several months may be required. The presence of other more readily degradable substrates can sometimes lengthen the acclimation period or slow the rate. Halogenated aliphatics are difficult to degrade aerobically. They cannot serve as a primary substrate with oxygen as the electron acceptor, but under anaerobic conditions, reductive dehalogenation can proceed. Polychlorinated biphenyls and pentachlorophenol can sometimes be dehalogenated under anaerobic conditions and then, if the compounds enter an aerobic environment, aerobic respiration proceeds to break the ring structure and to form catechols for eventual complete mineralization.

An interesting case is chlorinated solvents such as trichloroethylene (TCE) and tetrachloroethylene (PCE), which cannot serve as primary substrates for aerobic microorganisms but which can be degraded via co-metabolism or co-oxidation.[22] Oxygenase enzymes such as toluene dioxygenase (TDO), toluene monooxygenase (TMO), phenol monooxygenase (PMO), and methane monooxygenase (MMO) are

Table 9.7 Biodegradation of Organics

$CH_2O + O_2 \Leftrightarrow CO_2 + H_2O$	$CH_2O + \frac{4}{5}NO_3^- + \frac{4}{5}H^+ \Leftrightarrow CO_2 + \frac{2}{5}N_2 + \frac{7}{5}H_2O$	$CH_2O + \frac{1}{2}SO_4^{-2} + \frac{1}{2}H^+ \Leftrightarrow \frac{1}{2}HS^- + H_2O + CO_2$	$CH_2O + \frac{1}{2}CO_2 \Leftrightarrow \frac{1}{2}CH_4 + CO_2$
Aerobic	Anoxic Denitrification	Anoxic Sulfate Respiration	Anaerobic Methanogenesis
1. Aromatics, e.g., chlorinated benzenes and BTEX	1. Bromodichloromethane	All C_1 and C_2 Halogenated	All C_1 and C_2 halogenated aliphatics as co-metabolites with toxic intermediates.
2. Xylenes (dimethyl benzene) 3. Ethyl benzene 4. PAHs, e.g., napthalene 5. Benzene	2. Dibromodichloromethane 3. Chloroform, CCl_4 4. Bromoform 5. Xylenes after long acclimation period 6. 1, 1, 1 -Trichloroethane (TCA)	Aliphatics	
Not Degraded 1. Tetrachloroethylene (PCE) 2. TCA, CCl_4 3. Chloroform 4. Bromoform 5. TCE but only as co-metabolized with ϕ-OH,ϕ-CH_3, or methane	*Not Degraded Readily* 1. TCE 2. Chlorobenzenes 3. EtBz, Bz	*Not Degraded* 1. EtBz 2. Chlorobenzenes 3. TCE is slow	*Not Degraded* 1. EtBz 2. Chlorobenzenes (only dehalogenation of polyhalogenated chemicals occurs)

potent enzyme catalysts that serve to accelerate the aerobic oxidations of toluene, phenol, and methane, respectively.[22,23] In the process, other organic compounds may be co-oxidized in parallel reactions.[24]

$$C_6H_5CH_3 + \tfrac{1}{2}\,O_2 + H_2O \xrightarrow{\text{TDO}} CH_3C_6H_5(H)(OH)(OH)(H) \tag{42a}$$

$$\text{ClHC}{=}\text{CCl}_2 + \tfrac{1}{2}\,O_2 \xrightarrow[\text{or PMO}]{\text{TMO, MMO}} \text{ClHC}\overset{O}{—}\text{CCl}_2 \tag{42b}$$

As a groundwater remediation scheme, small amounts of toluene or phenol could be added to aerobic groundwater to induce the dioxygenase enzymes, and then a number of toxic organics could be oxidized. One difficulty with the scheme is that intermediates such as the epoxide product in equation (42b) is toxic to the bacterial biomass. If too much epoxide is produced, it will kill the biomass. On the other hand, if too much primary substrate is added (toluene), then the microorganisms will oxidize only the toluene and ignore the TCE. Optimization of such a remediation scheme would be needed. Other toxic intermediates can be produced under anaerobic conditions such as vinyl chloride, a carcinogen and an acute toxin implicated in liver disease.[23,24]

Because bacterial biomass and activity are so difficult to measure in the subsurface, engineers and scientists have resorted to using simplified expressions for chemical biodegradation such as the first-order (rate $= -kC$) and second-order reaction rate (rate $= -k'X_TC$, where k' is the second-order rate constant and X_T is the total biomass from acridine orange direct counts or other methods). X_T is usually assumed constant, so the second-order expression results in pseudo-first-order kinetics anyway.

If one could measure the enzyme activity (or viable biomass) of specific microorganisms for specific substrates, it would be possible to formulate two simultaneous Monod equations for microorganism–substrate interactions. For i microorganisms and j substrates:

$$\frac{dX_i}{dt} = +\sum_{j=1}^{m}\left(\frac{\mu_{i,j}S_jX_i}{K_{s_{i,j}} + S_j}\right) - b_iX_i \tag{43a}$$

$$\frac{dS_j}{dt} = \sum_{i=1}^{\ell}\frac{-\mu_{i,j}S_jX_i}{K_{s_{i,j}} + S_j}\,\frac{1}{Y_{i,j}} \tag{43b}$$

where X_i = viable biomasses, ML^{-3}

t = time, T

$Y_{i,j}$ = yield coefficients for the ith organism on the jth substrate, MM^{-1}
$\mu_{i,j}$ = maximum biomass growth rate constants, T^{-1}
S_j = substrate concentrations, ML^{-3}
$K_{s_{i,j}}$ = half-saturation constants, ML^{-3}
b_i = death or endogenous decay rate constants, T^{-1}

Even three microorganisms utilizing three substrates each would generate 36 constants that would be difficult to derive. A more realistic approach might be to assume that all of the bacterial biomass is active in degrading each substrate (or at least a constant fraction of the biomass is degrading each substrate). This obviates the need to identify the "active" microorganism biomass for each substrate. For i substrates and a total biomass X_T,

$$\frac{dX_T}{dt} = +\left(\sum_{i=1}^{n} \frac{\mu'_i S_i X_T}{K_{s_i} + S_i}\right) - \bar{b}X_T \tag{44a}$$

$$\frac{dS_i}{dt} = \frac{\mu'_i S_i X_T}{K_{s_i} + S_i}\frac{1}{Y_i} \tag{44b}$$

where $\bar{b}$ = the average biomass decay rate constant, T^{-1}

In some cases, it is possible to simplify equations (44a) and (44b) further if the chemical contaminants of interest can be lumped (such as total aromatic hydrocarbons, or methylene blue active substances, or dissolved organic carbon).

$$\frac{dS_b}{dt} = -\frac{\mu_b S_b X_T}{K_{s_b} + S_b}\frac{1}{Y_b} \tag{45a}$$

$$\frac{dX_T}{dt} = +\left(\frac{\mu_b S_b X_T}{K_{s_b} + S_b}\right) - \bar{b}X_T \tag{45b}$$

where S_b = bulk substrate concentration, ML^{-3}
μ_b = maximum cell growth rate constant, T^{-1}
K_{s_b} = half-saturation constant, ML^{-3}
Y_b = yield coefficient, MM^{-1}

Still we have not included effects of inhibition, toxic intermediates, co-metabolism, substrate–substrate interactions, or changes in electron acceptor condition. Even though biological transformations are critical to understanding the fate and transport of chemicals in the subsurface, we do not have a fundamental approach to quantifying the relationship. That is why first-order rate constants and pseudo-first-order rate constants will continue to be used, and it is also a challenge for the student to find paradigms for these complex relationships. Until now, we have ignored the microbial biomass (activity) because we did not have good methods for counting, but a better understanding of the role of subsurface microorganisms is evolving.

9.7 REDOX REACTIONS

Redox reactions in groundwater are crucial. They affect the fate and transport of both organic contaminants and metals in the subsurface. As redox changes, the indigenous microorganisms also go through an ecological succession from aerobic heterotrophs, to denitrifiers, to sulfate reducers and methanogens. As such, it is important for the modeler to appreciate the importance of redox potential and the changes that it causes in contaminant fate. In Section 9.6, we learned how certain organic chemicals can be degraded under aerobic, but not anaerobic, conditions. Likewise, Fe(II) is mobile under anaerobic conditions, but it is oxidized to Fe(III) under more oxygenated conditions and it becomes very insoluble and immobile.

Redox electrode potential (E_H in volts or pε) is defined under equilibrium conditions although groundwater reactions are not, in general, at equilibrium. Also, the electrode only measures redox couples (reactions) that are active at the specific electrode. Groundwater redox reactions can be very slow even on geological time scales, although they are often microbially mediated (enzyme catalyzed) over shorter periods. Bacteria cannot bring about reactions that are thermodynamically impossible. They can only mediate (catalyze) the rate of reaction, and they can use some of the free energy released in the redox reaction. Because there are no free electrons, every oxidation is accompanied by a reduction.

Figure 9.11 is a bar graph showing the redox sequence.[25] In groundwater, oxygen is the first electron acceptor to be utilized, followed by nitrate, followed by Mn(IV), Fe(III), SO_4^{2-}, and finally $CO_2 \rightarrow CH_4$. The order is not perfectly sequential in nature. Several electron acceptors (reduction reactions) can be utilized simultaneously or in microzones within the porous medium. To obtain a valid reading of electrode potential in the field, several conditions must be met: (1) species such as sulfide must not be adsorbed onto the Pt electrode, (2) the redox couple must be electroactive (electron transfer is rapid and reversible to attain chemical equilibrium), and (3) both members of the redox couple must be present at appreciable concentrations ($>10^{-5}$ M).

If we oxidize "typical" organic matter CH_2O with the sequence of electron acceptors, we can construct balanced redox reactions, Table 9.8. Note that all of the reactions have negative free energy $\Delta G^{\circ}(W)$, indicating spontaneity, but they may not all occur at a significant rate until all electron acceptors above it on Table 9.8 have been consumed. For example, reactions (1) and (2) may proceed in parallel due to microzones of low dissolved oxygen where denitrification occurs. Reactions (4)–(6) may occur simultaneously once the redox potential has been lowered to less than 0.0 V.

A two-dimensional, horizontal groundwater contaminant plume is depicted in Figure 9.12. The aquifer was originally aerobic, but the high concentration of organics from leachate has consumed the dissolved oxygen in the interior of the plume.

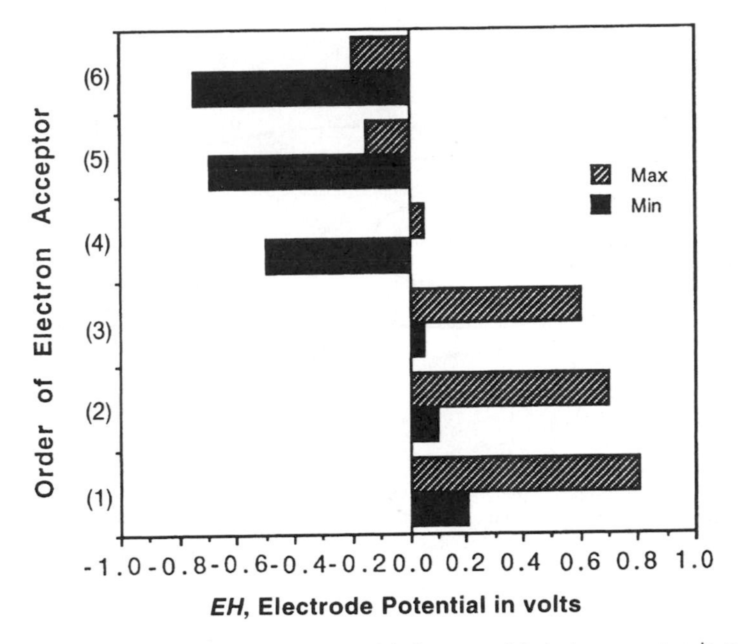

Figure 9.11 Range of measured electrode potentials for sequential electron acceptors in groundwater with organic reductants CH_2O. (1) $O_2 \rightarrow H_2O$. (2) $NO_3^- \rightarrow N_2$. (3) $MnO_2 \rightarrow Mn^{2+}$. (4) $FeOOH \rightarrow Fe^{2+}$. (5) $SO_4^{2-} \rightarrow H_2S$. (6) $CO_2 \rightarrow CH_4$.

Table 9.8 Progressive Reduction of Redox Intensity by Organic Substances in Groundwater: Sequence of Reactions at pH 7 and 25 °C

Reaction		$\Delta G^\circ(W)$, kJ equiv^{-1}
Oxygen consumption		
(1) $\frac{1}{4}CH_2O + \frac{1}{4}O_2$	$= \frac{1}{4}CO_2 + \frac{1}{4}H_2O$	–125
Denitrification		
(2) $\frac{1}{4}CH_2O + \frac{1}{5}NO_3^- + H^+$	$= \frac{1}{4}CO_2 + \frac{1}{10}N_2 + \frac{7}{20}H_2O$	–119
Formation of soluble Mn(II) by reduction of Mn(IV) oxides		
(3) $\frac{1}{4}CH_2O + \frac{1}{2}MnO_2(s) + H^+$	$= \frac{1}{4}CO_2 + \frac{1}{2}Mn^{2+} + \frac{3}{4}H_2O$	–85
Formation of soluble Fe(II) by reduction of Fe(III) (hydr)oxides		
(4) $\frac{1}{4}CH_2O + FeOOH(s) + 2\,H^+$	$= \frac{1}{4}CO_2 + \frac{7}{4}H_2O + Fe^{2+}$	–26.8
Sulfate reduction, formation of hydrogen sulfide		
(5) $\frac{1}{4}CH_2O + \frac{1}{8}SO_4^{2-} + \frac{1}{8}H^+$	$= \frac{1}{8}HS^- + \frac{1}{4}CO_2 + \frac{1}{4}H_2O$	–26
Methane fermentation		
(6) $\frac{1}{4}CH_2O + \frac{1}{8}CO_2$	$= \frac{1}{8}CH_4 + \frac{1}{4}CO_2$	–23.2

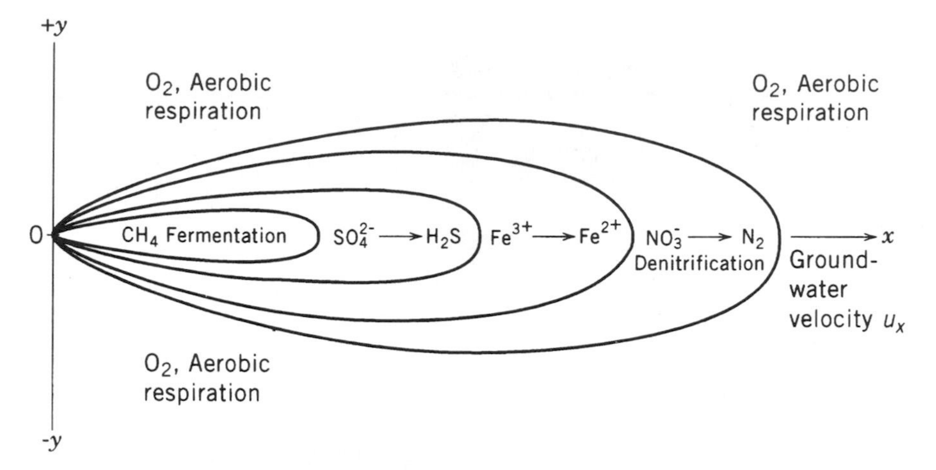

Figure 9.12 Two-dimensional plume of organics contamination and oxidation in groundwater. Dissolved oxygen is completely consumed on the interior of the plume, but aerobic oxidation occurs at the edges.

The isocontour lines correspond to zones where progressively more reducing conditions exist as you move into the interior of the plume. The example has implications for contaminants that only biodegrade under aerobic or anaerobic conditions. For example, aromatics that are degraded quickly under aerobic conditions may be transformed only at the edges of the plume, where oxygen is present. The EPA model BIOPLUMEII includes a mass balance for dissolved oxygen in the groundwater as well as a contaminant mass balance, and it accounts explicitly for this situation. The model uses a multiplicative Michaelis–Menton kinetic expression to account for decreased rates of organic biodegradation under low oxygen conditions.

$$\frac{dS}{dt} = \text{Rxn rate} = \frac{-1}{Y}\frac{\mu SX}{K_s + S}\left(\frac{[O_2]}{K_{O_2} + [O_2]}\right) \tag{46}$$

where S = substrate concentration of organics, ML^{-3}
t = time, T
Y = biomass yield, mass cells/mass substrate
μ = maximum biomass growth rate, T^{-1}
K_s = half-saturation constant, ML^{-3}

Aromatics such as BTEX chemicals (benzene, toluene, ethylbenzene, and xylenes) often cease to be degraded as $O_{2(aq)}$ concentrations approach zero.

As shown by Table 9.7, some organics can be degraded only under aerobic and some only under anaerobic conditions. Microorganisms can use xenobiotic organics as primary substrates, secondary substrates, or co-metabolically. As we recall from Chapter 7, primary substrates supply carbon and energy to the microorganisms; secondary substrates do not provide sufficient carbon and energy to maintain growth alone at the low concentration supplied. Co-metabolism involves a subset of secondary substrates, which can never be used as a primary energy source, but they can be transformed by an inducible enzyme system when primary substrate is available.

Organic chemicals can serve as electron acceptors in microbially mediated reactions. Sawyer et al.[24] discuss this possibility. Table 9.9 gives thermodynamic data from a number of half-reactions for chlorinated organics that can serve as electron acceptors, particularly under sulfate reducing or methanogenic conditions. Instead of carbon dioxide serving as the electron acceptor in methane fermentation, carbon tetrachloride could serve the purpose, provided that toxic intermediates did not develop to inhibit microbial mediation.

			$\Delta G^\circ(W)$
Methane fermentation	$\frac{1}{8}CO_2 + H^+ + e^-$	$= \frac{1}{8}CH_4 + \frac{1}{4}H_2O$	+5.76 kcal
	$\frac{1}{4}CH_2O + \frac{1}{4}H_2O$	$= \frac{1}{4}CO_2 + H^+ + e^-$	−10.00
Overall Rxn	$\frac{1}{4}CH_2O$	$= \frac{1}{8}CO_2 + \frac{1}{8}CH_4$	−4.24 kcal
CCl_4 as electron acceptor	$\frac{1}{2}CCl_4 + \frac{1}{2}H^+ + e^-$	$= \frac{1}{2}CHCl_3 + \frac{1}{2}Cl^-$	−15.45 kcal
	$\frac{1}{4}CH_2O + \frac{1}{4}H_2O$	$= \frac{1}{4}CO_2 + H^+ + e^-$	−10.00
Overall Rxn	$\frac{1}{2}CCl_4 + \frac{1}{4}CH_2O + \frac{1}{4}H_2O$	$= \frac{1}{2}CHCl_3 + \frac{1}{4}CO_2 + \frac{1}{2}H^+ + \frac{1}{2}Cl^-$	−25.45 kcal

It is an interesting exercise to put various electron acceptors from Table 9.9 with various substrates from Table 7.5 and to consider the possibility of the overall reaction. Vogel et al.[23] reported the possibility of trichloroethylene (TCE, $CHClCCl_2$) degrading anaerobically to form *trans*-dichloroethylene and then vinyl chloride (CH_2CHCl), a potent carcinogen. Not all TCE degradation would proceed through this pathway, but regulatory agencies must assume the worst case, so in risk assessments TCE is treated as if it produced an equal molar amount of vinyl chloride.

When the concentration of organic chemicals is not so large in groundwater to cause anaerobic conditions at the center of the plume, a more typical sequence of redox reactions may be observed. The following is an example of how electron acceptors can change as groundwater flows through an aquifer system.

Example 9.6 Sequence of Redox Reactions in Groundwater Below a Landfill

Jackson and Patterson[26] provide a nice example of a leachate from a landfill migrating through a 1-D water table aquifer. Shown below are data from a transect of monitoring wells. Interpret the water quality data in view of what we have learned about redox. Water velocity is approximately 10 cm d^{-1}, and the profile view of the lower

Table 9.9 Half-Reaction Potentials and Free Energies of Chlorinated Organic Chemicals and Some Electron Acceptors

		E°, volts[a]	$\Delta G°(W)$, kcal[b]
$\frac{1}{2}CCl_3CCl_3 + e^-$	$= \frac{1}{2}CCl_2CCl_2 + Cl^-$	1.14	–26.29
$\frac{1}{4}CH_2BrCH_2Br + \frac{1}{2}H^+ + e^-$	$= \frac{1}{4}CH_3CH_3 + \frac{1}{2}Br^-$	0.90	–20.75
$\frac{1}{4}O_2 + H^+ + e^-$	$= \frac{1}{2}H_2O$	0.81	–18.68
$Fe^{3+} + e^-$	$= Fe^{2+}$	0.771	–17.78
$\frac{1}{5}NO_3^- + \frac{6}{5}H^+ + e^-$	$= \frac{1}{10}N_2 + \frac{3}{5}H_2O$	0.743	–17.13
$\frac{1}{2}CCl_4 + \frac{1}{2}H^+ + e^-$	$= \frac{1}{2}CHCl_3 + \frac{1}{2}Cl^-$	0.67	–15.45
$\frac{1}{2}CCl_3CCl_3 + \frac{1}{2}H^+ + e^-$	$= \frac{1}{2}CHCl_2CCl_3 + \frac{1}{2}Cl^-$	0.66	–15.22
$\frac{1}{2}CCl_2CCl_2 + \frac{1}{2}H^+ + e^-$	$= \frac{1}{2}CHClCCl_2 + \frac{1}{2}Cl^-$	0.58	–13.38
$\frac{1}{2}CH_3CCl_3 + \frac{1}{2}H^+ + e^-$	$= \frac{1}{2}CH_3CHCl_2 + \frac{1}{2}Cl^-$	0.57	–13.15
$\frac{1}{2}CHCl_3 + \frac{1}{2}H^+ + e^-$	$= \frac{1}{2}CH_2Cl_2 + \frac{1}{2}Cl^-$	0.56	–12.91
$\frac{1}{2}CHClCCl_2 + \frac{1}{2}H^+ + e^-$	$= \frac{1}{2}CHClCHCl$ (*trans*) $+ \frac{1}{2}Cl^-$	0.53	–12.23
$\frac{1}{2}CH_3CHCl_2 + \frac{1}{2}H^+ + e^-$	$= \frac{1}{2}CH_3CH_2Cl + \frac{1}{2}Cl^-$	0.51	–11.76
$\frac{1}{2}NO_3^- + H^+ + e^-$	$= \frac{1}{2}NO_2^- \frac{1}{2}H_2O$	0.42	–9.60
$\frac{1}{2}CHClCHCl$ (*trans*) $+ \frac{1}{2}H^+ + e^-$	$= \frac{1}{2}CH_2CHCl + \frac{1}{2}Cl^-$	0.37	–8.53
$\frac{1}{8}SO_4^{2-} + \frac{19}{16}H^+ + e^-$	$= \frac{1}{16}H_2S + \frac{1}{16}HS^- + \frac{1}{2}H_2O$	–0.22	+5.08
$\frac{1}{8}CO_2 + H^+ + e^-$	$= \frac{1}{8}CH_4 + \frac{1}{4}H_2O$	–0.25	+5.76

[a] $E° = -96.485$ kJ V^{-1} equiv; 1 kcal = 4.184 kJ.
[b] $\Delta G°(W)$ is for unit activities and pH 7.
Source: Modified from Vogel et al.[23]

sand aquifer is shown in Figure 9.13. In the following data table, the electrode potentials were modified to reflect the actual E_H value in the aquifer based on redox couples.

Distance from Landfill, m	E_H, volts	pH	O_2, mg L^{-1}	$Fe(II)_{(aq)}$, mg L^{-1}	$S(-II)_{(aq)}$, mg L^{-1}
0	+ 0.52	5.6	4.0	< 0.1	< 0.01
100	+ 0.48	6.0	3.0	< 0.1	< 0.01
300	+ 0.05	6.8	0.4	2.0	< 0.01
500	- 0.10	7.3	0.0	6.0	0.02
700	- 0.15	7.8	0.0	2.5	0.07
900	- 0.20	8.3	0.0	0.2	0.12

Solution: First, dissolved oxygen is being consumed by aerobic respiration of organics. Then, iron is being reduced and subsequently reprecipitated as the sulfide and/or pyrite (after aging). Sulfate is reduced and sulfide is then precipitated.

$$Fe(OH)_3 \text{ or } FeOOH_{(s)} \rightarrow Fe(II)_{(aq)} \rightarrow FeS_{(s)} \text{ or } FeS_{2(s)}$$

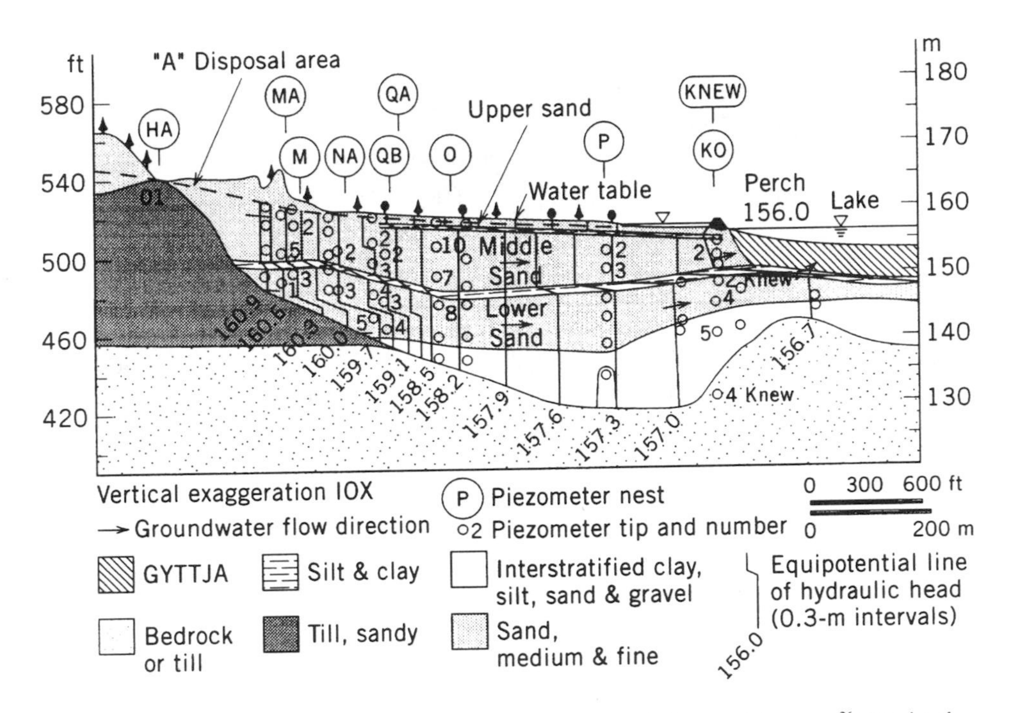

Figure 9.13 Landfill and observation wells in 2-D aquifer. (From Jackson and Patterson.[26]) Reprinted with permission of the American Geophysical Union. Copyright (1982).

The pH increases because of proton consumption by reduction reactions and mineral weathering. The following sequence of reactions is occurring:

O_2 consumption	$CH_2O + O_2 \rightleftharpoons CO_2 + H_2O$
Iron reduction	$CH_2O + 8H^+ + 4\,Fe(OH)_3 \rightleftharpoons 4\,Fe^{2+} + 11\,H_2O + CO_2$
Sulfate reduction	$CH_2O + \frac{1}{2}SO_4^{2-} + \frac{1}{2}H^+ \rightleftharpoons \frac{1}{2}HS^- + H_2O + CO_2$
Sulfide precipitation	$Fe^{2+} + HS^- \rightleftharpoons FeS_{(s)} + H^+$
Overall reaction to pyrite	$Fe(OH)_3 + 2\,SO_4^{2-} + 4\,H^+ \rightleftharpoons FeS_{2(s)} + \frac{15}{4}O_2 + \frac{7}{2}H_2O$

Oxygen that is produced by the overall reaction is immediately consumed by aerobic respiration.

It is important to remember that not only organic chemicals are influenced by changes in redox intensity in groundwater. Toxic metals are affected by redox potential directly, in the case of those metals with multiple valence states, or indirectly. Decreases in redox potential from aerobic to anaerobic conditions can:

- Change the valence state of metal ion (reduce it).
- Dissolve MnO_2 and FeOOH, thus releasing other metal ions bound at the surface into the groundwater.
- Change the pH, which affects metal sorption.
- Produce new solutes or ligands [S^{2-}, Fe(II), Mn(II)] in solution that can react with metal ions to complex, precipitate, or react with them.

Table 9.10 is a compilation of some important metals and metalloids (As, Se) that are known groundwater contaminants and their potential for changes in valence state as groundwater becomes more highly reduced. Fe(II) and Mn(II) are strong reductants. Could they, for example, reduce arsenic from arsenate As(V) to arsenite As(III) and render it more mobile in the groundwater?

Example 9.6 As(V) – As(III) – Equilibria[25]

a. Under what conditions is arsenate reduced to arsenite As(III)?
b. Can Fe^{2+} or Mn^{2+} reduce arsenate As(V) under conditions encountered in groundwater or soil waters (TOT As = 10^{-4} M, pH = 5, pH = 8)?

Solution: The key redox equilibrium is the reduction of arsenate to arsenite. Equilibrium constants can be found in Stumm and Morgan.[25]

$$H_2AsO_4^- + 3\ H^+ + 2\ e^- = H_3AsO_3 + H_2O; \quad \log K = 21.7\ (25\ °C) \tag{i}$$

Recall that $\log K = n(\text{p}\varepsilon)$ and $\text{p}\varepsilon = 16.9\ E°$, where n is the number of electrons transferred. The arsenate(V) species are from an acid/base point of view similar to the phosphate species; K_2 of H_3AsO_4 is the second acid dissociation constant

$$H_2AsO_4^- = H^+ + HAsO_4^{2-}; \quad \log K_2 = -6.8 \tag{ii}$$

The equilibrium matrix for all pertinent As(V) ($H_2AsO_4^-$, $HAsO_4^{2-}$) and As(III) (H_3AsO_3) is given below.

Species	Components: $H_2AsO_4^-$	H^+	e^-	log K
(i) $H_2AsO_4^-$	1	0	0	0
(ii) $HAsO_4^{2-}$	1	1	0	–6.8
(iii) H_3AsO_3	1	3	2	21.7
(iv) H^+	0	1	0	0
(v) e^-	0	0	1	0

$$\text{TOT As} = [(H_2AsO_4^-] + [(HAsO_4^{2-}] + [H_3AsO_3] = 10^{-4}\ \text{M}$$

Table 9.10 Potential Effects of a Reduction in Redox Intensity in a Groundwater for Heavy Metal and Metalloid Contaminants

Atomic Number	Element	Valence States	Fate Under Reducing Conditions
24	Cr	Cr(VI) → Cr(III)	Reduction by organics and some metals to immobilize Cr(III)
25	Mn	Mn(IV) → Mn(II)	MnO_2 dissolution and solubilization
26	Fe	Fe(III) → Fe(II)	FeOOH, Fe_2O_3 dissolution and solubilization, reprecipitation with sulfides
27	Co	Co(II)	One valence state, more mobile as a result of Fe-oxide dissolution
28	Ni	Ni(II)	One valence state, more mobile as a result of Fe-oxide dissolution but precipitates with sulfide
29	Cu	Cu(II) → Cu(I)	More mobile under reducing conditions but precipitates with sulfide
30	Zn	Zn(II)	One valence state, more mobile under reducing conditions but precipitates with sulfide
33	As	As(V) → As(III)	Anion reduced by organics $AsO_4^{3-} \rightarrow AsO_3^{3-}$, increased solubility and mobility
34	Se	Se(VI) → Se(IV)	Anion reduced by organics $SeO_4^{2-} \rightarrow SeO_3^{2-}$, increased solubility and mobility
42	Mo	Mo(VI) → Mo(III)	Anion reduced Mo(III) and increased mobility
48	Cd	Cd(II)	One valence state, relatively more mobile as a result of Fe-oxide dissolution but precipitates with sulfide
80	Hg	Hg(II) → Hg°	Can be reduced to elemental Hg or methylated under sulfate reducing conditions or precipitated with sulfide
82	Pb	Pb(II)	One valence state, relatively immobile, high K_d but precipitates with sulfide

The results are presented in Figure 9.14. Two pH values are considered that bracket the range of groundwater pH from 5 to 8. At pH 5, one needs a pε of +3.3. At pH 8, it must be a highly reducing pε of less than −1.8 (−0.1 volts, E_H).

In order to answer question (b) we have to consider under what condition Fe^{2+} and Mn^{2+} is oxidized to $Fe(OH)_{3(s)}$ or $MnO_{2(s)}$. The equilibria are

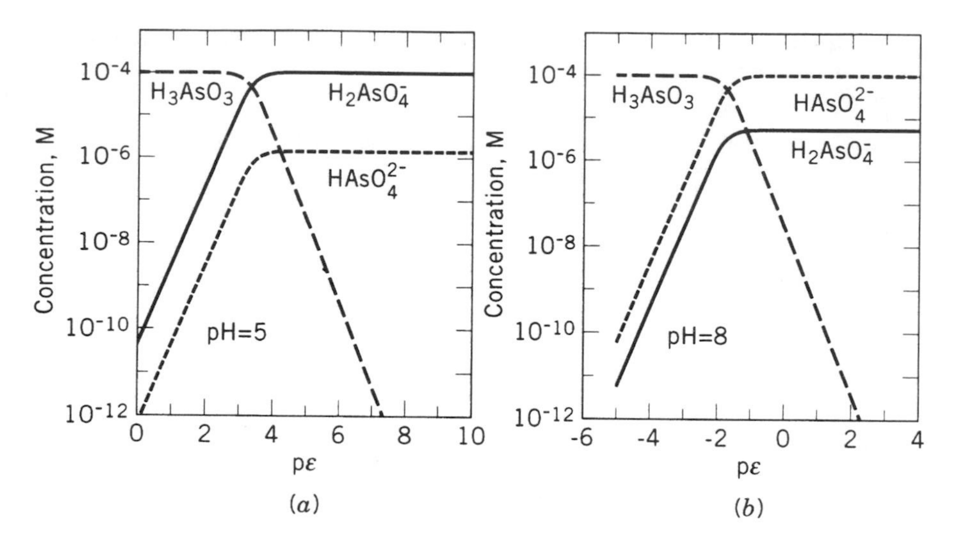

Figure 9.14 Arsenic speciation (log *C* versus pε) at (a) pH 5 and (b) pH 8, bracketing the conditions found in most groundwater.

$$Fe(OH)_3 + 3\ H^+ + e^- = Fe^{2+} + 3\ H_2O; \quad \log K = 16.0 \qquad \text{(iii)}$$

and

$$MnO_2 + 4\ H^+ + 2\ e^- = Mn^{2+} + 2\ H_2O; \quad \log K = 43.6 \qquad \text{(iv)}$$

We can consider these equations in the matrix below.

		Components				
	Species	$Fe(OH)_{3(s)}$	$MnO_{2(s)}$	H^+	e^-	log *K*
(i)	Fe^{2+}	1	0	3	1	16.0
(ii)	Mn^{2+}	0	1	4	2	43.6
(iii)	H^+	0	0	1	0	0
(iv)	e^-	0	0	0	1	0

The results are given in Figure 9.15. Obviously, Mn^{2+} cannot reduce arsenate(V) because typical concentrations (10^{-4} to 10^{-5} M) occur at much too high a pε value. In the case of Fe^{2+}, larger concentrations of Fe^{2+} ($Fe^{2+} > 10^{-4}$ M) at pH = 8 are, from a thermodynamic point of view, able to reduce $HAsO_4^{2-}$; at pH = 5 this reduction cannot be accomplished with Fe^{2+} concentrations typically encountered in natural waters. Note, however, that the equilibrium constant of reaction (iii) depends on stability of the solid Fe(III) (hydr)oxide phase considered. Interestingly, the thermodynamic data show that the reduction of As(V) and of $Fe(OH)_{3(s)}$ to As(III) and Fe(II),

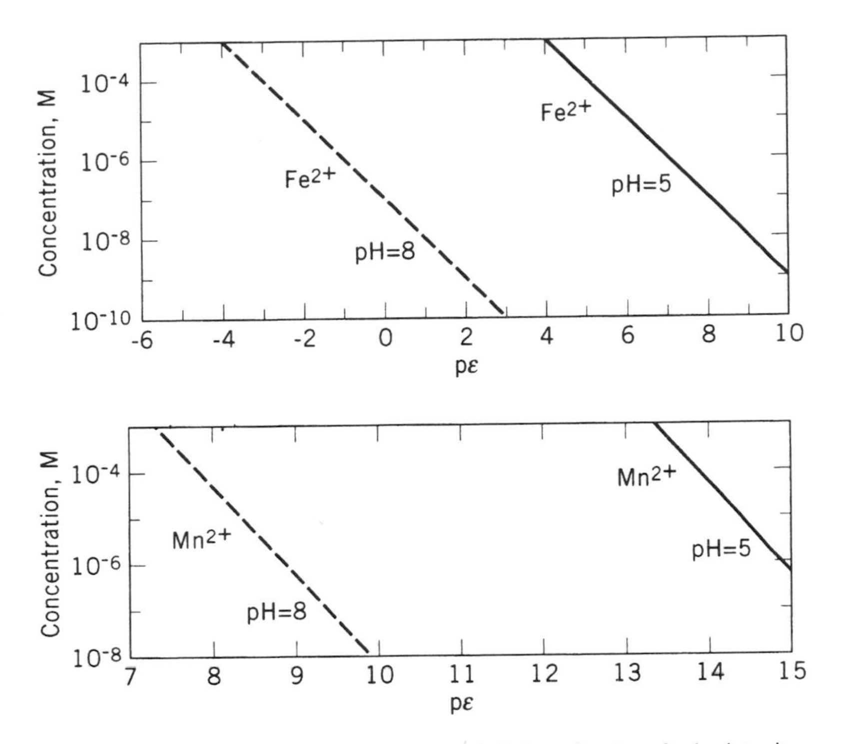

Figure 9.15 Fe^{2+} and Mn^{2+} concentrations at pH 5 and pH 8 as a function of redox intensity, pε.

respectively, occur at similar pε values, that is, at similar reducing conditions.

9.8 NONAQUEOUS PHASE LIQUIDS

9.8.1 LNAPLs and DNAPLs

Nonaqueous phase liquids (NAPLs) are immiscible in water, and they present another phase of concern in groundwater contamination problems. They can be classified as either LNAPLs (lighter-than-water nonaqueous phase liquids) or DNAPLs (denser-than-water nonaqueous phase liquids). They present special problems for the modeler because fate and transport are difficult to simulate (NAPL, dissolved contaminant, sorbed contaminant, gas phase) and because they may follow irregular flow paths in heterogeneous porous media. Leaking underground storage tanks of gasoline are a major problem with tens of thousands estimated to be leaking in the United States alone. Table 9.11 gives a brief summary of NAPL densities and solubilities. For a complete table at 25 °C refer to Schwarzenbach et al.[27].

Heterogeneities add complexity in groundwater modeling. Figure 9.16 shows how even dissolved constituents may tend to follow flow paths that are difficult to discern from the surface while taking samples. "Fingering" of plumes is common in

Table 9.11 Densities and Solubilities of NAPLs (Nonaqueous Phase Liquids)

Liquids	Density at 15 °C, g cm^{-3}	Solubility at 10 °C, mg L^{-1}
LNAPL		
Medium distillates (fuel oil)	0.82–0.86	3–8
Petroleum distillates (jet fuel)	0.77–0.83	10–150
Gasoline	0.72–0.78	150–300
Crude oil	0.80–0.88	3–25
DNAPL		
Trichloroethylene (TCE)	1.46	1,070
Tetrachloroethylene (PCE)	1.62	160
1,1,1-Trichloroethane (TCA)	1.32	1,700
Dichloromethane (CH_2Cl_2)	1.33	13,200
Chloroform ($CHCl_3$)	1.49	8,200
Carbon tetrachloride (CCl_4)	1.59	785
Creosote (napthalene, phenanthrene, etc.)	1.11	20

sedimentary deposits or alluvium that has been layered in its stratigraphy. Careful analysis of the well log records or core borings is necessary to understand the geology and the potential migration pathways of pollutants. Vertical cross-section diagrams must be obtained, such as Figure 9.16. Cracked clay, shale, fractured rock, or chalk aquifers are difficult to simulate. The saturated hydraulic conductivity of the bulk geologic material may be small ($K_z = 10^{-7}$ cm s^{-1}), but K_z is perhaps 10^{-1} cm s^{-1} in the crack. Sometimes modelers can average the cracks and the bulk material for an apparent "average" K_z on the order of 10^{-2}–10^{-5} cm s^{-1}. The exact magnitude of the hydraulic conductivity depends on the aperture openings, the connectedness of the cracks, and the number of cracks.

LNAPLs form a floating pool of material on the surface of the groundwater table

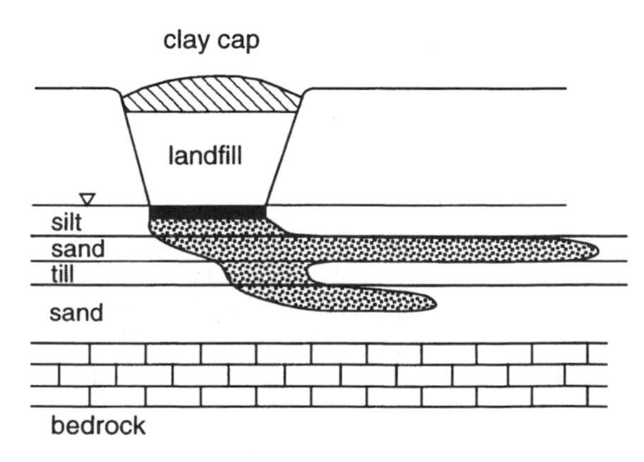

Figure 9.16 "Fingers" of dissolved contaminant plume beneath a leaking landfill. The plume follows sand units where K_x values are higher.

(Figure 9.17). Soluble constituents of the NAPL then dissolve into the groundwater and migrate in the direction of groundwater flow. Before the LNAPL reaches the groundwater table, it must percolate through the unsaturated zone. Soil retention capacities (SRT) for both DNAPLs and LNAPLs are substantial.

$$\text{SRT} = 3\text{–}5 \text{ L m}^{-3} \text{ soil in high permeability soil, sand} \tag{47a}$$

$$\text{SRT} = 30\text{–}50 \text{ L m}^{-3} \text{ soil in low permeability soil, silt–clay} \tag{47b}$$

Thus a leak of 10,000 gallons from a gasoline filling station tank might be completely retained within a low permeability silt–clay soil (~ 40 liters NAPL per m^3 soil) within a cube of soil 10 meters on each side. Only after 10,000 gallons was spilled would the gasoline begin to reach the saturated zone. It would be retained by surface tension (capillary forces) and by absorption into the micropores of the porous granules. These micropores provide an internal porosity for bound water and for NAPL. Of course, sandy soil does not retain as much NAPL as silt–clay soils [equations (47a) and (47b)].

If a significant portion of the LNAPL is retained in the unsaturated zone, volatilization will be an important fate pathway. Often, LNAPL spills are first detected by someone smelling the vapor in their yard or basement! In the groundwater, benzene, toluene, and napthalene are important dissolved components for which to analyze. Benzene is usually the most toxic and carcinogenic, and its concentration determines the remediation action plan for fuel spills.

Dense nonaqueous phase liquids (DNAPLs) are transported by gravity through the unsaturated zone, although a portion is retained according to equations (47a)

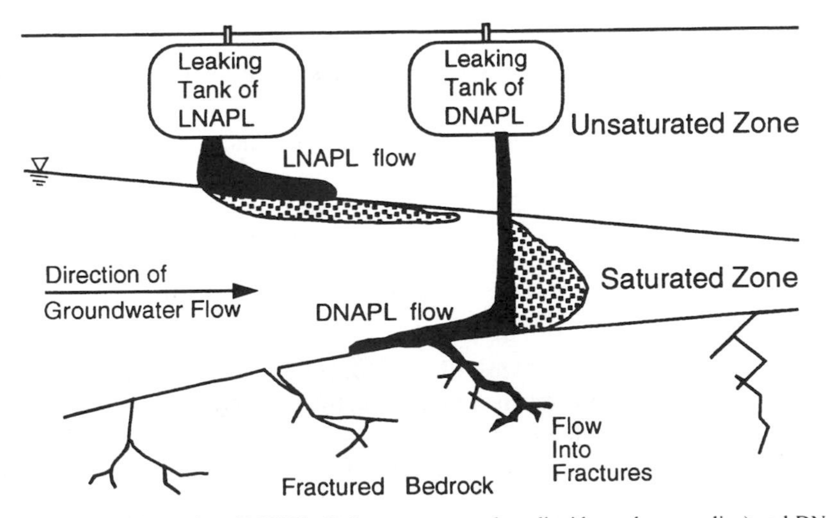

Figure 9.17 Leaking tanks of LNAPL (light nonaqueous phase liquids, such as gasoline) and DNAPL (dense nonaqueous phase liquids, such as CCl_4). Pure product is shown by dark area and dissolved constituents are shown by dotted area. (Modified from Hemond and Fechner.[6])

and (47b). Once it reaches the saturated zone, a plume can develop of dissolved chemical as depicted in Figure 9.17. DNAPLs can descend to the lowest rock unit and form a pool of pure organic liquid.

The first step in remediation of NAPL sites is to attempt to recover pure product (gasoline, trichloroethylene, etc.) from the subsurface by putting a submersible pump into a collection well and separating the mixture at the surface into water and NAPL phases. As shown in Figure 9.17, DNAPLs can actually move in different directions than the groundwater, so finding pools of pure liquid is sometimes difficult. The difficulty with collecting NAPL is (1) finding where it is located, (2) overcoming surface tension forces and getting it to move to a collection point, and (3) recovering small volumes of pure product that can contaminate large volumes of water.

Example 9.7 DNAPL Spill to the Unsaturated Zone and Groundwater

There has been a spill of 2000 gallons of tetrachloroethylene (PCE) to the soil. The groundwater table is 5 m deep and the soil is of low permeability. The area of the spill encompasses about 25 m^2. Answer the following questions.

a. Do you expect significant degradation of tetrachloroethylene?
b. Approximately how much will be retained in the unsaturated zone?
c. What will be the fate of the material once it reaches the groundwater table?
d. How many liters of groundwater can the remaining pool of NAPL contaminate above the MCL? (*Note:* The MCL for tetrachloroethylene is 5 $\mu g\ L^{-1}$.)

Solution:

a. Tetrachloroethylene will not biodegrade under aerobic conditions (Table 9.7). However, some volatilization/evaporation should occur from the unsaturated zone.
b. If we assume that the soil can retain 40 L m^{-3}, we find

$$(3.79\text{L gal}^{-1}) \times 2000\text{ gal} = 7580\text{ L of PCE}$$

$$\frac{(7580\text{ L})}{40\text{ L m}^{-3}} = 189.5\text{ m}^3\text{ of soil required to retain it}$$

But we only have 5 m × 25 m^2 = 125 m^3 of soil, so approximately 66% of the spill (1319 gallons) is retained in the unsaturated zone and 681 gallons leak through to the saturated zone.

c. PCE is more dense than water (1.62 g cm^{-3}), so it will sink to lower units (see DNAPL in Figure 9.17).
d. It will create a plume of dissolved PCE because it is soluble up to 160 mg L^{-1}. We could model the plume with equation (35) for a continuous input of source concentration 160 mg L^{-1}.

e. Mass of PCE in satd. zone = (2581 L) (1.62 kg L^{-1})
= 4181 kg

Mass = $C \times V$

$$4181 \text{ kg} = \frac{(5 \ \mu\text{g L}^{-1}) (V \text{ liters})}{10^9 \ \mu\text{g kg}^{-1}}$$

$V = 8.36 \times 10^{11}$ liters! theoretically

This is roughly enough water to supply all of New York City for one year (7 million people, 100 gallons per person per day).

It is the maximum volume of water that could be contaminated to the MCL, but it illustrates the point that a small volume of pure organic chemical can cause contamination problems for a long time if it is not remediated.

Nonaqueous phase liquids may become entrapped in cracks or fine pores in the subsurface, creating a long-term source of contamination to groundwater. It is difficult for hydraulic forces of groundwater flow to overcome the capillary forces that retain blobs of NAPL trapped in pore spaces.[6,28,29] Thus they tend to remain in place and slowly dissolve. Powers et al.[28] have shown that dissolution rates are highly dependent on the range and size distribution of NAPL blobs. Net observed mass transfer rates are slower for large or branched ganglia entrapped in coarse or graded sands due to the decreased surfacc area per unit volume. Small spherical blobs dissolve more quickly, and it is more likely to remediate such sites with soil flushing (Figure 9.18).

We may assume that mass transfer controls dissolution rates of NAPL blobs according to the development by Powers et al.[30] for one-dimensional saturated aquifers.

$$\theta_w \frac{\partial C}{\partial t} = \frac{\partial}{\partial x}\left(\theta_w D_x \frac{\partial C}{\partial x}\right) - v_x \frac{\partial C}{\partial x} + k_f a_0 (C_s - C) \tag{48}$$

where θ_w = aqueous phase volumetric fraction, dimensionless
C = aqueous phase concentration, ML^{-3}

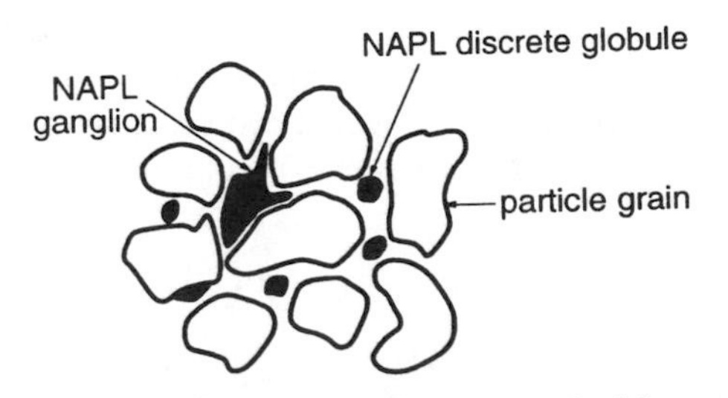

Figure 9.18 Small discrete globules of NAPL are more easily dissolved from porous media than large ganglia. Mass transfer limitations cause slow dissolution and remediation when flushing the porous medium with water. It is not feasible in most cases.

t = time, T
x = longitudinal distance, L
D_x = hydrodynamic dispersion coefficient, L^2T^{-1}
v_x = specific discharge, LT^{-1}
k_f = mass transfer coefficient between NAPL phase and water, LT^{-1}
a_0 = specific surface area, $L^2\ L^{-3}$
C_s = equilibrium concentration of the contaminant in water in contact with pure NAPL phase, ML^{-3}

Chemical engineering correlations have been used to estimate the mass transfer coefficient k_f from column studies with porous media and trichloroethylene, styrene, and napthalene dissolution.[28] Rewriting equation (48) in terms of dimensionless variables yields

$$\theta_{w'} \frac{\partial \beta}{\partial \tau} = \frac{\partial}{\partial \zeta}\left(\theta_{w'} \frac{1}{\text{Pe}} \frac{\partial \beta}{\partial \zeta}\right) \frac{\partial \beta}{\partial \zeta} + \frac{\text{Sh}}{\text{Re} \cdot \text{Sc}}\, \alpha\,(1-\beta) \tag{49}$$

where Re = $v_x \rho_{w'} \ell_c / \mu_{w'}$ = Reynolds number, dimensionless
Sc = $\mu_{w'}/D_L \rho_{w'}$ = Schmidt number, dimensionless
Sh = $77.6\ \text{Re}^{0.658}$ = Sherwood number, dimensionless
Pe = $v_x L/D_x$ = Peclet number, dimensionless
β = dimensionless concentration = C/C_s
ζ = dimensionless distance = x/L
τ = dimensionless time = $v_x t/L$
α = dimensionless surface area = $a_0 L$
L = length scale (simulation distance), L
$\rho_{w'}$ = density of water, ML^{-3}
ℓ_c = characteristic mixing scale (usually $\frac{1}{2}\ d_{50}$, particle diameter), L
D_L = molecular diffusivity of contaminant in water, L^2T^{-1}
$\mu_{w'}$ = viscosity of water, $ML^{-1}T^{-1}$

Equation (49) was solved using a 1-D finite element numerical method and Galerkin's method for the spatial derivatives and a backward finite-difference approximation of the time derivative for column studies.[30] In an actual groundwater contamination problem, equation (48) would be used and $k_f a_0$ would be a lumped parameter obtained from model calibration of field data.

9.8.2 Slow Processes

There are several "slow" processes at sites with large concentrations of organic contaminants in addition to NAPL dissolution. These processes cause groundwater remediation to be expensive, to be difficult to predict, and to require long recovery times.

- NAPL dissolution and ganglion globules.
- Intraparticle diffusion and slow desorption.
- Immobile water and slow mass transfer.

We may lump the latter two phenomena into one for the purpose of mathematical modeling. Heuristically, we can consider the slow processes as: (1) a local sorption equilibrium in the pore water with the outside surface of the particles; (2) a local sorption equilibrium inside the particles between immobile water and inner pore surfaces; and (3) a mass transfer limited diffusion between the two locations. Young and Ball[31] have measured the internal porosity of Borden sand aquifer material, which, normally, one would consider as "hard spheres" without any internal porosity. But calcite grains and organic coatings are believed to cause a measurable internal porosity and slow response even in sand aquifers. Internal porosity can be measured with BET analysis. Internal K_d values may differ from external (pore water) K_d values.

$$C \underset{\text{fast}}{\overset{K_d}{\longleftrightarrow}} q_s \underset{\text{slow}}{\overset{\gamma}{\longleftrightarrow}} q_{\text{im}} \underset{\text{fast}}{\overset{K_d}{\longleftrightarrow}} C_{\text{im}} \tag{50}$$

where C = dissolved pore water concentration, ML^{-3}
q_s = surface adsorbed contaminant, ML^{-2}
q_{im} = inner surface adsorbed contaminant, ML^{-2}
C_{im} = immobile internal pore water concentration, ML^{-3}
K_d = equilibrium distribution coefficient, water volume per mass of solids, L^3M^{-1}
γ = slow mass transfer rate coefficient, LT^{-1}

The total mass in a unit volume of aquifer consists of four reservoirs: (1) the mass in the bulk pore water, (2) the mass sorbed onto outer surfaces of the porous media, (3) the mass sorbed onto inner surfaces (internal pores) of the porous media, and (4) the mass of the immobile water in the internal pores. A total mass balance equation can be written.

$$\text{Total mass} = V_T C_T = n V_T C + \varepsilon_{\text{im}} V_T C_{\text{im}} + a_s V_T q_s + a_{\text{im}} V_T q_{\text{im}} \tag{51}$$

or

$$C_T = nC + \varepsilon_{\text{im}} C_{\text{im}} + a_s q_s + a_{\text{im}} q_{\text{im}} \tag{52}$$

where C_T = total concentration, mass per unit total volume, ML^{-3}
V_T = total volume, L^3
n = effective porosity
ε_{im} = internal porosity of immobile water
a_s = external specific surface area per unit total volume, L^2L^{-3}
a_{im} = internal specific surface area per unit total volume, L^2L^{-3}

Mass flux from the pore water to the immobile water would follow Fick's first law of diffusion.

$$J = -\frac{D_L A_s}{r}(C - C_{\mathrm{im}}) \tag{53}$$

where J = mass flux rate, MT^{-1}

D_L = molecular diffusion coefficient of contaminant in water, L^2T^{-1}

A_s = external surface area, L^2

r = length scale (d_{50} particle radius), L

The mass flux per unit volume (dC/dt) would be

$$\frac{dC}{dt} = -\frac{D_{\mathrm{app}}\, a_s}{rn}(C - C_{\mathrm{im}}) \tag{54}$$

in which D_{app} is the apparent diffusion coefficient between immobile water and bulk pore water chemical. Local equilibrium of the surface processes should take the form of a Langmuir or Freundlich isotherm, but if concentrations are in the linear portion of the curve, then the slope of the line can be described by an equilibrium distribution coefficient.

$$q_s = K_d\, C/\Omega_s \tag{55a}$$

$$q_{\mathrm{im}} = K_d\, C_{\mathrm{im}}/\Omega_{\mathrm{im}} \tag{55b}$$

where Ω_s and Ω_{im} are the BET surface area per mass of solids, L^2M^{-1}. The model equation for a 1-D saturated zone would be identical to equation (48) for NAPL dissolution except that the last term would be the apparent diffusion mass transfer limitation.

$$n\frac{\partial C}{\partial t} = \frac{\partial}{\partial x}\left(nD_x\frac{\partial C}{\partial x}\right) - v_x\frac{\partial C}{\partial x} - \frac{D_{\mathrm{app}}\, a_s}{r}(C - C_{\mathrm{im}}) \tag{56}$$

Batch slurry tests or short columns could be used to estimate the overall mass transfer rate constant, $D_{\mathrm{app}}a_s/rn$. Solving simultaneously equations (52), (54), and (55), one obtains the "slow concentration response," equation (57), for batch kinetics.

$$C = C_0\, e^{-\gamma(1+\alpha/\beta)t} + \frac{C_T}{\alpha+\beta}\left(1 - e^{-\gamma(1+\alpha/\beta)t}\right) \tag{57}$$

where C_0 = initial aqueous pore water chemical concentration, ML^{-3}

$\gamma = D_{\mathrm{app}}a_s/rn$ = the overall mass transfer rate constant, T^{-1} (57a)

$\alpha = n + a_sK_d/\Omega_s = n + \rho_bK_d/n$ = outer surface constant (57b)

$\beta = \varepsilon_{\mathrm{im}} + a_{\mathrm{im}}K_d/\Omega_{\mathrm{im}} = \varepsilon_{\mathrm{im}} + \rho_bK_d/\varepsilon_{\mathrm{im}}$ = inner surface constant (57c) where ρ_b is the dry bulk density of the porous medium. Equation (57) admits an asymptotic solution that approaches steady state, $C_{ss} = C_t/(\alpha + \beta)$, with a time constant of $1/\{\gamma(1 + \alpha/\beta)\}$. Apparent diffusion coefficients on the order of 1×10^{-8} to 1×10^{-10} cm^2 s^{-1} should be normal for mass transfer limitation in the aqueous phase (pore diffusion) considering molecular diffusion, tortuosity, constrictivity, and retardation. If diffusion through a solid film or phase is relevant, D_{app} will be much smaller.

Ball and Roberts[32,33] used a more mechanistic and detailed spherical intraparticle pore diffusion model to simulate batch isotherm data. In column studies, Young and Ball[31] found that sorption of trichloroethylene and 1,2,4,5-tetrachlorobenzene took years to attain full equilibrium compared with batch measured K_d values under realistic groundwater conditions, but pulverization of solid samples was shown to be an expedient alternative to excessively long equilibrium times in batch testing.[33] Response times in columns have been shown to depend on the intraparticle diffusion coefficient divided by the median particle radius squared.

$$\tau_c = \frac{D_p}{r^2}\left(\frac{L}{\nu_x}\,\varepsilon_{\text{im}}\right) \tag{58}$$

where τ_c = dimensionless time for column response
D_p = intraparticle pore water diffusion coefficient, L^2T^{-1}
r = particle radius, L
L = length of the column, L
ν_x = specific discharge, LT^{-1}
ε_{im} = immobile water pore volume ratio, dimensionless

9.9 BIOFILMS AND BIOAVAILABILITY

9.9.1 Biofilms

Rittmann and McCarty[34,35] pioneered the use of biofilm modeling for subsurface applications. When organic concentrations become large, bacteria can emit polysaccharides (Figure 9.10b) and growth occurs in films around porous media grains. It is especially a problem when biofilms cause plugging in the vicinity of wells where oxygen concentrations and organic loading rates are high. Most of the time in groundwater, dissolved organic concentrations are not sufficient to support biofilm growth except in cases where contaminants serve as the primary substrate for microorganisms. Even for cases where organic concentrations are not sufficient to support continuous biofilm growth around particles, biofilm model equations are still valid, but mass transfer limitation through the "film" is zero. Whether bacteria grow in the subsurface as continuous films around particles that restrict the pore size or in aggregates that accumulate in pore throats does not appreciably affect biotransformation rates, but it does affect permeability loss due to plugging.[36]

Figure 9.19 is a schematic of biofilms and discrete particles or aggregates in groundwater.

Assuming Monod kinetics, the coupled ordinary differential equations for substrate concentration and bacterial biomass are given by equations (59) and (60).

$$\frac{dS}{dt} = -\frac{q_m XS}{K_s + S} \tag{59}$$

$$\frac{dX}{dt} = Y\frac{q_m XS}{K_s + S} - b'X \tag{60}$$

where S = substrate concentration, ML^{-3}

t = time, T

q_m = maximum rate of substrate utilization, $M_S M_X^{-1} T^{-1}$

X = biomass concentration, ML^{-3}

K_s = half-saturation concentration, ML^{-3}

Y = yield coefficient, $M_X M_S^{-1}$

b' = overall loss rate of biomass, T^{-1}

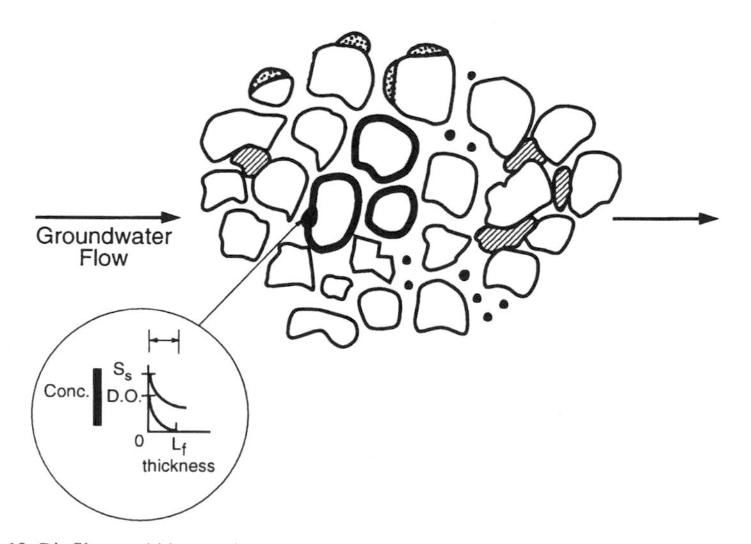

Figure 9.19 Biofilms and biomass in porous media. Schematic shows the types of biomass in saturated groundwater including continuous biofilm (black areas), discontinuous biofilm (dotted areas), aggregate flocs that clog pore throats (dashed areas) but are not attached, and discrete cells or colloids (e.g., virus particles) shown as small black spheres. Mass transfer limitations of oxygen and substrate can limit growth of continuous biofilms and aggregates and cause the interior of the film to be anoxic or even anaerobic. Concentration gradients of substrate(s) and oxygen exist through the biofilm.

For conditions of steady state ($dX/dt = 0$), equation (60) can be rearranged and solved algebraically to give

$$S_{\min} = \frac{K_s b'}{Yq_m - b'} \tag{61}$$

in which $S_{\min}$ is the steady-state substrate concentration below which the biofilm cannot sustain itself. For a steady-state biofilm, the rate at which substrate is removed from the water is[37]

$$r_{\text{bf}} = -(X_f L_f a)\,\eta\,\frac{q_m S_s}{K_s + S_s} \tag{62}$$

where r_{bf} = rate of substrate consumption due to biofilm uptake, $ML^{-3}T^{-1}$
X_f = attached biofilm biomass, ML^{-3}
L_f = biofilm thickness, L
a = specific surface area of the biofilm (the biofilm surface area per unit of reactor volume), L^2L^{-3}
η = fractional effectiveness factor due to mass transport through biofilm, dimensionless
S_s = substrate concentration at the water–biofilm interface after mass transfer through the liquid film resistance, ML^{-3}

$$S_s = S - r_{\text{bf}}\,(L/Da) \tag{63}$$

where L = thickness of biofilm layer that causes mass transfer limitation, L
D = molecular diffusion coefficient of the substrate in the biofilm, L^2T^{-1}

The product $X_f L_f$ equals the accumulation of biomass per unit surface area (ML^{-2}), and $X_f L_f a$ equals the accumulation of attached biomass per unit of reactor volume (ML^{-3}). For substrate concentrations greater than $S_{\min}$, eventually a steady-state biofilm exists with a surface accumulation

$$X_f L_f = YJb' \tag{64}$$

where J is the steady-state substrate flux ($ML^{-2}T^{-1}$), and it is equal to r_{bf}/a.[36]

Placing the biofilm kinetics into a simple 1-D transport model for the saturated zone, we have the coupled set of equations

$$\frac{d(nX_f)}{dt} = +Y(X_f L_f a)\,\eta\,\frac{q_m S_s}{K_s + S_s} - b' X_f \tag{65}$$

$$\frac{\partial(nS)}{\partial t} = \frac{\partial}{\partial x}\left(nD_x\frac{\partial S}{\partial x}\right) - \frac{\partial(nu_x S)}{\partial x} - (X_f L_f a)\,\eta\,\frac{q_m S}{K_s + S} \tag{66}$$

where u_x = actual velocity of the groundwater, LT^{-1}
D_x = hydrodynamic dispersion coefficient, L^2T^{-1}
n = effective porosity, dimensionless

All other parameters have been defined previously by equations (59)–(64).

The equations do not account for retardation due to sorption of the substrate to the biomass or porous media, nor for other degradation reactions other than utilization of the organic chemical as a primary substrate. Note that equation (65) is an ordinary differential equation because the biomass film is not transported (unless we allow for shear and readhesion at a later distance). Equation (66) is a transport equation for dissolved substrate, which is most easily solved by an operator splitting scheme. The transport routine can be solved by a Galerkin finite element scheme for the spatial integration and a central finite difference scheme for the temporal integration. Reaction terms are solved with an iterative predictor corrector numerical method.[38]

9.9.2 Secondary Substrate Utilization

Microbial biodegradation of organic chemicals in groundwater does not always provide energy for biofilm growth. Often organic chemical concentrations are below the S_{min} value and serve as secondary substrates (i.e., the concentration is too low for bacteria to survive but the chemical is nevertheless degraded because the enzyme system is active that can do it). In these cases the relevant microbial kinetics follow from equations (59) and (60), except that the secondary substrate does not provide energy for biomass growth.

$$\frac{dS_1}{dt} = -\frac{q_{m1}XS_1}{K_{s1} + S_1} \tag{67a}$$

$$\frac{dS_2}{dt} = -\frac{q_{m2}XS_2}{K_{s2} + S_2} = -\frac{q_{m2}XS_2}{K_{s2}} \quad \text{because } S_2 < S_{min} \tag{67b}$$

$$\frac{dX}{dt} = Y\frac{q_{m1}XS_1}{K_{s1} + S_1} - b'X \tag{67c}$$

where S_1, K_{s1}, and q_{m1}; and S_2, K_{s2}, and q_{m2} are Monod parameters for the primary and secondary substrate, respectively, as defined for equations (59) and (60). These kinetics [equation (67)] may be used in a transport model similar to equations (65) and (66). Reference Table 9.12 gives some primary and secondary substrates with their second-order utilization rate constants q_m/K_s, which applies when $S \ll K_s$. These rate constants were compiled by Fry and Istok[40] from groundwater field and column studies. Of course, rate constants are empirical and vary with groundwater chemistry and microbiology.

Table 9.12 Second-Order Rate Constants for Biodegradation of Primary and Secondary Substrates in Groundwater

Compound	q_m/K_s, L mg^{-1}d^{-1}	$q_m X/K_s$, day^{-1}, Pseudo-First-Order Rate Constant for $X = 1$ mg L^{-1}
	Bouwer and McCarty[39]: Aerobic Conditions	
Primary substrate		
Acetate	3.8	3.8
Secondary substrate		
Chlorobenzene	2.5	2.5
1,2-Dichlorobenzene	10	10
1,4-Dichlorobenzene	11	11
1,2,4-Trichlorobenzene	5	5
Ethylbenzene	35	35
Styrene	50	50
Naphthalene	40	40
	Bouwer and McCarty[39]: Anaerobic Conditions	
Primary		
Acetate	0.63	0.63
Secondary		
Chloroform	0.85	0.85
Carbon tetrachloride	2.1	2.1
1,1,1-Trichloroethane	0.46	0.46
Tetrachloroethylene	0.08	0.08
	Alvarez et al.[41]	
Benzene	1.33	1.33
Toluene	12	12
Toluene	0.01	0.01
Toluene	25.5	25.5
Toluene	0.38	0.38
Toluene	7.7	7.7
Toluene	0.75	0.75
Toluene	28.8	28.8
	Oldenhuis et al.[42]	
Methane	0.23	0.23
Dichloromethane	0.48	0.48
Chloroform	0.96	0.96
1,2-Dichloroethane	0.06	0.06
1,1,1-Trichloroethane	0.006	0.006
1,1-Dichloroethylene	0.06	0.06
trans-1,2-Dichloroethylene	0.12	0.12

(continued)

Table 9.12 *(continued)*

Compound	q_m/K_s, L mg^{-1}d^{-1}	$q_m X/K_s$, day^{-1}, Pseudo-First Order Rate Constant for $X = 1$ mg L^{-1}
cis-1,2-Dichloroethylene	0.42	0.42
Trichloroethylene	0.12	0.12
	Godsy et al.[42]	
Benzothiophene	0.15	0.15
Oxindole	0.15	0.15
2(1H)-Quinolinine	0.01	0.01
1(2H)-Isoquinolinone	0.02	0.02

Source: Fry and Istok.[40] Excerpted with permission from the American Geophysical Union. Copyright (1994).

Example 9.8 Biofilm Kinetics in Groundwater

a. From the following data for acetate and chlorobenzene in aerobic saturated groundwater, estimate the $S_{\min}$ values and determine if there is a continuous biofilm using acetate as the primary substrate. An acetate disappearance rate of 0.38 mg L^{-1}d^{-1} was obtained from column studies (r_{bf}) and you may assume $b' = 0.01$ day^{-1}.

	q_m/K_s, L mg^{-1} d^{-1}	S, mg L^{-1}	Y, mg X/mg S
Acetate	3.8	0.10	0.35
Chlorobenzene	2.5	0.010	0.15

The BET surface area was 1.0 m^2 g^{-1} and the dry bulk density was 1.7 kg L^{-1}, which allows estimation of the specific surface area (a) for a continuous biofilm.

b. What is the value of S_s? Is the biofilm continuous if it is 60 μm thick?

Solution:

a. Assume the groundwater concentrations are below the K_s values for acetate and chlorobenzene. Therefore

$$S_{\min} = \frac{K_s}{q_m}\frac{b'}{Y}$$

$S_{\min}$ for acetate is 0.0075 mg L^{-1} and for chlorobenzene is 0.027 mg L^{-1}, respectively. Acetate is above the $S_{\min}$ value in the groundwater so it can serve as a primary substrate; chlorobenzene is not.

b. The concentration S_s at the biofilm–water interface after diffusion through the liquid film is given by equation (63). Assume a molecular diffusion coefficient of 1×10^{-6} cm^2 s^{-1}, and a would be approximately 1.7×10^6 m^2 m^{-3}.

$$S_s = S - r_{\text{bf}}\left(\frac{L}{Da}\right) = 0.1 - 0.38\ \text{mg}\ \frac{\text{S}}{\text{L d}} \times$$

$$\times \left(\frac{6 \times 10^{-4}\ \text{cm}}{\frac{1 \times 10^{-6}\ \text{cm}^2}{\text{s}} \cdot \frac{1.7 \times 10^6\ \text{m}^2}{\text{m}^3} \cdot \frac{1\ \text{m}}{100\ \text{cm}} \cdot \frac{86{,}400\ \text{s}}{\text{d}}} \right)$$

$$S_s = 0.1\ \text{mg L}^{-1} - 1.55 \times 10^{-7}\ \text{mg L}^{-1} \simeq 0.1\ \text{mg L}^{-1}$$

At these low concentrations, there is not a mass transfer limitation, and a continuous biofilm likely does not exist.

9.9.3 Colloids

Zysset et al.[38] have extended biofilm kinetics in porous media to include discrete suspended bacteria that are not attached to the porous media but which move with the water. Bacteria tend to adhere to particles in the subsurface but not always. This raises the important issue of colloid transport in groundwater. Colloid transport (< 2-μm particles) can be a critical pathway for movement of chemical contaminants that would otherwise be highly immobile and retarded in groundwater. Example 9.4 showed that PCB is so highly sorbed in groundwater aquifers that it should move scarcely at all. Yet, if biomass or clay colloids sorb PCB (they do) and the particles are transported in the aquifer, our calculations will have been in error.

Colloids can include virus particles, clay fragments, fine precipitates, or bacterial cells. They are generally negatively charged at groundwater near-neutral pH values, and they are transported until they adhere or collide with grains of the porous medium and become entrapped. They may exhibit electrostatic as well as chemical surface coordination effects that influence their collision frequency and collision effectiveness factor. Transient flow conditions, especially near well fields, may produce colloids.

Ryan et al.[44] have shown that a simple consideration of colloid chemical transport can be estimated by modifying the retardation factor defined previously by equation (32). Now, three phases are included: chemical in H_2O, porous media, and colloids.

$$R = 1 + \frac{\rho_b K_d / n}{1 + K_d\,[\text{coll}]} \tag{68}$$

where R = retardation factor for the chemical contaminant
ρ_b = dry bulk density, ML^{-3}

K_d = distribution coefficient, L^3M^{-1}
n = effective porosity
[coll] = colloid concentration suspended in aqueous phase, ML^{-3}

Equation (68) shows that one needs a very high K_d value, on the order of 10^6 L kg^{-1}, before colloid transport affects the retardation factor of a chemical significantly. Colloid concentrations are thought to vary from 0.1 to 10.0 mg L^{-1} in groundwater; however, more research is needed to characterize them in natural groundwater and at landfill sites. Colloids are critical at radioactive waste sites, where the movement of radionuclides such as plutonium must be kept to a minimum. Ionic strength and complexing ligands must be considered, and irreversible deposition should not be assumed.

9.9.4 Bioavailability

Strong sorption of organic substrates to the porous medium can reduce biotransformation rates. Fry and Istok[40] have shown that when the linear desorption rate constant (k_2) is small relative to the pseudo-first-order degradation rate constant, one cannot effectively "bioremediate" a site. "Rebound" is the problem, whereby the aqueous phase is cleaned up only temporarily, and desorption results in recontamination at a later time.

$$C \underset{k_2}{\overset{k_1}{\rightleftharpoons}} q \qquad (69)$$

where C is the dissolved contaminant concentration and q is the mass adsorbed per mass of solids. For these cases, the slow process of desorption controls the remediation, and the organic chemical is not available for biotransformation until after it desorbs. Hydrophobic chemical contaminants exhibit slow desorption kinetics.

One approach is to modify the pseudo-first-order rate constant for biotransformation to include a bioavailability factor considering mass transfer limitations and a spherical particle back-diffusion process.

$$\frac{dq}{dt} = -B_f k_b q \qquad (70)$$

where q = absorbed concentration, MM^{-1}
t = time, T
B_f = bioavailability factor, dimensionless
k_b = biodegradation rate constant, T^{-1}

On the other hand, Bouwer et al. (*personal communication,* 1994) have demonstrated in laboratory columns that bacteria can aid NAPL dissolution via a "bioenhancement effect." In this case, the bacteria overcome mass transfer limitations of

NAPL dissolution and literally "pull components out" of the immobile phase for degradation. The effect can be significant when the Damkohler number becomes greater than 0.1.

$$D_{a2} = \frac{k_b L}{u_x} \tag{71}$$

where D_{a2} = the Damkohler dimensionless number
k_b = first-order biodegradation rate constant
L = length of the column
u_x = water velocity

The bioenhancement effect was evident for the dissolution and degradation of toluene from dodecane, components of jet fuel.

9.10 UNSATURATED ZONE

9.10.1 Measuring Pressure Heads

Water flows from areas of high head (energy per unit volume) to areas of low head in the unsaturated zone, just as in the saturated zonc. Usually it means that water percolates down through soil due to gravity, but this is not always the case. If the surface of the soil is fine textured, and if it becomes very dry, water will move vertically upward due to capillary forces, the "wick effect." Such vertical upward movement is typical in the arid southwestern United States, where evaporite or calichy deposits of salts may form over time, due to solute transport upward with the water, and ensuing evaporation leaves behind only the salts.

The equation for head in the unsaturated zone is identical to equation (11). Kinetic energy terms can be neglected, but the pressure head (ψ) is negative in the unsaturated zone.

$$h = z + \frac{p}{\rho g} = z + \psi$$

where h = hydraulic head, L
z = elevation head above datum, L
p = gage pressure (above atmospheric), $ML^{-1}T^{-2}$
ρ = water density, ML^{-3}
g = acceleration due to gravity, LT^{-2}
ψ = pressure head, L

Hydrogeologists term the negative pressure head in the unsaturated zone as the "suction head." It is measured by use of a tensiometer probe, a device akin to sticking a straw filled with water into the subsurface with your finger over the top, where the capillary forces at the bottom of the straw will draw out some water. The vacuum pressure created at the top of the straw, between your finger and the water, is a measure of the negative pressure head. Figure 9.20, from Freeze and Cherry,[1] is a schematic of pressure head and total hydraulic head with depth in an unsaturated and saturated zone.

The unsaturated zone becomes important to model accurately because it determines the boundary condition for development of a plume in the groundwater, which defines contaminant remediation strategies (Figure 9.21). The unsaturated

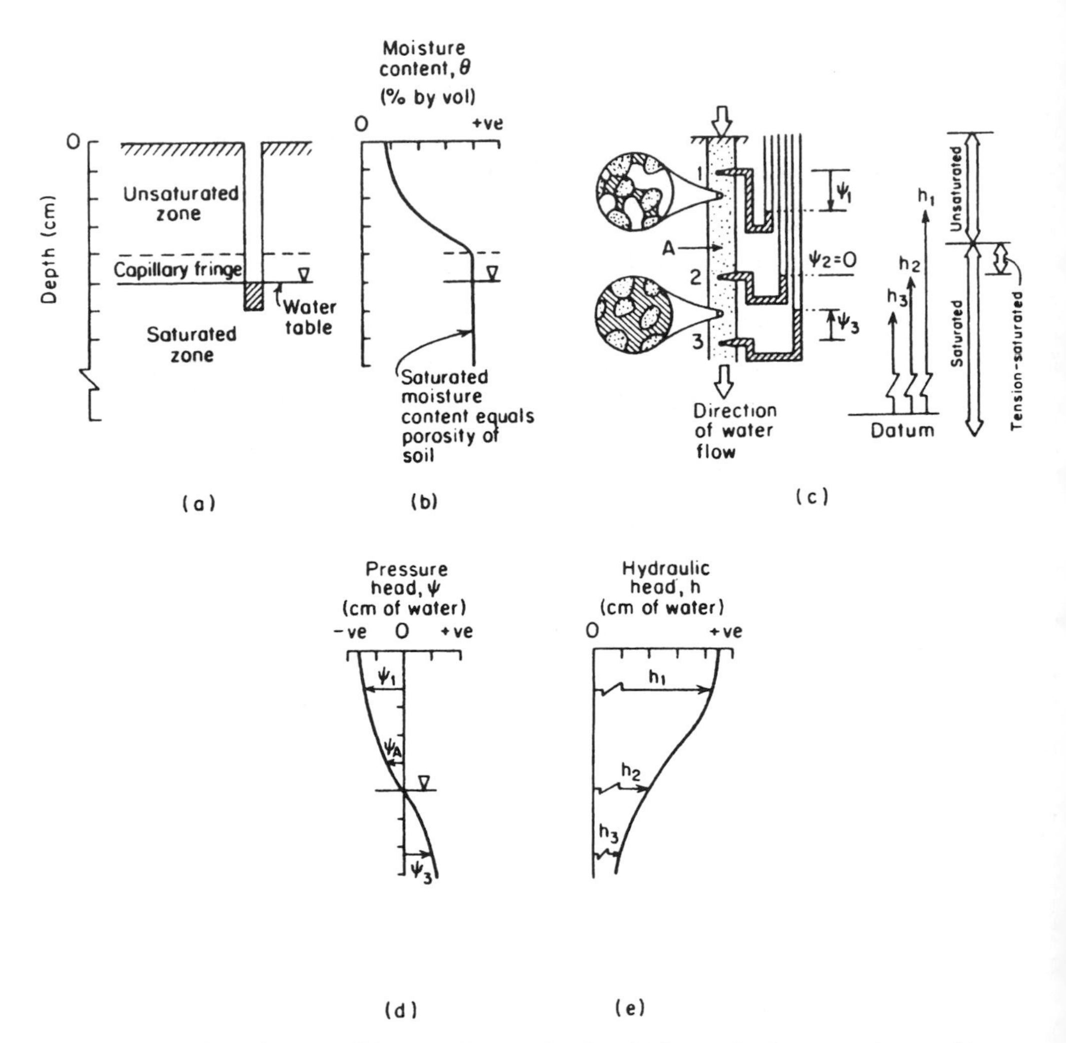

Figure 9.20 Groundwater conditions near the ground surface. (a) Saturated and unsaturated zones; (b) profile of moisture content versus depth; (c) pressure-head and hydraulic-head relationships; *insets:* water retention under pressure heads less than (*top*) and greater than (*bottom*) atmospheric; (d) profile of pressure head versus depth; (e) profile of hydraulic head versus depth. (From Freeze and Cherry.[1])

zone is the source of most saturated zone problems. Precipitation percolates through the unsaturated zone and carries the contaminant with it. Complication arises because soil gas can volatilize organic contaminants with a high Henry's constant as shown in Figure 9.21. So we must be concerned with four phases in the unsaturated zone:

- NAPL (if present).
- Sorbed contaminant on the soil.
- Aqueous phase contaminant in soil moisture.
- Gas phase contaminant (in soil gas).

Further complicating the picture is flow hysteresis that occurs with sequential wetting and drying cycles of the soil. The moisture content of the soil (θ = fractional volume) is greater during a drying cycle for a given suction head than during the subsequent wetting cycle, and the unsaturated hydraulic conductivity is a strong function of moisture content. As the soil becomes drier, the unsaturated hydraulic conductivity decreases substantially (Figure 9.22).

9.10.2 Flow and Contaminant Transport Equations

Shutter et al.[45] have provided a practical methodology for modeling contaminants in the 1-D vertical unsaturated zone. They use the van Genuchten[46] equation (73) that relates soil moisture to the measured pressure heads (ignoring hysteresis):

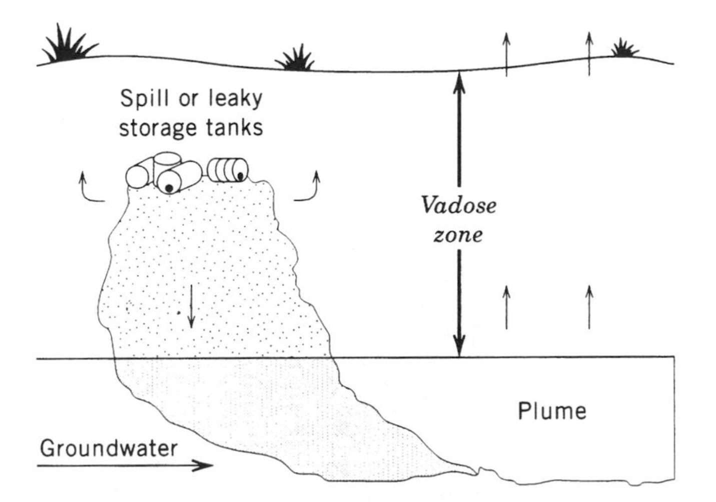

Figure 9.21 Contaminant movement from the unsaturated zone (vadose) to the groundwater. Percolation carries contaminants downward and volatilization may move them upward.

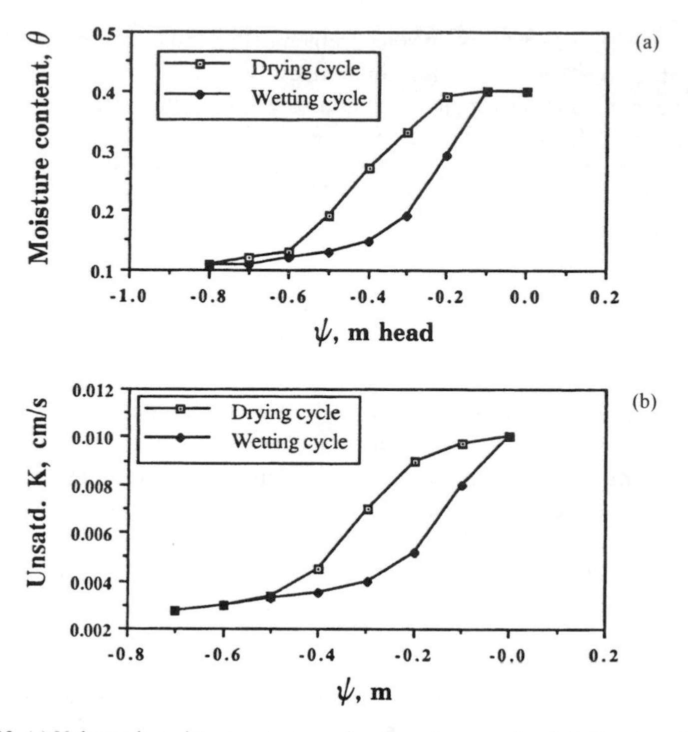

Figure 9.22 (a) Volumetric moisture content as a function of pressure head, ψ, for a hypothetical unsaturated zone soil. (b) Unsaturated hydraulic conductivity as a function of pressure head (and moisture content). Note the hysteresis between the drying and wetting cycles.

$$\theta = \theta_{res} + \frac{\theta_{sat} - \theta_{res}}{[1 + (\alpha |\psi - \psi_{air}|^{\beta}]^{\gamma}} \quad \text{for } \psi < \psi_{air} \tag{73}$$

where θ_{sat} = saturated moisture content = porosity

θ_{res} = residual moisture content under dry conditions (volume basis)

ψ_{air} = air-entry pressure head, L

ψ = pressure head in unsaturated zone, L

α, β, and γ are curve-fitting parameters that are obtained from nonlinear least-squares regression of data such as in Figure 9.22a. Then $K(\theta)$ can be determined from the following empirical equation.[46]:

$$K(\theta) = K_H \, k_{rw} = K_H \, S_e^{1/2} [1 - (1 - S_e^{1/\gamma})^{\gamma}]^2 \tag{74}$$

where S_e = effective saturation = $\dfrac{\theta - \theta_{res}}{\theta_{sat} - \theta_{res}}$

K_H = saturated hydraulic conductivity, LT^{-1}

k_{rw} = relative permeability with respect to saturated conditions, dimensionless

Moisture content and suction heads (ψ) must be measured in the field in order to fit the model parameters for a given location.

The equations for variably saturated flow through porous media and contaminant transport can solved sequentially.[45]

$$\frac{\partial}{\partial x_i}\left(K_{S_{ij}}\,k_{rw}\left[\frac{\partial\psi}{\partial x_j}+\frac{\partial z}{\partial x_j}\right]\right)=\left(S_w S_s + n\frac{\partial S_w}{\partial \psi}\right)\frac{\partial\psi}{dt} \tag{75a}$$

where ψ is the pressure head, $K_{S_{ij}}$ is the saturated hydraulic conductivity tensor, and $\partial z/\partial x_j$ represents the unit vector in the z direction. On the right side of the equation, S_w, equal to θ/θ_{sat}, is the relative saturation of the water phase; S_s is the specific storage coefficient; n is the effective porosity, defined as the fraction of the porous medium at which the groundwater is mobile; and t is time. The contaminant transport equation is

$$\frac{\partial}{\partial x_i}\left(nS_w\,D_{ij}\frac{\partial C}{\partial x_j}\right)-nS_w v_i\frac{\partial C}{\partial x_i}=nS_w R\left(\frac{\partial C}{\partial t}+\lambda C\right) \tag{75b}$$

where C is the solute concentration in the water, v_i is the fluid velocity, D_{ij} is the hydrodynamic dispersion tensor, R is the retardation factor, and λ is the first-order decay constant.

The retardation factor, R, in the unsaturated zone must be expressed in terms of the relative degree of saturation of the effective porosity.

$$R = 1 + \frac{\rho_b K_d}{nS_w} \tag{76}$$

If the relative saturation is only 5%, then the retardation factor of a hydrophobic chemical increases ~20 times for the unsaturated zone, although, macropore flow through old root channels or colloid transport can speed the vertical movement of contaminant. Caution in the use of retardation factors is warranted.

9.11 REMEDIATION

Pump-and-treat remediation of hazardous waste sites is still the preferred method of treatment in 70% of all cases.[47] The "rebound" effect remains a problem whereby pumping stops when regulatory standards have been achieved, but subsequent desorption and/or dissolution of contaminant causes groundwater concentrations to increase once again. Slow desorption and mass transfer limitations are a constant problem in the subsurface. Low permeability zones and subsurface heterogeneities "hide" contaminants, which are only slowly released. In essence, it is difficult to make water and nutrients flow to where the contaminants are located in the subsurface environment.

Figure 9.23 shows common methods for remediation of the unsaturated zone: soil vapor extraction, steam stripping, and bioventing. In soil vapor extraction, air is injected just above the groundwater table and it is collected with volatilized organic contaminants in the extraction wells. The only problem is that the extraction wells typically have a radius of influence of only 5 ft, and many wells are required.[47] Still, it is a proven and a popular technology. Bioventing is a similar technique, but the mode of treatment is to get oxygen to aerobic bacteria for in situ treatment. Steam injection aids in stripping contaminants from the unsaturated zone, but it is relatively expensive.

For the saturated groundwater, sites are usually isolated hydraulically using slurry walls or interceptor wells, as shown in Figure 9.24. In recirculating water, aerobic bacteria can increase their population and enhance bioremediation. To model the treatment scheme, a flow balance is needed based on the pumping rate of the interceptor well [see equation (18)]. Then a biofilm model [such as equations (65) and (66)] or a mass transfer limited NAPL dissolution model [such as equation (48)] may be used. If mass transfer limits the remediation, recovery times can be long. They are on the order of the intraparticle diffusion coefficient divided by the median particle radius squared, as in equation (58). Just flushing the system with water is not practical. Because of the slow velocity of groundwater, any more than two detention times is not feasible. Surfactant addition shows some promise in remediation of organic contaminants in the subsurface theoretically, but it has not been proved in the field yet.

Bioremediation is an attractive alternative for cleaning hazardous waste sites because of its low cost compared to excavation and disposal in drums, incineration of soil, or ex situ soil washing. *Intrinsic bioremediation* refers to leaving pollutants in place but estimating the time required for biological transformation of organics to innocuous end-products. Mathematical modeling is required for assessment of in-

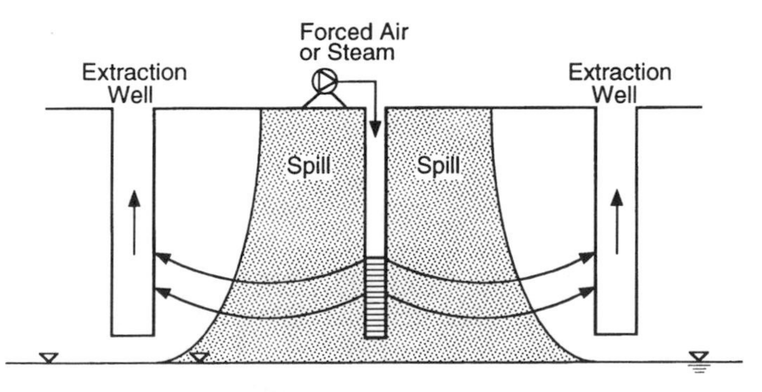

Figure 9.23 Soil vapor extraction, steam stripping, and bioventing all make use of injection of air or steam into the unsaturated zone to volatilize contaminants or enhance aerobic biological degradation reactions. Off-gases may require treatment. Nutrients can be added (N and P) by surface irrigation. Contaminated zone is between the injection well and the extraction well.

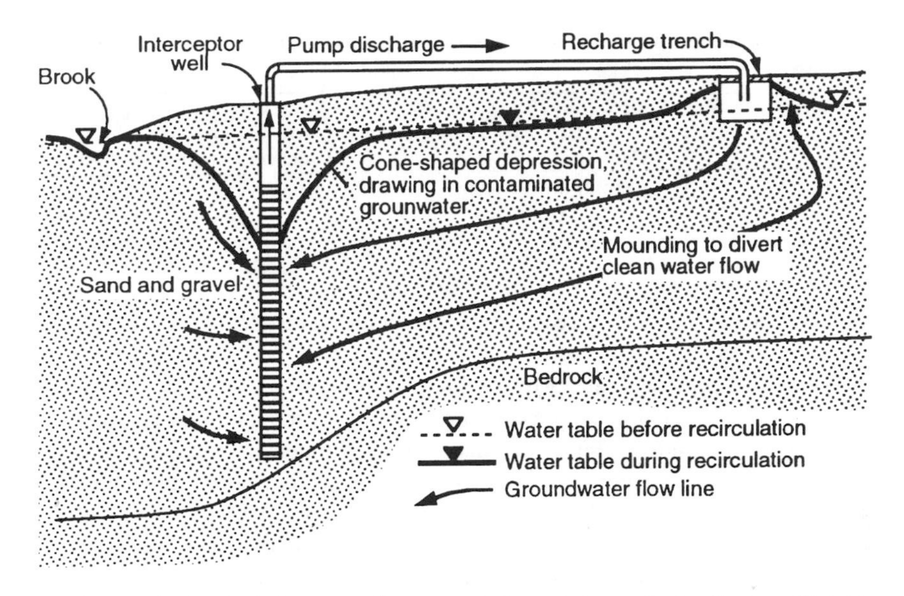

Figure 9.24 Typical recirculation system for treatment of saturated groundwater. Nutrients (N & P) can be added to recharge trench or oxidants (e.g., H_2O_2) can be injected. Often the intercepted water must be treated above ground before recharge.

trinsic bioremediation (the natural restoration alternative), and bioavailability and NAPL dissolution are important considerations in such modeling efforts.

9.12 NUMERICAL METHODS

Solving the partial differential flow equation (15) and mass balance equations for the contaminant [such as equation (48)] requires numerical methods and a computer. Finite element methods are the most popular for one-dimensional and two-dimensional problems. They often make use of Galerkin's method of weighted residuals, and complex geometries are easily handled by polygons of node points. (See Table 9.13.)

Finite difference techniques are also useful in solving the equations directly and in solving the reaction root-mean-squares of split operator methods. Careful attention to the boundary conditions and initial condition is necessary to be certain that an accurate solution is obtained. Figure 9.25 shows how a central differencing method is set up for a two-dimensional groundwater problem. As in all numerical methods, there is an error term (residual) due to approximating derivatives at a point by a finite difference scheme. The trick is to ensure that the errors do not grow.

Finite element techniques are useful in keeping numerical dispersion at a minimum, which is important because the reaction terms are concentration dependent. Large concentration gradients arise in subsurface remediation problems due to sharp boundaries of contamination. Also, the equations are complicated by non-

Table 9.13 Numerical Methods for Groundwater Contaminant Transport and Reaction Equations

Dimension	Finite Difference	Finite Element
1-D	Central or backward	Galerkin's method/split operator
2-D	Central differencing	Galerkin's method/split operator
3-D	ADI	Split operator

linearities and stiffness (time scales for reaction and transport may be quite different).

Wood et al.[48] provide a review of solving two-dimensional contaminant transport and biodegradation equations in a multilayered porous medium. An operator-splitting technique is used. The approach decouples the transport portion of the equations from the reaction portions, by first solving the transport problem with the source/sink term set to zero. The transport equations are solved by a finite-element modified method of characteristics (MMOC), which has the desirable attribute to handle large concentration gradients.[49] The concentrations obtained from this step are then used as the initial concentrations to solve the reaction equations, which are treated as ordinary differential equations and are solved with a second-order, explicit Runge–Kutta method with time steps that are generally much smaller than those used for transport.

Once the numerical technique has been developed and coded, it is important to test the numerical solution. Numerical solutions are exact solutions to an approximate equation. They contain round-off errors and potential propagation errors that must be tested.

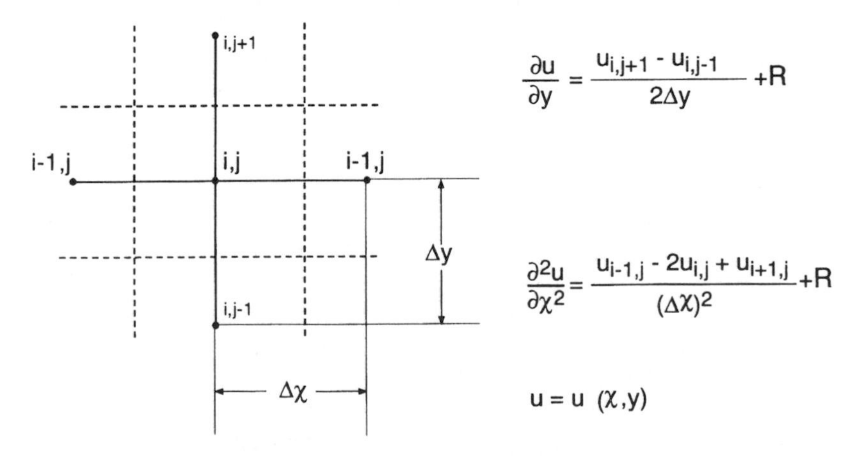

Figure 9.25 Numerical grid schemes.

The best way to test your numerical method is to run a simulation of a simple problem that has an analytical solution (exact solution). If this is not possible, you should devise a series of tests to gain confidence in the numerical accuracy.

- Change step sizes (Δx, Δy, Δt). (Does your result change?)
- Simulate steep concentration fronts.[49] (Look for numerical dispersion.)
- Check all mass balances. (Make sure mass is conserved.)
- Sensitivity analysis. (Does a small change in the parameter cause a large change in the results?)
- Test against another algorithm and code by running the identical problem.

9.13 REFERENCES

1. Freeze, R.A., and Cherry, J.A., *Groundwater,* Prentice-Hall, Englewood Cliffs, NJ (1979).
2. Bear, J., *Dynamics of Fluids in Porous Media,* American Elsevier, New York (1972).
3. Jacob, C.E., in *Engineering Hydraulics,* H. Rouse, Ed., Wiley, New York (1950).
4. Strack, O.D.L., *Groundwater Mechanics,* Prentice-Hall, Englewood Cliffs, NJ (1989).
5. Domenico, P.A., and Schwartz, F.W., *Physical and Chemical Hydrogeology,* Wiley, New York (1990).
6. Hemond, H.F., and Fechner, E.J., *Chemical Fate and Transport in the Environment,* Academic Press, San Diego, CA (1994).
7. Scheidegger, A., *Physics of Flow Through Porous Media,* University of Toronto Press, Toronto, Canada (1960).
8. McCarty, P.L., Reinhard, M., and Rittmann, B.E., Environ. Sci. Technol., 15, 40 (1981).
9. Javandel, I., Doughty, C., and Tsang, C.F., *Groundwater Transport: Handbook of Mathematical Models,* American Geophys. Union, Water Res. Monograph 10, Washington, DC (1994).
10. Lyman, W.J., Reehl, W.F., and Rosenbaltt, D.H., *Handbook of Chemical Property Estimation Methods,* Americal Chemical Society, Washington, DC (1990).
11. Schwarzenbach, R.P., and Westall, J., *Environ. Sci. Technol.,* 15, 1360 (1981).
12. Schnoor J.L., et al., *Processes, Coefficients, and Models for Simulating Toxic Organics and Heavy Metals in Surface Waters,* EPA/600/3-87-015, U.S. Environmental Protection Agency, Athens Environmental Research Laboratory, Athens, GA (1987).
13. Schwarzenbach, R.P., and Giger, W., in *Ground Water Quality,* C.H. Ward, W. Giger, and P.L. McCarty, Eds., Wiley, New York (1985).
14. Ghiorse, W.C., and Balkwill, D.L., in *Ground Water Quality,* C.H. Ward, W. Giger, and P.L. McCarty, Eds., Wiley, New York (1985).
15. Kuhn, E.P., et al., *Environ. Sci. Technol.,* 19, 961 (1985).
16. Barry, R.C., Ph.D. Dissertation, University of Iowa, Iowa City (1993).
17. Borden R.C., and Bedient, P.B., *Water Resour. Res.,* 22, 1973 (1982).
18. Bouwer, E.J., McCarty, P.L., and Lance, J.C., *Water Res.,* 15, 151 (1981).

19. Bouwer, E.J., Rittman, B.E., and McCarty, P.L., *Environ. Sci. Technol.,* 15, 596 (1981).
20. Nair, D.R., and Schnoor, J.L., *Environ. Sci. Technol.,* 26, 2298 (1992).
21. Nair, D.R., and Schnoor, J.L., *Water Res.,* 28, 1199 (1994).
22. Zylstra, G.J., Wackett, L.P., and Gibson, D.T., *Appl. Environ. Micro.,* 55, 3162 (1989).
23. Vogel, T.M., Criddle, C.S., and McCarty, P.L., *Environ. Sci. Technol.,* 21, 722 (1987).
24. Sawyer, C.N., McCarty, P.L., and Parkin, G.F., *Chemistry for Environmental Engineering,* 4th ed., McGraw-Hill, New York (1994).
25. Stumm, W., and Morgan, J.J., Aquatic Chemistry, 2nd ed., WIley, New York (1981).
26. Jackson, R.E., and Patterson, J.R., *Water Resour. Res.,* 18, 1255 (1982).
27. Schwarzenbach, R.P., Gschwend, P.M., and Imboden, D.M., *Environmental Organic Chemistry,* Wiley, New York (1993).
28. Powers, S.E., Abriola, L.M., Dunkin, J.S., and Weber, W.J., Jr., *J. Contaminant Hydrol.,* 16, 1 (1994).
29. Wilson, J.L., and Conrad, S.H., in *Proceedings Petroleum Hydrocarbons and Organic Chemicals in the Subsurface: Prevention, Detection, Restoration,* National Water Well Association, Worthington, OH (1988).
30. Powers, S.E., Loueiro, C.O., Abriola, L.M., and Weber, W.J., Jr., *Water Resour. Res.,* 27, 463 (1991).
31. Young, D.F., and Ball, W.P., *Environ. Prog.,* 13:1, 9 (1994).
32. Ball, W.P., and Roberts, P.V., *Environ. Sci. Technol.,* 25, 1223 (1991).
33. Ball, W.P., and Roberts, P.V., *Environ. Sci. Technol.,* 25, 1237 (1991).
34. Rittmann, B.E., and McCarty, P.L., *Biotechnol. Bioeng.,* 22, 2343 (1980).
35. Rittmann, B.E., and McCarty, P.L., *Biotechnol. Bioeng.,* 22, 2359 (1980).
36. Rittmann, B.E., *Water Resour. Res.,* 29, 2195 (1993).
37. Rittmann, B.E., *J. Environ. Eng.,* 107, 831 (1981).
38. Zysset, A., Stauffer, F., and Dracos, T., *Water Resour. Res.,* 30 2423 (1994).
39. Bouwer, E.J., and McCarty, P.L., *Biotechnol. Bioeng.,* 27, 1564 (1985).
40. Fry, V.A., and Istok, J.D., *Water Resour. Res.,* 30, 2413 (1994).
41. Alvarez, P.J.J., Anid, P.J., and Vogel, T.M., *Biodegradation,* 2, 43 (1991).
42. Oldenhuis, R., Oedzes, J.Y., vanderWaarde, J.J., and Janssen, D.B., *Appl. Environ. Microbiol.,* 57, 7 (1991).
43. Godsy, E.M., Goerlitz, D.F., and Grbic-Galic, D., Symposium on Bioremediation of Hazardous Wastes, EPA/600/R-93/054, U.S. Environmental Progection Agency, Washington, DC (1993).
44. Ryan, J.N., *Environ. Sci. Technol.,* 28, 1717 (1994).
45. Shutter, S.B., Sudicky, E.A., and Robertson, W.D., *Environ. Toxicol. Chem.,* 13, 223 (1994).
46. Nielsen, D.R., van Genuchten, M.Th., and Biggar, J.W., *Water Resour. Res.,* 22, 895 (1986).
47. MacDonald, J.A., and Kavanaugh, M.C., *Environ. Sci. Technol.,* 28, 362A (1994).
48. Wood, B.D., Dawson, C.N., Szecsody, J.E., and Streile, G.P., *Water Resour. Res.,* 30, 1833 (1994).
49. Huyakorn, P.S., Wu, Y.S., and Park, N.S., *J. Contaminant Hydrol.,* 16, 203 (1994).

9.14 PROBLEMS

1. Following is a sieve analysis for an unconsolidated sand aquifer material.

a. Using Hazen's empirical formula, estimate the saturated hydraulic conductivity.

b. The porosity of the sand is measured as 0.32. Estimate the saturated hydraulic conductivity using the Kozeny–Carmen equation.

U.S. Standard Sieve Size	Percentage of Mass Finer than
10	99.5
18	96.0
35	65.3
60	5.7
125	0.15
230	0.02

2. A transect of piezometers have been drilled into a deep, confined aquifer composed of limestone. For a porosity of 0.25 and an estimated hydraulic conductivity of 10^{-4} cm s^{-1}, estimate the specific discharge, actual velocity, and time of travel for the water to move 10 km from its recharge zone.

Piezometer	Head, m	Distance, km
1	50	0
2	45	5
3	40	10

3. A permeameter study has been completed in the laboratory on a coarse sand aquifer material. The sample was tamped-down into the permeameter tube to simulate field conditions. The cross-sectional area of the tube was 32.15 cm^2. For the data given below, estimate the saturated hydraulic conductivity from the volumetric flowrate and hydraulic gradient measurements. The length between manometer tubes was 100 mm.

Δh, mm	Q, cm^3 s^{-1}
2	0.032
10	0.16
20	0.33
50	0.80
100	1.59
200	3.22

4. Referring to the nested piezometers in Figure 9.4 (not to scale), suppose the new head measurements are given below.
   a. Interpret the hydrogeologic setting. In which direction(s) is water moving and why? The nests are each 100 m from each other along a transect.
   b. Estimate the specific discharge and the actual velocity between nested piezometers A1 and A2. K_z sand = 0.0001 cm s^{-1}, K_z clay = 10^{-7} cm s^{-1}.

Piezometer	Piezometer Elevation, m	Measured Head, m
A1	220	455
A2	245	471
A3	255	471
B1	220	460
B2	245	470
B3	255	470
C1	220	465
C2	245	469
C3	255	469
D1	160	474

5. In the development of the retardation factor, suppose the sorption of contaminant follows a Freundlich rather than a linear isotherm with a slope (1/m) less than 1.0.

$$S = K_F C^{1/m}$$

where K_F = Freundlich constant
$1/m$ = slope of the log–log isotherm plot

Can you define a new retardation factor? If not, why not, and how would you get around this problem?

6. A landfill is leaking CCl_4 and hexachlorobenzene into the surficial aquifer. The landfill sits on a flood plain above a nearby river 100 meters away. The hydraulic gradient of the water table is 0.005 m m^{-1}, and the saturated hydraulic conductivity is 0.001 cm s^{-1} with a porosity of 0.3. Considering retardation factors only, how long should it take for H_2O, CCl_4, and hexachlorobenzene to reach the river? Assume the bulk density of aquifer material is 1.5 g cm^{-3} and the f_{oc} is 0.001.

7. For the conditions given in Example 9.5, solve for the concentration (C versus x) for an impulse release of 100 kg of toluene into the 1-D aquifer. Assume that the aquifer is 10 m deep and the contamination is 10 m wide. Solve for the concentration after 1, 10, and 100 years assuming toluene does not biodegrade ($k = 0$, perhaps it doesn't under anaerobic conditions) and that it does biodegrade

($k = 0.03$ day^{-1}) aerobically. Make six plots total. $\rho_b = 1.5$ g cm^{-3}, $n = 0.4$, $f_{oc} = 0.001$.

8. Repeat problem 7 assuming a horizontal 1-D aquifer and a continuous input of 100 mg L^{-1}. Use equation (35) with $k = 0.001$ day^{-1}. Plot your results for toluene fate and transport (concentration versus distance at three different times.

9. a. Formulate a couple of simultaneous partial differential equations for an organic chemical that sorbs, and biodegrades in a shallow 1-D groundwater, similar to equation (33) but including substrate and microorganism growth as in equations (45a) and (45b). How might you solve the coupled set of equations?

b. Going one step farther with your equations from above, suppose S_b is the primary substrate, and a second chemical S_c is being co-metabolized. Postulate a third partial differential mass balance equation for the co-metabolite S_c. Do not solve.

10. Derive equation (57) for batch experiments with aquifer media that have mass transfer limitations between sorption to the outer surface of a particle and intraparticle diffusion.

11. Derive the $S_{\min}$ minimum substrate concentrations that will sustain growth for benzene ($q_m/K_s = 1.33$ L mg^{-1} d^{-1}) and toluene ($q_m/K_s = 0.38$ L mg^{-1} d^{-1}) from the Alvarez et al.[41] aerobic groundwater study. Assume that the loss rate b' is 0.01 day^{-1}, and the yield is 0.15 mg Y/mg S.

12. There is a continuous input of 800 gal d^{-1} from a septic tank to groundwater. The household chemicals, tartrate monosuccinate (TMS), and tartrate disuccinate (TDS) are the main chemicals in the effluent. The effluent concentration of TMS and TDS is approximately 6.35 mg L^{-1} and 1.22 mg L^{-1}, respectively. Sorption and biotransformation are two major decomposition processes for the chemicals in the subsurface region. The physical and hydrogeological characteristics are given below. The distribution coefficients (K_d) for TMS and TDS are 1.4 L kg^{-1} and 1.0 L kg^{-1}, respectively. The ranges of the biological reaction rate for TMS and TDS are 0.06–0.0015 day^{-1} and 0.1–0.0005 day^{-1}, respectively. The estimated longitudinal hydrodynamic dispersion coefficient (D) is 0.01 m^2 d^{-1}.

a. Develop a mathematical model to describe and predict the behavior and fate of TMS and TDS in the unsaturated zone (1-D model, vertical) and saturated zone (1-D model, longitudinal, eq. 35). Equation (75b) may be used for the unsaturated zone with simplifying assumptions of steady state and negligible dispersion.

b. Simulate the model with the information below and compare the results with the field data given. (No data are presented for the saturated zone.)

c. What is the estimated value of the biotransformation rate for TMS and TDS in the subsurface region?

Physical and Hydrogeological Characteristics of the Site

Septic tank volume	500 gal
Tile bed area	100 m^2
Bulk density	1.8 kg L^{-1}
Effective porosity	0.3
Relative saturation in unsatd. zone	0.2
Depth of unsaturated zone	2.0 m
Hydraulic conductivity, saturated	20.0 m d^{-1}
Hydraulic gradient	0.001
Depth of saturated zone	3.0 m
Flow velocity in unsaturated zone	15 cm d^{-1}

Lysimeter Monitoring Results in the Vadose Zone at the Site

Depth Beneath Tile, cm	TMS Average Concentration mg L^{-1}	TMS Standard Deviation of Concentration, mg L^{-1}	TDS Average Concentration, mg L^{-1}	TDS Standard Deviation of Concentration mg L^{-1}
0	4.94	3.14	0.89	0.65
10	4.53	0.95	0.87	0.26
20	2.20	1.77	0.66	0.36
25	1.84	1.12	0.87	0.35
30	2.89	1.97	0.87	0.50
38	2.54	1.49	0.64	0.27
40	2.11	0.00	0.34	0.00
51	0.69	0.52	0.91	0.61
91	0.19	0.00	0.42	0.00

ANSWERS TO SELECTED PROBLEMS—CHAPTER 9

2. 0.00864 cm d^{-1}, 0.0346 cm d^{-1}, 79,200 years

4b. -6.4×10^{-8} cm s^{-1} (flow is up)

6. 19 yr, 43 yr, 150,000 yr

7. for $k = 0$: C_{max} ($x = 4$ m, $t = 1$ yr) = 147 mg L^{-1}
C_{max} ($x = 37$ m, $t = 10$ yr) = 46 mg L^{-1}
C_{max} ($x = 365$ m, $t = 100$ yr) = 15 mg L^{-1}

10

ATMOSPHERIC DEPOSITION AND BIOGEOCHEMISTRY

For in the end we will conserve only what we love.
We will love only what we understand.
And we will understand only what we are taught.
—Baba Dioum,
African Conservationist

Atmosphere, land, and water are interconnected compartments. It makes no sense to speak solely of water pollution because of intermedia transfers to air or land. If one removes pollutants from a wastewater discharge in order to improve water quality, residuals are deposited onto land or, if incinerated, into the air. The atmosphere transfers pollutants to land and water via atmospheric deposition, that is, the transport of pollutants, both gaseous and particulate, from the air to land and water. Acid precipitation is the most common form of atmospheric deposition, and it affects elemental cycling in the environment (biogeochemistry) and heavy metals transport.

10.1 GENESIS OF ACID DEPOSITION

The oxidation of carbon, sulfur, and nitrogen, resulting from fossil fuel burning, disturbs redox conditions in the atmosphere. The atmosphere is more susceptible to anthropogenic emissions than are the terrestrial or aqueous environments because, from a quantitative point of view, the atmosphere is much smaller than the other reservoirs. Furthermore, the time constants concerning atmospheric alterations are small in comparison with those of the seas and the lithosphere.[1,2]

In oxidation–reduction reactions, electron transfers (e^-) are coupled with the transfer of protons (H^+) to maintain a charge balance. A modification of the redox balance corresponds to a modification of the acid–base balance.[3] The net reactions of the oxidation of C, S, and N exceed reduction reactions in these elemental cycles. A net production of H^+ ions in atmospheric precipitation is a necessary consequence. The disturbance is transferred to the terrestrial and aquatic environments, and it can impair terrestrial and aquatic ecosystems.

Figure 10.1 shows the various reactions that involve atmospheric pollutants and natural components in the atmosphere. The following reactions are of particular importance in the formation of acid precipitation: oxidative reactions, either in the

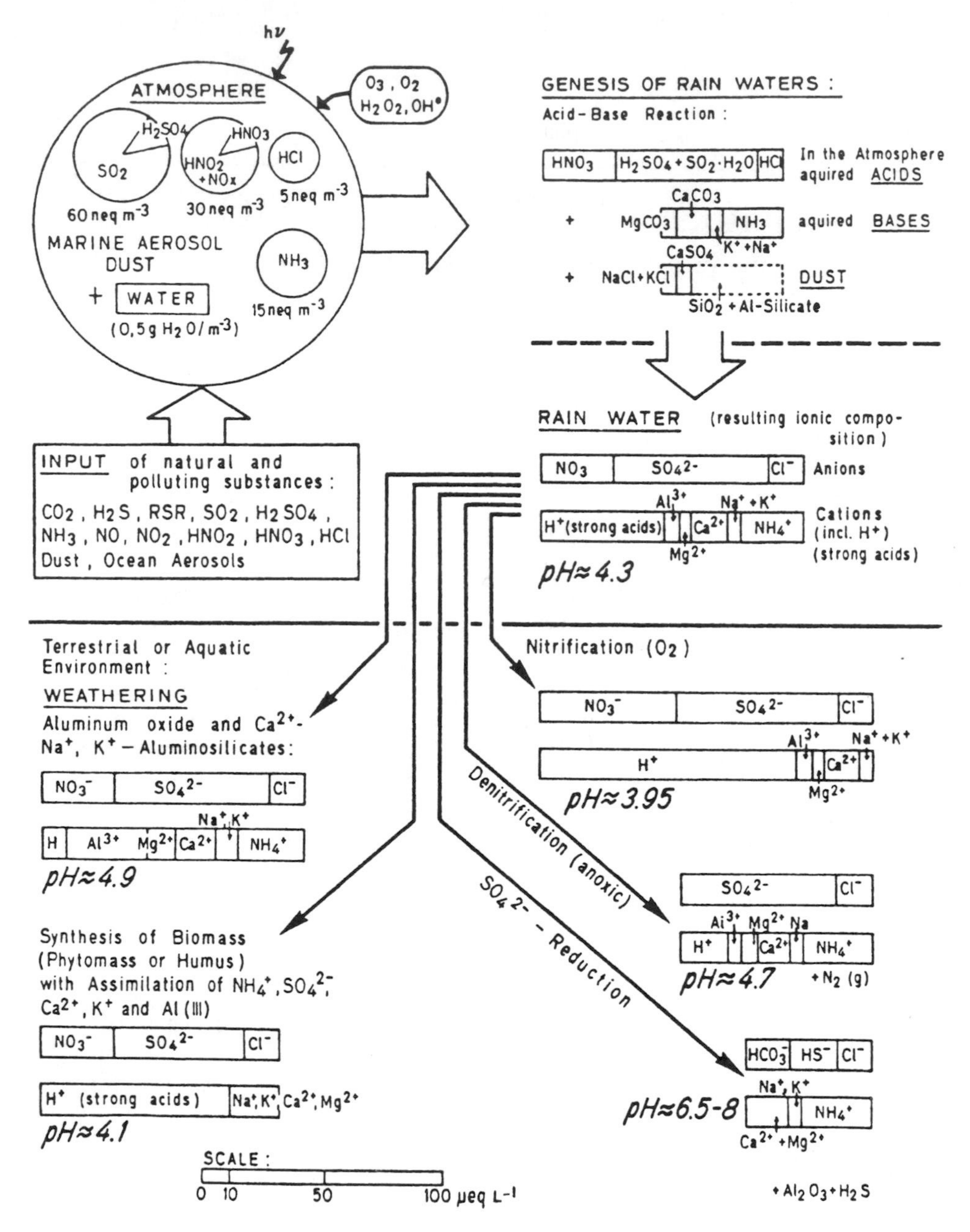

Figure 10.1 Depiction of the genesis of acid rain. From the oxidation of S and N during the combustion of fossil fuels, there is a buildup in the atmosphere (in the gas phase, aerosol particles, raindrops, snowflakes, and fog) of CO_2 and the oxides of S and N, which leads to acid–base interaction. The importance of absorption of gases into the various phases of gas, aerosol, and atmospheric water depends on a number of factors. (In this figure, we assume a yield of 50% for SO_4^{2-} and NO_3^-, of 80% for NH_3, and of 100% for HCl.) The genesis of acid rain is shown on the upper right as an acid–base titration. The data given are representative of the situation encountered in Zurich, Switzerland. Various interactions with the terrestrial and aquatic environment are shown in the lower part of the figure. (From Schnoor and Stumm.[27])

gaseous phase or in the aqueous phase, leading to the formation of oxides of C, S, and N (CO_2; SO_2, SO_3, H_2SO_4; NO, NO_2, HNO_2, HNO_3); absorption of gases into water (cloud droplets, falling raindrops, or fog) and interaction of the resulting acids ($SO_2 \cdot H_2O$, H_2SO_4, HNO_3) with ammonia (NH_3) and the carbonates of airborne dust; and the scavenging and partial dissolution of aerosols into water. In this case, aerosols are produced from the interaction of vapors and airborne particles (maritime and dust); they often contain $(NH_4)_2SO_4$ and NH_4NO_3.

The products of the various chemical and physical reactions are eventually returned to the earth's surface. Usually, one distinguishes between wet and dry deposition. Wet deposition (rainout and washout) includes the flux of all those components that are carried to the earth's surface by rain or snow, that is, those dissolved and particulate substances contained in rain or snow. Dry deposition is the flux of particles and gases (especially SO_2, HNO_3, and NH_3) to the receptor surface during the absence of rain or snow. Deposition also can occur through fog aerosols and droplets, which can be deposited on trees, plants, or the ground. With forests, approximately half of the deposition of SO_4^{2-}, NO_3^-, and H^+ occurs as dry deposition.

Three elementary chemical concepts are prerequisites to understanding the genesis and modeling of acid deposition. First of the three concepts is a simple stoichiometric model, which explains on a mass balance basis that the composition of the rain results primarily from a titration of the acids formed from atmospheric pollutants with the bases (NH_3- and CO_3^{2-}-bearing dust particles) introduced into the atmosphere. Next is an illustration of the absorption equilibria of such gases as SO_2 and NH_3 into water, which represents their interaction with cloud water, raindrops, fog droplets, or surface waters. Finally, a quantitative interpretation of the acid–base balance is presented. This requires a rigorous definition of acid-neutralizing capacity (alkalinity) and base-neutralizing capacity (acidity) to measure the residual acidity of acid precipitation and to interpret the interaction of acid deposition with the terrestrial and aquatic environment.

10.1.1 Stoichiometric Model

The rainwater shown in Figure 10.1 contains an excess of strong acids, most of which originate from the oxidation of sulfur during fossil fuel combustion and from the fixation of atmospheric nitrogen to NO and NO_2 (e.g., during combustion of gasoline by motor vehicles). In addition, there are natural sources of acidity, resulting from volcanic activity, from H_2S from anaerobic sediments, and from dimethyl sulfide and carbonyl sulfide that originate in the ocean.[4] HCl results from the combustion and decomposition of organochlorine compounds such as polyvinyl chloride. Bases originate in the atmosphere as the carbonate of wind-blown dust and from NH_3, generally of natural origin. The NH_3 comes from NH_4^+ and from the decomposition of urea in soil and agricultural environments.

Reaction rates for the oxidation of atmospheric SO_2 (0.05–0.5 day^{-1}) yield a sulfur residence time of several days at the most; this corresponds to a transport distance of several hundred to 1000 km.[5] The formation of HNO_3 by oxidation is more rapid and, compared with H_2SO_4, results in a shorter travel distance from the emission source. H_2SO_4 also can react with NH_3 to form NH_4HSO_4 or $(NH_4)_2SO_4$ aerosols. In addition, the NH_4NO_3 aerosols are in equilibrium with $NH_{3(g)}$ and $HNO_{3(g)}$.[6]

The flux of dry deposition is usually assumed to be a product of its concentration adjacent to the surface and the deposition velocity. Deposition velocity depends on the nature of the pollutant (type of gas, particle size), the turbulence of the atmosphere, and the characteristics of the receptor surface (water, ice, snow, vegetation, trees, rocks). Dry deposition is usually collected for study in open buckets during dry periods, but this method underestimates the flux of pollutants, particularly gases (SO_x, NO_x), to a foliar canopy or lake surface. (Fog can be collected with special devices, but it is not measured quantitatively in a wet or dry bucket collector.)

The foliar canopy receives much of its dry deposition in the form of sulfate, nitrate, and hydrogen ions, which occur primarily as SO_2, HNO_3, and NH_3 vapors. Dry deposition of coarse particles has been shown to be an important source of calcium and potassium ion deposition on deciduous forests in the eastern United States (Table 10.1).[7]

In addition to the acid–base components shown in Figure 10.1, various organic acids are often found. Many of these acids are by-products of the atmospheric oxidation of organic matter released into the atmosphere.[8] Of special interest are formic, acetic, oxalic, and benzoic acids, which have been found in rainwater in concentrations occasionally exceeding a few micromoles per liter. Thus they must

Table 10.1 Total Annual Atmospheric Deposition of Major Ions to an Oak Forest at Walker Branch Watershed, Tennessee.[a]

	Atmospheric Deposition, meq m^{-2} y^{-1}					
Process	SO_4^{2-}	NO_3^-	H^+	NH_4^+	Ca^{2+}	K^+
Precipitation	70 ± 5	20 ± 2	69 ± 5	12 ± 1	12 ± 2	0.9 ± 0.1
Dry deposition						
Fine particles	7 ± 2	0.1 ± 0.02	2.0 ± 0.9	3.6 ± 1.3	1.0 ± 0.2	0.1 ± 0.05
Coarse particles	19 ± 2	8.3 ± 0.8	0.5 ± 0.2	0.8 ± 0.3	30 ± 3	1.2 ± 0.2
Vapors[b]	62 ± 7	26 ± 4	85 ± 8	1.3	0	0
Total deposition	160 ± 9	54 ± 4	160 ± 9	18 ± 2	43 ± 4	2.2 ± 0.3

[a]Values are means ± standard errors for two years of data. Numbers of observations range from 15 (HNO_3) to 26 (particles) to 128 (precipitation) to 730 (SO_2). In comparing these deposition rates it must be recalled that any such estimates are subject to considerable uncertainty. The standard errors given provide only a measure of uncertainty in the calculated sample means relative to the population means; hence, additional uncertainties in analytical results, hydrologic measurements, scaling factors, and deposition velocities must be included. The overall uncertainty for wet deposition fluxes is about 20% and that for dry deposition fluxes is approximately 50% for SO_4^{2-}, Ca^{2+}, K^+, and NH_4^+ and approximately 75% for NO_3^- and H^+.

[b]Includes SO_2, HNO_3 and NH_3. Complete conversion of deposited SO_2 to H_2SO_4 and of NH_3 to NH_4^+ was assumed in determining the vapor input of H^+. NH_3 deposition was estimated from the literature.

Source: Lindberg et at.[7] Reprinted with permission. Copyright (1986). American Association for the Advancement of Science.

be included in the calculated acidity of rainwater, and their presence in larger concentrations has an influence on the pH of rain and fog samples.

Figure 10.2 illustrates the inorganic composition of representative rain samples.[9] The ratio of the cations (H^+, NH_4^+, Ca^{2+}, Na^+, and K^+) and the anions (SO_4^{2-}, NO_3^-, Cl^-) reflects the acid–base titration that occurs in the atmosphere and in rain droplets. Total concentrations (the sum of cations or anions) typically vary from 20 $\mu eq\ L^{-1}$ to 500 $\mu eq\ L^{-1}$. Dilution effects, such as washout by atmospheric precipitation, can in part explain the differences observed. For example, the highest concentrations are observed after long, dry periods; the lowest concentrations are typically recorded at the end of prolonged rain.

When fog is formed from water-saturated air, water droplets condense on aerosol particles. In addition to components of the aerosols, the fog droplets can absorb such gases as NO_x, SO_2, NH_3, and HCl; they form a favorable milieu for various oxidation processes, especially the formation of H_2SO_4. Fog droplets (10–50 μm in

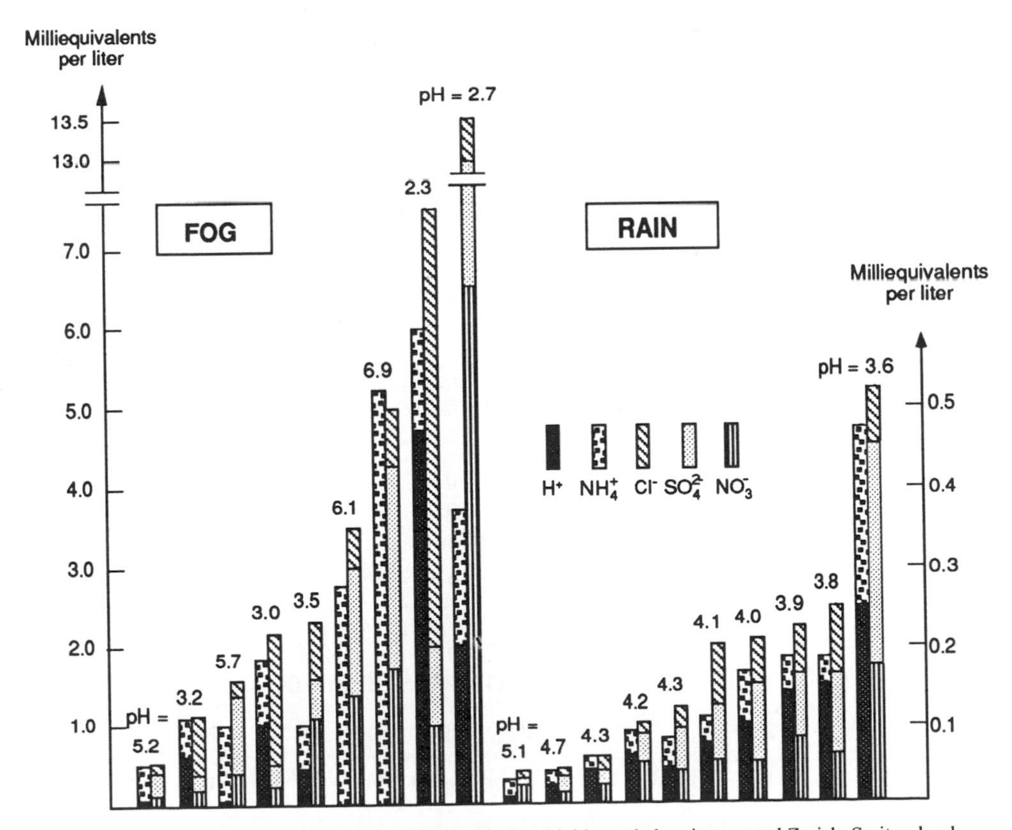

Figure 10.2 Composition of fog and rain samples, in a highly settled region around Zurich, Switzerland. The composition of fog varies widely and reflects to a larger extent than rain the influence of local emissions close to the ground. The fog concentration increases with decreasing liquid water content. (From Stumm et al.[1])

diameter) are much smaller than rain droplets; the liquid water content of fog is often in the range of 1×10^{-4} L m^{-3} air, so that acids in fog are typically 10–50 times more concentrated than are those found in rain (Figure 10.2).[9] Typical water contents in atmospheric systems are 5×10^{-5} to 5×10^{-3} L m^{-3} for fog and 10^{-4} to 10^{-3} L m^{-3} for clouds.

Rain clouds process a considerable volume of air over relatively large distances and thus are able to absorb gases and aerosols from a large region. Because fog is formed in the lower air masses, fog droplets are efficient collectors of pollutants close to the earth's surface. The influence of local emissions (such as NH_3 in agricultural regions or HCl near refuse incinerators) is reflected in the fog composition. Fog waters typically contain total ionic concentrations of 0.5-15 meq L^{-1}, as shown in Figure 10.2. Remarkably different pH values can be observed in fog. In addition to neutral fogs (pH 5–7)—some of which have very high anion concentrations—other fogs contain extreme acidity (pH 1.8).[10]

Table 10.2 is a summary of K_H (Henry's constant and other equilibrium constants at 25 °C (for Henry's law $C_{aq} = K_H p_{atm}$) for most gases of importance in atmospheric deposition to lakes and forests. Henry's law constants, as for other thermodynamic constants, are valid for ideal solutions. Ideally, they should be written in terms of activities and fugacities. Since activity coefficients for neutral molecules in aqueous solution become larger than 1.0 (salting-out effect), the solubility of gases is smaller in salt solution than in dilute aqueous medium (expressed in concentration units). We may use Henry's constant for *dilute* solutions without corrections as shown in Example 10.1. Henry's constants for absorption of gases in cloudwater (5 °C) must be corrected for temperature using the van't Hoff relationship.

Example 10.1 Solubility of SO_2 in Water

a. What is the solubility of SO_2 ($P_{atm} = 2 \times 10^{-9}$ atm) in water at 5 °C?
b. What is the solubility of CO_2 ($P_{atm} = 3.3 \times 10^{-4}$ atm) in water at 5 °C?
c. What is the composition of rain in equilibrium with both SO_2 and CO_2 in water under the conditions specified?

Solution:

a. The following constants are valid at 5 °C after correction using the van't Hoff relationship:

$$SO_{2(g)} + H_2O = SO_2 \cdot H_2O \qquad K_H = 3.0 \text{ M atm}^{-1}$$

$$SO_2 + H_2O = H^+ + HSO_3^- \qquad K_1 = 2.13 \times 10^{-2}$$

$$HSO_3^- = H^+ + SO_3^{2-} \qquad K_2 = 9.1 \times 10^{-8}$$

$$CO_2 + H_2O = H_2CO_3^* \qquad K_H = 6.3 \times 10^{-2} \text{ M atm}^{-1}$$

$$H_2CO_3^* = H^+ + HCO_3^- \qquad K_1 = 3.0 \times 10^{-7}$$

$$HCO_3^- = H^+ + CO_3^{2-} \qquad K_2 = 2.75 \times 10^{-11}$$

Table 10.2 Equilibrium Constants of Importance in Fog–Water Equilibrium

		$K_{25\,°C}$
1. $CO_{2(g)} + H_2O_{(l)}$	$= H_2CO^*_{3(aq)}$	3.39×10^{-2} [a]
2. $H_2CO_3^*$	$= H^+ + HCO_3^-$	4.45×10^{-7}
3. HCO_3^-	$= H^+ + CO_3^{2-}$	4.69×10^{-11}
4. $SO_{2(g)} + H_2O_{(l)}$	$= SO_2 \cdot H_2O_{(aq)}$	1.25[a]
5. $SO_2 \cdot H_2O_{(aq)}$	$= H^+ + HSO_3^-$	1.29×10^{-2}
6. HSO_3^-	$= H^+ + SO_3^{2-}$	6.25×10^{-8}
7. $NH_{3(g)}$	$= NH_{3(aq)}$	57[a]
8. $NH_{3(aq)} + H_2O$	$= NH_4^+ + OH^-$	1.77×10^{-5}
9. $HNO_{3(g)}$	$= H^+ + NO_3^-$	3.46×10^{6}
10. $HCl_{(g)}$	$= H^+ + Cl^-$	2.00×10^{6}
11. $HNO_{2(g)}$	$= HNO_{2(aq)}$	49[a]
12. $HNO_{2(aq)}$	$= H^+ + NO_2^-$	5.13×10^{-4}
13. $H_2S_{(g)}$	$= H_2S_{(aq)}$	1.05×10^{-1} [a]
14. $H_2S_{(aq)}$	$= H^+ + HS^-$	9.77×10^{-8}
15. HS^-	$= H^+ + S^{2-}$	1.00×10^{-19}
16. $NO_{(g)} + NO_{2(g)} + H_2O_{(l)}$	$= 2\ HNO_{2(aq)}$	1.24×10^{2}
17. $CH_3COOH_{(g)}$	$= CH_3COOH_{(aq)}$	7.66×10^{2} [a]
18. $CH_3COOH_{(aq)}$	$= H^+ + CH_3COO^-$	1.75×10^{-5}
19. $CH_2O_{(g)}$	$= CH_2O_{(aq)}$	6.3×10^{-3} [a]
20. $N_{2(g)}$	$= N_{2(aq)}$	6.61×10^{-4} [a]
21. $O_{2(g)}$	$= O_{2(aq)}$	1.26×10^{-3} [a]
22. $CO_{(g)}$	$= CO_{(aq)}$	9.55×10^{-4} [a]
23. $CH_{4(g)}$	$= CH_{4(aq)}$	1.29×10^{-3} [a]
24. $NO_{2(g)}$	$= NO_{2(aq)}$	1.00×10^{-2} [a]
25. $NO_{(g)}$	$= NO_{(aq)}$	1.9×10^{-3} [a]
26. $N_2O_{(g)}$	$= N_2O_{(aq)}$	2.57×10^{-2} [a]
27. $H_2O_{2(g)}$	$= H_2O_{2(aq)}$	1.0×10^{5} [a]
28. $O_{3(g)}$	$= O_{3(aq)}$	9.4×10^{-3} [a]

[a]Henry's constants are in units of M atm^{-1}
Source: Modified from Sigg and Stumm.[10]

The solubility of SO_2 can be calculated the same way as that of CO_2. The calculation for SO_2 solubility is as a function of pH. If SO_2 alone (no acids or bases added) comes into contact with water droplets, the composition is given approximately at a pH where $[H^+] \simeq [HSO_3^-]$ (see Figure 10.3a). The exact proton condition or charge balance condition is TOT H = $[H^+] - [HSO_3^-] - 2[SO_3^{2-}] - [OH^-] = 0$. This condition applies to the following composition: pH = 4.9, $[HSO_3^-] = 1.2 \times 10^{-6}$ M, $[SO_2 \cdot H_2O] = 3.7 \times 10^{-8}$ M, $[SO_3^{2-}] = 8 \times 10^{-8}$ M.

b. For this part we have pH = 5.65, $[H_2CO_3^*] = 2.1 \times 10^{-5}$ M, $[HCO_3^-] = 2.2 \times 10^{-6}$ M, $[CO_3^{2-}] = 2.7 \times 10^{-11}$ M. The answer to question (b) is obtained by plotting the corresponding diagram for CO_2 where $[H^+] = [HCO_3^-]$.

c. The answer to question (c) is obtained by superimposing the plots for SO_2 and

Table 10.3 Matrix for the Solubility of SO_2 and of CO_2

	Species	Components			log K (5 °C)
		$SO_{2(g)}$	$CO_{2(g)}$	H^+	
(i)	$SO_2 \cdot H_2O$	1	0	0	0.48
(ii)	HSO_3^-	1	0	−1	−1.2
(iii)	SO_3^{2-}	1	0	−2	−8.24
(iv)	$H_2CO_3^*$	0	1	0	−1.2
(v)	HCO_3^-	0	1	−1	−7.72
(vi)	CO_3^{2-}	0	1	−2	−18.3
(vii)	H^+	0	0	1	0

(viii) $p_{SO_2} = 2 \times 10^{-9}$ atm; $p_{CO_2} = 3.3 \times 10^{-4}$ atm
(ix) TOT H $= [H^+] - [HSO_3^-] - 2[SO_3^{2-}] - [OH^-]$

for CO_2. The matrix for solution of the chemical equilibrium problem is given by Table 10.3.

Electroneutrality [equation (ix)] specifies the condition for water in equilibrium with the given partial pressures of CO_2 and SO_2 (no acid or base added). This condition is fulfilled where $[H^+] \approx [HSO_3^-] + [HCO_3^-] + 2[SO_3^{2-}]$.

SO_2, even at small concentrations, has an influence on the pH of the water droplets. It has shifted from pH $\simeq$ 5.6 (where $[H^+] = [HCO_3^-]$) to pH = 4.75. The exact answer for this composition is in logarithmic units (Table 10.4).

The effect of CO_2 on the pH of the system is very small compared to that of SO_2.

The units for Henry's law constants in Table 10.2 are expressed as M atm^{-1}, but oftentimes they are given in inverse units in the literature, so one must be careful. Here, we will use K_H in M atm^{-1} and $R = 0.08206$ atm M^{-1} K^{-1} (at 25 °C, $RT = 24.5$ atm M^{-1}).

Table 10.4 Equilibrium Composition[a] for Example 10.1c

Species	log C
$SO_2 + H_2O$	−8.2
HSO_3^-	−5.15
SO_3^{2-}	−7.45
$H_2CO_3^*$	−4.7
HCO_3^-	−6.45
CO_3^{2-}	−12.3
H^+	−4.75

[a] Where $[H^+] = [HSO_3^-] + [HCO_3^-] + 2[SO_3^{2-}]$.

10.1.2 SO_2 and NH_3 Absorption

The distribution of gas molecules between the gas phase and the water phase depends on the Henry's law equilibrium distribution. In the case of CO_2, SO_2, and NH_3, the dissolution equilibrium is pH dependent because the components in the water phase—$CO_{2(aq)}$, H_2CO_3, $SO_2 \cdot H_2O_{(aq)}$, $NH_{3(aq)}$—undergo acid–base reactions.

Two varieties of chemical equilibrium modeling are possible. In an open system model, a constant partial pressure of the gas component is maintained. In a closed system, an initial partial pressure of a component is given, for example, for a cloud before rain droplets are formed or for a package of air before fog droplets condense. In this case, the system is considered closed from then on, the total concentration in the gas phase and in the solution phase is constant (Figure 10.3).[2,11,12]

To model the equilibrium of CO_2, SO_2, and NH_3 between the gas phase and the liquid phase of atmospheric water, it is first necessary to define the systems and to write equilibrium equations. For equilibrium at 25 °C (infinite dilution) the CO_2 system equations are as follows[2]:

$$[H_2CO_3^*]/P_{CO_2} = 10^{-1.5} \text{ M atm}^{-1} \quad (1)$$

$$[HCO_3^-]\,[H^+]/[H_2CO_3^*] = 10^{-6.3} \quad (2)$$

$$[CO_3^{2-}]\,[H^+]/[CO_3^{2-}] = 10^{-10.2} \quad (3)$$

Where P is partial pressure and $[H_2CO_3^*] = [CO_2 \cdot (aq)] + [H_2CO_3]$. The SO_2 system equations, also valid for equilibrium at 25 °C (infinite dilution), are written as follows:

$$[SO_2 \cdot H_2O]/P_{SO_2} = 10^{0.1} \text{ M atm}^{-1} \quad (4)$$

$$[HSO_3^-]\,[H^+]/[SO_2 \cdot H_2O] = 10^{-1.9} \quad (5)$$

$$[SO_3^{2-}]\,[H^+]/[HSO_3^-] = 10^{-7.2} \quad (6)$$

Finally, the NH_3 system equations are written as follows:

$$[NH_{3(aq)}]/P_{NH_3} = 10^{1.8} \text{ M atm}^{-1} \quad (7)$$

$$[NH_{3(aq)}]\,[H^+]/[NH_4^+] = 10^{-9.2} \quad (8)$$

The distribution of the species in an open system (constant partial pressure) is best obtained by plotting all the pertinent equations in a double logarithmic diagram (log molar concentration of the individual species in water versus pH). For example, in Figure 10.3a we obtain expressions for $P_{CO_2} = 10^{-3.5}$ atm (composition of the atmosphere) by combining equations (1)–(3) to arrive at

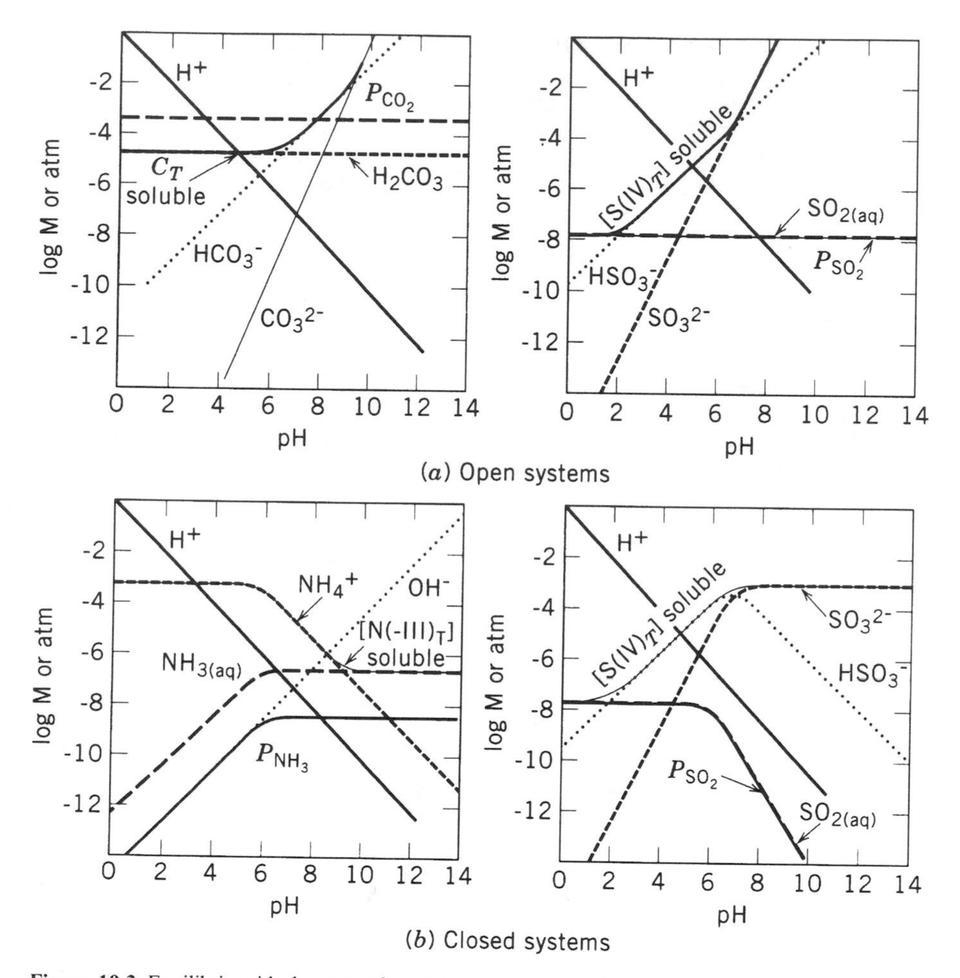

Figure 10.3 Equilibria with the atmosphere (atmospheric water droplets) for the conditions given. (a) Open systems: atmospheric CO_2 with water, $P_{CO_2} = 10^{-3.5}$ atm; $P_{SO_2} = 2 \times 10^{-8}$ atm. (b) Closed systems: atmospheric NH_3 with water, liquid water content $= 5 \times 10^{-4}$ L m^{-3}; total $NH_3 = 3 \times 10^{-7}$ mol m^{-3}; total $SO_2 = 8 \times 10^{-7}$ mol m^{-3}. (Example by Stumm et al.[1])

$$[H_2CO_3^*] = P_{CO_2} \cdot 10^{-1.5} = 10^{-5}\ M \tag{9}$$

$$[HCO_3^-] = 10^{-6.3} \cdot 10^{-1.5} \cdot P_{CO_2}/[H^+] = 10^{-11.3}\ [H^+]^{-1} \tag{10}$$

$$[CO_3^{2-}] = 10^{-6.3} \cdot 10^{-10.2} \cdot 10^{-1.5} \cdot P_{CO_2}/[H^+]^2 = 10^{-21.6}/[H^+]^2 \tag{11}$$

These equations are plotted in Figure 10.3a. Similarly, for an open SO_2 system, $P_{SO_2} = 2 \times 10^{-8}$ atm (constant), we obtain the distribution by combining equations (4)–(6) (Figure 10.3a).

$$[SO_2 \cdot H_2O] = P_{SO_2} \cdot 10^{0.1} = 10^{-7.6}\ M \tag{12}$$

$$[HSO_3^-] = 10^{-1.9} \cdot 10^{0.1} \cdot P_{SO_2}/[H^+] = 10^{-9.5}\ [H^+]^{-1} \tag{13}$$

$$[SO_3^{2-}] = 10^{-1.9} \cdot 10^{-7.2} \cdot 10^{0.1} \cdot P_{SO_2}/[H^+]^2 = 10^{-16.7}/[H^+]^2 \tag{14}$$

The closed system model can often be used expeditiously when a predominant fraction of the species is absorbed in the water phase. In a closed system, the total concentration is constant.

In a closed system, we can make the calculation on a mass balance basis for a given liquid water content of the atmosphere, q (in units of L m^{-3}). The distribution of atmospheric NH_3 also is a strong function of pH. If we assume a total concentration of NH_3 in gas and water of 3×10^{-7} mol m^{-3}, corresponding to an initial P_{NH_3} (in the absence of water) of 7.3×10^{-9} atm and a liquid water content of $q = 5 \times 10^{-4}$ L m^{-3}, the mass balance for total NH_3 in the gas and liquid water phases is

$$\begin{aligned} 3 \times 10^{-7}\ \text{mol m}^{-3} &= NH_{3(g)} + ([NH_{3(aq)}] + [NH_4^+])\,q[\text{mol m}^{-3}] \\ &= P_{NH_3}/RT + [P_{NH_3} \cdot 10^{-1.8}(1 + [H^+]/10^{-9.2})]q \end{aligned} \tag{15}$$

where RT (at 25 °C) = 2.446×10^{-2} m^3 atm mol^{-1}. The partial pressure of ammonia (P_{NH_3}) and then the other species can be calculated as a function of pH (Figure 10.3b). The calculation can be simplified if one realizes that at high pH nearly all NH_3 is in the gas phase, whereas at low pH nearly all of it is dissolved as NH_4^+. At low pH, water vapor is an efficient sorbent for NH_3 gas, but it decreases at higher pH. Thus pollution control efforts to remove SO_2 gas and decrease the acidity of acid rain may not be as efficient as hoped if less NH_3 is absorbed, which would have served to neutralize acid precipitation.

For SO_2 and an assumed total concentration of 8×10^{-7} mol m^{-3} (an initial P_{SO_2} of 2×10^{-8} atm) and a liquid water content of $q = 5 \times 10^{-4}$ L m^{-3}, the overall mass balance is given by the following equation:

$$\begin{aligned} SO_2(\text{total}) &= 8 \times 10^{-7}\ \text{mol m}^{-3} \\ &= P_{SO_2}/RT + [P_{SO_2} \cdot 10^{0.1}(1 + 10^{-1.9}/[H^+] + 10^{-9.1}/[H^+]^2)]q \end{aligned} \tag{16}$$

At low pH, nearly all SO_2 is in the gas phase; at high pH most of it is dissolved as SO_3^{2-} (Figure 10.3b).

These examples illustrate the strong pH dependence of the solubility of these atmospheric gases. They relate to the question of linearity of emission controls for SO_2. In some instances, other factors may affect the solubility. For example, the presence of formaldehyde can enhance the absorption of (total) SO_2 in water, because CH_2O reacts with HSO_3^- to form $CH_2OHSO_3^-$. Hydroxymethane–sulfonic acid is a very strong acid and oxidation of S(IV) bound to it is inhibited.[13]

10.2 ACIDITY AND ALKALINITY; NEUTRALIZING CAPACITIES

One has to distinguish between the H^+ concentration (or activity) as an intensity factor and the availability of H^+, that is, the H^+-ion reservoir as given by the base-neutralizing capacity, BNC. The BNC relates to the alkalinity [Alk] or acid-neutralizing capacity, ANC, by

$$[\mathrm{BNC}] = -[\mathrm{ANC}] \tag{17}$$

The BNC can be defined by a net proton balance with regard to a reference level: the sum of the concentrations of all the species containing protons in excess of the reference level, less the concentrations of the species containing protons in deficiency of the proton reference level. For natural waters, a convenient reference level (corresponding to an equivalence point in alkalimetric titrations) includes H_2O and H_2CO_3:

$$[\mathrm{H\text{-}Acy}] = [\mathrm{H^+}] - [\mathrm{HCO_3^-}] - 2[\mathrm{CO_3^{2-}}] - [\mathrm{OH^-}] \tag{18}$$

The acid-neutralizing capacity, ANC, or alkalinity [Alk] is related to [H-Acy] by

$$-[\mathrm{H\text{-}Acy}] = [\mathrm{Alk}] = [\mathrm{HCO_3^-}] + 2[\mathrm{CO_3^{2-}}] + [\mathrm{OH^-}] - [\mathrm{H^+}] \tag{19}$$

Considering a charge balance for a typical natural water (Figure 10.4), we realize that [Alk] and [H-Acy] also can be expressed by a charge balance: the equivalent sum of conservative cations, less the sum of conservative anions ([Alk] = $a - b$). The conservative cations are the base cations of the strong bases $Ca(OH)_2$, KOH, and the like; the conservative anions are those that are the conjugate bases of strong acids (SO_4^{2-}, NO_3^-, and Cl^-).

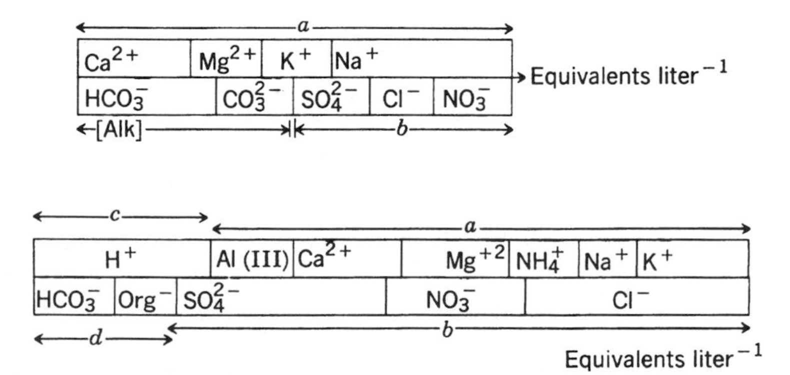

Figure 10.4 Natural water charge balance for an alkaline system (Alk = $a - b$) and an acid system (Alk = $a - b = d - c$).

$$[Alk] = [HCO_3^-] + 2[CO_3^{2-}] + [OH^-] - [H^+]$$
$$= [Na^+] + [K^+] + 2[Ca^{2+}] + 2[Mg^{2+}] - [Cl^-] - 2[SO_4^{2-}] - [NO_3^-] \quad (20)$$

The [H-Acy] for this particular water, obviously negative, is defined ([H-Acy] = $b - a$) as

$$[H\text{-}Acy] = [Cl^-] + 2[SO_4^{2-}] + [NO_3^-] - [Na^+] - [K^+] - 2[Ca^{2+}] - 2[Mg^{2+}] \quad (21)$$

These definitions can be used to interpret interactions of acid precipitation with the environment.

A simple accounting can be made: every base cationic charge unit (e.g., Ca^{2+} or K^+) that is removed from the water by whatever process is equivalent to a proton added to the water, and every conservative anionic charge unit (from anions of strong acids—NO_3^-, SO_4^{2-}, or Cl^-) removed from the water corresponds to a proton removed from the water; or generally,

$$\Sigma\,\Delta\,[\text{base cations}] - \Sigma\,\Delta\,[\text{conservative anions}] = +\Delta[Alk] = -\Delta[H\text{-}Acy] \quad (22)$$

If the water under consideration contains other acid- or base-consuming species, the proton reference level must be extended to the other components. In addition to the species given, if a natural water contains organic molecules, such as a carboxylic acid (H-Org) and ammonium ions, the reference level is extended to these species.

One usually chooses the ammonium species NH_4^+ and H-Org as a zero-level reference condition. In operation, we wish to distinguish between the acidity caused by strong acids (mineral acids and organic acids with $pK < 6$) typically called *mineral acidity* or *free acidity*, which often is nearly the same as the free-H^+ concentration, and the total acidity given by the BNC of the sum of strong and weak acids.[10,14]

The distinction is possible by careful alkalimetric titrations of rain and fog samples. Gran titrations have found wide acceptance in this area. The strong acidity, [H-Acy], is expressed with respect to the reference conditions: H_2O, H_2CO_3, SO_4^{2-}, NO_3^-, Cl^-, NO_2^-, F^-, HSO_3^-, NH_4^+, H_4SiO_4, and $\Sigma H_n Org$ (organic acids, $pK_a < 10$).

$$[H\text{-}Acy] = [H^+] + [HSO_4^-] + [HNO_2] + [HF] + [H_2SO_3] - [OH^-]$$
$$- [HCO_3^-] - 2[CO_3^{2-}] - [NH_3] - [H_3SiO_4^-] - \Sigma n\,[Org^{n-}] \quad (23)$$

Components such as HSO_4^-, HNO_2, HF, H_2SO_3,CO_3^{2-}, NH_3, and $H_3SiO_4^-$ are in negligible concentrations in typical rainwater. Thus the equation may be simplified to

$$[H\text{-}Acy] = [H^+] - [OH^-] - [HCO_3^-] - \Sigma n\,[Org^{n-}] \quad (24)$$

For most rain samples of pH 4–4.5, [H-Acy] is equal to $[H^+]$, but in highly concentrated fog waters (in extreme cases, $pH < 2.5$) HSO_4^- and $SO_2 \cdot H_2O$ become important species contributing to the strong acidity.[15]

The reference conditions pertaining to the determination of total acidity $[Acy_T]$ are H_2O, CO_3^{2-}, SO_4^{2-}, NO_3^-, Cl^-, NO_2^-, F^-, SO_3^{2-}, NH_3, $H_3SiO_4^-$, ΣOrg^{n-}, and $Al(OH)_3$. (When aluminum is titrated from pH 4 to pH 10, it will hydrolyze from Al^{3+} to $Al(OH)_3$, consuming three equivalents of base for every mole of Al^{3+}.)

$$[Acy_T] = [H^+] + 2[H_2CO_3] + [HSO_3^-] + [NH_4^+] + [H_4SiO_4] + 3[Al^{3+}] + 2[Al(OH)^{2+}] + [Al(OH)_2^+] + \Sigma n[H_nOrg] - [OH^-] \quad (25)$$

For most samples this equation can be simplified as

$$[Acy_T] = [H^+] + [NH_4^+] + 2[H_2CO_3] + \Sigma n[H_nOrg] \quad (26)$$

A Gran titration of the strong acidity usually gives a good approximation of the acidity [H-Acy], as defined above, but one must be aware that organic acids with pK_a 3.5–5 are partly included in this titration and may affect the resulting Gran functions. For the determination of total acidity, H_2CO_3 is usually stripped (with N_2) out of the samples, because when H_2CO_3 is present it is more difficult to reach the end point of the titration and to calculate the concentration of the weak acids.

10.2.1 Atmospheric Acidity and Alkalinity

In a (hypothetically closed) large system of the environment consisting of the reservoirs atmosphere, hydrosphere, and lithosphere, a proton and electron balance is maintained. Temporal and spatial inhomogeneities between and within individual reservoirs cause significant shifts in electron and proton balance, so that subsystems contain differences in acidity or alkalinity. Any transfer of an oxidant or reductant, of an acid or base, or of ions from one system to another (however caused, by transport, chemical reaction, or redox process) causes a corresponding transfer of acidity or alkalinity.

It is essential to consider the acid–base neutralization process in the entire atmospheric system. Morgan,[16] Liljestrand,[12] and Jacob et al.[17] introduced the concept of atmospheric acidity and alkalinity to interpret the interactions of NH_3 with strong acids emitted into and/or produced within the atmosphere. A net atmospheric acidity can be defined by summing over all potential acids and bases in the gas, aerosol, and liquid phases, using reference conditions for specific redox conditions, and assuming the presence of liquid water.

Figure 10.5 exemplifies the concept of alkalinity, Alk, and acidity, Acy, for a gas–water environment and defines the relevant reference conditions. In Figures 10.5a and 10.5b, it is shown how the gases NH_3, SO_2, NO_x, HNO_3, HCl, and CO_2 (potential bases or acids, respectively), subsequent to their dissolution in water and the oxidation of SO_2 to H_2SO_4 and of NO_x to HNO_3, become alkalinity or acidity components. Thus $\{Alk\}_{(gas)}$ and $\{Acy\}_{(gas)}$, for the gas phase, is defined by the following relation[18]:

$$\{\text{Alk}\}_{(\text{gas})} = \{\text{NH}_3\}_{(\text{gas})} - 2\,\{\text{SO}_2\}_{(\text{gas})} - \{\text{NO}_x\}_{(\text{gas})} - \{\text{HNO}_3\}_{(\text{gas})} - \{\text{HCl}\}_{(\text{gas})} - \{\text{H-Org}\}_{(\text{gas})} \tag{27}$$

and

$$\{\text{Alk}\}_{(\text{gas})} = -\{\text{Acy}\}_{(\text{gas})} \tag{28}$$

where { } indicate mol m^{-3} and H-Org is the sum of volatile organic acids. Figure 10.5b shows that these potential acids and bases, subsequent to their dissolution in cloud or fog water with $q = 10^{-4}$ L H_2O per m^3 atmosphere, give a water with the equivalent acidity.

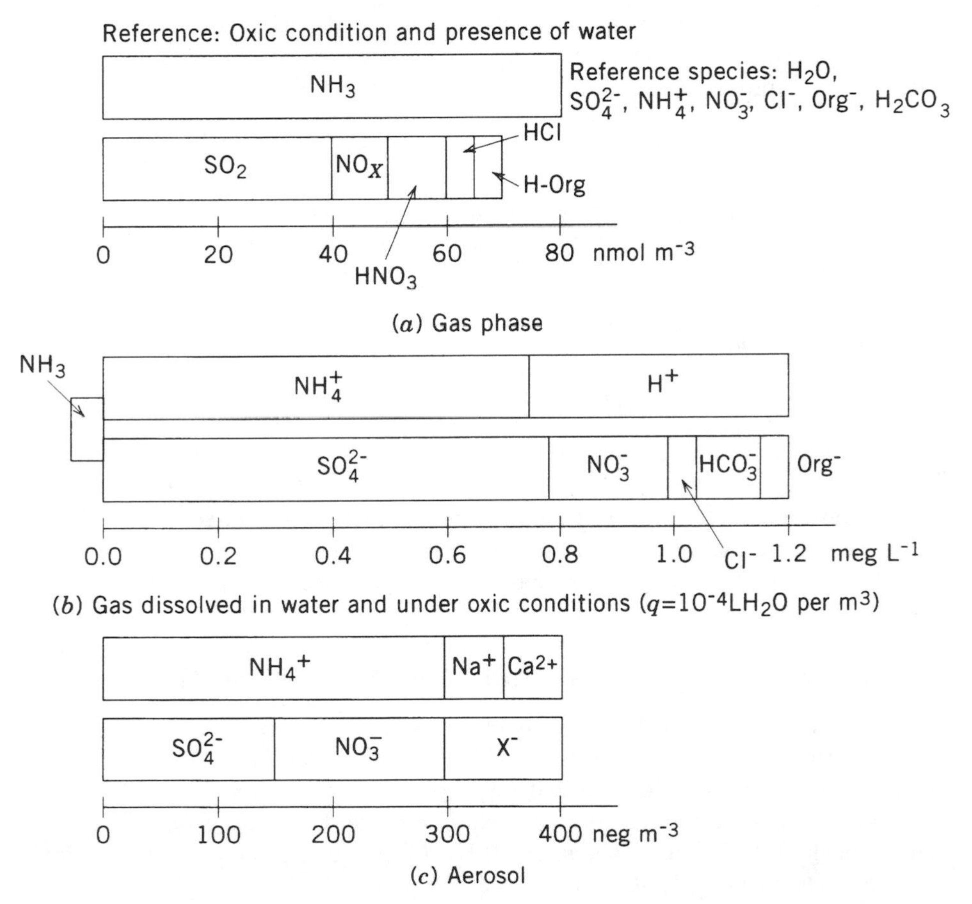

Figure 10.5 Alkalinity/acidity in atmosphere, aerosols, and atmospheric water. Alkalinity and acidity can be defined for the atmosphere using a reference state valid for oxide conditions (SO_2 and NO_x oxidized to H_2SO_4 and HNO_3) and in the presence of water. The neutralization of atmospheric acidity by NH_3 is a major driving force in atmospheric deposition. (From Behra et al.[18])

In the case of *aerosols*, we can define the alkalinity by a charge balance of the sum of conservative cations, $\Sigma n\{Cat.^{n+}\}_{(ae)}$, of NH_4^+, $\{NH_4^+\}_{(ae)}$, and of the sum of conservative anions, $\Sigma m\{An.^{m-}\}_{(ae)}$ (Figure 10.5c):

$$\{Alk\}_{(ae)} = \Sigma n\{Cat.^{n+}\}_{(ae)} + \{NH_4^+\}_{(ae)} - \Sigma m\{An.^{m-}\}_{(ae)} \quad (29)$$

NH_3 essentially "titrates" the inputs of the potential strong acids in the atmosphere.[17,18] The oxidation of SO_2 in atmospheric water (cloud, rain, liquid aerosol, and fog) is influenced by the presence of NH_3. The enhancing effect of NH_3 is especially pronounced if the oxidation occurs with an oxidant such as O_3 for which the reaction rate increases strongly with increasing pH, because NH_3 (1) codetermines the pH of the water and thus, in turn, the solubility of SO_2 and (2) provides acid-neutralizing capacity as well as buffer intensity to the heterogeneous atmosphere–water system in counteracting the acidity produced by the oxidation of SO_2.[18] At low buffer intensity, for example, in the case of residual atmospheric acidity, the acidity production alleviates further SO_2 oxidation, if

$$\{NH_3\}_{(gas)} < [(2\{SO_2\}_{(gas)} + \{NO_x\}_{(gas)} + \{HNO_3\}_{(gas)} + \{HCl\}_{(gas)}] \quad (30)$$

Thus, while NH_3 introduced into the atmosphere reduces its acidity, it enhances the oxidation of SO_2 by ozone, participates in the formation of ammonium sulfate and ammonium nitrate aerosols, and accelerates the deposition of SO_4^{2-}. Furthermore, any NH_3 or NH_4^+ that is returned to the earth's surface becomes HNO_3 as a consequence of nitrification and/or H^+ ions as a consequence of plant uptake,

$$NH_3 + 2\,O_2 = H^+ + NO_3^- + H_2O$$
$$NH_4^+ + 2\,O_2 = 2\,H^+ + NO_3^- + H_2O \quad (31)$$

and it may aggravate the acidification of soils and lakes. This effect is not sufficiently considered in the assessment of NH_3 emissions (e.g., agriculture, feed lots) and the use of excess NH_3 in air pollution control processes to reduce nitrogen oxides.

Example 10.2 Mixing of Waters with Different Acid-Neutralizing Capacities

The effluent from an acid lake with [H-Acy] = 5×10^{-5} eq L^{-1} and a pH of 4.3 mixes with a river containing an [Alk] = 1.5×10^{-4} eq L^{-1} and a pH of 7.4 in a 1:1 volumetric ratio. What is the alkalinity and the pH of the mixed waters? You may assume that the mixed water is in equilibrium with the CO_2 of the atmosphere (3.5×10^{-4} atm) and at 10 °C. The acidity constant of $H_2CO_3^*$ is 3×10^{-7} and Henry's constant for the reaction $CO_{2(g)} + H_2O = H_2CO_3^*$ is $K_H = 0.050$ M atm^{-1}.

Solution: Alkalinity (ANC) is a conservative quantity that is unaffected by $CO_{2(g)}$ sorption. We may calculate the alkalinity of the mixture by volume-weighted averaging.

$$[\text{Alk}]_{\text{mix}} = \tfrac{1}{2}([\text{Alk}]_{\text{river}} - [\text{H-Acy}]_{\text{lake}})$$

$$= \tfrac{1}{2}(1.5 \times 10^{-4} - 5 \times 10^{-5}) = 10^{-4} \text{ eq L}^{-1}.$$

The concentration of $H_2CO_3^*$ in equilibrium with the atmosphere is given by

$$[H_2CO_3^*] = K_H P_{CO_2} = (0.05)(3.5 \times 10^{-4})$$

$$= 1.75 \times 10^{-5} \text{ M}$$

At a pH of 7.4, most of the alkalinity is due to HCO_3^- so that $[\text{Alk}] = [HCO_3^-] = 10^{-4}$ M. Then, $[H^+]$ is given by the equilibrium expression at 10 °C:

$$\frac{[H^+][HCO_3^-]}{[H_2CO_3^*]} = 3 \times 10^{-7}$$

and $[H^+] = 5.2 \times 10^{-8}$; pH = 7.3, close to that of the original river.

This example illustrates that (1) [Alk] = –[H-Acy], (2) [Alk] and [H-Acy] are conservative parameters and can be used directly in mixing calculations, and (3) $[H^+]$ and pH are not conservative parameters. The river was well buffered by the bicarbonate system despite an equal volume of acid input at low concentration.

10.3 WET AND DRY DEPOSITION

10.3.1 Wet Deposition

Wet deposition occurs when pollutants fall to the ground or sea by rainfall, snowfall, or hail/sleet. *Dry deposition* is when gases and aerosol particles are intercepted by the earth's surface in the absence of precipitation. Let us first discuss wet deposition. Wet deposition to the surface of the earth is directly proportional to the concentration of pollutant in the rain, snow, or ice phase.

The wet deposition flux is defined by equation (32):

$$F_{\text{wet}} = IC_w \tag{32}$$

where F_{wet} is the areal wet deposition flux in μg cm^{-2} s^{-1}, I is the precipitation rate in cm s^{-1} (as liquid H_2O), and C_w is the concentration of the pollutant associated with the precipitation in μg cm^{-3}. Wet deposition is measured with a bucket collector and a rain gage. The rain gage is placed at the receptor site and provides an accurate measure of precipitation rate, I. The wet bucket collector is open only during the precipitation event, and its contents are analyzed for pollutant concentration, C_w.

The concentration of pollutants in wet deposition is due to two important effects with quite different physical mechanisms:

- Aerosol particle scavenging.
- Gas scavenging.

Aerosols begin their life cycle after nucleation and formation of a submicron hygroscopic particle, for example, $(NH_4)_2 SO_4$, which hydrates and grows very quickly due to condensation of water around the particle. At this stage, it is neither solid nor liquid, but merely a stable aerosol with a density between 1.0 and 1.1 g cm^{-3}. Mass quantities of air are processed by clouds, creating updrafts that take up surface layer pollutants. Cloud droplets are very small, on the order of 10 μm in diameter. Typically 1 million cloud droplets are needed to comprise a 1 mm diameter raindrop. Assuming an average spacing of 1-mm between cloud droplets, condensation of 10^6 cloud droplets into a 1-mm raindrop would scavenge enough air for a washout ratio of 10^6.

$$W = \frac{C_{w'}}{C_{ae}} \tag{33}$$

where $C_{w'}$ is the concentration of the pollutant in precipitation water in μg cm^{-3}, C_{ae} is the concentration of the pollutant associated with aerosol droplets in air in μg cm^{-3}, and W is the washout ratio for aerosols, dimensionless (cm^3 air/cm^3 precipitation).

Details of the physics of the scavenging process are beyond the scope of this book, but reference texts include Schwartz and Slinn,[19] Pruppacher et al.,[20] and Eisenreich.[21]

Table 10.5 provides a few values of washout ratios for metals associated with particles, and they are typically on the order of 10^5–10^6. *Rainout* sometimes refers to below-cloud processes, whereby pollutants are scavenged as raindrops fall through polluted air. Because clouds process such large quantities of air and pull up polluted air from the surface, washout is caused predominantly by in-cloud processes. Therefore we will not distinguish between washout and rainout.

Washout ratios for gas scavenging operate by a different mechanism than aerosol particle scavenging. Here, Henry's law is applicable because chemical equilibrium for absorption processes in the atmosphere is on the time scale of 1 second. Gas

Table 10.5 Some Measured Values for Size and Washout Ratio of Metals as Aerosols in the Atmosphere

Element	Mass Median Diameter (MMD), μm	Washout Ratio ($\times 10^6$)
Mg	5.7	0.38
K	5.7	0.46
Ca	5.7	0.29
Fe	1.7, 3.0	0.21
Mn	1.4, 2.0	0.31
Zn	0.8, 1.0	0.15
Pb	0.5	0.06

Source: Gatz.[22]

scavenging, therefore, is reversible, while aerosol scavenging is an irreversible process.

If we express Henry's constant K_H in units of M atm^{-1}, the following equations apply for Henry's law and the washout ratio:

$$C_{w} = K_H \, p_{\text{atm}} \tag{34}$$

$$W = \frac{C_{w}}{C_g} = K_H RT \tag{35}$$

where C_{w} is the concentration in the water phase (M), p_{atm} is the atmospheric partial pressure (atm), W is the washout ratio (dimensionless, i.e., L H_2O/L gas), C_g is the concentration in the gas (mol L^{-1} gas), and RT is the universal gas law constant times temperature (24.46 atm M^{-1} at 25 °C).

Some estimates for washout ratios of selected pesticides are presented in Table 10.6. Henry's constants are taken from Schwarzenbach et al.[23] In general, washout ratios are large for soluble and polar compounds, intermediate for semivolatile chemicals (such as DDT, dieldrin, dioxin, and PCBs), and low for volatile organic chemicals. Semivolatile pollutants are an interesting case because these gases can be transported long distances and recycled many times before being deposited in polar regions by a "cold-trap" effect. Although washout ratios of gases by snow are smaller than by rain, there can be appreciable liquid water contained in snow that absorbs gases. Adsorption of gases to snowflake surfaces can also be significant.

Table 10.6 Estimates of Washout Ratios for Selected Gases, 25 °C

Chemical	Henry's Constant K_H, M atm^{-1}	Washout Ratio $W = K_H RT$ [a]
Aldrin	100	2,450
Benzene	0.18	4.4
Benzo[a]pyrene	830	20,300
CCl_4	0.042	1.0
Dioxin	20	490
DDT	105	2,570
Dieldrin	89	2,200
Di-*n*-butyl phthalate	780	19,000
Methane	0.0015	0.037
Naphthalene	2.3	56
Parathion	2,630	64,000
Trichloroethene	0.093	2.3
Toluene	0.15	3.7
2,2′,5,5′-PCB	3.5	86

[a] $RT = 24.46$ atm M^{-1} at 25 °C.

The washout ratio does not give enough information to calculate the mass of pollutant in a column of air that is actually "washed-out" by rain. Example 10.3 shows how to estimate the mass contained in the rain assuming gas sorption equilibria.

Example 10.3 Washout of Pollutants from the Atmosphere

To what extent are atmospheric pollutants washed out by rain? We can try to answer this question by considering the gas absorption equilibria. Our estimate is based on the following assumptions and mass balance considerations. For example, calculate the mass fraction that is washed out (f_{water}) for the pesticide lindane (γ-hexachlorocyclohexane, $C_6H_6Cl_6$) with a Henry's constant K_H of 309 M atm^{-1}.

Solution: Assume the height of the air column is 5×10^3 m. This column is "washed out" by a rain of 25 mm (corresponding to 25 L m^{-2}). In other words,

gas volume $V_g = 5 \times 10^3 \text{ m}^3$
water volume $V_w = 0.025 \text{ m}^3$

$$\frac{V_g}{V_w} = 2 \times 10^5$$

The total quantity of the pollutant is

$$A_{\text{tot}} = ((A)_g \times V_g) + ((A)_w \times V_w)$$

where $(A)_g$ and $(A)_w$ are the concentrations in the gas phase in atm and the water phase in M.

The fraction of pollutants in the water phase, f_{water}, is given by

$$f_{\text{water}} = \frac{(A)_w \times V_w}{(A)_g V_g + (A)_w V_w} = \frac{1}{\left[\frac{(A)_g}{(A)_w} \times \frac{V_g}{V_w}\right] + 1}$$

$$f_{\text{water}} = \frac{1}{\frac{1}{K_H RT}\frac{V_g}{V_w} + 1} = \frac{1}{\left[\frac{1}{(309 \text{ M atm}^{-1})(24.46 \text{ atm M}^{-1})} \cdot 2 \times 10^5\right] + 1}$$

$$f_{\text{water}} = 0.0365$$

Lindane is quite soluble, relatively speaking, but only about 3.65% of it is washed out by the rainfall.

10.3.2 Dry Deposition

Both wet and dry deposition are important transport mechanisms. For total sulfur deposition in the United States, they are roughly of equal magnitude. Dry deposition takes place (in the absence of rain) by two pathways:

- Aerosol and particle deposition.
- Gas deposition.

There are three resistances to aerosol and gas deposition: (1) aerodynamic resistance, (2) boundary layer resistance, and (3) surface resistance. Aerodynamic resistance involves turbulent mixing and transport from the atmosphere (~1-km elevation) to the laminar boundary layer in the quiescent zone above the earth's surface. Boundary resistance refers to the difficulty of pollutant transport through the laminar boundary layer, and surface resistance involves the physical and chemical reactions that may occur at the surface of the receptor (sea surface, vegetation, snow surface, etc.). Dry deposition velocity encompasses the electrical analog of these three resistances in series:

$$V_d = \frac{1}{r_a + r_b + r_s} \tag{36}$$

where V_d is defined as the dry deposition velocity (cm s^{-1}), r_a is the aerodynamic resistance, r_b is the boundary layer resistance, and r_s is the resistance at the surface.

The deposition velocity is affected by a number of factors including relative humidity, type of aerosol or gas, aerosol particle size, wind velocity profile, type of surface receptor, roughness factor, atmospheric stability, and temperature. V_d increases with wind speed because sheer stress at the surface causes increased vertical turbulence and eddies. An excellent summary of dry deposition measurements and a comparison of collector surfaces is given by Davidson and Wu.[24]

For aerosol particles, the deposition velocity is dependent on particle diameter as shown in Figure 10.6.[25] In reality, aerosols change constantly due to changes in relative humidity; they evaporate or condense water continually. We often use the mass median diameter (MMD) as a measure of the particle size distribution. Milford and Davidson[26] showed a general power-law correlation for the dependence of V_d on particle size:

$$V_d = 0.388 \text{ MMD}^{0.76} \tag{37}$$

where V_d is the deposition velocity in cm s^{-1} and MMD is the mass median diameter of the particle in μm. Table 10.7 is a compilation of dry deposition velocities for chemicals of interest from Davidson and Wu.[24]

In general, gases that react at the surface (e.g., SO_2, HNO_3, HCl, and O_3) tend to have slightly higher deposition velocities, on the order of 1.0 cm s^{-1}. HNO_3 vapor has a very large deposition velocity because there is no surface resistance—it is im-

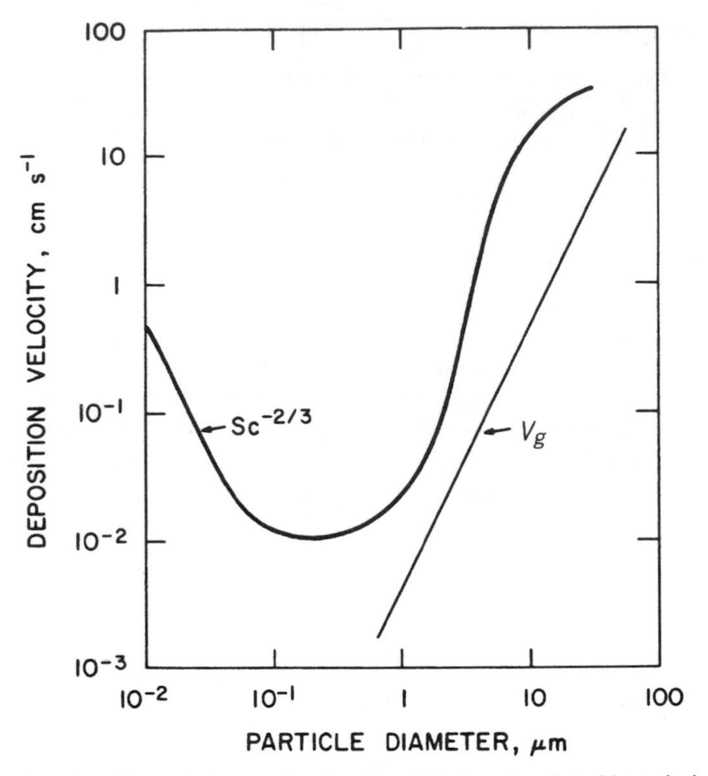

Figure 10.6 Dry deposition velocity as a function of particle diameter. Deposition velocity is always greater than the Stokes law discrete particle settling velocity (V_g) because of turbulent mixing and reaction at the surface. For very fine aerosols (less than 0.1 μm), the curve follows mass transfer correlations of the Schmidt number $Sc^{-2/3}$. (From Schmel and Sutter.[25])

mediately absorbed and neutralized by vegetation and/or water. Some gases such as NO_x display higher V_d values in daylight because vegetation transpires at that time, and gas exchange through the stomata serves to increase the concentration gradient and the flux at the leaf surface.

The receptor surface is critical. Deposition velocities in Table 10.7 are mostly to natural earth surfaces. Surrogate surfaces tend to underestimate the actual dry deposition because of differences in reactivity at the surface, differences in surface area, and aerodynamic differences around the collector. Natural vegetation and trees are relatively efficient interceptors of gases and particles based on specific surface areas. SO_2 dry deposition velocity for a coniferous forest may be several times higher than for an open field or a snow field.

Metals associated with wind-blown dust and coarse particles (Ca, Mg, K, Fe, Mn) tend to have higher deposition velocities due to the effect of particle size. Anthropogenic metals from fossil fuel combustion tend to be associated with fine particles and enriched relative to background concentrations of the earth's crust. Aluminum is used as a reference element for the earth's crust.

Table 10.7 Dry Deposition Velocities for a Number of Aerosol Particles and Gases

Pollutant	V_d, cm s^{-1}, Typical Range	V_d, cm s^{-1}, Typical Median	Conditions
$SO_{2(g)}$	0.3–1.6	0.95	To natural vegetation
	0.04–0.22	0.13	To snow field
SO_4^{2-}	0.01–1.2	0.55	Submicron aerosols in field (micrometeorological)
	0.01–0.5	0.26	To surrogate surfaces
NO_3^-	0.1–2.0	0.7	Aerosol particle deposition
NH_4^+	0.05–2.0	0.8	Aerosol particle deposition
$HNO_{3(g)}$	1–3	1.4	Gas, no surface resistance
$NO_{x(g)}$	0.01–0.5	0.05	Night, closed stomata
	0.1–1.7	0.6	Day, open stomata
Cl^-	1–5	2	Particles, MMD = 1-4 μm
$HCl_{(g)}$	0.6–0.8	0.7	Sorption by dew
$O_{3(g)}$	0.01–1.5	0.4	By measured gradients
Pb	0.1–1.0	0.26	Aerosol particle deposition from autos, MMD < 1 μm
Crustal metals (Ca, Mg, K, Fe, Mn)	0.3–3.0	1.5	Associated with coarse particles, MMD = 1–4 μm
Enriched (anthropogenic metals—Ag, As, Cd, Cu, Zn, Pb, Ni)	0.1–1.0	0.3	Associated with fine particles, enriched MMD < 1 μm
Fine particles	0.1–1.2	0.4	Submicron particles

Source: Summarized from Davidson and Wu.[24]

$$EF = \frac{X_{air}/Al_{air}}{X_{crust}/Al_{crust}} \tag{38}$$

where X_{air} and Al_{air} represent the airborne concentrations of any element X and aluminum, respectively, and X_{crust} and Al_{crust} are the concentrations in the earth's crust. Ag, As, Cd, Cu, Zn, Pb, and Ni tend to be enriched relative to aluminum, indicating anthropogenic origin in the atmosphere.

10.4 PROCESSES THAT MODIFY THE ANC OF SOILS AND WATERS

Subsequent to acid deposition, there are various processes in the terrestrial and aquatic environment that can neutralize or enhance the acidity of the precipitation. Of the processes depicted in Figure 10.1, denitrification, sulfate sorption, sulfate reduction, and chemical weathering decrease acidity, while photosynthetic assimilation, sulfur oxidation, and nitrification increase the acidity of the waters (Table 10.8).

Table 10.8 Some Processes that Modify the H^+ Balance in Waters[27]

Processes		Changes in Alkalinity $\Delta[\text{Alk}]^a = -\Delta[\text{H-Acy}]^b$ [equivalents per mole reacted (reactant is underlined)]
1. Weathering reactions (calcite, anorthite, orthoclase, gibbsite):		
	$\underline{CaCO_{3(s)}} + 2H^+ \rightleftharpoons Ca^{2+} + CO_2 + H_2O$	+2
	$\underline{CaAl_2Si_2O_{8(s)}} + H_2O + 2H^+ \rightleftharpoons Ca^{2+} + Al_2Si_2O_5(OH)_{4(s)}$	+2
	$\underline{KAlSi_3O_{8(s)}} + H^+ + 4\frac{1}{2}H_2O \rightleftharpoons K^+ + 2H_4SiO_4 + \frac{1}{2}Al_2Si_2O_5(OH)_{4(s)}$	+1
	$\underline{Al_2O_3 \cdot 3H_2O} + 6H^+ \rightleftharpoons 2Al^{3+} + 6H_2O$	+6
2. Ion exchange and sulfate sorption:		
	$2ROH + \underline{SO_4^{2-}} \rightleftharpoons R_2SO_4 + 2OH^-$	+2
	$NaX + \underline{H^+} \rightleftharpoons HX + Na^+$	+1
3. Redox processes (microbial mediation):		
Nitrification	$\underline{NH_4^+} + 2O_2 \rightleftharpoons NO_3^- + H_2O + 2H^+$	-2
Denitrification	$1\frac{1}{4}CH_2O + \underline{NO_3^-} + H^+ \rightarrow 1\frac{1}{4}CO_2 + \frac{1}{2}N_2 + 1\frac{3}{4}H_2O$	+1
Oxidation of H_2S	$\underline{H_2S} + 2O_2 \rightarrow SO_4^{2-} + 2H^+$	-2
SO_4^{2-} reduction	$\underline{SO_4^{2-}} + 2CH_2O + 2H^+ \rightarrow 2CO_2 + H_2S + 2H_2O$	+2
Pyrite oxidation	$\underline{FeS_{2(s)}} + 3\frac{3}{4}O_2 + 3\frac{1}{2}H_2O \rightarrow Fe(OH)_3 + 2SO_4^{2-} + 4H^+$	-4
4. Synthesis ($\rightarrow$) and decomposition ($\leftarrow$) of biomass and of humus[c]:		
Photosynthesis with NO_3 assimilation ($\rightarrow$), aerobic respiration ($\leftarrow$):		
	$a\,CO_{2(g)} + b\,NO_3^- + c\,HPO_4^{2-} + \cdots + g\,Ca^{2+} + h\,Mg^{2+} + i\,K^+ + f\,Na^+ + x\,H_2O \rightleftharpoons \{C_aN_bP_cS_d \cdots Ca_gMg_hK_iNa_fH_2O_m\}_{\text{biomass}} + (a+2b)\,O_{2(g)}$	$+b+2c+2d$ $-(2g+2h+i+f)$
NH_4 assimilation ($\rightarrow$); aerobic mineralization (back reaction):		
	$a\,CO_{2(g)} + b\,NH_4^+ + \cdots \rightleftharpoons \{C_a \cdots N_b\}_{\text{biomass}} + a\,O_{2(g)}$	$-b$
N_2 fixation (half-reaction):		
	$\cdots + \frac{1}{2}b\,N_{2(g)} + \cdots \rightarrow \{\cdots N_b \cdots\}_{\text{biomass}}$	0

[a][Alk] = alkalinity = acid-neutralizing capacity.

[b][H-Acy] = mineral acidity.

[c]In the buildup of biomass (or the exchange of ions), the uptake of each equivalent of conservative anions causes an equivalent increase in alkalinity in the environment, and each equivalent of base cations that is taken up results in an equivalent decrease of alkalinity. One comes to the same conclusion as long as the biomass or humus is formed of neutral components $\{(CH_2O)_a(NH_3)_b(H_3PO_4)_c(H_2SO_4)_d + \cdots + (Ca(OH)_2)_g(Mg(OH)_2)_h(KOH)_i(NaOH)_f(H_2O)_m\}$ and provided that one has written the corresponding stoichiometric equations for formation and decomposition.

10.4.1 ANC of Soil

In weathering reactions, alkalinity is added from the soil–rock system to the water:

$$\Delta[\mathrm{Alk}]_{\mathrm{water}} = -\Delta[\mathrm{Alk}]_{\mathrm{soil,rock}} \tag{39}$$

The acid-neutralizing capacity of a soil is given by the bases, carbonates, silicates, and oxides of the soil system. In a similar way as for an aquatic system, alkalinity and acidity can be defined. If the composition of the soil is not known but its elemental analysis is given in oxide components, the following kind of accounting is equivalent to that given by equation (20) for natural waters:

$$\begin{aligned} \mathrm{ANC} = {} & [\mathrm{Alk}] = 6(\mathrm{Al_2O_3}) + 2(\mathrm{CaO}) + 2(\mathrm{MgO}) + 2(\mathrm{K_2O}) \\ & + 2(\mathrm{Na_2O}) + 4(\mathrm{MnO_2}) + 2(\mathrm{FeO}) + (\text{“}\mathrm{NH_4OH}\text{”}) \\ & -2(\mathrm{SO_3}) - (\mathrm{HCl}) - 2(\mathrm{N_2O_5}) \end{aligned} \tag{40}$$

Equation (40) is expressed in oxide equivalents of each element in soil. Sulfates, nitrates, and chlorides incorporated or adsorbed are subtracted from ANC. On the basis of acid balance alone, one cannot distinguish whether uptake by the soil of H^+ and the release of cations is caused by weathering or by ion exchange. Similarly, the loss of SO_4^{2-} from the water accompanied by OH^- or [Alk] release may be caused by sulfate reduction (see Table 10.8) or by ion exchange of sulfate.

10.4.2 Chemical Weathering

Figure 10.7 shows some processes that affect the acid-neutralizing capacity of soils. Ion exchange occurs at the surface of clays and organic humus in various soil horizons. The net effect of ion exchange processes is identical to chemical weathering (and alkalinity); that is, hydrogen ions are consumed and basic cations (Ca^{2+}, Mg^{2+}, Na^+, K^+) are released. However, the kinetics of ion exchange are rapid relative to those of chemical weathering (taking minutes compared to hours or even days). In addition, the pool of exchangeable bases is small compared to the total ANC of the soil [equation (40)]. Thus there exists two pools of bases in soils—a small pool of exchangeable bases with relatively rapid kinetics and a large pool of mineral bases with the slow kinetics of chemical weathering. If chemical weathering did not replace exchangeable bases in acid soils of temperate regions receiving acidic deposition, the base exchange capacity would be completely diminished over a period of decades to centuries.[27]

In the long run, chemical weathering is the rate-limiting step in the supply of basic cations for export from watersheds. Surface waters usually are not at chemical equilibrium with regard to common minerals in soils or sediments with the possible exception of amorphous aluminum hydroxide. The chemistry of natural waters is predominantly kinetically controlled.

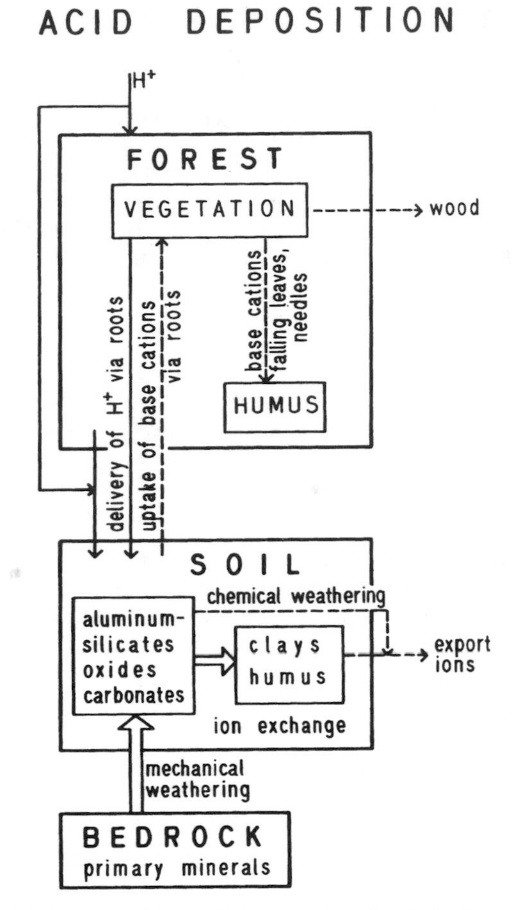

Figure 10.7 Processes affecting the acid-neutralizing capacity of soils (including the exchangeable bases, cation exchange, and mineral bases). H^+ ions from acid precipitation and from release by the roots react by weathering carbonates, aluminum silicates, and oxides and by surface complexation and ion exchange on clays and humus. Mechanical weathering resupplies weatherable minerals. Lines drawn out indicate flux of protons; dashed lines show flux of base cations (alkalinity). The trees (plants) act like a base pump. (From Schnoor and Stumm.[29])

The genesis of podzolic acid soils stems from a chromatographic effect whereby bases are depleted from the upper soil profile and iron and aluminum are mobilized. Soil profiles become stratified into horizons, for example, O, A, B, and C. In podzolic soils the organic "O" horizon is typically 1–10 cm deep and contains the humus from leaf and litter decomposition. The "A" horizon is the uppermost "inorganic" soil horizon. In podzolic soils, it is depleted of basic cations and appears as a gray-to-white leached layer ("E" horizon). The "B" horizon is classified as "spodic." It appears somewhat reddish and contains much of the iron and aluminum that were mobilized from above due to acidic conditions. As soil water percolates, it is neutralized by exchange reactions and chemical weathering, and the iron and aluminum precipitate or adsorb from solution.

Even in acid podzolic soils, there exist significant amounts of "fresh" minerals or weatherable bases available for dissolution. One reason is mechanical weathering, a physical process. Solifluction (the process of bedrock crack and surfacing on steep slopes) is an important mechanism in mountainous terrain. Freezing, thawing, and cracking of rocks—a form of mechanical weathering—also avails fresh minerals. Especially in pit and mound topographies, tree throw provides a significant amount of fresh minerals from the B and C soil horizons. Burrowing animals can do likewise. Finally, forest canopies provide a "base pump" whereby basic cations taken from lower soil horizons are deposited in a more weatherable form on the surface of the soil during litter fall.

There are several factors that affect the rate of chemical weathering in soil solution. These include:

- Hydrogen ion activity of the solution.
- Ligand activities in solution.
- Dissolved CO_2 activity in solution.
- Temperature of the soil solution.
- Mineralogy of the soil.
- Flowrate through the soil.
- Grain size of the soil particles.

For a given silicate mineral, the hydrogen ion activity contributes to the formation of surface-activated complexes, which determine the rate of mineral dissolution at $pH < 6$. Also, since chemical weathering is a surface reaction-controlled phenomenon, organic and inorganic ligands (e.g., oxalate, formate, succinate, humic and fulvic acids, fluoride, and sulfate) may form other surface-activated complexes that enhance dissolution.[27–31] (Some organics are known to form surface complexes that "block" dissolution.) Dissolved carbon dioxide accelerates chemical weathering presumably due to its effect on soil pH and the aggression of H_2CO_3. Increases in temperature generally increase the rate of chemical weathering. Because mineral dissolution is kinetically controlled, increases in the rate of flow and decreases in the particle size can increase the rate of chemical weathering.[27]

Mineralogy is of prime importance. The Goldich dissolution series, which is roughly the reverse of the Bowen crystallization series, indicates that chemical weathering rates should decrease as we go from carbonates → olivine → pyroxenes → Ca, Na plagioclase → amphiboles → K-feldspars → muscovite → quartz. If the pH of the soil solution is low enough ($pH < 4.5$), aluminum oxides provide some measure of neutralization to the aqueous phase along with an input of monomeric inorganic aluminum. The key parameter that affects the activated complex and determines the rate of dissolution of the silicates is the Si/O ratio. The less the ratio of Si/O is, the greater its chemical weathering rate is. Anorthite and forsterite have Si/O of 1:4, while quartz has Si/O of 1:2, the slowest to dissolve in acid.

Chemical weathering requires the mass transport of solutes to a mineral surface to form a coordination complex. In a series of steps involving geometric and chemical principles at surface defects (kinks, edges, and screw dislocations), a detachment occurs and transport of the reactants into bulk solution follows. In surface-controlled dissolution reactions, detachment of an activated surface-coordinated complex is the rate-determining step. The general rate law may be expressed as[32]

$$R = k X_a P_j S_{\text{sites}} \tag{41}$$

where R is the proton or ligand-promoted dissolution rate (mol m^{-2} s^{-1}), k is the rate constant (s^{-1}), X_a denotes the mole fraction of dissolution active sites (dimensionless), P_j represents the probability of finding a specific site in the coordinative arrangement of the activated precursor complex, and S_{sites} is the total surface concentration of sites (mol m^{-2}). The rate expression in equation (41) is essentially a first-order reaction in the concentration of activated surface complex, C_j (mol m^{-2}):

$$C_j = X_a P_j S_{\text{sites}} \tag{42}$$

Formulation of equation (42) is consistent with transition state theory (Chapter 3), where the rate of the reaction far from equilibrium depends solely on the activity of the activated transition state complex.

The surfaces of oxides and aluminosilicate minerals in the presence of water are characterized by amphoteric surface hydroxyl groups. The surface OH^- group has a complex-forming O-donor atom that coordinates with H^+ and metal ions.[33,34] The underlying aluminum or metal ion in the surface tetrahedra of the mineral and other cations are subject to coordination with OH^- groups, which, in turn, weakens the bond to their structural oxygen atoms. Detachment of an activated complex removes the coordination complex and renews the surface for further hydrolysis, protonation, and dissolution. Also, the underlying central metal ions can exchange their structural OH^- ions for other ligands (e.g., organic anions) to form a strong surface coordination complex that can facilitate dissolution.

A very important result of laboratory studies has been the fractional order dependence of mineral dissolution on bulk phase hydrogen ion activity. If the dissolution reaction is controlled by hydrogen ion diffusion through a thin liquid film or residue layer, one would expect a first-order dependence on $\{H^+\}$. If the dissolution reaction is controlled by some other factor such as surface area alone, then the dependence on hydrogen ion activity should be zero-order. Rather, the dependence has been fractional order in a wide variety of studies, indicating a surface reaction-controlled dissolution.[35–40] For a description of the kinetic equations, see Chapter 3 for fractional order reactions. The rate of the chemical weathering can be written

$$R = k[H^+]^m = k'[H^+]^n_{\text{ads}} \tag{43}$$

where R is the rate of the dissolution reaction, k is the rate constant for H^+ ion attack, m is the fractional order dependence on hydrogen ion concentration in bulk solution, $[H^+]_{ads}$ is the proton concentration sorbed to the surface of the mineral, and n is the valence of the central metal ion under attack.

Figure 10.8 summarizes laboratory data on the rate of weathering of some minerals.[41] Dissolution rates of different minerals vary by orders of magnitude; they are strongly pH dependent. Obviously, carbonate minerals dissolve much faster than oxides or aluminum silicates. The reactivity of the surface, that is, its tendency to dissolve, depends on the type of surface species present. Surface protonation tends to increase the dissolution rate, because it leads to highly polarized interatomic bonds in the immediate proximity of the surface central ions and thus facilitates the detachment of a cationic surface group into the solution. On the other hand, a surface coordinated metal ion (e.g., Cu^{2+} or Al^{3+}) may block a surface group and thus retard dissolution. An outer-sphere surface complex has little effect on the dissolu-

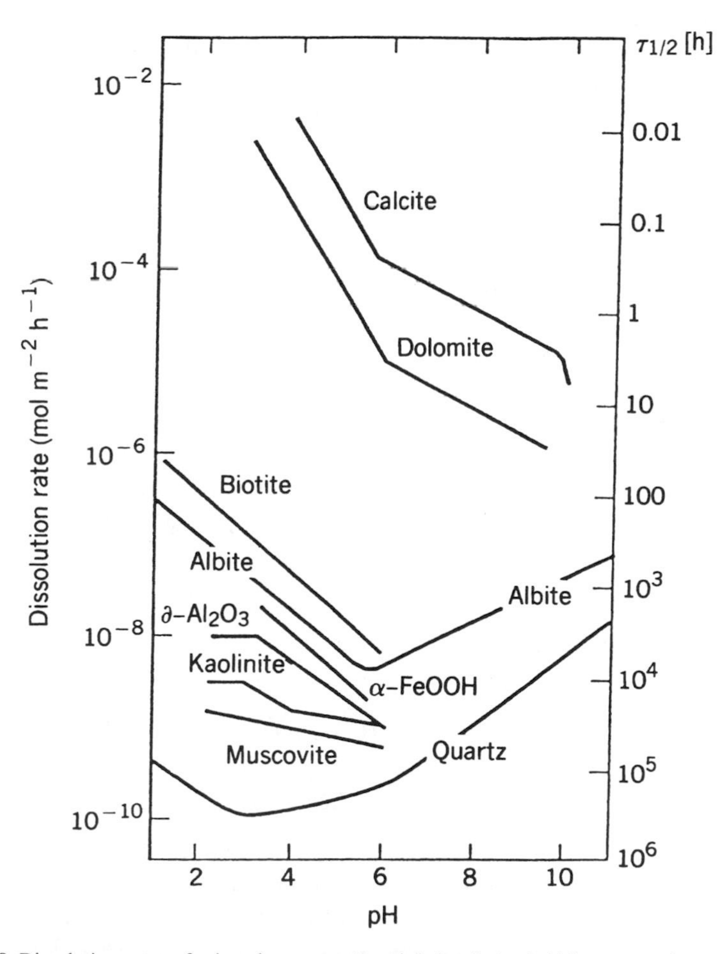

Figure 10.8 Dissolution rates of minerals versus pH and their relative half-lives assuming 10 sites or central metal atoms per nm^2 surface area. (From Sigg and Stumm.[41])

tion rate. Changes in the oxidation state of surface central ions have a pronounced effect on the dissolution rate. Inner-sphere complexes form with a ligand attack on a central metal ion such as that shown for oxalate,

$$\begin{matrix} \diagdown & OH_2 \\ Me & \\ \diagup & OH \end{matrix} + C_2O_4^{2-} + H+ \rightleftharpoons \begin{matrix} \diagdown & O-C=O^- \\ Me & | \\ \diagup & O-C=O \end{matrix} + 2\,H_2O \tag{44}$$

or other dicarboxylates, dihydroxides, or hydroxy-carboxylic acids.

$$R\begin{matrix} COOH \\ COOH \end{matrix} \quad R\begin{matrix} OH \\ OH \end{matrix} \quad \text{or} \quad R\begin{matrix} COOH \\ OH \end{matrix}$$

It facilitates the detachment of a central metal ion and enhances the dissolution. This is readily understandable, because the ligands shift electron density toward the central metal ion at the surface and bring negative charge into the coordination sphere of the Lewis acid center. It enhances simultaneously the surface protonation and can labilize the critical Me–oxygen lattice bonds, thus enabling the detachment of the central metal ion into solution.

Oxalate is sometimes used as a surrogate for natural organic matter because it is known to exist in soils from plant exudates at levels of 1–100 μM. It forms strong complexes at mineral surfaces and can accelerate dissolution at high concentrations near the root–mineral interface, the rhizosphere.

The ideas developed here are largely based on the concept of the coordination at the (hydr)oxide interface; the ideas apply equally well to aluminum silicates.[31] Somewhat modified concepts for the surface chemistry of carbonate, phosphate, sulfide, and disulfide minerals have to be developed.

In the dissolution reaction of an oxide mineral, the coordinative environment of the metal changes; for example, in dissolving an aluminum oxide layer, the Al^{3+} in the crystalline lattice exchanges its O_2^- ligand for H_2O or another ligand L. The most important reactants participating in the dissolution of a solid mineral are H_2O, H^+, OH^-, ligands (surface complex building), and reductants and oxidants (in the case of reducible or oxidizable minerals).

Thus the reaction occurs schematically in two sequences:

$$\text{Surface sites} + \text{Reactants } (H^+, OH^-, \text{or ligands}) \xrightarrow{\text{fast}} \text{Surface species} \tag{45}$$

$$\text{Surface species} \xrightarrow[\text{detachment of Me}]{\text{slow}} Me_{(aq)} \tag{46}$$

where Me stands for the metal ion. Although each sequence may consist of a series of smaller reaction steps, the rate law of surface-controlled dissolution is based on the idea that the attachment of reactants to the surfaces sites is fast and that the subsequent detachment of the metal species from the surface of the crystalline lattice into the solution is slow and thus rate limiting. An example of Na-feldspar dissolution by H^+ ions is given in Figure 10.9.

In the first sequence the dissolution reaction is initiated by the surface coordination with H^+, OH^-, and ligands, which polarize, weaken, and tend to break the metal–oxygen bonds in the lattice of the surface. Since reaction (46) is rate limiting and by using a steady-state approach, the rate law on the dissolution reaction will show a dependence on the concentration (activity) of the particular surface species, C_j (mol m^{-2}):

$$\text{Dissolution rate} \propto (\text{Surface species}) \tag{47}$$

We reach the same conclusion if we treat the reaction sequence according to transition state theory. The particular surface species that has formed from the interaction of H^+, OH^-, or ligands with surface sites is the precursor of the activated complex.

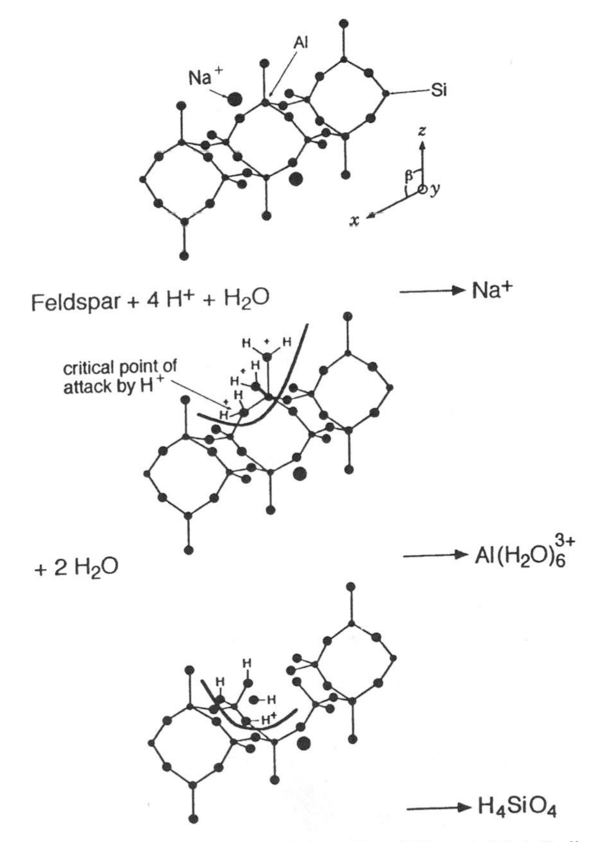

Figure 10.9 Hydrogen ion attack and initial dissolution of Na-feldspar (albite). Sodium ions, monomeric aluminum ions, and dissolved silica are produced. Atoms at the vertices of the ring structures are alternately Si and Al atoms. All other atoms are oxygen. Critical point of H^+ attack is on the oxygen atom in the lattice next to an Al atom.

$$\text{Dissolution rate} \propto (\text{Precursor of the activated complex}) \tag{48}$$

The overall rate of dissolution is given by mixed kinetics:

$$R = k_{\mathrm{H}} (C^{s}_{\mathrm{H}})^{j} + k_{\mathrm{OH}} (C^{s}_{\mathrm{OH}})^{i} + k_{\mathrm{L}} (C^{s}_{\mathrm{L}}) + k_{\mathrm{H_2O}} \tag{49}$$

the sum of the individual reaction rates, assuming that the dissolution occurs in parallel at different metal centers. The last term in equation (49) is due to the effect of hydration and reflects the pH-independent portion of the dissolution rate in Figure 10.8. In equation (49), R is the rate of the reaction, k_{H} is the rate constant of dissolution due to hydrogen ions coordinated at the surface C^{s}_{H}, k_{OH} is the rate constant of dissolution due to hydroxide ions coordinated at the surface C^{s}_{OH}, k_{L} is the rate constant of dissolution due to ligands coordinated at the surface C^{s}_{L}, and $k_{\mathrm{H_2O}}$ is the pH-independent portion of the dissolution rate.

Scanning electron microscopy and energy dispersive x-ray spectroscopy (SEM/EDX) have allowed observations of mineral surfaces during dissolution. While secondary mineral formation is sometimes observed, a coating of the primary mineral surface does not occur. Dissolution reactions occur preferentially along crystal defects, screw dislocations, and etch pits. These results have been reported for a wide range of minerals including microcline, feldspars, augite, diopside, hypersthene, and hornblende.[28,42–45]

Figure 10.10 shows photomicrographs of weathered sediments from Lake Piccolo Naret in the Maggia Valley of Ticino, Switzerland, on the southern face of the Alps at 2500-m elevation. The area receives moderate acidic deposition (pH 4.6–4.9). Because of its crystalline rock geology (granitic gneiss bedrock) and flashy hydrograph, the lake is slightly acidic (pH 5.5–6.5). Plagioclase feldspars and biotite are the primary weathering minerals based on scanning electron microscopy and energy dispersive x-ray spectrometry. The chemical equation for biotite weathering in the presence of strong acid is

$$\begin{aligned} &\mathrm{KMg}_{x}\mathrm{Fe}_{(3-x)}\mathrm{AlSi_3O_{10}(OH)_{2(s)}} + (1+2x)\mathrm{H^+} + (9-2x)\mathrm{H_2O} \\ &\rightarrow \mathrm{K^+} + x\mathrm{Mg^{2+}} + 3\mathrm{H_4SiO_4} + \mathrm{Al(OH)_3} + (3-x)\mathrm{Fe(OH)_2} \end{aligned} \tag{50}$$

where x is the mole fraction of magnesium. On reaction with dissolved oxygen in water, ferrous hydroxide becomes oxidized to ferric hydroxide:

$$(3-x)\mathrm{Fe(OH)_{2(aq)}} + [(3-x)/2]\mathrm{O_{2(aq)}} \rightleftharpoons (3-x)\mathrm{Fe(OH)_{3(s)}} \tag{51}$$

Dissolution of plagioclase feldspar in the presence of strong acids may be written

$$\begin{aligned} &\mathrm{Na}_{(1-x)}\mathrm{Ca}_{x}\mathrm{Al}_{(1+x)}\mathrm{Si}_{(3-x)}\mathrm{O_{(s)}} + (1+x)\mathrm{H^+} + (7-x)\mathrm{H_2O} \\ &\rightarrow (1-x)\mathrm{Na^+} + x\mathrm{Ca^{2+}} + (3-x)\mathrm{H_4SiO_4} + (1+x)\mathrm{Al(OH)_3} \end{aligned} \tag{52}$$

where x is the mole fraction of calcium.

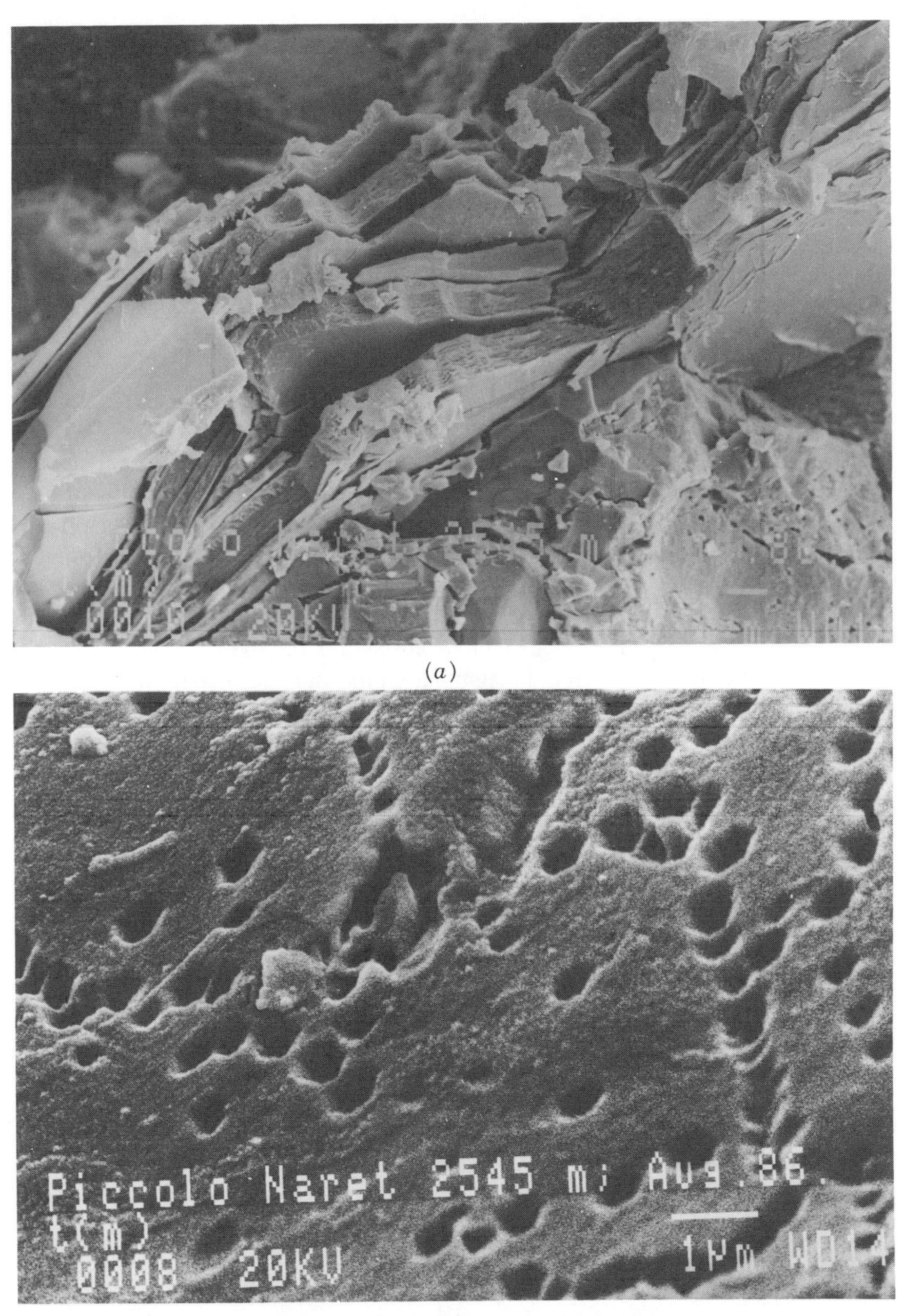

Figure 10.10 SEM/EDX images of (a) chemical weathering of biotite platelets and (b) etch pits on a plagioclase mineral grain. All grains were taken from surface sediments of Lake Piccolo Naret, Switzerland. Photographs are courtesy of Professor Rudolf Giovanoli, Laboratory of Electron Microscopy, University of Bern.

Dissolved silica H_4SiO_4 is one of the best "tracers" to estimate chemical weathering because it is roughly conservative in upland streams and watersheds. Field and laboratory studies were compiled for aluminosilicate minerals with reference to their weathering rates and flowrates or discharge measurements.[30] Table 10.9 shows the results for eight plots, five of which were soils from Bear Brook Watershed (BBW), Maine (a stream with near zero ANC and 70 meq m^{-2} yr^{-1} acid deposition). Site 1 refers to silica export measured at the discharge from East Bear Brook. Site 2 was a small (1.4×1.4-m^2) weathering plot experiment at BBW with HCl applications of pH 2, 2.5, and 3.0. Sites 3–5 were all laboratory experiments at pH 3–4 on Bear Brook size-fractionated soils. Site 6 was Coweeta Watershed 27 in the Southern Blue Ridge mountains of North Carolina.[46] Coweeta soils were composed of three primary weatherable minerals: plagioclase, garnet, and biotite. Site 7 was Filson Creek in northern Minnesota, a large watershed of 25 km^2 with waters of pH 6 and plagioclase and olivine as the predominant weatherable minerals in the till.[47] Site 8 was Lake Cristallina in the Swiss Alps with plagioclase, biotite, and epidote as predominant weathering minerals.[28] These examples ranged over 10 orders of magnitude in dissolved silica export and flowrate.

When silica export is correlated against flowrate on a log–log scale, most points lie near the line with a slope of 1.0, corresponding to a concentration of approxiamtely 60 μM dissolved silica. A strong correlation exists in part because flowrate is a multiplier on both the ordinate and the abscissa. Even though the sites and experiments differ over 10 orders of magnitude in the amount of water flowing, they are all within about one order of magnitude in dissolved silica concentration (25–300 μM), except for the fluidized-bed experiment by Mast and Drever.[48] Thus one conclusion is that experiments can be scaled to field conditions by ensuring that the dissolved silica concentration is of the same order of magnitude as the prototype ecosystem. Weathering export rates are correlated to flowrate in log–log plots in

Table 10.9 Laboratory and Field Studies of Silicon Export and Release Rates (Si RR) and Flowrates

Site	Si Concentration, μM	Si Export, mol s^{-1}	Flowrate, L s^{-1}	Si RR, mol m^{-2} s^{-1}	Flowrate/ Mass Ratio, L d^{-1} g^{-1}
1. Bear Brook, ME	70	2.9×10^{-4}	4.2	9×10^{-15}	2×10^{-6}
2. Small plot, ME	300	2.9×10^{-8}	9.7×10^{-5}	3×10^{-13}	1×10^{-5}
3. Soil column, ME	30	2.8×10^{-10}	9.2×10^{-6}	6×10^{-12}	3×10^{-3}
4. pH-stat, ME	100	1.2×10^{-11}	1.2×10^{-7}	6×10^{-11}	1×10^{-3}
5. Fluidized bed, ME	3	5.4×10^{-12}	1.7×10^{-6}	3×10^{-12}	4×10^{-2}
6. Coweeta, NC	42	1.4×10^{-3}	33	7×10^{-13}	2×10^{-5}
7. Filson Creek, MN	167	4.2×10^{-2}	250	5×10^{-15}	1×10^{-6}
8. Cristallina, Switzerland	25	1.6×10^{-4}	6.5	6×10^{-14}	2×10^{-5}

Source: Schnoor.[30]

many natural systems. This implies that a surface coordination reaction could not be controlling the rate of weathering in natural systems because there should not be an increase in ion export rates with flowrate. Indeed, in laboratory experiments 4 and 5 of Table 10.9 there was no change in weathering rate with changing flowrate or stirring rate, but these systems were saturated with water (the flowrate/mass ratio was large).

If the ratio of flowrate to mass of wetted soil ($L\ d^{-1}\ g^{-1}$) is plotted on the abscissa, and silica release rate [silica export per square meter of wetted mineral ($mol\ m^{-2}\ s^{-1}$)] is plotted on the ordinate, an asymptotic relationship for silica release rate is determined (Figure 10.11). Silica release rates reach a maximum of 10^{-11} to 10^{-12} at a flowrate/mass ratio greater than $10^{-3.4}$ $L\ d^{-1}\ g^{-1}$. This corresponds to the flow regime and weathering rates most frequently reported in laboratory studies of weathering of pure minerals. The flowrate/mass ratio of $10^{-3.4}$ is a flushing rate of approximately 1.0 volume of water per volume of soil per day, and it represents a limit above which surface reactions control mineral dissolution rates and below which hydrology controls (limits) the rate of weathering. Hydrologic control may exist because of unsaturated macropore flow through soils, which results in insufficient flow to wet all the available minerals and to carry away the dissolved solutes. Uncertainty in weathering rates measured in the laboratory is one order of magnitude, but limitations of hydrology can result in two orders of magnitude lower estimates of weathering rates.

The silica release rate and flowrate/mass ratio of Figure 10.11 depend on estimations of the wetted surface area of reacting minerals and the mass of wetted soil. As discussed by Paces,[49] there is much uncertainty in estimating wetted surface areas of reacting minerals. In Table 10.9, it was assumed that 40% of the surface area of

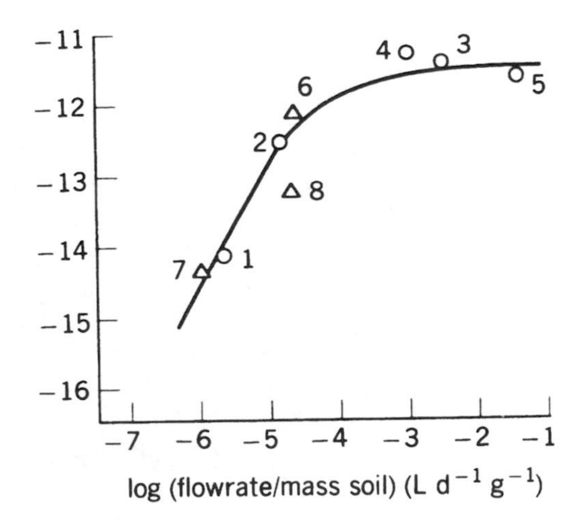

Figure 10.11 Dissolved silica release rate (weathering rate) versus flowrate/mass ratio. Circles represent BBW soils and triangles are other field sites. Numbers refer to site numbers listed in Table 10.9.

soil was weatherable minerals; the soil surface area was 0.5 $m^2\ g^{-1}$ (except at Coweeta, where Velbel's[46] procedure was used), the bulk density of the soil was 1.5 kg L^{-1}, and the equivalent saturated water depth was 0.5 m (except at Coweeta and Cristallina, where it was taken to be 0.1 m). All of these parameters were measured at Bear Brook Watershed and applied to the other sites.

10.4.3 Ion Exchange and Aluminum Dissolution

Ion exchange in soils serves to neutralize acid deposition in many cases. Through geologic time, soils are formed by chemical weathering, vegetational uptake, and biomineralization of organic matter. Eventually, a large pool of exchangeable base cations (Ca^{2+}, Mg^{2+}, K^+, and Na^+) are accumulated in the upper soil horizons. These cations can be exchanged for H^+ ions or other cations, as in the following example.

$$2H^+ + CaX_2 \rightleftharpoons 2HX + Ca^{2+} \tag{53}$$

CaX_2 represents calcium ions on soil exchange sites. Calcium, magnesium, sodium, and potassium ions are termed "base cations" because their oxides (CaO, MgO, Na_2O, and K_2O) are capable of neutralizing protons much like the exchange equation (53).

Forested soils in areas with crystalline bedrock are especially sensitive to acid deposition. These soils are already somewhat acidic (pH $\leq$ 5 in 1:1 vol with H_2O) because of the slow production of base cations by rock-forming aluminosilicate minerals. Plants accumulate base cations in vegetation but gradually acidify the soil. Organic acids are released as exudates and decomposition products from vegetation, which then complex base cations and deplete them from the upper soil horizons.

Acid deposition on acidic soils can result in acidic surface waters. Although forest soils in crystalline rock areas tend to be somewhat acidic, under natural conditions, they have a small positive ANC in soil solution. When acid deposition decreases the ANC, it may cause the soil solution to become negative, resulting in release of H^+ ions and Al^{3+}. Soil water tends to be supersaturated with carbon dioxide partial pressures of 0.01–0.06. When it contains positive ANC, the pH of the water increases significantly upon degassing to $P_{CO_2} = 10^{-3.5}$ atm. As soil water becomes acidic due to acid deposition and the "salt effect" (in which ion exchange processes increase the H^+-ion activity in solution), the ANC of soil water becomes negative, and the pH does not increase upon degassing.[50] Then lakes and streams become acidic also.

Cation exchange capacity (CEC) of soils includes all the cations on the exchange complex of the soil.

$$\text{CEC} = AlX_3 + CaX_2 + MgX_2 + NaX + KX\ (+HX) \tag{54}$$

where AlX_3, CaX_2, MgX_2, NaX, KX, and HX denote cation concentrations sorbed on the soil in meq/100 g (a traditional soil science measurement unit). HX is usually neglected in definitions of the cation exchange capacity by soil scientists. Base exchange capacity (BEC) is the CEC minus the acidic cations: H^+, a strong acid, and Al^{3+}, a Lewis acid.

$$\mathrm{BEC} = \mathrm{CaX_2} + \mathrm{MgX_2} + \mathrm{NaX} + \mathrm{KX} \tag{55}$$

Dividing BEC by CEC yields the percent base saturation, BS:

$$\mathrm{BS} = \frac{\mathrm{BEC}}{\mathrm{CEC}}$$

Exchangeable acidity ($HX + AlX_3$) does not play a significant role in soil solution chemistry until the base saturation becomes small (< 0.20). Reuss[51] showed that for a simple three-component system (H^+, Ca^{2+}, Al^{3+}), aluminum ions are released into solution only when the base saturation (in this case, the exchangeable-Ca fraction) was less than 0.1. H^+ ions in soil solution do not increase very much even when $E_{Ca} < 0.1$; mostly Al^{3+} ions are released as base saturation decreases (Figure 10.12).

As acid deposition is added to soils, the pH change of soil solution is relatively

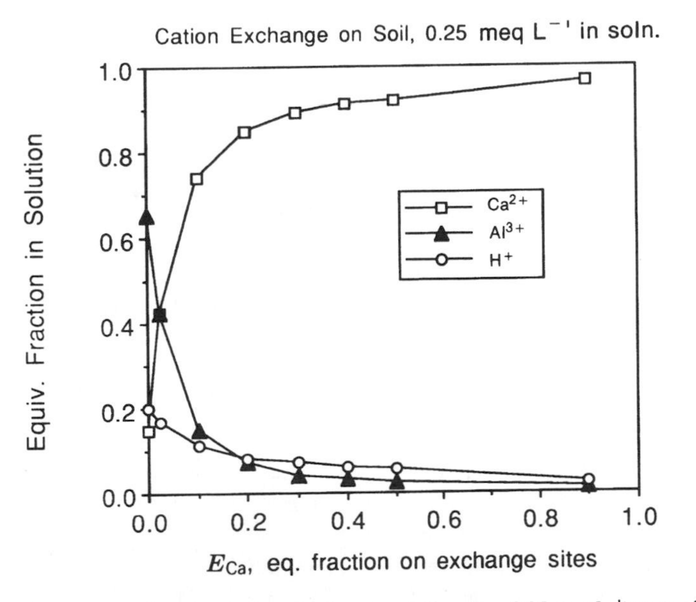

Figure 10.12 Ion concentrations (equivalent fractions in solution, $C_T = 0.25$ meq L^{-1}) versus base saturation (equivalent fraction of Ca on soil exchange sites). $K_{s0} = 10^{8.5}$ for $Al(OH)_3$ solubility and $K_{gt} = 10 = (E_{Al}^2 \{Ca^{2+}\}^3 / E_{Ca}^3 \{Al^{3+}\}^2)$. Modified from Reuss.[51]

small. Initially, it is buffered by exchange of base cations, especially Ca^{2+}. Eventually, if chemical weathering does not supply sufficient base cations to resupply the exchange complex, then base saturation will become depleted and aluminum ions and H^+ ions will be released. Because dissolved aluminum is so toxic to fish and vegetation, acid deposition is a serious concern in the special circumstances where acid deposition falls on acid-sensitive soils. It is also a problem in cases where water runs quickly through sensitive watersheds with little time for contact with the soil (hydrologically flashy catchments) and for seepage lakes in which a large fraction of the total volume comes directly from acid precipitation with little buffering capacity in sediments.

Cation exchange is a complex process that lumps a variety of mechanisms:

- Exchange of ions with the structural or permanent charge of interlayer clays.
- Exchange of ions with nonstructural sites in clays and oxide minerals.
- Exchange of ions with organic matter and its coordinated metal ions.

Four exchange reactions can be used to summarize ion exchange in soils.[52] All other exchange reactions between pairs of cations can be written as algebraic combinations of these four equations. Exchange with hydrogen ions in soil solution is neglected—it is generally small, but the assumption can introduce errors under some circumstances as pointed out by McBride.[53]

$$2Al^{3+} + 3\ CaX_2 \rightleftharpoons 3\ Ca^{2+} + 2\ AlX_3 \qquad (56)$$

$$Ca^{2+} + 2\ NaX \rightleftharpoons 2\ Na^+ + CaX_2 \qquad (57)$$

$$Mg^{2+} + 2\ NaX \rightleftharpoons 2\ Na^+ + MgX_2 \qquad (58)$$

$$K^+ + NaX \rightleftharpoons Na^+ + KX \qquad (59)$$

Soil scientists have long recognized that aluminum solubility in soil water is ultimately controlled not by equation (56) but rather by a solubility relationship between gibbsite or amorphous $Al(OH)_3$ and pH.[53–55]

$$Al(OH)_3 + 3\ H^+ \rightleftharpoons Al^{3+} + 3\ H_2O \qquad (60)$$

where $K_{s0} = 10^{8.04}$ for crystalline gibbsite and as high as $10^{9.66}$ for amorphous $Al(OH)_3$.[51] As soil pH becomes less than 5.0, aluminum ions are liberated from exchange sites, but solubility is controlled by equation (60).

Filtered aluminum concentrations in Lake Cristallina, Switzerland, are shown in Figure 10.13. The lake receives acid deposition and displays a wide range of pH values seasonally. It follows amorphous aluminum hydroxide control on aluminum solubility ($\log K_{s0} = 8.9$). Generally, aluminum concentrations in lakes and streams are controlled by gibbsite or amorphous $Al(OH)_3$ solubility.[41,56–58] Exceptions to the

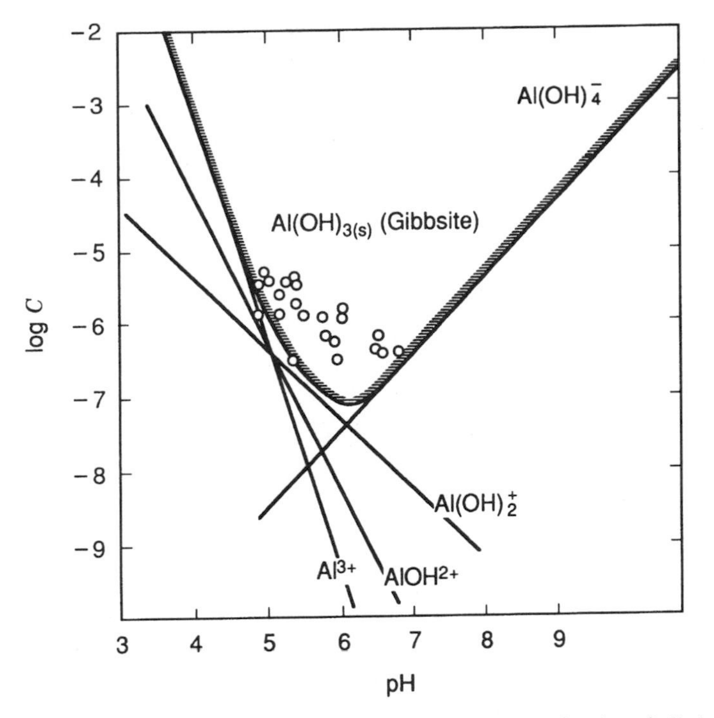

Figure 10.13 Al concentrations from Lake Cristallina, Switzerland, as a function of pH. A pC–pH diagram for gibbsite solubility log K_{s0} = 8.9 is superimposed, suggesting that mineral phases have some control on aluminum solubility in natural waters. From Sigg and Stumm.[41]

rule are likely when soil is frozen and snowmelt water recharges streams with very little soil contact. In most cases, there is an excess of amorphous $Al(OH)_3$ available in sediments for dissolution and reprecipitation.[28] An average log K_{s0} of 8.5 is often applicable to many soil waters and streams.

$$\frac{\{Al^{3+}\}}{\{H^+\}^3} = 10^{8.5} \tag{61}$$

Ion exchange includes the fast pool of cations available for neutralizing acid deposition in upper soil horizons, but chemical weathering of minerals sets up the exchanger. As acid deposition increases and the base exchange complex is depleted, chemical weathering increases but not sufficiently to neutralize all of the acidity. In such cases, a slow drawdown of the exchanger may occur and soils and surface waters will become progressively more acidic over a time scale of decades.

10.4.4 Biomass Synthesis

Assimilation by vegetation of an excess of cations can have acidifying influences in the watershed, which may rival acid loading from the atmosphere. The synthesis of a terrestrial biomass, for example, on the forest and forest floor, could be written with the following approximate stoichiometry:

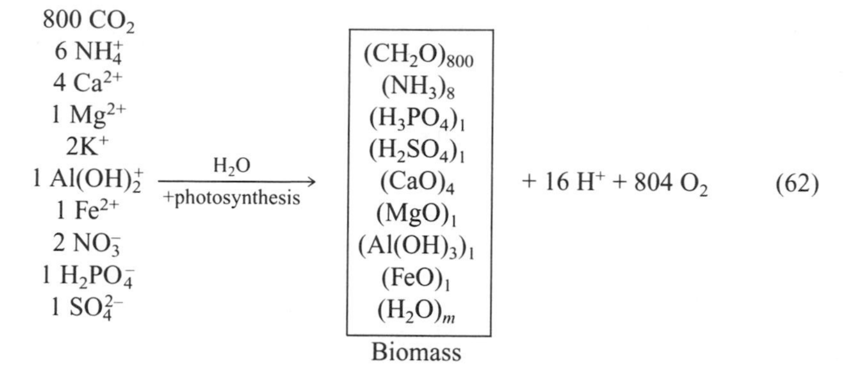

From the point of view of acid-neutralizing capacity, the biomass may also be represented in terms of corresponding cations, anions, and bases (Table 10.8). It is evident from this equation that the aggrading production of terrestrial vegetation is accompanied by an acidification of the surrounding waters; correspondingly, the ANC of the biomass increases. The ashes of trees are alkaline. Every temporal or spatial decoupling of the production and mineralization of biomass leads to a modification of the H^+ balance in the environment. This is the result of intensive agricultural and forestry practices and seasonal fluctuations. The release of H^+ by intensive growing vegetation (especially by forests used for wood exploitation) leads to leaching of base cations from the soil.

The interaction between acidification by an aggrading forest and the leaching (weathering) of the soil is schematically depicted in Figure 10.7. If the weathering rate equals or exceeds the rate of H^+ release by the biota, such as would be the case in a calcareous soil, the soil will maintain a buffer in base cations and residual alkalinity. On the other hand, in noncalcareous "acid" soils, the rate of H^+ release by the biomass may exceed the rate of H^+ consumption by weathering and cause a progressive acidification of the soil. In some instances, the acidic atmospheric deposition may be sufficient to disturb an existing H^+ balance between aggrading vegetation and weathering reactions.

Humus and peat can likewise become very acid and deliver some humic or fulvic acids to the water. Note that the release of humic or fulvic material (H-Org, Org^-) to the water in itself is not the cause of resulting acidity, but rather the aggrading humus and net production of base-neutralizing capacity. The aerobic decomposition of organic biomass creates organic acids, which help to leach aluminum, iron, and base cations, but overall they do not contribute to acidification when fully oxidized. For example, decomposition and oxidation of a simple sugar is shown in equation (63):

$$C_6H_{12}0_6 \xrightarrow[\text{in soil}]{\text{bacteria}} 3\ CH_3COOH \qquad (63a)$$

$$3\ CH_3COOH + 6\ O_2 \xrightarrow{\text{bacteria}} 6\ CO_2 + 6\ H_2O \qquad (63b)$$

Carbon dioxide degasses to the atmosphere, and ANC is conservative with respect to CO_2 exchange.

10.4.5 Role of Sulfate and Nitrate

Table 10.8 lists some changes in the proton balance resulting from redox processes. Alkalinity or acidity changes can be computed as before: any addition of NO_3^- or SO_4^{2-} to the water as in nitrification or sulfur oxidation increases acidity, while NO_3^- reduction (denitrification) and SO_4^{2-} reduction cause an increase in alkalinity. As shown by Schindler[59] in experimental lakes, the nitrate removal caused by denitrification is linearly related to an increase in alkalinity. Incipient decreases of pH resulting from the addition of sulfuric acid or nitric acid to a lake is reversed by subsequent denitrification or SO_4^{2-} reduction.

$$SO_4^{2-} \rightarrow H_2S \rightarrow FeS_2 \tag{64a}$$

$$NO_3^- \rightarrow N_{2(g)} \tag{64b}$$

Sediments in a water–sediment system are usually highly reducing environments. Electrons, delivered to the sediments by the "reducing" settling biological debris, and H^+ are consumed; thus in the sediments (including pore water) alkalinity increases. Results from an acidified lake are shown in Figure 10.14. Sulfate reduction in the sediment generates considerable bicarbonate alkalinity, some of which enters the water column by vertical eddy diffusion. A small amount of nitrate is also reduced in the sediments of this acidified lake system.

In eutrophic lakes, large amounts of nitrate may be taken up by algae (an alkalizing process for the lake) or reduced in the sediments. That is why eutrophic lakes tend to be higher pH than oligotrophic lakes, all other variables being similar. Sulfate reduction to H_2S and pyrite can be enhanced due to significant inputs from organic seston to sediments. Urban[60] has shown that sulfate flux to sediments and reduction follow first-order kinetics, but the ultimate accumulation of sulfur in sediments (as FeS, FeS_2, Org^- S) is complicated and depends on a number of biogeochemical factors.

$$R = \frac{d[SO_4^{2-}]}{dt} = -k[SO_4^{2-}] \tag{65}$$

where R is the rate of bacterially mediated sulfate reduction and k is the first order rate constant dependent on temperature and microbial activity. *Desulfivibrio* spp. are known to be effective in using sulfate as an electron acceptor in lake sediments.

Sulfate sorption is another reaction that can produce alkalinity in watersheds. Sulfate can sorb to iron and aluminum oxides in B-horizon soils, as shown in Table 10.8. This is especially important in older, unglaciated soils of the southeastern United States. Sulfate sorption provides a hysteresis effect that slows recovery of acidified watersheds as acid deposition is curtailed.

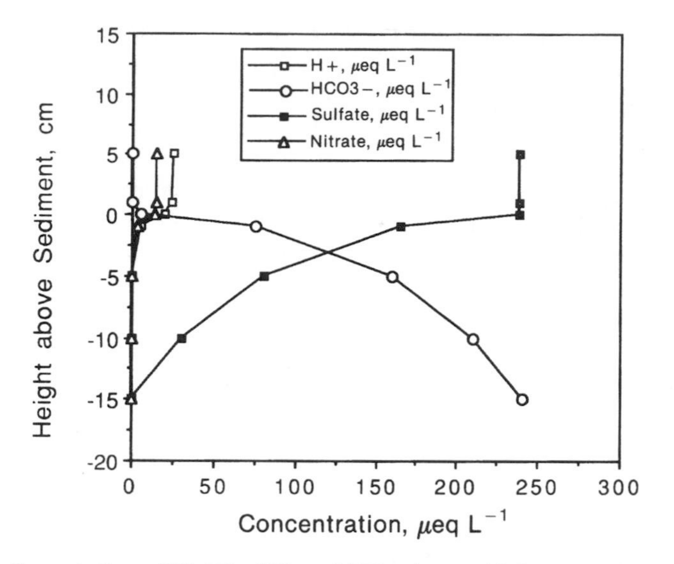

Figure 10.14 Concentrations of H^+, NO_3^-, SO_4^{2-}, and HCO_3^- above and below the sediment–water interface in an acidified lake. Overlying waters are a source of acid and reduced organic matter (electrons) to the sediment. Sediment flux upward contributes bicarbonate alkalinity to the overlying water as sulfate and nitrate are reduced.

10.5 BIOGEOCHEMICAL MODELS

Several biogeochemical models have been used to estimate the effects of acid deposition on forested watersheds, lakes, and streams.[61–67] The issue first received attention in Scandinavia where scientists believed that lakes had become increasingly acidified since the 1960s. Henriksen[61] developed an empirical charge balance model based on data from a survey of lakes in Norway. He noticed that lakes deficient in calcium ions relative to sulfate ions tended to be acidic. The rationale was that lakes with high sulfate concentrations had received considerable acid deposition in the form of H_2SO_4 and SO_2. If watersheds could not supply sufficient calcium ions (the major cation in most lakes) by ion exchange and chemical weathering, then lakes would become more acidic. Figure 10.15 shows an application of Henriksen's model to lakes of the Eastern Lake Survey. Some lakes are misclassified by the two empirical dividing lines, but the correlation is reasonably strong.

Christophersen et al.[67] took the charge balance idea one step further and developed a numerical model with a simple two-compartment hydrologic model and a three-equation chemistry model that included gibbsite equilibrium, a modified Gapon equation for ion exchange, and a charge balance. An independent mass balance was used for sulfate (assumed to be conservative).

$$\frac{[Al^{3+}]}{[H^+]^3} = K_{s0} = 10^{8.1} \tag{66}$$

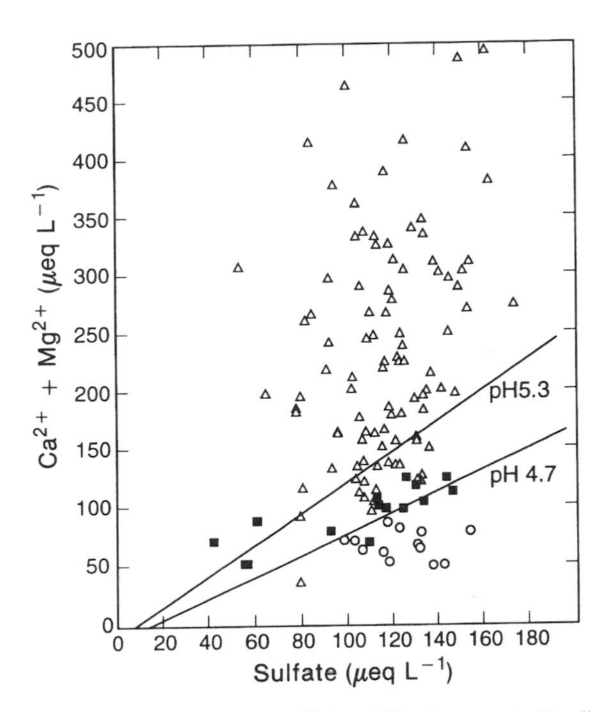

Figure 10.15 Classification of U.S. Eastern Lakes[56] for acidification status by Henriksen plot.[61] Triangle data points are lakes with pH > 5.3 (some ANC); black squares are for lakes with 4.7 ≤ pH < 5.3 (near zero ANC); and circles are for acidic lakes with pH < 4.7. Some lakes are misclassified by the Henriksen approach, based on a simple charge balance.

$$\frac{[H^+]}{[M^{2+}]^{1/2}} = K_G = 10^{-2.2} \tag{67}$$

$$[H^+] + 2[M^{2+}] + 3[Al^{3+}] = 2[SO_4^{2-}] \tag{68}$$

The model was applied to the Birkenes stream catchment in Norway, which was quite acidic (pH 4.5), so the charge balance using tri-protic aluminum ions was appropriate. Aluminum was a major cation, and fish could not survive at such high Al concentrations (0.1–1.0 mg L^{-1}).

Cosby et al.[66] improved on the charge balance concept by introduction of a detailed ion exchange complex in equilibrium with amorphous aluminum trihydroxide ($\log K_{s0} = 8.5$) after Reuss.[50,51] Fluoride, sulfate, and hydroxide complexes of aluminum were also considered, and carbon dioxide effects on pH were included. The MAGIC model was run successfully for many streams and lakes in the United States and Scandinavia when coupled with a hydrologic code. Table 10.10 shows all the equilibrium reactions. Selectivity coefficients for ion exchange and the aluminum solubility constants were calibrated by fitting model results to field data. A simple zero-order input of cations for chemical weathering was added.

Table 10.10 Summary of Equations Included in the MAGIC Model[52,66]

Equations
Soil–Water Cation Exchange Reactions
1. $E_{Al} + E_{Ca} + E_{Mg} + E_K + E_{Na} = 1$
2. $BS = E_{Ca} + E_{Mg} + E_K + E_{Na} = 1 - E_{Al}$
3. $\dfrac{\{Ca^{2+}\}^3 E_{Al}^2}{\{Al^{3+}\}^2 E_{Ca}^3} = S_{AlCa}$
4. $\dfrac{\{Na^+\}^2 E_{Ca}}{\{Ca^{2+}\} E_{Na}^2} = S_{CaNa}$
5. $\dfrac{\{Na^+\}^2 E_{Mg}}{\{Mg^{2+}\}^2 E_{Na}^2} = S_{MgNa}$
6. $\dfrac{\{Na^+\} E_K}{\{K^+\} E_{Na}} = S_{KNa}$
Inorganic Aluminum Reactions
7. $\dfrac{\{Al^{3+}\}}{\{H^+\}^3} = K_{Al}$
8. $\dfrac{\{Al(OH)^{2+}\}\{H^+\}^2}{\{Al^{3+}\}} = K_{Al1}$
9. $\dfrac{\{Al(OH)_2^+\}\{H^+\}^2}{\{Al^{3+}\}} = K_{Al2}$
10. $\dfrac{\{Al(OH)_3^0\}\{H^+\}^3}{\{Al^{3+}\}} = K_{Al3}$
11. $\dfrac{\{Al(OH)_4^-\}\{H^+\}^4}{\{Al^{3+}\}} = K_{Al4}$
12. $\dfrac{\{AlF^{2+}\}}{\{Al^{3+}\}\{F^-\}} = K_{Al5}$
13. $\dfrac{\{AlF_2^+\}}{\{Al^{3+}\}\{F^-\}^2} = K_{Al6}$
14. $\dfrac{\{AlF_3^0\}}{\{Al^{3+}\}\{F^-\}^3} = K_{Al7}$
15. $\dfrac{\{AlF_4^-\}}{\{Al^{3+}\}\{F^-\}^4} = K_{Al8}$
16. $\dfrac{\{AlF_5^{2-}\}}{\{Al^{3+}\}\{F^-\}^5} = K_{Al9}$
17. $\dfrac{\{AlF_6^{3-}\}}{\{Al^{3+}\}\{F^-\}^4} = K_{Al10}$
18. $\dfrac{\{Al(SO_4)^+\}}{\{Al^{3+}\}\{SO_4^{2-}\}} = K_{Al11}$
19. $\dfrac{\{Al(SO_4)_2^-\}}{\{Al^{3+}\}\{SO_4^{2-}\}^2} = K_{Al12}$
Inorganic Carbon Reactions
20. $\dfrac{\{CO_{2(aq)}\}}{P_{CO_2}} = K_{CO21}$
21. $\dfrac{\{HCO_3^-\}\{H^+\}}{\{CO_{2(aq)}\}} = K_{CO22}$
22. $\dfrac{\{CO_3^{2-}\}\{H^+\}}{\{HCO_3^-\}} = K_{CO23}$
23. $\{H^+\}\{OH^-\} = K_w$
Ionic Balance
24. $(H^+) - (OH^-) + 2(Ca^{2+}) + 2(Mg^{2+}) + (K^+) + (Na^+) + 3(Al^{3+}) + 2(Al(OH)^{2+}) + (Al(OH)_2^+)$ $- (Al(OH)_4^-) + 2(AlF^{2+}) + (AlF_2^+) - (AlF_4^-) - 2(AlF_5^{2-}) - 3(AlF_6^{3-}) + (Al(SO_4)^+)$ $- (Al(SO_4)_2^-) - (Cl^-) - (F^-) - (NO_3^-) - 2(SO_4^{2-}) - (HCO_3^-) - 2(CO_3^{2-}) = 0$

Assumptions of chemical equilibrium in the exchange complex seemed to work well.

Gherini et al.[62] made a very detailed model for watershed and lake response to acidification called ILWAS, the Integrated Lake Watershed Acidification Study. It is the most complex of all the biogeochemical models used today; it is also the most mechanistic. ILWAS includes processes neglected by the other models that may be important in certain applications. These processes include nitrogen dynamics, organic carbon mineralization and effects on pH and binding, mineral geochemistry, uptake of cations by vegetation, and sulfate sorption as a function of pH. Because ILWAS is so large, with such a great number of adjustable parameters and input data requirements, it is more difficult to utilize than the other models, but it does account for almost all processes that have been identified as influencing the acid–base balance of watersheds.

Schnoor et al.[63] Lin et al.[64] and Nikolaidis et al.[65] took a different approach, focusing instead on alkalinity or acid-neutralizing capacity (ANC) as a master variable, which is the equivalent sum of all the base cations (Ca, Mg, Na, K) minus the mobile anions (chloride, sulfate, nitrate), equation (20). In this case, all the various cations and anions do not need to be simulated, only their summation expressed as ANC.

$$\text{ANC} = \Sigma \text{ Base cations} - \Sigma \text{ Acid anions} \tag{69}$$

The model was called the Enhanced Trickle Down (ETD) Model, after an emphasis on hydrology (Figure 10.16) as a key factor in whether lakes become acidic or not. Deep flow paths through B- and C-horizon soils increase chemical weathering and ion exchange neutralization reactions. Results for two lakes with quite different flow paths are presented in Table 10.11. The groundwater table had alkalinity of ~400 μeq L^{-1}, which fed Lake Panther and caused it to be protected from acidic deposition. Lake Woods was acidic (−10 μeq L^{-1}) due to little inflow from the groundwater table.

There are lakes with two types of hydrologic sensitivity: lakes in mountainous regions with flashy hydrology and little contact time between acidic runoff and soil minerals (Figure 10.17a), and seepage lakes with no tributary inlets or outlets that receive most of their water directly from precipitation onto the surface of the lake (Figure 10.17b).

The ETD model computed sulfate and ANC mass balances on a daily basis and summed them to arrive at annual average mass balance, such as shown in Table 10.11. A summary of the mass balance differential equations for the ith compartment of Figure 10.16 is given by equations (70)–(72).

$$\frac{dh_i}{dt} = \Sigma Q_i^I - \Sigma Q_i^0 \tag{70}$$

$$\frac{dS_i}{dt} = \frac{1}{h_i + \beta_i}\left[\Sigma(Q_i^I S_i) - \Sigma(Q_i^0 S_i) - S_i \frac{dh_i}{dt}\right] \tag{71}$$

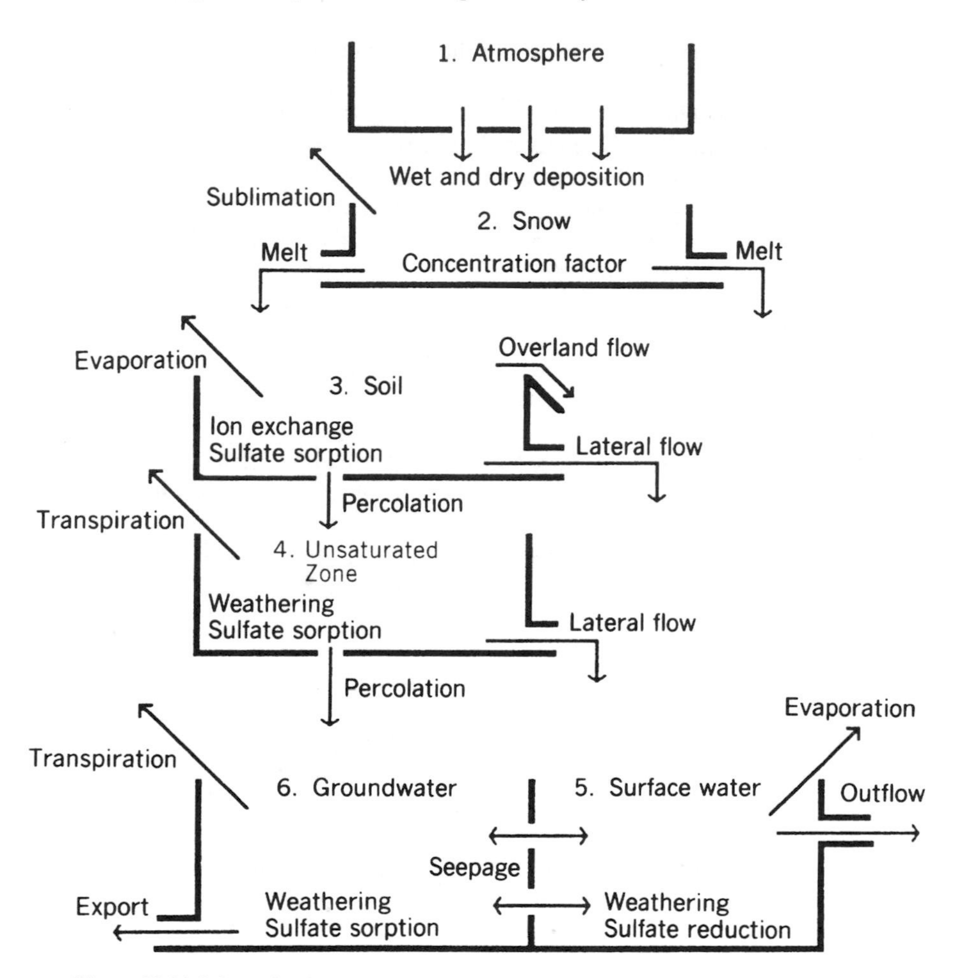

Figure 10.16 Schematic of the hydrologic flow paths in the ETD model by Nikolaidis et al.[65]

Table 10.11 Comparison of ANC Simulation Results for Acidic Lake Woods (with a Flashy Hydrograph) and Lake Panther (with Deeper Flow Paths) in Similar Geologic Areas Within Adirondack Park, New York

	ANC, eq ha^{-1} yr^{-1}	
	Lake Woods	Lake Panther
Atmospheric inputs	− 1140	− 1169
+ Sulfate sorption	+ 121	+ 89
+ Sulfate reduction	+ 116	+ 129
+ Ion Exchange & Weathering	+ 1126	+ 2637
= Outflow TOTAL	− 106	+ 1250

Source: Modified from Nikolaidis et al.[65]

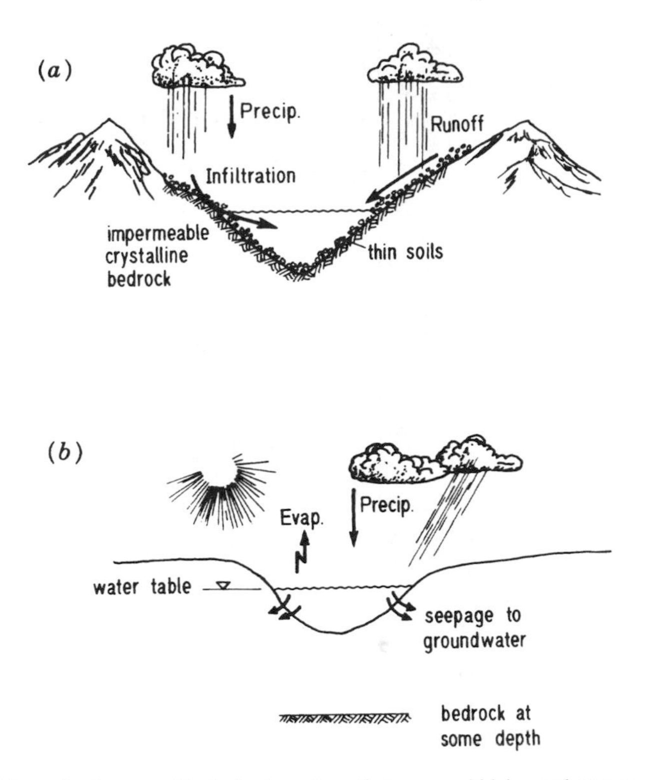

Figure 10.17 Schematic diagram of hydrologic systems that cause acid lakes and streams in areas with crystalline bedrock and sensitive geology:.(a) steep, rocky catchments with thin soils where water moves quickly with little contact time for acid neutralization; and (b) seepage lakes with no inlets or outlets—most of the water in the lake comes directly from acid precipitation.

$$\frac{dA_i}{dt} = \frac{1}{h_i}\left[\Sigma(Q_i^I A_i) - \Sigma(Q_i^0 A_i) - A_i\frac{dh_i}{dt} + \frac{dM_{Si}}{dt} + WR_i\right] \qquad (72)$$

where i = the ith compartment of Figure 10.16

h_i = area normalized water depth, L

S_i = dissolved sulfate concentration, ML^{-3}

A_i = alkalinity concentration, ML^{-3}

M_{Si} = sulfate mass sorbed per unit area, ML^{-2}

Q_i^I = area normalized compartment inflowrate, LT^{-1}

Q_i^0 = area normalized compartment outflowrate, LT^{-1}

WR_i = ion exchange and weathering rate, $ML^{-2}T^{-1}$

β_i = sulfate sorption retardation factor, L

Table 10.12 Comparison of Selected Biogeochemical Models for Acid Deposition Assessments

Model	Principle	Cations and Anions	Ion Exchange	Al Speciation	H-Org	Chemical Weathering	Sulfate Sorption	Hydrologic Balance
Henricksen[61]	Charge balance	Ca^{2+}, SO_4^{2-}	No	No	No	No	No	No
Christophersen et al.[67]	Charge + mass balance	Al^{3+}, M^{2+}, SO_4^{2-}	Yes	Yes	No	No	No	Yes
Cosby et al.[52,66]	I.E., charge + mass balance	All	Yes	Yes	No	Zero order	No	No
Gherini et al.[62]	Include all processes	All	Yes	Yes	Yes	Fractional order	Yes	Yes
Schnoor et al.[63–65]	ANC mass balance + kinetics	ANC, SO_4^{2-}, Cl^-	Kinetic	No	No	Fractional order	Yes	Yes

A summary comparison of model formulations is presented in Table 10.12. Overall model results indicate that a small number of U.S. lakes have been acidified by acid deposition. Some lakes were already naturally acidic (pH 5–6) due to crystalline bedrock geology and organic acids from flooded soils and peat. Lake chemistry in these acid systems is buffered somewhat by aluminum and organic weak acids. Thus only a fraction of currently acidic lakes can be expected to recover as a result of controls on sulfur dioxide emissions.[68]

10.6 ECOLOGICAL EFFECTS

Regions where acid deposition has been reported to affect lakes (e.g., Adirondacks, New York; southern Norway; and Sweden) also have acid soils. If chemical weathering cannot replace exchangeable bases in soils rapidly enough, base cations become depleted from the upper soil profile and iron and aluminum are mobilized. Such chromatographic effects can lead to podzolic acid soils. Aluminum is not only liberated from the soil in the rooting zone, but it is at high concentrations as Al^{3+} or $AlOH^{2+}$ on the fine roots and mycorrhizal fungal associations. The possible toxic effects of Al^{3+}, its uptake by trees in acid soils, and sorbed Al effects on active and passive transport of nutrients to the tree are the subject of much debate.[69,70] Lower pH values accentuate the potential negative effects of heavy metals.

It has been pointed out by Schindler[71] that sensitivities of terrestrial and aquatic ecosystems to atmospheric pollutants are remarkably different. Primary production seems to be reduced at a much earlier stage of air pollution stress in the terrestrial ecosystem than in the aquatic ecosystem. Soils like lake sediments tend to be sinks for pollutants; this may protect the pelagic regions of lakes from influxes of toxic substances that would occur if watersheds and sediments were unreactive.[71] The careful studies by Schindler et al.[72] of experimental acidification on a small lake (Lake 223)—over a period of 8 years the pH of the lake was gradually decreased from 6.8 to 5.0 by the addition of H_2SO_4—have provided an invaluable documentation and interpretation of response of a lake to long-term ecosystem stress. The acidification produced dramatic changes in the lake's food web, above all, in phytoplankton species, in the disappearance of benthic crustaceans, and the cessation of fish reproduction. Schindler et al.[72] report that key organisms in the food web leading to lake trout were eliminated from the lake at pH values as high as 5.8; they interpret this as an indication that irreversible stresses on aquatic ecosystems occur earlier in the acidification process than was heretofore believed. These changes are caused in Lake 223 by H^+ ion alone and not by the secondary effect of aluminum toxicity.

Humic substances are adsorbed to oxide and other soil minerals; the adsorption is pH dependent and decreases with increasing pH. Thus the acidification of soil systems reduces the drainage of humic substances into receiving waters. Furthermore, the increased dissolved Al(III) forms complexes with residual humic acids. All of these effects [lowering of pH, increase of Al(III) and decrease in concentration of humic acids] increase the activity of free heavy metals. Increased free metal ion activity can have an impact on ecological structure of phytoplankton. There has been speculation on whether toxic algal blooms in the North Sea and Adriatic may have been accentuated by these indirect consequences of increased acid deposition during the last decades.

Nitrogen in the form of ammonium and nitrate is a fertilizer for forest growth. In most forests of the northern temperate and boreal zones, nitrogen is considered to be the limiting nutrient for forest growth. However, nitric acid deposition and other forms of nitrogen are increasing worldwide. Human capacity to fix nitrogen from the atmosphere in the form of ammonia for fertilizer and NO_x from fossil fuel combustion now rivals the total nitrogen fixed by natural processes such as nitrogen-fixing plants and by lightning. While sulfur deposition has been decreasing over the past 5–10 years in developed countries, nitrogen deposition continues to increase and it is becoming a larger fraction of total acid deposition. Nitrogen saturation may be defined as a surplus of mineral nitrogen in forests beyond the capacity for plant uptake or soil immobilization. Increasingly saturated with nitrogen, forest growth could become limited by other factors such as phosphate, magnesium, water, light, or temperature.

When forests lie in hydrogeographic settings that are sensitive to acid deposition, the leaching of nitrate from the forest floor during snowmelt or runoff events becomes an acidifying influence on soils, streams, and lakes. Organic nitrogen is mineralized to ammonium in forest soils, and the ammonium undergoes nitrification, releasing hydrogen ions. Similarly, hydrogen ions are released from root cells when ammonium is taken up by vegetation, the process is necessary to achieve a charge balance in the plant tissue. Leaching of nitrate from forest soils has been correlated with low ANC and acid pulses in many stream ecosystems in the northeastern and southeastern United States.[73,74] It has also been linked to liberation of aluminum in soils, root toxicity, and/or forest nutrition problems. Leakage of nitrate can become a significant contribution to eutrophication of lakes and estuaries, particularly if they are nitrogen-limited, such as Chesapeake Bay.

Forests in southern Sweden and Denmark are currently receiving more than 20 (kg N) ha^{-1} yr^{-1}, and nitrogen deposition in northernmost Scandinavia, which probably reflects background levels, is about 5 (kg N) ha^{-1} yr^{-1}.[75] Figure 10.18 is a schematic of the flow of nitrogen through a forest ecosystem. Typically 30–50 (kg N) ha^{-1} yr^{-1} is mineralized in the forest floor in Scandinavia, so deposition in excess of these quantities is required to sustain vegetation. Nitrate export from the watershed has been termed *nitrogen saturation*. Rosen et al.[75] have published critical load maps for nitrogen deposition in Nordic countries based on a simple mass balance approach that ranges from 35 (kg N) ha^{-1} yr^{-1} in the south to 1.3 (kg N) ha^{-1} yr^{-1} in the north of the Nordic countries. These levels are exceeded in southern and southwestern parts of the Nordic countries, with few or no exceedances in the north.

Figure 10.19 is a hypothetical plot of nitrogen accumulation in forests and export to streams as a function of nitrogen deposition. It is an idealized graph because it

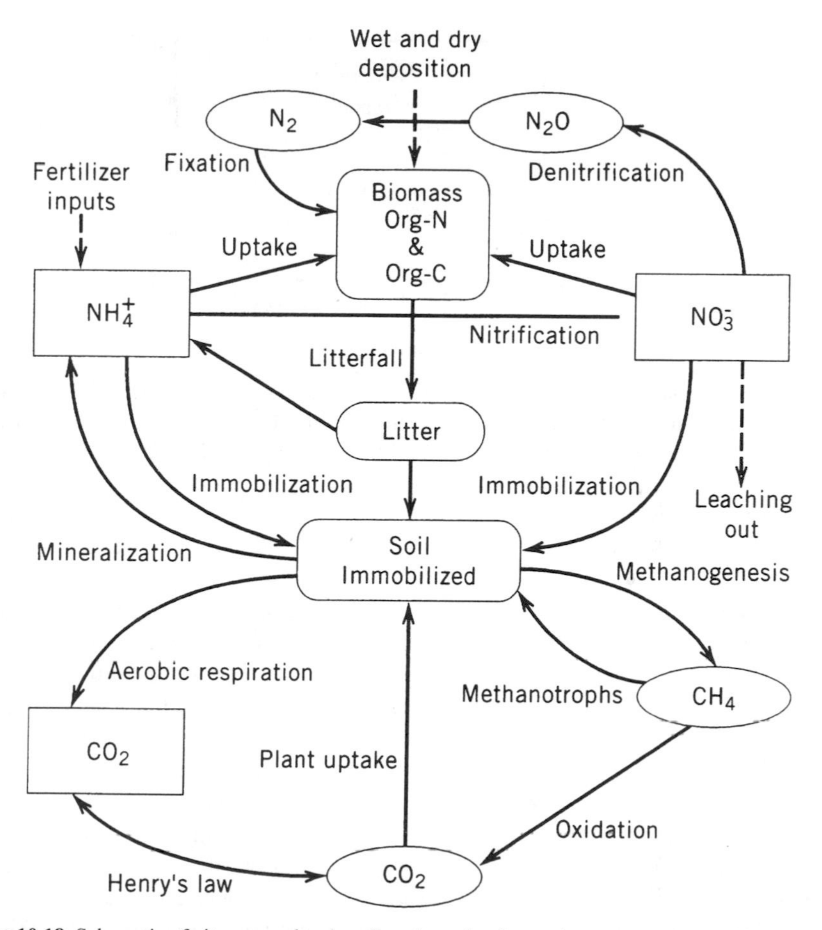

Figure 10.18 Schematic of nitrogen and carbon flow through a forested ecosystem showing the immobilization of ammonium and nitrate into the organic soil pool and subsequent mineralization and nitrification of ammonium that results in the leakage of nitrate to drainage waters. Ovals represent gases; rectangular boxes are aqueous phase concentrations. (From Rees and Schnoor.[68])

does not take into account the age of the forest nor the time horizon for nitrogen saturation to occur. Low levels of nitrogen deposition (0–25 kg ha^{-1} yr^{-1}) are immobilized into the organic pool and taken up by vegetation, a slight fertilization effect resulting from increased primary production. In sensitive forested ecosystems, N-deposition in the range of 25–50 (kg N) ha^{-1} yr^{-1} results in initial export of nitrate to drainage water. Generally nitrate-nitrogen concentrations are low (<0.2 mg L^{-1}), but as N-deposition increases they may increase to 0.3–0.7 mg L^{-1}, the first signs of N-saturation and potential forest effects. The mass of nitrate exported via the stream is still small compared to the total deposited, but it increases as deposition increases to become >50 (kg N) ha^{-1} yr^{-1}, and N-saturation becomes more serious as shown in Figure 10.19.

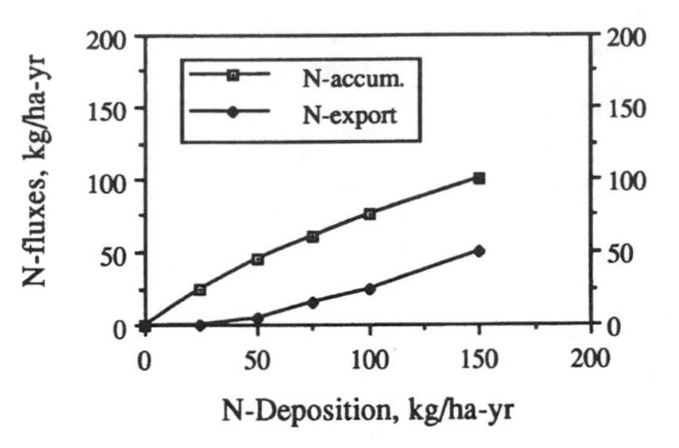

Figure 10.19 Hypothetical plot of nitrogen accumulation in an ecosystem (organic-N in canopy, bole, roots, and soil) and nitrogen export (nitrate predominantly) from forested catchments as a function of atmospheric deposition in (kg N) ha^{-1} yr^{-1}.

10.7 CRITICAL LOADS

Critical loads have been defined as "*a quantitative estimate of an exposure to one or more pollutants below which significant harmful effects on specified sensitive elements of the environment do not occur according to our present knowledge.*"[76] The interest in critical loads has increased over the past few years because of its appeal to both scientists and policy-makers. It is appealing to some scientists because the concept of critical loads considers biological effects of pollution in a quantitative framework. Biological effects are really the end point that the scientist and the public are seeking, for example, preservation of a fishery rather than placing an arbitrary effluent limit on industry that may or may not produce the desired effect. In practice, the environmental scientist resorts to chemical indicators of biological effects. The concept of critical loads appeals to some policy-makers as well because there is an implicit assumption that the critical load will truly protect the natural resource, and the decision-maker can take comfort in that fact with constituencies.

In 1985, the Convention on Long Range Transboundary Air Pollution of the United Nations Economic Commission for Europe was signed and ratified by 21 European countries, the United States, and Canada. It formed the so-called *30% club*, countries that pledged to decrease their sulfur emissions by 30% from 1980 levels by 1993. The agreement has been largely successful, but it left people wondering whether 30% was enough. What reduction in emissions would be required to prevent future acidification of lakes or forest decline? *Critical loads* have been the management tool of choice in Europe ever since. In 1988, the NO_x protocol was signed by 25 European countries, the United States, and Canada, promising that NO_x emissions should not increase from 1980–85 levels. It included the use of critical load estimates as the protocol for future pollution control agreements.[76]

Estimates of critical loads are fraught with many scientific ambiguities, however. Turning back to the definition, let us examine a few key nouns that point up the difficulties in estimating critical pollutant loads for a given region or country.

- Pollutants—Which pollutants are the most critical and how do the various pollutants interact to cause biological effects?
- Harmful effects—Which effects are *most* harmful?
- Sensitive elements of the environment—Which elements of the ecosystem are we trying to protect and which one is the most sensitive on what spatial scale?

It may not be possible to protect the most sensitive lake in a region, so how much of the total resource can be protected? Maintaining pollution levels below the critical load is supposed to ensure that harmful effects do not occur, but what is the appropriate time horizon on which to base an estimate of the critical load? Perhaps the most vexing problem from an implementation standpoint is the difficulty of defining a source–receptor relationship for allocation of transboundary pollutant emissions. It must be both scientifically defensible and politically palatable.

The estimation of critical loads is viewed as a scientific problem. Once the critical loads have been established, it is up to each country to determine *target loads*. Target loads take into account technical, social, economic, and political considerations, including what is achievable and what is desirable for the individual country. The relationship between critical loads and target loads is roughly analogous to that of water quality criteria and water quality standards in the United States. Target loads, like water quality standards, would be enforceable by law. Target loads could be set more stringently than critical loads in order to ensure a safety factor and to prevent significant deterioration of resource quality. On the other hand, target loads could be set less stringently than the critical load, thus admitting that some resources would be affected but, perhaps, allowing for more stringent target loads to be set at some point in the future when the country or countries are better able to respond.

10.7.1 Critical Load Maps

Pollutants that should be considered for critical loads and target loads to protect lakes and forest soils include sulfur deposition (both wet and dry, SO_4^{2-} and SO_2) for lakes and hydrogen ion deposition for forest soils (because it includes the effects of both sulfuric acid and nitric acid inputs to soils). Smith et al.[77] have shown that atmospheric deposition of hydrogen and calcium ions to dystrophic peat in Great Britain is capable of producing pH changes in soil solution of 0.2–0.6 units more readily than previously reported. In addition, forests and forest soils may be sensitive to total nitrogen inputs (N-saturation), so critical loads for nitrogen are being explored.[78] Sulfate concentration in lakes has also been utilized to set critical loads in lieu of sulfur deposition,[79] and acid-neutralizing capacities (ANCs) have been used as the water quality parameter that protects against effects (fish mortality) in lakes. ANC values of 20 μeq L^{-1},[80] 25 μeq L^{-1}, and 0 μeq L^{-1} (ref. 81) have been cited for use in critical loads assessments. Hydrogen ion deposition to forest soils is a surrogate parameter to estimate the ratio of base cations to aluminum ion in soil moisture. Sverdrup et al.[82] assume that the critical molar ratio $(Ca^{2+} + Mg^{2+})/Al$ is

1:1; above this ratio, roots are protected from aluminum toxicity and magnesium deficiency. The critical molar ratio is used together with a gibbsite solubility constant ($[Al^{3+}]/[H^+]^3) = 10^{8.3}$) to estimate the alkalinity (or base cations) that need to be leached from soils in order to prevent aluminum buildup in soil solution.

Kamari et al.[83] have produced a map of critical loads expressed in terms of acidity (from both wet and dry deposition) deposited in Europe. The pollutant designated for control is therefore hydrogen ion deposition, and the element of the ecosystem intended for protection is forests. The biological effect in consideration is forest decline stemming from acidification of soil over a decade–centuries time horizon. Only a fraction of the resource can be protected—the most sensitive forest soil cannot—so the decision was made to include 95% of the forest soils on an areal basis (i.e., 5% of the forest soils would require critical load values more stringent than published[83]). The Steady-State Mass Balance Method (SMB) by Hettlingh et al.[84] was used. Considerable concern exists for forest soils due to the reported die-back in Germany, Poland, the Czech Republic, and the Slovak Republic, and because there is evidence that soils have acidified during the 20th century in southern Sweden.[85] But Henriksen et al.[80] report that lakes are the most sensitive receptors in Nordic countries. Yet, 82% of the forest soils in Nordic countries receive deposition exceeding the critical loads given by Kamari et al.[83] Thus target loads must be set more stringently than the critical loads in order to affect recovery.[82,83]

Preliminary target loads have been published for the Nordic countries of Sweden, Norway, Finland, and Denmark.[76] They are 0. 5 (g S) m^{-2} yr^{-1}, except in southern Sweden where the target load has been designated as 0.3 (g S) m^{-2} yr^{-1}, and in central and northern Finland where the target loads are 0.4 and 0.2 (g S) m^{-2} yr^{-1}, respectively. These are very stringent targets for enforcement, and most of the area will require further emission reductions to be able to meet the target loads. A European Monitoring Network (EMEP) plot of total sulfur deposition for Europe is given in Figure 10.20 (top panel), demonstrating that most of Europe received deposition greater than 0.5 g m^{-2} yr^{-1} in 1985. Alcamo et al.[86] have developed a regional assessment model (RAINS) to incorporate transfer functions from the EMEP model and to relate emissions of a country to deposition in 150-km by 150-km grid cells. Models such as RAINS will be needed to arrive at estimates for target loads.

The bottom panel of Figure 10.20 shows acid precipitation pH contours for the continental United States from the National Atmospheric Deposition Program (Illinois State Water Survey) superimposed on cross-hatched areas designating sensitive geology from James Omernik at the U.S. Environmental Protection Agency (Corvallis Environmental Research Laboratory).

10.7.2 Models for Critical Load Evaluation

Simple steady-state models have been used to determine critical loads for lakes and forest soils. The basic principle is that primary mineral weathering in the watershed is the ultimate supplier of base cations, which are required elements for vegetation and lake water to ensure adequate acid-neutralizing capacity (see previous definitions of ANC as the sum of base cations minus acid anions). If more acid is deposit-

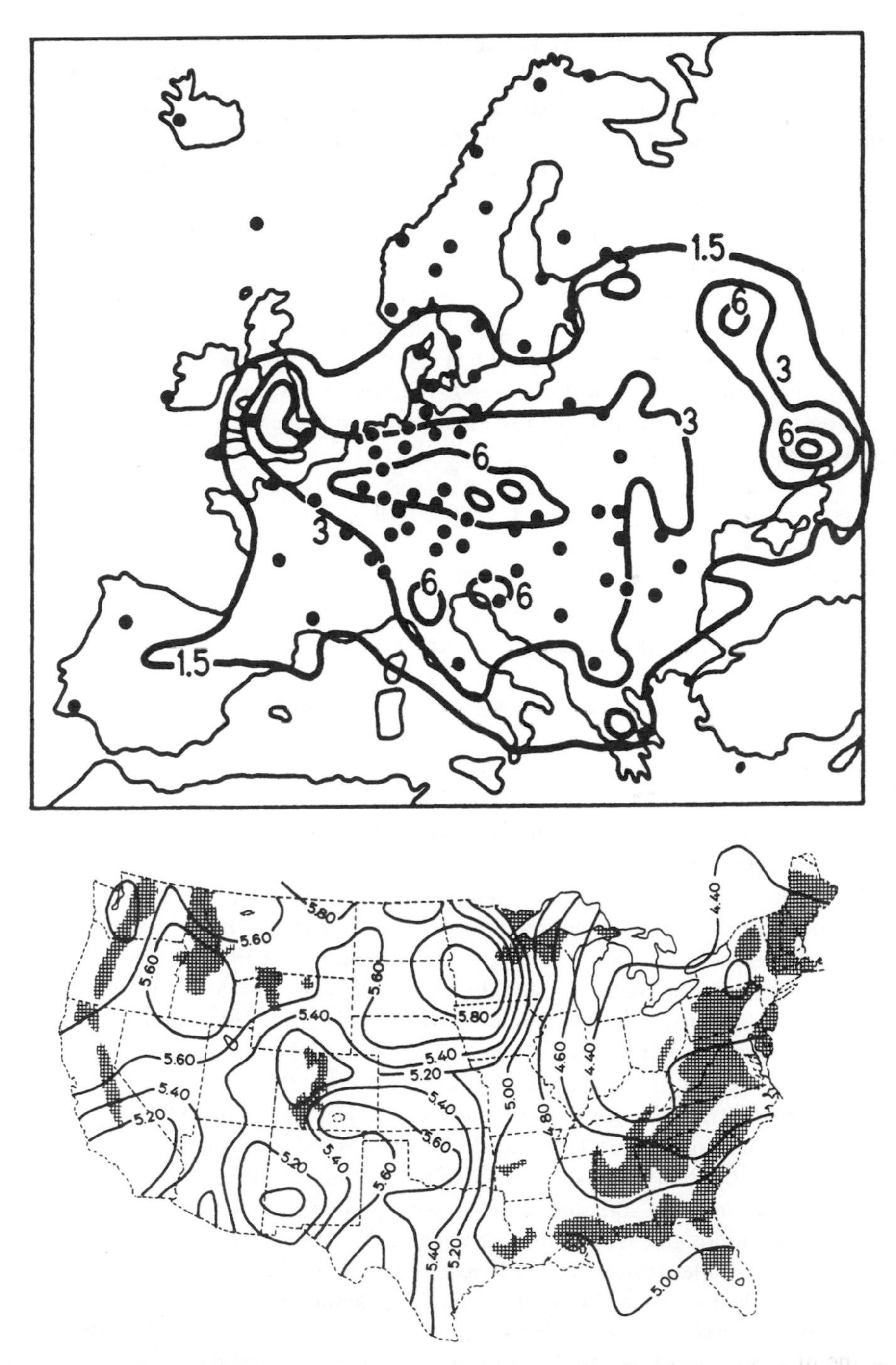

Figure 10.20 *Top*: European Monitoring Network (EMEP) map of total sulfur deposition in (g S) m^{-2} yr^{-1} for 1985. Target loads were exceeded by S-deposition over much of Europe, but emissions are improving dramatically. *Bottom*: Volume-weighted annual average pH contours (1980–1984) and sensitive geologic regions in the continental United States (cross-hatched areas).

ed in a watershed than chemical weathering can neutralize, acidification of the soils and water will eventually occur. It may not happen immediately; there could be a delayed response because ion exchange reactions can supply base cations to vegetation and runoff water for a period, but eventually the exchange capacity of the soil will become depleted and acidification will result. The delay period can be estimated as the base cation exchange capacity (exchangeable Ca, Mg, Na, and K ions) of the soil divided by the cation denudation rate, typically on the order of a few decades in poor, acidic soils. The exchanger (cation exchange complex in soil) is "set up" by the mineralogy and lithology of the parent material, and this determines the base exchange capacity and even the chemistry of the soil.

Sverdrup et al.[82] developed the PROFILE model, which is based on the principle of continuity of alkalinity or ANC in soil. The critical load is defined as the allowable acid loading that will not acidify forest soils and cause the release of aluminum and hydrogen ions to soil solution.

$$\mathrm{CL} = BC_w - \mathrm{Alk}_L \tag{73}$$

where CL = the critical load, meq m^{-2} yr^{-1}

BC_w = weathering rate, meq m^{-2} yr^{-1}

Alk_L = alkalinity leaching, meq m^{-2} yr^{-1}

The amount of alkalinity that is necessary to leach from the soil can be estimated from the ratio of base cations to aluminum, (Ca + Mg)/Al, assuming a critical threshold value for biological effects at 1:1. Then, based on gibbsite solubility and a minimum required base cation concentration for forest nutrition of 5 meq m^{-2} yr^{-1}, it is possible to estimate the required alkalinity leaching for healthy soils. The SMB model includes a third term on the right-hand side of equation (73) for cation uptake required by vegetation, which is another acidifying influence on forest soils that must be subtracted from the weathering rate.[79] Furrer et al.[87] have a more detailed steady-state model than SMB or PROFILE. It includes weathering, ion exchange, base cation uptake by trees, nitrogen transformation reactions, and equilibrium soil chemistry under steady-state conditions. The model has not been used for critical load assessments, but it represents a more comprehensive approach.

Methods to calculate critical loads from steady-state models do not take time scales for acidification and recovery into account. Detailed nonsteady-state models have been used to estimate target loads for three watersheds in Norway and Finland.[88] Three different models were used in the predictive mode such that deposition was adjusted until the soil solution chemistry produced the desired 1:1 ratio between base cations and aluminum by the year 2037 (50-year simulations). All three models gave similar results for catchments with shallow soils, but the differences were great when soils were deep, and hydrologic assumptions in the models varied as to how water should be routed through the soil and to the stream. The Birkenes catchment in southern Norway[67] has become so acidified that the target load must be reduced to less than 0 meq m^{-2} yr^{-1} of S deposition in order for it to recover by the year 2037.

Holdren et al.[81] found that results of dynamic models gave quite different results for target loads in a population of 762 lakes from northeastern Pennsylvania to Maine. This is the population of lakes that was selected for the Eastern Lake Survey (ELS) in a stratified random sample of 7150 lakes in the region. Three models were used to assess the projected recovery of these lakes under a scenario of a 30% decline in S-deposition, as a result of the 1990 Clean Air Act Amendments in the United States. The three models included the ILWAS model by Gherini et al.,[62] the MAGIC model by Cosby et al.,[66] and the Enhanced Trickle-Down model by Nikolaidis et al.[65]

Each data point in Figure 10.21 represents a lake selected in the stratified random sample with different weighting factors; each point represents approximately 12–27 lakes in the region of similar hydrogeochemical characteristics. These are among the most sensitive lakes in the region, generally with crystalline bedrock that is slow to weather and flashy hydrographs. Figure 10.21 shows a slight correlation suggesting that sulfate is an important variable in the ANC concentration of lakes in the region. In lakes with high sulfate concentrations (in areas receiving greater acid deposition), the ANC concentration generally decreases. This offers some evidence of sulfuric acid deposition as being responsible for acidification of lakes among a class of lakes with the greatest sensitivity in the northeastern United States. The linear least-squares regression line through the data in Figure 10.21 defines an F-factor, which is the change in the sum of base cation concentrations in the lakes divided by the change in the sulfate anion concentration:

$$F = \{\Delta \Sigma \text{Base cations}\} / \Delta \{SO_4^{2-}\} \tag{74}$$

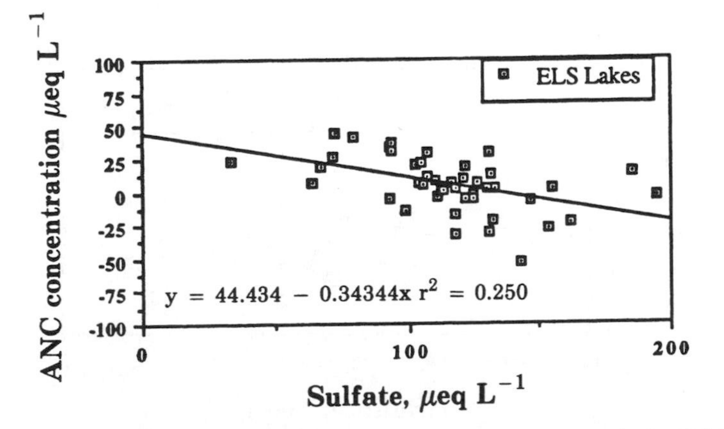

Figure 10.21 Acid-neutralizing capacity versus sulfate concentration (in μeq L^{-1}) for 43 lakes in the northeastern United States, part of the Eastern Lakes Survey database. These lakes were the most sensitive lakes out of 762 lakes sampled representing the entire database of 7150 lakes from northeastern Pennsylvania to Maine.

Equation (74) gives a measure of the degree of buffering that a watershed can provide to increased sulfur deposition. For example, if the sulfate anion concentration in a lake increased by 50 μeq L^{-1} due to acid deposition, increased ion exchange and weathering might be able to provide an additional release of 25 μeq L^{-1} of base cations, resulting in an *F*-factor of 0.5. Alkalinity is a surrogate variable for base cations in Figure 10.21 because for every release of a base cation from weathering or exchange, an equivalent amount of alkalinity is generated. In the example, the release of 25 μeq L^{-1} of base cations would provide 25 μeq L^{-1} of alkalinity, and the change in alkalinity in the lake due to 50 μeq L^{-1} of strong acid would be -25μeq L^{-1}. The slope of the regression line in Figure 10.19 for lakes in the Eastern Lake Survey indicates an *F*-factor of 0.66.

Results of the ETD model for 81 representative lakes in the ELS survey are shown in Figure 10.22. A 30% decrease in total S-deposition was used as input to the model; the decrease was input as a declining ramp function over a 15-year period, and the results shown in Figure 10.22 are the changes in ANC and sulfate ion concentrations in the lake after 50 years (~steady-state). The surprising result was that the lakes did not respond to the declining deposition to a large extent. The slope of the regression line indicates an *F*-factor of 0.9. The lakes were relatively well buffered due to a variety of mechanisms including ion exchange, chemical weathering, and sulfate sorption—it is difficult to change the ANC and pH of the lakes dramatically either via increased or decreased deposition. In this case, the prospect for increasing ANC in lakes by decreasing atmospheric emissions was modest—six lakes out of a stratified random sample of 81 lakes recovered from an ANC of less than zero to a positive ANC. Initially, there were 15 acid lakes with ANC concentrations less than zero. The average lake increased its ANC concentration by only 6 μeq L^{-1}.

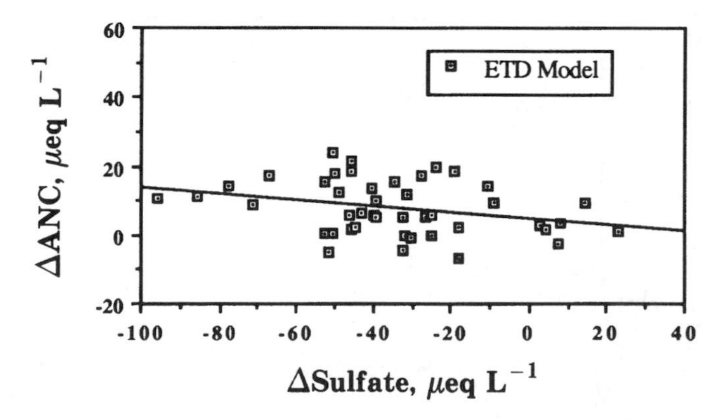

Figure 10.22 Change in acid-neutralizing capacity (ANC) versus the change in sulfate concentration following a 30% ramp function decrease in sulfur deposition, 50-year simulation, using the Enhanced Trickle Down Model by Lee et al.[89]

10.8 CASE STUDIES

Two case studies are given to illustrate lake and stream acidification. In the first example, acidification of a lake in a sensitive small watershed alpine area with sparse vegetation and thin soils has occurred at relatively low levels of acidic deposition. Here, emphasis is on explaining the weathering of crystalline rocks in acidic catchments. The second example gives data from a watershed study in a forested area, where a whole catchment of 40 acres has been artificially acidified. The result is consistent with other watershed studies such as that of Likens et al.,[90] and it illustrates the importance of nitrogen status of a forested ecosystem to increased deposition of nitric acid or ammonium salts. Many stream systems are poised to become acidic, especially during snowmelt or runoff events if acid deposition increases.

10.8.1 Chemical Weathering of Crystalline Rocks in the Catchment Area of Acidic Ticino Lakes, Switzerland

Figure 10.23 gives the water composition of four lakes at the top of the Maggia valley.[28] Although these lakes are less than 10 km apart, they differ markedly in their water composition as influenced by different bedrocks in their catchments. All lakes are at an elevation of 2100–2550 m. The small catchments are characterized by sparse vegetation (no trees), thin soils, and steep slopes.

The compositions of Lakes Cristallina and Zota, situated within a drainage area characterized by the preponderance of gneissic rocks and the absence of calcite and dolomite, and Lake Val Sabbia, the catchment area of which contains dolomite, are markedly different. The waters of Lake Cristallina and Lake Zota exhibit mineral acidity (i.e., caused by mineral acids and HNO_3), their calcium concentrations are 10–15 μmol L^{-1} and their pH is <5.3. On the other hand, the water of Lake Val Sabbia is characterized by an alkalinity of 130 μmol L^{-1} and a calcium concentration of 85 μmol L^{-1}. The water of Lake Piccolo Naret is intermediate; its alkalinity is <50 μmol L^{-1} and appears to have been influenced by the presence of some calcite or dolomite in its catchment area.

Weathering processes regulate the chemical water composition. On the basis of simple mass balance considerations, plausible reconstructions were attempted for the contribution of the various weathering processes responsible for the residual water composition of acidic Lake Cristallina. Stoichiometric reactions were developed for the interaction of acidic deposition (snowmelt) with weatherable minerals in the drainage area.[28] The composition of atmospheric deposition results generally from the interactions of strong acids (H_2SO_4, HNO_3), bases (NH_3) and wind-blown dust and aerosols ($CaCO_3$, $MgCO_3$, NH_4NO_3, NaCl, KCl).

In Table 10.13 the contributions of the individual weathering reactions were assigned and combined in such a way as to yield the concentrations of Ca^{2+}, Mg^{2+}, Na^+, K^+, and H^+ measured in these lakes; the amounts of silicic acid and aluminum hydroxide produced and the hydrogen ions consumed were calculated stoichiometrically from the quantity of minerals assumed to have reacted. Corrections must be

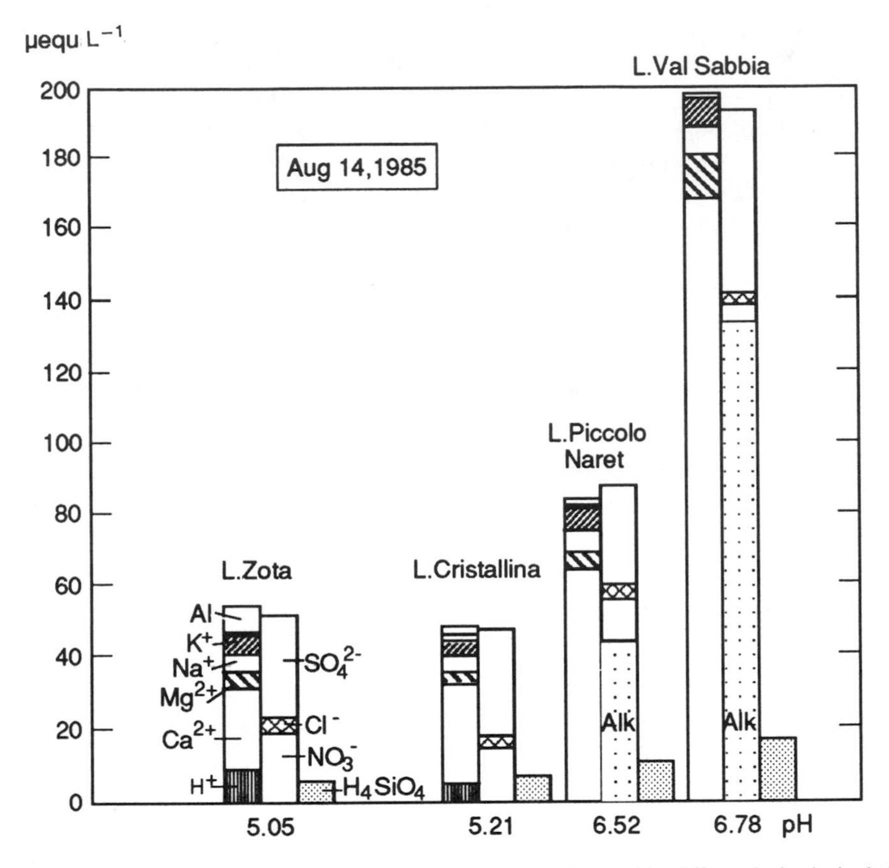

Figure 10.23 Comparison of water composition of four lakes influenced by different bedrocks in their catchments. Drainage areas of Lake Zota and Lake Cristallina contain only gneiss and granitic gneiss; that of Lake Piccolo Naret contains small amounts of calcareous schist; that of Lake Val Sabbia exhibits a higher proportion of schist.[28]

made for biological processes, such as ammonium assimilation and nitrification and the uptake of silicic acid by diatoms. Some of the H_4SiO_4 was apparently lost by adsorption on aluminum hydroxide and Fe(III) (hydr)oxides, but the extent of these reactions was difficult to assess.

Although an unequivocal quantitative mass balance could not be obtained, a plausible reaction sequence was deduced that accounts reasonably well for the residual chemical water composition. The amount of $CaCO_3$ that had to be dissolved to establish the residual water composition is about what can be accounted for by wind-blown calcite dust. The neutralization of the acidic precipitation by NH_3 was, subsequent to its deposition, largely annulled by the H^+ ions produced by nitrification and NH_4^+ assimilation.

Based on Table 10.13, in the drainage area of Lake Cristallina, about 4 μmol of plagioclase, 2 μmol of epidote, 1 μmol of biotite, and 1 μmol of K-feldspar per liter

Table 10.13 Minerals that Undergo Chemical Weathering in the Lake Cristallina (Switzerland) Catchment Area and Estimated Rates of Weathering and H^+-Ion Neutralization Based on Water Chemistry

Rock-Forming Mineral	Chemical Weathering Reaction, μmol L^{-1}	Estimated Weathering Rate, mmol m^{-2} yr^{-1}	H^+-Ion Neutralization, meq m^{-2} yr^{-1}
Calcite	$CaCO_3 + 2\ H^+ \rightarrow Ca^{2+} + H_2CO_3$	1.2	2.4
Plagioclase	$4\ Na_{0.75}Ca_{0.25}Al_{1.25}Si_{2.75}O_8 + 5\ H^+ + 27\ H_2O$ $\rightarrow 3\ Na^+ + Ca^{2+} + 11\ H_4SiO_4 + 5\ Al(OH)_3$	4.8	6.0
Epidote	$2\ Ca_2Al_{2.5}Fe_{0.5}Si_3O_{12}\ (OH) + 8\ H^+ + 16\ H_2O$ $\rightarrow 4\ Ca^{2+} + 6\ H_4SiO_4 + 5\ Al(OH)_3 + Fe(OH)_3$	2.4	9.6
Biotite	$KMg_{0.5}Fe_{2.5}AlSi_3O_{10}\ (OH)_2 + 2\ H^+ + 8\ H_2O$ $\rightarrow K^+ + 0.5\ Mg^{2+} + 3\ H_4SiO_4 + Al(OH)_3 + 2.5\ Fe(OH)_2$	1.2	2.4
Orthoclase (K-feldspar)	$KAlSi_3O_8 + H^+ + 7\ H_2O$ $\rightarrow K^+ + 3\ H_4SiO_4 + Al(OH)_3$	1.2	1.2
		10.8	21.6

of runoff water are weathered per year. This amounts to 21.6 meq of cations per m^2 per year.

10.8.2 Watershed Manipulation Project (WMP) at Bear Brooks, Maine

The site of Bear Brooks Watershed at Lead Mountain, Maine (latitude 44°51′75″, longitude 68°6′25″), contains two almost identical streams, each draining 15 hectares, with a southern exposure from the top of Lead Mountain, 450 m above mean sea level. It is a part of the U.S. Environmental Protection Agency's (EPA's) Watershed Manipulation Project, a long-term experimental manipulation in which the western catchment has been artificially acidified with applications of ammonium sulfate, and the eastern catchment has remained as a reference.

Bedrock at the site consists of metamorphosed quartz sandstones with granitic sills and dikes, overlain by glacial till 4–5 m thick at the lower elevations. Exposed bedrock is visible at the summit of Lead Mountain. Mineralogy by x-ray diffraction of the soils (B and C horizons) consists of quartz as the major mineral, followed by feldspar as a significant mineral [$X(Si, Al)_4O_8$, X = Na, K, Ca, Ba], and minor minerals muscovite [$KAl_3Si_3O_{10}(OH)_2$], chlorite [$(Mg, Fe, Al)_6(Si, Al)_4O_{10}(OH)_8$], and hornblende [$X_2Y_5(Si,Al)_8O_{22}(OH)_2$, X = Na, K, Ca, Mn, Fe, and Y = Mg, Al, Ti, Mn, Fe]. The parent material and soils are quite deficient in calcium (0.8–1.3% by weight as oxide equivalents). In the fine size fractions of the upper B-horizon soil, there is considerable iron oxide (4.1% by mass) and organic carbon (15% as measured by loss on ignition at 1050 °C). Little chemical differences exist among the various size fractions otherwise. The B-horizon soils have more Fe and Al than do the C-horizon soils, as expected for acid, podzolic soils.

Manipulation at Bear Brook forested catchment using stable isotopic $(^{15}NH_4)_2SO_4$ applications have allowed a tracer experiment to document the fertilization effect of ammonium ions and the degree of nitrogen saturation. Steve Norton at the University of Maine, in cooperation with Aber and Nadelhofer, applied approximately 16.8 (kg N) ha^{-1} yr^{-1} of dry ammonium sulfate to a 15-ha catchment, West Bear Brook. It was expected that most of the ammonium would be taken up by the mixed hardwood forest in the middle of succession and/or immobilized into the forest floor in a stable organic-humate phase. The forest is owned by a paper company, which allowed the experiment to be performed. It was clear-cut less than 40 years ago, so the system was thought to be in a rather immature state of succession, and it was not expected that it would leak much nitrogen.

Most of the sulfate applied was adsorbed in the first year after treatment, but SO_4^{2-} has been increasingly mobile in subsequent years, as shown by Figure 10.24. Treatment of ammonium sulfate began as quarterly applications in November 1989 (approximately day 730 in Figure 10.24). The reference catchment (East Bear Brook) did not demonstrate an increase in sulfate concentration in the Brook, but the treated catchment (West Bear Brook) showed an increase from an average concentration of 105 to 135 μeq L^{-1}. In the third year of applications, about one-third of the applied sulfate was exported from the catchment in West Bear Brook. In effect, the uptake of ammonium by vegetation caused a small release of sulfuric acid

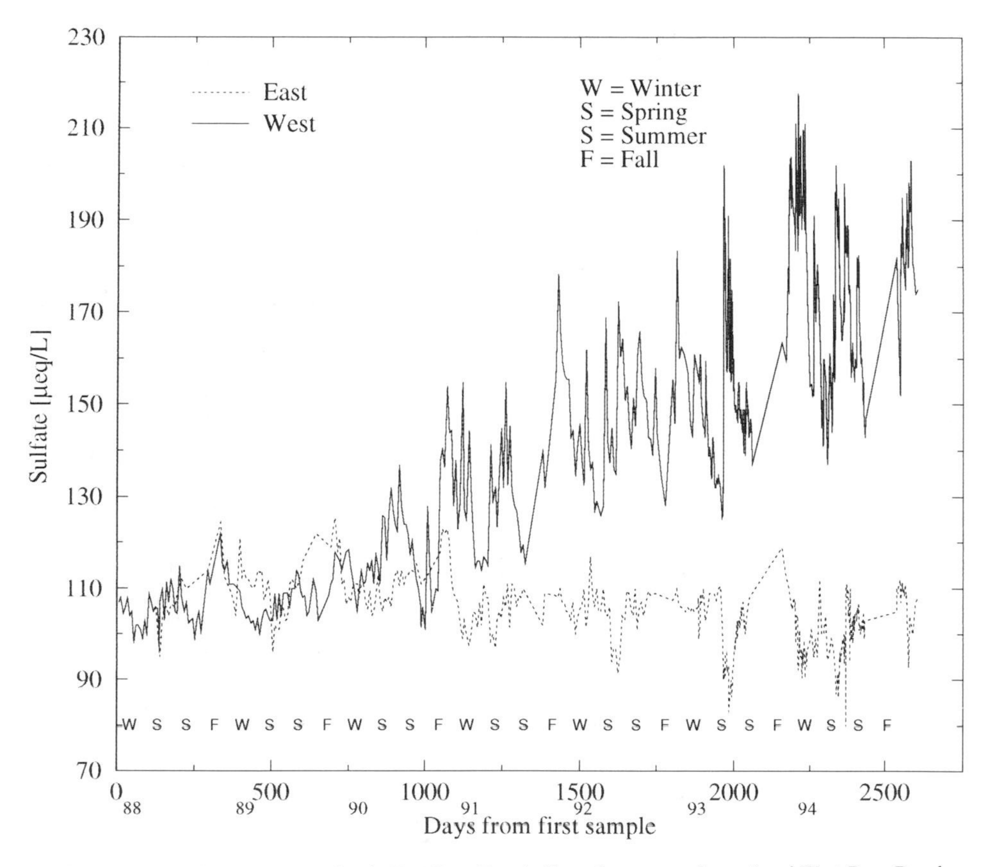

Figure 10.24 Sulfate concentration in East Bear Brook (the reference catchment) and West Bear Brook [receiving 19.2 (kg S) ha^{-1} yr^{-1} in the form of ammonium sulfate additions]. Ammonium sulfate additions began on day 730 in November 1989. The sulfate concentration in West Bear Brook (*solid line*) has increased significantly compared to the reference (*dashed line*).

to Bear Brook, acidifying the soil and the Brook and releasing inorganic monomeric aluminum (Figure 10.25). Total aluminum concentrations doubled from the preapplication period, and high concentrations of inorganic monomeric aluminum were evident during spring snowmelt and runoff events.

A significant increase in the export of nitrate-nitrogen entering Bear Brook was observed after applications of ammonium sulfate began (Figure 10.26), indicating that excess nitrogen had been nitrified and that the soils were somewhat nitrogen saturated. Approximately 2.5 (kg N) ha^{-1} yr^{-1}, 15% of the application rate, was measured in West Bear Brook in 1991–92 (Figure 10.26). Vegetation could not take up all of the applied nitrogen, and most of it was incorporated into the immobilized organic fraction of the soil, ~12–13 (kg N) ha^{-1} yr^{-1}. The chemistry and the fate of this immobilized organic nitrogen in the upper soil are not very well understood, yet, it accounts for a very large pool that could supply ammonium (via mineraliza-

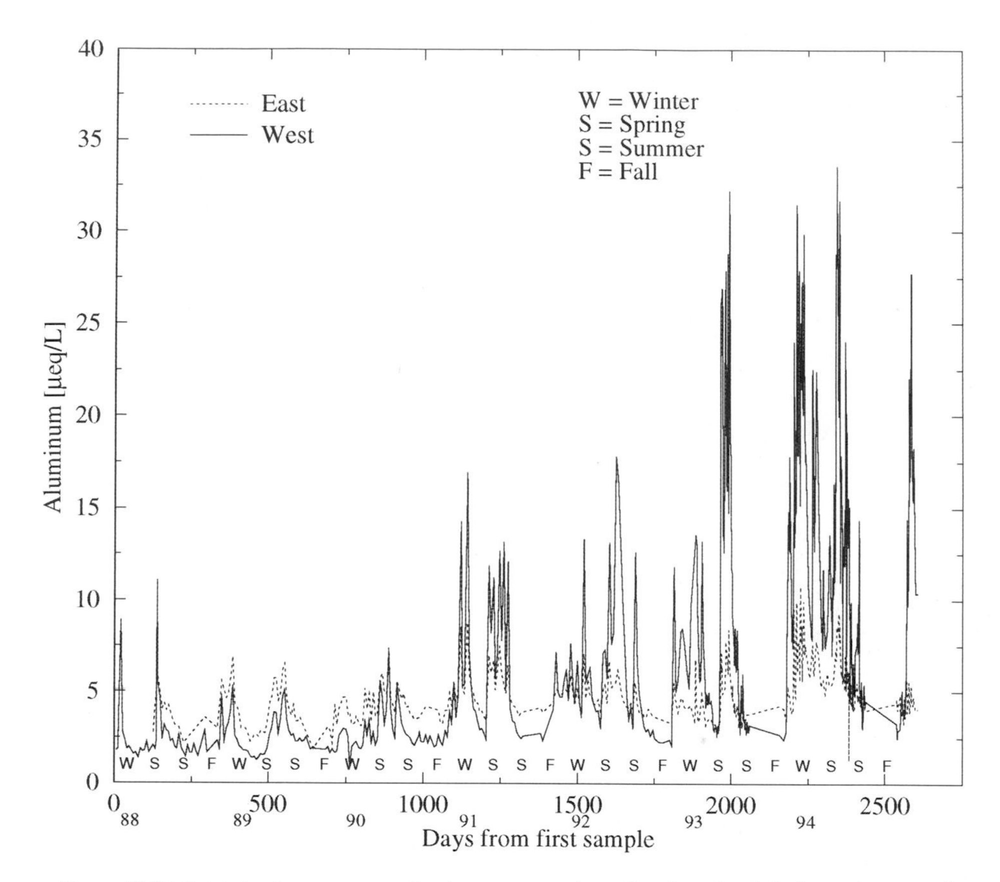

Figure 10.25 Total aluminum concentration in stream samples at East Bear Brook (reference) compared to West Bear Brook, which was acidified via the addition of ammonium sulfate. Total aluminum concentrations in West Bear Brook (*solid line*) have increased since the manipulation began on day 730.

tion) for the annual increment of the forest for many years. Uptake by vegetation only accounted for 1–2 (kg ^{15}N) ha^{-1} yr^{-1} of the 16.8 (kg N) ha^{-1} yr^{-1} applied, so the forest was fertilized but not as much as might have been expected. The organic immobile pool in the soil is large, 350–800 (kg N) ha^{-1}, and it is equivalent to about 20 years of nitrogen application.

The manipulation at West Bear Brook testifies to the importance of long-term, large-scale field experiments to test models and hypotheses. It appears that the forest is leaking nitrogen more readily than was expected, even though it is a small fraction of the total amount that was applied. A better understanding is needed of the large immobile organic pool that accumulates in forests and the effects of excess nitrate on ecosystems. The application rate of 16.8 (kg N) ha^{-1} yr^{-1} is not a large loading, and critical load estimates for the region would probably range ~20–30 (kg N) ha^{-1} yr^{-1}. Because the background levels of nitrogen and acid deposition to West Bear Brook are ~10–12 (kg N) ha^{-1} yr^{-1}, the total nitrogen deposition to West Bear

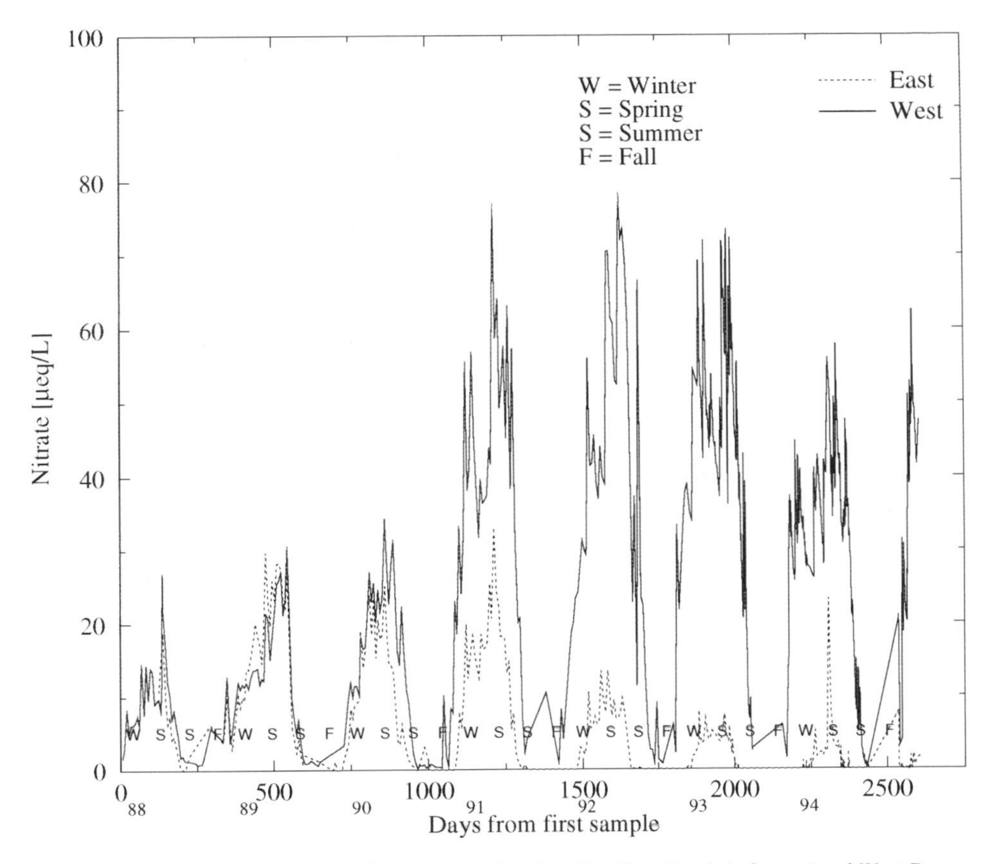

Figure 10.26 Nitrate concentration in stream samples from East Bear Brook (reference) and West Bear Brook, amended with 16.8 (kg N) ha^{-1} yr^{-1} beginning on day 730 (November 1989). Nitrate concentrations in West Bear Brook (*solid line*) increased steadily since the applications began, and the export of nitrate-nitrogen amounts to ~15% of the ammonium sulfate additions.

Brook is in the range of the critical loading, making the manipulation a particularly pertinent experiment. So far, long-term harmful effects on the forest have not been observed.

10.9 METALS DEPOSITION

Two major pollution problems in the world today—acid deposition and metal pollution—are both manifestations of acid–base disturbances in the most general sense. As we have seen, deposition of acidity results from intensive oxidative production of Brønsted acids (H_2SO_4, HNO_3), causing increased BNC in ecosystems. Metal pollution introduces excessive quantities of certain Lewis acid metals to ecosystems.[91] The biological consequences of metal pollution strongly depend on the resulting chemical speciation, which is a function of the kinds and amounts of Lewis bases, the redox intensity (pϵ), and the acidity–alkalinity (pH) characteristics of particular environments. In the biota, the key Lewis bases include ligands (in solution as well as on particle surfaces) containing oxygen (O) donors (e.g.,—OH, RCOOH), sulfur (S) donors (e.g.,—SH), or nitrogen donors (e.g.,—NH,—NH_2), which can become coordinated to *essential trace metals*, (e.g., Mg, Mn, Fe, Co, Zn, Cu).

In biochemistry, geochemistry, environmental chemistry, and the chemistry of metal emission control, it is important to recognize and make use of the general pattern of chemical affinities between Lewis acids and bases, that is the energetics of equilibria for all reactions of interest of metal coordination

$$M_i + L_j \rightleftharpoons M_i L_j \tag{75}$$

including aqueous, solid, and surface complex species and extending to multiple metal (e.g., polynuclear) or multiple ligand (e.g., mixed ligand) forms. The *aqueous proton* is one of the most influential M_i, and the hydroxide ion one of the more influential ligands. The redox state of the electron acceptors, M_i, and of the donors, L_j, *strongly* affect the affinities (e.g., Fe^{3+} versus Fe^{2+} and SO_4^{2-} versus S^{2-}) as shown by Morgan and Stumm.[91]

The proportion of free metal ions and thus potential ecologically adverse effects increase markedly with lower pH because H^+ ions compete successfully with metal ions for the available ligands:

$$CO_3^{2-} + H^+ = HCO_3^-$$

$$RCOO^- + H^+ = RCOOH \tag{76}$$

$${=}MeO^- + H^+ = {=}MeOH$$

$$Al(OH)_3 + H^+ = Al(OH)_2^+$$

$$CuCO_{3(aq)} + H^+ = Cu^{2+} + HCO_3^-$$

$$RCOO\,Zn^+ + H^+ = RCOOH + Zn^{2+} \tag{77}$$

$${=}MeOPb^+ + H^+ = {=}MeOH + Pb^{2+}$$

A schematic of metal deposition, fate, and transport is provided in Figure 10.27. For example, Sigg[92] has shown that heavy metal concentrations measured in acidic mountain lakes despite significantly smaller pollutional metal input are much larger than those measured in lakes in eutrophic carbonate-bearing watersheds because metal ion binding by sorption to particle surfaces decreases with decreasing pH. The scavenging of metals by sinking particles is less in acid lakes; thus the concen-

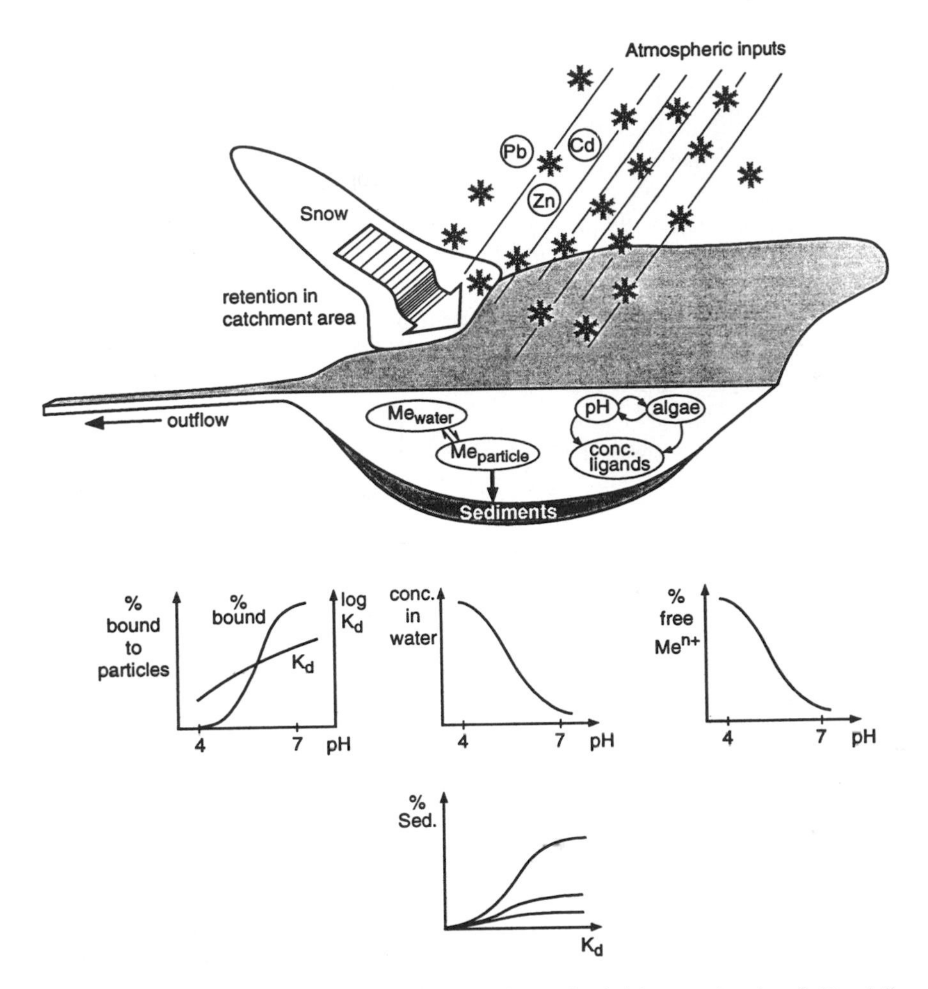

Figure 10.27 Schematic of metal ion deposition and sedimentation in lakes as a function of pH and distribution coefficient K_d (mL g^{-1}). Acidification of lakes increases metal concentration, free metal ion activity, and toxicity to aquatic biota.[92]

tration of dissolved metal ions in acid lakes is often larger in the hypolimnetic waters than in the sedimentary pore waters, especially if redox conditions are conducive to metal sulfide formation; steep concentration gradients develop at the sediment–water interface.[93]

A comparison of typical metals essential for life with the potentially *hazardous* metals is of interest (metalloids are underlined):

Essential metals: Na, K, Mg, Ca, Cr, Mn, Fe, Co, Ni, Cu, Zn, Mo
Hazardous metals: Cr, Cu, Zn, As, Se, Ag, Cd, In, Sn, Sb, Hg, Tl, Pb, Bi

10.9.1 Cycling of Metals

The dispersion of metals into the atmosphere appears to rival and sometimes exceed natural mobilization. Despite many uncertainties, the conclusion emerges from Sposito's review[94] that the trace metals Cu, Zn, Ag, Sb, Sn, Hg, and Pb are the most potentially hazardous on a global or regional scale. Lead is of acute concern on the global scale, because of its prominent showing in all the enrichment factors and transfer rates considered. These conclusions are in accord with those of Andreae and Byrd.[95]

The B-metals (i.e., the soft Lewis acids) are not only enriched in the natural environment, but they are also, because of their toxic effect (i.e., their tendency to react with soft bases, e.g., to SH and NH groups in enzymes), potentially hazardous to ecology and human health.

Settling particles, especially bioorganic particles, play a dominating role in binding heavy metals and transferring them to the deeper portion of oceans and lakes, where they are partially mineralized and transformed into the sediments. Biological surfaces are especially efficient scavengers—probably (for equal specific surface area) better ones than the mineral surfaces—for heavy metals.[96–98]

Lakes, despite being polluted with metal ions 10–100 times as much as oceans from riverine and atmospheric inputs, are often nearly as much depleted in these trace metals as are the oceans. Therefore the elimination mechanisms in lakes must be more efficient than those in oceans. Larger productivities and higher sedimentation rates for particles are primarily responsible for the more efficient scavenging in lakes through adsorption of metals on phytoplankton and to a lesser extent by other particles. The input of phosphorus into a lake, influencing the production of biogenic particles, is a major factor controlling the sedimentation rate of biogenic particles and in removing "biophile" heavy metals.[97] Settle and Patterson[99] have used data on the memory record of sediments in the Pacific to compare prehistoric and present-day eolian inputs. These data suggest that the present Pb(II) input is two orders of magnitude larger than that of prehistoric time. What does this mean with regard to the ecology? We really do not know.

10.9.2 Lead in Soils

Acid deposition is accompanied by toxic metals from industrial point sources and mobile sources. One of the most serious metals from a health perspective has been lead, a primary anti-knock compound in gasoline in the United States until 1974. Deposition of acids may affect the fate and transport of metals as they move through the soil horizons to groundwater or surface water. In the case of lead, it is important to understand factors controlling the recovery time for ecosystems following a major management action, such as the removal of lead from gasoline. Friedland et al.[100] reported the rapid decrease of lead in the forest floor of the northeastern United States between 1980 and 1990, a 12–17% change in lead concentrations. The estimated residence time for lead in the organic horizon of forest soils is approximately 150–500 years, so a change of 12–17% in the lead concentration in only ten years is, indeed, unexpected. The forest floor is defined as the top few centimeters of soil that is rich in organic carbon (50–75% by mass) and that overlies the mineral soil in forests.

A case study for lead has been conducted in the Hubbard Brook Experimental Forest (HBEF), U.S. Department of Agriculture, in central New Hampshire. Figure 10.28 summarizes the research of Driscoll et al.[101] at Hubbard Brook and the Coun-

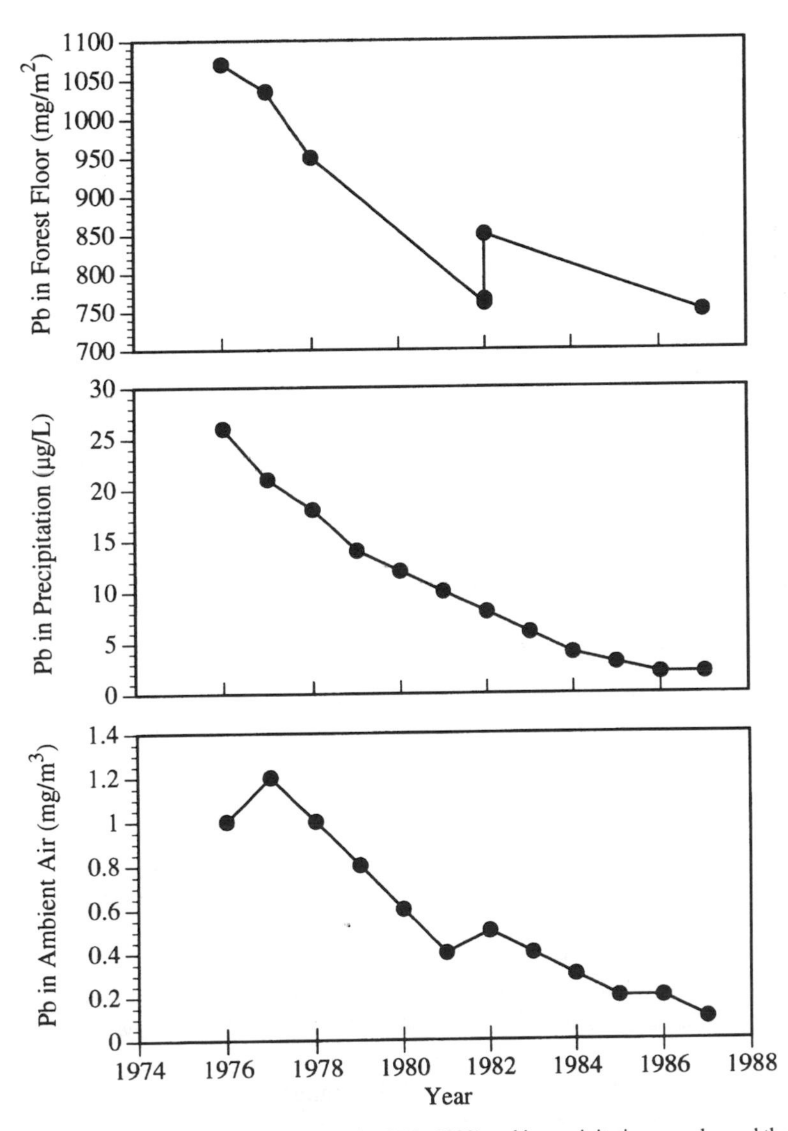

Figure 10.28 Lead concentrations in ambient air (CEQ, 1990) and in precipitation samples and the forest floor at the Hubbard Brook Experimental Forest, New Hampshire, 1976–1987. (From Driscoll et al.[101])

cil on Environmental Quality[102] ambient air quality monitoring network in the Northeast. Concentrations of lead in ambient air decreased rapidly after 1974, from ~1.2 mg m^{-3} to ~0.1 mg m^{-3}. The dramatic improvement in air quality was corroborated by the decrease of Pb in precipitation samples collected at HBEF between 1976 and 1988. The data at HBEF demonstrate a dramatic decrease of Pb in the forest floor. Although Pb has been leaving the organic soil horizon, it has not been measured in streams draining the forests, and it would seem that it is adsorbing in the deeper mineral soils. It is likely that there is a large labile pool of Pb in the forest floor that responds to changes in deposition and biomineralization very rapidly, but the exact mechanism of mobilization has yet to be determined.[100]

10.10 REFERENCES

1. Stumm, W., Sigg, L., and Schnoor, J.L., *Environ. Sci. Technol.,* 21, 8 (1987).
2. Stumm, W., and Morgan, J.J., *Aquatic Chemistry,* 2nd ed., Wiley-Interscience, New York (1981).
3. Stumm, W., Morgan, J.J., and Schnoor, J.L. *Naturwissenschaften,* 70, 216 (1983).
4. Schlesinger, W.H., *Global Biogeochemistry,* Academic Press, New York (1991).
5. Samson, P.J., and Small, M.J., in *Modeling of Total Acid Precipitation Impacts.* J.L. Schnoor, Ed., Butterworth, Boston (1984).
6. Seinfeld, J.H., *Atmospheric Chemistry and Physics of Air Pollution,* Wiley-Interscience, New York, p. 738 (1985).
7. Lindberg, S.E., et al. *Science,* 231, 141 (1986).
8. Keene, W.C., and Galloway, J.N., *Atmos. Environ.,* 18, 2491 (1984).
9. Johnson, C.A., et al. *EAWAG News,* 18/19, 13 (1985).
10. Sigg, L., and Stumm, W., *Aquatische Chemie,* VDF Publishers, Zürich (1989).
11. Charlson, R.J., and Rodhe, H., *Nature* 295, 683 (1982).
12. Liljestrand, H.M., Atmos. *Environ.,* 19, 487 (1985).
13. Hoffmann, M.R., and Boyce, S.D., *J. Phys. Chem.,* 88, 4740 (1984).
14. Johnson, C.A., and Sigg, L., *Chimia,* 39, 59 (1985).
15. Waldman, J.M., *Science,* 218, 677 (1982).
16. Morgan, J.J., in *Atmospheric Chemistry,* E.D. Goldberg, Ed., Springer-Verlag, Berlin (1982).
17. Jacob, D.J., Munger, J.W., Waldman, J.M., and Hoffman, M.R., *J. Geophys. Res.,* 91, 1073 (1986).
18. Behra, P., Sigg, L., and Stumm, W., *Atmos. Environ.,* 23, 2691 (1989).
19. Schwartz, S.E., and Slinn, W.G.N., Eds., *Precipitation Scavenging and Atmosphere–Surface Exchange,* Vols. 1, 2, and 3, Hemisphere Publishing, Washington, DC (1992).
20. Pruppacher, H.R., Semonin, R.G., and Slinn, W.G.N., Eds., *Precipitation Scavenging, Dry Deposition, and Resuspension,* Vols. 1 and 2, Elsevier, New York (1983).
21. Eisenreich, S.J., Ed., *Atmospheric Pollutants in Natural Waters,* Ann Arbor Science, Ann Arbor, MI (1981).

22. Gatz, D.F., in *Precipitation Scavenging—1974,* R.W. Beadle and R.G. Semonin, Eds., CONF-741014 ERDA, NTIS, Springfield, VA, pp. 71–87 (1975).
23. Schwarzenbach, R.P., Gschwend, P.M., and Imboden, D.M., *Environmental Organic Chemistry,* Wiley, New York (1993).
24. Davidson, C.I., and Wu, Y.L., *Acidic Precipitation—Sources, Deposition, and Canopy Interactions,* Vol. 3, S.E. Lindberg, A.L. Page, and S.A. Norton, Eds., Springer-Verlag, New York, pp. 103–216 (1990).
25. Schmel, G.A., and Sutter, S.L., *Particle Deposition Rates on a Water Surface as a Function of Particle Diameter and Velocity,* BNWL-1850, Battelle Pacific Northwest Laboratories, Richland, WA (1974).
26. Milford, J.B., and Davidson, C.I., *J. Air Pollut. Control Assoc.,* 35, 1249 (1985).
27. Schnoor, J.L., and Stumm, W., in *Chemical Processes in Lakes,* W. Stumm, Ed., Wiley Interscience, New York (1985).
28. Giovanoli, R., Schnoor, J.L., Sigg, L., Stumm, W., and Zobrist, J., *Clays Clay Miner.,* 36:6, 521 (1988).
29. Schnoor, J.L. and Stumm, W,. *Schweiz. Z. Hydrol.,* 48, 171 (1986).
30. Schnoor, J.L., in *Aquatic Chemical Kinetics,* W. Stumm, Ed., Wiley Interscience, New York (1990).
31. Stumm, W., *Chemistry of the Solid–Water Interface*, Wiley-Interscience, New York (1992).
32. Wieland, E., Wehrli, B., and Stumm, W., *Geochim. Cosmochim. Acta,* 52, 1969 (1988).
33. Schindler, P.W., in *Adsorption of Inorganics at Oxide Surfaces,* M.A. Anderson and A. Rubin, Eds., Ann Arbor Science, Ann Arbor, MI (1982).
34. Sigg, L., and Stumm, W., *Colloids Surf.,* 2, 101 (1980).
35. Schott, J., Berner, R.A., and Sjöberg, E.L., *Geochim. Cosmochim. Acta,* 45, 2123 (1981).
36. Busenberg, E., and Clemency, C.V., *Geochim. Cosmochim. Acta,* 41, 41 (1976).
37. Busenberg, E., and Plummer, L.N., *Am. J. Sci.* 282, 45 (1982).
38. Furrer, G., and Stumm, W., *Chimia* 37, 338 (1983).
39. Furuichi, R., Sato, N., and Okamoto, G., *Chimia* 23, 455 (1969).
40. Grandstaff, D.E., *Geochim. Cosmochim. Acta,* 41, 1097 (1977).
41. Sigg, L., and Stumm, W., *Aquatische Chemie,* 4th ed., Teubner Publishers, Stuttgart (1996).
42. Berner, R.A., and Holdren, G.R. Jr., *Geology,* 5, 369 (1977).
43. Berner, R.A., and Holdren, G.R. Jr., *Geochim. Cosmochim. Acta,* 43, 1173 (1979).
44. Berner, R.A., Sjöberg, E.L., Velbel, M.A., and Krom, M.D., *Science,* 207, 1205 (1980).
45. Schott, J., Berner, R.A., and Sjöberg, E.L., *Geochim. Cosmochim. Acta,* 43, 1161 (1979).
46. Velbel, M.A., *Am. J. Sci.,* 285, 904 (1985).
47. Siegel, D.I., and Pfannkuch, H.O., *Geol. Soc. Am. Bull.,* 95, 1446 (1984).
48. Mast, M.A., and Drever, J.I., *Geochim. Cosmochim. Acta,* 51, 2559 (1987).
49. Paces, T., *Geochim. Cosmochim. Acta,* 47, 1855 (1983).
50. Reuss, J.O., and Johnson, D.W., *J. Environ. Qual.,* 14, 26 (1985).

51. Reuss, J.O., *J. Environ. Qual.*, 12, 591 (1983).
52. Cosby, B.J., Hornberger, G.M., and Galloway, J.N., *Water Resour. Res.*, 21, 51 (1985).
53. McBride, M.B., *Environmental Chemistry of Soils,* Oxford University Press, Oxford, England (1994).
54. Lindsay, W.L., *Chemical Equilibria in Soils,* Wiley, New York (1979).
55. Sposito, G., *The Chemistry of Soils,* Oxford University Press, Oxford, England (1989).
56. Linhurst, R.A., Landers, D.H., Eilers, J.M., Brakke, D.F., Overton, W.S., Meier, E.P. and Crowe, R.E., *Characteristics of Lakes in the Eastern United States,* EPA/600/4-86/007a, U.S. Environmental Protection Agency, Washington, DC (1988).
57. Lee, S., and Schnoor, J.L., *Environ. Sci. Technol.*, 22, 190 (1988).
58. Driscoll, C.T., Baker, J.P., Bisogni, J.J., and Schofield, C.L., *Nature,* 284, 161 (1980).
59. Schindler, D.W., in *Chemical Processes in Lakes,* W. Stumm, Ed., Wiley, New York (1985).
60. Urban, N.R., in *Environmental Chemistry of Lakes and Reservoirs,* American Chemical Society, New York (1994), pp. 323–369.
61. Henriksen, A., *Regional Survey of Lakewater Chemistry of Large Lakes in South Norway,* TN 50/79 SNSF, Norwegian Institute for Water Research, Oslo (1979).
62. Gherini, S.A., Mok, L., Hudson, R.J.M., Davis, G.F., Chen, C.W., and Goldstein, R.A., *Water Air Soil Pollut.*, 26, 425 (1985).
63. Schnoor, J.L., Palmer, W.D. Jr., and Glass, G.E., in *Modeling of Total Acid Precipitation Impacts,* J.L. Schnoor, Ed., Butterworth, Stoneham, MA (1984).
64. Lin, J.C., and Schnoor, J.L., *J. Environ. Eng.*, 114, 677 (1986).
65. Nikolaidis, N.P., Rajaram, H., Schnoor, J.L., and Georgakakos, K.P., *Water Resour. Res.*, 24, 1983 (1988).
66. Cosby, B.J., Wright, R.F., Hornberger, G.M., and Galloway, J.N. *Water Resour. Res.*, 21, 1591 (1985).
67. Christophersen, N., Seip, H.M., and Wright, R.F., *Water Resour. Res.*, 18, 977 (1982).
68. Rees, T.F., and Schnoor, J.L., *J. Environ. Eng.* 120, 291 (1994).
69. Ulrich, B., *Fortswiss. Centralbl.*, 100, 228 (1981).
70. Van Breeman, N., Driscoll, C.T., and Mulder, J., *Nature,* 307, 599 (1983).
71. Schindler, D.W., *Can. J. Fish. Aquat. Sci.*, 44, 6 (1987).
72. Schindler, D.W., Mills, K.H., Malley, D.F., Findlay, D.L., Shearer, J.A., Davies, I.E., Turner, M.A., Lindsey, G.A., and Cruikshank, D.R., *Science,* 228, 1395 (1985).
73. Murdock, P.S., and Stoddard, J.L., *Water Resour. Res.*, 28, 2707 (1992).
74. Herlihy, A.T., Kaufmann, P.R., Church, M.R., Wigington, P.J., Webb, J.R., and Sale, M.J., *Water Resour. Res.*, 29, 2687 (1993).
75. Rosen, K., Gundersen, P., Tegnhammer, L., Johansson, M., and Frogner, T., *Ambio,* 21, 364 (1992).
76. Brodin, Y.W., and Kuylenstierna, J.C.I., *Ambio* 21, 332 (1992).
77. Smith, C.M.S., Cresser, M.S., and Mitchell, R.D.J., *Ambio,* 22, 22 (1993).
78. Kamari, J., Forsius, M., and Posch, M., *Water Air Soil Pollut.*, 66, 77 (1993).
79. Posch, M., Forsius, M., and Kamari, J. *Water Air Soil Pollut.*, 66, 173 (1993).
80. Henriksen, A., Kamari, J., Posch, M., and Wilander, A., *Ambio,* 21, 356 (1992).

81. Holdren, G.R. Jr., Strickland, T.C., Shaffer, P.W., Ryan, P.F., Ringold, P.L., and Turner, R.S., *J. Environ. Qual.,* 22, 279 (1993).
82. Sverdrup, H., Warfvinge, P., Frogner, T., Haoya, A.O., Johansson, M., and Andersen, B., *Ambio,* 21, 348 (1992).
83. Kamari, J., Amann, M., Brodin, Y.W., Chadwick, M.J., Henriksen, A., Hettlingh, J.P., Kuylenstierna, J., Posch, M., and Sverdrup, H., *Ambio,* 21, 377 (1992).
84. Hettlingh, J.P., Downing, R.J., and deSmet, P.A.M., Eds., *Mapping Critical Loads for Europe,* CCE Technical Report No. 1, National Institute of Public Health and Environmental Protection, Bilthoven, The Netherlands (1991).
85. Eriksson, E., Karitun, E., and Lundmark, J.E., *Ambio,* 21, 150 (1992).
86. Alcamo, J., Shaw, R., and Hordijk, L., *The RAINS Model of Acidification: Science and Strategies in Europe,* Kluwer, Dordrecht, The Netherlands (1990).
87. Furrer, G., Wetall, J., and Sollins, P., *Geochim. Cosmochim. Acta,* 53, 595 (1989).
88. Warfvinge, P., Holmberg, M., Posch, M., and Wright, R.F., *Ambio,* 21, 369 (1992).
89. Lee, S., Georgakakos, K.P., and Schnoor, J.L., *Water Resour. Res.,* 26, 459 (1990).
90. Liken, G.E., Bormann, F.H., Pierce, R.S., Eaton, J.S., and Johnson, N.M., *Biogeochemistry of a Forested Ecosystem,* Springer-Verlag, New York (1977).
91. Morgan, J.J., and Stumm, W., in *Metals and Their Compounds in the Environment,* E. Merian, Ed., VCH Verlag, Weinham, Germany (1989).
92. Sigg, L., in *Chemical Processes in Lakes,* W. Stumm, Ed., Wiley, New York (1985).
93. Tessier, A., Carignan, R., Dubreul, B., and Rapin, F., *Geochim. Cosmochim. Acta,* 53, 1511 (1989).
94. Sposito, G., *CRC Crit. Rev. Environ. Control,* 16, 193 (1986).
95. Andreae, M.O., and Byrd, J.T., *Anal. Chim. Acta,* 156, 147 (1984).
96. Morel, F.M.M., and Hudson, R.J.M., in *Chemical Processes in Lakes,* W. Stumm, Ed., Wiley, New York (1985).
97. Sigg, L., in *Aquatic Surface Chemistry,* W. Stumm, Ed., Wiley, New York (1987).
98. Whitfield, M., and Turner, D.R., in *Aquatic Surface Chemistry,* W. Stumm, Ed., Wiley, New York (1987).
99. Settle, D.M., and Patterson, C.C., *Science,* 207, 1167 (1980).
100. Friedland, A.J., Craig, B.W., Miller, E.K., Herrick, G.T., Siccama, T.G., and Johnson, A.H., *Ambio,* 21, 400 (1992).
101. Driscoll, C.T., Iverfeldt, A., and Otton, J.K., in *Biogeochemistry of Small Catchments,* J. Cerny and B. Moldan, Eds., SCOPE Series, Wiley, New York (1994).
102. Council on Environmental Quality, *Environmental Quality,* Twentieth Annual Report, U.S. Govternment Printing Office, Washington, DC (1990).

11

GLOBAL CHANGE AND GLOBAL CYCLES

Sustainable development meets the needs of the present without compromising the ability of future generations to meet their own needs.

—Gro Brundtland, *Our Common Future* (1987)

11.1 INTRODUCTION

Humans have always been in tension with their environment as they seek out a better standard of living. Land has been cleared for agriculture or commerce, and animal populations have been exploited. What is different about the situation that we find ourselves in today is the magnitude of our impacts. We are more numerous: 5.7 billion people on earth seek out an existence, and they are multiplying. Every six months, there is another contingent of people equal in population to the country of France, almost 50 million. Imagine, every six months another France for whom to provide food, housing, shelter, and jobs. Every ten years, there is another population nearly the size of China. And coupled with population is the problem of an ever-increasing per capita consumption among developed countries, especially. We are powerful. Twin juggernauts—burgeoning population and per capita consumption—are driving global change.

In this chapter, we will see how mathematical models can help us to analyze where we are going from where we have been with respect to our global environs. Environmental modeling should aid us in decision-making about our collective future, as dynamics of our environmental and social systems become quite complex.

A projection of global population from 1750 to 2100 is shown in Figure 11.1.[1] Following the industrial revolution, population in developed countries increased rapidly from 1850 to 1950. But more recently, developing countries in Asia, Africa, and South America have increased population as the industrializing countries did 100 years ago. Global population is projected to rise rapidly until the end of the 21st century and then stabilize at 10–11 billion. Population increase is the first driving force of global change.

The second driving force of global change is per capita consumption. It has

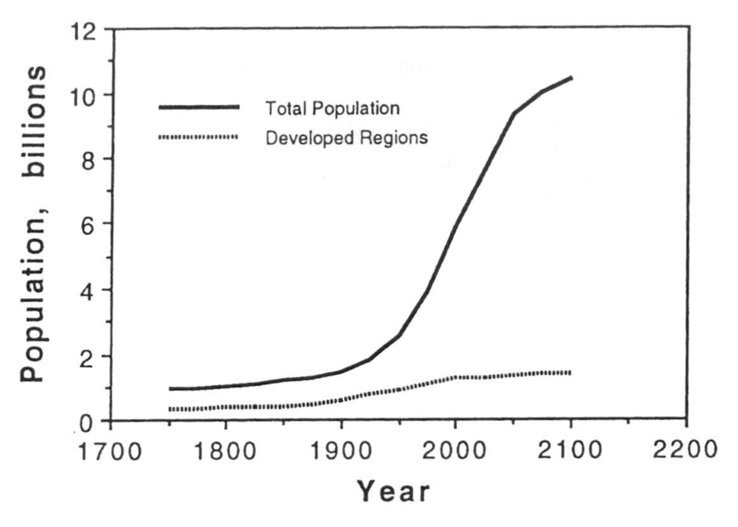

Figure 11.1 Historical and projected world population.

grown as fast or faster than population. From 1950 to 1986, the population doubled from 2.5 billion to 5.0 billion, a rate of 2.0% per year. But during that same period, the annual growth in global primary energy consumption was 4.0% per year, about double that of population. The example shows that per capita consumption, especially among developed countries, has contributed significantly to our energy and pollution (emission) problems. Global primary energy consumption is an indicator of global emissions to water, soil, and atmosphere (Figure 11.2).

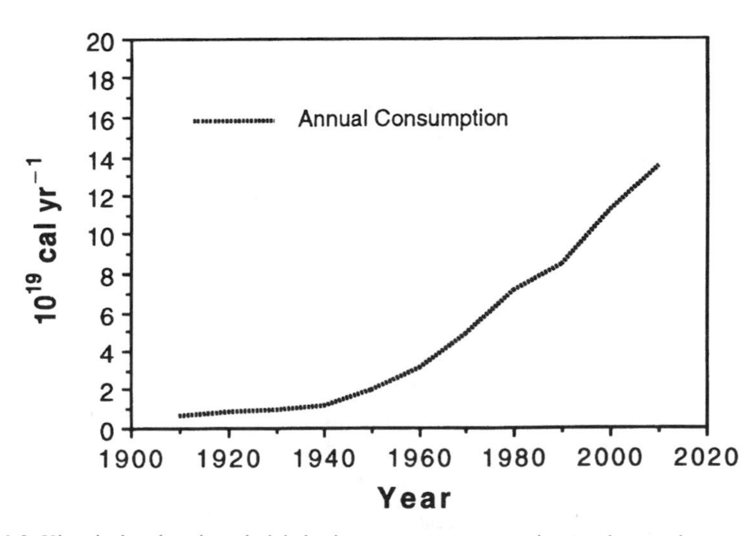

Figure 11.2 Historical and projected global primary energy consumption (coal, natural gas, petroleum, wood, nuclear, hydro).

Atmosphere is a relatively small compartment in the earth's movement of environmental pollutants. It contains a mass of 5×10^{18} kg of air that is dwarfed by the ocean (1.4×10^{21} kg for 3800-m deep ocean). Furthermore, residence times of pollutants in the atmosphere are short compared to the ocean. It is relatively easy to pollute the atmosphere. It responds quickly to global emissions, but it also cleanses itself quickly when emissions are abated.

Greenhouse gases are increasing in the atmosphere. Carbon dioxide from fossil fuel emissions is increasing at 0.4–0.5% per year (Figure 11.3). A little more than half of the anthropogenic greenhouse gas effect is due to carbon dioxide (Table 11.1).[2] Methane concentrations in the atmosphere, another potent greenhouse gas, are increasing at ~0.7% per year, mainly due to flooded agriculture (rice production) and animal husbandry required to feed an expanding population (Figure 11.4).[3] But the *rate of increase* in methane concentrations is beginning to decline, perhaps because leaks at natural gas pipelines have been controlled.

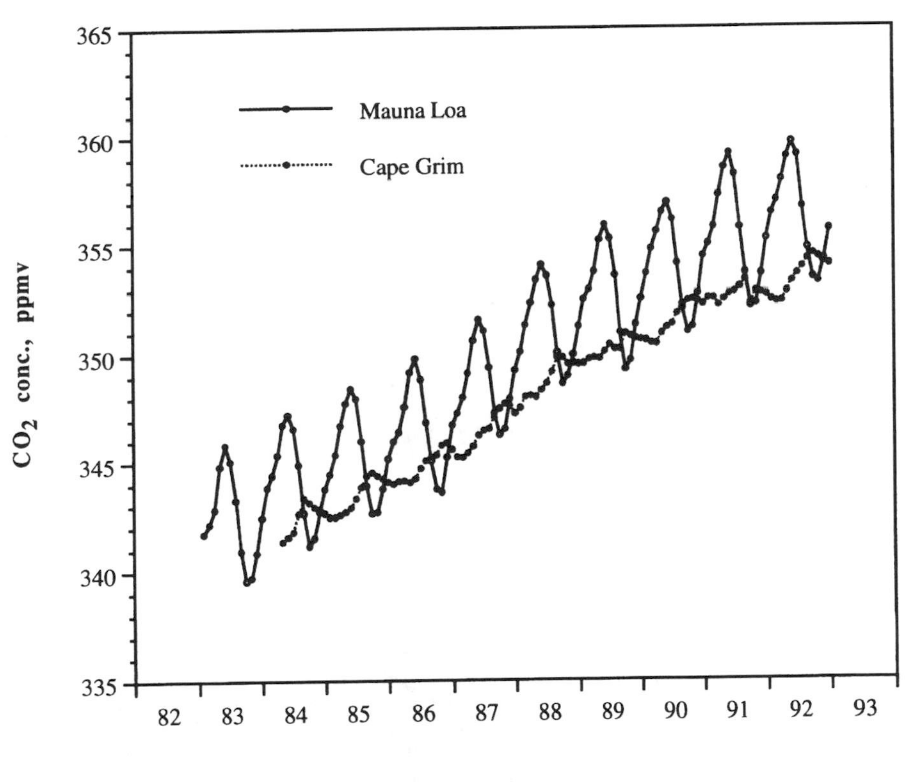

Figure 11.3 Trend of carbon dioxide concentration at two stations: in the northern hemisphere, Mauna Loa, Hawaii, and the southern hemisphere at Cape Grim, showing the exponential increase in CO_2 at 0.5% per year.[3]

Table 11.1 Summary of Major Anthropogenic Greenhouse Gases

Gas	Concentration,[a] ppmv	Greenhouse Contribution	Rate of Increase % yr^{-1}	Half-life, yr	Relative Greenhouse Effect	
					per kg	per mol
CO_2	355	60	0.4–0.5	150	1	1
CH_4	1.7	15–20	0.7–0.9	7–10	70	25
N_2O	0.31	5	0.2	150	200	200
O_3	0.01–0.05	8	0.1–0.5	0.01	1800	2000
$CFCl_3$	0.28 ppbv	4	0–4	65	4000	12,000
CF_2Cl_2	0.48 ppbv	8	0–4	120	6000	15,000

[a]Concentrations are on a volumetric basis (mixing ratios)
Source: Modified from Rodhe.[2]

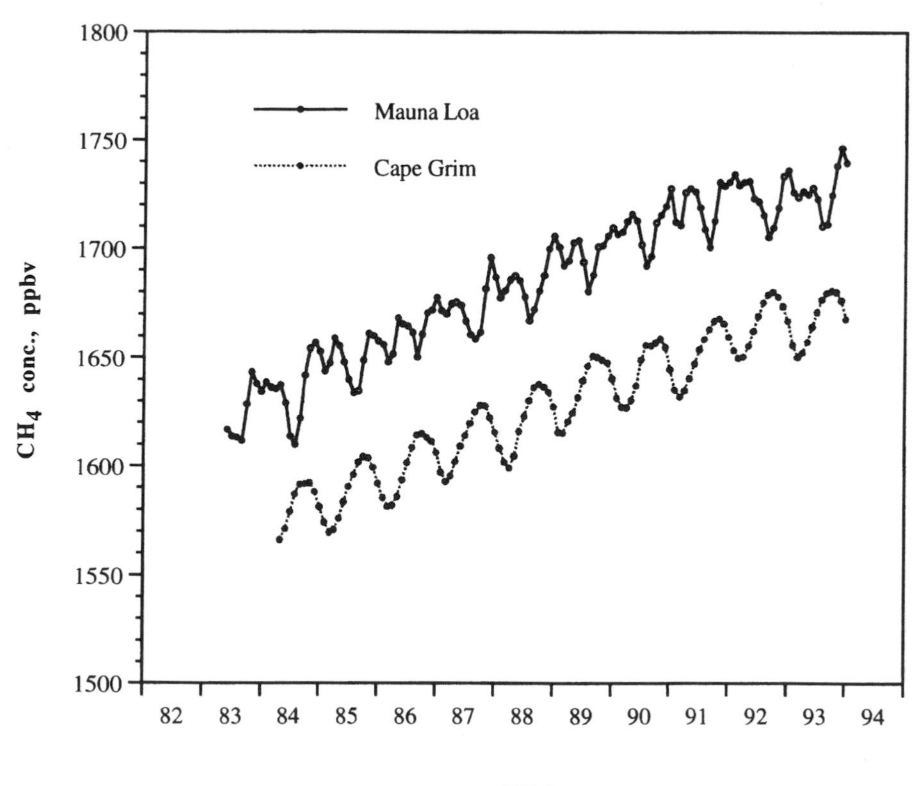

Figure 11.4 Trend of methane concentrations at two stations: in the northern hemisphere, Mauna Loa, Hawaii, and the southern hemisphere at Cape Grim, showing the exponential increase of methane at 0.9% per year but concentrations appear to be leveling off.[3]

Nitrous oxide (N_2O) is the result of increasing fossil fuel emissions, biomass burning, and nitrogen fertilizer applications worldwide. As ammonia-nitrogen is oxidized in the soil and nitrate is denitrified, nitrous oxide represents an intermediate oxidation state, which is volatilized to the atmosphere. Figure 11.5 shows that N_2O is increasing rapidly in the atmosphere and, even though it is only ~5% of the anthropogenic greenhouse gas effect, it will be a difficult gas to control. Fertilizer applications of ammonia, ammonium nitrate, and ammonium sulfate are likely to increase worldwide.

Chlorofluorocarbon (CFC) concentrations are also increasing, and they have a long half-life in the atmosphere (Table 11.1). They are used as refrigerants, blowing agents, and cleaning chemicals in the microelectronics industry. Two of the most widely used are CFC-11 ($CFCl_3$, trichlorofluoromethane) and CFC-12 (CF_2Cl_2, dichloro-difluoromethane). Due to a slow reaction rate with hydroxyl radical, ·OH, in the lower atmosphere, they are transported into the stratosphere (upper atmosphere), where they deplete the ozone layer that shields the earth from harmful ultra-

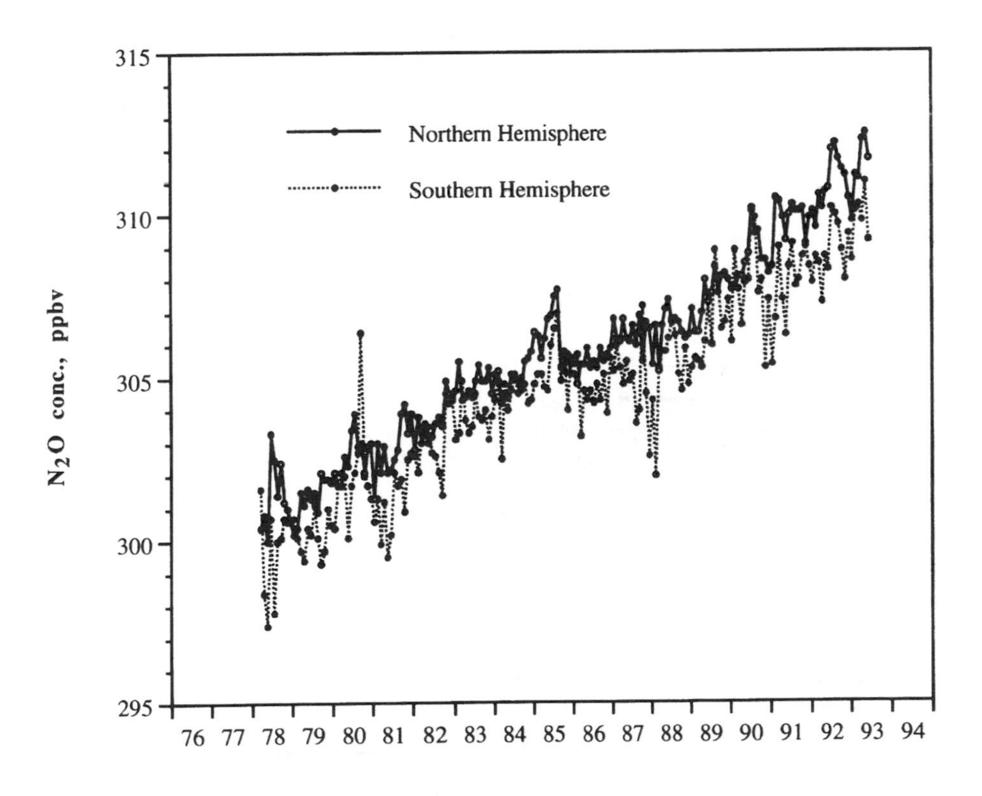

Figure 11.5 Trend of nitrous oxide concentrations at two stations in the northern and the southern hemispheres, showing the exponential increase of N_2O at 0.2% per year.[3]

violet radiation. The Montreal Protocol of 1987 and subsequent amendments (London, 1990; Copenhagen, 1992) have banned the production of CFCs by the year 2000. From conception of the problem by Mario Molina and Sherwood Rowland[4] in 1974, to measurement of the Antarctic ozone hole in 1985, and acceptance by industry and the scientific community in 1987, the control of CFC chemicals is one of the major success stories of international environmental agreements.[5,6] Methyl chloroform, a potent ozone-destroying chemical, is now decreasing in the atmosphere[6] as a result of the Montreal Protocol (Figure 11.6). Other CFC chemicals will do likewise within the next decade or so. For their vision, Molina and Rowland were awarded the Nobel Prize in Chemistry for 1995. Still, due to long half-lives of the chemicals, ozone depletion will continue for some years. In addition, CFC chemicals account for a significant fraction of the anthropogenic greenhouse gas effect.

Globally, the clearing of land for agriculture and commerce amounts to ~50,000 mi^2 per year or roughly the area of the State of Iowa. It results in destruction of habitats for plant, animal, and microbial species. Approximately 100,000 species are thought to become extinct per year, from a total of perhaps 4–40 million species.[7] Half of that biodiversity is in the tropics and developing countries, so the clearing of rain forests is of particular concern. Nations are clearing about 0.5% of the forests each year, and if this causes a loss of half of the natural species present, then the rate of extinction would be ~0.25% per year, a very grave long-term problem.

Global poverty is perhaps the greatest environmental problem of all: 1.3 billion people lack safe drinking water; 2.3 billion lack access to sanitation facilities; 1.5 billion do not have enough firewood or fuel for cooking and heating; and 13 million children die from diarrhea, dysentery, and hunger each year.[8] Until the developed

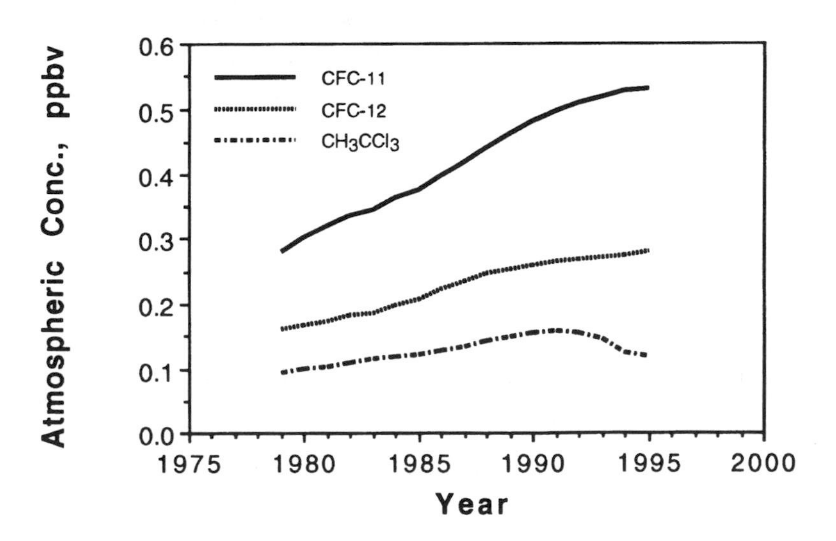

Figure 11.6 Chlorofluorocarbon (CFC) and methyl chloroform (CH_3CCl_3) concentrations (mixing ratios in ppb by volume) in the atmosphere.

world finds its way clear to invest in developing countries, it will be difficult to make collaborative progress on forest preservation, biodiversity, and CO_2 emissions. Developed countries must restrain their own per capita consumption (a major part of the problem) while helping other countries with their development. Much of the global atmospheric problems were created by emissions from the developed world, and they are most responsible for controlling them. One responsibility is to analyze global problems rationally using environmental models and by engineers contributing to international policy discussions and decisions.

Example 11.1 Mass of the Earth's Compartments

What is the mass of the atmosphere compared to that of the oceans and "living soil" (pedosphere)?

Solution: The column of air above earth exerts an atmospheric pressure that is a measure of the integrated mass. We can use pressure to estimate the mass of the atmosphere.

$$P_A = \text{Atmospheric pressure} = 101 \text{ kPa} = 101 \text{ kN m}^{-2}$$

$$S_E = \text{Surface area of earth} = 5.1 \times 10^{14} \text{ m}^2$$

$$M_A = \text{Mass of atmosphere} = \frac{P_A}{g} S_E = \left(\frac{101{,}000 \text{ N m}^{-2}}{9.807 \text{ m s}^{-2}}\right) (5.1 \times 10^{14} \text{ m}^2)$$

$$= 5.25 \times 10^{18} \text{ kg}$$

Earth's lower atmosphere (~10-km elevation) is termed the troposphere. The barometric pressure at that elevation is 220 millibars or 22 kPa. Thus the troposphere contains 78% of the mass of the total atmosphere. The earth's upper atmosphere or stratosphere (10–100-km elevation) contains the remainder.

$$\text{Mass (\%) in troposphere} = \frac{(101 - 22) \text{ kPa}}{101 \text{ kPa}}$$

$$= 78\%$$

The mass of the oceans is huge. Assuming a density of 1.03 g cm^{-3}, a depth of 3800 m, and a surface area of 70% of the earth's surface, the following calculations apply:

$$\text{Ocean mass} = S_0\, d\rho = (0.7 \times 5.1 \times 10^{14} \text{ m}^2)(3800 \text{ m})(1.03 \text{ g cm}^{-3})$$

$$= 1.40 \times 10^{21} \text{ kg}$$

The mass of living soil is small by comparison. Assuming a depth of 2 m, a bulk density of 1.5 kg L^{-1}, and a surface area of 25.5% of the earth's surface area, the mass of the soil is much smaller than the oceans and one order-of-magnitude smaller than the atmosphere.

$$\text{Soil mass} = S_s\, d\rho_b = (0.255 \times 5.1 \times 10^{14}\ \text{m}^2)(2\ \text{m})(1.5\ \text{kg L}^{-1})$$
$$= 3.9 \times 10^{17}\ \text{kg}$$

The large mass of the oceans indicates that residence times of pollutants could be very long compared to the atmosphere and soil. Later in the chapter, we will make estimates of atmospheric residence times.

11.2 CLIMATE CHANGE AND GENERAL CIRCULATION MODELS

11.2.1 Greenhouse Gases and Climate

The increase in "greenhouse gases" in the atmosphere has led to a concern regarding the potential for global warming and climate change. While it is certain that greenhouse gases that absorb long-wave radiation are increasing (Table 11.1), it is not certain whether climate has been affected. The natural structure (temperature and pressure) of the earth's atmosphere is shown by Figure 11.7.[9] Divisions of troposphere, stratosphere, mesosphere, and thermosphere are based on temperature variations.

Figure 11.8 is a schematic of the energy budget for the earth and its atmosphere.[10] Of the incoming short-wave solar radiation, only about 51% eventually strikes the earth and is absorbed. Twenty-six percent of the incoming solar radiation is reflected by clouds, dust, and haze. Once radiant energy strikes the earth and is absorbed, warming occurs and energy is emitted back to the atmosphere according to the Stefan–Boltzmann law of blackbody radiation. Eventually, 70% of all short-wave solar energy that is received by the earth's atmosphere is radiated back in the form of long-wave radiation. But water vapor, clouds, CO_2, and other greenhouse gases absorb back-radiation and cause the "greenhouse" effect.

The vast majority (~98%) of the greenhouse effect is natural, caused by water vapor, preindustrial CO_2, and clouds. Without these gases, the earth would be so cold as to be uninhabitable, 33 °C cooler than it is presently. The annual global mean surface temperature of the earth is 15 °C, and the global temperature in the absence of natural greenhouse gases would be –18 °C. The "greenhouse effect" is natural and it is beneficial to life on earth.[10]

Concern arises because of the increase of anthropogenic greenhouse gases including excess CO_2, CH_4, CFCs, and N_2O (Table 11.1). They are increasing at a rapid rate, and they have the potential to change climate if left unabated. Molecules with more than two atoms absorb long-wave radiation, and when they do, they increase their vibrational and rotational energies. There is a "window" in the absorption wavelength spectrum of the atmosphere that is particularly important to the planetary radiation budget, 8–20-μm wavelength, where H_2O, CO_2, O_3, N_2O, CH_4,

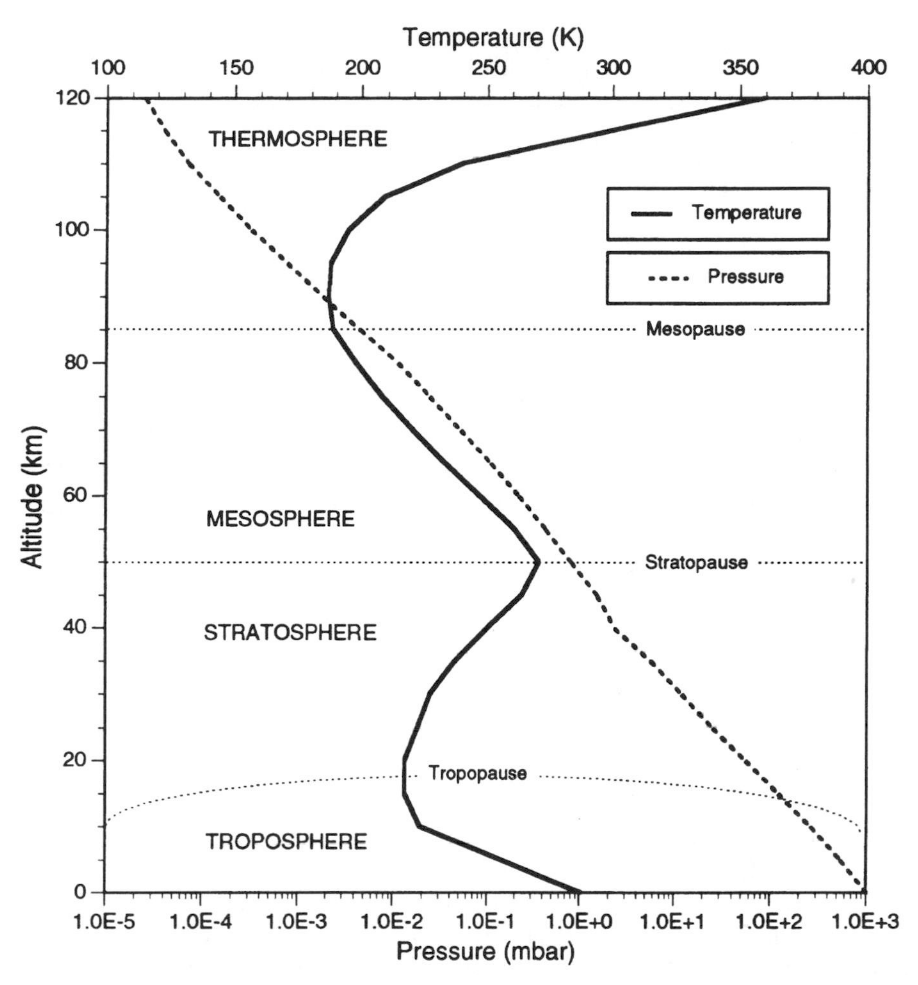

Figure 11.7 Typical vertical thermal and pressure structure of earth's atmosphere. From the U.S. Standard Atmosphere.[9] Reprinted by permission of Kluwer Academic Publishers.

$CFCl_3$, and CF_2Cl_2 are efficient absorbers. The relative effectiveness of each of the gases is not equal. Table 11.1 shows that $CFCl_3$ (CFC-11) and CF_2Cl_2 (CFC-12) are 12,000 and 15,000 times more effective, respectively, at absorbing long-wave radiation compared to carbon dioxide on a molar basis.[2]

11.2.2 Climate Projections

All of the information on greenhouse gases can be included in a model for planetary heat, mass, and momentum transfer. Global models of coupled atmospheric, terrestrial, and ocean processes have been summarized by the Intergovernmental Panel on Climate Change (IPCC).[11,12]

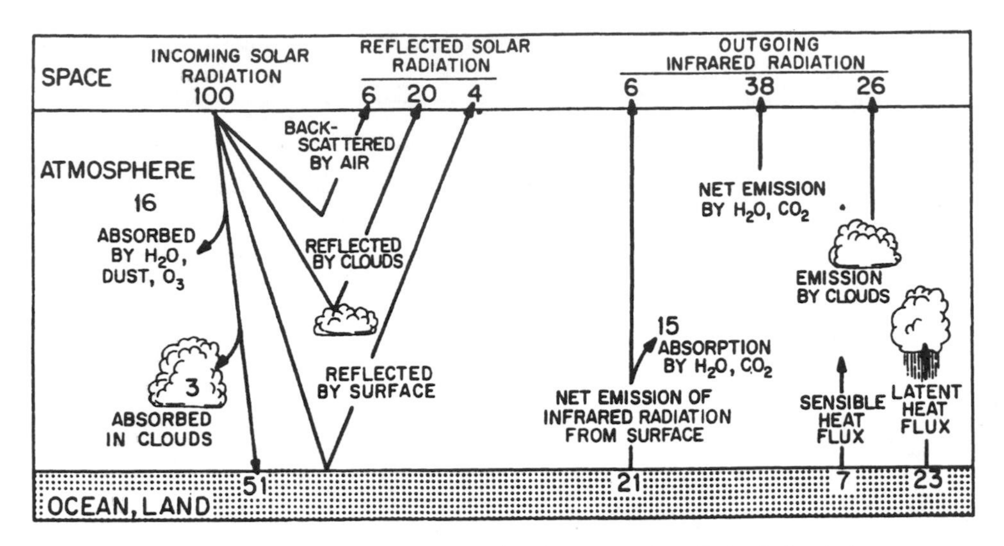

Figure 11.8 Earth atmospheric heat balance. Incoming short wave solar radiation is absorbed and warms the atmosphere. Some is back-scattered, and the remainder is eventually emitted as outgoing infrared radiation (steady state condition).[13]

Calculated changes in the earth's radiation budget as a result of increasing greenhouse gases are shown in Figure 11.9. Carbon dioxide has increased 27% from 280 ppm in 1765 to 355 ppm in 1990. A total change of 2.5 watts m^{-2} is attributed to anthropogenic greenhouse gases. To put the number into perspective, a standard Christmas tree light bulb puts out ~4 watts of energy, so if there was one additional light bulb of this kind shining continually on every square meter of the earth's surface, it would result in a climate forcing of 4 W m^{-2}. Nevertheless, the climate forcing is small relative to the total input of energy from the sun (the solar irradiance or solar constant) of 1372 W m^{-2}.[10] The solar constant is not constant—it varies with sunspot activities and other factors (± 0.1%)—and it may have increased a little in recent times (Figure 11.9). Also, humans have put more sulfur dioxide into the atmosphere (from coal combustion), which results in the formation of sulfate aerosols and brighter clouds (albedo increases) that causes "global cooling" (Figure 11.9).[11,12] This cooling effect is on the same order of magnitude, but somewhat less than, the estimated warming effect due to greenhouse gases. There is considerable uncertainty in all of these effects. They are estimated by use of global dynamic models, General Circulation Models or GCMs.

One of the "feedback effects" of greenhouse gases in GCMs is the potential for a slight surface warming to be amplified by clouds and water vapor. As the earth's surface becomes warmer, evaporation increases. As evaporation increases, water vapor increases and clouds are formed, which are potent greenhouse gases. This climate feedback can multiply the effect of CO_2 by four times in GCMs. We do not know how many new clouds would form or where they would be located in the atmosphere. Thus it is difficult to parameterize the models to determine the exact

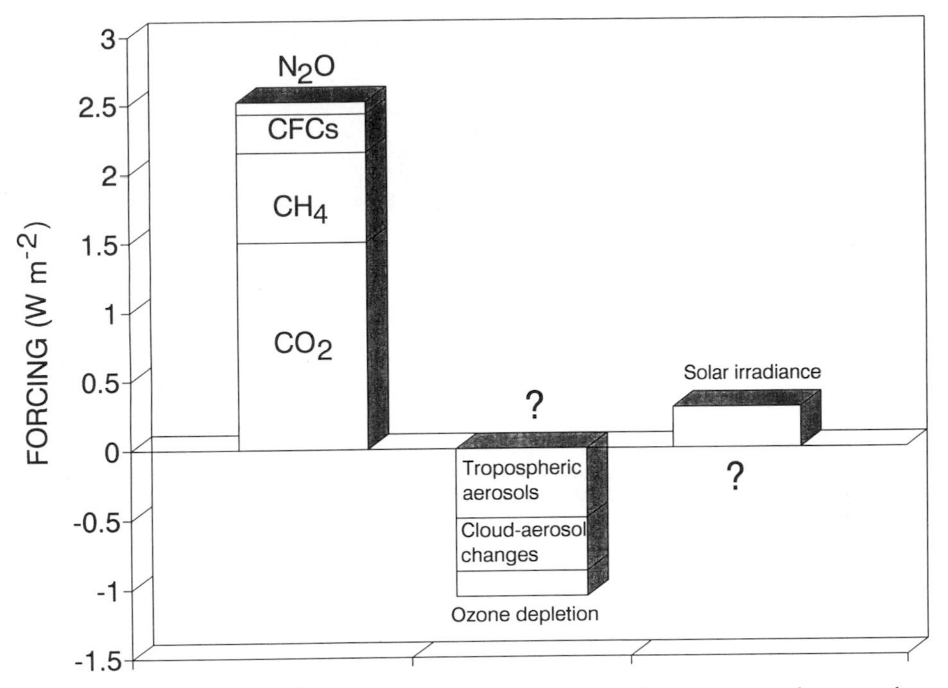

Figure 11.9 Changes in contributions (in W m^{-2}) to global warming of various gases and processes between 1765 and 1990 (IPCC).[11,12]

feedback effect. For now, one can only parameterize models in logical ways and compare the results of future scenarios and simulations from various models. At issue is the extent of warming, not the presence of warming. The relative degree of warming will be strongly affected by cloud and water vapor feedback mechanisms.

We are currently in a relatively warm climate pattern (Figure 11.10).[11,12] The global average surface temperature of the earth is ~0.5 °C warmer than the earliest period of record, 1860. But this temperature change is still within the interannual variability of temperature (±0.7 °C) due to large-scale circulation patterns (El Niño Southern Oscillation events, etc.) that we do not fully understand. Seven out of eight of the warmest years on record have occurred since 1980 (1980–1995) in a 130-year period. The year 1995 was the warmest on record and 1990 was the second warmest year. The years 1992 and 1993 were relatively cool because SO_2 and ash particles were blown high into the atmosphere (> 35,000 ft) by the eruption of Mt. Pinatubo volcano. All of these factors have been modeled with some successes and failures using GCMs.

11.2.3 Potential for Global Warming

General circulation models are driven by an assumed 1% yr^{-1} increase in CO_2 concentrations. This is done to mimic the ~0.5% yr^{-1} increase in CO_2 plus a similar increase in all other greenhouse gases. The best estimate of warming is 1.0–3.5 °C

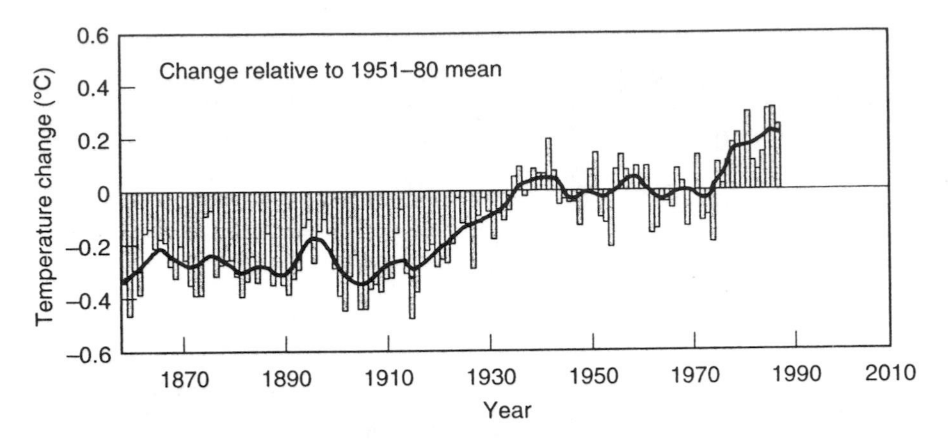

Figure 11.10 Annual deviation of the global mean (land and sea) temperature change over the past century relative to the average for 1951–1980. The curve shows the results of a smoothing filter applied to the annual values.[11]

with a most probable estimate of 2.0 °C by the middle of the 21st century.[12] If carbon dioxide and other greenhouse gases continue to increase, all models show a warming trend in the 21st century.

Global average surface air temperatures are currently about 0.5 °C higher than average temperatures of the 19th century. However, this change cannot yet be unambiguously attributed to an increased concentration of greenhouse gases. GCMs have hind-casted a larger change during the past 130 years than has been observed, and this is perhaps due to the cooling effects of anthropogenic aerosols.[13]

GCMs predict global precipitation to increase in the 21st century under a scenario of 1% yr^{-1} increasing greenhouse gases, but the geographic distribution is uncertain. The global scale of GCMs does not lend itself very well to mesoscale (regional) predictions, and the various models differ on the locations where precipitation will change the most.[13]

Northern hemisphere sea-ice will be reduced by warming in the 21st century according to GCMs. The models suggest polar amplification of global warming, and Arctic land areas will experience wintertime warming.

Mean sea level is projected to rise at an accelerating rate in the 21st century, at first due to thermal expansion of the ocean as its surface warms, and then due to the melting of polar ice and the breaking off and melting of the Ross ice shelf in Antarctica. The sea level has risen 3.9 ± 0.8 mm yr^{-1} in 1993 and 1994.[14] Most models estimate the rate of sea level rise as 15–95 cm by the year 2100.[12] With predictions of such importance for humans everywhere, we might ask ourselves how these models are formulated and how great their uncertainty is.

11.2.4 General Circulation Models

There are 30 or more General Circulation Models (GCMs) in use today for analyzing climate change, and five of the original models were evaluated by the IPCC

(1990).[11] These include the Geophysical Fluids Dynamics Laboratory (GFDL) model of the National Oceanic and Atmospheric Administration,[15] the Goddard Institute for Space Studies (GISS) model of the National Atmospheric and Space Administration,[16] the National Center for Atmospheric Research (NCAR) model,[17] the Oregon State University (OSU) model,[18] and the United Kingdom Meteorological Office (UKMO) model.[19] The models are run in a steady-state mode with a doubling of CO_2 and also in a transient mode in which the concentration of CO_2 is increased in a time-variable fashion (~1% per year). Some GCMs are directly coupled to the oceans with a full, dynamical treatment of ocean mixing and dynamics, but in early simulations the oceans were simply a constant temperature heat sink for atmospheric warming.

Each model divides the atmosphere into layers and the earth into a grid along lines of latitude and longitude. The grid cells are as large as 4.5 degrees latitude × 7.5 degrees longitude (450 × 750 km^2) or as small as 1° × 3°. Mesoscale (regional) models are sometimes embedded within the GCM for smaller scale model testing with resolutions as fine as 10 × 10 km^2 to 50 × 50 km^2. Atmospheric layers usually number between 10 and 30, and they rise at least 30 km into the stratosphere. The time step in most GCMs is ~30 minutes and the number of equations is on the order of 150,000, so the computing power of a supercomputer is required. A 100-year simulation of the NCAR ocean-coupled model requires 1000 processor hours on a Cray Y-MP supercomputer. However, coarse grids and uncoupled ocean models can be run on a workstation. A detailed description of GCMs is beyond the scope of this text, and the reader is referred to the book *Climate System Modeling* by K.E. Trenberth.[20]

GCMs rely on first principles to simulate atmospheric and oceanic large-scale circulation and heat content. This amounts to the principle of continuity for heat, momentum, and mass transfer. The basic equations are derived from Newton's laws: the second law of motion (momentum transport); conservation of mass; conservation of energy (the first law of thermodynamics); an equation of state that relates pressure, density, and temperature; and an equation of state that relates temperature, density, and salinity of the oceans.[21] The change in heat content of a layer of atmosphere is equal to the change in internal energy and the pressure–volume work that it does to the surroundings (the first law of thermodynamics for an ideal gas in a closed system).

$$dq = du + dw = du + P\,dV \tag{1}$$

where dq is the change in heat, du is the change in internal energy, dw is the change in work, P is the pressure, and dV is the change in volume of the parcel of air. The change in internal energy depends on the change in enthalpy minus the work performed.

$$dq = (dh - P\,dV - V\,dP) + P\,dV \tag{2}$$

$$dq = dh - V\,dP \tag{3}$$

$$dq = C_p\, dT - \frac{1}{\rho}\, dP \tag{4}$$

where C_p is the heat capacity of the volume of air at constant pressure, dT is the change in absolute temperature, ρ is the density of the air, and dP is the change in pressure. Equation (4) shows the interplay between pressure, density, and temperature for a parcel of air. As the parcel of air increases in temperature or decreases in pressure at a constant density, the heat content (dq) increases. Changes in pressure cause forces to be created that move the mass of air:

$$\Delta F = \frac{\partial P}{\partial x}\, dx\, dy\, dz \tag{5}$$

where $\partial P/\partial x$ is the pressure gradient in the longitudinal dimension and dx, dy, dz are the incremental distances in three-dimensions. Equation (5) leads to the momentum equation (due to the net pressure gradient force). It allows atmospheric circulation as air moves from areas of high pressure to areas of low pressure. Pressure gradients are created by differential heating and cooling.

$$\Delta M_0 = \left(\frac{\partial P}{\partial x}\, dx\, dy\, dz \right) \Delta t \tag{6}$$

$$\Delta M_0 = m\, \Delta v \tag{7}$$

where ΔM_0 is the change in momentum of the parcel of air, Δv is the change in velocity, and Δt is the time increment.

In the vertical dimension, the air is approximately adiabatic as it expands into the upper atmosphere ($dq = 0$). We can rewrite equation (4) as

$$C_p\, dT = \frac{1}{\rho}\, dP \tag{8}$$

Taking the derivative in the vertical direction, and substituting the hydrostatic pressure equation $dP = -\rho g\, dz$ into equation (8), we have

$$\left(-\frac{dT}{dz} \right)_{\text{adia}} = \frac{g}{C_p} \tag{9}$$

where g is the gravitational constant. If the changes in g and C_p are assumed to be negligible over small vertical elevations, then the change in temperature with elevation under adiabatic conditions is a constant, independent of elevation! The following units apply in the SI system for dry air at 25 °C:

$$C_p = 1.005 \text{ kJ kg}^{-1}\, °\text{C}^{-1}$$

$$g = 9.806 \text{ m s}^{-2}$$

Equation (9) is known as the dry adiabatic lapse rate, and it is equal to –9.8 °C km^{-1}. Air cools as it rises about 10 °C per km; this is especially applicable in the troposphere, the lower atmosphere. However, equation (9) does not include the effect of water vapor condensation as the parcel of air cools (latent heat). The actual standard temperature profile for air as it rises in the troposphere is –6.6 °C km^{-1} due to water vapor condensation.

Equations (1)–(9) are simplifications of the numerical calculations in a GCM that generate atmospheric circulation and temperature change, but they help to illustrate the principles. In addition to heat and momentum transfer, we have mass transfer, including effects of carbon dioxide sorption by oceans and the biosphere. In the next section, we will examine the biogeochemistry of carbon cycling.

11.3 GLOBAL CARBON BOX MODEL

11.3.1 Model Equations

As an example of a global biogeochemical model, a Simple Global Carbon Model (SGCM) for carbon flows through the ocean, atmosphere, and terrestrial biosphere is presented in this section. A box model is a compartmentalized approach to solution of mass balance transport equations. Bulk dispersion and flows between boxes allow simplification of the problem and result in a numerical solution of the sets of ordinary differential equations rather than partial differential equations. A schematic of a global carbon box model is presented in Figure 11.11, together with carbon reservoirs (numbers inside boxes) and fluxes (numbers by arrows) based on reported values.[22] Units are Gt C (gigatons carbon, 1×10^{15} g C) in reservoirs and Gt C per year in fluxes. It is composed of three large carbon pools that are the atmosphere, the terrestrial biosphere, and the ocean. The terrestrial biosphere is compartmentalized into the land biota, soil, and detritus. The ocean is compartmentalized into the warm ocean surface water, the cold ocean surface water, the intermediate and deep waters, and two ocean biota compartments in both warm and cold surface waters. Therefore the box model consists of eight compartments and 19 fluxes, as shown in Figure 11.11. Table 11.2 is the physical description of the eight compartments. The volume of the atmosphere was calculated from the total carbon mass divided by the carbon concentration of the atmosphere, and surface areas of the land and the ocean were taken from *Medallion World Atlas*.[23] The depth of ocean surface water is normally taken to be 75–100 m, and the total average ocean depth ~3800 m.[22] Most processes are formulated with first-order kinetics except for CO_2 fertilization of terrestrial biota, which follow a Michaelis–Menton formulation.

Ninety-five percent of the land biota compartment is composed of C_3 plants, and the remainder consists of C_4 plants.[24] C_3 plants have different photosynthetic mechanisms such that photosynthesis increases with increases of atmospheric CO_2 concentration, while that of C_4 plants, mainly tropical grasses, does not increase at high

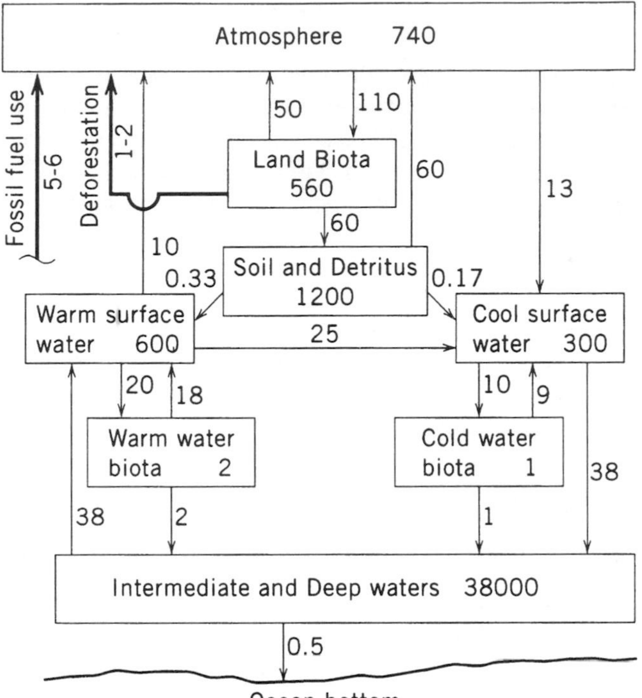

Figure 11.11 Compartmentalized simple global carbon model (SGCM). Units are gigatons (GT) as carbon in reservoirs (numbers inside boxes) and gigatons as carbon per year in fluxes (numbers by arrows).

Table 11.2 Physical Descriptions: Areas and Volumes of Global Box Model

Description	Size
Volume of atmosphere (V_1)	3.99×10^{18} m^3
Area of land biota (A_2)	1.33×10^{14} m^2
Volume of soil and detritus (V_3)	6.65×10^{14} m^3
Depth of ocean surface water	100 m
Depth of average ocean	3800 m
Area of warm ocean surface water (A_4)	2.42×10^{14} m^2
Volume of warm ocean surface water (V_4)	2.42×10^{16} m^3
Area of cold ocean surface (A_6)	1.21×10^{14} m^2
Volume of cold ocean surface water (V_6)	1.21×10^{16} m^3
Area of intermediate and deep waters (A_8)	3.63×10^{14} m^2
Volume of intermediate and deep waters (V_8)	1.318×10^{18} m^3

CO_2 concentrations.[25] Here we assume all plants are C_3. Plants fix CO_2 from the atmosphere and store it as organic carbon (photosynthesis). They also release organic carbon in the form of CO_2 (respiration). In addition, organic carbon in the stems, leaves, and roots is incorporated into the soil compartment seasonally and as plants die (litterfall). The CO_2 fertilization effect is assumed to follow a Michaelis–Menton kinetic expression similar to the hyperbolic relationship used by Raich et al.[26]; and temperature feedback follows a Q_{10}-type exponential function. Thus the CO_2 fertilization effect on the land biota is included in the following rate expression:

$$k = k_0 \cdot \frac{C_{\text{atm}}}{K_M + C_{\text{atm}}} \tag{10}$$

where k = the rate of CO_2 uptake
k_0 = the maximum rate of uptake
C_{atm} = the CO_2 concentration in the atmosphere
K_{M} = half-saturation velocity

A similar Michaelis–Menton kinetic expression is applied to photosynthesis, biota respiration, and litterfall. Temperature feedback is expressed by the relationship

$$k = k_0 \cdot (Q_{10})^{(T-T_0)/10} \tag{11}$$

in which Q_{10} is the increasing rate for every 10 °C increase of temperature, T is temperature, and T_0 is the initial temperature. Rearranging equation (11) results in the following equation:

$$k = k_0 \cdot \theta^{\Delta T} \tag{12}$$

in which $\theta(= Q_{10}^{0.1})$ is the temperature-variable coefficient, $\Delta T = T - T_0$. The temperature feedback is applied to photosynthesis and biota respiration to which the temperature is sensitive, while it is not applied to litterfall. Combined with deforestation as an efflux, the mass balance for the land biota compartment is written as[27]

$$A_2 \cdot \frac{dC_2}{dt} = \left\{k_p \cdot \frac{C_1}{(K_M + C_1)} \cdot C_2 \cdot A_2 \cdot \theta_p^{\Delta T}\right\} - \left\{k_{br} \cdot \frac{C_1}{(K_M + C_1)} \cdot C_2 \cdot A_2 \cdot \theta_{br}^{\Delta T}\right\}$$

$$- \left\{k_e \cdot \frac{C_1}{(K_M + C_1)} \cdot C_2 \cdot A_2\right\} - D(t) \tag{13}$$

where A_2 = land surface area, L^2
C_1 = carbon concentration in the atmosphere, $M\ L^{-3}$ (converted from ppm)
C_2 = carbon concentration in land biota, ML^{-2}
K_M = half-saturation velocity, ML^{-3}

k_p = maximum photosynthesis rate constant, T^{-1}
k_{br} = maximum land biota respiration rate constant, T^{-1}
k_e = maximum litterfall rate constant, T^{-1}
θ_p = temperature-variable coefficient of photosynthesis, dimensionless
θ_{br} = temperature-variable coefficient of land biota respiration, dimensionless
ΔT = temperature change from steady-state (preindustrial time) temperature, °C
$D(t)$ = deforestation, MT^{-1}

Most organic carbon exists in the upper layer of the land surface that is assumed to be 5-m depth global average. Organic carbon is added to the soil by litterfall, where it is decomposed by soil microbes and emitted to the atmosphere in the form of CO_2 (soil respiration). Because of a strong influence of temperature on microbial activity, equation (12) is applied to soil respiration to account for the effect of temperature. In addition, a small amount of carbon is washed out and transported to the ocean surface water (runoff). During this process, organic carbon is assumed to be completely oxidized to inorganic carbon and contributes to the increase of total inorganic carbon of the ocean surface water. A mass balance of the soil compartment is[27]

$$V_3 \cdot \frac{dC_3}{dt} = \left\{k_e \cdot \frac{C_1}{(K_M + C_1)} \cdot C_2 \cdot A_2\right\} - \{k_{sr} \cdot C_3 \cdot V_3 \cdot \theta_{sr}^{\Delta t}\} - \{k_w \cdot C_3 \cdot V_3\} \tag{14}$$

where V_3 = volume of the soil compartment, L^3
C_3 = carbon concentration of the soil compartment, ML^{-3}
k_{sr} = soil respiration rate constant, T^{-1}
k_w = runoff rate constant, T^{-1}
θ_{sr} = temperature-variable coefficient of soil respiration, dimensionless.

As depicted in Figure 11.11, the ocean is composed of three compartments, the warm ocean surface water, the cold ocean surface water, and the intermediate and deep waters. Carbon is internally cycled by way of ocean circulation.[28] Denser water is formed owing to lower temperature and increased salinity in the cold surface water compartment. It is primarily composed of water from two regions, Antarctic cold water and water near Greenland, and it sinks to the intermediate and deep water compartment (cold surface water sinking); whereas a part of the intermediate and deep water is upwelling to the warm surface water compartment, primarily in the equatorial area (deep water upwelling). To balance the water mass, water is transported from the warm surface water compartment to the cold surface water compartment (surface water advection). In addition to ocean circulation, ocean biota, which is composed mainly of organic carbon, is also involved in the carbon cycle through biological, chemical, and physical processes. Biota photosynmthesize and incorporate dissolved CO_2 and nutrients such as nitrogen and phosphorus (ocean biota primary production). It is generally believed that primary production is not rate limited by dissolved CO_2 but by nutrients. In this chapter, ocean primary production follows first-order kinetics and is linear with respect to the dissolved CO_2 concentration of the surface water. Roughly 90% of primary production is remineralized to dissolved CO_2 within the surface water (ocean biota respiration), with the remaining 10% sinking to the intermediate and deep waters, where most of it is remineralized, and a small amount is buried into the ocean bottom as organic carbon (organic carbon burial).[22] Even though the activity of ocean biota strongly controls alkalinity and CO_2 partial pressure in the ocean, SGCM does not include diurnal and seasonal oscillations because the model deals with an annually averaged carbon cycle. Organic carbon is also transported into the ocean surface water from the soil compartment (runoff). Two-thirds of it is assumed to be transported into the warm ocean surface water and one-third of it into the cold ocean surface water. Carbon in the form of CO_2 is exchanged between the atmosphere and the ocean surface waters (gas exchange). The driving force of CO_2 gas exchange is the difference of CO_2 partial pressure between the atmosphere and the ocean surface water. CO_2 partial pressure in the ocean is a function of alkalinity, temperature, salinity, and total inorganic carbon and is computed with the method proposed by Peng et al.[29] A mass balance of carbon of each compartment is as follows.[27]

For the warm ocean surface water,

$$V_4 \cdot \frac{dC_4}{dt} = -\{k_\ell \cdot S_{wo} \cdot (P_{wo} - P_{\mathrm{atm}}) \cdot A_4\} - \{k_{pwo} \cdot C_4 \cdot V_5\} + \{k_{rwo} \cdot C_5 \cdot V_5) + \{k_{\mathrm{up}} \cdot C_8 \cdot A_4\} - \{Q_{46} \cdot C_4\} + \{\tfrac{2}{3} \cdot k_w \cdot C_3 \cdot V_3\} \tag{15}$$

where V_4 = volume of warm ocean surface water, L^3

V_5 = volume of warm ocean biota, L^3

C_4 = carbon concentration of warm ocean surface water, ML^{-3}

k_ℓ = gas transfer velocity, LT^{-1}

S_{wo} = solubility of warm ocean surface water, ML^{-3} ppm^{-1}

P_{wo} = CO_2 partial pressure of warm ocean surface water, ppm

P_{atm} = CO_2 partial pressure of the atmosphere, ppm

k_{pwo} = warm ocean biota primary production rate constant, T^{-1}

k_{rwo} = warm ocean biota respiration rate constant, T^{-1}

C_5 = carbon concentration of warm ocean biota, ML^{-3}

k_{up} = intermediate and deep waters upwelling velocity, LT^{-1}

C_8 = carbon concentration of intermediate and deep waters, ML^{-3}

A_4 = area of warm ocean surface water, L^2

Q_{46} = advection between surface waters, L^3T^{-1}

For the cold ocean surface water,

$$V_6 \cdot \frac{dC_6}{dt} = \{k_\ell \cdot S_{co} \cdot (P_{\text{atm}} - P_{co}) \cdot A_6\} - \{k_{pco} \cdot C_6 \cdot V_7\} + \{k_{rco} \cdot C_7 \cdot V_7)$$

$$- \{k_{dc} \cdot C_6 \cdot A_6\} + \{Q_{46} \cdot C_4\} + \{\tfrac{1}{3} \cdot k_w \cdot C_3 \cdot V_3\} \quad (16)$$

where V_6 = volume of cold ocean surface water, L^3
V_7 = volume of cold ocean biota, L^3
C_6 = carbon concentration of cold ocean surface water, ML^{-3}
S_{co} = solubility of cold ocean surface water, ML^{-3}/ppm
P_{co} = CO_2 partial pressure of cold ocean surface water, ppm
k_{pco} = cold ocean biota primary production rate constant, T^{-1}
k_{rco} = cold ocean biota respiration rate constant, T^{-1}
C_7 = carbon concentration of cold ocean biota, ML^{-3}
k_{dc} = cold ocean surface water sinking velocity, LT^{-1}
A_6 = area of cold ocean surface water, L_2

For the warm ocean biota,

$$V_5 \cdot \frac{dC_5}{dt} = \{k_{pwo} \cdot C_4 \cdot V_5\} - \{k_{rwo} \cdot C_5 \cdot V_5\} - \{k_{swo} \cdot C_5 \cdot A_4\} \quad (17)$$

where k_{swo} = warm ocean biota sinking rate constant, LT^{-1}

For the cold ocean biota,

$$V_7 \cdot \frac{dC_7}{dt} = \{k_{pco} \cdot C_6 \cdot V_7\} - \{k_{rco} \cdot C_7 \cdot V_7\} - \{k_{sco} \cdot C_7 \cdot A_6\} \quad (18)$$

where k_{sco} = cold ocean biota sinking rate constant, LT^{-1}

For the intermediate and deep waters,

$$V_8 \cdot \frac{dC_8}{dt} = \{k_{dc} \cdot C_6 \cdot A_6\} - \{k_{\text{up}} \cdot C_8 \cdot A_4\} + \{k_{swo} \cdot C_5 \cdot A_4\} + \{k_{sco} \cdot C_7 \cdot A_6\}$$

$$-\{k_{\text{sed}} \cdot C_8 \cdot A_8\} \quad (19)$$

where V_8 = volume of intermediate and deep waters, L^3
A_8 = area of intermediate and deep waters, L^2
k_{sed} = organic carbon burial rate constant, T^{-1}

Carbon mass fluxes to and from the atmosphere include land biota photosynthesis, land biota respiration, soil respiration, and CO_2 gas exchange between the atmosphere and the ocean surface water. Combined with anthropogenic emissions including fossil fuel combustion and deforestation as inputs (forcing functions in this model), a mass balance for the atmosphere compartment is

$$V_1 \cdot \frac{dC_1}{dt} = F(t) + D(t) - \left\{k_p \cdot \frac{C_1}{K_M} + C_1 \cdot C_2 \cdot A_2 \cdot \theta_p^{\Delta T}\right\}$$
$$+ \left\{k_{br} \cdot \frac{C_1}{K_M + C_1} \cdot C_2 \cdot A_2 \cdot \theta_{br}^{\Delta T}\right\} + \left\{k_{sr} \cdot C_3 \cdot V_3 \cdot \theta_{sr}^{\Delta T}\right\} \qquad (20)$$
$$+ \{k_\ell \cdot S_{wo} \cdot (P_{wo} - P_{atm}) \cdot A_4\} - \{k_\ell \cdot S_{co} \cdot (P_{atm} - P_{co}) \cdot A_6\}$$

where V_1 = volume of the atmosphere, L^3
$F(t)$ = fossil fuel combustion, MT^{-1}

Eight mass balance equations (13 through 20) form a set of ordinary differential equations that were solved numerically with the time step of 1 day via a fourth-order Runge–Kutta method.[30]

11.3.2 Model Calibration

SGCM was calibrated with a 32-year record (1958–1989) of atmospheric CO_2 concentrations measured at Mauna Loa (Hawaii, United States),[3] which was assumed to represent the global average CO_2 concentration for the purpose of this study (it is slightly higher than the true global average, but the calibration of the model is consistent with Mauna Loa data). An oceanic sink constraint was set as 0.6–2.6 Gt C yr^{-1} in the 1980s, based on literature values.[11,12,31] Annual CO_2 emissions from fossil fuel combustion was used as input data.[3] The other input, CO_2 emissions from deforestation, was estimated at 0.4–2.6 Gt C in 1980.[32,33] Since it included a wide range of uncertainty, however, it was treated as an adjustable parameter. For the temperature input the annual mean global surface air temperature (Figure 11.12) was taken from observed data reported by Hansen et al.[34] The temperature change ΔT that appears in temperature-related flux terms is a difference between the temperature at time t and the preindustrial temperature that was assumed to be –0.4 °C based on historical data.[35]

Calibrated parameter values including deforestation are presented in Table 11.3. Most rate constants were derived from the carbon reservoirs and fluxes (Figure 11.11) and adjusted. A half-saturation constant K_M of 400 ppm for C_3 plants was used.[36] Temperature-variable coefficients θ_p, θ_{br}, and θ_{sr} were derived from model calibrations, and their values were 1.032, 1.066, and 1.069, respectively. The conversion of those to Q_{10} results in 1.37 for photosynthesis, 1.89 for biota respiration,

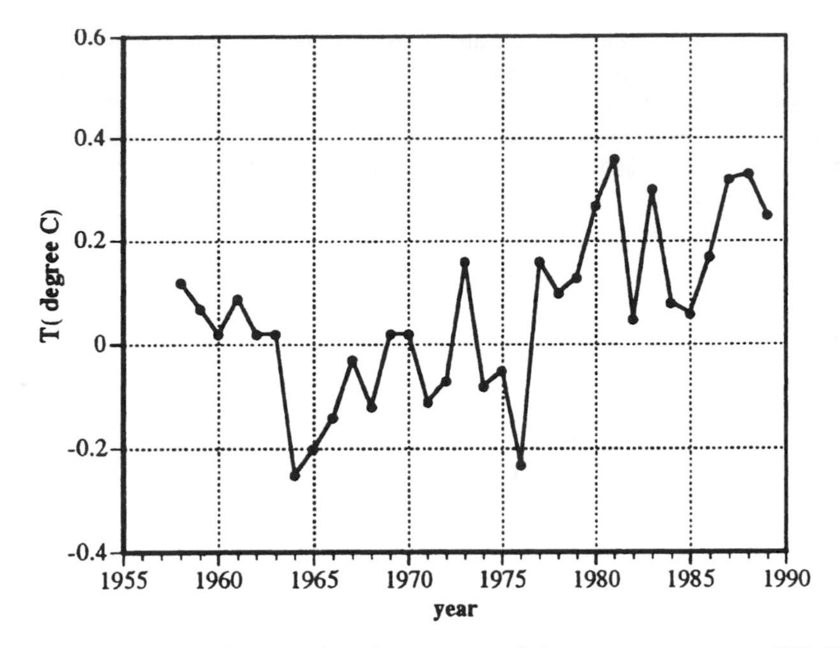

Figure 11.12 Global annual mean surface air temperature relative to mean temperature, 1951–1980. (Data from Hansen et al.[34])

Table 11.3 Parameter Values Including Deforestation ($D(t)$)

Parameter	Value	Parameter	Value
$D(t)$	1.21 Gt C yr^{-1}	k_{rco}	9.0 yr^{-1}
k_p	0.44286 yr^{-1}	k_{swo}	100 m yr^{-1}
k_{br}	0.19232 yr^{-1}	k_{sco}	100 m yr^{-1}
k_e	0.24911 yr^{-1}	k_{sed}	0.051 m yr^{-1}
k_{sr}	0.05146 yr^{-1}	θ_p	1.032
k_w	0.00042 yr^{-1}	θ_{br}	1.066
K_M	400 ppm	θ_{sr}	1.069
k_ℓ	1752 m yr^{-1}	T_{wo}	25 °C
k_{up}	5.55856 m yr^{-1}	T_{co}	8 °C
k_{dc}	12.4502 m yr^{-1}	SA_{wo}	35.51‰
Q_{46}	1.046×10^{16} m^3 yr^{-1}	SA_{co}	34.17‰
k_{pwo}	0.03448 yr^{-1}	TA_{wo}	2272 μmol kg^{-1}
k_{rwo}	9.0 yr^{-1}	TA_{co}	2237 μmol kg^{-1}
k_{pco}	0.03448 yr^{-1}		

and 1.95 for soil respiration. These are in agreement with reported values in the literature, which range from 1.03 to 2.0 for photosynthesis,[37,38] 1.4 to 3.0 for biota respiration,[38–42] and 0.77 to 3.3 for soil respiration[38,43–47] (except for one high value of 4.3).[48] The model calibrated value[49] for deforestation was 1.21 Gt C yr^{-1} and that is within the range of 0.4–2.6 Gt C yr^{-1} reported by Moore and Bolin and by Houghton et al.[22,32] and 1.6 ± 1.0 Gt C yr^{-1} by IPCC.[11] Gas transfer velocity k_ℓ was taken from the study using natural and bomb-produced ^{14}C methods.[50] Ocean CO_2 partial pressures (P_{wo} and P_{co}) as well as solubilities S_{wo} and S_{co} were calculated with the method proposed by Peng et al.[29] To compute them, total alkalinities (TA_{wo} and TA_{co}), salinities (SA_{wo} and SA_{co}), and surface water temperatures (T_{wo} and T_{co}) that were assumed constant were based on the Geochemical Ocean Sections Study data[51] but were slightly adjusted (within ±5%) to satisfy the oceanic sink constraint.

11.3.3 Global Carbon Budget

Figure 11.13 represents the best fit of the box model with the measured atmospheric CO_2 concentration. The standard error (std. dev./$\sqrt{n}$) was 0.08 ppm. Figure 11.14 represents the model-derived carbon budget during the calibration period. Positive values are sources, and negative values are sinks. Annual average sources in the 1980s were 5.42 Gt C yr^{-1} from fossil fuel use and 1.21 Gt C yr^{-1} from deforestation; sinks were 3.35 Gt C yr^{-1} in the atmosphere, 2.29 Gt C yr^{-1} in the ocean, and 0.99 Gt C yr^{-1} in the terrestrial biosphere. The fluctuation of the atmospheric carbon

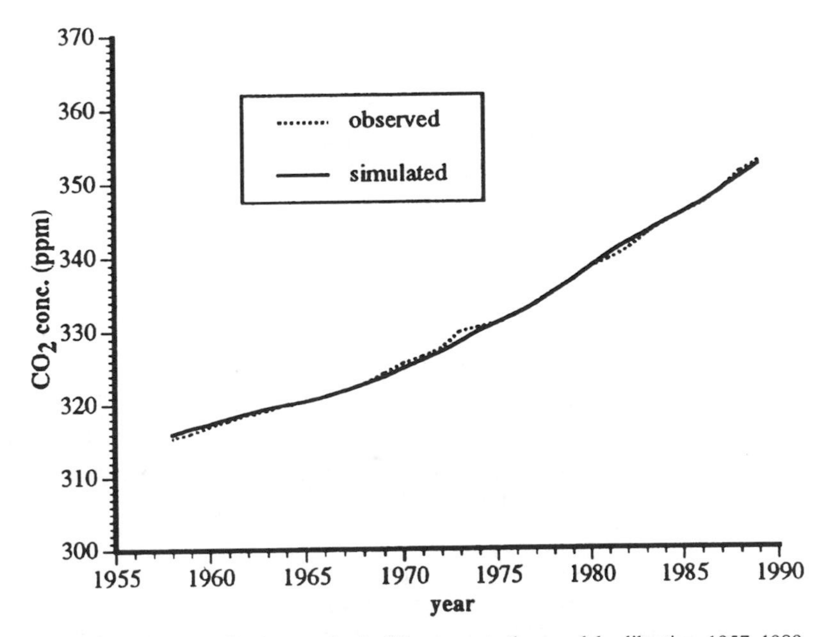

Figure 11.13 Best fit of atmospheric CO_2 concentration, model calibration, 1957–1989.

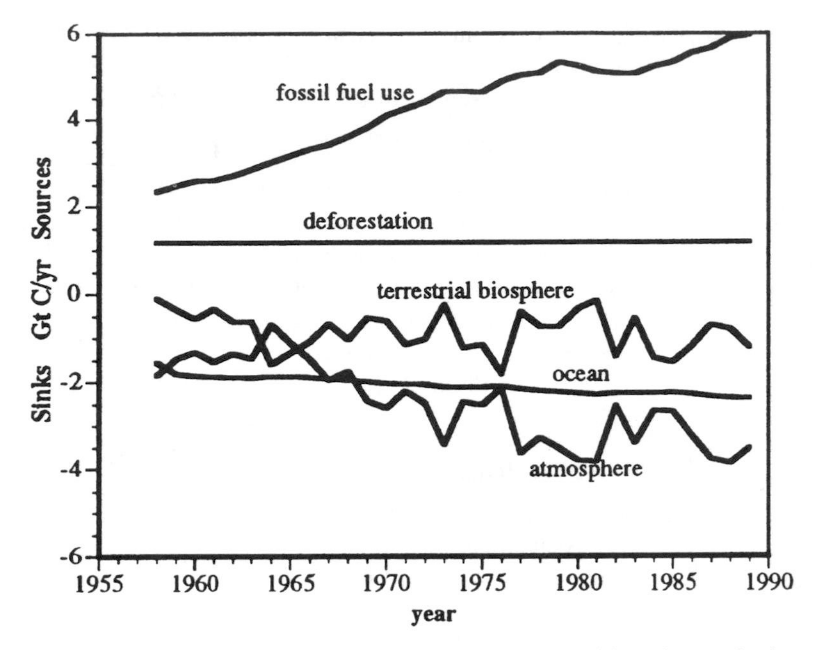

Figure 11.14 Model results of carbon budget for sources (pluses) and sinks (minuses), in gigatons of carbon per year.[49].

sink shows the reverse pattern and was correlated with that of the terrestrial biospheric carbon sink, which was due to the temperature effect. As seen in Figure 11.14, during the calibration period, the terrestrial biospheric sink offset deforestation (approximately ± 1.2 Gt C yr^{-1}). The atmospheric sink and the oceanic sink increased with an increase of CO_2 emissions from fossil fuel use. However, the atmosphere accumulated carbon faster than the ocean. This is due to the so called "buffering capacity" of the ocean. The buffer factor (β) is written as

$$\beta = \frac{\Delta pCO_2/pCO_2}{\Delta TIC/TIC} \tag{21}$$

where ΔpCO_2 = change of atmospheric CO_2 concentration
pCO_2 = atmospheric CO_2 concentration at preindustrial level
ΔTIC = change of total inorganic carbon concentration in the surface water
TIC = total inorganic carbon concentration in the surface water at the preindustrial level

The reported buffer factor β for the present ocean surface waters are on the order of 10.[52] This means that a relative increase of pCO_2 in the atmosphere of 10% corresponds to a relative increase of TIC in surface water of only 1%. The ocean buffering capacity will decrease with an increase of the atmospheric CO_2 concentration because $\beta > 1.0$.

11.3.4 Parameter Sensitivity Analysis

A conventional sensitivity analysis was performed to understand where more research is needed (priority setting). Sensitivity of the atmospheric CO_2 concentration was analyzed for ±10% change of each parameter value, independently. The atmospheric CO_2 concentration of 1989 (end of the calibration period) was used for the sensitivity comparison. The sensitivity parameter SP is expressed as

$$SP = \frac{\Delta pCO_2/pCO_2}{|\Delta Z/Z|} \times 100\% \qquad (22)$$

where ΔpCO_2 = change of atmospheric CO_2 concentration resulting from parameter change
pCO_2 = atmospheric CO_2 concentration in year 1989
ΔZ = change of parameter value
Z = parameter value

Results of the sensitivity analysis are summarized in Table 11.4. Photosynthesis, respiration, and litterfall of land biota were the most sensitive parameters. Soil respiration, half-saturation velocity, and deep water upwelling and cold surface water sinking in the ocean were the next group. Others were not sensitive, as indicated by $|SP| \leq 10\%$. Temperature-variable coefficients θ_p, θ_{br}, and θ_{sr} were not included in the sensitivity analysis because of the difficulty of direct comparison with other parameters, but results are quite sensitive to these coefficients. Sensitivity parameter results show that terrestrial biospheric activities and carbon exchange processes be-

Table 11.4 Sensitivity of CO_2 Concentration Due to a ±10% Change in Model Parameters

	Percent Change in CO_2 SP (%)			Percent Change in CO_2 SP (%)	
Parameter	+ 10%	−10%	Parameter	+ 10%	−10%
k_p	−516	+649	Q_{46}	−4	+5
k_{br}	+292	−266	k_{pwo}	−5	+5
k_e	+289	−275	k_{rwo}	+4	−5
k_{sr}	+67	−74	k_{pco}	−2	+2
K_M	+37	−46	k_{rco}	+2	−2
k_ℓ	−2	+3	k_{swo}	−4	−4
k_{up}	+85	−99	k_{sco}	−2	+2
k_{dc}	−86	+80	k_{sed}	0	0

tween the surface water and the intermediate and deep waters are most important in controlling the atmospheric CO_2 concentration.

11.3.5 Model Scenario Analysis

Prior to scenario analysis, the feedback temperature due to global warming by CO_2 was determined by model calibration. Since historical data show strong evidence of a linear relationship between global average surface temperature and the atmospheric CO_2 concentration,[53] the global temperature was assumed to be a linear function of the atmospheric CO_2 concentration. To estimate the feedback temperature, the global temperature at preindustrial CO_2 levels (here 280 ppm) was assumed to be –0.4 °C (relative temperature referred to in Figure 11.12). Setting the feedback temperature as χ °C ppm^{-1}, the temperature at a given atmospheric CO_2 concentration was expressed as

$$\text{Temp}(t) = -0.4 + [pCO_2(t) - 280] \times \chi \tag{23}$$

in which Temp(t) is global temperature at time t and $pCO_2(t)$ is the atmospheric CO_2 concentration at time t. Test results are presented in Figure 11.15. The model with 0.01 °C ppm^{-1} of feedback temperature resulted in the best fit with the measured values (thick line). This is a reasonable result because a global average surface temperature increase of roughly 0.5–0.7 °C and an increase of 60 ppm CO_2 have been measured since the mid-19th century.

Scenario analysis with a selected feedback temperature (0.01 °C ppm^{-1}) was performed for three CO_2 emission scenarios from fossil fuel use with a constant emission from deforestation (1.21 Gt C yr^{-1}). Case A is the 1.5% increase per year; case B, the constant emission; and case C, the recovery case, that is, 5% decrease per year beginning in 1989. Figure 11.16 illustrates the resulting concentrations and masses of atmospheric CO_2, the land biota carbon reservoir, and the soil carbon reservoir. The atmospheric CO_2 concentration doubled from preindustrial levels (280 ppm) in the year 2060 in case A; it increased but did not double within the mid-22nd century in case B; and it stabilized to 350 ppm in case C. Land biota carbon decreased and then increased until a sudden drop occurred in case A because fertilization effects were dominated ultimately by respiration effects. Concentrations of CO_2 slowly decreased to a steady state in case B, but they continued to decrease in case C. Changes in the mass of carbon in land biota resulted due to the competition between allowed deforestation and land biota growth stimulated by CO_2 fertilization and temperature effects. The decrease of land biota carbon mass in Figure 11.16 (case C) is not due to the decrease of net primary productivity NPP (photosynthesis minus respiration) but rather due to a constant rate of deforestation. Model results show that in case A, NPP increased by ~24% when the CO_2 concentration in the atmosphere was doubled. Even though an enhanced temperature was favorable to the biota growth, a relatively high temperature (+5.5 °C and 900 ppm) resulted in a sudden drop, as shown in case A. This was because the increased biota respiration (Q_{10} = 1.89) combined with increased litterfall exceeded the enhanced

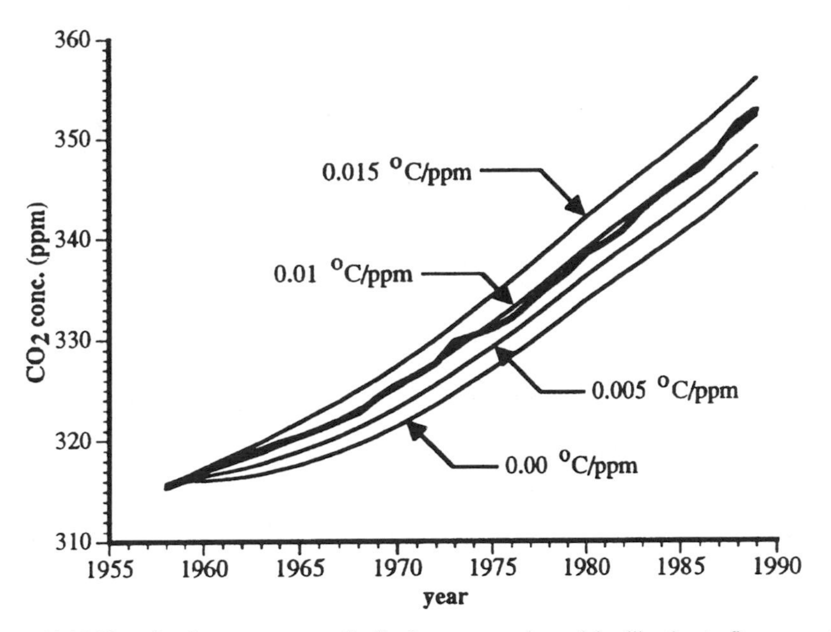

Figure 11.15 Use of various temperature feedback parameters in model calibration to fit measured atmospheric CO_2 data.

photosynthesis ($Q_{10} = 1.37$). The soil carbon reservoir increased until a sudden drop occurred in case A; it maintained nearly constant value in case B; and it decreased in case C. Change in the soil carbon reservoir occurred, mainly due to competition between carbon accumulation by litterfall that was influenced by a direct fertilization effect of CO_2, and carbon depletion by soil respiration influenced by a direct temperature effect. High temperatures resulted in a sudden decrease of the soil carbon reservoir similar to that of the land biota carbon reservoir (Figure 11.16). The high temperature reduced the land biota rapidly and thereby decreased litterfall, and soil respiration was enhanced by high temperature, which accelerated the soil carbon depletion.

Figure 11.17 illustrates the scenarios of both ocean surface waters and the intermediate and deep waters. The carbon concentration time series in both surface waters showed a very similar pattern to that of the atmospheric CO_2 concentration. The difference is that the rate of increase was slower in the surface waters than in the atmosphere; that is, the increase of buffer factor (β) occurred with the increase of the atmospheric CO_2 concentrations. Using equation (21) with an assumed steady-state concentration for the surface waters (1980 μmol kg^{-1} in the cold ocean surface water and 1955 μmol kg^{-1} in the warm ocean surface water) satisfied the buffer factor of 10.0 in year 1980, but buffer factors of the warm and cold ocean surface waters increased to 16.4 and 13.9 in the year 2035 (CO_2 450 ppm), and 21.6 and 18.5 in the year 2070 (600 ppm), respectively. This results in a decreased capacity for oceans to uptake atmospheric carbon dioxide. The carbon concentration of the intermediate

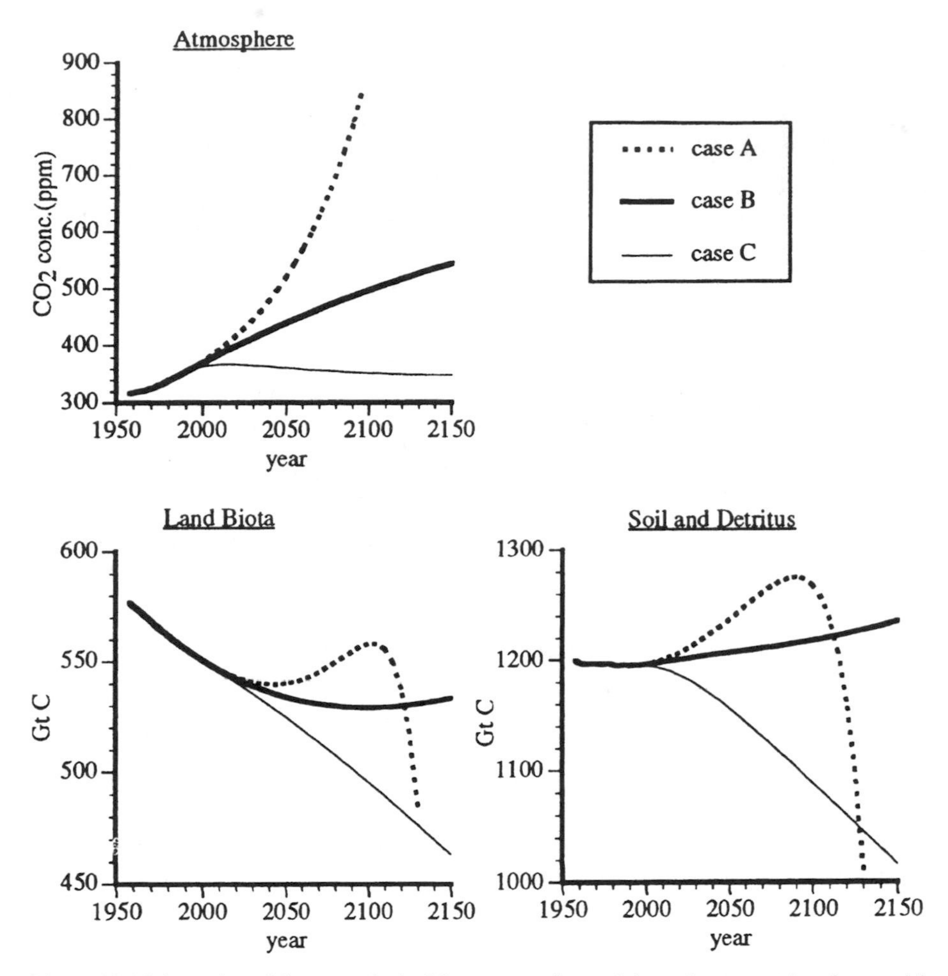

Figure 11.16 Scenarios of the atmospheric CO_2 concentration and the carbon reservoirs of terrestrial biosphere. Case A is the 1.5% increase per year, case B is the constant emission, and case C is the 5% decrease per year from year 1989.[49]

and deep waters increased for all three cases because of increased biota sinking to the intermediate and deep waters. Cold water sinking into intermediate and deep waters was a major flux of carbon in scenarios of the SGCM model.

11.3.6 Summary of Global Carbon Box Model

In this chapter, a Simple Global Carbon Model (SGCM) with an atmosphere–terrestrial biosphere–ocean interaction has been discussed. Even though it is a rather simple model in its spatial and temporal scale, it includes CO_2 fertilization and temperature effects for the terrestrial biosphere and inorganic chemistry that with ocean

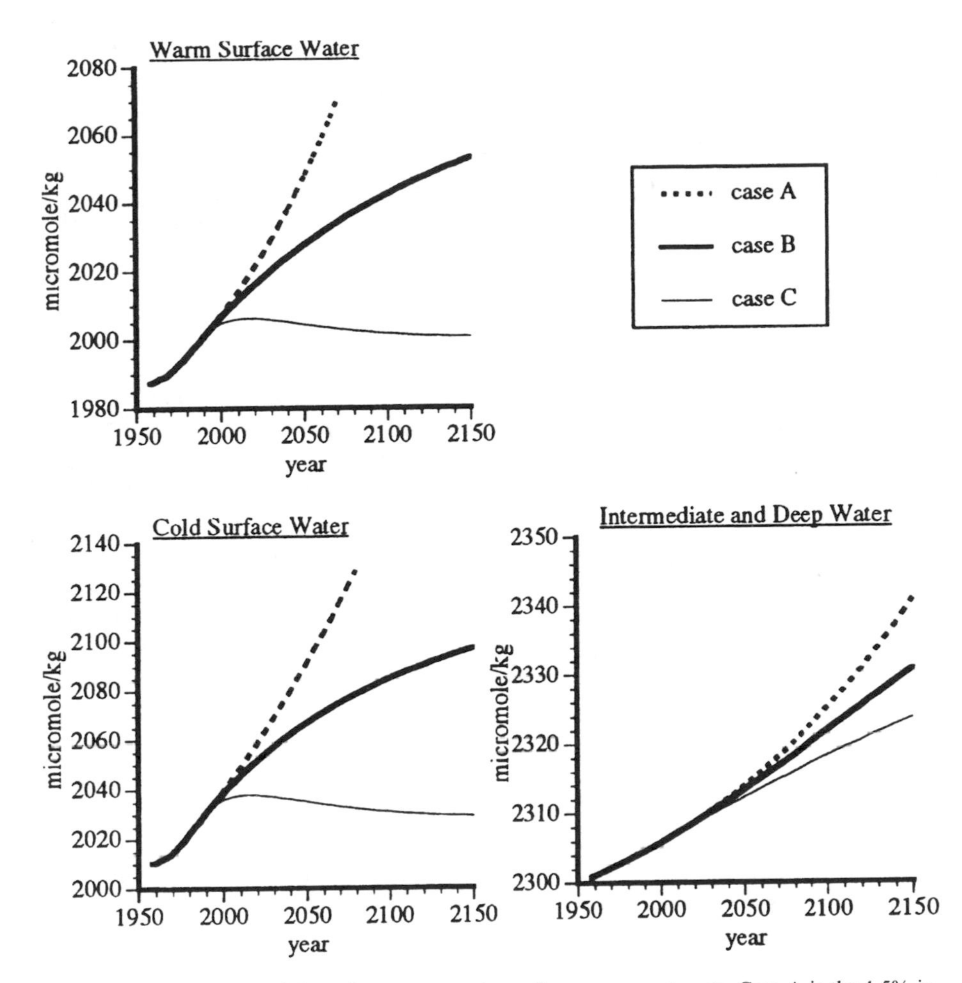

Figure 11.17 Scenarios of the carbon concentrations of ocean compartments. Case A is the 1.5% increase per year, case B is the constant emission, and case C is the 5% decrease per year from year 1989.[49]

water circulation enables the calculation of time-variable oceanic carbon uptake. It illustrates that simple box mass balance models can yield realistic and rich dynamics of global biogeochemical cycles. Model-derived Q_{10} values for land biota photosynthesis (1.37), land biota respiration (1.89), and soil respiration (1.95) are in good agreement with reported values.

Using a model-derived feedback temperature of 0.01 °C ppm^{-1}, a scenario analysis was performed. On the basis of the model results, a CO_2 doubling from preindustrial level (280 ppm) will appear in the mid-21st century if CO_2 emissions from fossil fuel continue at the present increasing rate of 1.5% yr^{-1}, whereas it will not occur until the mid-22nd century if CO_2 emissions are held constant at the present level. The ocean buffering capacity to take up excess CO_2 will decrease with an in-

Table 11.5 Carbon Sources and Sinks Based on Box Model in the Year 1989

	Carbon Sources, Gt C yr^{-1}	Carbon Sinks, Gt C yr^{-1}	
Fossil fuel emissions	6.0	1.2	Biosphere uptake
Deforestation	1.2	2.5	Oceanic uptake
		3.5	Atmosphere accumulation
TOTAL	7.2	7.2	

crease of atmospheric CO_2 concentrations. A climate feedback temperature of 0.01 °C ppm^{-1} is quite sensitive to the terrestrial biospheric carbon reservoirs, indicating an increase of land biota and a decrease of soil carbon due to the global warming. Models such as these are useful for prioritizing research needs and understanding global cycles.

Prospects for the future indicate that CO_2 concentrations in the atmosphere will continue to increase. Table 11.5 shows a global carbon budget for 1989. If fossil fuel emissions continue to increase as they are projected by the IPCC (Figure 11.18), then the atmosphere will be the main sink for CO_2. The Intergovernmental Panel on Climate Change examined many carbon models of the kind discussed in this section, and projections of CO_2 concentrations in the atmosphere follow those of the SGCM quite closely. Three scenarios were analyzed: (1) a high scenario of ~1.5% yr^{-1} increase in emissions; (2) a mid-scenario of slowly increasing emissions, ~0.95% yr^{-1}; and (3) gradually decreasing emissions. In all cases, carbon dioxide concentrations in the atmosphere are projected to increase between 1990 and the year 2000. Comparing results of similar mass balance models, but with different assumptions and formulations, is one method to gain confidence in model results and greater understanding of processes.

Example 11.2 Mass of Carbon in Earth's Compartments

Based on the following information, estimate the mass of carbon in the atmosphere, land biota, soil, oceans, and plankton. Assume the mass of atmosphere, oceans, and soil from Example 11.1.

Atmosphere	5.25×10^{18} kg air
	355 ppmv CO_2
	44 g mol^{-1} MW of CO_2
	12 g mol^{-1} MW of C
	29.2 g mol^{-1} MW of air
Land biota	100 metric tons per ha dry biomass
	1.33×10^{14} m^2 land surface area
	0.4 g C per g biomass (dry)

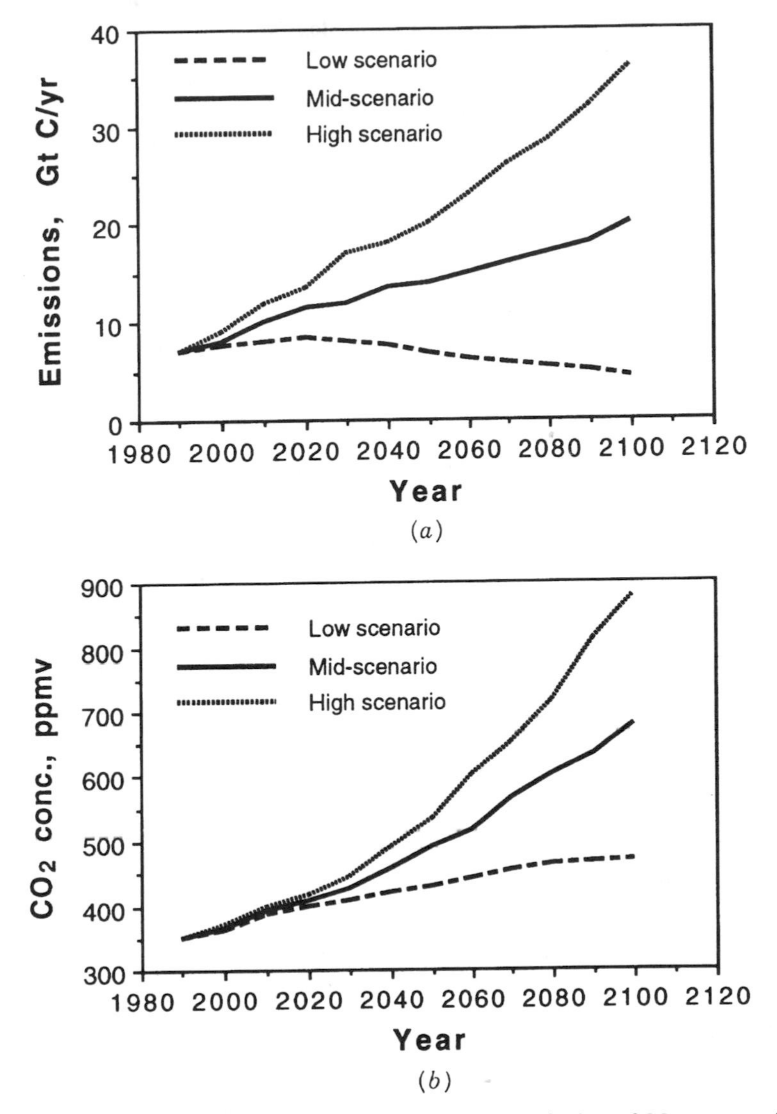

Figure 11.18 Intergovernmental Panel on Climate Change (IPCC) projections of CO_2 concentrations in the atmosphere for three scenarios. Results are a best estimate from several models.[54] (a) Emissions scenarios. (b) CO_2 concentrations.

Soil	3.9×10^{17} kg
	0.5% organic carbon (avg.)
Ocean	1.4×10^{21} kg ocean mass
	2250 μeq kg^{-1} HCO_3^- alkalinity
Plankton	80 μg L^{-1} planktonic organic carbon in ocean
	3.6×10^{16} m^3 surface ocean volume

Solution: Dimensional analysis allows calculation of the mass of carbon in each compartment. The answers are:

Atmosphere = 7.66×10^{17} g C (766 Gt C)
Land biota = 5.32×10^{17} g C (532 Gt C)
Soil = 1.95×10^{18} g C (1950 Gt C)
Ocean = 3.78×10^{19} g C (37,800 Gt C)
Plankton = 2.88×10^{15} g C (2.9 Gt C)

These results compare favorably with those estimated by Moore and Bolin[22] and shown in Figure 11.11.

Example 11.3 The Simplest Climate Model

Back-radiation from the Earth must be equal to solar energy inputs to the earth's surface under steady-state conditions. Estimate the average temperature of the earth based on a solar constant of 1367 W m^{-2} and the Stefan–Boltzmann law.

Solution

$$\text{Back radiation:} \quad E_e = \sigma T_e^4 \cdot (4\pi r^2) \quad \text{Stefan–Boltzmann law} \qquad \text{(i)}$$

where σ = Stefan–Boltzmann constant = 5.5597×10^{-8} W m^{-2} K^{-4}
T_e = temperature of the earth, K (kelvin units)
E_e = energy emitted, W
$4\pi r^2$ = surface area of the earth, m^2

$$\text{Energy input:} \quad E_i = (\pi r^2) \cdot S(1 - A) \qquad \text{(ii)}$$

where πr^2 = circular disk (earth) receiving solar energy
S = solar constant = 1367 W m^{-2}
A = albedo or reflectivity of the earth ~ 0.31

Equations (i) and (ii) can be equated, and one can solve for T_e, the average temperature of the earth.

$$\text{Energy emitted} = \text{Energy input from sun}$$

$$\sigma T_e^4 = S(1 - A)/4 \qquad \text{(iii)}$$

$$\therefore \; T_e = 255 \text{ K}$$

The average temperature of the earth is calculated to be 255 K or –18 °C. This is very cold! It reflects what the temperature would be in the absence of a natural greenhouse effect. However, the calculation is in error because it fails to consider the fraction of back-radiation (long-wave, infrared) that is absorbed by water vapor (λ = 12 μm, where H_2O molecules rotate) and CO_2 (λ = 15 μm where molecular vibration occurs). If we include a factor for emissivity, the opacity or fraction of back-radiation that escapes the earth's atmosphere, we can modify equation (iii) to account for the "natural greenhouse effect." The emissivity (e) is ~0.615.)

$$e\sigma T^4 = S(1 - A)/4 \quad \text{(iv)}$$

$$\therefore \quad T = 288 \text{ K}$$

The global average surface temperature of the earth is 288 K or 15 °C, which is verified by statistics from monitoring stations throughout the world.

One can use equation (iv) as a simple climate model to examine the effect of increasing cloud cover (decreased emissivity by 1%) or increasing the albedo of the earth due to sulfate aerosols by 1%. In the case of decreased emissivity, the global average surface temperature increases by 0.7 °C. For increased albedo of 1%, the global average surface temperature decreases by 0.3 °C.

11.4 NITROGEN CYCLE

The nitrogen cycle is critical to life because it provides major nutrients for plants (NH_4^+, NO_3^-) and animals (protein N, amino acids) alike. In a natural ecosystem, biological N_2-fixation is the crucial mechanism that provides nutrients for photosynthesis; however, human activities of fertilizer production and fossil fuel combustion now rival natural rates of N-fixation.

Figure 11.19 is a schematic of the N-cycle in a natural ecosystem. Major processes include N-fixation by legumes and bacteria on roots; N-fixation from the atmosphere by lightning to form NO; redox reactions in soils and biomass burning that form N_2O, NO_x, and NO_2; and uptake of ammonium and nitrate by vegetation (Table 11.6). The problem is that human activities are fixing amounts of nitrogen equal to those of nature, which results in disturbances to ecosystems. Human activities include fertilizer production of anhydrous ammonia from natural gas; NO_x production from internal combustion engines that convert $N_2 \rightarrow NO_x$ during fossil fuel burning; NO_x production from organic nitrogen in coal and other fossil fuels when they are combusted; and N_2O and other N_xO_y compounds from biomass burning (Table 11.7). As a result of these activities, air pollution and smog are formed from NO_x, N_2O is increasing in the atmosphere, and forests are showing symptoms of nitrogen saturation and decline due to acid deposition (Chapter 10). Natural cycles are being disrupted, the consequence of human activity, intensive agriculture, and fossil fuel combustion.

Galloway et al.[55] have estimated the global nitrogen cycle in preindustrial and anthropogenic times (Figure 11.20). The accumulation of N_2O in the atmosphere amounts to ~3.4 Tg N yr^{-1} (teragram = 1×10^{12} g). It emanates from fertilized soil and biomass burning. Fluxes of NO (and NO_x) to the atmosphere have increased

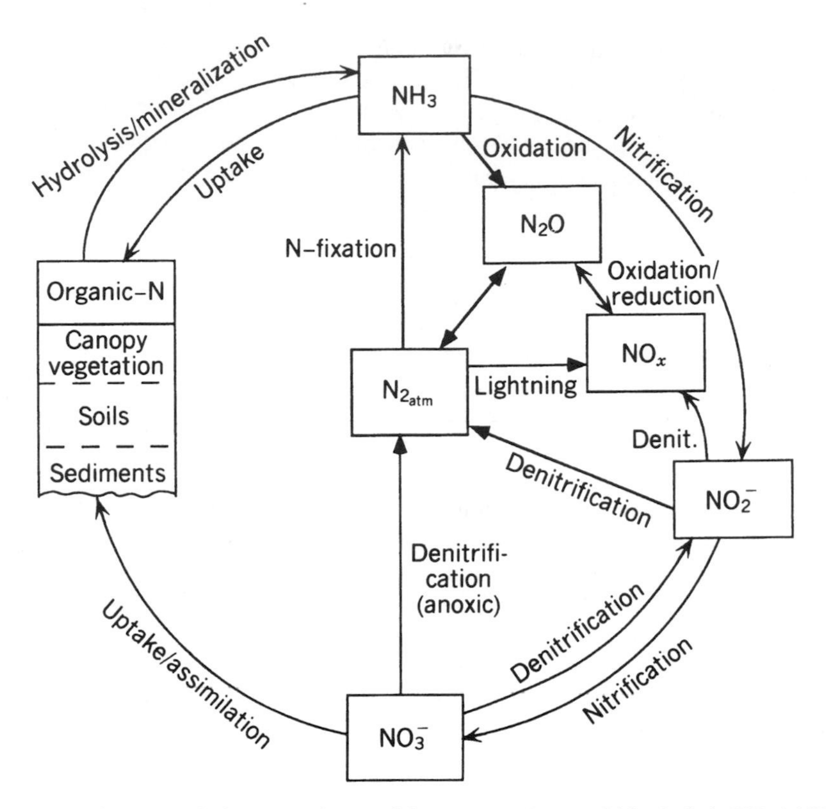

Figure 11.19 Nitrogen cycle in a natural terrestrial ecosystem. State variables include NH_3 ($+NH_4^+$), N_2O, NO_3^-, and organic-N.

Table 11.6 Some Important Reactions of the Nitrogen Cycle

N_2-Fixation by Plants

$$N_2 + CH_2O + 2H_2O \rightarrow 2NH_3 + CO_2 + \tfrac{1}{2}\,O_2$$

N_2-Fixation by Lightning

$$N_2 + O_2 \rightarrow 2\,NO$$

Nitrification

$$NH_3 + 2\,O_2 \rightarrow NO_3^- + H_2O + H^+$$

Dentrification

$$NO_3^- + \tfrac{5}{4}\,CH_2O + H^+ \rightarrow \tfrac{1}{2}\,N_2 + \tfrac{5}{4}\,CO_2 + \tfrac{7}{4}\,H_2O$$

Biomass Synthesis:

$$106\,CO_2 + 16\,NO_3^- + HPO_4^{2-} + 122\,H_2O + 18\,H^+ \underset{R}{\overset{P}{\rightleftharpoons}} C_{106}H_{263}O_{110}N_{16}P_1 + 138\,O_2$$

Table 11.7 Anthropogenic Reactions Now Entering the Nitrogen Cycle

Fertilizer Production (Haber Process)

$$3\ CH_4 + 6H_2O + 4N_2 \rightarrow 3\ CO_2 + 8\ NH_3$$

NO_x Production from Nitrogen Gas During Fossil Fuel Combustion

$$N_2 + O_2 \xrightarrow{2000\ °F} 2\ NO$$

NO_x Production from Fossil Fuel Organic Nitrogen

$$C_{10}\,H_{20}\,N_{0.04} + (15 + 0.02)\ O_2 \rightarrow 10\ CO_2 + 10\ H_2O + 0.04\ NO$$

N_2O Production from Biomass Burning (and fertilized soils)

$$C_{10}\,H_{20}\,N_{0.2} + 15.05\ O_2 \rightarrow 0.1\ N_2O + 10\ CO_2 + 10\ H_2O$$

sixfold due to fossil fuel combustion and biomass burning. Nitrate deposition to terrestrial and oceanic ecosystems has increased, resulting in a fertilization effect. Fertilization constitutes a disruption of natural cycles, and, in some instances, it has resulted in problems of eutrophication and forest decline.

The sources and sinks of reactive forms of nitrogen. The sources and sinks of reactive forms of nitrogen (NH_3, NO_x, N_2O) that are caused by human activities is given by Table 11.8.[55] Nitrogen is accumulating globally, but many of the sinks have not been measured, including the accumulation of nitrogen in soils, the accumulation of nitrate in groundwater, and denitrification to N_2 gas. A large fraction of reactive nitrogen runs off to oceans as nitrate or it is deposited by atmospheric deposition as $NO_{x(g)}$ and $NO^-_{3(aq)}$.

Figure 11.21 shows results of the Geophysical Fluid Dynamics Laboratory Chemical Transport GCM for global NO_y deposition. Preindustrialized NO_y deposition was less than 10 (mmol N) $m^{-2}\ yr^{-1}$. It is apparent that a large fertilization and deposition experiment is underway for nitrogen species to terrestrial and oceanic environments. (NO_y refers to natural plus anthropogenic sources.)

There is a linkage between the global carbon model of Section 11.3 and the global nitrogen deposition model shown by Figure 11.21. Nitrogen could act as the "trigger," the most important variable controlling net primary photosynthesis. If biomass production in boreal and temperate forests is increasing at 1.0 Gt C yr^{-1}, it is possible that N-fertilization, and not CO_2-fertilization, has been responsible for the increased growth rate of forests (the so-called missing carbon sink in many global carbon models). Hudson et al.[56] estimate that anthropogenic deposition of nitrogen species has resulted in an increased biomass synthesis of 26–30 Tg N yr^{-1}. Table 11.8 includes a conservative estimate of ~20 Tg N yr^{-1} due to biomass uptake.

Example 11.4 Nitrogen–Carbon Coupling and the Missing Carbon Sink

Estimates of CO_2 emissions, deforestation, and biomass burning generally exceed measurements of CO_2 accumulation in the atmosphere, ocean uptake, and terrestrial uptake by at least 1.0 Gt C yr^{-1}. The so-called missing sink for carbon dioxide may

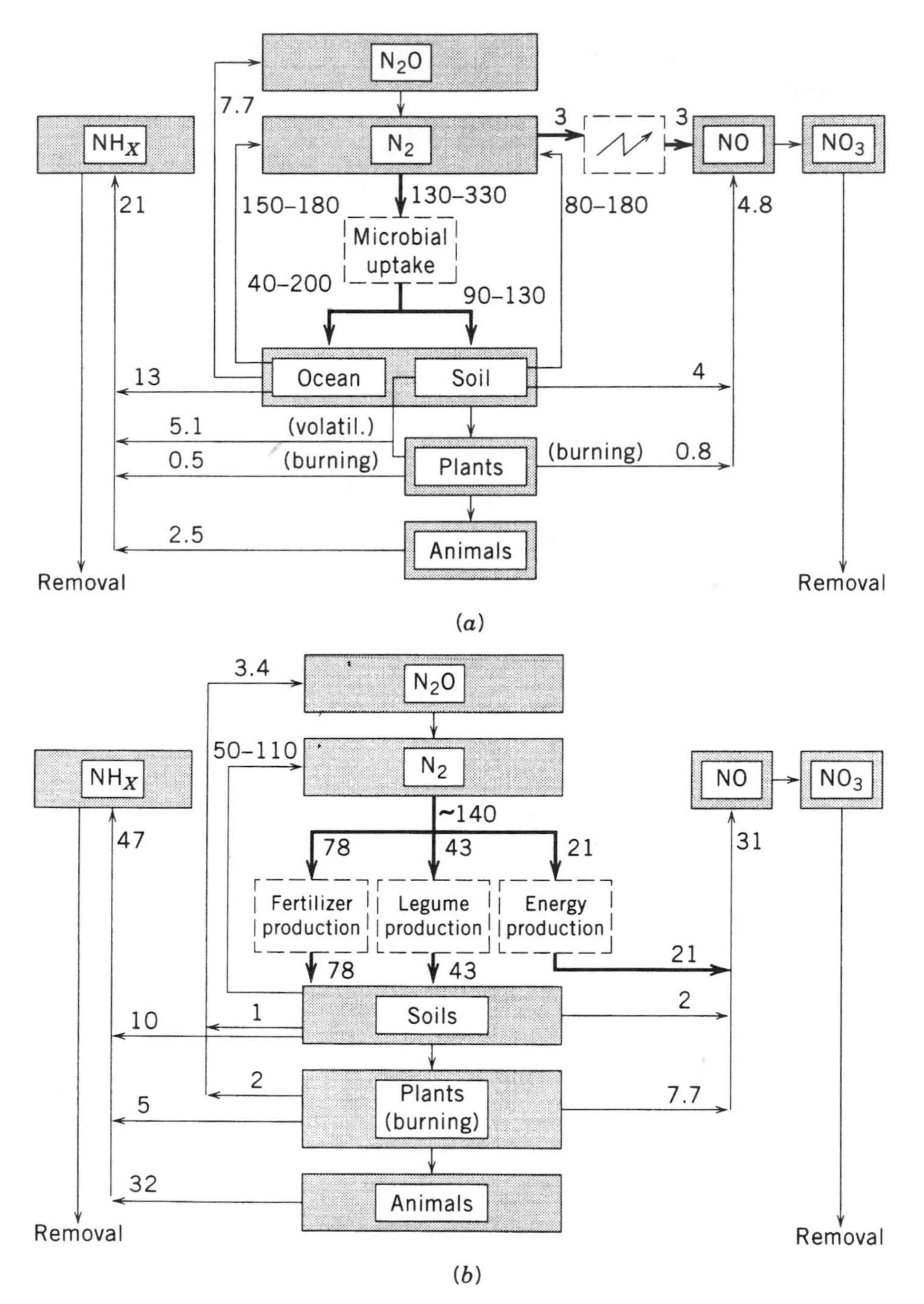

Figure 11.20 Global nitrogen cycle. A) Biogenic or preindustrial conditions where microbial uptake is the principal form of N-fixation and the fluxes to the atmosphere of NH_x and NO are relatively small. B) Additional fluxes due to the anthropogenic factors that rival or exceed natural fluxes of reactive nitrogen. (From Galloway et al.[55]) Reprinted with permission from the American Geophysical Union. Copyright (1995).

Table 11.8 Estimated Sources and Sinks of Global Reactive Nitrogen (NH_3, NO_x, N_2O) from Human Activities[55]

Sources of Reactive Nitrogen	Tg N yr^{-1}
Energy production (fossil fuel combustion)	20
Fertilizer production	80
Legume crop agriculture (land use changes)	40
TOTAL	140

Sinks of Reactive Nitrogen	Tg N yr^{-1}
N_2O in atmosphere	3
Riverine transport to oceans	41
Oceanic atmospheric deposition	18
Biomass uptake (fertilization)	20
Remaining sinks (?)	58
Denitrification to N_2	
Soil accumulation	
Groundwater accumulation	
TOTAL	140

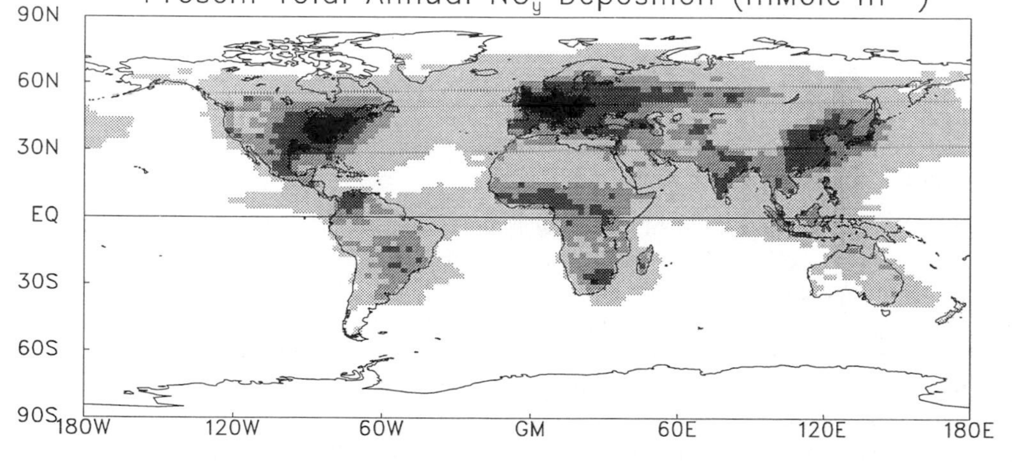

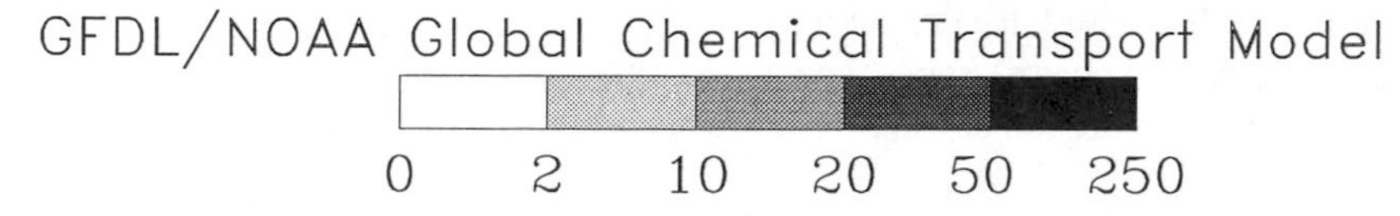

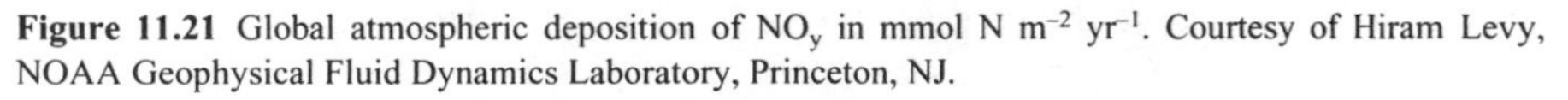

Figure 11.21 Global atmospheric deposition of NO_y in mmol N m^{-2} yr^{-1}. Courtesy of Hiram Levy, NOAA Geophysical Fluid Dynamics Laboratory, Princeton, NJ.

be in forests of temperate and boreal zones that were clear-cut by settlers at the turn of the century and are now in a regrowth phase. Alternatively, the fertilization of forests by HNO_3, NH_3, and excess CO_2 may have increased net primary production. In either case, the result would be an increased standing crop of forests and, eventually, more organic carbon stored in upper soil horizons. Reported ranges for carbon/nitrogen ratios in woody biomass range from 25:1 (C/N) to 50:1 (C/N). In soils, C/N ratios are typically 25:1.

Based on a potential carbon sink of 1 Gt C yr^{-1} (1.0×10^{15} g C yr^{-1}), estimate a range for the nitrogen sink that would be required for incremental forest growth to explain the missing carbon sink.

Solution: Stoichiometry allows us to couple carbon uptake to nitrogen uptake in forests.

$$\frac{1\text{ N}}{25\text{ C}} \times \frac{1 \times 10^{15}\text{ g C}}{\text{yr}} = 4 \times 10^{13}\,\frac{\text{g N}}{\text{yr}} = 40\text{ Tg N yr}^{-1}$$

$$\frac{1\text{ N}}{50\text{ C}} \times \frac{1 \times 10^{15}\text{ g C}}{\text{yr}} = 2 \times 10^{13}\text{ g N yr}^{-1} = 20\text{ Tg N yr}^{-1}$$

The latter estimate is consistent with the global nitrogen balance in Table 11.8. All of these global estimates are very crude, and they will be refined as measurements improve.

11.5 GLOBAL SULFUR CYCLE

11.5.1 Species and Reactions

It is difficult to make a global mas balance for sulfur, in part, because there are so many species involved. Anthropogenic emissions of sulfur dioxide (SO_2) are greater than biogenic emissions (natural reactions involving microorganisms and plants), but biogenic sources are also important. Anthropogenic emissions emanate from coal combustion (60%), petroleum refining (28%), smelting of ores (~10%), and the pulp and paper industry (2%).[57] Biogenic emissions of dimethyl sulfide or DMS (CH_3SCH_3) come from oceanic algae, while other reduced sulfur species are emitted from wetlands and coastal regions that include hydrogen sulfide (H_2S), carbonyl sulfide (COS), carbon disulfide (CS_2), and DMS. These reduced sulfur species are volatile, and they eventually oxidize to form $SO_{2(g)}$ and sulfate aerosols that are deposited by both wet and dry deposition.

In wetlands and coastal zones, anaerobic microbial processes cause sulfate ions to be reduced to hydrogen sulfide (or more accurately as bisulfide anion, HS^-).

$$SO_4^{2-} + 2\,CH_2O \xrightarrow{\text{Desulfivibrio}} H_2S + 2\,HCO_3^-$$

CH_2O is indicative of organic carbon compounds that are present as a source of substrate for sulfate-reducing bacteria, while sulfate ions serve as the electron acceptor. H_2S is a volatile species, but most of it oxidizes in aerobic waters before it can escape the water surface. In soils, H_2S is reoxidized or it is precipitated as FeS before being emitted. Aerobic soils can actually absorb H_2S from the atmosphere. However, in swamps and coastal zones, significant biogenic emissions of H_2S and DMS occur.

Dimethyl sulfide (DMS) is produced by ocean phytoplankton, and it is released when phytoplankton are grazed upon by zooplankton

$$2\ CO_2 + 2\ H_2O \xrightarrow{\text{sun}} 2\ CH_2O + 2\ O_2$$

$$SO_4^{2-} + 2\ CH_2O + 2\ H^+ \xrightarrow{\text{algae}} CH_3SCH_3 + 3\ O_2$$

DMS is quite volatile, and it accounts for a large fraction of biogenic sulfur emissions from oceans.

Reduced forms of sulfur in the atmosphere react with hydroxyl radicals. For H_2S and DMS, the reaction is quite fast with lifetimes in the troposphere of 4.4 days and 0.6 days, respectively.[57] A series of free radical reactions occur for H_2S.

$$\cdot OH + H_2S \rightarrow H_2O + HS\cdot$$

$$HS\cdot + HCHO \rightarrow H_2S + HCO\cdot$$

or

$$HS\cdot + O_3 \rightarrow HSO + O_2$$

The HS· radical does not build up in the atmosphere, so it is thought that formaldehyde and ozone react with HS·. Likewise, DMS undergoes a series of free radical reactions.

$$\cdot OH + CH_3SCH_3 \rightarrow CH_3S(OH)CH_3$$

$$CH_3S(OH)CH_3 + O_2 \rightarrow (O_2\text{–DMS–OH})\ \text{complex}$$

$$\text{complex} \rightarrow \underbrace{CH_3\ (SO_3)\ H}_{\text{MSA}} + CH_3\cdot$$

Methane sulfonic acid (MSA) is a major product of the reaction between hydroxyl radicals and DMS. It is a condensable product that helps to form aerosol particles in the atmosphere (cloud condensation nuclei). James Lovelock has sugested that the formation of MSA is a negative feedback loop to global warming, consistent with his Gaia hypothesis. The hypothesis states that the earth is something akin to a huge organism capable of homeostasis and feedback controlling processes. As the earth heats up under greater CO_2 concentrations, algal production will increase causing DMS release and conversion to MSA, resulting in more aerosols, more clouds, and more albedo for the earth (a negative feedback effect). Many scientists feel that Gaia is a fascinating hypothesis and that feedback loops are, indeed, very important in earth surface processes, but that there is no need to invoke teleological reasoning or to attribute a "persona" to the earth. Still, it makes a good story.

Carbon disulfide (CS_2) also reacts rapidly with hydroxyl radicals in the atmosphere with a lifetime of ~12 days.[57]

$$\cdot OH + CS_2 \rightarrow SCSOH$$

$$SCSOH + O_2 \rightarrow COS + SO_2 + H$$

But carbonyl sulfide (COS) is slow to be oxidized. Its lifetime in the atmosphere is on the order of 44 years.[57]

$$\cdot OH + COS \rightarrow CO_2 + H_2S$$

When it is impossible to estimate a residence time for a chemical in the atmosphere because of a lack of information on reservoir pools or flux rates, it is often convenient to use the concept of "lifetime." Lifetime is defined as the reciprocal of the pseudo-first-order rate constant. It is the time that it takes to reach one-third of its original concentration ($0.368\ C_0$) in a batch reactor, which is equal to $1/k'$.

$$\text{Lifetime} = \frac{1}{k'} \tag{24}$$

where k' is the pseudo-first-order rate constant.

Kinetics of reactions of reduced sulfur species with hydroxyl radicals are not truly first order; they are generally second order. For example, the oxidation of hydrogen sulfide with ·OH is first order with respect to ·OH, first order with respect to H_2S, and second order overall.

$$\frac{d[H_2S]}{dt} = \text{Rxn rate} = -k[\cdot OH][H_2S] \tag{25}$$

A pseudo-first-order reaction can be utilized if the hydroxyl radical concentration is roughly constant.

$$\frac{d[H_2S]}{dt} = -\,k'[H_2S] \tag{26}$$

where $k' = k[\cdot OH]$, the pseudo-first-order rate constant.

As discussed in Chapter 10, both gas and aqueous phase reactions are important in the oxidation of sulfur dioxide in the atmosphere. Oxygen and hydroxyl radicals can oxidize SO_2 to SO_4^{2-} in the gas phsae, and humidity in the air (H_2O) can convert SO_3 to acidic aerosol particles.

$$SO_2 + \cdot OH \rightarrow HOSO_2 \xrightarrow{O_2} SO_3 + HO_2\text{(superoxide)}$$

$$SO_3 + H_2O \rightarrow H_2SO_4 \text{ (aerosol)}$$

Aqueous phase reactions for SO_2 include reaction with O_3 and H_2O_2; both can be important depending on the concentrations of ozone and hydrogen peroxide in clouds.

$$SO_2 + H_2O \rightarrow SO_3^{2-} + 2H^+$$

$$SO_3^{2-} + O_{3(aq)} \rightarrow SO_4^{2-} + O_2$$

$$SO_3^{2-} + H_2O_{2(aq)} \rightarrow SO_4^{2-} + H_2O$$

Lifetimes ($1/k'$) for the reactions in clouds are on the order of 1–50 days. Clouds process a tremendous amount of air and water vapor. They serve as a concentrating vortex for particles and gases that react with SO_2.

Figure 11.22 is a schematic of the reactive sulfur species in the atmosphere, a summary of the reactions in this section. Most emissions (both biogenic and anthropogenic) are in the form of reduced sulfur species and SO_2. Sulfur falls back to earth (continents and oceans) in the form of $SO_{2(g)}$ (dry deposition), sulfate aerosols (H_2SO_4, $MgSO_4$, $CaSO_4$ in dry deposition), and sulfate ions (H_2SO_4 and $CaSO_4$ in wet deposition). Sulfate aerosols and cloud condensation nucleii play an important role as a negative feedback effect to global warming by increasing the earth's albedo.

11.5.2 Global Sulfur Budget

Major emissions and transport mechanisms of the global sulfur cycle are shown in Figure 11.23. Anthropogenic emissions of SO_2 on land are 93 Tg S yr^{-1} (93×10^{12} g S yr^{-1}), which are greater than emissions due to volcanoes, wind-blown dust, and biogenic sources. Most sulfur is deposited over the oceans as SO_2 (dry deposition) and sulfate (dry + wet deposition). Emissions follow those of Schlesinger,[58] and they are estimated such that inputs are equal to outputs in the atmosphere compartment. We assume that sulfur is not accumulating in the atmosphere because removal mechanisms have such fast reaction rates. If we sum all the inputs and outputs to the land compartment and the oceans compartment in Figure 11.23, we see that the land is emitting slightly more sulfur than is coming-in, and the ocean is accumulating sulfur species. But the reservoirs of sulfur in the land (reduced S and sulfates) and in the ocean (sulfates) are enormous.

The atmosphere is a small reservoir for sulfur species, only 4.6 Tg S, resulting in a mean residence time in the atmosphere, only 4.9 days. SO_2 travels 500–2000 km by long-range transport, but it does not accumulate in the atmosphere.

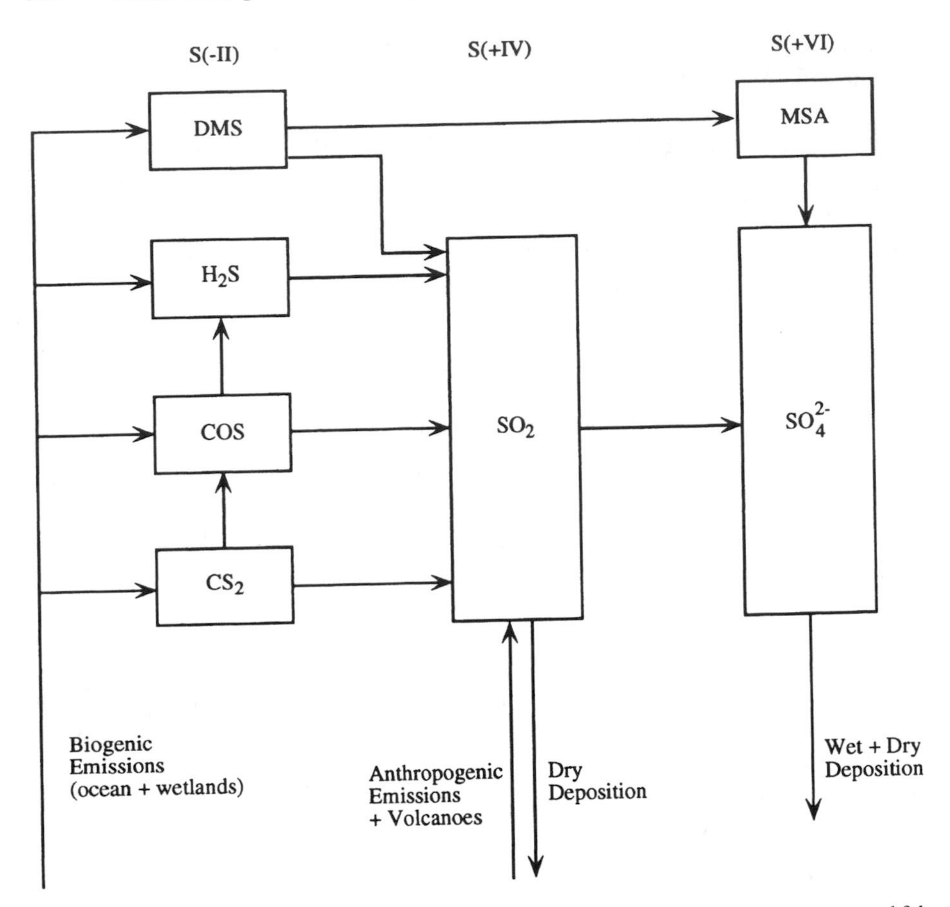

Figure 11.22 Reactive sulfur species in the atmosphere (reduced species in the left-hand column and fully oxidized species in the right-hand column).

$$\tau_s = \frac{\text{Mass of S in the atmosphere}}{\text{Mass flux of S out of the atmosphere}} \tag{27}$$

A summary of the atmospheric global sulfur budget is given by Table 11.9. Estimates have a large uncertainty and each year can vary significantly. Volcanic activity is highly variable. There are approximately 500 active volcanoes on earth, 90 of which are active in an average year. The Mt. Pinatubo eruption in 1991 was the largest in 100 years, and it emitted 15–20 Tg S, (on the same order of magnitude as U.S. electric utilities in that year, 14 Tg S). But Mt. Pinatubo blasted SO_2 into the lower level of the stratosphere where sulfate aerosols have a longer lifetime. The albedo effect caused global cooling (0.3–0.5 °C) for two years after the event. The Kuwait oil fires following the Gulf War emitted 3–10 Tg S, but the plume rose only

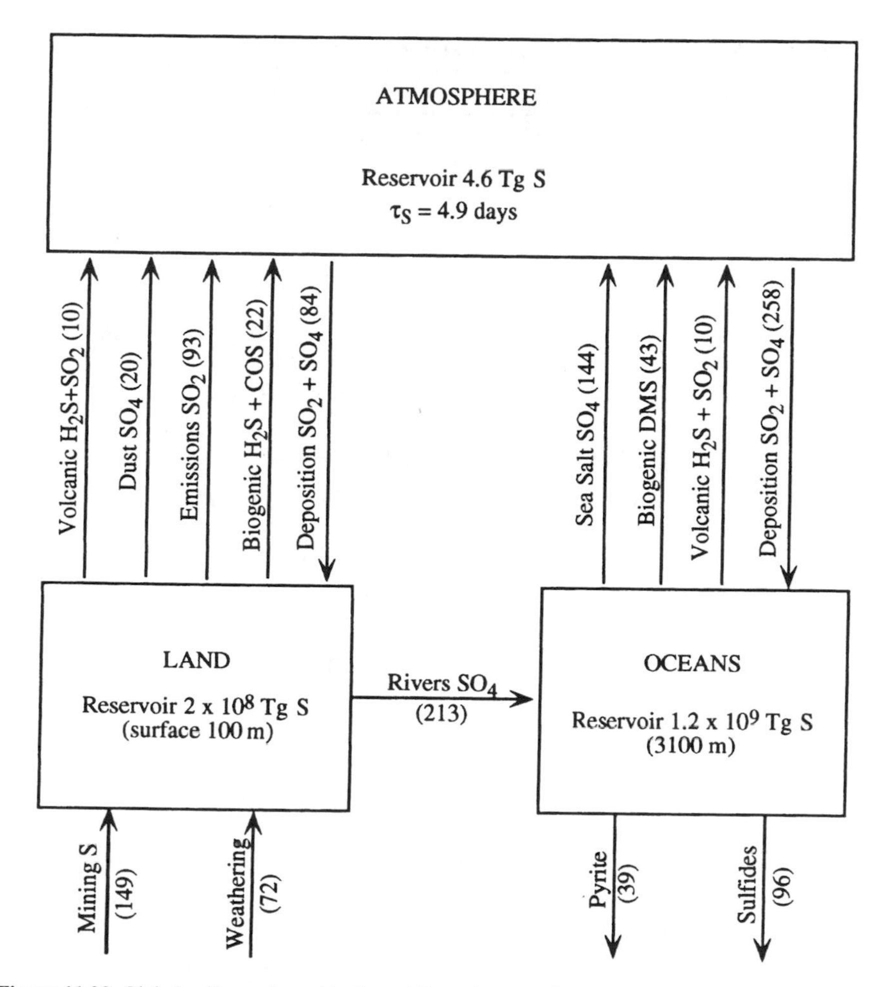

Figure 11.23 Global sulfur cycle and budget. All numbers are fluxes expressed in Tg S yr^{-1}, except for the reservoir masses, which are expressed in units of Tg S. (Modified from Schlesinger.[58])

about 3 km into the troposphere and it resulted in regional sulfur pollution, which did not produce a global effect.

Example 11.5 Characteristic Times for SO_2 in the Atmosphere

Calculate the half-life, lifetime, and residence time of SO_2 species in the atmosphere.

$SO_{2(g)}$ = 2 Tg S mass reservoir in the atmosphere
SO_2 deposition = 150 Tg S yr^{-1} total
$k' = 0.02\ h^{-1}$ pseudo-first-order rate constant for oxidation reactions

Table 11.9 Global Sulfur Budget[a] to and from the Atmosphere

Sources and Sinks	Tg S yr^{-1}
Sulfur Sources to Atmosphere	
Volcanoes ($SO_2 + H_2S$)	20
Dust ($CaSO_4$)	20
Emissions (SO_2)	93
Soil and wetlands ($H_2S + COS$)	22
Sea salt (Na_2SO_4	144
Ocean flux (DMS)	43
TOTAL	342
Sulfur Sinks from the Atmosphere	
Wet and dry deposition (terrestrial)	84
Deposition (oceanic)	258
TOTAL	342

[a] 10^{12} g S yr^{-1} = 1 million metric tonnes.

Solution: Each characteristic time has a different meaning and definition. It is important not to confuse the terms. Half-life and lifetime are based on the assumption of pseudo-first-order kinetics in a completely mixed batch reactor (50% remaining and 36.8% remaining, respectively). Residence time assumes that steady state conditions are applicable in a flow-through system.

Half-life:

$$C = C_0 e^{-k't}$$

$$\tfrac{1}{2} = e^{-k't}$$

$$t_{1/2} = \frac{0.693}{k'}$$

$$t_{1/2} = \frac{0.693}{0.02} = 35 \text{ hours}$$

Lifetime:

$$t_{\text{life}} = \frac{1}{k'} = 50 \text{ hours}$$

Residence time:

$$\tau_s = \frac{\text{Mass}}{\text{Flux out}} = \frac{2}{150} = 0.13 \text{ yr}$$

$$\tau_s = 4.9 \text{ days}$$

Response time is another characteristic time in model simulations. It is based on the eigenvalue of the dynamic mass balance equation, and it predicts how quickly the environmental compartment will change after a step-function change in inputs.

11.6 TRACE GASES

Trace gases in the atmosphere include methane (CH_4), nitrous oxide (N_2O), methyl chloroform, and chlorofluorocarbons (CFCs). Methane and nitrous oxide are anthropogenic greenhouse gases. The global budget for N_2O is not well known, especially regarding its sources. Sources include industrial emissions and incomplete combustion of fossil fuels, 1 Tg N_2O yr^{-1}, and biomass burning ~1 Tg N_2O yr^{-1}. Natural ecosystems emit 3–9 Tg N_2O yr^{-1} as an intermediate oxidation state (leakage) from the nitrogen cycle. Fertilized fields are thought to emit up to ten times more N_2O per m^2 than without fertilizer. Deforestation and the opening of the soil nitrogen cycle after clear-cutting may account for another large source, and surface oceans waters are ~4% supersaturated with N_2O. Since N_2O is increasing in the atmosphere at only 0.2% yr^{-1}, there must be a large sink in the atmosphere. Atmospheric sinks for N_2O include photolysis and a slow oxidation with monoatomic oxygen to form NO and N_2 (Table 11.11).

Methane is another trace greenhouse gas that occupies 1.7 ppm by volume in the atmosphere. Anthropogenic sources of methane rival natural sources with flooded rice agriculture and ruminant animals as the largest sources (Table 11.10). Wetlands emit large amounts of natural methane due to methanogenesis in anaerobic sediments and soils.[59] Methane reacts with hydroxyl radicals in the atmosphere as the principal sink. Eventually methane reacts all the way to form CO_2, but the reactions are slow. Table 11.11 is a compilation of some trace gas reactions for carbon species in the atmosphere including methane assuming pseudo-steady state approximations. Methane reacts with ·OH to form formaldehyde, HCHO; and formaldehyde undergoes photolytic oxidation to form carbon monoxide, which eventually gives carbon dioxide. Natural emissions of non-methane hydrocarbons (NMHC) are also important sources of formaldehyde and carbon monoxide to the atmosphere. They enter the photolytic cycle and participate in the formation of ozone and smog. NMHCs are primarily C_{10} and higher alkenes that are emitted by vegetation, such as terpene and isoprene (Table 11.12). They are responsible for the haze found in the Smokey Mountains of Appalachia.

Trace gas models indicate that CH_4 is the trace gas that would be most simple to control. If we could control emissions at 1990 levels (constant anthropogenic emissions), the concentration in the atmosphere would level off at 1.85 ppmv by the year 2005.[60] This is because methane has a relatively short residence time in the atmosphere (5–10 years). Indeed, the rate of increase of CH_4 concentrations in the atmosphere is already decreasing. Control of leaky gas pipelines may be sufficient to cause CH_4 concentrations to remain constant in the 21st century. But nitrous oxide concentrations are projected to increase even if we can keep emissions at a constant level. N_2O has a long residence time in the atmosphere that will cause it to increase over the next few decades, even assuming constant emission rates.

Table 11.10 Methane Balance[a] for the Atmosphere Adapted from Paterson[59]

Sources and Sinks	Tg CH_4 yr^{-1}
Sources	
Anthropogenic	
Biomass burning	44
Coal extraction	37
Waste systems	52
Natural gas losses	51
Rice production	99
Ruminant animals	82
Subtotal	365
Natural	
Biomass burning	10
Freshwater	5
Hydrates–clathrates	5
Oceans	10
Termites	21
Wetlands	109
Subtotal	160
TOTAL	525
Sinks	
Oxidation with hydroxyl radical	436
Oxidation with chlorine in stratosphere	26
Accumulation in atmosphere	26
Oxidation by soil microorganisms	37
TOTAL	525

[a] 1 Tg $CH_4 = 10^{12}$ g CH_4.

Table 11.11 Simplified Reactions for Selected Trace Gas Species in the Atmosphere (not Balanced)[60]

	Reactions	
(A)	$CO + HO\cdot$	$\rightarrow CO_2 + HO_2\cdot$
(B)	$CH_4 + OH\cdot$	$\rightarrow HCHO + HO_2\cdot + H_2O$
(C)	$CH_4 + Cl$	$\rightarrow HCHO + HO_2\cdot + HCl$
(D)	$HCHO + h\nu$	$\rightarrow H_2 + CO + 2HO_2\cdot$
(E)	$HCHO + HO\cdot$	$\rightarrow CO + HO_2\cdot + H_2O$
(F)	$HCHO + Cl$	$\rightarrow CO + HO_2\cdot + HCl$
(G)	$NMHC + HO\cdot$	$\rightarrow 1.23HCHO + 1.68CO_2 + 0.09CO + H_2O$
(H)	$NMHC + NO_3$	$\rightarrow 1.23HCHO + 1.68CO_2 + 0.09CO + HNO_3$
(I)	$N_2O + h\nu$	$\rightarrow N_2 + O('D)$
(J)	$N_2O + O('D)$	$\rightarrow N_2 + O_2$
(K)	$N_2O + O('D)$	$\rightarrow 2\,NO$

Table 11.12 Trace Gas Emissions[a] from Anthropogenic and Natural Sources[60]

		Natural Sources		
	Anthropogenic Sources	Ocean	Continents	Total
N_2O	6.5	1	6	13
CH_4	274	11	109	394
CO	500	71	29	600
NMHC	125	90	920	1135

[a]Units: Tg C yr^{-1} for carbon species; Tg N yr^{-1} for N_2O.

Example 11.6 Mean Residence Time of N_2O in the Atmosphere

Find the mean residence time of N_2O in the atmosphere. Its concentration is currently 0.31 ppmv and emissions are given in Table 11.12.

Solution: The concentration in the atmosphere is 0.31 ppmv, and the sources (or sinks at steady state) are 13 Tg N yr^{-1} (20.4 Tg N_2O yr^{-1}).

$$0.31 \times 10^{-6}\ N_2O \times \frac{(44\ g\ mol^{-1}\ N_2O)}{(29\ g\ mol^{-1}\ air)} \times 5.25 \times 10^{18}\ kg\ air = 2.47 \times 10^{12}\ kg\ N_2O$$

$$\tau_{N_2O} = \frac{\text{Mass}}{\text{Flux out}} = \frac{2470\ Tg\ N_2O}{20.4\ Tg\ N_2O\ yr^{-1}} = 121\ yr$$

The residence time of N_2O is quite long, indicating that the concentration in the atmosphere is likely to continue to increase for several decades even if emissions are controlled at constant levels. In this example, we have assumed steady state, which is not the case for N_2O. It is likely that the "Flux out" must be greater to achieve steady state.

Stratospheric ozone depletion is caused primarily by CFC chemicals and methyl chloroform. These chemicals react with hydroxyl radicals in the troposphere before they are transported by vertical mixing to the stratosphere, but the rate of reaction is slow compared to transport times. Thus they accumulate in the stratosphere, where they react with ozone and deplete the ozone layer.

Hydrochlorofluorocarbon (HCFC) and hydrofluorocarbon (HFC) chemicals have been mandated to replace CFCs. These substitutes have shorter lifetimes ($1/k'$) in the troposphere (Table 11.13). The reaction rate is second order:

$$\frac{dC}{dt} = \text{Rxn rate} = k[\cdot OH][C] \qquad (28)$$

where k is the second-order reaction rate constants.

By examining Table 11.13, it is easy to see that the CFC substitutes are much more rapidly degraded than CFC-12, but they are still likely to accumulate in the stratosphere similar to that of methyl chloroform (CH_3CCl_3). The good news is that methyl chloroform has a short enough lifetime that it is now decreasing in the atmosphere as a result of the Montreal Protocol and subsequent amendments. Eventually, there will be a need to phase-out of HCFC and HFC chemicals as well. There is an additional concern regarding HFC-134a, HCFC-123, and HCFC-124. Their reaction products include trifluoroacetic acid (CF_3COOH), which is quite persistent and toxic in aquatic and terrestrial ecosystems.

11.7 Lifetime and Reaction Rate of HCFC-22

Estimate the reaction rate and the lifetime of HCFC-22 from the following information. Assume standard temperature and pressure.

$[\cdot OH] = 5 \times 10^5$ molecules cm^{-3} (average tropospheric concentration)
$[CHF_2Cl] = 0.1$ ppbv (average tropospheric concentration)
$k = 1.44 \times 10^{-11}$ cm^3 molecule^{-1} h^{-1} (second-order rate constant)

Solution: Use a second-order kinetic expression to calculate the reaction rate. The lifetime is defined as the reciprocal of the pseudo-first-order reaction rate constant.

Table 11.13 Atmospheric Lifetime of CFCs, HCFCs, and HFCs for the Reaction with Tropospheric Hydroxyl Radical

HCFC or HFC	Total Lifetime, yr
CFC-11 ($CFCl_3$)	22.8
CFC-12 (CF_2Cl_2)	114
CFC-14 (CF_3Cl)	680
CH_3CCl_3 (methyl chloroform)	2.9
HFC-32 (CH_2F_2)	6.7
HCFC-22 (CHF_2Cl)	15.8
HFC-152a (CH_3CHF_2)	1.6
HFC-134a (CH_2FCF_3)	14.3
HFC-125 (CHF_2CF_3)	30
HCFC-141b (CH_3CFCl_2)	10.8
HCFC-142b (CH_3CF_2Cl)	22.4
HCFC-123 ($CHCl_2CF_3$)	1.7
HCFC-124 ($CHFClCF_3$)	6.9
HCFC-225ca ($CHCl_2CF_2CF_3$)	2.8
HCFC-225cb ($CHClFCF_2CClF_2$)	8.0

Source: David.[61]

$$\frac{dC}{dt} = -k[\cdot OH][C] = -1.44 \times 10^{-11}(5 \times 10^{5})(C)$$

$$C = \frac{0.1 \times 10^{-9}\text{ mol}}{\text{mol}} \left| \frac{6.02 \times 10^{23}\text{ molecules}}{\text{mol}} \right| \frac{\text{mol}}{22.4\text{ L}} \left| \frac{\text{L}}{1000\text{ cm}^3} \right.$$

$$= 2.69 \times 10^{9} \frac{\text{molecules}}{\text{cm}^3}$$

$$-\frac{dC}{dt} = +1.94 \times 10^{4} \frac{\text{molecules}}{\text{cm}^3\text{ h}}$$

$$\frac{1}{k'} = \frac{1}{k[\cdot OH]} = 15.8\text{ yr lifetime}$$

The result is equal to the estimate of David[61] in Table 11.13.

Example11.8 Concentration of Products in Global Rainwater: HCFC-141b

Find the concentration of HF and HCl in global rainwater as a result of HCFC-141b reaction with hydroxyl radicals in the tropopshere. Global annual rainfall is 4.8×10^{17} L yr^{-1}, and the future annual production of HCFC-141b is estimated as 527,000 metric tons of $CF_2Cl_2CH_3$. (This example is from David.[61])

Solution: Use stoichiometry to calculate the yield of HF and HCl.

$$CFCl_2CH_3 + 2\,O_2 \rightarrow HF + 2\,HCl + 2\,CO_2$$

$$\text{M W HCFC-141b} = 116.9\text{ g mol}^{-1}$$

$$\therefore \quad \text{Production of HCFC-141b} \simeq 5 \times 10^{9}\text{ mol yr}^{-1}$$

Thus there are 2 moles of HCl and 1 mole of HF formed for every mole of HCFC-141b reacted. Assuming steady state, global annual rainfall would contain the total annual production.

HF: $5 \times 10^{9}(20\text{ g mol}^{-1}) = 1 \times 10^{11}\text{ g}$

$(1 \times 10^{11}\text{ g})/(4.8 \times 10^{17}\text{ L}) = 2.1 \times 10^{-7}\text{ g L}^{-1}$

$= 0.21\ \mu\text{g L}^{-1}$

HCl: $(10 \times 10^{9})(36.5\text{ g mol}^{-1}) = 3.7 \times 10^{11}\text{ g}$

$(3.7 \times 10^{11}\text{ g})/(4.8 \times 10^{17}\text{ L}) = 7.6 \times 10^{-7}\text{ g L}^{-1}$

$= 0.76\ \mu\text{g L}^{-1}$

Once CFC, HCFC, and HFC chemicals reach the stratosphere they react in free radical reactions to yield free chlorine. It is the free chlorine in the stratosphere that reacts with ozone. Ozone is continuously formed in the stratosphere by a photodissociation of molecular oxygen and recombination to form O_3, but it is also decomposed by the back-reaction[10]:

$$CFCl_3 + O_2 \xrightarrow{h\nu} CO_2 + F + 3Cl$$

$$2\,O_3 \overset{h\nu}{\rightleftharpoons} 3\,O_2$$

Chlorine combines with ozone to form chlorine monoxide radicals. ClO reaches a peak concentration over Antarctica during the months of October and November when there is early spring sunlight, stable vortex conditions, and extremely cold air with polar stratospheric clouds for heterogeneous reactions that catalyze the following reactions:

$$Cl + O_3 \rightarrow ClO + O_2$$

$$O_3 + h\nu \rightarrow O + O_2$$

$$ClO + O \rightarrow Cl + O_2$$

$$\text{Net:}\quad 2\,O_3 + h\nu \rightarrow 3\,O_2$$

Note that Cl is a catalyst in the reaction that is regenerated. One chlorine atom can recycle ~20,000 times before it is terminated in the series of free radical reactions, destroying considerable O_3 molecules.

Some chlorine atoms would reach the stratosphere naturally, ~0.6 ppbv, by volcanic eruptions and the transport of methyl chloride (CH_3Cl) from seaweed.[62] But the current concentration due to CFC chemicals is five times greater than background, and it is responsible for destruction of ozone in the stratosphere.

Mathematical models of global cycles and trace gas constituents can be used to understand global change, to set research priorities, and to predict the results of policy decisions. In this chapter, we have examined the role that greenhouse gases (CO_2, CH_4, N_2O) play in global warming. Major nutrient cycles of sulfur and nitrogen are coupled to the carbon cycle through common sources of emissions and fertilization effects on biota. CFCs will decrease in concentration in the atmosphere as emissions are curtailed, but long lifetimes of the chemicals and some of their substitutes will remain a concern for stratospheric ozone depletion for years. Because reaction times are slow in the atmosphere, some trace gases are accumulating, and their long-term effects are not fully known. It will be difficult to curtail emissions of CO_2, SO_x, and NO_x because of the twin driving forces: burgeoning population and increasing per capita consumption of energy and mineral resources. Sustainable development will require increased use of renewable resources, pollution prevention, recycling, reuse, energy efficiency, and a stable population.

11.7 REFERENCES

1. Soule, M.E., *Science*, 253, 744 (1991).
2. Rodhe, H., *Science*, 248, 1217 (1990).
3. Boden, T.A., Kanciruk, P., and Farrel, M.P., *Trends '90: A Compendium of Data on Global Change*, Oak Ridge National Laboratory, U.S. Department of Energy, Oak Ridge, TN (1990).
4. Molina, M.J., and Rowland, F.S., *Nature*, 249, 810 (1974).
5. World Meteorological Organization (WMO), Scientific Assessment of Ozone Depletion: 1991, Report No. 25, Geneva, Switzerland (1992).
6. Ravishankara, A.R., and Albritton, D.L., *Science*, 269, 183 (1995).
7. Mann, C.C., *Science*, 253, 736 (1991).
8. United Nations Development Report, New York (1992).
9. Brasseur, G., and Solomon, S., *Aeronomy of the Middle Atmosphere*, Reidel Publishers, New York (1986).
10. Graedel, T.E., and Crutzen, P.J., *Atmospheric Change—An Earth System Perspective*, W.H. Freeman and Company, New York (1993).
11. Intergovernmental Panel on Climate Change (IPCC), *Climate Change, The IPCC Scientific Assessment WMO/UNEP*, Cambridge University Press, Cambridge, United Kingdom (1990).
12. Intergovernmental Panel on Climate Change (IPCC), *Climate Change, The Second IPCC Scientific Assessment WMO/UNEP*, Cambridge University Press, Cambridge, United Kingdom (1996).
13. Barron, E.J., *EOS Trans. Am. Geophys. Union*, 76, 185 (1995).
14. Nerem, R.S., *Science*, 268, 667 (1995).
15. Manabe, S., and Wetherald, R.T., *J. Atmos. Sci.*, 32, 3 (1985).
16. Hansen, J., Lacis, A., Rind, D., Russell, G., Stone, P., Fung, I., Ruedy, R., and Lerner, J., in *Climate Processes and Climate Sensitivity*, J.E. Hansen and T. Takahashi, Eds., Maurice Ewing Series Vol. 5, p. 130 (1985).
17. Washington, W.M., and Meehl, G.A., *J. Geophys. Res.*, 89, 9475 (1984).
18. Schlesinger, M.E., and Jiang, X., *Climate Dynamics*, 3, 1 (1988).
19. Wilson, C.A., and Mitchell, J.F.B., *J. Geophys. Res.*, 92, 13315 (1987).
20. Trenberth, K.E., Ed., *Climate System Modeling*, Cambridge University Press, Cambridge, United Kingdom (1992).
21. Perry, T.S., *IEEE Spectrum*, 30: 7, 33 (1993).
22. Moore, B. III, and Bolin, B., *Oceanus*, 29, 16–29 (1986/87).
23. Hammond, *Medallion World Atlas*, Hammond Incorporated, Maplewood, NJ (1985).
24. Warrick, R., Gifford, R., and Parry, M., in *The Greenhouse Effect, Climate Change, and Ecosystems, SCOPE 29*, B. Bolin et al., Eds., Wiley, New York, pp. 393–473 (1986).
25. Acock, B., in *Impact of Carbon Dioxide, Trace Gases, and Climate Change on Global Agriculture, ASA Spec. Publ. 53*, pp. 45–60, (1990).
26. Raich, J.W., Rastetter, E.B., Melillo, J.M., Kicklighter, D.W., Steudler, P.A., Peterson, B.J., Grace, A.L., Moore, B. III, and Vörösmarty, C.J., *Ecol. Appl.*, 1, 399–429 (1991).
27. Kwon, O. Yul, Ph.D. Dissertation, The University of Iowa, Iowa City (1994).

28. Murray, J.W., in *Global Biogeochemical Cycles,* S.S. Butcher, R.J. Charlson, G.H. Orians, and G.V. Wolfe, Eds., Academic Press, San Diego, CA, pp. 175–211, (1992).
29. Peng, T.H., Takahashi, T., Broecker, W.S., and Olafsson, J., *Tellus Ser. B*, 39B, 439–458 (1987).
30. Griffiths, D.V., and Smith, I.M., Eds., *Numerical Methods for Engineers*, CRC Press, Boca Raton, FL (1991).
31. Quay, P. D., Tilbrook, B., and Wong, C.S., *Science*, 256, 74–79 (1992).
32. Houghton, R., et al., *Tellus Ser. B*, 39B, 122–139 (1987).
33. Detwiler, R.P., and Hall, C.A.S., *Science*, 239, 42–50 (1988).
34. Hansen, J., Lacis, A., Ruedy, R., and Sato, M., *Geophys. Res. Lett.*, 19, 215–218 (1992).
35. Jones, P.D., Wigley, T.M.L., and Wright, P.B., *Nature*, 322, 430–434 (1986).
36. Budyko, M.I., and Izrael, Y.A., Eds., *Anthropogenic Climate Change*, University of Arizona Press, Tucson (1991).
37. Larcher, W., Ed., *Physiological Plant Ecology*, Springer-Verlag, New York (1980).
38. Harvey, L.D.D., *Global Biogeochem. Cycles*, 3, 137–153 (1989).
39. Enoch, H.Z., and Sacks, J.M., *Photosynthetica*, 12, 150–157 (1978).
40. Webb, W.L., Lauenroth, W.K., Szarek, S.R., and Kinerson, R.S., *Ecology*, 64, 134–151 (1983).
41. Woodwell, G.M., in *Changing Climate, Carbon Dioxide Assessment Committee*, National Academy of Science Press, Washington, DC pp. 216–251, (1983).
42. Running, S.W., and Coughlan, J.C., *Ecol. Modelling*, 42, 125–154 (1988).
43. Reuss, J.O., and Innis, G.S., *Ecology*, 58, 379–388 (1977).
44. Jansson, P.E., and Berg, B. *Can. J. Bot.*, 63, 1008–1016 (1985).
45. Carlyle, J.C. and Than, U.B.A., *J. Ecol.*, 76, 654–662 (1988).
46. Crill, P.M., *Global Biogeochem. Cycles*, 5, 319–334 (1991).
47. Raich, J.W., and Schlesinger, W.H. *Tellus Ser. B*, 44B, 81–99 (1992).
48. Born, M., Dörr, H. and Levin, I., *Tellus Ser. B*, 42B, 2–8 (1990).
49. Kwon, O. Yul, and Schnoor, J.L., *Global Biogeochem. Cycles*, 8, 295–305 (1994).
50. Broecker, W.S., and Peng, T.H., in *Gas Transfer at Water Surfaces*, W. Brutsaert et al., Eds., D. Reidel, Norwell, MA, pp. 479–493 (1984).
51. Takahashi, T., Broecker, W.S., and Bainbridge, A.E., in *Carbon Cycle Modelling, SCOPE 16*, B. Bolin, Ed., Wiley, New York, pp. 159–199 (1981).
52. Oeschger, H., in *Carbon Dioxide and Other Greenhouse Gases: Climate and Associated Impacts*, R. Fantechi and A. Ghazi, Eds., Kluwer Academic, Norwell, MA, pp. 40–54 (1989).
53. Barnola, J.M., Raynaud, D., Korotkevich, Y.S., and Lorius, C., *Nature*, 329, 408–414 (1987).
54. Intergovernmental Panel on Climate Change (IPCC), *Climate Change 1994 Working Groups I and III of the IPCC*, Cambridge University Press, Cambridge, United Kingdom (1994).
55. Galloway, J.N., Schlesinger, W.H., Levy, H. II, Michaels, A., and Schnoor, J.L., *Global Biogeochem. Cycles*, 9:2, 235–252 (1995).
56. Hudson, R.J.M., Gherini, S.A., and Goldstein, R.A., *Global Biogeochem. Cycles*, 8:3, 307–333 (1994).

57. Warnek, P., *Chemistry of the Natural Atmosphere*, Academic Press, London (1988).
58. Schlesinger, W., *Biogeochemistry: An Analysis of Global Change*, Academic Press, New York (1991).
59. Paterson, K.G., Ph.D. Dissertation, The University of Iowa, Iowa City (1993).
60. Yan, Y.L., Ph.D. Dissertation, The University of Iowa, Iowa City (1993).
61. David, M., *Estimation of the Yields of Tropospheric Degradation Products of HCFCs and HFCs*, EAWAG No. 1800, Swiss Federal Institute for Environmental Science and Technology, Duebendorf, Switzerland (1993).

11.8 PROBLEMS

1. Prove that the "lifetime" of a chemical in a batch system with first-order decay is the time that it takes to reach $1/e$ of its original concentration. The definition of lifetime is the reciprocal pseudo-first-order reaction rate constant.

2. What is the mean residence time of methane (CH_4) in the atmosphere if its global average concentration is 1.7 ppmv, and the sources/sinks are ~525 Tg CH_4 yr^{-1} (from Table 11.10)? Assume steady state.

3. Find the mean residence time of carbon (carbon dioxide) in the atmosphere based on Figure 11.11. (The atmosphere is accumulating CO_2 at ~0.5% yr^{-1}, so the assumption of steady state is not appropriate in this case, but the calculation does indicate that CO_2 should respond rapidaly to changes in emissions which it does on a seasonal basis.) Discuss why the simulations in Figures 11.16 and 11.17 have the shape and response time indicated. (An eigenvalue analysis of the box model equations could be used to explain quantitatively the response curve.)

$$\tau_{CO_2} = \frac{\text{Mass in the Atmosphere}}{\text{Mass flux out}}$$

4. What is the residence time of carbon in terrestrial biota, soil and detritus, cold surface water, warm surface water, and intermediate and deep ocean water?

5. Assume the present concentration of CO_2 in the atmosphere is 360 ppmv (mL m^{-3}), and the average concentration in the surface layer of the ocean is 0.868 mg L^{-1} of $CO_{2(aq)}$. Calculate the flux rate in Gt C yr^{-1} (1 gigaton = 10^{15} g) to the ocean from the atmosphere. Do you understand why it is so difficult to get accurate estimates of CO_2 flux to the ocean?

$A = 3.6 \times 10^{18}$ cm^2 ocean area
H = Henry's constant = 0.77 (mg L^{-1} air)/(mg L^{-1} H_2O) at 15 °C
K_L = overall mass transfer coefficient = 20.0 cm h^{-1}

6. Solve the following mass balance differential equation for the mass of carbon dioxide (as C) in the atmosphere. You may use the data given in the previous problem. Plot the mass in the atmosphere versus time.

$$\frac{d(\text{Mass})}{dt} = D + F + (R - P) + R_s - K_L A\,(C_g/H - C_\ell)$$

C_g = CO_2 atmospheric gas phase concentration, mg L^{-1} air CO_2

C_ℓ = liquid film concentration of CO_2 in oceans, 0.868 mg L^{-1}

D = deforestation and biomass burning, 2.0 Gt C yr^{-1}

F = fossil fuel combustion, 6.0 Gt C yr^{-1}

$(R - P)$ = (negative net primary production) = –60 Gt C yr^{-1}

R_s = soil degassing and respiration = 60 Gt C yr^{-1}

Mass = 740 Gt C yr^{-1} initial CO_2 (as carbon)

What is the change in the model results if emissions are curtailed? If emissions increase at 2% per year?

7. You need to assess the global environmental impact of large-scale fertilization of forests with "acid rain" NH_4HSO_4. Develop a model of the nitrogen cycle for a forest floor. Assume nitrogen-nitrate is the limiting nutrient. Write a set of mass balance differential equations to comprise your model. Include the following reactions (k_1, k_2, and k_3 are first order rate constants):

$$NH_4^+ \xrightarrow{k_1} NO_2^- \xrightarrow{k_2} NO_3^-$$

$$NO_3^- \xrightarrow{\mu, K_M} \text{plant organic–N } (C_5H_7NO_2)$$

$$C_5H_7NO_2 \xrightarrow{k_3} NH_4^+ \text{ (mineralization, aerobic)}$$

$$NH_4^+ \overset{K_{eq}}{\rightleftharpoons} NH_{3(aq)} + H^+$$

$$NH_{3(aq)} \overset{K_H}{\rightleftharpoons} NH_{3(g)} \xrightarrow{k_L} 0.$$

8. HFC-134a has a projected emission rate of 663,000 metric tonnes per year. It is predicted that its degradation will produce 400,000 metric tonnes of HF and some CF_3COOH. Estimate the concentration of HF and CF_3COOH in global average rainfall..

9. If the lifetime in the atmosphere of HFC-134a is 14.3 years, estimate the reaction rate and rate constant for its reaction with ·OH in the troposphere. Assume an average ·OH concentration of 5×10^5 molecules cm^{-3} and HFC-134a concentration of 0.2 ppbv.

ANSWERS TO SELECTED PROBLEMS—CHAPTER 11

2. Mass of atmosphere $\simeq 5.25 \times 10^{18}$ kg; 4950 Tg CH_4 in the atmosphere; $\tau_{CH4} =$ 9.4 yr.

3. 6.0 yr; slow equilibration of the atmosphere with the oceans and their long response times.

5. $CO_2 = 0.671$ mg L^{-1} at 15 °C in the atmosphere; 2.3 Gt C yr^{-1} to the ocean.

9. $k = 1.6 \times 10^{-11}$ cm^3 $molecule^{-1}$ hr^{-1}
rxn rate = 0.014 ppbv yr^{-1}

APPENDIX A

DISSOLVED OXYGEN AS A FUNCTION OF SALINITY AND TEMPERATURE

	Chloride Concentration in Water, mg L^{-1}				
Temperature, °C	0	5,000	10,000	15,000	20,000
0	14.6	13.8	13.0	12.1	11.3
1	14.2	13.4	12.6	11.8	11.0
2	13.8	13.1	12.3	11.5	10.8
3	13.5	12.7	12.0	11.2	10.5
4	13.1	12.4	11.7	11.0	10.3
5	12.8	12.1	11.4	10.7	10.0
6	12.5	11.8	11.1	10.5	9.8
7	12.2	11.5	10.9	10.2	9.6
8	11.9	11.2	10.6	10.0	9.4
9	11.6	11.0	10.4	9.8	9.2
10	11.3	10.7	10.1	9.6	9.0
11	11.1	10.5	9.9	9.4	8.8
12	10.8	10.3	9.7	9.2	8.6
13	10.6	10.1	9.5	9.0	8.5
14	10.4	9.9	9.3	8.8	8.3
15	10.2	9.7	9.1	8.6	8.1
16	10.0	9.5	9.0	8.5	8.0
17	9.7	9.3	8.8	8.3	7.8
18	9.5	9.1	8.6	8.2	7.7
19	9.4	8.9	8.5	8.0	7.6
20	9.2	8.7	8.3	7.9	7.4
21	9.0	8.6	8.1	7.7	7.3
22	8.8	8.4	8.0	7.6	7.1
23	8.7	8.3	7.9	7.4	7.0
24	8.5	8.1	7.7	7.3	6.9
25	8.4	8.0	7.6	7.2	6.7

Source: American Public Health Association (1960).

APPENDIX B

DIMENSIONS, UNITS, CONVERSIONS

Definitions, Dimensions, and SI Units for Basic Mechanical Properties

					Dimension of Unit	
Property	Symbol	Definition	SI Unit	SI Symbol	Derived	Basic
Mass	M		kilogram	kg		kg
Length	L		meter	m		m
Time	T		second	s		s
Area	A	$A = l^2$				m^2
Volume	V	$V = l^3$				m^3
Velocity	v	$v = l/t$				$m\ s^{-1}$
Acceleration	a	$a = l/t^2$				$m\ s^{-2}$
Force	F	$F = Ma$	newton	N	N	$kg\ m\ s^{-2}$
Weight	w	$w = Mg$	newton	N	N	$kg\ m\ s^{-2}$
Pressure	p	$p = F/A$	pascal	Pa	$N\ m^{-2}$	$kg\ m^{-1}\ s^{-2}$
Work	W	$W = Fl$	joule	J	N m	$kg\ m^2\ s^{-2}$
Energy		Work done	joule	J	N m	$kg\ m^2\ s^{-2}$
Mass density	ρ	$\rho = M/V$				$kg\ m^{-3}$
Weight density	γ	$\gamma = w/V$			$N\ m^{-3}$	$kg\ m^{-2}\ s^{-2}$
Stress	σ, τ	Internal response to external p	pascal	Pa	$N\ m^{-2}$	$kg\ m^{-1}\ s^{-2}$
Strain	ϵ	$\epsilon = \Delta V/V$				Dimensionless
Young's modulus	E	Hooke's law			$N\ m^{-2}$	$kg\ m^{-1}\ s^{-2}$

Definitions, Dimensions, and SI Units for Fluid Properties and Groundwater Terms

Property	Symbol	Definition	SI Unit	SI Symbol	Dimension of Unit	
					Derived	Basic
Volume	V	$V = l^3$	liter (= $m^3 \times 10^{-3}$)		L	m^3
Discharge	Q	$Q = l^3/t$			$L\ s^{-1}$	$m^3\ s^{-1}$
Fluid pressure	p	$p = F/A$	pascal	Pa	$N\ m^{-2}$	$kg\ m^{-1}\ s^{-2}$
Head	h					m
Mass density	ρ	$\rho = M/V$				$kg\ m^{-3}$
Dynamic viscosity	μ	Newton's law	centipoise (= $N\ s\ m^{-2} \times 10^{-3}$)	cP	cP, $N\ s\ m^{-3}$	$kg\ m^{-1}\ s^{-1}$
Kinematic viscosity	v	$v = \mu/\rho$	centistoke (= $m^2\ s^{-1} \times 10^{-6}$)	cSt	cSt	$m^2\ s^{-1}$
Compressibility	α, β	$\alpha = 1/E$			$m^2\ N^{-1}$	$m\ s^2\ kg^{-1}$
Hydraulic conductivity	K	Darcy's law			$cm\ s^{-1}$	$m\ s^{-1}$
Permeability	k	$k = K\mu/pg$			cm^2	m^2
Porosity	n					Dimensionless
Specific storage	S_s	$S_s = pg(\alpha + n\beta)$				m^{-1}
Storativity	S	$S = S_s b$ [a]				Dimensionless
Transmissivity	T	$T = Kb$ [a]				$m^2\ s^{-1}$

[a] b, thickness of confined aquifer (see Section 2.10).

APPENDIX C

COMPLEMENTARY ERROR FUNCTION

x	erfc(x)	x	erfc(x)
0	1.0		
0.05	0.943628	1.1	0.119795
0.1	0.887537	1.2	0.089686
0.15	0.832004	1.3	0.065992
0.2	0.777297	1.4	0.047715
0.25	0.723674	1.5	0.033895
0.3	0.671373	1.6	0.023652
0.35	0.620618	1.7	0.016210
0.4	0.571608	1.8	0.010909
0.45	0.524518	1.9	0.007210
0.5	0.479500	2.0	0.004678
0.55	0.436677	2.1	0.002979
0.6	0.396144	2.2	0.001863
0.65	0.357971	2.3	0.001143
0.7	0.322199	2.4	0.000689
0.75	0.288844	2.5	0.000407
0.8	0.257899	2.6	0.000236
0.85	0.229332	2.7	0.000134
0.9	0.203092	2.8	0.000075
0.95	0.179109	2.9	0.000041
1.0	0.157299	3.0	0.000022

$$\text{erfc}(x) = 1 - (2/\pi) \int_0^x e^{-\epsilon^2}\, d\epsilon$$

$$\text{erfc}(x) = 1 - \text{erf}(x)$$

$$\text{erfc}(-x) = 2 - \text{erfc}(x)$$

Source: Modified from R.A. Freeze and J.A. Cherry, *Groundwater*, Prentice-Hall, Englewood Cliffs, NJ (1979)

APPENDIX D

RUNGE KUTTA FOURTH ORDER ACCURATE NUMERICAL MODEL FOR PCBs IN THE GREAT LAKES

```
      Program gr_lakes
* This program solves a model of PCB concentration in the Great
Lakes
* Completely mixed steady flow,instantaneous sorption
*
* Define & Initialize Variables
*
      Real wao(5),wqo(5),ks(5),kv(5),fp(5),fd(5),tau(5),v(5),
     & ct(5,101),dt,tn,step,error
      Integer i,f,j
*
* Link variables to routines
      Common /runge/dt
      Common /calc/wao,wqo,ks,fp,fd,tau,v,kv
*
* Read in data values for each lake
      Open(unit=1,file='lake.dat',status='old')
      Do 5 f=1,5
            Read(1,300) wao(f),wqo(f)(f,1),ks(f),kv(f),fp(f),
     &      fd(f),tau(f),v(f)
C           write(*,*) wao(f),wqo(f),ct(f,1),ks(f),kv(f),fp(f),
C    &      fd(f),tau(f),v(f)

5     Continue
      Close(unit=1)
      open (unit=3,file='gll.out', status='unknown')
      tn=0.0
      dt=0.25
      n=80
* Loop to solve ODE-'s (five for five lakes)
      Do 25 i=1,n
            tn=tn+dt
```

```
C         write(*,*)tn
          Do 20 j=1,5
                 Call rkutta(j,i,tn,ct,step,error)
                 ct(j,i+1)=ct(j,i)+step
C                write(*,*) ct(j,i),ct(j,i+1)
20        Continue
25    Continue
*
* Write output to file
      Open(unit=2,file='gl.out',status='unknown')
      Do 50 j=1,5
           count=0
           write(2,340) j
           write(2,350) 1980,ct(j,1),fp(j)*ct(j,1),fd(j)*ct(j,1)
        Do 45 i=1,n
                 count=count+1
             if(count.eq.4)then
                 write(2,350) 1980+i*dt,ct(j,i),fp(j)*ct(i,i),
    &            fd(j)*ct(j,i)
                 count=0
             endif
45         Continue
50    Continue
300   Format (f7.2,2x,f7.2,2x,f4.2,2x,f5.2,2x,f4.2,2x,f4.2,
    &             2x,f4.2,f6.2,2x,f8.2)
340   Format (1x,i1)
350   Format(1x,i4,3(3x,f4.2))
*
      close(unit=2)
      close(unit=3)
*
      End
*********Subroutine to calculate Runge-Kutta scheme***********
*
      Subroutine rkutta(j,i,tn,ct,step,error)
* Define Variables
      Real tn,ct(5,101),step,error,dt,k1,k2,k3,k4,k5,c
      Integer i,j
      Common /runge/dt
*
* Calculate Runge k's
*
      c=ct(j,i)
      k1=dt*funct(j,tn,c)
      k2=dt*funct(j,tn+dt/3.,c+k1/3.)
      k3=dt*funct(j,tn+dt/2.,c+(k1+k2)/6.)
      k4=dt*funct(j,tn+dt/2.,c+(k1+3.*k3)/6.)
      k5=dt*funct(j,tn+dt,c+(k1-3.*k3+4.*k4)/2.)
      step=(k1+4*k4+k5)/6.
```

```
      error=(k1-4.5*k3+4.*k4-.5*k5)/15.
C     write(*,*)k1,k2,k3,k4,k5,step
C     write(*,*)c
      Return
    end
*
*
*********** Function to calculate f(t,Ct)*******************
*
23456789012345678980123456789012345678901234567890123456789012345678
9012
      Real Function funct(j,t,cdum)
* Define variables
      Real wao(5),wqo(5),wa,wq,ks(5),fp(5),fd(5),tau(5),
    & v(5),kv(5),t,cdum
      Integer j
      Common /calc/wao,wqo,ks,fp,fd,tau,v,kv
*
* calculate loading rates
*
* comments toggle for different loading rates
*
C     wa=wao(j)*exp(-0.02*t)
C     wq=wqo(j)*exp(-0.03*t)
      wq=wqo(j)
      wa=wao(j)
      funct=(wa+wq)/v(j)-cdum*(1./tau(j)+ks(j)*fp(j)+kv(j)*fd(j))
C     write(*,*)t,j,cdum,v(j),tau(j),ks(j),fp(j),kv(j),fd(j),
C   & funct
      if(t.eq.20.) write(3,20)j,wa,wq,v(j)*cdum*1./tau(j)
    & v(j)*cdum*ks(j)*fp(j),v(j)*kv(j)*fd(j)
20    Format(1x,i1,5(1x,f9.2))
      /END/
```

APPENDIX E

CHEMICAL EQUILIBRIUM FORTRAN PROGRAM FOR ALUMINUM SPECIATION IN NATURAL WATERS USING A NEWTON-RAPHSON ITERATIVE TECHNIQUE

```
      IMPLICIT DOUBLE PRECISION (A-H, O-Z)
      CHARACTER*1 ANS
      CHARACTER*3 CUNIT
      DIMENSION XO(4),TO(4)
    1 TO(3)=0.0
      WRITE(1,100)
      READ(1,33)CUNIT
      WRITE(1,101)
      READ(1,*)TO(1)
      WRITE(1,102)
      READ(1,*)TO(2)
      WRITE(1,122)
      READ(1,*)TO(3)
      WRITE(6,106)TO(1),CUNIT
      WRITE(6,107)TO(2),CUNIT
      WRITE(6,108)TO(3),CUNIT
      IF(CUNIT .EQ. 'MOL') GO TO 2
      TO(1)=TO(1)/26981.5
      TO(2)=TO(2)/18998.4
      TO(3)=TO(3)/9606.16
    2 TO(4)=0.0
      XO(4)=0.0
      XO(1)=DLOG10(TO(1)/10.0)
      XO(2)=DLOG10(TO(2)/10.0)
      XO(3)=DLOG10(TO(3)/10.0)
      CALL ALEQ(XO,TO)
      WRITE(1,105)
      READ(1,88)ANS
```

```
      IF(ANS .EQ. 'Y') GO TO 1
  33  FORMAT(A3)
  44  FORMAT(20X,A1)
  55  FORMAT(16X,F6.3)
  88  FORMAT(A1)
 100  FORMAT(12X, 'Enter the unit for conc >>',/
    > ,15X, 'PPM or MOL')
 101  FORMAT(12X, 'Enter total conc. for AI >>')
 102  FORMAT(12X, 'Enter total conc. for F >>')
 122  FORMAT(12X, 'Enter total conc. for SO4 >>')
 105  FORMAT(12X, 'Would you like to run a new job ?',/,
    > 15X,' >> Y or N')
 106  FORMAT(12X, 'TOTAL CONC. FOR AL+++ IS ',F10.5, ' ',A3)
 107  FORMAT(12X, 'TOTAL CONC. FOR F- IS ',F10.5, ' ',A3)
 108  FORMAT(12X, 'TOTAL CONC. FOR S04-2 IS ',F10.5, ' ',A3)
      STOP
      END
      SUBROUTINE ALEQ(X,T)
      IMPLICIT DOUBLE PRECISION(A-H,O-Z)
      CHARACTER*10 ID(19)
      DIMENSION X(4),T(4),C(19),YO(4),D0(4),E(4),Y(4),RK(19),
    > IA(19,4),Z0(4,4),Z(4,4),Z1(4,4)
      DATA ID /'AL+++    ','ALOH++   ','AL(OH)2+  ','AL(OH)3 AQ',
    > 'AL(OH)4-  ','ALF++     ','ALF2+  ','ALF3 AQ  ',
    > 'ALF4-     ','HF AQ     ','HF2-   ','H2F2 AQ  ',
    > 'H+        ','OH-       ','F-     ','HSO4-    ',
    > 'ALSO4+    ','AL(SO4)2- ','SO4-2  '/
      DATA IA/9*1,7*0,2*1,6*0,1,2,3,4,1,2*2,2*0,1,19*0,2*1,2,1,
    > 0,-1,-2,-3,-4,4*0,2*1,2,1,-1,0,1,3*0/
      DATA RK/0.0,-4.99,-10.1,-16,0,-23.0,7.01,12.75,17.02,19.72,
    > 3.169,3.749,6.768,0.0,-13.998,0.0,1.987,3.02,4.92,0.0/
      M1=19
      KNUM=4
      N1=4
      EPI=1.0D-4
      ITMAX=120
      I8=0
 998  DO 2 I=1,M1
   2  C(I) =0.0
      DO 3 I= 1,M1
      DO 3 J=1,N1
   3  C(I)=IA(I,J)*X(J)+C(I)
      DO 4 I=1,M1
   4  C(I)=C(I)+RK(I)
      DO 5 I=1,M1
   5  C(I)=10.0**C(I)
      DO 6 J=I,N1
   6  E(J)=10.0**X(J)
      I7=0
```

```
      DO 9 J=I,N1
      V8=-T(J)
      V9=DABS(T(J))
      DO 7 I=1,M1
      V8=V8+IA(I,J)*C(I)
    7 V9=V9+DABS(DBLE(FLOAT(IA(I,J))))*C(I)
      IF (J .GT. KNUM) GO TO 8
      IF (DABS(V8)/V9 .GT. EPI) I7=1
    8 Y(J)=V8
    9 CONTINUE
      DO 150 J=1,KNUM
      DO 150 K=1,KNUM
      V9=0.0
      DO 140 I=1,M1
  140 V9=V9+IA(I,J)*IA(I,K)*C(I)/E(K)
  150 Z(J,K)=V9
      IF (I7 .EQ. 0) GO TO 191
      I8=I8+1
      IF (I8 .GT. ITMAX) GO TO 190
      DO 170 J=1,KNUM
      DO 160 K=1,KNUM
  160 ZO(J,K)=Z(J,K)
  170 YO(J)=Y(J)
      CALL MATINV(ZO,KNUM)
      DO 171 I=1,KNUM
  171 DO (I)=0.0
      DO 172 I=1,KNUM
      DO 172 J=1,KNUM
  172 DO(1)=ZO(I,J)*YO(J)+DO(I)
      DO 180 J=1,KNUM
      E(J)=E(J)-DO(J)
      IF (E(J) .LE. 0.0) E(J)=(E(J)+DO(J))/10.0
  180 X(J)=DLOG10(E(J))
      GO TO 998
  190 WRITE(6,444)
  191 WRITE(6,555)
      DO 192 I=1,M1
      WRITE(6,666) ID(I),C(I)
  192 CONTINUE
  444 FORMAT(12X, 'AFTER ALLOWABLE MAX. ITERATIONS'/)
  555 FORMAT(12X, 'THE RESULT IS '/)
  666 FORMAT(/10X,A10, ' CONC. = ',1PD15.5)
      RETURN
      END
      SUBROUTINE MATINV(A,N)
      IMPLIClT DOUBLE PRECISION(A-H,O-Z)
      DIMENSION A(4,8)
      DATA EPS/1.D-8/
      L=N+1
```

```
      M=2*N
      DO 2 I=1,N
      DO 2 J=L,M
      A(I,J)=0.0
      IF (I+N-J) 2, 1,2
    1 A(I,J)=1.
    2 CONTINUE
      DO 13 I=1,N
      K=I
      IF(I-N)3,8,3
    3 IF(A(I,I)-EPS)4,5,8
    4 IF(-A(I,I)-EPS)5,5,8
    5 K=K+1
      DO 7 J=1,M
    7 A(I,J)=A(I,J)+A(K,J)
      GO TO 3
    8 DIV=A(I,I)
      DO 9 J=1,M
    9 A(I,J)=A(I,J)/DIV
      DO 13 K=1,N
      DELT=A(K,I)
      IF(DABS(DELT)-EPS)13,13,10
   10 IF(K-I)11,13,11
   11 DO 12 J=1,M
   12 A(K,J)=A(K,J)-A(I,J)*DELT
   13 CONTINUE
      DO 14 I=1,N
      DO 14 J=1,N
      A(I,J)=A(I,N+J)
   14 CONTINUE
      RETURN
      END
```

EXAMPLE FOR RUNNING FORTRAN VERSION

```
      OK, O EQ.AL.OUT 2 2
      OK, SEG #EQ.AL.FTN
              Enter the unit for conc >>
                 PPM or MOL
      MOL
              Enter total conc. for Al >>
      0.0001
              Enter total conc. for F >>
      0.0001
              Enter total conc. for SO4 >>
      0.0001
              Would you like to run a new job?
                 >> Y or N
      N
```

```
 **** STOP

 OK, SLIST EQ.AL.OUT
        TOTAL CONC. FOR AL+++ IS   0.00010 MOL
        TOTAL CONC. FOR F- IS      0.00010 MOL
        TOTAL CONC. FOR SO4-2 IS   0.00010 MOL
        THE RESULT IS

        AL+++         CONC. =    9.42872E-06
        ALOH++        CONC. =    6.64306E-06
        AL(OH)2+      CONC. =    3.55045E-06
        AL(OH)3 AQ    CONC. =    3.07751E-07
        AL(OH)4-      CONC. =    2.11892E-09
        ALF++         CONC. =    5.90184E-05
        ALF2+         CONC. =    1.98391E-05
        ALF3 AQ       CONC. =    2.25973E-07
        ALF4-         CONC. =    6.92773E-11
        HF AQ         CONC. =    1.31105E-08
        HF2-          CONC. =    3.04897E-14
        H2F2 AQ       CONC. =    4.62634E-16
        H+            CONC. =    1.45239E-05
        OH-           CONC. =    6.91697E-10
        F-            CONC. =    6.11695E-07
        HSO4-         CONC. =    1.39362E-07
        ALSO4+        CONC. =    9.76143E-07
        AL(SO4)2-     CONC. =    7.66610E-09
        SO4-2         CONC. =    9.88692E-05
OK, COMO -E
```

APPENDIX F

IMPLICIT FINITE DIFFERENCE NUMERICAL TECHNIQUE FOR ADVECTIVE-DISPERSION EQUATION IN A VERTICALLY STRATIFIED LAKE OR RESERVOIR

The generalized mass balance for a one-dimensional reservoir with vertical mixing and flow between compartments (i compartment) is given by equation (1).

$$\frac{\partial V_i C_i}{\partial t} = \sum_1^n Q_i C_{\text{in}} + Q_{v_{i-1}} C_{i-1} - EA_i \frac{\partial C_{i-1}}{\partial z}$$

$$- Q_i C_i - Q_{v_i} C_i + EA_{i+1} \frac{\partial C_i}{\partial z} \tag{1}$$

IMPLICIT SOLUTION TECHNIQUE

Equation (1) represents the generalized mass balance equation for water quality. It is a partial differential equation including derivatives with respect to depth and time. Approximating the $\partial C / \partial z$ terms with a one-step forward difference produces an ordinary differential equation below that is more easily soluble.

$$\frac{\partial C_{i-1}}{\partial z} \simeq \frac{C_i - C_{i-1}}{\Delta z} \tag{2}$$

$$\frac{\partial C_i}{\partial z} \simeq \frac{C_{i+1} - C_i}{\Delta z} \tag{3}$$

$$\frac{dV_iC_i}{dt} = \sum_1^n (Q_iC)_{\text{in}} + Q_{v_{i-1}}C_{i-1} - \frac{EA_iC_i}{\Delta z}$$

$$+ \frac{EA_iC_{i-1}}{\Delta z} - (Q_iC_i)_{\text{out}} - Q_{v_i}C_i$$

$$+ \frac{EA_{i+1}C_{i+1}}{\Delta z} - \frac{EA_{i+1}C_i}{\Delta z} \tag{4}$$

From the approximation equations (2) and (3) we may expect errors of order Δz, so the grid spacing should be kept fairly tight to minimize error. The error approaches zero as $\Delta z \to 0$.

Notice that the mass flow rate of equation (4) is dependent on the concentration from only three adjacent slices in the reservoir, at $i + 1$, i, and $i - 1$. We may rewrite equation (4) grouping the like terms:

$$V_i \frac{dC_i}{dt} = x_{i-1}\, C_{i-1} + x_i\, C_i + x_{i+1}\, C_{i+1} + \{\mathrm{p}'\}. \tag{5}$$

The x terms include the turbulent diffusion, vertical advection, and output terms of equation (4), while p′ simply is equal to $\Sigma_1^n\, Q_i\, C_{in}$. For reactive substances and external boundary condition problems, the reaction term may be included in the x_i term, while the boundary conditions are put into p′.

Rewriting equation (5) in matrix notation gives

$$[V]\left\{\frac{dC}{dt}\right\} = [x]\,\{C\} + \{\mathrm{p}'\}. \tag{6}$$

Where $[V]$ is the diagonal volume matrix and $[x]$ is tridiagonal.

Using the implicit trapezoid rule to approximate C at $(t + \Delta t)$ we may write:

$$C_i(t + \Delta t) = C_i(t) + \frac{\Delta t}{2}\left(\frac{dC_i(t)}{dt} + \frac{dC_i(t + \Delta t)}{dt}\right) \tag{7}$$

or in vector notation

$$\{C\,(t + \Delta t)\} = \left\{C(t) + \frac{\Delta t}{2}\frac{dC(t)}{dt}\right\} + \left\{\frac{\Delta t}{2}\frac{dC(t + \Delta t)}{dt}\right\} \tag{8}$$

There are several advantages of this method. The error is of $\theta(\Delta t^2, \Delta z^2)$ and the stability or amplification factor |G| is identically equal to one.

By substituting equation (8) into equation (6) taken at $(t + \Delta t)$, we have:

$$\left[V - \frac{\Delta t}{2}x\right]\left\{\frac{dC(t + \Delta t)}{dt}\right\} = [x]\left\{C(t) + \frac{\Delta t}{2}\frac{dC(t)}{dt}\right\} + \{\mathrm{p}'\} \tag{9}$$

or

$$[S(t+\Delta t)] = \left\{\frac{dC(t+\Delta t)}{dt}\right\} = \{F(t+\Delta t)\} \tag{10}$$

where

$$[S(t+\Delta t)] = \left[V - \frac{\Delta t}{2}x\right] \tag{11}$$

and

$$[F(t+\Delta t)] = [x]\left\{C(t) + \frac{\Delta t}{2}\frac{dC(t)}{dt}\right\} + \{p'\}. \tag{12}$$

The solution of equation (9) proceeds as a recursion relationship to be performed in sequence at each time step.

Step 1. Solve equation (10) for $\{dC(t+\Delta t)/dt\}$ via the Thomas algorithm.
Step 2. Define $\{\alpha(t)\} = \{C(t)\} + \{\Delta t/2\; dC(t)/dt\}$ and set $\{\alpha(o)\} = \{C(o)\}$ to initialize α. Then solve equation (8) for $\{C(t+\Delta t)\}$.
Step 3. Find the next α value in time from the definition given in Step 2.

$$\{\alpha(t+\Delta t)\} = \{C(t+\Delta t)\} + \frac{\Delta t}{2}\left\{\frac{dC}{dt}(t+\Delta t)\right\}.$$

Repeat

The Thomas algorithm mentioned in Step 1 is used to solve equation (10). It is a modified form of Gaussian elimination, and a synopsis of the algorithm follows

THOMAS ALGORITHM

$$[S]\,\{\dot{C}\} = \{F\}$$

$$\begin{bmatrix} S_{12} & S_{13} & & & \\ S_{21} & S_{22} & S_{23} & & \\ & S_{31} & S_{32} & S_{33} & \\ & & & \vdots & \\ & & & S_{n1} & S_{n2} \end{bmatrix} \begin{Bmatrix} \dot{C}_1 \\ \dot{C}_2 \\ \dot{C}_3 \\ \vdots \\ \dot{C}_n \end{Bmatrix} = \begin{Bmatrix} F_1 \\ F_2 \\ F_3 \\ \vdots \\ F_n \end{Bmatrix}$$

1. Set $\overline{F_1} = \dfrac{F_1}{S_{1,2}}$ and $\overline{S_{1,3}} = \dfrac{S_{1,3}}{S_{1,2}}$

2. Operate from the second row to the last row, n, using the relations given below:

$$\overline{F_j} = \frac{F_j - S_{j,1}\,\overline{F_{j-1}}}{S_{j,2} - S_{j,1}\,S_{j-1,3}}$$

$$\overline{S_{j,3}} = \frac{S_{j,3}}{S_{j,2} - \overline{S_{j-1,3}}}$$

3. For the final row, n

$$\dot{C}_n = \overline{F_n}$$

4. Operate from row n – 1 to row 1 to find $\dot{C}$, by:

$$\dot{C}_j = \overline{F_j} - \overline{S_{j,3}} = \dot{C}_{j+1}$$

INDEX

ENVIRONMENTAL SCIENCE AND TECHNOLOGY

A Wiley-Interscience Series of Texts and Monographs

Edited by JERALD L. SCHNOOR, *University of Iowa*
ALEXANDER ZEHNDER, *Swiss Federal Institute for Water Resources and Water Pollution Control*

PHYSIOCHEMICAL PROCESSES FOR WATER QUALITY CONTROL
Walter J. Weber, Jr., Editor

pH AND pION CONTROL IN PROCESS AND WASTE STREAMS
F. G. Shinskey

AQUATIC POLLUTION: An Introductory Text
Edward A. Laws

INDOOR AIR POLLUTION: Characterization, Prediction, and Control
Richard A. Wadden and Peter A. Scheff

PRINCIPLES OF ANIMAL EXTRAPOLATION
Edward J. Calabrese

SYSTEMS ECOLOGY: An Introduction
Howard T. Odum

INTEGRATED MANAGEMENT OF INSECT PESTS OF POME AND STONE FRUITS
B. A. Croft and S. C. Hoyt, Editors

WATER RESOURCES: Distribution, Use and Management
John R. Mather

ECOGENETICS: Genetic Variation in Susceptibility to Environmental Agents
Edward J. Calabrese

GROUNDWATER POLLUTION MICROBIOLOGY
Gabriel Bitton and Charles P. Gerba, Editors

CHEMISTRY AND ECOTOXICOLOGY OF POLLUTION
Des W. Connell and Gregory J. Miller

SALINITY TOLERANCE IN PLANTS: Strategies for Crop Improvement
Richard C. Staples and Gary H. Toenniessen, Editors

ECOLOGY, IMPACT ASSESSMENT, AND ENVIRONMENTAL PLANNING
Walter E. Westman

CHEMICAL PROCESSES IN LAKES
Werner Stumm, Editor

INTEGRATED PEST MANAGEMENT IN PINE-BARK BEETLE ECOSYSTEMS
William E. Waters, Ronald W. Stark, and David L. Wood, Editors

PALEOCLIMATE ANALYSIS AND MODELING
Alan D. Hecht, Editor

BLACK CARBON IN THE ENVIRONMENT: Properties and Distribution
E. D. Goldberg

GROUND WATER QUALITY
C. H. Ward, W. Giger, and P. L. McCarty, Editors

TOXIC SUSCEPTIBILITY: Male/Female Differences
Edward J. Calabrese

ENERGY AND RESOURCE QUALITY: The Ecology of the Economic Process
Charles A. S. Hall, Cutler J. Cleveland, and Robert Kaufmann

AGE AND SUSCEPTIBILITY TO TOXIC SUBSTANCES
Edward J. Calabrese

ENVIRONMENTAL SCIENCE AND TECHNOLOGY

List of Titles (*Continued)*

ECOLOGICAL THEORY AND INTEGRATED PEST MANAGEMENT PRACTICE
Marcos Kogan, Editor

AQUATIC SURFACE CHEMISTRY: Chemical Processes at the Particle Water Interface
Werner Stumm, Editor

RADON AND ITS DECAY PRODUCTS IN INDOOR AIR
William W. Nazaroff and Anthony V. Nero, Jr., Editors

PLANT STRESS–INSECT INTERACTIONS
E. A. Heinrichs, Editor

INTEGRATED PEST MANAGEMENT SYSTEMS AND COTTON PRODUCTION
Ray Frisbie, Kamal El-Zik, and L. Ted Wilson, Editors

ECOLOGICAL ENGINEERING: An Introduction to Ecotechnology
William J. Mitsch and Sven Erik Jorgensen, Editors

ARTHROPOD BIOLOGICAL CONTROL AGENTS AND PESTICIDES
Brian A. Croft

AQUATIC CHEMICAL KINETICS: Reaction Rates of Processes in Natural Waters
Werner Stumm, Editor

GENERAL ENERGETICS: Energy in the Biosphere and Civilization
Vaclav Smil

FATE OF PESTICIDES AND CHEMICALS IN THE ENVIRONMENT
J. L. Schnoor, Editor

ENVIRONMENTAL ENGINEERING AND SANITATION, Fourth Edition
Joseph A. Salvato

TOXIC SUBSTANCES IN THE ENVIRONMENT
B. Magnus Francis

CLIMATE-BIOSPHERE INTERACTIONS
Richard G. Zepp, Editor

AQUATIC CHEMISTRY: Chemical Equilibria and Rates in Natural Waters, Third Edition
Werner Stumm and James J. Morgan

PROCESS DYNAMICS IN ENVIRONMENTAL SYSTEMS
Walter J. Weber, Jr., and Francis A. DiGiano

ENVIRONMENTAL CHEMODYNAMICS: Movement of Chemicals in Air, Water, and Soil, Second Edition
Louis J. Thibodeaux

ENVIRONMENTAL MODELING: Fate and Transport of Pollutants in Water, Air, and Soil
Jerald L. Schnoor

TRANSPORT MODELING FOR ENVIRONMENTAL ENGINEERS AND SCIENTISTS
Mark M. Clark